ARTH SCIENCES
w.cengage.com/earthscience

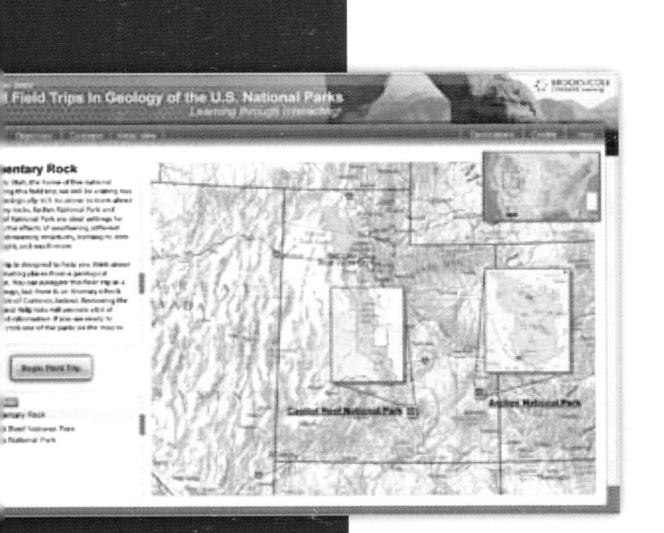

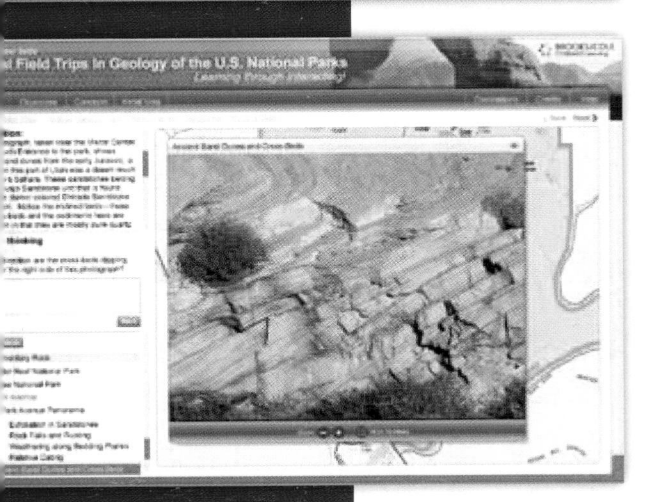

VIRTUAL FIELD TRIPS IN GEOLOGY

Explore some of America's most awe-inspiring natural features without ever leaving campus.

The **Virtual Field Trips in Geology** are concept-based modules that teach students geology by using famous locations throughout the United States. Sideling Hill Syncline, Grand Canyon, and Hawaii Volcanoes National Parks are included, as well as many others. Designed to be used as homework assignments or lab work, the modules use a rich array of multimedia to demonstrate concepts. High definition videos, images, animations, quizzes, and Google Earth layers work together in **Virtual Field Trips** to bring the concepts to life.

These modules are available as a discounted bundle with any Brooks/Cole Earth Science text or as a stand-alone product.

Available for Fall 2009

- Sedimentary Rocks: Formation and Correlation
- Geologic Time
- Desert Environment
- Running Water
- Groundwater

Coming Soon

- Metamorphism and Metamorphic Rocks
- Mass Wasting Processes
- Glaciers and Glaciation
- Igneous Rock Textures
- Earthquakes & Seismicity
- Volcano Types
- Plate Tectonics
- Mineral Resources
- Hydrothermal Activity
- Shorelines and Shoreline Processes

Students, ask your instructor how you can get access.

Instructors, contact your local Brooks/Cole representative for more information.

Visit the Earth Science Community Site to

www.cengage.com/community/

Essentials of Physical Geology

FIFTH EDITION

Reed Wicander
Central Michigan University

James S. Monroe
Professor Emeritus
Central Michigan University

Australia • Brazil • Japan • Korea • Mexico • Singapore • Spain • United Kingdom • United States

Essentials of Physical Geology, Fifth Edition
Reed Wicander and James S. Monroe

Development Editor: Amy K. Collins

Assistant Editor: Liana Monari

Technology Project Managers: Melinda Newfarmer, Alexandria Brady

Marketing Manager: Joe Rogove

Marketing Assistant: Ashley Pickering

Marketing Communications Manager: Belinda Krohmer

Project Manager, Editorial Production: Hal Humphrey

Art Director: Vernon Boes

Print Buyer: Karen Hunt

Permissions Editor: Bob Kauser

Production Service: Kevin Shea, Pre-Press PMG

Text Designer: Lisa Buckley

Photo Researcher: Terri Wright

Copy Editor: Patricia A. Onufrak

Illustrator: Precision Graphics, Pre-Press PMG

Cover Designer: Denise Davidson

Cover Image: Dr. Parvinder Sethi

Lava flows meeting the Pacific Ocean near the Laeapuki region on the south coast of Hawaii Volcanoes National Park. Such lava flows begin their journey some 7–8 km upslope from the flanks of the Kilauea-Pu u O o Summit and travel towards the ocean mostly through a network of underground lava tubes. Occasionally, some flows will "break out" on the surface, such as these ones. As the molten lava collides with the cold Pacific waters it causes chemical reactions creating plumes of highly toxic, hydrochloric acid—seen here as white gas.

Compositor: Pre-Press PMG

Library of Congress Control Number: 2007940303

ISBN-13: 978-0-538-79790-0

ISBN-10: 0-538-79790-8

Brooks/Cole
10 Davis Drive
Belmont, CA 94002-3098
USA

Cengage Learning is a leading provider of customized learning solutions with office locations around the globe, including Singapore, the United Kingdom, Australia, Mexico, Brazil, and Japan. Locate your local office at **international.cengage.com/region**

Cengage Learning products are represented in Canada by Nelson Education, Ltd.

For your course and learning solutions, visit **academic.cengage.com**

Purchase any of our products at your local college store or at our preferred online store **www.ichapters.com**

Printed in the United States
1 2 3 4 5 6 7 12 11 10 09

BRIEF CONTENTS

CONTENTS

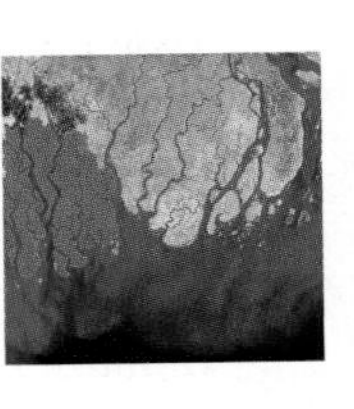

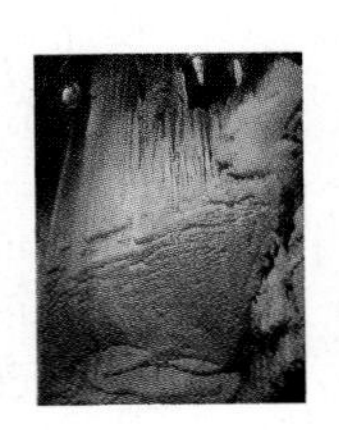

PREFACE

Earth is a dynamic planet that has changed continuously during its 4.6 billion years of existence. The size, shape, and geographic distribution of the continents and ocean basins have changed through time, as have the atmosphere and biota. As scientists and concerned citizens, we have become increasingly aware of how fragile our planet is and, more importantly, how interdependent all of its various systems and subsystems are.

We have also learned that we cannot continually pollute our environment and that our natural resources are limited and, in most cases, nonrenewable. Furthermore, we are coming to realize how central geology is to our everyday lives. For example, a 9.0-magnitude earthquake in the Indian Ocean on December 26, 2004 generated a tsunami that killed more than 220,000 people in Indonesia, Sri Lanka, India, Thailand, Somalia, Myanmar, Malaysia, and the Maldives, and caused billions of dollars in damage. Two hurricanes in 2005, Katrina and Rita, damaged many offshore oil platforms and oil refineries in Texas and Louisiana, demonstrating how fragile our energy network is, from production to the finished refined product, and how dependent we are on petroleum to run our economy. For these and other reasons, geology is one of the most important college or university courses a student can take.

Essentials of Physical Geology, Fifth Edition, is designed for a one-semester introductory course in geology and the earth sciences. One of the problems with any introductory science course is that students are overwhelmed by the amount of material that must be learned. Furthermore, most of the material does not seem to be linked by any unifying theme and does not always appear to be relevant to their lives. This book, however, is written to address that problem in that it shows, in its easy-to-read style, that geology is an exciting and ever-changing science, and one in which new discoveries and insights are continually being made.

The goals of this book are to provide students with a basic understanding of geology and its processes and, most importantly, with an understanding of how geology relates to the human experience—that is, how geology affects not only individuals, but society in general. With these goals in mind, we introduce the major themes of the book in the first chapter to provide students with an overview of the subject and to enable them to see how the various systems and subsystems of Earth are interrelated. We then cover the unifying theme of geology—plate tectonics, in the second chapter. Plate tectonic theory is central to the study of geology, because it links many aspects of geology together. It is a theme that is woven throughout this edition. We also discuss the economic and environmental aspects of geology throughout the book rather than treating these topics in separate chapters. In this way, students can see, through relevant and interesting examples, how geology impacts our lives.

NEW AND RETAINED FEATURES IN THE FIFTH EDITION

Just as Earth is dynamic and evolving, so too is *Essentials of Physical Geology*. The fifth edition has undergone considerable rewriting and updating, resulting in a volume that is still easy to read and has a high level of current information. Drawing on the comments and suggestions of reviewers, we have retained our most popular features and incorporated a number of new features into this edition.

- The *Chapter Objectives* outline at the beginning of each chapter has been retained to alert students to the key points that the chapter will address.
- Chapter content has been rewritten to help clarify concepts and make the material more exploratory.
- Current events and integration of topics can be found throughout the text.
- Many of the popular *Geo-Focus* features contain new topics or have been updated.
- The successful *What Would You Do?* boxes in each chapter continue to encourage students to think critically about what they're learning by asking open-ended questions related to the chapter material. Many of these boxes are either new or have been rewritten to reflect current topics in the news.
- A new feature, *Geo-inSight*, has been added to each chapter. These two-page, highly illustrated pieces are designed to enhance students' interest in the chapter material.
- The art program has undergone revision to better illustrate the material covered in the text. In addition, the figure captions have been expanded and improved to help explain what the student is seeing.
- Many photographs in the fourth edition have been replaced, including many of the chapter-opening photographs.
- The *Review Questions* section at the end of each chapter has been revised to include 10 multiple-choice questions and 10 short-answer questions. Answers to all of the multiple-choice questions as well as two of the short-answer questions per chapter are provided at the back of the book.

It is our strong belief that the rewriting and updating done in the text as well as the addition of new photographs and newly rendered art greatly improve the fifth edition of *Essentials of Physical Geology*. We think that these changes and enhancements make this textbook easier to read and comprehend, as well as a more effective teaching tool.

TEXT ORGANIZATION

Plate tectonic theory is the unifying theme of geology and this book. This theory has revolutionized geology because it provides a global perspective of Earth and allows geologists to treat many seemingly unrelated geologic phenomena as part of a total planetary system. Because plate tectonic theory is so important, it is covered in Chapter 2 and is discussed in most subsequent chapters in terms of the subject matter of that chapter.

Another theme of this book is that Earth is a complex, dynamic planet that has changed continuously since its origins some 4.6 billion years ago. We can better understand this complexity by using a systems approach to the study of Earth and emphasizing this approach throughout the book.

We have organized *Essentials of Physical Geology*, Fifth Edition, into the following informal categories.

- Chapter 1 is an introduction to geology and Earth systems, geology's relevance to the human experience, and the origin of the solar system and Earth's place in it.
- Chapter 2 deals with plate tectonics in detail and sets the stage for its integration throughout the rest of the book.
- Chapters 3–7 examine Earth's materials (minerals and igneous, sedimentary, and metamorphic rocks) and the geologic processes associated with them, including the role of plate tectonics in their origin and distribution.
- Chapters 8–10 deal with the related topics of Earth's interior, the seafloor, earthquakes, and deformation and mountain building.
- Chapters 11–16 cover Earth's surface processes.
- Chapter 17 discusses geologic time.

We have found that presenting the material in the order discussed above works well for most students. We know, however, that many instructors prefer an entirely different order of topics, depending on the emphasis in their course. We have therefore written this book so that instructors can present the chapters in any order that suits the needs of a particular course.

CHAPTER ORGANIZATION

All chapters have the same organizational format that follows.

- Each chapter opens with a photograph relating to the chapter material, an Outline of the topics covered, and an Objectives list that alerts students to the learning outcome objectives of the chapter.
- An *Introduction* follows that is intended to stimulate interest in the chapter and show how the chapter material fits into the larger geologic perspective.
- The text is written in a clear, informal style, making it easy for students to comprehend.
- Numerous newly rendered color diagrams and new photographs complement the text and provide a visual representation of the concepts and information presented.
- Each chapter contains one *Geo-Focus* feature that presents a brief discussion of an interesting aspect of geology or geologic research.
- A *Geo-inSight* art spread is now in every chapter. These two-page features are designed to enhance students' interest in the chapter material through visual learning.
- Two *What Would You Do?* boxes per chapter encourage students to engage in critical thinking by solving a hypothetical problem or issue that is related to the chapter material.
- Topics relating to environmental and economic geology are discussed throughout the text. Integrating economic and environmental geology with the chapter material helps students relate the importance and relevance of geology to their lives.
- The end-of-chapter *Geo-Recap* begins with a concise review of important concepts and ideas presented in the chapter.
- The *Important Terms*, which are printed in boldface type in the chapter text, are listed at the end of each chapter for easy review along with the page numbers on which they are first defined. A full *Glossary* of important terms appears at the end of the text.
- The *Review Questions* are another important feature of this book and include multiple-choice questions with answers as well as short-answer questions, two of which have the answers provided at the end of the book. Many new questions have been added to each chapter of the fifth edition.

ANCILLARY MATERIALS

FOR INSTRUCTORS

We are pleased to offer a full suite of text and multimedia products to accompany *Essentials of Physical Geology*, Fifth Edition.

Geology Resource Center *academic.cengage.com/earthscience*

Book Companion Website *academic.cengage.com/earthscience*

The Geology Resource Center and the Book Companion Website feature a rich array of learning resources for your students. The text-specific companion website includes quizzing and other web-based activities that will help students explore the concepts presented in the text.

PowerLecture with JoinIn™ on Turning Point® A complete all-in-one reference for instructors, the PowerLecture CD contains PowerPoint slides of images from the text, stepped art from the text, zoomable art figures from the text, Active Figures that interactively demonstrate concepts, and lectures that outline the main points of each chapter. In addition to providing you with fantastic course presentation material, the PowerLecture CD also contains electronic files of the Test Bank and Instructor's Manual as well as JoinIn, the easiest Audience Response System to use, featuring instant classroom assessment and learning.

ISBN-10: 0495564958 | ISBN-13: 978-0495564959

Online Instructor's Manual with Test Bank This invaluable guide contains resources designed to streamline and maximize the effectiveness of your course preparation, including a complete test item file in Microsoft Word™ format, ideal for homework, group work or laboratory exercises.

ISBN-10: 0495555010 | ISBN-13: 978-0495555018

ExamView Create, deliver, and customize tests and study guides (both print and online) in minutes with this easy-to-use assessment and tutorial system. ExamView offers both a Quick Test Wizard and an Online Test Wizard that guide you step by step through the process of creating tests, whereas its "what you see is what you get" interface allows you to see the test you are creating on the screen exactly as it will print or display online. You can build tests of up to 250 questions using up to 12 question types. Using ExamView's complete word processing capabilities, you can enter an unlimited number of new questions or edit existing questions.

ISBN-10: 0495555045 | ISBN-13: 978-0495555049

Active Earth Collection CD The Active Earth Collection allows you to pick and choose from over 120 earth science animations and active figures, ABC natural hazards video clips, and in-depth Google Earth lecture activities. Grab your students' attention by creating your lectures using these dynamic tools.

ISBN-10: 0495555320 | ISBN-13: 978-0495555322

FOR STUDENTS

Geology Resource Center *academic.cengage.com/earthscience*

Book Companion Website *academic.cengage.com/earthscience*

This website features a rich array of learning resources, including quizzing and other web-based activities that will help you explore the concepts that are presented in the text.

ACKNOWLEDGMENTS

As the authors, we are, of course, responsible for the organization, style, and accuracy of the text, and any mistakes, omissions, or errors are our responsibility. The finished product is the culmination of many years of work during which we received numerous comments and advice from many geologists who reviewed all or parts of the previous editions. They are: Lawrence Balthaser, *California Polytechnic State University, San Luis Obispo*; Mary Lou Bevier, *University of British Columbia*; Richard Bonnett, *Marshall University*; Dale H. Easley, *University of New Orleans*; Terry Engelder, *Pennsylvania State University*; Normal Fox, *Hocking Technical College*; Joan Fryxell, *California State University, San Bernardino*; David Gibson, *University of Maine*; Fredric R. Goldstein, *Trenton State College*; Brian Grant, *Brock University*; Patrick Hicks, *La Grange College*; William F. Kean, *University of Wisconsin, Milwaukee*, Paul Morgan, *Northern Arizona University*; Susan Morgan, *Utah State University*; Louis Pinto, *Monroe Community College*; Robert Reynolds, *Central Oregon Community College*; John Ritter, *Wittenberg University*; Robert J. Smith, *Seattle University*; James L. Talbot, *Western Washington University*; Dale Valentino, *State University of New York, Oswego*; William H. Wright, *Sonoma State University*.

We wish to express our sincere appreciation to the reviewers who reviewed the Fourth Edition. They are: Michael Handke, *University of Kentucky*; Brian Kirchner, *Henry Ford Community College*; Ronald Endris, *Indiana University Southeast*; Xenia V. Conquy, *Broward Community College*; John Dassinger, *Chandler-Gilbert Community College*; John Hickey, *Inver Hills Community College*; Eric R. Swanson, *University of Texas, San Antonio*. We also want to extend our gratitude to the reviewers of the physical chapters of the Fourth Edition of *The Changing Earth* and made many helpful and useful comments that led to the many improvements seen in this Fifth Edition. They are: David Berry, California State Polytechnic University, Pomona; Wesley A. Brown, Stephen F. Austin State University; David Cordero, Lower Columbia College; Kathleen Devaney, El Paso Community College; Yongli Gao, East Tennessee State University; Jorg Maletz, University at Buffalo-SUNY; Kevin McCartney, University of Maine - Presque Isle; Thomas J. Weiland, Georgia Southwestern State University; and William N. Mode, University of Wisconsin Oshkosh.

We also wish to thank Kathy Benison, Richard V. Dietrich (Professor Emeritus), David J. Matty, Jane M. Matty, Wayne E. Moore (Professor Emeritus), and Sven Morgan of the Geology Department, and Bruce M. C. Pape (Emeritus) of the Geography Department of Central Michigan University, as well as Eric Johnson (Hartwick College, New York), and Stephen D. Stahl (St. Bonaventure, New York) for providing us with photographs and answering our questions concerning various topics. We are also grateful for the generosity of the various agencies and individuals from many countries who provided photographs.

Special thanks must go to Peter Adams, Executive Editor Health/Nutrition/Earth Sciences at Brooks/Cole, Cengage Learning, who initiated this fifth edition, and to our Development Editor, Amy Collins of WriteWorks, who edited and managed the content for this edition. We are indebted to our production manager at Pre-Press PMG, Kevin Shea, for his attention to detail and consistency. We would also like to thank Lisa Buckley for the fresh design, and

Patricia A. Onufrak for her copyediting skills. We thank Parvinder Sethi for his help in locating appropriate photographs. We would also like to recognize Hal Humphrey, Cengage Learning Production Project Manager; Melinda Newfarmer and Alexandria Brady, Technology Project Managers, for developing the media program; Joe Rogove, Executive Marketing Manager; Talia Wise and Belinda Krohmer, Marketing Communications Managers; and Liana Monari, Assistant Editor, who managed the accompanying ancillary package. We also extend thanks to the artists at Pre-Press PMG and Precision Graphics, who are responsible for updating much of the art program.

As always, our families were very patient and encouraging when much of our spare time and energy were devoted to this book. We again thank them for their continued support and understanding.

Reed Wicander
James S. Monroe

Essentials of Physical Geology

CHAPTER 1

Understanding Earth: A Dynamic and Evolving Planet

OUTLINE

OBJECTIVES

At the end of this chapter, you will have learned that

- Geology is the study of Earth.
- Earth is a complex, integrated system of interconnected components that interact and affect one another in various ways.
- Theories are based on the scientific method and can be tested by observation and/or experiment.
- Geology plays an important role in the human experience and affects us as individuals and members of society and nation-states.
- The universe is thought to have originated approximately 14 billion years ago with a big bang. The solar system and

Satellite-based image of Earth. North America is visible in the center of this view, as well as Central America and South America. The present locations of continents and ocean basins are the result of plate movements. The interaction of plates through time has affected the physical and biological history of Earth.

planets evolved from a turbulent, rotating cloud of material surrounding the embryonic Sun.

- Earth consists of three concentric layers—core, mantle, and crust—and this orderly division formed during Earth's early history.
- Plate tectonics is the unifying theory of geology and this theory revolutionized the science.
- The rock cycle illustrates the interrelationships between Earth's internal and external processes and shows how and why the three major rock groups are related.
- The theory of organic evolution provides the conceptual framework for understanding the history of life.
- An appreciation of geologic time and the principle of uniformitarianism is central to understanding the evolution of Earth and its biota.
- Geology is an integral part of our lives.

INTRODUCTION

A major benefit of the space age has been the ability to look back from space and view our planet in its entirety. Every astronaut has remarked in one way or another on how Earth stands out as an inviting oasis in the otherwise black void of space (see this chapter's opening photograph). We are able to see not only the beauty of our planet, but also its fragility. We can also decipher Earth's long and frequently turbulent history by reading the clues preserved in the geologic record.

A major theme of this book is that Earth is a complex, dynamic planet that has changed continuously since its origin some 4.6 billion years ago. These changes and the present-day features we observe result from the interactions among Earth's internal and external systems, subsystems, and cycles. Earth is unique among the planets of our solar system in that it supports life and has oceans of water, a hospitable atmosphere, and a variety of climates. It is ideally suited for life as we know it because of a combination of factors, including its distance from the Sun and the evolution of its interior, crust, oceans, and atmosphere. Life processes have, over time, influenced the evolution of Earth's atmosphere, oceans, and, to some extent, its crust. In turn, these physical changes have affected the evolution of life.

By viewing Earth as a whole—that is, thinking of it as a system—we not only see how its various components are interconnected, but also better appreciate its complex and dynamic nature. The system concept makes it easier for us to study a complex subject such as Earth because it divides the whole into smaller components that we can easily understand, without losing sight of how the components fit together as a whole.

A **system** is a combination of related parts that interact in an organized manner. An automobile is a good example of a system. Its various components or subsystems, such as the engine, transmission, steering, and brakes, are all interconnected in such a way that a change in any one of them affects the others.

We can examine Earth in the same way we view an automobile—that is, as a system of interconnected components that interact and affect each other in many ways. The principal subsystems of Earth are the *atmosphere, biosphere, hydrosphere, lithosphere, mantle*, and *core* (Figure 1.1). The complex interactions among these subsystems result in a dynamically changing planet in which matter and energy is continuously recycled into different forms (Table 1.1). For example, the movement of plates has profoundly affected the formation of landscapes, the distribution of mineral resources, and atmospheric and oceanic circulation patterns, which, in turn, have affected global climate changes.

We must also not forget that humans are part of the Earth system, and our activities can produce changes with potentially wide-ranging consequences. When people discuss and debate such environmental issues as acid rain, the greenhouse effect and global warming, and the depleted ozone layer, it is important to remember that these are not isolated issues, but are part of the larger Earth system. Furthermore, remember that Earth goes through time cycles that are much longer than humans are used to. Although they may have disastrous short-term effects on the human species, global warming and cooling are also part of a longer-term cycle that has resulted in many glacial advances and retreats during the past 1.8 million years.

Accordingly, we must understand that actions we take can produce changes with wide-ranging consequences that we might not initially be aware of. For this reason, an understanding of geology, and science in general, is of paramount importance. If the human species is to survive, we must understand how the various Earth systems work and interact and, more importantly, how our actions affect the delicate balance between these systems.

As you study the various topics covered in this book, keep in mind the themes discussed in this chapter and how, like the parts of a system, they are interrelated. By relating each chapter's topic to its place in the entire Earth system, you will gain a greater appreciation of why geology is so integral to our lives.

WHAT IS GEOLOGY?

Geology, from the Greek *geo* and *logos*, is defined as the study of Earth, but now must also include the study of the planets and moons in our solar system. It is generally divided into two broad areas—physical geology and historical geology. *Physical geology* is the study of Earth materials, such as minerals and rocks, as well as the processes operating within Earth and on its surface. *Historical geology* examines the origin and evolution of Earth, its continents, oceans, atmosphere, and life.

The discipline of geology is so broad that it is subdivided into numerous fields or specialties. Table 1.2 shows many of the diverse fields of geology and their relationship to the sciences of astronomy, biology, chemistry, and physics.

Nearly every aspect of geology has some economic or environmental relevance. Many geologists are involved in

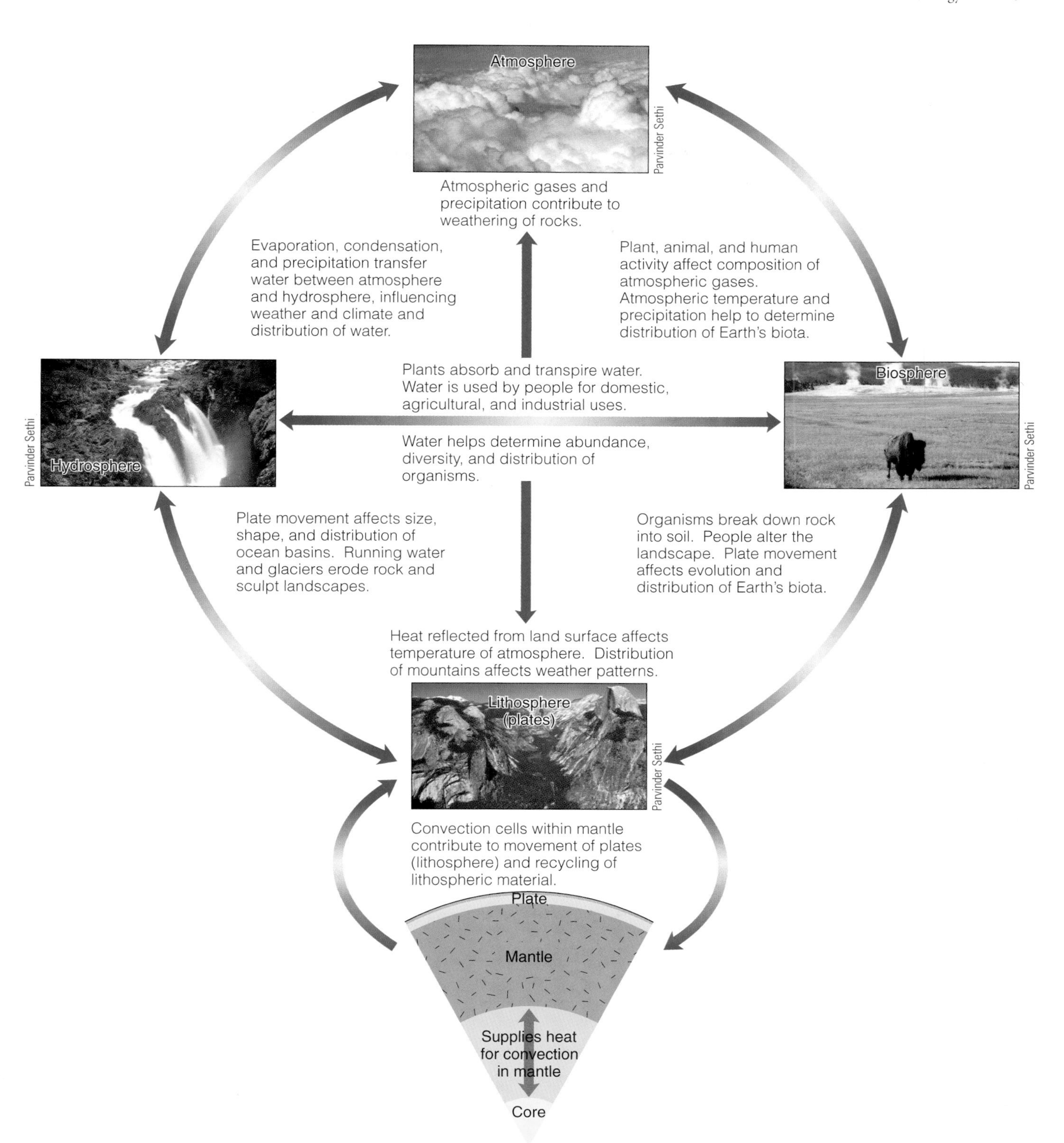

Figure 1.1 Subsystems of Earth The atmosphere, hydrosphere, biosphere, lithosphere, mantle, and core are all subsystems of Earth. This simplified diagram shows how these subsystems interact, with some examples of how materials and energy are cycled throughout the Earth system. The interactions between these subsystems make Earth a dynamic planet that has evolved and changed since its origin 4.6 billion years ago.

TABLE 1.1 Interactions Among Earth's Principal Subsystems

	Atmosphere	Hydrosphere	Biosphere	Lithosphere
Atmosphere	Interaction among various air masses	Surface currents driven by wind; evaporation	Gases for respiration; dispersal of spores, pollen, and seeds by wind	Weathering by wind erosion; transport of water vapor for precipitation of rain and snow
Hydrosphere	Input of water vapor and stored solar heat	Hydrologic cycle	Water for life	Precipitation; weathering and erosion
Biosphere	Gases from respiration	Removal of dissolved materials by organisms	Global ecosystems; food cycles	Modification of weathering and erosion processes; formation of soil
Lithosphere	Input of stored solar heat; landscapes affect air movements	Source of solid and dissolved materials	Source of mineral nutrients; modification of ecosystems by plate movements	Plate tectonics

exploration for mineral and energy resources, using their specialized knowledge to locate the natural resources on which our industrialized society is based. As the demand for these nonrenewable resources increases, geologists apply the basic principles of geology in increasingly sophisticated ways to focus their attention on areas that have a high potential for economic success.

Whereas some geologists work on locating mineral and energy resources, other geologists use their expertise to help solve environmental problems. Finding adequate sources of groundwater for the ever-burgeoning needs of communities and industries is becoming increasingly important, as is the monitoring of surface and underground water pollution and its cleanup. Geologic engineers help find safe locations for dams, waste-disposal sites, and power plants, as well as designing earthquake-resistant buildings.

Geologists are also engaged in making short- and long-range predictions about earthquakes and volcanic eruptions, and the potential destruction that may result. Following the tragic events in Indonesia in 2004, geologists are now more involved than ever in working with various governmental agencies and civil defense planners to ensure that timely warnings are given to potentially affected regions when natural disasters such as tsunami occur, and that contingency plans are in place.

GEOLOGY AND THE FORMULATION OF THEORIES

The term **theory** has various meanings. In colloquial usage, it means a speculative or conjectural view of something—hence, the widespread belief that scientific theories are little more than unsubstantiated wild guesses. In scientific usage, however, a theory is a coherent explanation for one or several related natural phenomena supported by a large body of objective evidence. From a theory, scientists derive predictive statements that can be tested by observations and/or experiments so that their validity can be assessed. The law of universal gravitation is an example of a theory that describes the attraction between masses (an apple and Earth in the popularized account of Newton and his discovery).

TABLE 1.2 Specialties of Geology and Their Broad Relationship to the Other Sciences

Specialty	Area of Study	Related Science
Geochronology	Time and history of Earth	Astronomy
Planetary geology	Geology of the planets	
Paleontology	Fossils	Biology
Economic geology	Mineral and energy resources	
Environmental geology	Environment	
Geochemistry	Chemistry of Earth	Chemistry
Hydrogeology	Water resources	
Mineralogy	Minerals	
Petrology	Rocks	
Geophysics	Earth's interior	Physics
Structural geology	Rock deformation	
Seismology	Earthquakes	
Geomorphology	Landforms	
Oceanography	Oceans	
Paleogeography	Ancient geographic features and locations	
Stratigraphy/sedimentology	Layered rocks and sediments	

Theories are formulated through the process known as the **scientific method**. This method is an orderly, logical approach that involves gathering and analyzing facts or data about the problem under consideration. Tentative explanations, or **hypotheses**, are then formulated to explain the observed phenomena. Next, the hypotheses are tested to see whether what was predicted actually occurs in a given situation. Finally, if one of the hypotheses is found, after repeated tests, to explain the phenomena, then the hypothesis is proposed as a theory. Remember, however, that in science, even a theory is still subject to further testing and refinement as new data become available.

The fact that a scientific theory can be tested and is subject to such testing separates it from other forms of human inquiry. Because scientific theories can be tested, they have the potential for being supported or even proven wrong. Accordingly, science must proceed without any appeal to beliefs or supernatural explanations, not because such beliefs or explanations are necessarily untrue, but because we have no way to investigate them. For this reason, science makes no claim about the existence or nonexistence of a supernatural or spiritual realm.

Each scientific discipline has certain theories that are of particular importance. In geology, the formulation of plate tectonic theory has changed the way geologists view Earth. Geologists now view Earth from a global perspective in which all of its subsystems and cycles are interconnected, and Earth history is seen to be a continuum of interrelated events that are part of a global pattern of change.

Figure 1.2 Geology and Art *Kindred Spirits* by Asher Brown Durand (1849) realistically depicts the layered rocks along gorges in the Catskill Mountains of New York State. Durand was one of numerous artists of the 19th-century Hudson River School, which was known for realistic landscapes. This painting shows Durand conversing with the recently deceased Thomas Cole, the original founding force of the Hudson River School.

HOW DOES GEOLOGY RELATE TO THE HUMAN EXPERIENCE?

You would probably be surprised at the extent to which geology pervades our everyday lives and the numerous references to geology in the arts, music, and literature. Many sketches and paintings depict rocks and landscapes realistically. Leonardo da Vinci's *Virgin of the Rocks* and *Virgin and Child with Saint Anne*, Giovanni Bellini's *Saint Francis in Ecstasy* and *Saint Jerome*, and Asher Brown Durand's *Kindred Spirits* (Figure 1.2) are just a few examples by famous painters.

In the field of music, Ferde Grofé's *Grand Canyon Suite* was no doubt inspired by the grandeur and timelessness of Arizona's Grand Canyon and its vast rock exposures. The rocks on the Island of Staffa in the Inner Hebrides provided the inspiration for Felix Mendelssohn's famous *Hebrides Overture*.

References to geology abound in *The German Legends of the Brothers Grimm*. Jules Verne's *Journey to the Center of the Earth* describes an expedition into Earth's interior. There is even a series of mystery books by Sarah Andrews that features the fictional geologist Em Hansen, who uses her knowledge of geology to solve crimes. On one level, the poem "Ozymandias" by Percy B. Shelley deals with the fact that nothing lasts forever and even solid rock eventually disintegrates under the ravages of time and weathering. Even comics contain references to geology. One of the best known is *The Far Side* by Gary Larson, which, over the years, has had many cartoons with a geological theme.

Geology has also played an important role in the history and culture of humankind. Empires throughout history have risen and fallen on the distribution and exploitation of natural resources. Wars have been fought for the control of such natural resources as oil and gas, and valuable minerals such as gold, silver, and diamonds. The configuration of Earth's surface, or its topography, which is shaped by geologic agents, played a critical role in military tactics. For example, Napoleon included two geologists in his expeditionary forces when he invaded Egypt in 1798, and the Russians used geologists as advisors in selecting fortification sites during the Russo-Japanese war of 1904–1905. Natural barriers such as mountain ranges and rivers have frequently served as political boundaries, and the shifting of river channels has sparked numerous border disputes. Deserts, which most people think of as inhospitable areas, have been the home to many people, such as the Bedouin, throughout history.

HOW DOES GEOLOGY AFFECT OUR EVERYDAY LIVES?

The most obvious connection between geology and our everyday lives is when natural disasters strike. Less apparent,

What Would You Do?

Because of budget shortfalls and in an effort to save money, your local school board is considering eliminating the required geology course that all students must pass to graduate from high school. As a concerned parent and citizen, you have rallied community support to keep geology in the curriculum, and you will be making your presentation to the school board next week. What will be your arguments to keep geology as part of the basic body of knowledge that all students should have when they graduate from high school? Why should students have a working knowledge of geology? What arguments would you expect from the school board in favor of eliminating geology from the curriculum and how will you counter their reasons?

but equally significant, are the connections between geology and economic, social, and political issues. Whereas most readers of this book will not become professional geologists, everyone should have a basic understanding of the geologic processes that ultimately affect all of us.

Natural Events

Destructive events such as volcanic eruptions, earthquakes, landslides, tsunami, floods, and droughts make headlines and affect many people in obvious ways. Although we cannot prevent most of these natural disasters from happening, the more we learn about what causes them, the better we will be able to predict and mitigate the severity of their impact.

Economics and Politics

Equally important, but not always as well understood or appreciated, is the connection between geology and economic and political power. Mineral and energy resources are not equally distributed and no country is self-sufficient in all of them. Throughout history, people have fought wars to secure these resources. The United States was involved in the 1990–1991 Gulf War largely because it needed to protect its oil interests in that region. Many foreign policies and treaties develop from the need to acquire and maintain adequate supplies of mineral and energy resources.

Our Role as Decision Makers

You may become involved in geologic decisions in various ways—for instance, as a member of a planning board or as a property owner with mineral rights. In such cases, you must have a basic knowledge of geology to make informed decisions. Many professionals must also deal with geologic issues as part of their jobs. Lawyers, for example, are becoming more involved in issues ranging from ownership of natural resources to how development activities affect the environment. As government plays a greater role in environmental issues and regulations, members of Congress have increased the number of staff devoted to studying issues related to the environment and geology.

Consumers and Citizens

If issues such as nonrenewable energy resources, waste disposal, and pollution seem simply too far removed or too complex to be fully appreciated, consider for a moment just how dependent we are on geology in our daily routines (Figure 1.3).

Much of the electricity for our appliances comes from the burning of coal, oil, natural gas, or uranium consumed in nuclear-generating plants. It is geologists who locate the coal, petroleum (oil and natural gas), and uranium. The copper or other metal wires through which electricity travels are manufactured from materials found as the result of mineral exploration. The concrete foundation (concrete is a mixture of clay, sand, or gravel, and limestone), drywall (made largely from the mineral gypsum), and windows (the mineral quartz is the principal ingredient in the manufacture of glass) of the buildings we live and work in owe their very existence to geologic resources.

When we go to work, the car or public transportation we use is powered and lubricated by some type of petroleum byproduct and is constructed of metal alloys and plastics. And the roads or rails we ride over come from geologic materials, such as gravel, asphalt, concrete, or steel. All of these items are the result of processing geologic resources.

As individuals and societies, we enjoy a standard of living that is obviously directly dependent on the consumption of geologic materials. We therefore need to be aware of how our use and misuse of geologic resources may affect the environment, and develop policies that not only encourage management of our natural resources, but also allow for continuing economic

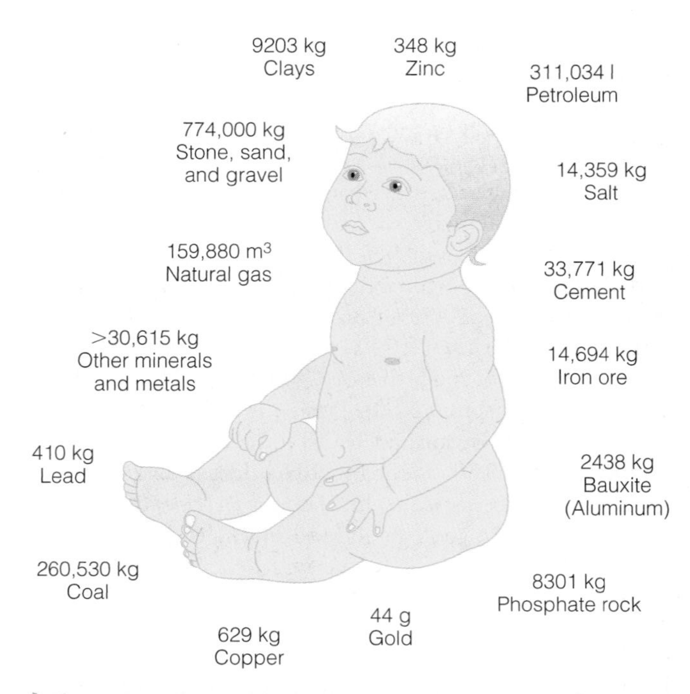

Figure 1.3 Lifetime Mineral Usage According to the Mineral Information Institute in Golden, Colorado, the average American born in 2006 has a life expectancy of 77.8 years and will need 1,672,393 kg of minerals, metals, and fuels to sustain his or her standard of living over a lifetime. That is an average of 21,496 kg of mineral and energy resources per year for every man, woman, and child in the United States.

development among all the world's nations. Geologists will continue to play an important role in meeting these demands by locating the needed resources and ensuring protection of the environment for the benefit of future generations.

GLOBAL GEOLOGIC AND ENVIRONMENTAL ISSUES FACING HUMANKIND

Most scientists would argue that overpopulation is the greatest environmental problem facing the world today. The world's population reached 6.7 billion in 2007, and projections indicate that this number will grow by at least another billion people during the next two decades, bringing Earth's human population to more than 7.7 billion. Although this may not seem to be a geologic problem, remember that these people must be fed, housed, and clothed, and all with a minimal impact on the environment. Much of this population growth will be in areas that are already at risk from such hazards as earthquakes, tsunami, volcanic eruptions, and floods. Adequate water supplies must be found and protected from pollution. Additional energy resources will be needed to help fuel the economies of nations with ever-increasing populations. New techniques must be developed to reduce the use of our dwindling nonrenewable resource base and to increase our recycling efforts so that we can decrease our dependence on new sources of these materials.

The problems of overpopulation and how it affects the global ecosystem vary from country to country. For many poor and non-industrialized countries, the problem is too many people and not enough food. For the more developed and industrialized countries, it is too many people rapidly depleting both the nonrenewable and renewable natural resource base.

What Would You Do?

An important environmental issue facing the world today is global warming. How can this problem be approached from a global systems perspective? What are the possible consequences of global warming, and can we really do anything about it? Are there ways to tell whether global warming occurred in the geologic past?

And in the most industrially developed countries, it is people producing more pollutants than the environment can safely recycle on a human timescale. The common thread tying these varied situations together is an environmental imbalance created by a human population exceeding Earth's short-term carrying capacity.

An excellent example of how Earth's various subsystems are interrelated is the relationship between the greenhouse effect and global warming (see Geo-Focus on pages 10 and 11). As a by-product of respiration and the burning of organic material, carbon dioxide is a component of the global ecosystem and is constantly being recycled as part of the carbon cycle. The concern in recent years over the increase in atmospheric carbon dioxide levels is related to its role in the greenhouse effect.

The recycling of carbon dioxide between Earth's crust and atmosphere is an important climate regulator because carbon dioxide and other gases, such as methane, nitrous oxide, chlorofluorocarbons, and water vapor, allow sunlight to pass through them, but trap the heat reflected back from Earth's surface. This retention of heat is called the *greenhouse effect.* It results in an increase in the temperature of Earth's surface and, more importantly, its atmosphere, thus producing global warming (Figure 1.4).

Figure 1.4 The Greenhouse Effect and Global Warming

a Short-wavelength radiation from the Sun that is not reflected back into space penetrates the atmosphere and warms Earth's surface.

b Earth's surface radiates heat in the form of long–wavelength radiation back into the atmosphere, where some of it escapes into space. The rest is absorbed by greenhouse gases and water vapor and reradiated back toward Earth.

c Increased concentrations of greenhouse gases trap more heat near Earth's surface, causing a general increase in surface and atmospheric temperatures, which leads to global warming.

Geo-Focus — Global Warming and Climate Change, and How They Affect You

The greenhouse effect, global warming, climate change—these headlines and topics are in the news all the time and are global issues that affect us all and the planet we live on. But just how will global warming and the resultant climate change personally affect you? Are they something that you should really be concerned about? After all, there are exams to worry about, graduation, finding a job, and that doesn't even include the everyday issues we all must deal with, not to mention your personal life. Yet, part of the college experience is examining and debating the "big picture" and issues facing society today. So what about global warming and you?

You may recall we talked about the greenhouse effect and its relationship to global warming. The greenhouse effect helps regulate Earth's temperature because as sunlight passes through the atmosphere, some of the heat is trapped in the lower atmosphere and not radiated back into space, thus effectively warming Earth's surface and atmosphere. The issue is not whether we have a greenhouse effect, because we do, but rather the degree to which human activity, such as the burning of fossil fuels, is increasing the greenhouse effect, and thus contributing to global warming.

Based on many studies using a variety of techniques, it is clear that carbon dioxide levels have increased since the Industrial Revolution in the 19th century. (Carbon dioxide is one of the greenhouse gases that allows short-wavelength solar radiation to pass through it, but traps some of the long-wavelength radiation reflected back from Earth's surface.) Furthermore, global surface temperatures have increased about 0.6°C since the late 1800s, and about 0.4°C during the past 25 years. However, this warming trend has not been globally uniform and some areas, such as the southwestern part of the United States, have actually cooled during this time period.

One thing we must be careful about is mistaking regional trends for global

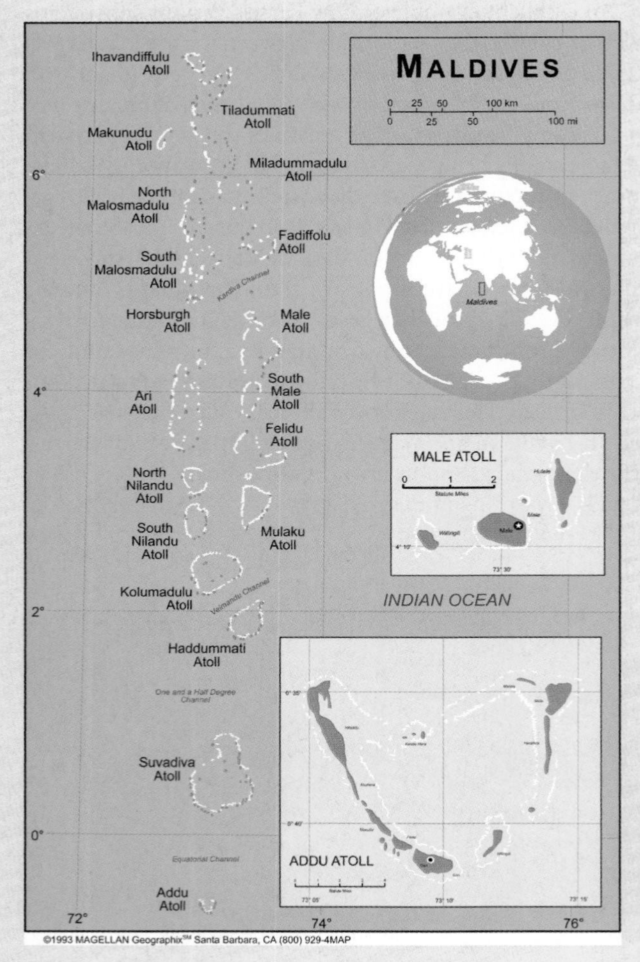

a Map showing the location of the Maldives Islands in the Indian Ocean.

b Aerial view of the Maldives Islands.

Figure 1 A rise in sea level due to global warming could easily submerge the Maldives islands

With industrialization and its accompanying burning of tremendous amounts of fossil fuels, carbon dioxide levels in the atmosphere have been steadily increasing since about 1880, causing many scientists to conclude that a global warming trend has already begun and will result in severe global climate shifts. Most computer models based on the current rate of increase in greenhouse gases show Earth warming by as much as 5°C during the next hundred years. Such a temperature change will be uneven, however, with the greatest warming occurring in the higher latitudes. As a

Parvinder Sethi

Figure 2 Withered corn due to drought conditions.

trends. For example, there is compelling evidence of climate variability or extremes on a regional scale, but on a global scale, there is currently little evidence of a sustained trend in climate variability or extremes. That doesn't mean that we can ignore the overall increase in average global surface temperatures, because if left unchecked, such changes can have significant environmental, ecological, and economic effects.

So what are some of the effects we should be worried about? For starters, there is the problem of rising sea level. During the past 100 years, global sea level has been rising at an average rate of 1 to 2 mm per year, and the projected rate of increase by 2100 is anywhere from 9 to 88 cm, depending on which climate model is used. What this means is that low-lying coastal areas will experience flooding and increased erosion along the coastline, endangering housing and communities. For instance, about 17 million people live less than one meter above sea level in Bangladesh, and are at risk due to rising sea levels. Furthermore, many major cities are just above sea level, and could also suffer from rising sea levels. A rise in sea level at the upper end could completely submerge such island nations as the Maldives in the Indian Ocean (Figure 1).

Based on various climatic computer models and taking into account the complexity and variability of the atmospheric–oceanic system, most predictions show Earth's average surface temperature increasing by 1.4 to 5.8°C during the period of 1990–2100. This will result in widely varying regional responses, such that land areas will warm more and faster than ocean areas, particularly in the high latitudes of the northern hemisphere. Expect to see more hot days and heat waves over nearly all the land areas, with increasing drought frequency within all continental interiors during the summer (Figure 2). There will also be increased precipitation during the 21st century, particularly in the northern middle to high latitudes. And expect to see a continuation of the retreat of glaciers and ice caps, with a decrease in northern hemisphere snow cover and sea ice.

What do all these predictions mean to you? With increasingly hotter summers and greater frequency of drought, expect to see higher food prices as crop yields decrease. There will also be an increased risk of wildfires, and deadly heat waves will result in more heat-related deaths, such as occurred in Europe during the summer of 2003.

As climates change, diseases such as malaria are easily spreading into areas of warmer, wetter climates. Disease-carrying mosquitoes are expanding their reach as climate changes allow them to survive in formerly inhospitable regions.

Higher temperatures will affect regional water supplies, creating potential water crises in the western United States within the next 20 years, as well as other areas such as Peru and western China. Just as many regions will experience longer and hotter summers, other areas will suffer from intense and increased rainfall, which will result in severe flooding and landslides.

Everyone is vulnerable to weather-related disasters; however, large-scale changes brought about by climate change will impact people in poor countries more than those in the more industrialized countries. However, whether these climate changes are part of a natural global cycle taking place over thousands or hundreds of thousands of years—that is, on a geological time scale—or are driven, in part, by human activities, is immaterial. The bottom line is that we already are, or eventually will be, affected in some way, be it economic or social, by the climate changes that are taking place.

consequence of this warming, rainfall patterns will shift dramatically, which will have a major effect on the largest grain-producing areas of the world, such as the American Midwest. Drier and hotter conditions will intensify the severity and frequency of droughts, leading to increased crop failure and higher food prices. With such shifts in climate, Earth's deserts may expand, with a resulting decrease in the amount of valuable crop and grazing lands.

Continued global warming will result in a rise in mean sea level as icecaps and glaciers melt and contribute their water to the world's oceans. It is predicted that at the current rate of glacial melting, sea level will rise 21 cm around 2050, thus increasing the number of people at risk from flooding in coastal areas by approximately 20 million!

We would be remiss, however, if we did not point out that many other scientists are not convinced that the global warming trend is the direct result of increased human activity related to industrialization. They point out that although the level of greenhouse gases has increased, we are still uncertain about their rate of generation and rate of removal, and whether the rise in global temperatures during the past century resulted from normal climatic variations through time or from human activity. Furthermore, these scientists point out that even if there is a general global warming during the next hundred years, it is not certain that the dire predictions made by proponents of global warming will come true.

Earth, as we know, is a remarkably complex system, with many feedback mechanisms and interconnections throughout its various subsystems and cycles. It is very difficult to predict all of the consequences that global warming would have for atmospheric and oceanic circulation patterns and its ultimate effect on Earth's biota.

ORIGIN OF THE UNIVERSE AND SOLAR SYSTEM, AND EARTH'S PLACE IN THEM

How did the universe begin? What has been its history? What is its eventual fate, or is it infinite? These are just some of the basic questions people have asked and wondered about since they first looked into the nighttime sky and saw the vastness of the universe beyond Earth.

Origin of the Universe—Did It Begin with a Big Bang?

Most scientists think that the universe originated about 14 billion years ago in what is popularly called the **Big Bang**. The Big Bang is a model for the evolution of the universe in which a dense, hot state was followed by expansion, cooling, and a less dense state.

According to modern *cosmology* (the study of the origin, evolution, and nature of the universe), the universe has no edge and therefore no center. Thus, when the universe began, all matter and energy were compressed into an infinitely small high-temperature and high-density state in which both time and space were set at zero. Therefore, there is no "before the Big Bang," only what occurred after it. As demonstrated by Einstein's theory of relativity, space and time are unalterably linked to form a space–time continuum, that is, without space, there can be no time.

How do we know that the Big Bang took place approximately 14 billion years ago? Why couldn't the universe have always existed as we know it today? Two fundamental phenomena indicate that the Big Bang occurred. First, the universe is expanding, and second, it is permeated by background radiation.

When astronomers look beyond our own solar system, they observe that everywhere in the universe galaxies are moving away from each other at tremendous speeds. Edwin Hubble first recognized this phenomenon in 1929. By measuring the optical spectra of distant galaxies, Hubble noted that the velocity at which a galaxy moves away from Earth increases proportionally to its distance from Earth. He observed that the spectral lines (wavelengths of light) of the galaxies are shifted toward the red end of the spectrum; that is, the lines are shifted toward longer wavelengths. Galaxies receding from each other at tremendous speeds would produce such a redshift. This is an example of the *Doppler effect,* which is a change in the frequency of a sound, light, or other wave caused by movement of the wave's source relative to the observer (▶ Figure 1.5).

One way to envision how velocity increases with increasing distance is by reference to the popular analogy of a rising loaf of raisin bread in which the raisins are uniformly distributed throughout the loaf (▶ Figure 1.6). As the dough rises, the raisins are uniformly pushed away from each other at velocities directly proportional to the distance between any two raisins. The farther away a given raisin is to begin with, the farther it must move to maintain the regular spacing during the expansion, and hence the greater its velocity must be. In the same way that raisins move apart in a rising loaf of bread, galaxies are receding from each other at a rate proportional to the distance between them, which is exactly what astronomers see when they observe the universe. By measuring this expansion rate, astronomers can calculate how long ago the galaxies were all together at a single point, which turns out to be about 14 billion years, the currently accepted age of the universe.

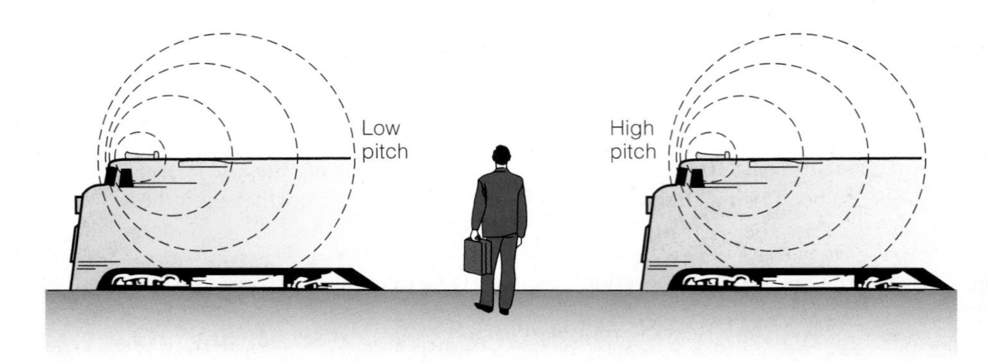

▶ **Figure 1.5 The Doppler Effect** The sound waves of an approaching whistle are slightly compressed so that the individual hears a shorter-wavelength, higher-pitched sound. As the whistle passes and recedes from the individual, the sound waves are slightly spread out, and a longer-wavelength, lower-pitched sound is heard.

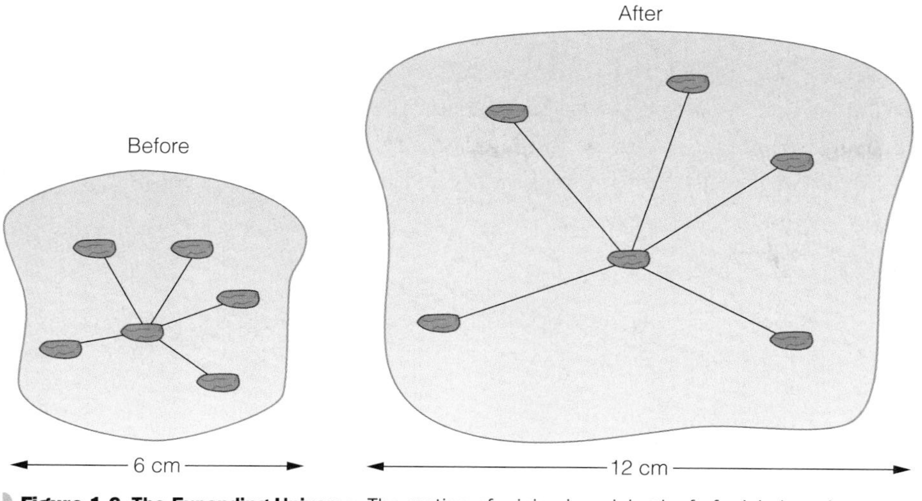

Figure 1.6 The Expanding Universe The motion of raisins in a rising loaf of raisin bread illustrates the relationship that exists between distance and speed and is analogous to an expanding universe. In this diagram, adjacent raisins are located 2 cm apart before the loaf rises. After one hour, any raisin is now 4 cm away from its nearest neighbor and 8 cm away from the next raisin over, and so on. Therefore, from the perspective of any raisin, its nearest neighbor has moved away from it at a speed of 2 cm per hour, and the next raisin over has moved away from it at a speed of 4 cm per hour. In the same way that raisins move apart in a rising loaf of bread, galaxies are receding from each other at a rate proportional to the distance between them.

Arno Penzias and Robert Wilson of Bell Telephone Laboratories made the second important observation that provided evidence of the Big Bang in 1965. They discovered that there is a pervasive background radiation of 2.7 Kelvin (K) above absolute zero (absolute zero equals –273°C; 2.7 K = –270.3°C) everywhere in the universe. This background radiation is thought to be the fading afterglow of the Big Bang.

Currently, cosmologists cannot say what it was like at time zero of the Big Bang because they do not understand the physics of matter and energy under such extreme conditions. However, it is thought that during the first second following the Big Bang, the four basic forces—*gravity* (the attraction of one body toward another), *electromagnetic force* (combines electricity and magnetism into one force and binds atoms into molecules), *strong nuclear force* (binds protons and neutrons together), and *weak nuclear force* (responsible for the breakdown of an atom's nucleus, producing radioactive decay)—separated and the universe experienced enormous expansion. By the end of the first three minutes following the Big Bang, the universe was cool enough that almost all nuclear reactions had ceased, and by the time it was 30 minutes old nuclear reactions had completely ended and the universe's mass consisted almost entirely of hydrogen and helium nuclei.

As the universe continued expanding and cooling, stars and galaxies began to form and the chemical makeup of the universe changed. Initially, the universe was 100% hydrogen and helium, whereas today it is 98% hydrogen and helium and 2% all other elements by weight. How did such a change in the universe's composition occur? Throughout their life cycle, stars undergo many nuclear reactions in which lighter elements are converted into heavier elements by nuclear fusion. When a star dies, often explosively, the heavier elements that were formed in its core are returned to interstellar space and are available for inclusion in new stars. In this way, the composition of the universe is gradually enhanced by heavier elements.

Our Solar System—Its Origin and Evolution

Our solar system, which is part of the Milky Way galaxy, consists of the Sun, eight planets, one dwarf planet (Pluto), 101 known moons or satellites (although this number keeps changing with the discovery of new moons and satellites surrounding the Jovian planets), a tremendous number of asteroids—most of which orbit the Sun in a zone between Mars and Jupiter—and millions of comets and meteorites, as well as interplanetary dust and gases (Figure 1.7). Any theory formulated to explain the origin and evolution of our solar system must therefore take into account its various features and characteristics.

Many scientific theories for the origin of the solar system have been proposed, modified, and discarded since the French scientist and philosopher René Descartes first proposed, in 1644, that the solar system formed from a gigantic whirlpool within a universal fluid. Today, the **solar nebula theory** for the origin of our solar system involves the condensation and collapse of interstellar material in a spiral arm of the Milky Way galaxy (Figure 1.8).

The collapse of this cloud of gases and small grains into a counterclockwise-rotating disk concentrated about 90% of the material in the central part of the disk and formed an embryonic Sun, around which swirled a rotating cloud of material called a *solar nebula*. Within this solar nebula were localized eddies in which gases and solid particles condensed. During the condensation process, gaseous, liquid, and solid particles began to accrete into ever-larger masses called *planetesimals*, which collided and grew in size and mass until they eventually became planets.

The composition and evolutionary history of the planets are a consequence, in part, of their distance from the Sun (see Geo-inSight on pages 14 and 15). The **terrestrial planets**—Mercury, Venus, Earth, and Mars—so named because they are similar to *terra*, Latin for "earth," are all small and composed of rock and metallic elements that condensed at the high temperatures of the inner nebula. The **Jovian planets**—Jupiter, Saturn, Uranus, and Neptune—so named because they resemble Jupiter (the Roman god was also called *Jove*), all have small rocky cores compared to their overall size, and are composed mostly of hydrogen, helium, ammonia, and methane, which condense at low temperatures.

Geo-inSight The Terrestrial and Jovian Planets

The planets of our solar system can be divided into two major groups that are quite different, indicating that the two underwent very different evolutionary histories. The four inner planets-Mercury, Venus, Earth, and Mars—are the terrestrial planets; they are small and dense (composed of a metallic core and silicate mantle-crust), ranging from no atmosphere (Mercury) to an oppressively thick one (Venus).

The outer four planets (Pluto, is now considered a dwarf Planet)—Jupiter, Saturn, Uranus, and Neptune—are the Jovian planets; they are large, ringed, low-density planets with liquid interiors cores surrounded by thick atmospheres.

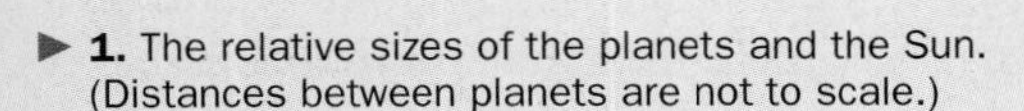

► **1.** The relative sizes of the planets and the Sun. (Distances between planets are not to scale.)

► **2. Venus** is surrounded by an oppressively thick atmosphere that completely obscures its surface. However, radar images from orbiting spacecraft reveal a wide variety of terrains, including volcanic features, folded mountain ranges, and a complex network of faults.

▲ **3.** The **Moon** is one-fourth the diameter of Earth, has a low density relative to the terrestrial planets, and is extremely dry. Its surface is divided into low-lying dark-colored plains and light-colored highlands that are heavily cratered, attesting to a period of massive meteorite bombardment in our solar system more than 4 billion years ago. The hypothesis that best accounts for the origin of the Moon has a giant planetesimal, the size of Mars or larger, crashing into Earth 4.6 to 4.4 billion years ago, causing ejection of a large quantity of hot material that cooled and formed the Moon.

◄ **4. Mercury** has a heavily cratered surface that has changed very little since its early history. Because Mercury is so small, its gravitational attraction is insufficient to retain atmospheric gases; any atmosphere that it may have held when it formed probably escaped into space quickly.

JPL/NASA

► **5. Earth** is unique among our solar system's planets in that it has a hospitable atmosphere, oceans of water, and a variety of climates, and it supports life.

▼ **6. Mars** has a thin atmosphere, little water, and distinct seasons. Its southern hemisphere is heavily cratered like the surfaces of Mercury and the Moon. The northern hemisphere has large smooth plains, fewer craters, and evidence of extensive volcanism. The largest volcano in the solar system is found in the northern hemisphere as are huge canyons, the largest of which, if present on Earth, would stretch from San Francisco to New York!

◄ **7. Jupiter** is the largest of the Jovian planets. With its moons, rings, strong magnetic field, and intense radiation belts, Jupiter is the most complex and varied planet in our solar system. Jupiter's cloudy and violent atmosphere is divided into a series of different colored bands and a variety of spots (the Great Red Spot) that interact in incredibly complex motions.

▼ **8. Saturn's** most conspicuous feature is its ring system, consisting of thousands of rippling, spiraling bands of countless particles. The width of Saturn's rings would just reach from Earth to the Moon.

► **9. Uranus** is the only planet that lies on its side; that is, its axis of rotation nearly parallels the plane in which the planets revolve around the Sun. Some scientists think that a collision with an Earth-sized body early in its history may have knocked Uranus on its side. Like the other Jovian planets, Uranus has a ring system, albeit a faint one.

► **10. Neptune** is a dynamic stormy planet with an atmosphere similar to those of the other Jovian planets. Winds up to 2000 km/hr blow over the planet, creating tremendous storms, the largest of which, the Great Dark Spot, seen in the center, is nearly as big as Earth and is similar to the Great Red Spot on Jupiter.

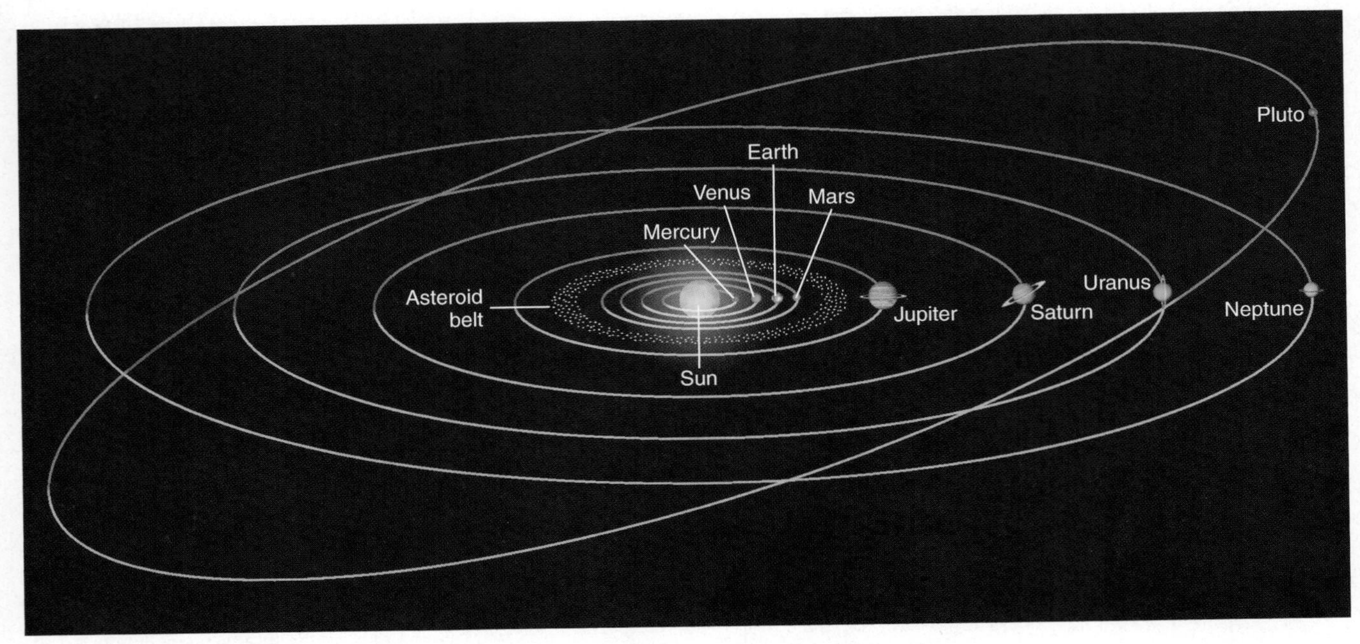

Figure 1.7 Diagrammatic Representation of the Solar System This representation of the solar system shows the planets and their orbits around the Sun. On August 24, 2006, the International Astronomical Union downgraded Pluto from a planet to a dwarf planet. A dwarf planet has the same characteristics as a planet, except that it does not clear the neighborhood around its orbit. Pluto orbits among the icy debris of the Kuiper Belt, and therefore does not meet the criteria for a true planet.

While the planets were accreting, material that had been pulled into the center of the nebula also condensed, collapsed, and was heated to several million degrees by gravitational compression. The result was the birth of a star, our Sun.

During the early accretionary phase of the solar system's history, collisions between various bodies were common, as indicated by the craters on many planets and moons. Asteroids probably formed as planetesimals in a localized eddy between

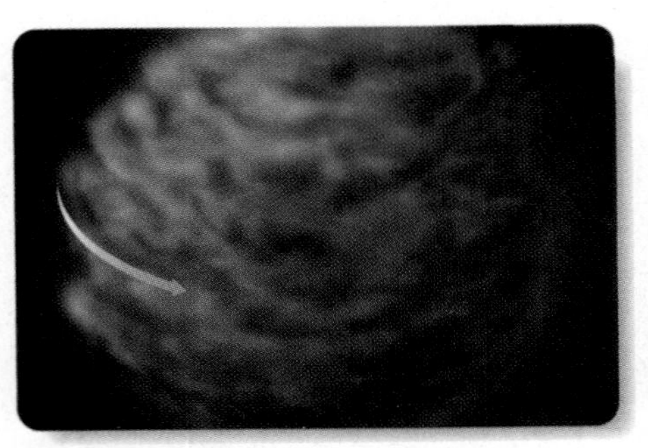

a A huge rotating cloud of gas contracts and flattens

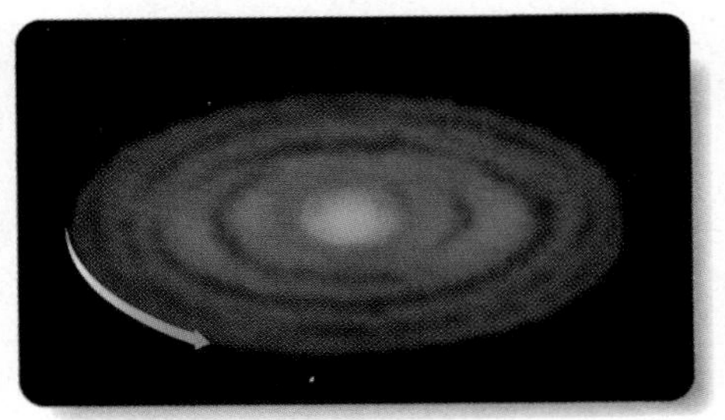

b to form a disk of gas and dust with the sun forming in the center,

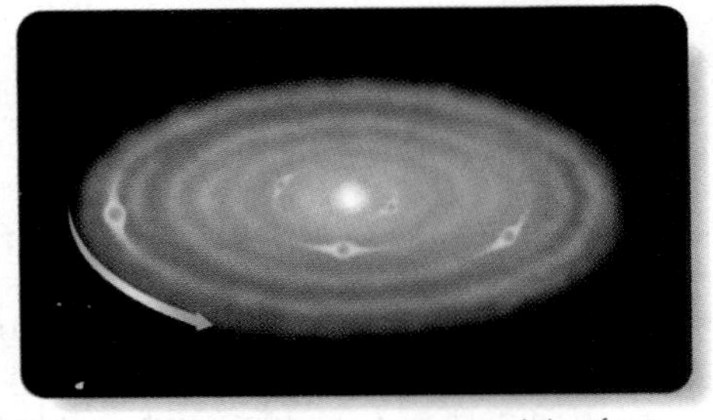

c and addies gathering up material to form planets.

Figure 1.8 Solar Nebula Theory According to the currently accepted theory for the origin of our solar system, the planets and the Sun formed from a rotating cloud of gas.

a Early Earth probably had a uniform composition and density throughout.

b The temperature of early Earth reached the melting point of iron and nickel, which, being denser than silicate minerals, settled to Earth's center. At the same time, the lighter silicates flowed upward to form the mantle and the crust.

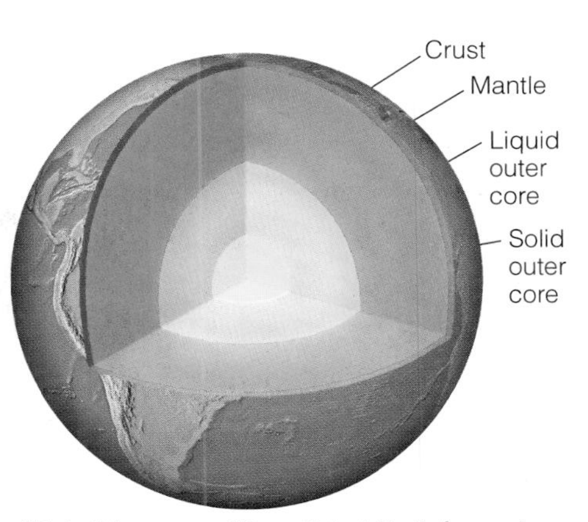

c In this way, a differentiated Earth formed, consisting of a dense iron–nickel core, an iron-rich silicate mantle, and a silicate crust with continents and ocean basis.

Figure 1.9 Homogeneous Accretion Theory for the Formation of a Differentiated Earth

what eventually became Mars and Jupiter in much the same way that other planetesimals formed the terrestrial planets. The tremendous gravitational field of Jupiter, however, prevented this material from ever accreting into a planet. Comets, which are interplanetary bodies composed of loosely bound rocky and icy materials, are thought to have condensed near the orbits of Uranus and Neptune.

The solar nebula theory for the formation of the solar system thus accounts for most of the characteristics of the planets and their moons, the differences in composition between the terrestrial and Jovian planets, and the presence of the asteroid belt. Based on the available data, the solar nebula theory best explains the features of the solar system and provides a logical explanation for its evolutionary history.

Earth—Its Place in Our Solar System

Some 4.6 billion years ago, various planetesimals in our solar system gathered enough material together to form Earth and the other planets. Scientists think that this early Earth was probably cool, of generally uniform composition and density throughout, and composed mostly of silicates, compounds consisting of silicon and oxygen, iron and magnesium oxides, and smaller amounts of all the other chemical elements (Figure 1.9a). Subsequently, when the combination of meteorite impacts, gravitational compression, and heat from radioactive decay increased the temperature of Earth enough to melt iron and nickel, this homogeneous composition disappeared (Figure 1.9b) and was replaced by a series of concentric layers of differing composition and density, resulting in a differentiated planet (Figure 1.9c).

This differentiation into a layered planet is probably the most significant event in Earth's history. Not only did it lead to the formation of a crust and eventually continents, but it also was probably responsible for the emission of gases from the interior that eventually led to the formation of the oceans and atmosphere.

WHY EARTH IS A DYNAMIC AND EVOLVING PLANET

Earth is a dynamic planet that has continuously changed during its 4.6-billion-year existence. The size, shape, and geographic distribution of continents and ocean basins have changed through time, the composition of the atmosphere has evolved, and life-forms existing today differ from those that lived during the past. Mountains and hills have been worn away by erosion, and the forces of wind, water, and ice have sculpted a diversity of landscapes. Volcanic eruptions and earthquakes reveal an active interior, and folded and fractured rocks are testimony to the tremendous power of Earth's internal forces.

Earth consists of three concentric layers: the core, the mantle, and the crust (Figure 1.10). This orderly division results from density differences between the layers as a function of variations in composition, temperature, and pressure.

The **core** has a calculated density of 10–13 grams per cubic centimeter (g/cm^3) and occupies about 16% of Earth's total volume. Seismic (earthquake) data indicate that the core consists of a small, solid inner region and a larger, apparently liquid, outer portion. Both are thought to consist mostly of iron and a small amount of nickel.

The **mantle** surrounds the core and comprises about 83% of Earth's volume. It is less dense than the core (3.3–5.7 g/cm^3) and is thought to be composed mostly of *peridotite*, a dark, dense igneous rock containing abundant iron and magnesium. The mantle can be divided into three distinct zones based on physical characteristics. The lower mantle is solid and forms most of the volume of Earth's interior. The **asthenosphere** surrounds the lower mantle. It has the same composition as the lower mantle, but behaves plastically and flows slowly. Partial melting within the asthenosphere generates *magma* (molten material), some of which rises to the surface because it is less dense than the rock from which it

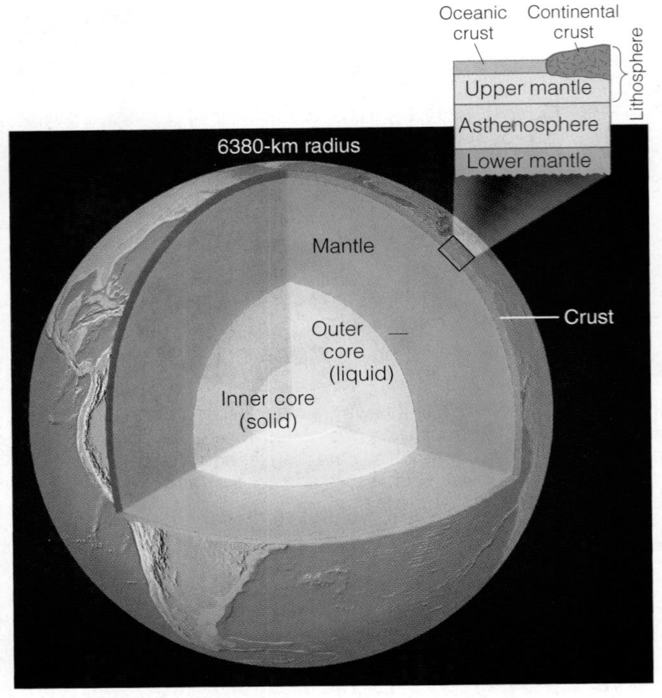

Figure 1.10 Cross Section of Earth Illustrating the Core, Mantle, and Crust The enlarged portion shows the relationship between the lithosphere (composed of the continental crust, oceanic crust, and solid upper mantle) and the underlying asthenosphere and lower mantle.

was derived. The upper mantle surrounds the asthenosphere. The solid upper mantle and the overlying crust constitute the **lithosphere**, which is broken into numerous individual pieces called **plates** that move over the asthenosphere, partially as a result of underlying *convection cells* (Figure 1.11). Interactions of these plates are responsible for such phenomena as earthquakes, volcanic eruptions, and the formation of mountain ranges and ocean basins.

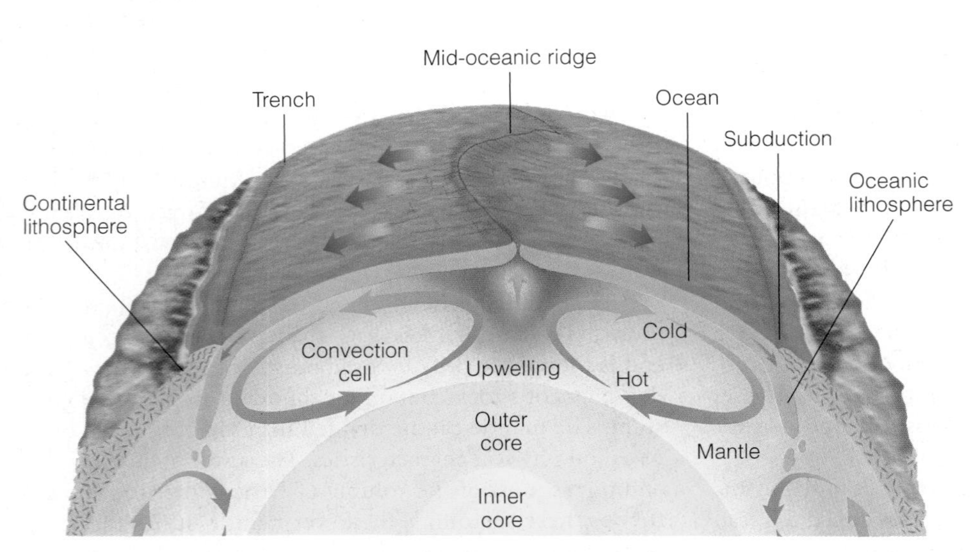

Figure 1.11 Movement of Earth's Plates Earth's plates are thought to move partially as a result of underlying mantle convection cells in which warm material from deep within Earth rises toward the surface, cools, and then upon losing heat, descends back into the interior as shown in this diagrammatic cross section.

The **crust**, Earth's outermost layer, consists of two types. *Continental crust* is thick (20–90 km), has an average density of 2.7 g/cm^3, and contains considerable silicon and aluminum. *Oceanic crust* is thin (5–10 km), denser than continental crust (3.0 g/cm^3), and is composed of the dark igneous rocks *basalt* and *gabbro*.

Plate Tectonic Theory

The recognition that the lithosphere is divided into rigid plates that move over the asthenosphere forms the foundation of **plate tectonic theory** (Figure 1.12). Zones of volcanic activity, earthquakes, or both mark most plate boundaries. Along these boundaries, plates separate (diverge), collide (converge), or slide sideways past each other (Figure 1.13).

The acceptance of plate tectonic theory is recognized as a major milestone in the geologic sciences, comparable to the revolution that Darwin's theory of evolution caused in biology. Plate tectonics has provided a framework for interpreting the composition, structure, and internal processes of Earth on a global scale. It has led to the realization that the continents and ocean basins are part of a lithosphere–atmosphere–hydrosphere system that evolved together with Earth's interior (Table 1.3).

A revolutionary concept when it was proposed in the 1960s, plate tectonic theory has had far-reaching consequences in all fields of geology because it provides the basis for relating many seemingly unrelated phenomena. Besides being responsible for the major features of Earth's crust, plate movements also affect the formation and occurrence of Earth's natural resources, as well as the distribution and evolution of the world's biota.

The impact of plate tectonic theory has been particularly notable in the interpretation of Earth's history. For example, the Appalachian Mountains in eastern North America and the mountain ranges of Greenland, Scotland, Norway, and Sweden are not the result of unrelated mountain-building episodes, but, rather, are part of a larger mountain-building event that involved the closing of an ancient Atlantic Ocean and the formation of the supercontinent Pangaea approximately 251 million years ago.

THE ROCK CYCLE

A **rock** is an aggregate of **minerals**, which are naturally occurring, inorganic, crystalline solids that have definite physical and chemical properties. Minerals are composed of elements such as oxygen, silicon, and aluminum, and elements are made up of atoms, the smallest particles

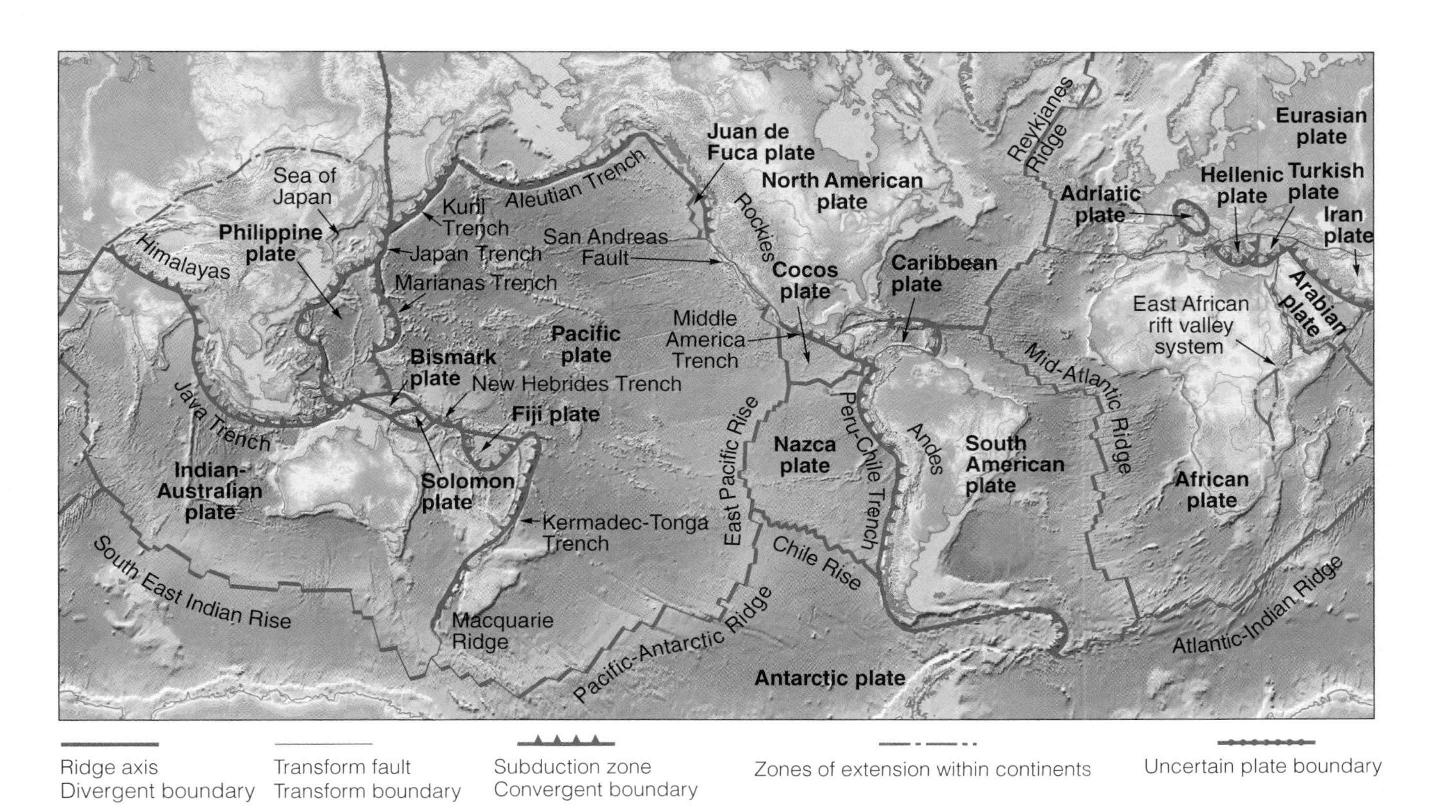

Figure 1.12 Earth's Plates Earth's lithosphere is divided into rigid plates of various sizes that move over the asthenosphere.

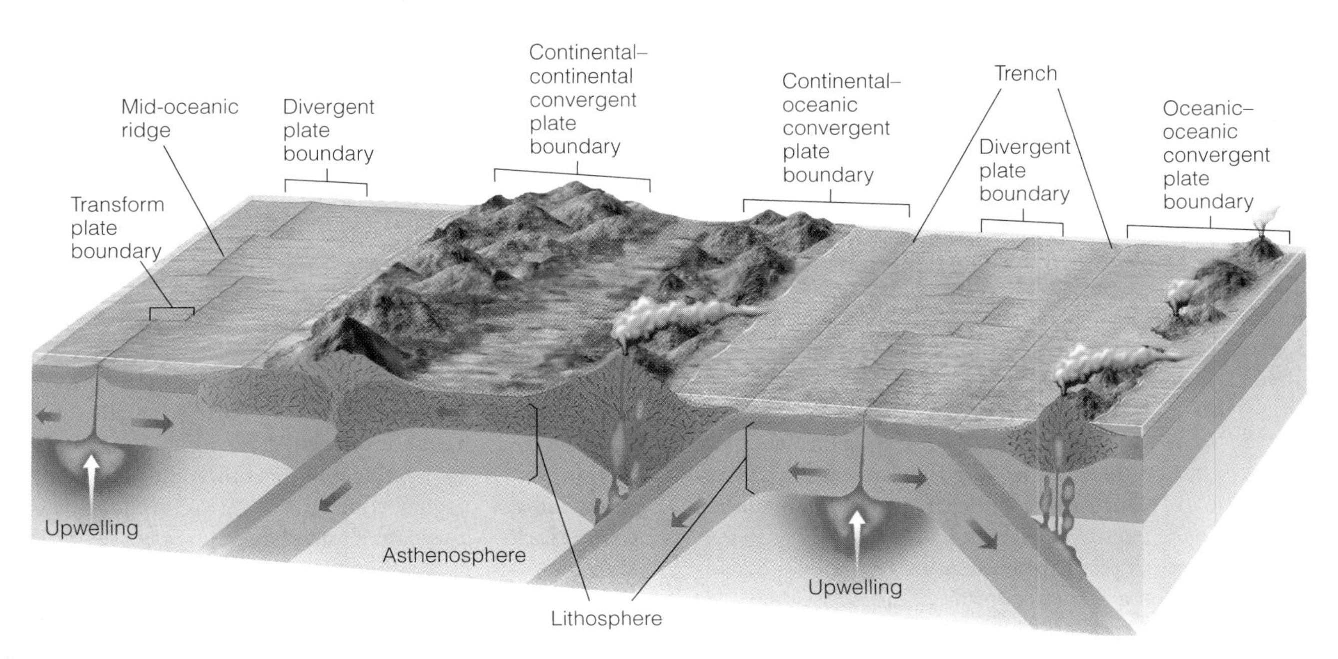

Figure 1.13 Relationship Between Lithosphere, Asthenosphere, and Plate Boundaries An idealized cross section illustrating the relationship between the lithosphere and the underlying asthenosphere and the three principal types of plate boundaries: divergent, convergent, and transform.

TABLE 1.3 Plate Tectonics and Earth Systems

Solid Earth

Plate tectonics is driven by convection in the mantle and in turn drives mountain building and associated igneous and metamorphic activity.

Atmosphere

Arrangement of continents affects solar heating and cooling, and thus winds and weather systems. Rapid plate spreading and hot-spot activity may release volcanic carbon dioxide and affect global climate.

Hydrosphere

Continental arrangement affects ocean currents. Rate of spreading affects volume of mid-oceanic ridges and hence sea level. Placement of continents may contribute to onset of ice ages.

Biosphere

Movement of continents creates corridors or barriers to migration, the creation of ecological niches, and the transport of habitats into more or less favorable climates.

Extraterrestrial

Arrangement of continents affects free circulation of ocean tides and influences tidal slowing of Earth's rotation.

Source: Adapted by permission from Stephen Dutch, James S. Monroe, and Joseph Moran, *Earth Science* (Minneapolis/St. Paul: West Publishing Co., 1997).

of matter that retain the characteristics of an element. More than 3500 minerals have been identified and described, but only about a dozen make up the bulk of the rocks in Earth's crust (see Table 3.3).

Geologists recognize three major groups of rocks—*igneous*, *sedimentary*, and *metamorphic*—each of which is characterized by its mode of formation. Each group contains a variety of individual rock types that differ from one another on the basis of their composition or texture (the size, shape, and arrangement of mineral grains).

The **rock cycle** is a pictorial representation of events leading to the origin, destruction and/or changes, and reformation of rocks as a consequence of Earth's internal and surface processes (Figure 1.14). Furthermore, it shows that the three major rock groups—igneous, sedimentary, and metamorphic—are interrelated; that is, any rock type can be derived from the others. Notice in Figure 1.14 that the ideal cycle involves those events depicted on the circle leading from magma to igneous rocks and so on. Notice also that the circle has several internal arrows indicating interruptions in the cycle.

Igneous rocks result when magma crystallizes or volcanic ejecta, such as ash, accumulate and consolidate. As magma cools, minerals crystallize and the resulting rock is characterized by interlocking mineral grains. Magma that cools slowly beneath the surface produces *intrusive igneous rocks* (Figure 1.15a); magma that cools at the surface produces *extrusive igneous rocks* (Figure 1.15b).

Rocks exposed at Earth's surface are broken into particles and dissolved by various weathering processes. The particles and dissolved materials may be transported by wind, water, or ice and eventually deposited as *sediment*. This sediment may then be compacted or cemented (lithified) into sedimentary rock.

Sedimentary rocks form in one of three ways: consolidation of mineral or rock fragments, precipitation of mineral matter from solution, or compaction of plant or animal remains (Figure 1.15c, d). Because sedimentary rocks form at or near Earth's surface, geologists can make inferences about the environment in which they were deposited, the transporting agent, and perhaps even something about the source from which the sediments were derived (see Chapter 6). Accordingly, sedimentary rocks are especially useful for interpreting Earth history.

Metamorphic rocks result from the alteration of other rocks, usually beneath the surface, by heat, pressure, and the chemical activity of fluids. For example, marble, a rock preferred by many sculptors and builders, is a metamorphic rock produced when the agents of metamorphism are applied to the sedimentary rocks limestone or dolostone. Metamorphic rocks are either *foliated* (Figure 1.15e) or *nonfoliated* (Figure 1.15f). Foliation, the parallel alignment of minerals due to pressure, gives the rock a layered or banded appearance.

How Are the Rock Cycle and Plate Tectonics Related?

Interactions between plates determine, to some extent, which of the three rock groups will form (Figure 1.16). For example, when plates converge, heat and pressure generated along the plate boundary may lead to igneous activity and metamorphism within the descending oceanic plate, thus producing various igneous and metamorphic rocks.

Some of the sediments and sedimentary rocks on the descending plate are melted, whereas other sediments and sedimentary rocks along the boundary of the nondescending plate are metamorphosed by the heat and pressure generated along the converging plate boundary. Later, the mountain range or chain of volcanic islands formed along the convergent plate boundary will be weathered and eroded, and the new sediments will be transported to the ocean to begin yet another cycle.

The interrelationship between the rock cycle and plate tectonics is just one example of how Earth's various subsystems and cycles are all interrelated. Heating within Earth's interior results in convection cells that power the movement of plates, and also magma, which forms intrusive and extrusive igneous rocks. Movement along plate boundaries may result in volcanic activity, earthquakes, and, in some cases, mountain building. The interaction between the atmosphere, hydrosphere, and biosphere contributes to the weathering of rocks exposed on Earth's surface. Plates descending back into Earth's interior are subjected to increasing heat and pressure, which may lead to metamorphism, as well as the generation of magma and yet another recycling of materials.

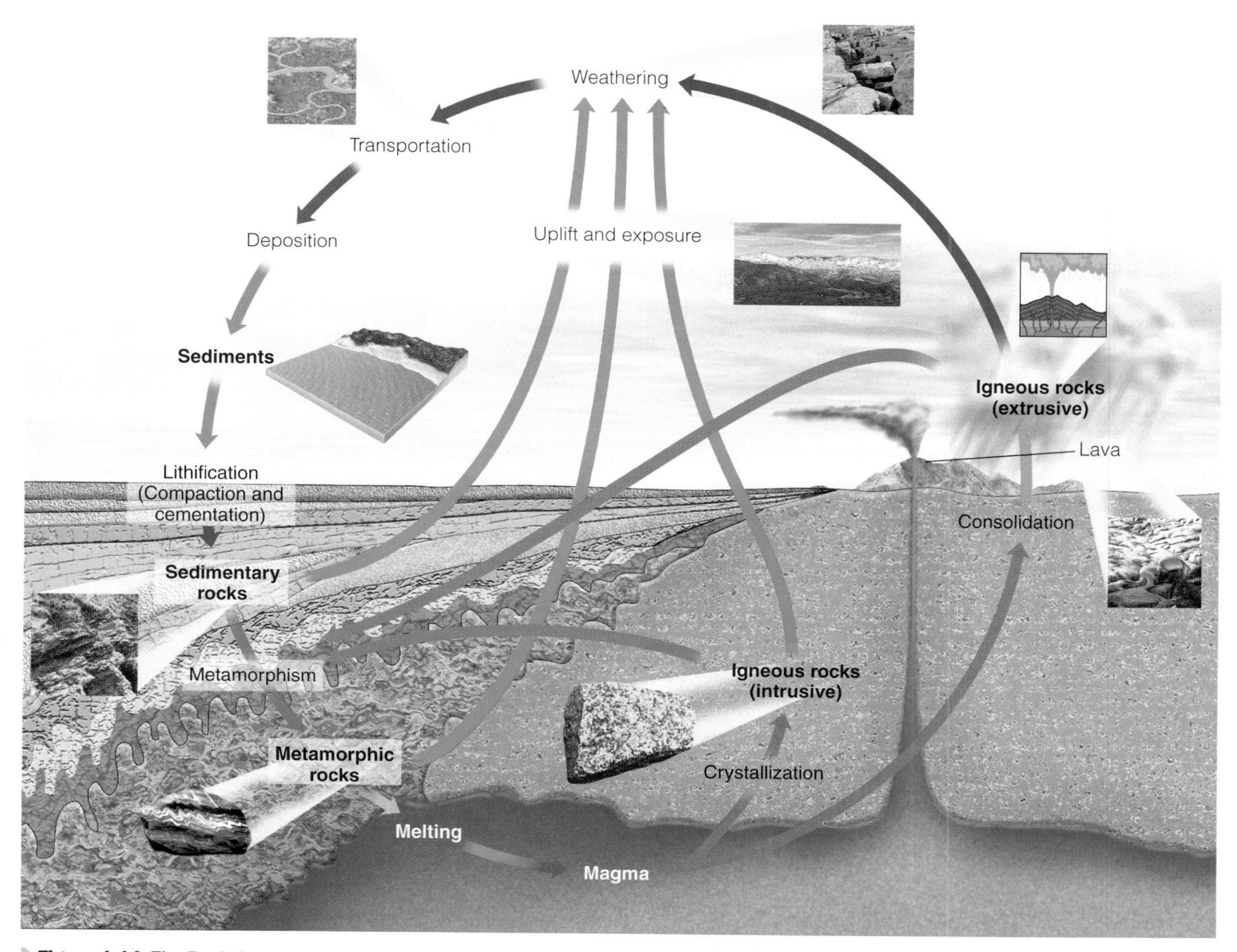

Figure 1.14 The Rock Cycle This cycle shows the interrelationships between Earth's internal and external processes and how the three major rock groups are related. An ideal cycle includes the events on the outer margin of the cycle, but interruptions, indicated by internal arrows, are common.

ORGANIC EVOLUTION AND THE HISTORY OF LIFE

Plate tectonic theory provides us with a model for understanding the internal workings of Earth and its effect on Earth's surface. The theory of **organic evolution** (whose central thesis is that all present-day organisms are related, and that they have descended with modifications from organisms that lived in the past) provides the conceptual framework for understanding the history of life. Together, the theories of plate tectonics and organic evolution have changed the way we view our planet, and we should not be surprised at the intimate association between them. Although the relationship between plate tectonic processes and the evolution of life is incredibly complex, paleontological data provide indisputable evidence of the influence of plate movement on the distribution of organisms.

The publication in 1859 of Darwin's *On the Origin of Species by Means of Natural Selection* revolutionized biology and marked the beginning of modern evolutionary biology. With its publication, most naturalists recognized that evolution provided a unifying theory that explained an otherwise encyclopedic collection of biologic facts.

When Darwin proposed his theory of organic evolution, he cited a wealth of supporting evidence, including the way organisms are classified, embryology, comparative anatomy, the geographic distribution of organisms, and, to a limited extent, the fossil record. Furthermore, Darwin proposed that *natural selection*, which results in the survival to reproductive age of those organisms best adapted to their environment, is the mechanism that accounts for evolution.

Perhaps the most compelling evidence in favor of evolution can be found in the fossil record. Just as the geologic

Parvinder Sethi

a **Granite**, an intrusive igneous rock.

Parvinder Sethi

b **Basalt**, an extrusive igneous rock.

Parvinder Sethi

c **Conglomerate**, a sedimentary rock formed by the consolidation of rounded rock fragments.

Parvinder Sethi

d **Limestone**, a sedimentary rock formed by the extraction of mineral matter from seawater by organisms or by the inorganic precipitation of the mineral calcite from seawater.

Parvinder Sethi

e **Gneiss**, a foliated metamorphic rock.

Parvinder Sethi

f **Quartzite**, a nonfoliated metamorphic rock.

Figure 1.15 Hand Specimens of Common Igneous, Sedimentary, and Metamorphic Rocks

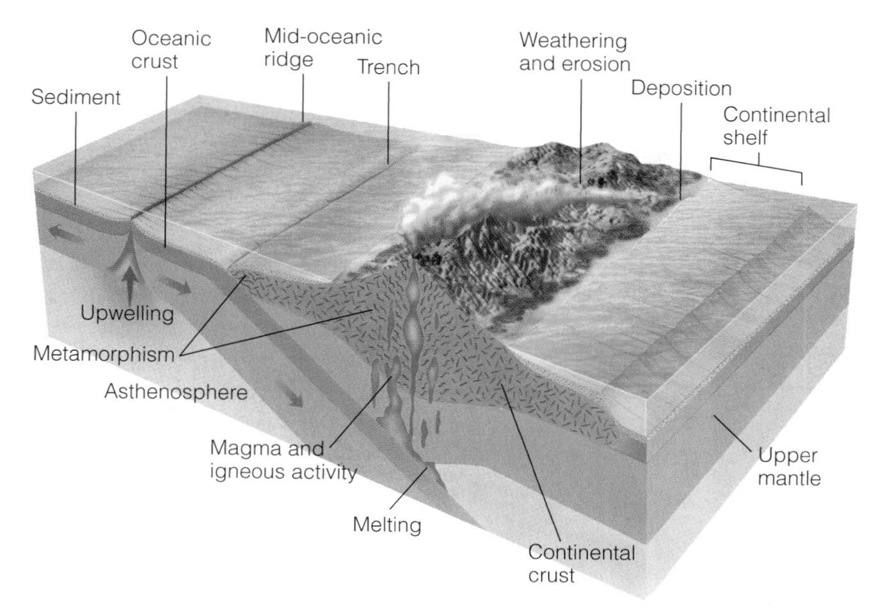

Figure 1.16 Plate Tectonics and the Rock Cycle Plate movement provides the driving mechanism that recycles Earth materials. The cross section shows how the three major rock groups—igneous, sedimentary, and metamorphic—are recycled through both the continental and oceanic regions. Subducting plates are partially melted to produce magma, which rises and either crystallizes beneath Earth's surface as intrusive igneous rock or spills out on the surface, solidifying as extrusive igneous rock. Rocks exposed at the surface are weathered and eroded to produce sediments that are transported and eventually lithified into sedimentary rocks. Metamorphic rocks result from pressure generated along converging plates or adjacent to rising magma.

record allows geologists to interpret physical events and conditions in the geologic past, **fossils**, which are the remains or traces of once-living organisms, not only provide evidence that evolution has occurred, but also demonstrate that Earth has a history extending beyond that recorded by humans. The succession of fossils in the rock record provides geologists with a means for dating rocks and allowed for a relative geologic time scale to be constructed in the 1800s.

GEOLOGIC TIME AND UNIFORMITARIANISM

An appreciation of the immensity of geologic time is central to understanding the evolution of Earth and its biota. Indeed, time is one of the main aspects that sets geology apart from the other sciences, except astronomy. Most people have difficulty comprehending geologic time because they tend to think in terms of the human perspective—seconds, hours, days, and years. Ancient history is what occurred hundreds or even thousands of years ago. When geologists talk of ancient geologic history, however, they are referring to events that happened hundreds of millions or even billions of years ago. To a geologist, recent geologic events are those that occurred within the last million years or so.

It is also important to remember that Earth goes through cycles of much longer duration than the human perspective of time. Although they may have disastrous effects on the human species, global warming and cooling are part of a larger cycle that has resulted in numerous glacial advances and retreats during the past 1.8 million years. Because of their geologic perspective on time and how the various Earth subsystems and cycles are interrelated, geologists can make valuable contributions to many of the current environmental debates such as those involving global warming and sea-level changes.

The **geologic time scale** subdivides geologic time into a hierarchy of increasingly shorter time intervals; each time subdivision has a specific name. The geologic time scale resulted from the work of many 19th-century geologists who pieced together information from numerous rock exposures and constructed a chronology based on changes in Earth's biota through time. Subsequently, with the discovery of radioactivity in 1895 and the development of various radiometric dating techniques, geologists have been able to assign numerical ages (also known as absolute ages) in years to the subdivisions of the geologic time scale (Figure 1.17).

One of the cornerstones of geology is the **principle of uniformitarianism**, which is based on the premise that present-day processes have operated throughout geologic time. Therefore, to understand and interpret geologic events from evidence preserved in rocks, we must first understand present-day processes and their results. In fact, uniformitarianism fits in completely with the system approach that we are following for the study of Earth.

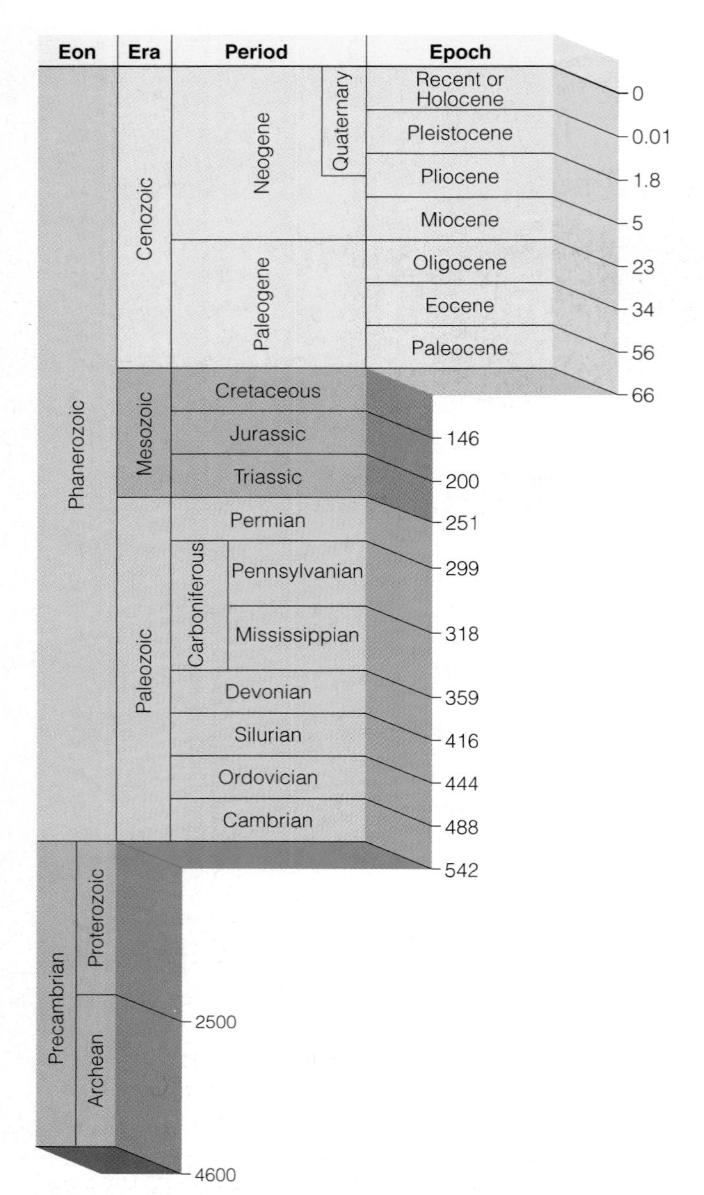

Figure 1.17 The Geologic Time Scale The numbers to the right of the columns are ages in millions of years before the present. Dates are from Gradstein, F., J. Ogg and A. Smith. *A Geologic Time Scale 2004* (Cambridge, UK: Cambridge University Press, 2005), Figure 1.2.

Uniformitarianism is a powerful principle that allows us to use present-day processes as the basis for interpreting the past and for predicting potential future events. We should keep in mind, however, that uniformitarianism does not exclude sudden or catastrophic events such as volcanic eruptions, earthquakes, tsunami, landslides, or floods. These are processes that shape our modern world, and some geologists view Earth history as a series of such short-term or punctuated events. This view is certainly in keeping with the modern principle of uniformitarianism.

Furthermore, uniformitarianism does not require that the rates and intensities of geologic processes be constant through time. We know that volcanic activity was more intense in North America 5 to 10 million years ago than it is today, and that glaciation has been more prevalent during the last several million years than in the previous 300 million years.

What uniformitarianism means is that even though the rates and intensities of geologic processes have varied during the past, the physical and chemical laws of nature have remained the same. Although Earth is in a dynamic state of change and has been ever since it formed, the processes that shaped it during the past are the same ones operating today.

HOW DOES THE STUDY OF GEOLOGY BENEFIT US?

The most meaningful lesson to learn from the study of geology is that Earth is an extremely complex planet in which interactions are taking place between its various subsystems and have been for the past 4.6 billion years. If we want to ensure the survival of the human species, we must understand how the various subsystems work and interact with each other and, more importantly, how our actions affect the delicate balance between these systems. We can do this, in part, by studying what has happened in the past, particularly on the global scale, and use that information to try to determine how our actions might affect the delicate balance between Earth's various subsystems in the future.

The study of geology goes beyond learning numerous facts about Earth. In fact, we don't just study geology—we *live* it. Geology is an integral part of our lives. Our standard of living depends directly on our consumption of natural resources, resources that formed millions and billions of years ago. However, the way we consume natural resources and interact with the environment, as individuals and as a society, also determines our ability to pass on this standard of living to the next generation.

As you study the various topics covered in this book, keep in mind the themes discussed in this chapter and how, like the parts of a system, they are interrelated and responsible for the 4.6-billion-year history of Earth. View each chapter's topic in the context of how it fits in the whole Earth system, and remember that Earth's history is a continuum and the result of interaction between its various subsystems. By relating each chapter's topic to its place in the Earth system, you will gain a greater appreciation of why geology is so integral to our lives.

Geo-Recap

Chapter Summary

- Earth can be viewed as a system of interconnected components that interact and affect one another. The principal subsystems of Earth are the atmosphere, hydrosphere, biosphere, lithosphere, mantle, and core. Earth is considered a dynamic planet that continually changes because of the interactions among its various subsystems and cycles.
- Geology, the study of Earth, is divided into two broad areas: Physical geology is the study of Earth materials, as well as the processes that operate within Earth and on its surface; historical geology examines the origin and evolution of Earth, its continents, oceans, atmosphere, and life.
- The scientific method is an orderly, logical approach that involves gathering and analyzing facts about a particular phenomenon, formulating hypotheses to explain the phenomenon, testing the hypotheses, and finally proposing a theory. A theory is a testable explanation for some natural phenomenon that has a large body of supporting evidence.
- Geology is part of the human experience. We can find examples of it in art, music, and literature. A basic understanding of geology is also important for dealing with the many environmental problems and issues facing society.
- Geologists engage in a variety of occupations, the main one being exploration for mineral and energy resources. They are also becoming increasingly involved in environmental issues and making short- and long-range predictions of the potential dangers from such natural disasters as volcanic eruptions, tsunami, and earthquakes.
- The universe began with a big bang approximately 14 billion years ago. Astronomers have deduced this age by observing that celestial objects are moving away from each other in an ever-expanding universe. Furthermore, the universe has a background radiation of 2.7 K above absolute zero (2.7 K = –270.3°C) which is thought to be the faint afterglow of the Big Bang.
- About 4.6 billion years ago, our solar system formed from a rotating cloud of interstellar matter. As this cloud condensed, it eventually collapsed under the influence of gravity and flattened into a counterclockwise-rotating disk. Within this rotating disk, the Sun, planets, and moons formed from the turbulent eddies of nebular gases and solids.
- Earth formed from a swirling eddy of nebular material 4.6 billion years ago, accreting as a solid body, and soon thereafter differentiated into a layered planet during a period of internal heating.
- Earth's outermost layer is the crust, which is divided into continental and oceanic portions. The crust and underlying solid part of the upper mantle, also known as the lithosphere, overlie the asthenosphere, a zone that behaves plastically and flows slowly. The asthenosphere is underlain by the solid lower mantle. Earth's core consists of an outer liquid portion and an inner solid portion.
- The lithosphere is broken into a series of plates that diverge, converge, and slide sideways past one another.
- Plate tectonic theory provides a unifying explanation for many geologic features and events. The interaction between plates is responsible for volcanic eruptions, earthquakes, the formation of mountain ranges and ocean basins, and the recycling of rock materials.
- The three major rock groups are igneous, sedimentary, and metamorphic. Igneous rocks result from the crystallization of magma or the consolidation of volcanic ejecta. Sedimentary rocks are typically formed by the consolidation of rock fragments, precipitation of mineral matter from solution, or compaction of plant or animal remains. Metamorphic rocks result from the alteration of other rocks, usually beneath Earth's surface, by heat, pressure, and chemically active fluids.
- The rock cycle illustrates the interactions between Earth's internal and external processes and how the three rock groups are interrelated.
- The central thesis of the theory of organic evolution is that all living organisms evolved (descended with modifications) from organisms that existed in the past.
- Time sets geology apart from the other sciences except astronomy, and an appreciation of the immensity of geologic time is central to understanding Earth's evolution. The geologic time scale is the calendar geologists use to date past events.
- The principle of uniformitarianism is basic to the interpretation of Earth history. This principle holds that the laws of nature have been constant through time and that the same processes operating today have operated in the past, although not necessarily at the same rates.
- Geology is an integral part of our lives. Our standard of living depends directly on our consumption of natural resources, resources that formed millions and billions of years ago.

Important Terms

asthenosphere (p. 17)
Big Bang (p. 12)
core (p. 17)
crust (p. 18)
fossil (p. 23)
geologic time scale (p. 23)
geology (p. 4)
hypothesis (p. 7)
igneous rock (p. 20)
Jovian planets (p. 13)
lithosphere (p. 18)
mantle (p. 17)
metamorphic rock (p. 20)
mineral (p. 18)
organic evolution (p. 21)
plate (p. 18)
plate tectonic theory (p. 18)
principle of uniformitarianism (p. 23)
rock (p. 18)
rock cycle (p. 20)
scientific method (p. 7)
sedimentary rock (p. 20)
solar nebula theory (p. 13)
system (p. 4)
terrestrial planets (p. 13)
theory (p. 6)

Review Questions

1. The change in frequency of a sound wave caused by movement of its source relative to an observer is known as the
   a. ______ Curie point;
   b. ______ Hubble shift;
   c. ______ Doppler effect;
   d. ______ Quasar force;
   e. ______ none of these.
2. Rocks that result from the alteration of other rocks, usually beneath the surface, by heat, pressure, and the chemical activity of fluids are
   a. ______ igneous;
   b. ______ sedimentary;
   c. ______ metamorphic;
   d. ______ volcanic;
   e. ______ answers a and d.
3. The study of the origin and evolution of Earth is
   a. ______ astronomy;
   b. ______ historical geology;
   c. ______ astrobiology;
   d. ______ physical geology;
   e. ______ paleontology.
4. That all living organisms are the descendents of different life-forms that existed in the past is the central claim of
   a. ______ the principle of fossil succession;
   b. ______ plate tectonics;
   c. ______ the principle of uniformitarianism;
   d. ______ organic evolution;
   e. ______ none of these.
5. The movement of plates is thought to result from
   a. ______ density differences between the inner and outer core;
   b. ______ rotation of the mantle around the core;
   c. ______ gravitational forces;
   d. ______ the Coriolis effect;
   e. ______ convection cells.
6. A combination of related parts interacting in an organized fashion is
   a. ______ a cycle;
   b. ______ a theory;
   c. ______ uniformitarianism;
   d. ______ a hypothesis;
   e. ______ a system.
7. Which of the following statements about a scientific theory is not true?
   a. ______ It is an explanation for some natural phenomenon;
   b. ______ Predictive statements can be derived from it;
   c. ______ It is a conjecture or guess;
   d. ______ It has a large body of supporting evidence;
   e. ______ It is testable.
8. What two observations lead scientists to conclude that the Big Bang occurred approximately 14 billion years ago?
   a. ______ A steady-state universe and 2.7 K background radiation;
   b. ______ A steady-state universe and opaque background radiation;
   c. ______ An expanding universe and opaque background radiation;
   d. ______ An expanding universe and 2.7 K background radiation;
   e. ______ A shrinking universe and opaque background radiation.
9. The concentric layer that makes up most of Earth's volume is the
   a. ______ inner core;
   b. ______ outer core;
   c. ______ mantle;
   d. ______ asthenosphere;
   e. ______ crust.
10. The premise that present-day processes have operated throughout geologic time is the principle of
    a. ______ organic evolution;
    b. ______ plate tectonics;
    c. ______ uniformitarianism;
    d. ______ geologic time;
    e. ______ scientific deduction.
11. Using plate movement as the driving mechanism of the rock cycle, explain how the three rock groups are related and how each rock group can be converted into a different rock group.

12. Why is viewing Earth as a system a good way to do it? Are humans a part of the Earth system? If so, what role, if any, do we play in Earth's evolution?
13. Discuss why an accurate geologic time scale is particularly important for geologists in examining changes in global temperatures during the past and how an understanding of geologic time is crucial to the current debate on global warming and its consequences.
14. In what ways does geology affect our everyday lives at the individual, local, and nation-state levels?
15. What is the Big Bang? What evidence do we have that the universe began approximately 14 billion years ago?
16. Why is Earth considered a dynamic planet? What are its three concentric layers and their characteristics?
17. Explain how the principle of uniformitarianism allows for catastrophic events.
18. Why is plate tectonic theory so important to geology? How does it fit into a systems approach to the study of Earth?
19. Why do most scientists think overpopulation is the greatest environmental problem facing the world today? Do you agree? If not, what do you think is the greatest threat to our existence as a species?
20. How does the solar nebula theory account for the formation of our solar system, its features, and evolutionary history?

CHAPTER 2

Plate Tectonics: A Unifying Theory

Dita Alangkara/AP/Wide World Photo

OUTLINE

OBJECTIVES

At the end of this chapter, you will have learned that

- Plate tectonics is the unifying theory of geology and has revolutionized geology.
- The hypothesis of continental drift was based on considerable geologic, paleontologic, and climatologic evidence.
- The hypothesis of seafloor spreading accounts for continental movement and the idea that thermal convection cells provide a mechanism for plate movement.
- The three types of plate boundaries are divergent, convergent, and transform. Along these boundaries new plates are formed, consumed, or slide past one another.
- Interaction along plate boundaries accounts for most of Earth's earthquake and volcanic activity.

An earthquake survivor walks among the ruins of houses in Bantul, central Indonesia, where a 6.3-magnitude earthquake on May 27, 2006, left more than 6200 dead and more than 200,000 people homeless. Devastating earthquakes such as this are the result of movement along plate boundaries. Such earthquakes are part of the interaction between plates and, unfortunately for humans, will continue to result in tremendous loss of life and property damage in seismically active areas.

- The rate of movement and motion of plates can be calculated in several ways.
- Some type of convective heat system is involved in plate movement.
- Plate movement affects the distribution of natural resources.
- Plate movement affects the distribution of the world's biota and has influenced evolution.

INTRODUCTION

Imagine it is the day after Christmas, December 26, 2004, and you are vacationing on a beautiful beach in Thailand. You look up from the book you're reading to see the sea suddenly retreat from the shoreline, exposing a vast expanse of seafloor that had moments before been underwater and teeming with exotic and colorful fish. It is hard to believe that within minutes of this unusual event, a powerful tsunami will sweep over your resort and everything in its path for several kilometers inland. Within hours, the coasts of Indonesia, Sri Lanka, India, Thailand, Somalia, Myanmar, Malaysia, and the Maldives will be inundated by the deadliest tsunami in history. More than 220,000 people will die, and the region will incur billions of dollars in damage.

One year earlier, on December 26, 2003, violent shaking from an earthquake awakened hundreds of thousands of people in the Bam area of southeastern Iran. When the magnitude-6.6 earthquake was over, an estimated 43,000 people were dead, at least 30,000 were injured, and approximately 75,000 survivors were left homeless. At least 85% of the structures in the Bam area were destroyed or damaged. Collapsed buildings were everywhere, streets were strewn with rubble, and all communications were knocked out.

Now go back another 12½ years to June 15, 1991, when Mount Pinatubo in the Philippines erupted violently, discharging huge quantities of ash and gases into the atmosphere. Fortunately, in this case, warnings of an impending eruption were broadcast and heeded, resulting in the evacuation of 200,000 people from areas around the volcano. Unfortunately, the eruption still caused at least 364 deaths not only from the eruption, but also from the ensuing mudflows.

What do these three recent tragic events have in common? They are part of the dynamic interactions involving Earth's plates. When two plates come together, one plate is pushed or pulled under the other plate, triggering large earthquakes such as the one that shook India in 2001, Iran in 2003, Pakistan in 2005, and Indonesia in 2006. If conditions are right, earthquakes can produce a tsunami such as the one in 2004 or the 1998 Papua New Guinea tsunami that killed more than 2200 people.

As the descending plate moves downward and is assimilated into Earth's interior, magma is generated. Being less dense than the surrounding material, the magma rises toward the surface, where it may erupt as a volcano such as Mount Pinatubo did in 1991 and others have since. It therefore should not be surprising that the distribution of volcanoes and earthquakes closely follows plate boundaries.

As we stated in Chapter 1, *plate tectonic theory* has had significant and far-reaching consequences in all fields of geology because it provides the basis for relating many seemingly unrelated phenomena. The interactions between moving plates determines the location of continents, ocean basins, and mountain systems, all of which, in turn, affect atmospheric and oceanic circulation patterns that ultimately determine global climate (see Table 1.3). Plate movements have also profoundly influenced the geographic distribution, evolution, and extinction of plants and animals. Furthermore, the formation and distribution of many geologic resources, such as metal ores, are related to plate tectonic processes, so geologists incorporate plate tectonic theory into their prospecting efforts.

If you're like most people, you probably have only a vague notion of what plate tectonic theory is. Yet plate tectonics affects all of us. Volcanic eruptions, earthquakes, and tsunami are the result of interactions between plates. Global weather patterns and oceanic currents are caused, in part, by the configuration of the continents and ocean basins. The formation and distribution of many natural resources are related to plate movement, and thus have an impact on the economic well-being and political decisions of nations. It is therefore important to understand this unifying theory, not only because it affects us as individuals and as citizens of nation-states, but also because it ties together many aspects of the geology you will be studying.

EARLY IDEAS ABOUT CONTINENTAL DRIFT

The idea that Earth's past geography was different from today is not new. The earliest maps showing the east coast of South America and the west coast of Africa probably provided people with the first evidence that continents may have once been joined together, then broken apart and moved to their present positions. As far back as 1620, Sir Francis Bacon commented on the similarity of the shorelines of western Africa and eastern South America. However, he did not make the connection that the Old and New Worlds might once have been joined together.

Antonio Snider-Pellegrini's 1858 book *Creation and Its Mysteries Revealed* is one of the earliest specific references to the idea of continental drift. Snider-Pellegrini suggested that all of the continents were linked together during the Pennsylvanian Period and later split apart. He based his conclusions on the resemblances between plant fossils in the Pennsylvanian-age coal beds of Europe and North America.

During the late 19th century, the Austrian geologist Edward Suess noted the similarities between the Late Paleozoic plant fossils of India, Australia, South Africa, and South America, as well as evidence of glaciation in the rock sequences of these continents. The plant fossils comprise a unique flora that occurs in the coal layers just above the glacial deposits of these southern continents. This flora is very different from the contemporaneous coal swamp flora of the northern continents, which Snider-Pellegrini noted earlier, and is collectively known as the ***Glossopteris* flora** after its most conspicuous genus (Figure 2.1).

Figure 2.1 Fossil *Glossopteris* Leaves Plant fossils, such as these *Glossopteris* leaves from the Upper Permian Dunedoo Formation in Australia, are found on all five of the Gondwana continents. Their presence on continents with widely varying climates today is evidence that the continents were at one time connected. The distribution of the plants at that time was in the same climatic latitudinal belt.

Figure 2.2 Alfred Wegener Alfred Wegener, a German meteorologist, proposed the continental drift hypothesis in 1912 based on a tremendous amount of geologic, paleontologic, and climatologic evidence. He is shown here waiting out the Arctic winter in an expedition hut in Greenland.

In his book, *The Face of the Earth*, published in 1885, Suess proposed the name *Gondwanaland* (or **Gondwana** as we will use here) for a supercontinent composed of the aforementioned southern continents. Abundant fossils of the *Glossopteris* flora are found in coal beds in Gondwana, a province in India. Suess thought these southern continents were at one time connected by land bridges over which plants and animals migrated. Thus, in his view, the similarities of fossils on these continents were due to the appearance and disappearance of the connecting land bridges.

The American geologist Frank Taylor published a pamphlet in 1910 presenting his own theory of continental drift. He explained the formation of mountain ranges as a result of the lateral movement of continents. He also envisioned the present-day continents as parts of larger polar continents that eventually broke apart and migrated toward the equator after Earth's rotation was supposedly slowed by gigantic tidal forces. According to Taylor, these tidal forces were generated when Earth captured the Moon approximately 100 million years ago.

Although we now know that Taylor's mechanism is incorrect, one of his most significant contributions was his suggestion that the Mid-Atlantic Ridge, discovered by the 1872–1876 British HMS *Challenger* expeditions, might mark the site along which an ancient continent broke apart to form the present-day Atlantic Ocean.

Alfred Wegener and the Continental Drift Hypothesis

Alfred Wegener, a German meteorologist (Figure 2.2), is generally credited with developing the hypothesis of **continental drift**. In his monumental book, *The Origin of Continents and Oceans* (first published in 1915), Wegener proposed that all landmasses were originally united in a single supercontinent that he named **Pangaea**, from the Greek meaning "all land." Wegener portrayed his grand concept of continental movement in a series of maps showing the breakup of Pangaea and the movement of the various continents to their present-day locations. Wegener amassed a tremendous amount of geologic, paleontologic, and climatologic evidence in support of continental drift; however, initial reaction of scientists to his then-heretical ideas can best be described as mixed.

Nevertheless, the eminent South African geologist Alexander du Toit further developed Wegener's arguments and gathered more geologic and paleontologic evidence in support of continental drift. In 1937, du Toit published *Our Wandering Continents*, in which he contrasted the glacial deposits of Gondwana with coal deposits of the same age found in the continents of the Northern Hemisphere. To resolve this apparent climatologic paradox, du Toit moved the Gondwana continents to the South Pole and brought the northern continents together such that the coal deposits were located at the equator. He named this northern landmass **Laurasia**. It consisted of present-day North America, Greenland, Europe, and Asia (except for India).

WHAT IS THE EVIDENCE FOR CONTINENTAL DRIFT?

What then was the evidence Wegener, du Toit, and others used to support the hypothesis of continental drift? It includes the fit of the shorelines of continents, the appearance of the same rock sequences and mountain ranges of the same age on continents now widely separated, the matching of glacial deposits and paleoclimatic zones, and the similarities of many extinct plant and animal groups whose fossil remains are found today on widely separated continents. Wegener and his supporters argued that this vast amount of evidence from a variety of sources surely indicated that the continents must have been close together in the past.

Continental Fit

Wegener, like some before him, was impressed by the close resemblance between the coastlines of continents on opposite sides of the Atlantic Ocean, particularly South America and Africa. He cited these similarities as partial evidence that the continents were at one time joined together as a supercontinent that subsequently split apart. As his critics pointed out, though, the configuration of coastlines results from erosional and depositional processes and therefore is continuously being modified. So, even if the continents had separated during the Mesozoic Era, as Wegener proposed, it is not likely that the coastlines would fit exactly.

A more realistic approach is to fit the continents together along the continental slope where erosion would be minimal. In 1965, Sir Edward Bullard, an English geophysicist, and two associates showed that the best fit between the continents occurs at a depth of about 2000 m (Figure 2.3). Since then, other reconstructions using the latest ocean basin data have confirmed the close fit between continents when they are reassembled to form Pangaea.

Figure 2.3 Continental Fit When continents are placed together based on their outlines, the best fit isn't along their present-day coastlines, but rather along the continental slope at a depth of about 2000 m, where erosion would be minimal.

Similarity of Rock Sequences and Mountain Ranges

If the continents were at one time joined, then the rocks and mountain ranges of the same age in adjoining locations on the opposite continents should closely match. Such is the case for the Gondwana continents (Figure 2.4). Marine, nonmarine,

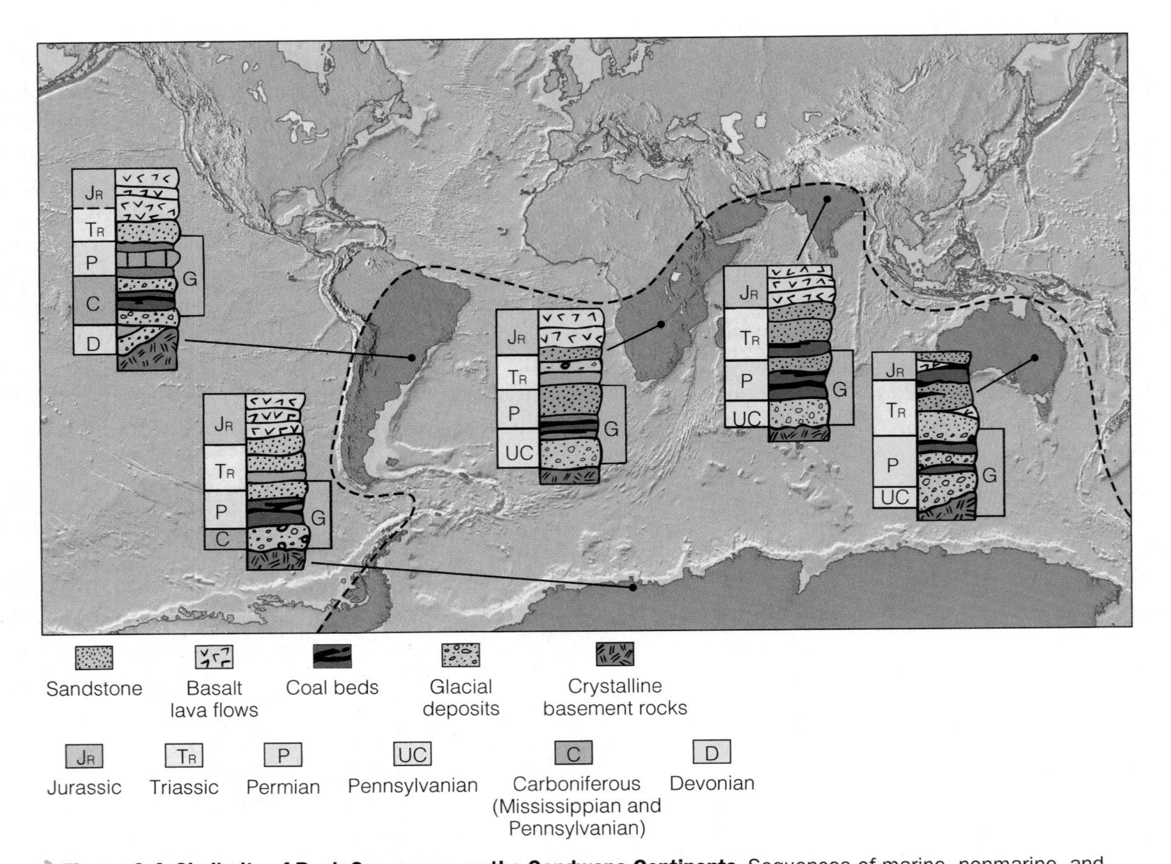

Figure 2.4 Similarity of Rock Sequences on the Gondwana Continents Sequences of marine, nonmarine, and glacial rocks of Pennsylvanian (UC) to Jurassic (JR) age are nearly the same on all five Gondwana continents (South America, Africa, India, Australia, and Antarctica). These continents are widely separated today and have different environments and climates ranging from tropical to polar. Thus, the rocks forming on each continent are very different. When the continents were all joined together in the past, however, the environments of adjacent continents were similar and the rocks forming in those areas were similar. The range indicated by G in each column is the age range (Carboniferous–Permian) of the *Glossopteris* flora.

and glacial rock sequences of the Pennsylvanian to Jurassic periods are almost identical on all five Gondwana continents, strongly indicating that they were joined at one time.

The trends of several major mountain ranges also support the hypothesis of continental drift. These mountain ranges seemingly end at the coastline of one continent only to apparently continue on another continent across the ocean. The folded Appalachian Mountains of North America, for example, trend northeastward through the eastern United States and Canada and terminate abruptly at the Newfoundland coastline. Mountain ranges of the same age and deformational style are found in eastern Greenland, Ireland, Great Britain, and Norway. Interestingly, the same red sandstones used in the construction of many English and Scottish castles are used in various buildings throughout New York. So, even though the Appalachian Mountains and their equivalent-age mountain ranges in Great Britain are currently separated by the Atlantic Ocean, they form an essentially continuous mountain range when the continents are positioned next to each other as they were during the Paleozoic Era.

Glacial Evidence

During the Late Paleozoic Era, massive glaciers covered large continental areas of the Southern Hemisphere. Evidence for this glaciation includes layers of till (sediments deposited by glaciers) and striations (scratch marks) in the bedrock beneath the till. Fossils and sedimentary rocks of the same age from the Northern Hemisphere, however, give no indication of glaciation. Fossil plants found in coals indicate that the Northern Hemisphere had a tropical climate during the time that the Southern Hemisphere was glaciated.

All of the Gondwana continents except Antarctica are currently located near the equator in subtropical to tropical climates. Mapping of glacial striations in bedrock in Australia, India, and South America indicates that the glaciers moved from the areas of the present-day oceans onto land. This would be highly unlikely because large continental glaciers (such as occurred on the Gondwana continents during the Late Paleozoic Era) flow outward from their central area of accumulation toward the sea.

If the continents did not move during the past, one would have to explain how glaciers moved from the oceans onto land and how large-scale continental glaciers formed near the equator. But if the continents are reassembled as a single landmass with South Africa located at the South Pole, the direction of movement of Late Paleozoic continental glaciers makes sense (Figure 2.5). Furthermore, this geographic arrangement places the northern continents nearer the tropics, which is consistent with the fossil and climatologic evidence from Laurasia.

Fossil Evidence

Some of the most compelling evidence for continental drift comes from the fossil record (Figure 2.6). Fossils of the *Glossopteris* flora are found in equivalent Pennsylvanian-

Figure 2.5 Glacial Evidence Indicating Continental Drift

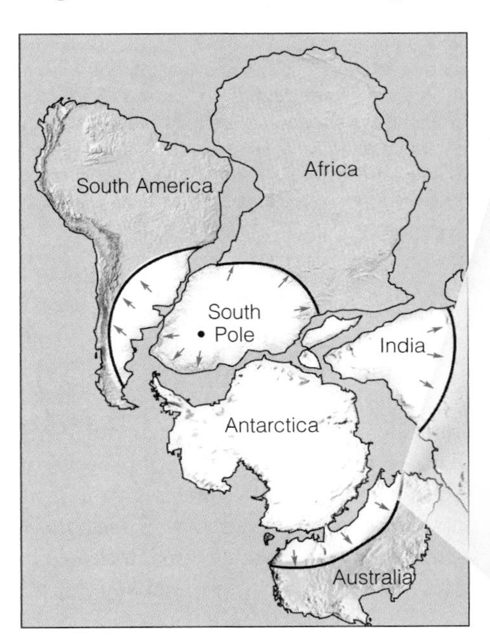

a When the Gondwana continents are placed together so that South Africa is located at the South Pole, the glacial movements indicated by striations (red arrows) found on rock outcrops on each continent make sense. In this situation, the glacier (white area) is located in a polar climate and has moved radially outward from its thick central area toward its periphery.

b Glacial striations (scratch marks) on an outcrop of Permian-age bedrock exposed at Hallet's Cove, Australia, indicate the general direction of glacial movement more than 200 million years ago. As a glacier moves over a continent's surface, it grinds and scratches the underlying rock. The scratch marks that are preserved on a rock's surface (glacial striations) thus provide evidence of the direction (red arrows) the glacier moved at that time.

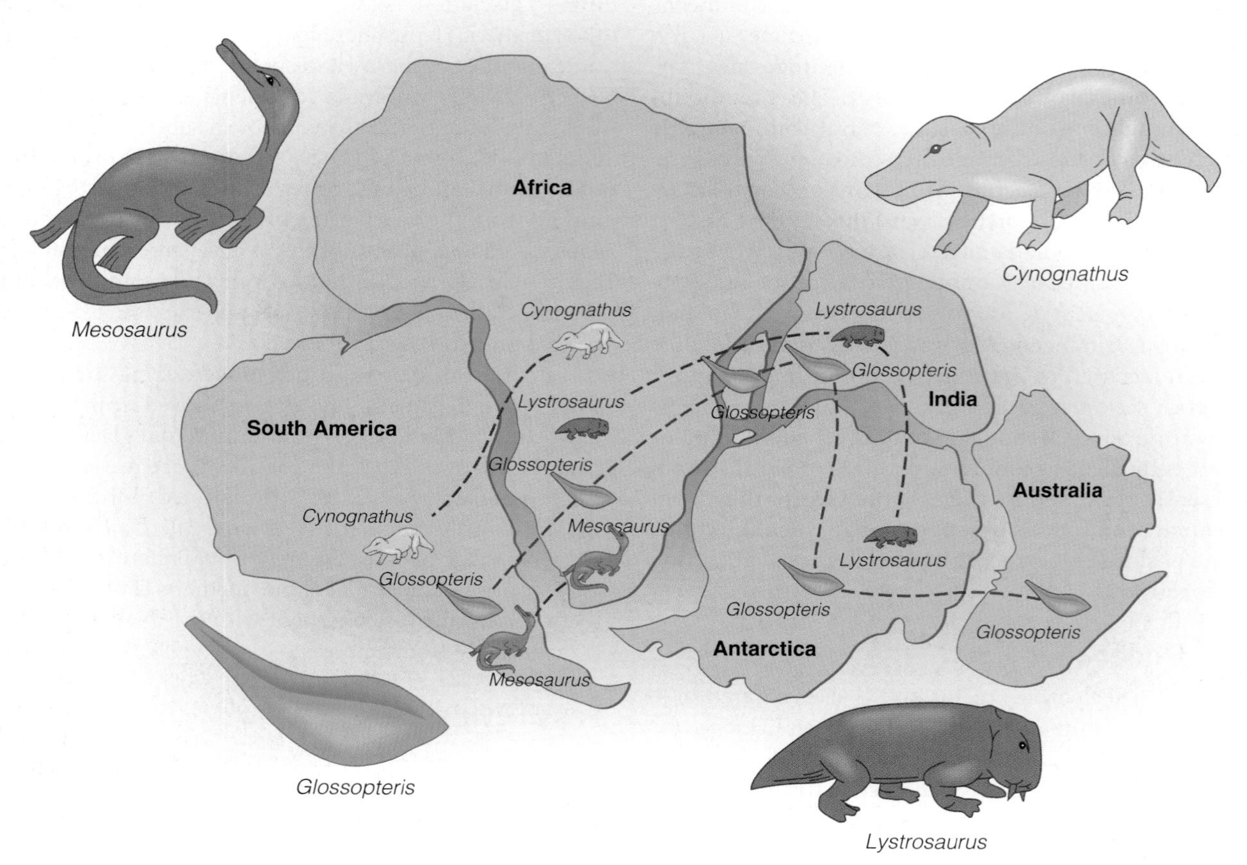

Figure 2.6 Fossil Evidence Supporting Continental Drift Some of the plants and animals whose fossils are found today on the widely separated continents of South America, Africa, India, Australia, and Antarctica. During the Late Paleozoic Era, these continents were joined together to form Gondwana, the southern landmass of Pangaea. Plants of the *Glossopteris* flora are found on all five continents, which today have widely different climates; however, during the Pennsylvanian and Permian periods, they were all located in the same general climatic belt. *Mesosaurus* is a freshwater reptile whose fossils are found only in similar nonmarine Permian-age rocks in Brazil and South Africa. *Cynognathus* and *Lystrosaurus* are land reptiles that lived during the Early Triassic Period. Fossils of *Cynognathus* are found in South America and Africa, whereas fossils of *Lystrosaurus* have been recovered from Africa, India, and Antarctica. It is hard to imagine how a freshwater reptile and land-dwelling reptiles could have swum across the wide oceans that presently separate these continents. It is more logical to assume that the continents were once connected.

and Permian-age coal deposits on all five Gondwana continents. The *Glossopteris* flora is characterized by the seed fern *Glossopteris* (Figure 2.1), as well as by many other distinctive and easily identifiable plants. Pollen and spores of plants can be dispersed over great distances by wind; however, *Glossopteris*-type plants produced seeds that are too large to have been carried by winds. Even if the seeds had floated across the ocean, they probably would not have remained viable for any length of time in saltwater.

The present-day climates of South America, Africa, India, Australia, and Antarctica range from tropical to polar and are much too diverse to support the type of plants in the *Glossopteris* flora. Wegener therefore reasoned that these continents must once have been joined so that these widely separated localities were all in the same latitudinal climatic belt (Figure 2.6).

The fossil remains of animals also provide strong evidence for continental drift. One of the best examples is *Mesosaurus*, a freshwater reptile whose fossils are found in Permian-age rocks in certain regions of Brazil and South Africa and nowhere else in the world (Figure 2.6). Because the physiologies of freshwater and marine animals are completely different, it is hard to imagine how a freshwater reptile could have swum across the Atlantic Ocean and found a freshwater environment nearly identical to its former habitat. Moreover, if *Mesosaurus* could have swum across the ocean, its fossil remains should be widely dispersed. It is more logical to assume that *Mesosaurus* lived in lakes in what are now adjacent areas of South America and Africa, but were once united into a single continent.

Lystrosaurus and *Cynognathus* are both land-dwelling reptiles that lived during the Triassic Period; their fossils are found only on the present-day continental fragments of Gondwana (Figure 2.6). Because they are both land animals, they certainly could not have swum across the oceans currently separating the Gondwana continents. Therefore, it is logical to assume that the continents must once have been connected. Recent discoveries of dinosaur fossils in the Gondwana continents further solidifies the argument that these landmasses were in close proximity during the Early Mesozoic Era.

Notwithstanding all of the empirical evidence presented by Wegener and later by du Toit and others, most geologists simply refused to entertain the idea that continents might have moved in the past. The geologists were not necessarily being obstinate about accepting new ideas; rather, they found the evidence for continental drift inadequate and unconvincing. In part, this was because no one could provide a suitable mechanism to explain how continents could move over Earth's surface.

Interest in continental drift waned until new evidence from oceanographic research and studies of Earth's magnetic field showed that the present-day ocean basins were not as old as the continents, but were geologically young features that resulted from the breakup of Pangaea.

EARTH'S MAGNETIC FIELD

What is magnetism and what is a magnetic field? **Magnetism** is a physical phenomenon resulting from the spin of electrons in some solids—particularly those of iron—and moving electricity. A **magnetic field** is an area in which magnetic substances such as iron are affected by lines of magnetic force emanating from a magnet (Figure 2.7). The magnetic field shown in Figure 2.7 is *dipolar,* meaning that it possesses two unlike magnetic poles referred to as the north and south poles.

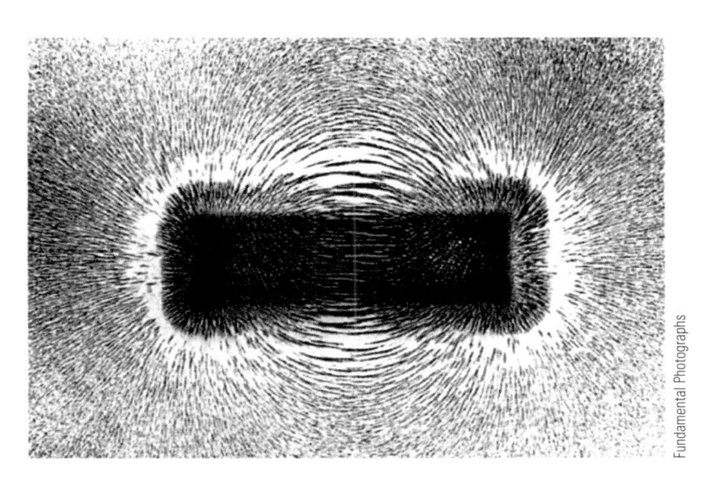

Figure 2.7 Magnetic Field Iron filings align along the lines of magnetic force radiating from a bar magnet.

Earth can be thought of as a giant dipole magnet in which the magnetic poles essentially coincide with the geographic poles (Figure 2.8). This arrangement means that the strength of the magnetic field is not constant, but varies. Notice in Figure 2.8 that the lines of magnetic force around Earth parallel its surface only near the equator. As the lines of force approach the poles, they are oriented at increasingly larger angles with respect to the surface, and the strength of the magnetic field increases; it is strongest at the poles and weakest at the equator.

Figure 2.8 Earth's Magnetic Field

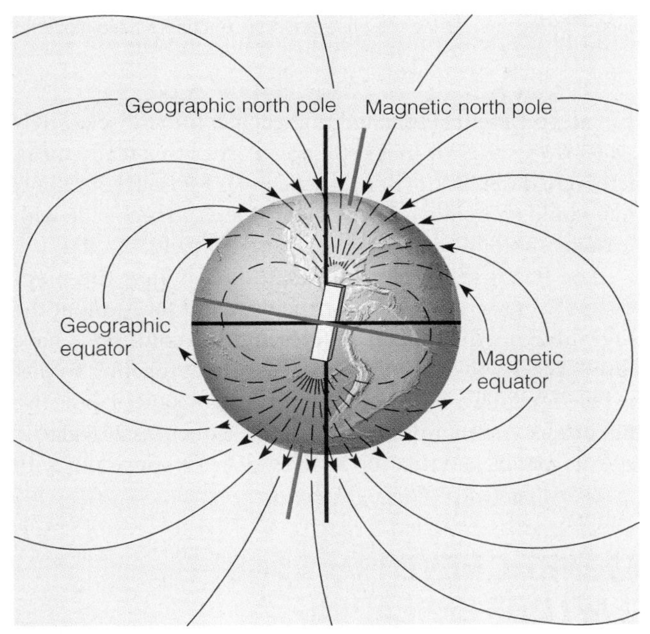

a Earth's magnetic field has lines of force like those of a bar magnet.

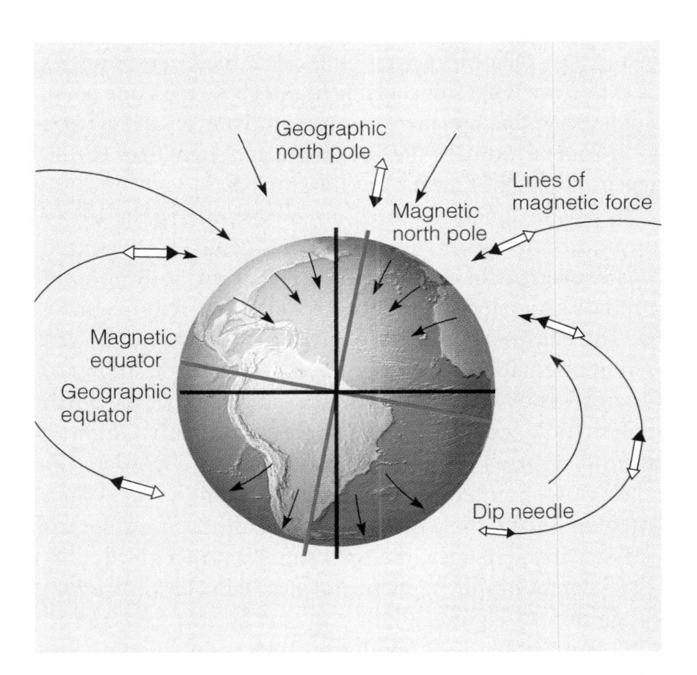

b The strength of the magnetic field changes from the magnetic equator to the magnetic poles. This change in strength causes a dip needle (a magnetic needle that is balanced on the tip of a support so that it can freely move vertically) to be parallel to Earth's surface only at the magnetic equator, where the strength of the magnetic north and south poles are equally balanced. Its inclination or dip with respect to Earth's surface increases as it moves toward the magnetic poles, until it is at 90 degrees or perpendicular to Earth's surface at the magnetic poles.

Another important aspect of the magnetic field is that the magnetic poles, where the lines of force leave and enter Earth, do not coincide with the geographic (rotational) poles. Currently, an 11.5° angle exists between the two (Figure 2.8). Studies of the Earth's magnetic field show that the locations of the magnetic poles vary slightly over time, but that they still correspond closely, on average, with the locations of the geographic poles.

Experts on magnetism do not fully understand all aspects of Earth's magnetic field, but most agree that electrical currents resulting from convection in the liquid outer core generate it. Furthermore, it must be generated continuously or it would decay and Earth would have no magnetic field in as little as 20,000 years. The model most widely accepted now is that thermal and compositional convection within the liquid outer core, coupled with Earth's rotation, produce complex electrical currents or a *self-exciting dynamo* that, in turn, generates the magnetic field.

PALEOMAGNETISM AND POLAR WANDERING

Interest in continental drift revived during the 1950s as a result of evidence from paleomagnetic studies, a relatively new discipline at the time. **Paleomagnetism** is the remanent magnetism in ancient rocks recording the direction and intensity of Earth's magnetic poles at the time of the rock's formation.

When magma cools, the magnetic iron-bearing minerals align themselves with Earth's magnetic field, recording both its direction and strength. The temperature at which iron-bearing minerals gain their magnetization is called the **Curie point**. As long as the rock is not subsequently heated above the Curie point, it will preserve that remanent magnetism. Thus, an ancient lava flow provides a record of the orientation and strength of Earth's magnetic field at the time the lava flow cooled.

As paleomagnetic research progressed during the 1950s, some unexpected results emerged. When geologists measured the paleomagnetism of geologically recent rocks, they found that it was generally consistent with Earth's current magnetic field. The paleomagnetism of ancient rocks, though, showed different orientations. For example, paleomagnetic studies of Silurian lava flows in North America indicated that the north magnetic pole was located in the western Pacific Ocean at that time, whereas the paleomagnetic evidence from Permian lava flows pointed to yet another location in Asia. When plotted on a map, the paleomagnetic readings of numerous lava flows from all ages in North America trace the apparent movement of the magnetic pole (called *polar wandering*) through time (Figure 2.9).

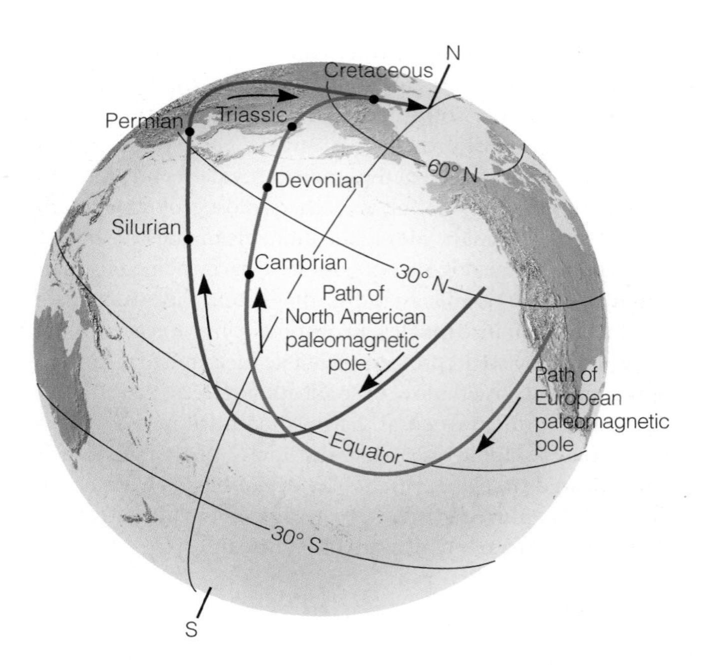

Figure 2.9 Polar Wandering The apparent paths of polar wandering for North America and Europe. The apparent location of the north magnetic pole is shown for different periods on each continent's polar wandering path. If the continents have not moved through time, and because Earth has only one magnetic north pole, the paleomagnetic readings for the same time in the past, taken on different continents, should all point to the same location. However, the north magnetic pole has different locations for the same time in the past when measured on different continents, indicating multiple north magnetic poles. The logical explanation for this dilemma is that the magnetic north pole has remained at the same approximate geographic location during the past, and the continents have moved.

This paleomagnetic evidence from a single continent could be interpreted in three ways: The continent remained fixed and the north magnetic pole moved; the north magnetic pole stood still and the continent moved; or both the continent and the north magnetic pole moved.

Upon analysis, magnetic minerals from European Silurian and Permian lava flows pointed to a different magnetic pole location from those of the same age from North America (Figure 2.9). Furthermore, analysis of lava flows from all continents indicated that each continent seemingly had its own series of magnetic poles. Does this really mean there were different north magnetic poles for each continent? That would be highly unlikely and difficult to reconcile with the theory accounting for Earth's magnetic field.

The best explanation for such data is that the magnetic poles have remained near their present locations at the geographic north and south poles and the continents have moved. When the continental margins are fit together so that the paleomagnetic data point to only one magnetic pole, we find, just as Wegener did, that the rock sequences and glacial deposits match, and that the fossil evidence is consistent with the reconstructed paleogeography.

MAGNETIC REVERSALS AND SEAFLOOR SPREADING

Geologists refer to Earth's present magnetic field as being normal—that is, with the north and south magnetic poles located approximately at the north and south geographic poles. At various times in the geologic past, however, Earth's magnetic field has completely reversed, that is, the magnetic north and south poles reverse positions, so that the magnetic north pole becomes the magnetic south pole, and the magnetic south pole becomes the magnetic north pole. During such a reversal, the magnetic field weakens until it temporarily disappears. When the magnetic field returns, the magnetic poles have reversed their position. The existence of such **magnetic reversals** was discovered by dating and determining the orientation of the remanent magnetism in lava flows on land (Figure 2.10). Although the cause of magnetic reversals is still uncertain, their occurrence in the geologic record is well documented.

A renewed interest in oceanographic research led to extensive mapping of the ocean basins during the 1960s. Such mapping revealed an oceanic ridge system more than 65,000 km long, constituting the most extensive mountain range in the world. Perhaps the best-known part of the ridge system is the Mid-Atlantic Ridge, which divides the Atlantic Ocean basin into two nearly equal parts (Figure 2.11).

As a result of oceanographic research conducted during the 1950s, Harry Hess of Princeton University proposed, in a 1962 landmark paper, the theory of **seafloor spreading** to account for continental movement. He suggested that continents do not move through oceanic crust as do ships plowing through sea ice, but rather that the continents and oceanic crust move together as a single unit. Thus, the theory of seafloor spreading answered a major objection of the opponents of continental drift—namely, how could continents move through oceanic crust? In fact, the continents moved with the oceanic crust as part of a lithospheric system.

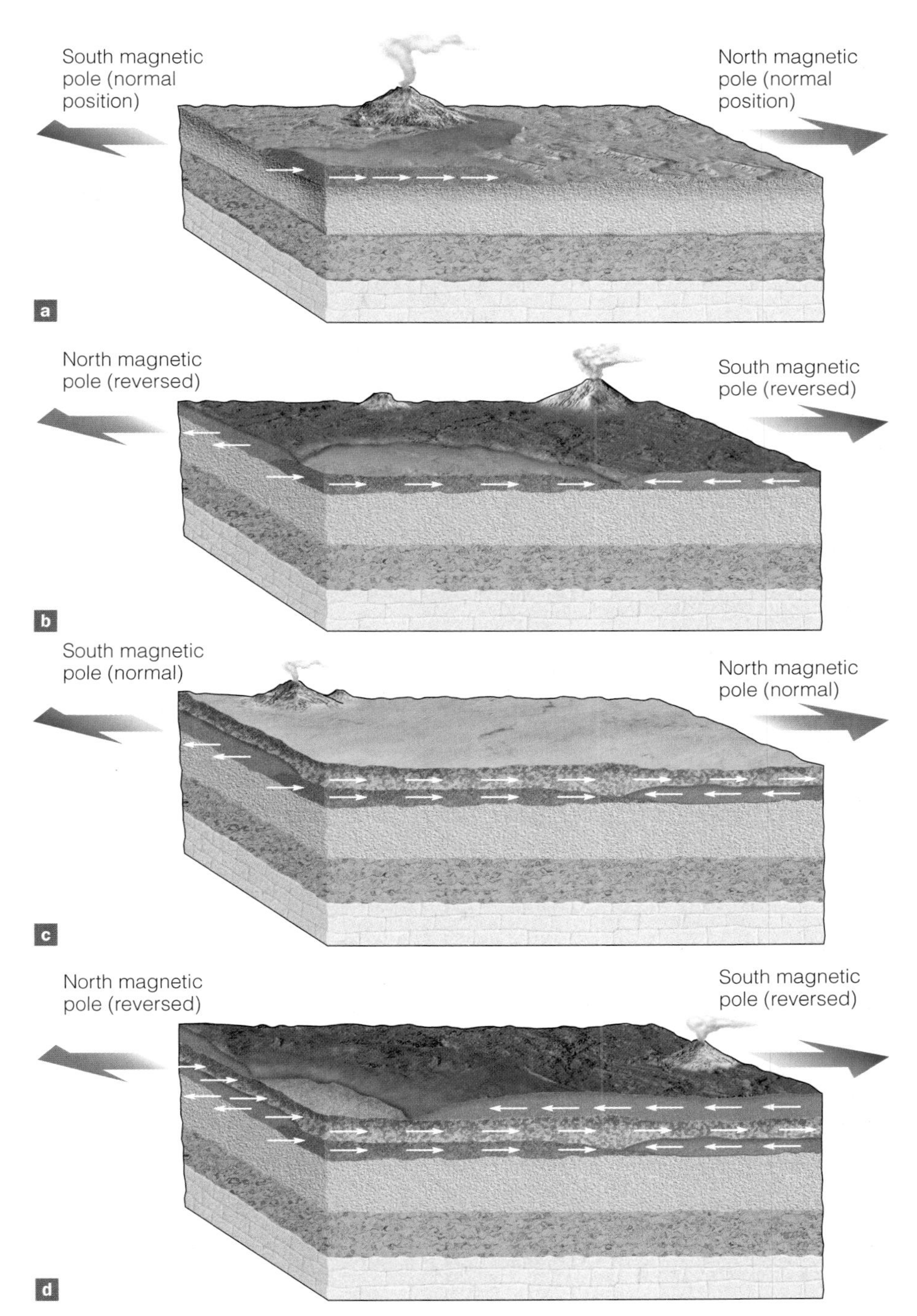

Figure 2.10 Magnetic Reversals During the time period shown (a–d), volcanic eruptions produced a succession of overlapping lava flows. At the time of these volcanic eruptions, Earth's magnetic field completely reversed—that is, the magnetic north pole moved to the geographic south pole, and the magnetic south pole moved to the geographic north pole. Thus, the end of the needle of a magnetic compass that today would point to the North Pole would point to the South Pole if the magnetic field should again reverse. We know that Earth's magnetic field has reversed numerous time in the past because when lava flows cool below the Curie point, magnetic minerals within the flow orient themselves parallel to the magnetic field at the time. They thus record whether the magnetic field was normal or reversed at that time. The white arrows in this diagram show the direction of the north magnetic pole for each individual lava flow, thus confirming that Earth's magnetic field has reversed in the past.

Hess postulated that the seafloor separates at oceanic ridges, where new crust is formed by upwelling magma. As the magma cools, the newly formed oceanic crust moves laterally away from the ridge.

As a mechanism to drive this system, Hess revived the idea (first proposed in the late 1920s by the British geologist Arthur Holmes) of a heat transfer system—or **thermal convection cells**—within the mantle as a mechanism to move the plates. According to Hess, hot magma rises from the mantle, intrudes along fractures defining oceanic ridges, and thus forms new crust. Cold crust is subducted back into the mantle at oceanic trenches, where it is heated and recycled, thus completing a thermal convection cell (see Figure 1.11).

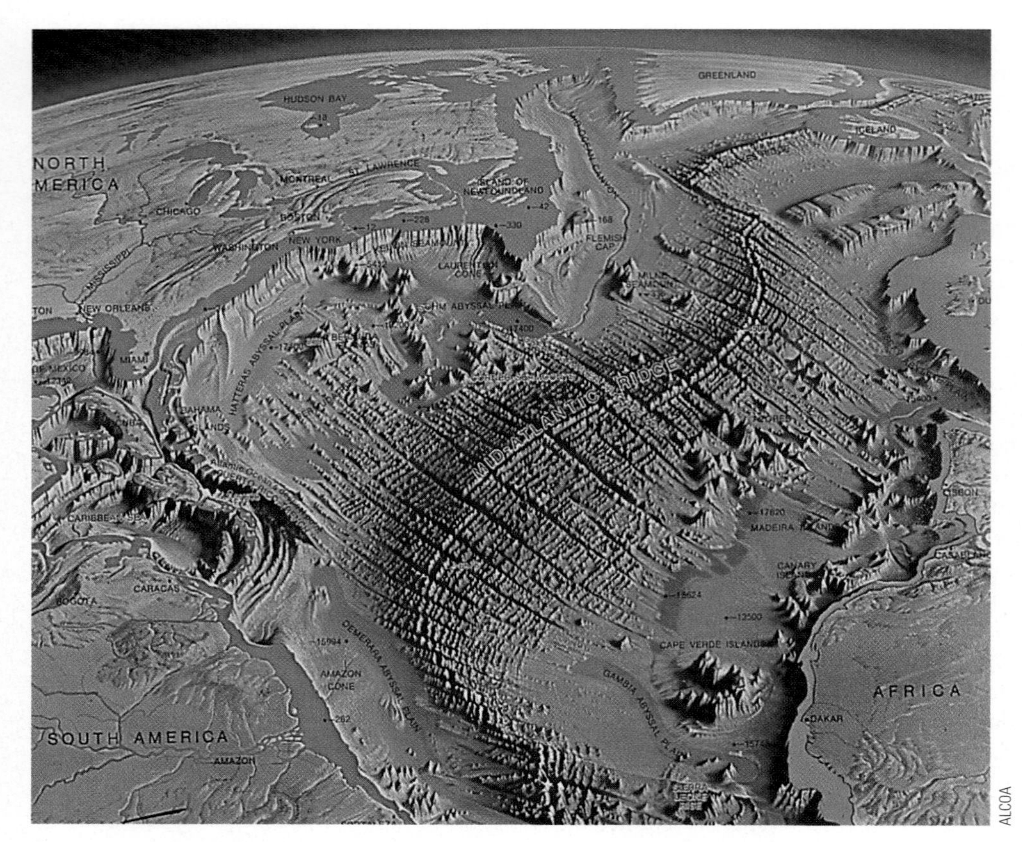

Figure 2.11 **Topography of the Atlantic Ocean Basin** Artistic view of what the Atlantic Ocean basin would look like without water. The major feature is the Mid-Atlantic Ridge, an oceanic ridge system that is longer than 65,000 km and divides the Atlantic Ocean basin in half. It is along such oceanic ridges that the seafloor is separating and new oceanic crust is forming from upwelling magma in Earth's interior.

How could Hess's hypothesis be confirmed? Magnetic surveys of the oceanic crust revealed a pattern of striped **magnetic anomalies** (deviations from the average strength of Earth's present-day magnetic field) in the rocks that are both parallel to and symmetric around the oceanic ridges (Figure 2.12). A positive magnetic anomaly results when Earth's magnetic field at the time of oceanic crust formation along an oceanic ridge summit was the same as today, thus yielding a stronger than normal (positive) magnetic signal. A negative magnetic anomaly results when Earth's magnetic field at the time of oceanic crust formation was reversed, therefore yielding a weaker than normal (negative) magnetic signal.

Thus, as new oceanic crust forms at oceanic ridge summits and records Earth's magnetic field at the time, the previously formed crust moves laterally away from the ridge. These magnetic stripes therefore represent times of normal and reversed polarity at oceanic ridges (where upwelling magma forms new oceanic crust), and conclusively confirm Hess's theory of seafloor spreading.

One of the consequences of the seafloor spreading theory is its confirmation that ocean basins are geologically young features whose openings and closings are partially responsible for continental movement (Figure 2.13). Radiometric dating reveals that the oldest oceanic crust is somewhat younger than 180 million years old, whereas the oldest continental crust is 3.96 billion years old. Although geologists do not universally accept the idea of thermal convection cells as a driving mechanism for plate movement, most accept that plates are created at oceanic ridges and destroyed at deep-sea trenches, regardless of the driving mechanism involved.

Deep-Sea Drilling and the Confirmation of Seafloor Spreading

For many geologists, the paleomagnetic data amassed in support of continental drift and seafloor spreading were convincing. Results from the Deep-Sea Drilling Project (see Chapter 9) confirmed the interpretations made from earlier paleomagnetic studies. Cores of deep-sea sediments and seismic profiles obtained by the *Glomar Challenger* and other research vessels have provided much of the data that support the seafloor spreading theory.

According to this theory, oceanic crust continuously forms at mid-oceanic ridges, moves away from these ridges by seafloor spreading, and is consumed at subduction zones. If this is the case, then oceanic crust should be youngest at the ridges and become progressively older with increasing distance away from them. Moreover, the age of the oceanic crust should be symmetrically distributed about the ridges. As we have just noted, paleomagnetic data confirm these statements. Furthermore, fossils from sediments overlying the oceanic crust and radiometric dating of rocks found on oceanic islands both substantiate this predicted age distribution.

Sediments in the open ocean accumulate, on average, at a rate of less than 0.3 cm in 1000 years. If the ocean basins were as old as the continents, we would expect deep-sea sediments to be several kilometers thick. However, data from numerous drill holes indicate that deep-sea sediments are, at most, only a few hundred meters thick and are thin or absent at oceanic ridges. Their near-absence at the ridges should come as no surprise because these are the areas where new crust is continuously produced by volcanism and seafloor spreading. Accordingly, sediments have had little time to accumulate at or very close to spreading ridges where the oceanic crust is young; however, their thickness increases with distance away from the ridges (Figure 2.14).

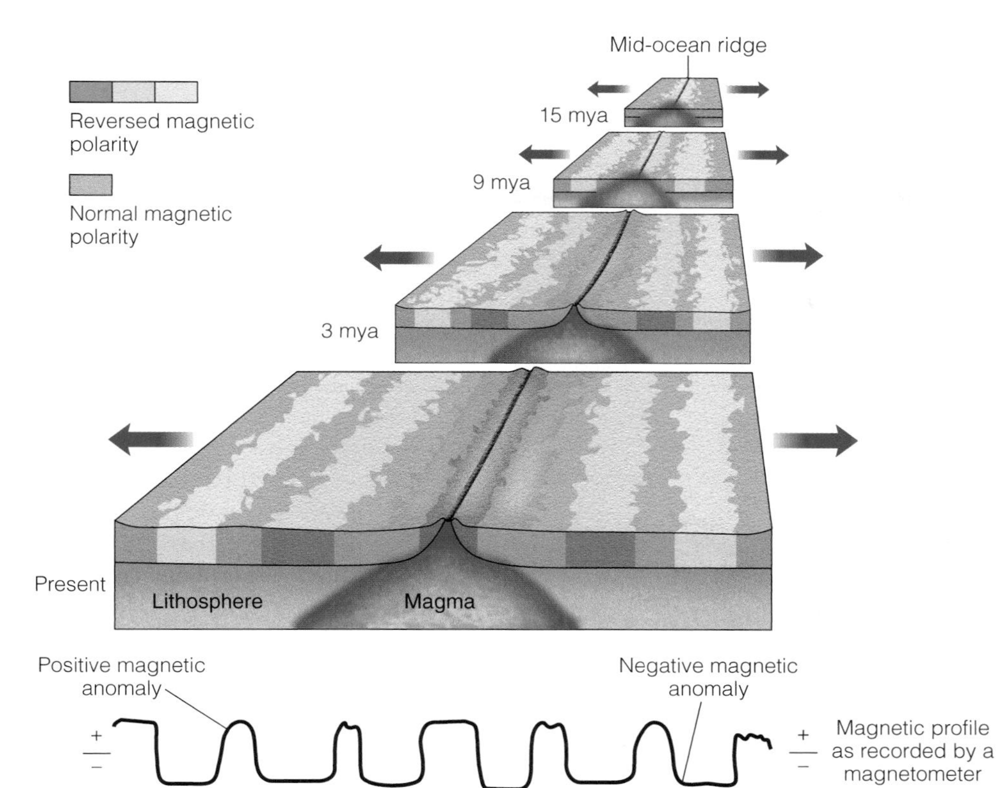

Figure 2.12 Magnetic Anomalies and Seafloor Spreading The sequence of magnetic anomalies preserved within the oceanic crust is both parallel to and symmetric around oceanic ridges. Basaltic lava intruding into an oceanic ridge today and spreading laterally away from the ridge records Earth's current magnetic field or polarity (considered by convention to be normal). Basaltic intrusions 3, 9, and 15 million years ago record Earth's reversed magnetic field at those times. This schematic diagram shows how the solidified basalt moves away from the oceanic ridge (or spreading ridge), carrying with it the magnetic anomalies that are preserved in the oceanic crust. Magnetic anomalies are magnetic readings that are either higher (positive magnetic anomalies) or lower (negative magnetic anomalies) than Earth's current magnetic field strength. The magnetic anomalies are recorded by a magnetometer, which measures the strength of the magnetic field. Modified from Kious and Tilling, USGS and Hyndman & Hyndman, *Natural Hazards and Disasters,* Brooks/Cole, 2006, p. 15, Fig. 2.6b.

An easy way to visualize plate movement is to think of a conveyer belt moving luggage from an airplane's cargo hold to a baggage cart. The conveyer belt represents convection currents within the mantle, and the luggage represents Earth's lithospheric plates. The luggage is moved along by the conveyer belt until it is dumped into the baggage cart in the same way that plates are moved by convection cells until they are subducted into Earth's interior. Although this analogy allows you to visualize how the mechanism of plate movement takes place, remember that this analogy is limited. The major limitation is that, unlike the luggage, plates consist of continental and oceanic crust, which have different densities, and only oceanic crust is subducted into Earth's interior.

Most geologists accept plate tectonic theory because the evidence for it is overwhelming and it ties together many seemingly unrelated geologic features and events and shows how they are interrelated. Consequently, geologists now view many geologic processes from the global perspective of plate tectonic theory in which plate interaction along plate margins is responsible for such phenomena as mountain building, earthquakes, and volcanism. Furthermore, because all of the inner planets have had a similar origin and early history, geologists are interested in determining whether plate tectonics is unique to Earth or whether it operates in the same way on other planets (see Geo-inSight on pages 42 and 43).

PLATE TECTONICS: A UNIFYING THEORY

Plate tectonic theory is based on a simple model of Earth. The rigid lithosphere, composed of both oceanic and continental crust, as well as the underlying upper mantle, consists of numerous variable-sized pieces called *plates* (Figure 2.15). There are seven major plates (Eurasian, Indian-Australian, Antarctic, North American, South American, Pacific, and African), and numerous smaller ones ranging from only a few tens to several hundreds of kilometers in width. Plates also vary in thickness; those composed of upper mantle and continental crust are as much as 250 km thick, whereas those of upper mantle and oceanic crust are up to 100 km thick.

The lithosphere overlies the hotter and weaker semiplastic asthenosphere. It is thought that movement resulting from some type of heat-transfer system within the asthenosphere causes the overlying plates to move. As plates move over the asthenosphere, they separate, mostly at oceanic ridges; in other areas, such as at oceanic trenches, they collide and are subducted back into the mantle.

THE THREE TYPES OF PLATE BOUNDARIES

Because it appears that plate tectonics has operated since at least the Proterozoic Eon, it is important that we understand how plates move and interact with each other and how ancient plate boundaries are recognized. After all, the movement of plates has profoundly affected the geologic and biologic history of this planet.

Geologists recognize three major types of plate boundaries: *divergent*, *convergent*, and *transform* (Table 2.1). Along these boundaries, new plates are formed, are consumed, or

What Would You Do?

You've been selected to be part of the first astronaut team to go to Mars. While your two fellow crew members descend to the Martian surface, you'll be staying in the command module and circling the Red Planet. As part of the geologic investigation of Mars, one of the crew members will be mapping the geology around the landing site and deciphering the geologic history of the area. Your job will be to observe and photograph the planet's surface and try to determine whether Mars had an active plate tectonic regime in the past and whether there is current plate movement. What features would you look for, and what evidence might reveal current or previous plate activity?

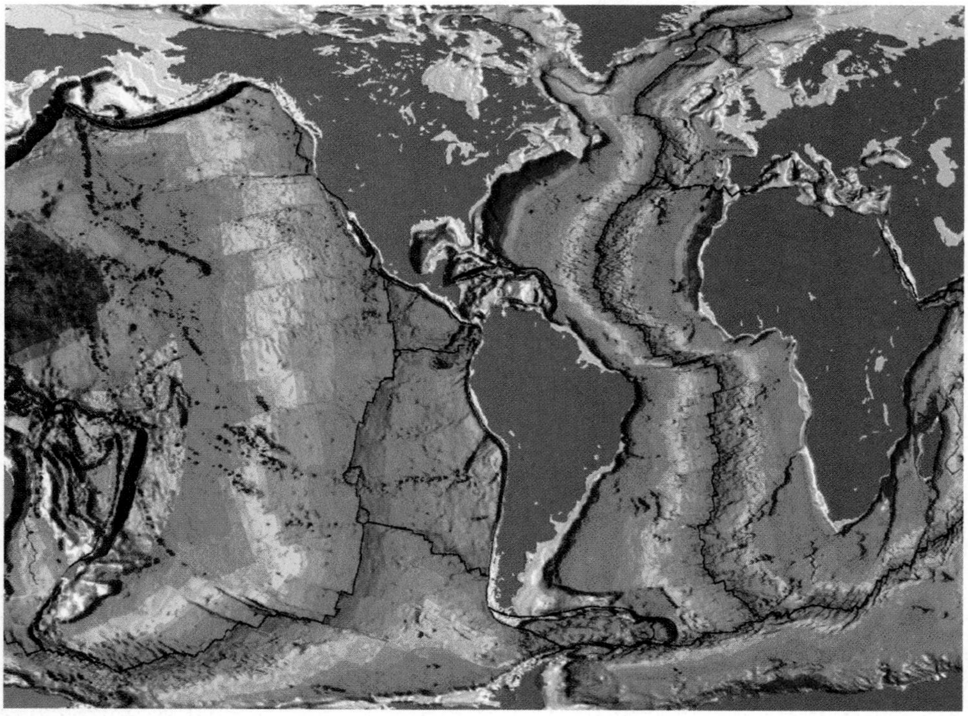

Figure 2.13 Age of the World's Ocean Basins The age of the world's ocean basins has been determined from magnetic anomalies preserved in oceanic crust. The red colors adjacent to the oceanic ridges are the youngest oceanic crust. Moving laterally away from the ridges, the red colors grade to yellow at 48 million years ago, to green at 68 million years ago, and to dark blue some 155 million years ago. The darkest blue color is adjacent to the continental margins and is just somewhat less than 180 million years old.

slide laterally past one another. Interaction of plates at their boundaries accounts for most of Earth's volcanic eruptions and earthquakes, as well as the formation and evolution of its mountain systems.

Divergent Boundaries

Divergent plate boundaries or *spreading ridges* occur where plates are separating and new oceanic lithosphere is forming. Divergent boundaries are places where the crust is extended, thinned, and fractured as magma, derived from the partial melting of the mantle, rises to the surface. The magma is almost entirely basaltic and intrudes into vertical fractures to form dikes and pillow lava flows (see Figure 5.7). As successive injections of magma cool and solidify, they form new oceanic crust and record the intensity and orientation of Earth's magnetic field (Figure 2.12). Divergent boundaries most commonly occur along the crests of oceanic ridges—for example, the Mid-Atlantic Ridge. Oceanic ridges are thus characterized by rugged topography with high relief resulting from displacement of rocks along large fractures, shallow-depth earthquakes, high heat flow, and basaltic flows or pillow lavas.

Divergent boundaries are also present under continents during the early stages of continental breakup. When magma wells up beneath a continent, the crust is initially elevated, stretched, and thinned, producing fractures, faults, rift valleys, and volcanic activity (Figure 2.16a). As magma intrudes into faults and fractures, it solidifies or flows out onto the surface as lava flows; the latter often covering the rift valley floor (Figure 2.16b). The East African Rift Valley is an excellent example of continental breakup at this stage (Figure 2.17a).

As spreading proceeds, some rift valleys continue to lengthen and deepen until the continental crust eventually breaks and a narrow linear sea is formed, separating two continental blocks (Figure 2.16c). The Red Sea separating the Arabian Peninsula from Africa (Figure 2.17b) and the Gulf of California, which separates Baja California from mainland Mexico, are good examples of this more advanced stage of rifting.

As a newly created narrow sea continues to enlarge, it may eventually become an expansive ocean basin such as the Atlantic Ocean basin is today, separating North and South America from Europe and Africa by thousands of kilometers (Figure 2.16d). The Mid-Atlantic Ridge is the boundary between these diverging plates (Figure 2.11); the American plates are moving westward, and the Eurasian and African plates are moving eastward.

An Example of Ancient Rifting What features in the geologic record can geologists use to recognize ancient rifting? Associated with regions of continental rifting are faults, dikes (vertical intrusive igneous bodies), sills (horizontal intrusive

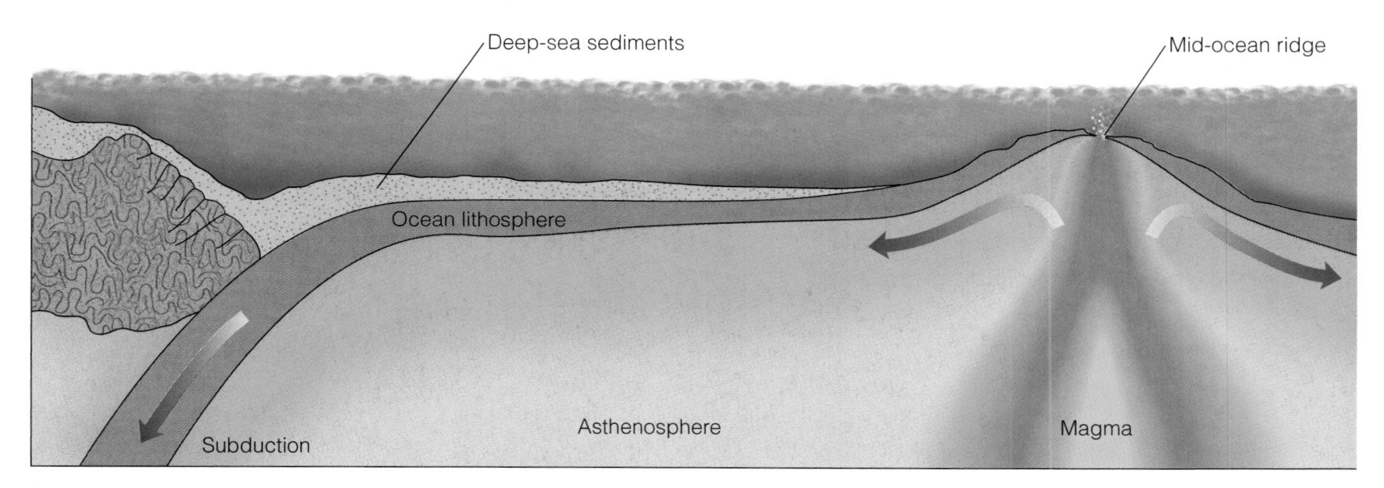

Figure 2.14 Deep-Sea Sediments and Seafloor Spreading The total thickness of deep-sea sediments increases away from oceanic ridges. This is because oceanic crust becomes older away from oceanic ridges, and there has been more time for sediment to accumulate.

igneous bodies), lava flows, and thick sedimentary sequences within rift valleys, all features that are preserved in the geologic record. The Triassic fault basins of the eastern United States are a good example of ancient continental rifting (see Figure 22.7). These fault basins mark the zone of rifting that occurred when North America split apart from Africa. The basins contain thousands of meters of continental sediment and are riddled with dikes and sills (see Chapter 22).

Pillow lavas, in association with deep-sea sediment, are also evidence of ancient rifting. The presence of pillow lavas marks the formation of a spreading ridge in a narrow linear sea. A narrow linear sea forms when the continental crust in the rift valley finally breaks apart, and the area is flooded with seawater. Magma, intruding into the sea along this newly formed spreading ridge, solidifies as pillow lavas, which are preserved in the geologic record, along with the sediment being deposited on them.

Convergent Boundaries

Whereas new crust forms at divergent plate boundaries, older crust must be destroyed and recycled in order for the entire

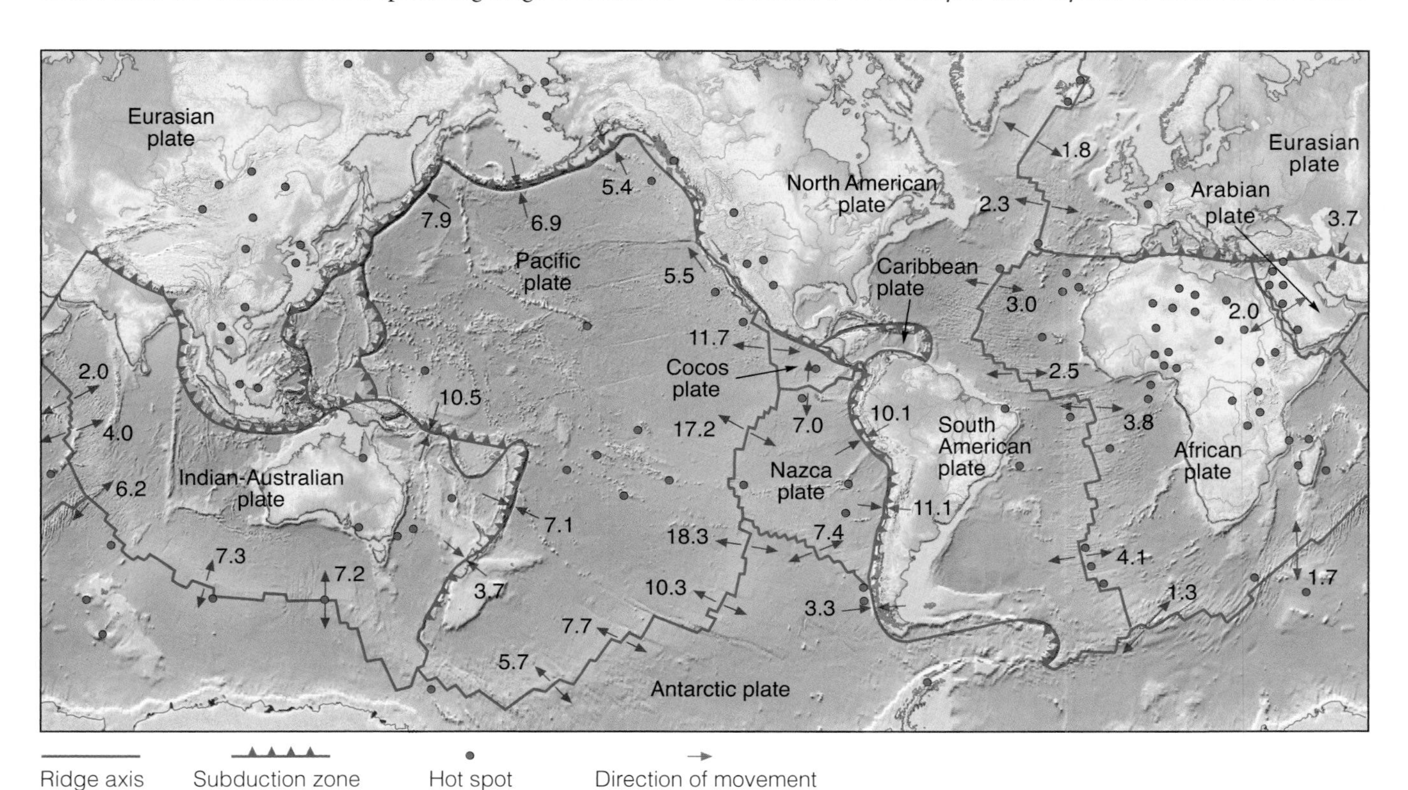

Figure 2.15 Earth's Plates A world map showing Earth's plates, their boundaries, their relative motion and rates of movement in centimeters per year, and hot spots.

Geo-inSight Tectonics of the Terrestrial Planets

The four inner, or terrestrial, planets—Mercury, Venus, Earth, and Mars—all had a similar early history involving accretion, differentiation into a metallic core and silicate mantle and crust, and formation of an early atmosphere by outgassing. Their early history was also marked by widespread volcanism and meteorite impacts, both of which helped modify their surfaces. Whereas the other three terrestrial planets as well as some of the Jovian moons display internal activity, Earth appears to be unique in that its surface is broken into a series of plates.

◀ **1.** A color-enhanced photomosaic of Mercury shows its heavily cratered surface, which has changed very little since its early history.

JPL/ NASA

Courtesy of Victor Royer

Images of **Mercury** sent back by *Mariner 10* show a heavily cratered surface with the largest impact basins filled with what appear to be lava flows similar to the lava plains on Earth's Moon. The lava plains are not deformed, however, indicating that there has been little or no tectonic activity.

Another feature of Mercury's surface is a large number of scarps, a feature usually associated with earthquake activity. Yet, some scientists think that these scarps formed when Mercury cooled and contracted.

▲ **2.** Seven scarps (indicated by arrows) can clearly be seen in this image. These scarps might have formed when Mercury cooled and contracted early in its history.

JPL/ NASA

Of all the planets, **Venus** is the most similar in size and mass to Earth, but it differs in most other respects. Whereas Earth is dominated by plate tectonics, volcanism seems to have been the dominant force in the evolution of the Venusian surface. Even though no active volcanism has been observed on Venus, the various-sized volcanic features and what appear to be folded mountains indicate a once-active planetary interior. All of these structures appear to be the products of rising convection currents of magma pushing up under the crust and then sinking back into the Venusian interior.

▲ **3.** A color-enhanced photomosaic of Venus based on radar images beamed back to Earth by the *Magellan* spacecraft. This image shows impact craters and volcanic features characteristic of the planet.

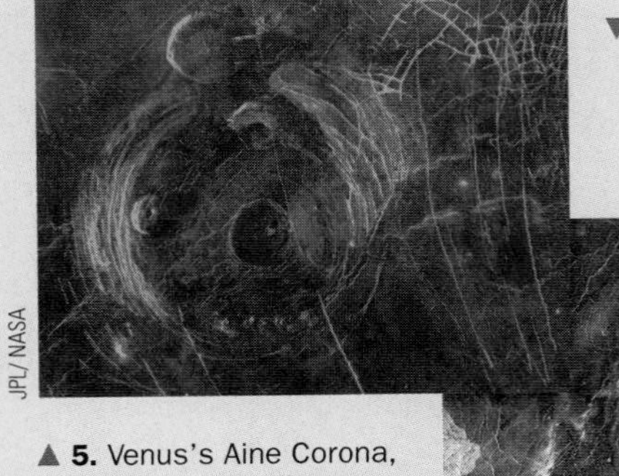

JPL/ NASA

▲ **5.** Venus's Aine Corona, about 200 km in diameter, is ringed by concentric faults, suggesting that it was pushed up by rising magma. A network of fractures is visible in the upper right of this image as well as a recent lava flow at the center of the corona, several volcanic domes in the lower portion of the image, and a large volcanic pancake dome in the upper left of the image.

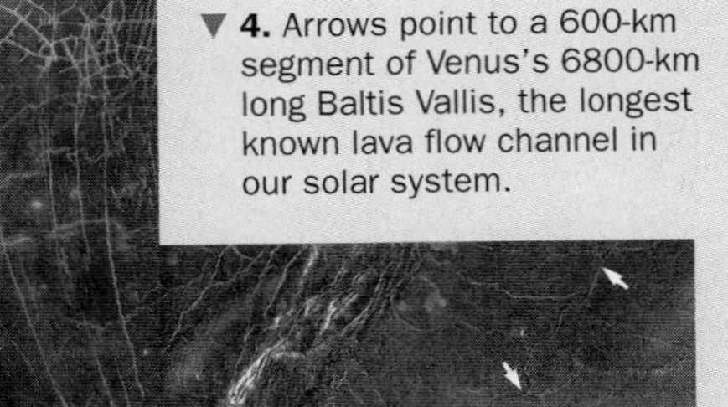

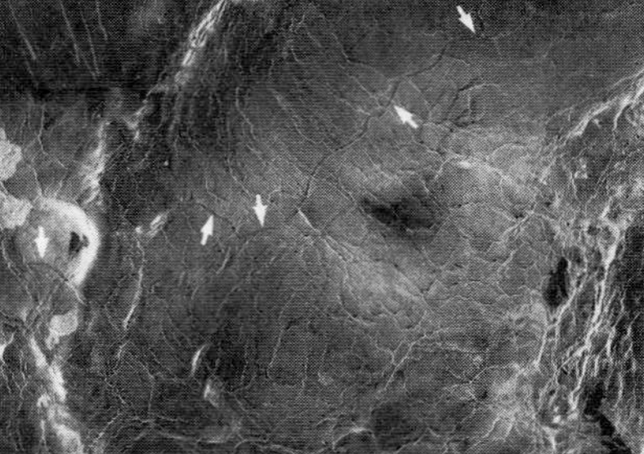

NASA

▼ **4.** Arrows point to a 600-km segment of Venus's 6800-km long Baltis Vallis, the longest known lava flow channel in our solar system.

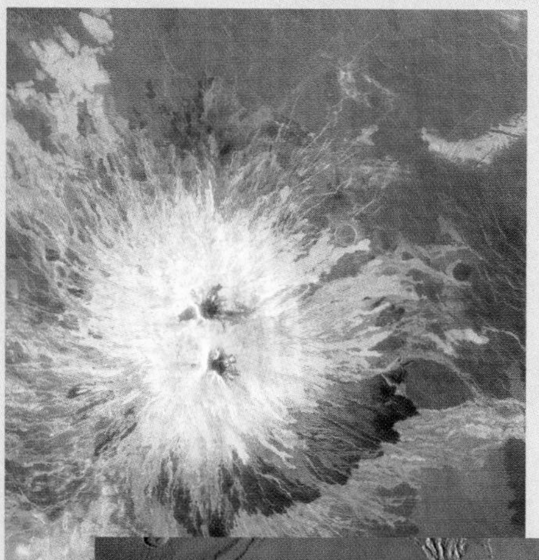

◀ **6.** Volcano Sapas Mons contains two lava-filled calderas and is flanked by lava flows, attesting to the volcanic activity that was once common on Venus.

Mars, the Red Planet, has numerous features that indicate an extensive early period of volcanism. These include Olympus Mons, the solar system's largest volcano, lava flows, and uplifted regions thought to have resulted from mantle convection. In addition to volcanic features, Mars displays abundant evidence of tensional tectonics, including numerous faults and large fault-produced valley structures. Whereas Mars was tectonically active during the past, no evidence indicates that plate tectonics comparable to that on Earth has ever occurred there.

▲ **7.** A vertical view of Olympus Mons, a shield volcano and the largest volcano in our solar system. The edge of the Olympus Mons caldera is marked by a cliff several kilometers high rather than a moat as in Mauna Loa, Earth's largest shield volcano.

USGS

▲ **8.** A photomosaic of Mars shows a variety of geologic structures, including the southern polar ice cap.

Although not a terrestrial planet, **Io**, the innermost of Jupiter's Galilean moons, must be mentioned. Images from the *Voyager* and *Galileo* spacecrafts show that Io has no impact craters. In fact, more than a hundred active volcanoes are visible on the moon's surface, and the sulfurous gas and ash erupted by these volcanoes bury any newly formed meteorite impact craters. Because of its proximity to Jupiter, the heat source of Io is probably tidal heating, in which the resulting friction is enough to at least partially melt Io's interior and drive its volcanoes.

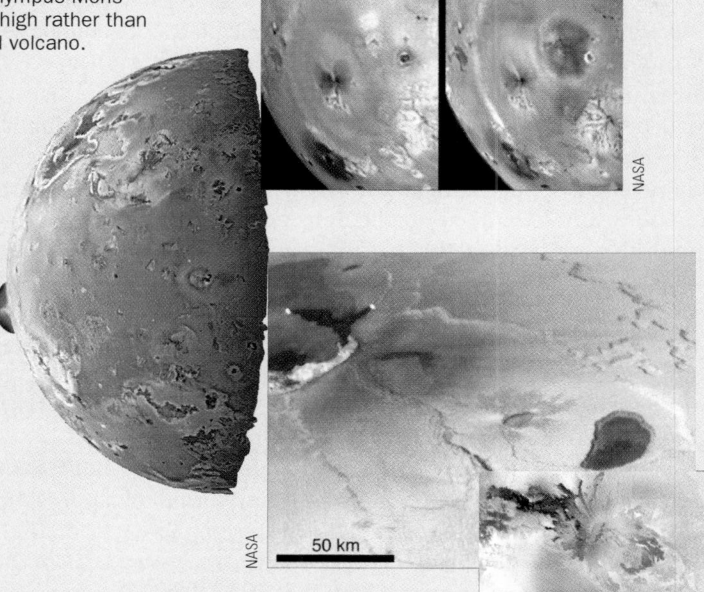

▶ **9.** Volcanic features of Io, the innermost moon of Jupiter. As shown in these digitally enhanced color images, Io is a very volcanically active moon.

TABLE 2.1 Types of Plate Boundaries

Type	Example	Landforms	Volcanism
Divergent			
Oceanic	Mid-Atlantic Ridge	Mid-oceanic ridge with axial rift valley	Basalt
Continental	East African Rift Valley	Rift valley	Basalt and rhyolite, no andesite
Convergent			
Oceanic–oceanic	Aleutian Islands	Volcanic island arc, offshore oceanic trench	Andesite
Oceanic–continental	Andes	Offshore oceanic trench, volcanic mountain chain, mountain belt	Andesite
Continental–continental	Himalayas	Mountain belt	Minor
Transform	San Andreas fault	Fault valley	Minor

surface area of Earth to remain the same. Otherwise, we would have an expanding Earth. Such plate destruction occurs at **convergent plate boundaries** (Figure 2.18), where two plates collide and the leading edge of one plate is subducted beneath the margin of the other plate and eventually incorporated into the asthenosphere. A dipping plane of earthquake foci, called a *Benioff zone,* defines subduction zones (see Figure 8.5). Most of these planes dip from oceanic trenches beneath adjacent island arcs or continents, marking the surface of slippage between the converging plates.

Deformation, volcanism, mountain building, metamorphism, earthquake activity, and deposits of valuable minerals characterize convergent boundaries. Three types of convergent plate boundaries are recognized: *oceanic–oceanic, oceanic–continental,* and *continental–continental.*

Oceanic–Oceanic Boundaries When two oceanic plates converge, one is subducted beneath the other along an **oceanic–oceanic plate boundary** (Figure 2.18a). The subducting plate bends downward to form the outer wall of an oceanic trench. A *subduction complex*, composed of wedge-shaped slices of highly folded and faulted marine sediments and oceanic lithosphere scraped off the descending plate, forms along the inner wall of the oceanic trench. As the subducting plate descends into the mantle, it is heated and partially melted, generating magma commonly of andesitic composition (see Chapter 4). This magma is less dense than the surrounding mantle rocks and rises to the surface of the nonsubducted plate to form a curved chain of volcanic islands called a *volcanic island arc* (any plane intersecting a sphere makes an arc). This arc is nearly parallel to the oceanic trench and is separated from it by a distance of up to several hundred kilometers—the distance depends on the angle of dip of the subducting plate (Figure 2.18a).

In those areas where the rate of subduction is faster than the forward movement of the overriding plate, the lithosphere on the landward side of the volcanic island arc may be subjected to tensional stress and stretched and thinned, resulting in the formation of a *back-arc basin.* This back-arc basin may grow by spreading if magma breaks through the thin crust and forms new oceanic crust (Figure 2.18a). A good example of a back-arc basin associated with an oceanic–oceanic plate boundary is the Sea of Japan between the Asian continent and the islands of Japan.

Most present-day active volcanic island arcs are in the Pacific Ocean basin and include the Aleutian Islands, the Kermadec–Tonga arc, and the Japanese (Figure 2.18a) and Philippine Islands. The Scotia and Antillean (Caribbean) island arcs are in the Atlantic Ocean basin.

Oceanic–Continental Boundaries When an oceanic and a continental plate converge, the denser oceanic plate is subducted under the continental plate along an **oceanic–continental plate boundary** (Figure 2.18b). Just as at oceanic–oceanic plate boundaries, the descending oceanic plate forms the outer wall of an oceanic trench.

The magma generated by subduction rises beneath the continent and either crystallizes as large intrusive bodies, called *plutons*, before reaching the surface or erupts at the surface to produce a chain of andesitic volcanoes, also called a *volcanic arc*. An excellent example of an oceanic–continental plate boundary is the Pacific Coast of South America where the oceanic Nazca plate is currently being subducted under South America (Figure 2.18b; see also Chapter 10). The Peru–Chile Trench marks the site of subduction, and the Andes Mountains are the resulting volcanic mountain chain on the nonsubducting plate.

Continental–Continental Boundaries Two continents approaching each other are initially separated by an ocean floor that is being subducted under one continent. The edge of that continent displays the features characteristic of oceanic–continental convergence. As the ocean floor continues to be subducted, the two continents come closer together until they eventually collide. Because continental lithosphere, which consists of continental crust and the upper mantle, is less dense

Figure 2.16 History of a Divergent Plate Boundary

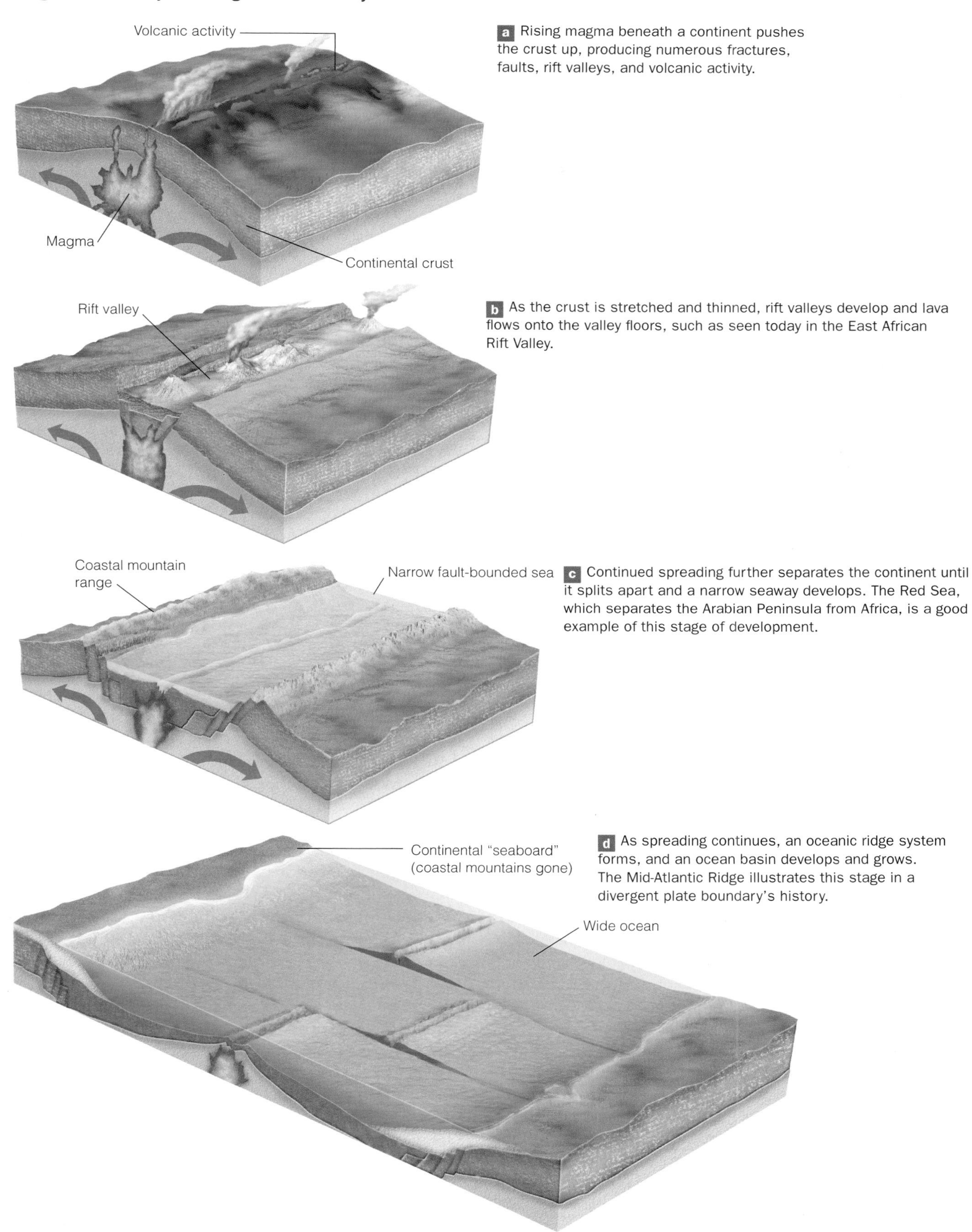

a Rising magma beneath a continent pushes the crust up, producing numerous fractures, faults, rift valleys, and volcanic activity.

b As the crust is stretched and thinned, rift valleys develop and lava flows onto the valley floors, such as seen today in the East African Rift Valley.

c Continued spreading further separates the continent until it splits apart and a narrow seaway develops. The Red Sea, which separates the Arabian Peninsula from Africa, is a good example of this stage of development.

d As spreading continues, an oceanic ridge system forms, and an ocean basin develops and grows. The Mid-Atlantic Ridge illustrates this stage in a divergent plate boundary's history.

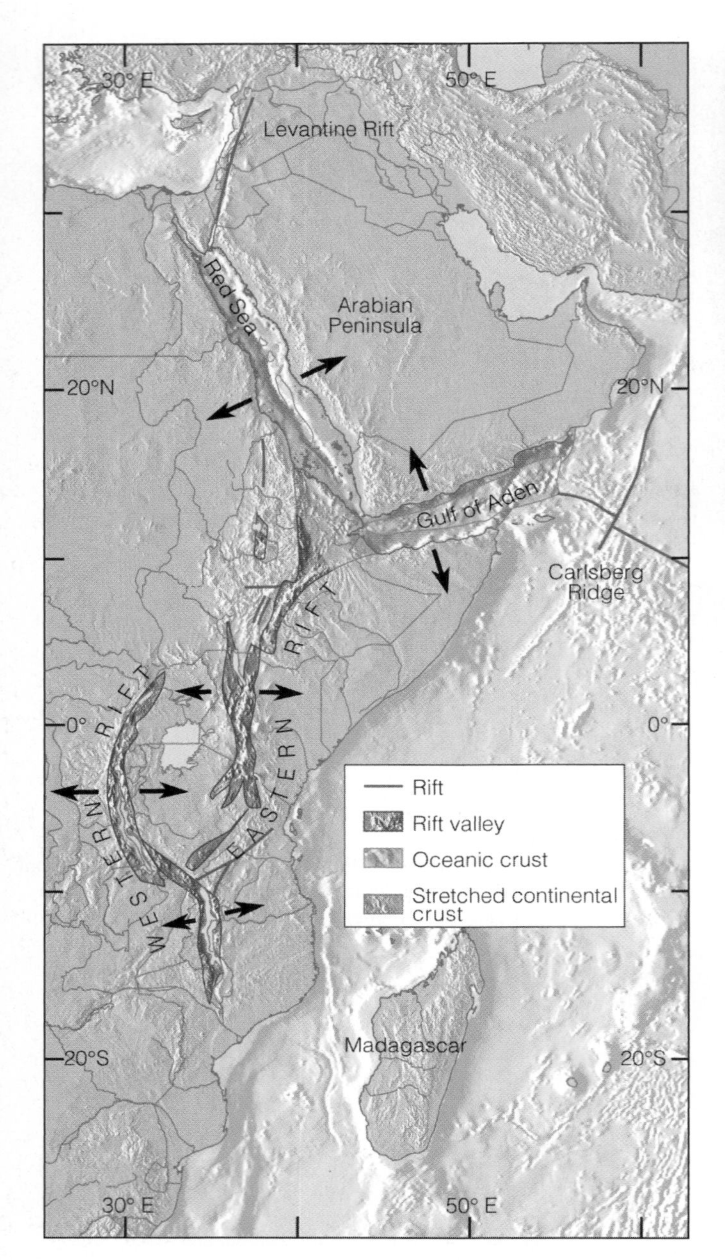

a The East African Rift Valley is being formed by the separation of eastern Africa from the rest of the continent along a divergent plate boundary.

than oceanic lithosphere (oceanic crust and upper mantle), it cannot sink into the asthenosphere. Although one continent may partially slide under the other, it cannot be pulled or pushed down into a subduction zone (Figure 2.18c).

When two continents collide, they are welded together along a zone marking the former site of subduction. At this **continental–continental plate boundary,** an interior mountain belt is formed consisting of deformed sediments and sedimentary rocks, igneous intrusions, metamorphic rocks, and fragments of oceanic crust. In addition, the entire region is subjected to numerous earthquakes. The Himalayas in central Asia, the world's youngest and highest mountain

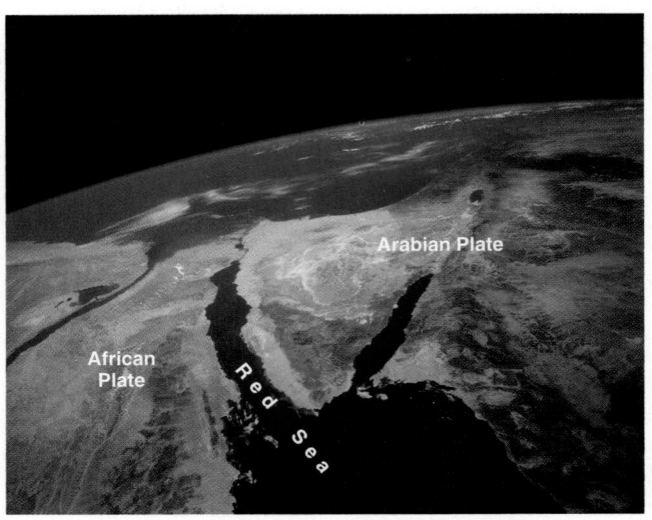

b The Red Sea represents a more advanced stage of rifting, in which two continental blocks (Africa and the Arabian Peninsula are separated by a narrow sea.

Figure 2.17 East African Rift Valley and the Red Sea—Present-Day Examples of Divergent Plate Boundaries The East African Rift Valley and the Red Sea represent different stages in the history of a divergent plate boundary.

system, resulted from the collision between India and Asia that began 40 to 50 million years ago and is still continuing (Figure 2.18c; see Chapter 10).

Recognizing Ancient Convergent Plate Boundaries How can former subduction zones be recognized in the geologic record? Igneous rocks provide one clue to ancient subduction zones. The magma erupted at the surface, forming island arc volcanoes and continental volcanoes, and is of andesitic composition. Another clue is the zone of intensely deformed rocks between the deep-sea trench where subduction is taking place and the area of igneous activity. Here, sediments and submarine rocks are folded, faulted, and metamorphosed into a chaotic mixture of rocks termed a *mélange.*

During subduction, pieces of oceanic lithosphere are sometimes incorporated into the mélange and accreted onto the edge of the continent. Such slices of oceanic crust and upper mantle are called *ophiolites* (Figure 2.19). They consist of a layer of deep-sea sediments that include graywackes (poorly sorted sandstones containing abundant feldspars and rock fragments, usually in a clay-rich matrix), black shales, and cherts (see Chapter 6). These deep-sea sediments are underlain by pillow lavas, a sheeted dike complex, massive gabbro (a dark intrusive igneous rock), and layered gabbro, all of which form the oceanic crust. Beneath the gabbro is peridotite (a dark intrusive igneous rock composed of the mineral olivine), which probably represents the upper mantle. The presence of ophiolite in an outcrop or drilling core is a key indicator of plate convergence along a subduction zone.

Figure 2.18 Three Types of Convergent Plate Boundaries

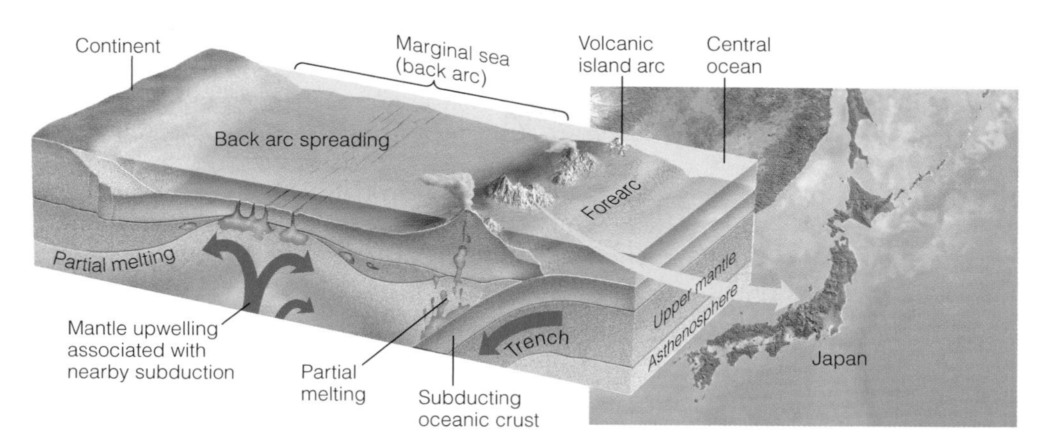

a Oceanic–oceanic plate boundary. An oceanic trench forms where one oceanic plate is subducted beneath another. On the nonsubducted plate, a volcanic island arc forms from the rising magma generated from the subducting plate. The Japanese Islands are a volcanic island arc resulting from the subduction of one oceanic plate beneath another oceanic plate.

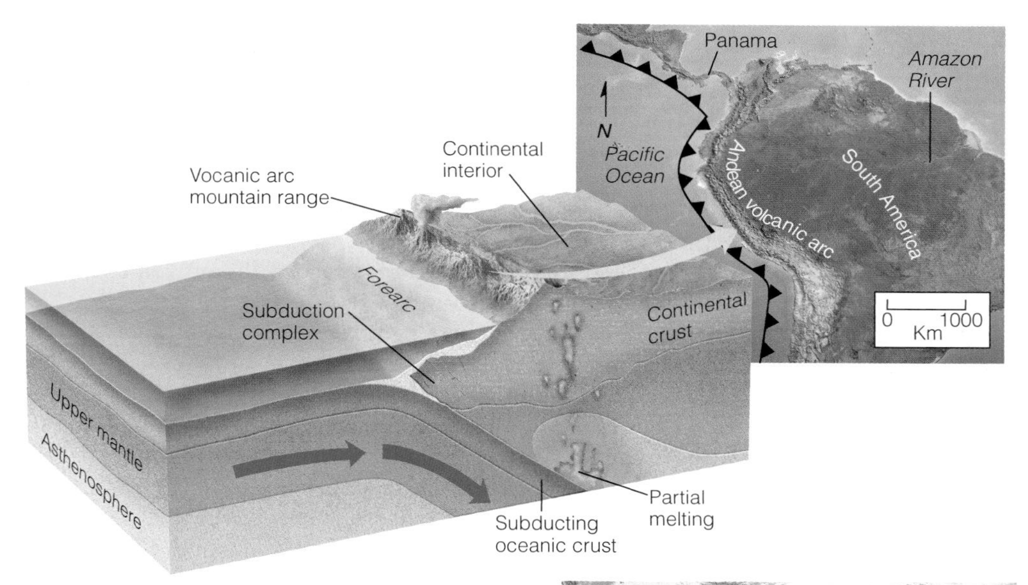

b Oceanic–continental plate boundary. When an oceanic plate is subducted beneath a continental plate, an andesitic volcanic mountain range is formed on the continental plate as a result of rising magma. The Andes Mountains in Peru are one of the best examples of continuing mountain building at an oceanic–continental plate boundary.

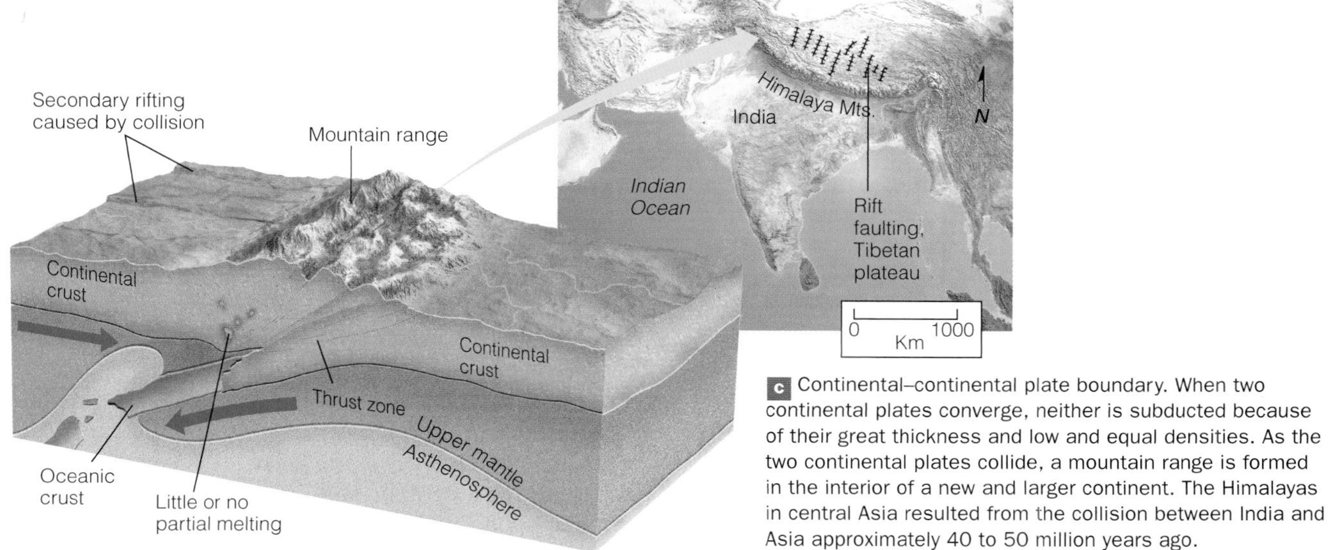

c Continental–continental plate boundary. When two continental plates converge, neither is subducted because of their great thickness and low and equal densities. As the two continental plates collide, a mountain range is formed in the interior of a new and larger continent. The Himalayas in central Asia resulted from the collision between India and Asia approximately 40 to 50 million years ago.

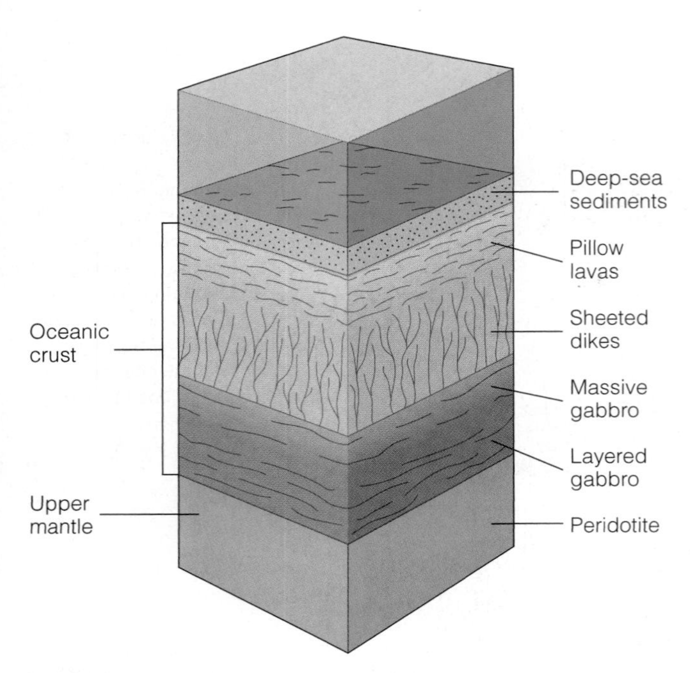

Figure 2.19 Ophiolites Ophiolites are sequences of rock on land consisting of deep-sea sediment, oceanic crust, and upper mantle. Ophiolites are one feature used to recognize ancient convergent plate boundaries.

Elongated belts of folded and faulted marine sedimentary rocks, andesites, and ophiolites are found in the Appalachians, Alps, Himalayas, and Andes mountains. The combination of such features is significant evidence that these mountain ranges resulted from deformation along convergent plate boundaries.

Transform Boundaries

The third type of plate boundary is a **transform plate boundary**. These mostly occur along fractures in the seafloor, known as *transform faults,* where plates slide laterally past one another roughly parallel to the direction of plate movement. Although lithosphere is neither created nor destroyed along a transform boundary, the movement between plates results in a zone of intensely shattered rock and numerous shallow-depth earthquakes.

Transform faults "transform" or change one type of motion between plates into another type of motion. Most commonly, transform faults connect two oceanic ridge segments; however, they can also connect ridges to trenches and trenches to trenches (Figure 2.20). Although the majority of transform faults are in oceanic crust and are marked by distinct fracture zones, they may also extend into continents.

One of the best-known transform faults is the San Andreas fault in California. It separates the Pacific plate from the North American plate and connects spreading ridges in the Gulf of California with the Juan de Fuca and Pacific plates off the coast of northern California (Figure 2.21). Many of the earthquakes that affect California are the result of movement along this fault (see Chapter 8).

Unfortunately, transform faults generally do not leave any characteristic or diagnostic features except for the obvious displacement of the rocks with which they are associated. This displacement is usually large, on the order of tens to hundreds of kilometers. Such large displacements in ancient rocks can sometimes be related to transform fault systems.

HOT SPOTS AND MANTLE PLUMES

Before leaving the topic of plate boundaries, we should mention an intraplate feature found beneath both oceanic and continental plates. A **hot spot** (Figure 2.15) is the location on Earth's surface where a stationary column of magma, originating deep within the mantle (*mantle plume*), has slowly risen to the surface and formed a volcano. Because the mantle plumes apparently remain stationary (although some evidence suggests that they might not) within the mantle while the plates move over them, the resulting hot spots leave a trail of extinct and progressively older volcanoes called *aseismic ridges* that record the movement of the plate.

One of the best examples of aseismic ridges and hot spots is the Emperor Seamount–Hawaiian Island chain (Figure 2.22). This chain of islands and seamounts (structures of volcanic origin rising higher than 1 km above the seafloor) extends from the island of Hawaii to the Aleutian Trench off Alaska, a distance of some 6000 km, and consists of more than 80 volcanic structures.

Currently, the only active volcanoes in this island chain are on the island of Hawaii and the Loihi Seamount. The rest of the islands are extinct volcanic structures that become progressively older toward the north and northwest. This means that the Emperor Seamount–Hawaiian Island chain records the direction that the Pacific plate traveled as it moved over an apparently stationary mantle plume. In this case, the Pacific plate first moved in a north-northwesterly direction and then, as indicated by the sharp bend in the chain, changed to a west-northwesterly direction approximately 43 million years ago. The reason that the Pacific plate changed directions is not known, but the shift might be related to the collision of India with the Asian continent at about the same time (see Figure 10.22).

Mantle plumes and hot spots help geologists explain some of the geologic activity occurring within plates as opposed to activity occurring at or near plate boundaries. In addition, if mantle plumes are essentially fixed with respect to Earth's rotational axis, they can be used to determine not only the direction of plate movement, but also the rate of movement. They can also provide

Figure 2.20 Transform Plate Boundaries Horizontal movement between plates occurs along transform faults. Extensions of transform faults on the seafloor form fracture zones.

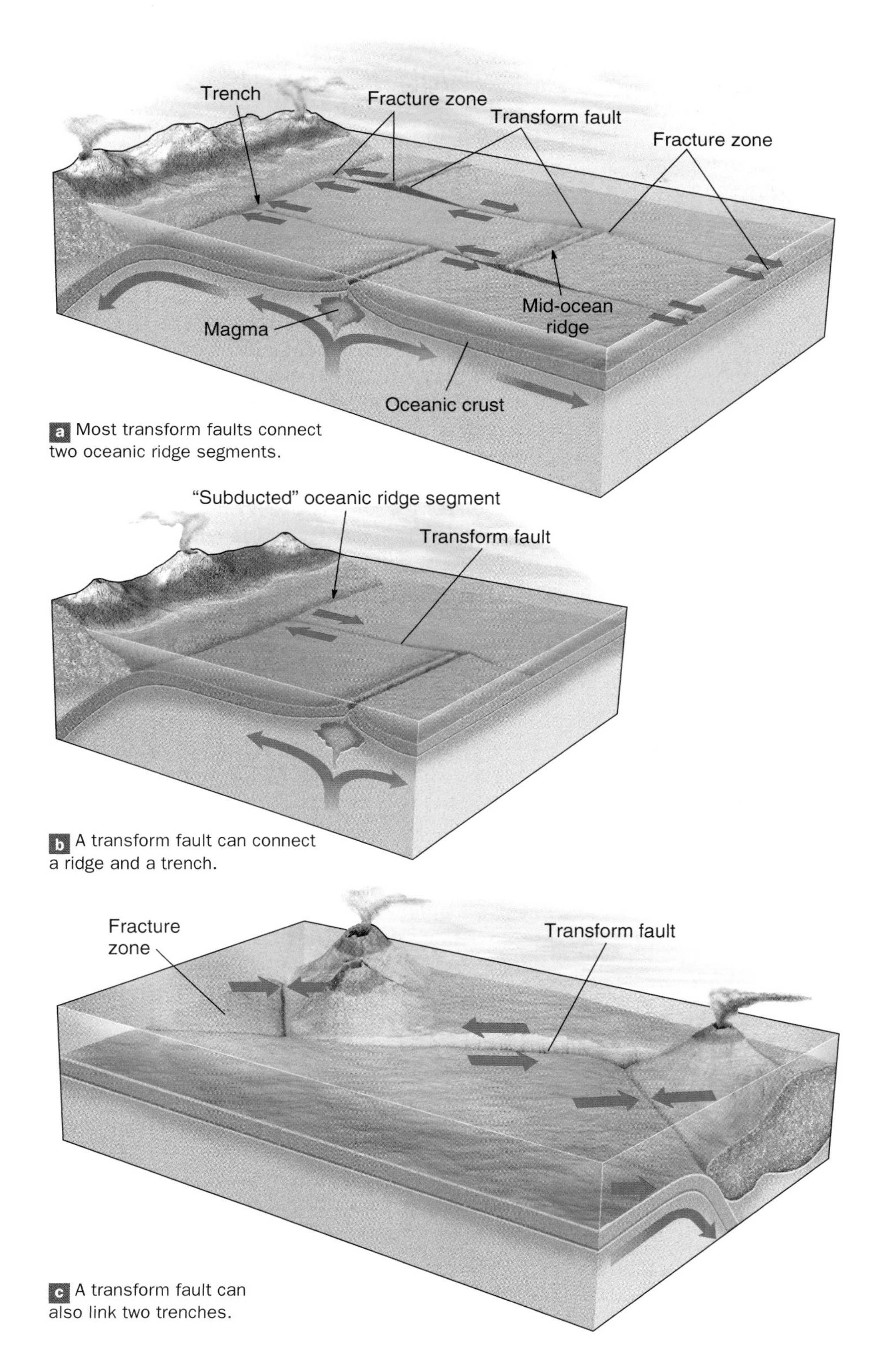

a Most transform faults connect two oceanic ridge segments.

b A transform fault can connect a ridge and a trench.

c A transform fault can also link two trenches.

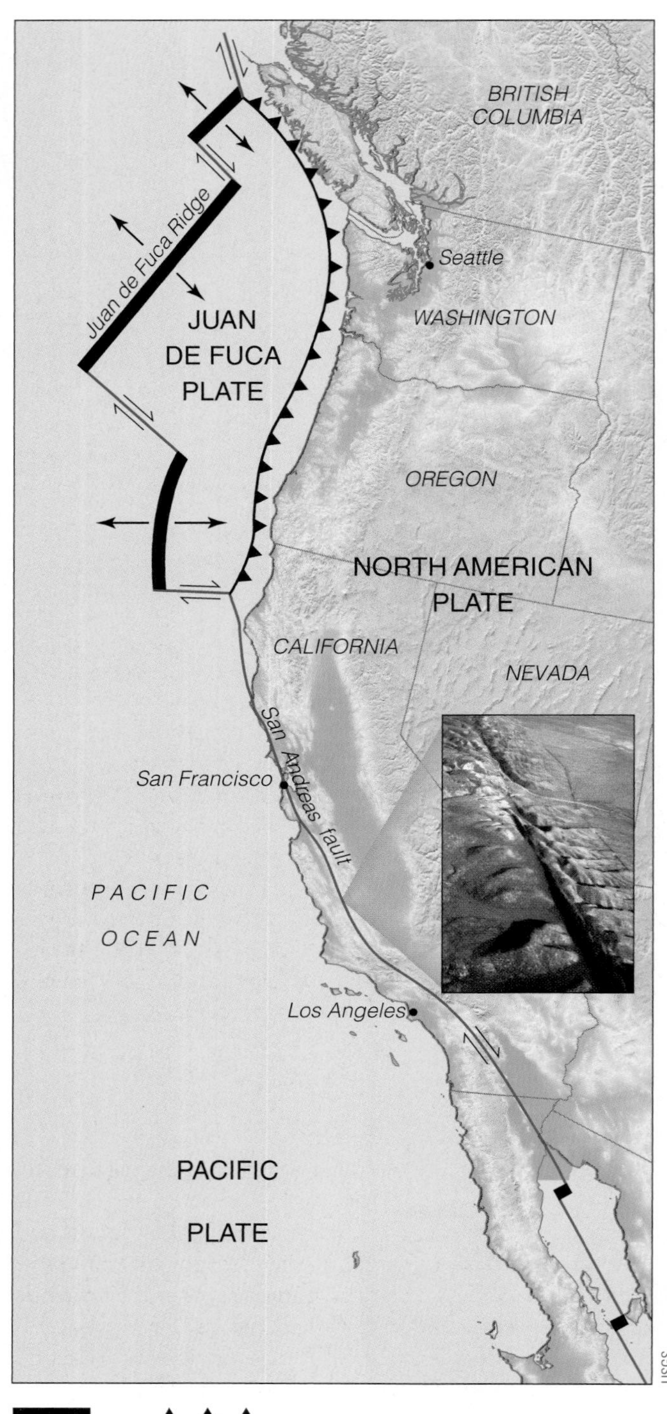

Figure 2.21 The San Andreas Fault—A Transform Plate Boundary The San Andreas fault is a transform fault separating the Pacific plate from the North American plate. It connects the spreading ridges in the Gulf of California with the Juan de Fuca and Pacific plates off the coast of northern California. Movement along the San Andreas fault has caused numerous earthquakes. The insert photograph shows a segment of the San Andreas fault as it cuts through the Carrizo Plain, California.

reference points for determining paleolatitude, an important tool when reconstructing the location of continents in the geologic past.

PLATE MOVEMENT AND MOTION

How fast and in what direction are Earth's plates moving? Do they all move at the same rate? Rates of plate movement can be calculated in several ways. The least accurate method is to determine the age of the sediments immediately above any portion of the oceanic crust and then divide the distance from the spreading ridge by that age. Such calculations give an average rate of movement.

A more accurate method of determining both the average rate of movement and relative motion is by dating the magnetic anomalies in the crust of the seafloor. The distance from an oceanic ridge axis to any magnetic anomaly indicates the width of new seafloor that formed during that time interval. For example, if the distance between the present-day Mid-Atlantic Ridge and anomaly 31 is 2010 km, and anomaly 31 formed 67 million years ago (Figure 2.23), then the average rate of movement during the past 67 million years has been 3 cm per year (2010 km, which equals 201 million cm divided by 67 million years; 201,000,000 cm/67,000,000 years = 3 cm/year). Thus, for a given interval of time, the wider the strip of seafloor, the faster the plate has moved. In this way, not only can the present average rate of movement and relative motion be determined (Figure 2.15), but the average rate of movement in the past can also be calculated by dividing the distance between anomalies by the amount of time elapsed between anomalies.

Geologists use magnetic anomalies not only to calculate the average rate of plate movement, but also to determine plate positions at various times in the past. Because magnetic anomalies are parallel and symmetric with respect to spreading ridges, all one must do to determine the position of continents when particular anomalies formed is to move the anomalies back to the spreading ridge, which will also move the continents with them (Figure 2.23). Unfortunately, subduction destroys oceanic crust and the magnetic record that it carries. Thus, we have an excellent record of plate movements since the breakup of Pangaea, but not as good an understanding of plate movement before that time.

The average rate of movement, as well as the relative motion between any two plates, can also be determined by satellite-laser ranging techniques. Laser beams from a station on one plate are bounced off a satellite (in geosynchronous orbit) and returned to a station on a different plate. As the plates move away from each other, the laser beam takes more time to go from the sending station to the stationary satellite and back to the receiving station. This difference in elapsed time is used to calculate the rate of movement and the relative motion between plates.

Plate motions derived from magnetic reversals and satellite-laser ranging techniques give only the relative motion of one plate with respect to another. Hot spots allow

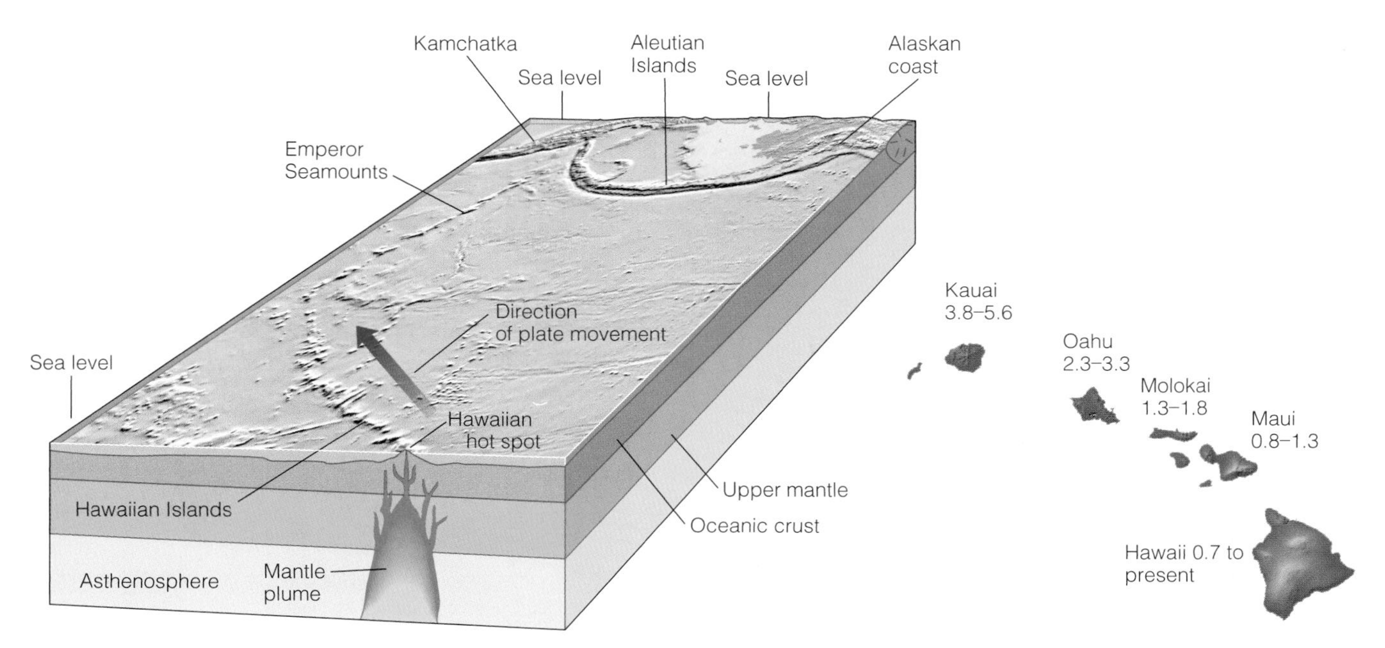

Figure 2.22 Hot Spots A hot spot is the location where a stationary mantle plume has risen to the surface and formed a volcano. The Emperor Seamount—Hawaiian Island chain formed as a result of the Pacific plate moving over a mantle plume, and the line of volcanic islands in this chain traces the direction of plate movement. The island of Hawaii and the Loihi Seamount are the only current hot spots of this island chain. The numbers indicate the ages of the Hawaiian Islands in millions of years.

Figure 2.23 Reconstructing Plate Positions Using Magnetic Anomalies

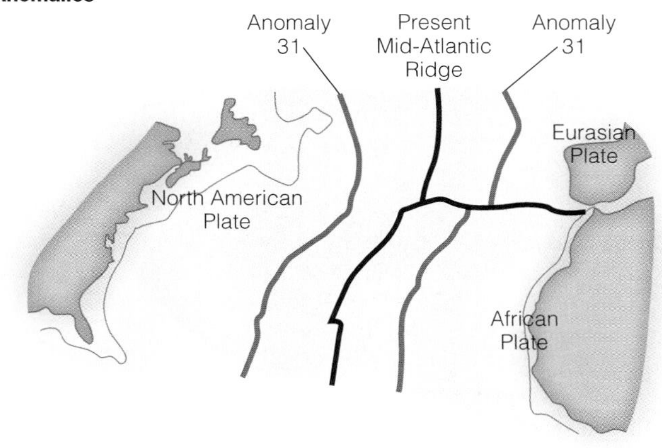

a The present North Atlantic, showing the Mid-Atlantic Ridge and magnetic anomaly 31, which formed 67 million years ago.

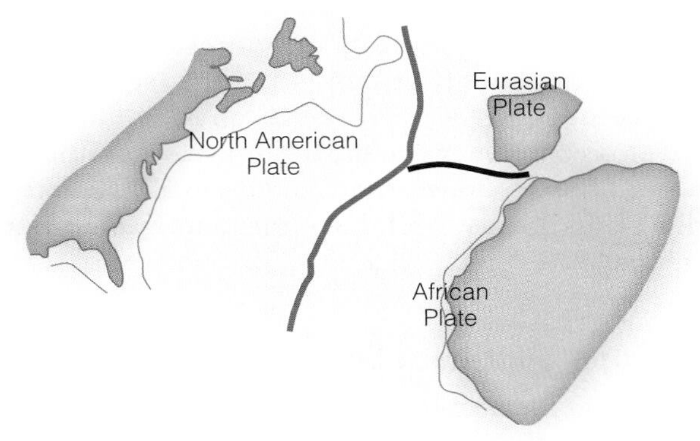

b The Atlantic 67 million years ago. Anomaly 31 marks the plate boundary 67 million years ago. By moving the anomalies back together, along with the plates they are on, we can reconstruct the former positions of the continents.

geologists to determine absolute motion because they provide an apparently fixed reference point from which the rate and direction of plate movement can be measured.

The previously mentioned Emperor Seamount–Hawaiian Island chain formed as a result of movement over a hot spot. Thus, the line of the volcanic islands traces the direction of plate movement, and dating the volcanoes enables geologists to determine the rate of movement.

THE DRIVING MECHANISM OF PLATE TECTONICS

A major obstacle to the acceptance of the continental drift hypothesis was the lack of a driving mechanism to explain continental movement. When it was shown that continents and ocean floors moved together, not separately, and that new crust is formed at spreading ridges by rising magma, most geologists accepted some type of convective heat system (convection cells) as the basic process responsible for plate motion. The question still remains, however: What exactly drives the plates?

Most of the heat from Earth's interior results from the decay of radioactive elements, such as uranium (see Chapter 17), in the core and lower mantle. The most efficient way for this heat to escape Earth's interior is through some type of slow convection of mantle rock in which hot rock from the interior rises toward the surface, loses heat to the overlying lithosphere, becomes denser as it cools, and then sinks back into the interior where it is heated, and the process repeats itself. This type of convective heat system is analogous to a pot of stew cooking on a stove (Figure 2.24).

Two models involving thermal convection cells have been proposed to explain plate movement (Figure 2.25). In one model, thermal convection cells are restricted to the

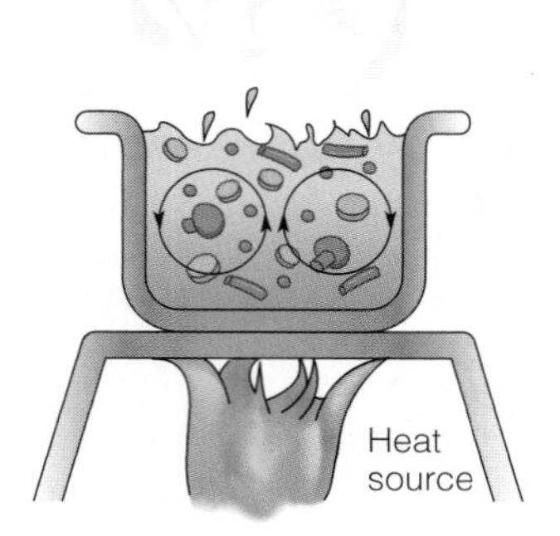

Figure 2.24 **Convection in a Pot of Stew** Heat from the stove is applied to the base of the stew pot causing the stew to heat up. As heat rises through the stew, pieces of the stew are carried to the surface, where the heat is dissipated, the pieces of stew cool, and then sink back to the bottom of the pot. The bubbling seen at the surface of the stew is the result of convection cells churning the stew. In the same manner, heat from the decay of radioactive elements produce convection cells within Earth's interior.

Figure 2.25 Thermal Convection Cells as the Driving Force of Plate Movement. Two models involving thermal convection cells have been proposed to explain plate movement.

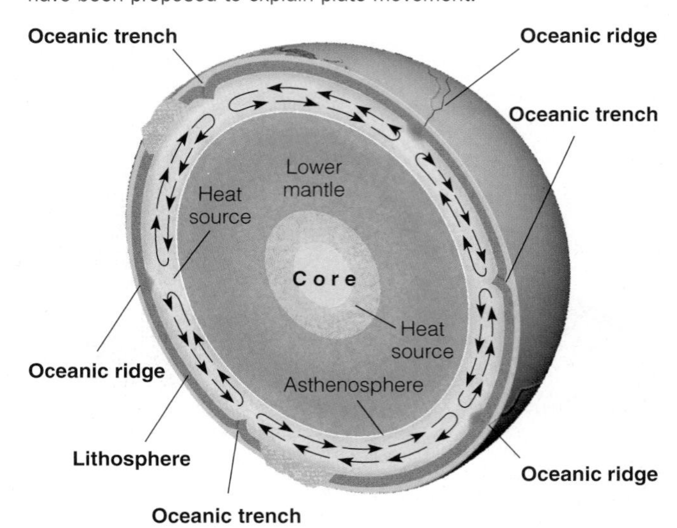

a In the first model, thermal convection cells are restricted to the asthenosphere.

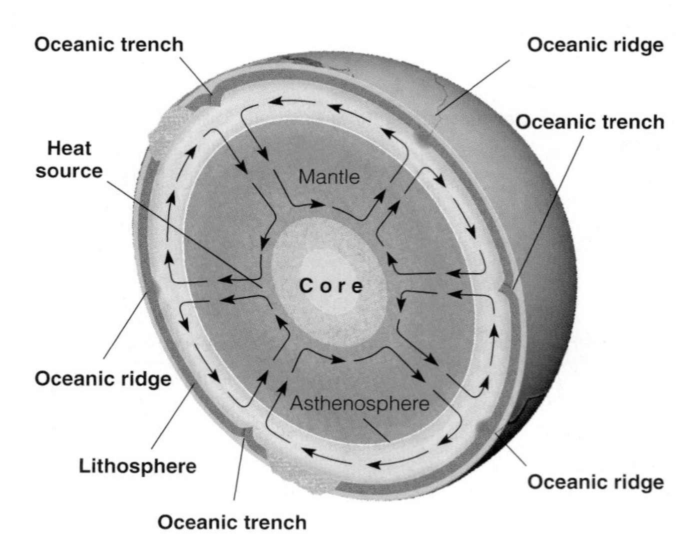

b In the second model, thermal convection cells involve the entire mantle.

asthenosphere; in the second model, the entire mantle is involved. In both models, spreading ridges mark the ascending limbs of adjacent convection cells, and trenches are present where convection cells descend back into Earth's interior. The convection cells therefore determine the location of spreading ridges and trenches, with the lithosphere lying above the thermal convection cells. Thus, each plate corresponds to a single convection cell and moves as a result of the convective movement of the cell itself.

Although most geologists agree that Earth's internal heat plays an important role in plate movement, there are problems with both models. The major problem associated with the first model is the difficulty of explaining the source of heat for the convection cells and why they are restricted to the asthenosphere. In the second model, the heat comes from the outer core, but it is still not known how heat is transferred from the outer core to the mantle. Nor is it clear how convection can involve both the lower mantle and the asthenosphere.

In addition to some type of thermal convection system driving plate movement, some geologists think that plate movement occurs because of a mechanism involving "slab-pull" or "ridge-push," both of which are gravity driven but still depend on thermal differences within Earth (Figure 2.26). In slab-pull, the subducting cold slab of lithosphere, being denser than the surrounding warmer asthenosphere, pulls the rest of the plate along as it descends into the asthenosphere. As the lithosphere moves downward, there is a corresponding upward flow back into the spreading ridge.

Operating in conjuction with slab-pull is the ridge-push mechanism. As a result of rising magma, the oceanic ridges are higher than the surrounding oceanic crust. It is thought that gravity pushes the oceanic lithosphere away from the higher spreading ridges and toward the trenches.

Currently, geologists are fairly certain that some type of convective system is involved in plate movement, but the extent to which other mechanisms, such as slab-pull and ridge-push, are involved is still unresolved. However, the fact that plates have moved in the past and are still moving today has been proven beyond a doubt. And although a comprehensive theory of plate movement has not yet been developed, more and more of the pieces are falling into place as geologists learn more about Earth's interior.

The Supercontinent Cycle

As a result of plate movement, all the continents came together to form the supercontinent Pangaea by the end of the Paleozoic Era. Pangaea began fragmenting during the Triassic Period and continues to do so, thus accounting for the present distribution of continents and ocean basins. It has been proposed that supercontinents consisting of all or most of Earth's landmasses form, break up, and come together again in a cycle spanning approximately 500 million years.

The *supercontinent cycle hypothesis* is an expansion on the ideas of the Canadian geologist J. Tuzo Wilson. During the early 1970s, Wilson proposed a cycle (now known as the Wilson cycle) that includes continental fragmentation, the opening and closing of an ocean basin, and reassembly of the continent. According to the supercontinent cycle hypothesis, heat accumulates beneath a supercontinent because the rocks of continents are poor conductors of heat. As a result of the heat accumulation, the supercontinent domes upward and fractures. Basaltic magma rising from below fills the fractures. As a basalt-filled fracture widens, it begins subsiding and forms a long narrow ocean, such as the present-day Red Sea. Continued rifting eventually forms an expansive ocean basin, such as the Atlantic.

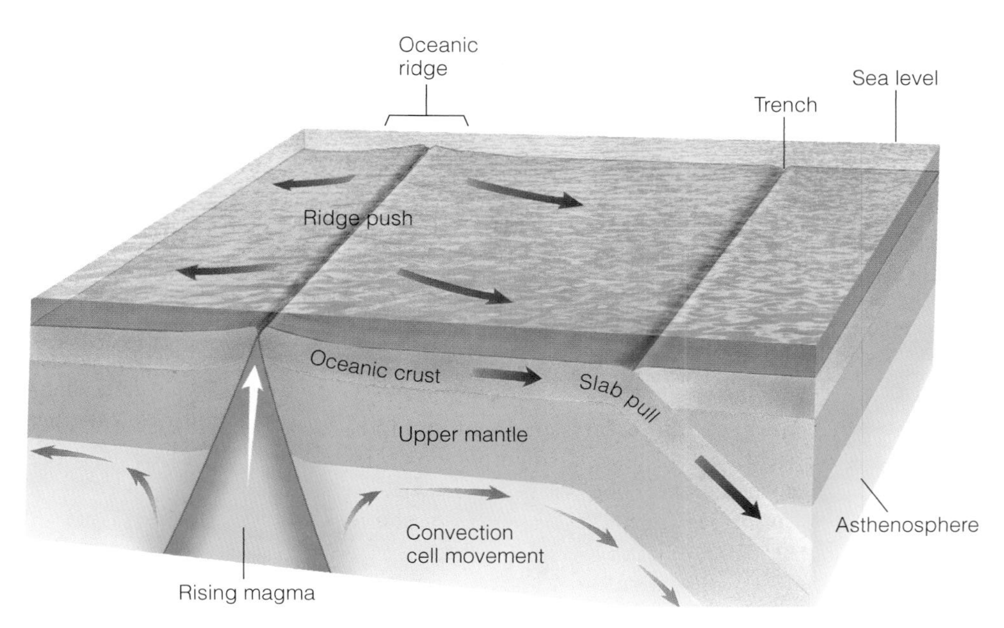

Figure 2.26 Plate Movement Resulting from Gravity-Driven Mechanisms Plate movement is also thought to result, at least partially, from gravity-driven "slab-pull" or "ridge-push" mechanisms. In slab-pull, the edge of the subducting plate descends into the interior, and the rest of the plate is pulled downward. In ridge-push, rising magma pushes the oceanic ridges higher than the rest of the oceanic crust. Gravity thus pushes the oceanic lithosphere away from the ridges and toward the trenches.

One of the most convincing arguments for proponents of the supercontinent cycle hypothesis is the "surprising regularity" of mountain building caused by compression during continental collisions. These mountain-building episodes occur about every 400 to 500 million years and are followed by an episode of rifting about 100 million years later. In other words, a supercontinent fragments and its individual plates disperse following a rifting episode, an interior ocean forms, and then the dispersed fragments reassemble to form another supercontinent.

The supercontinent cycle is yet another example of how interrelated the various systems and subsystems of Earth are and how they operate over vast periods of geologic time.

PLATE TECTONICS AND THE DISTRIBUTION OF NATURAL RESOURCES

In addition to being responsible for the major features of Earth's crust and influencing the distribution and evolution of the world's biota, plate movement also affects the formation and distribution of some natural resources. Consequently, geologists are using plate tectonic theory in their search for petroleum (see Geo-Focus on page 55) and mineral deposits and in explaining the occurrence of these natural resources. It is becoming increasingly clear that if we are to keep up with the continuing demands of a global industrialized society, the application of plate tectonic theory to the origin and distribution of natural resources is essential.

Although large concentrations of petroleum occur in many areas of the world, more than 50% of all proven reserves are in the Persian Gulf region. The reason for this is paleogeography and plate movement. Elsewhere in the world, plate tectonics is also responsible for concentrations of petroleum. The formation of the Appalachians, for example, resulted from the compressive forces generated along a convergent plate boundary and provided the traps necessary for petroleum to accumulate.

Mineral Deposits

Many metallic mineral deposits such as copper, gold, lead, silver, tin, and zinc are related to igneous and associated hydrothermal (hot water) activity. So it is not surprising that a close relationship exists between plate boundaries and the occurrence of these valuable deposits.

The magma generated by partial melting of a subducting plate rises toward the surface, and as it cools, it precipitates and concentrates various metallic ores. Many of the world's major metallic ore deposits are associated with convergent plate boundaries, including those in the Andes of South America, the Coast Ranges and Rockies of North America, Japan, the Philippines, Russia, and a zone extending from the

What Would You Do?

You are part of a mining exploration team that is exploring a promising and remote area of central Asia. You know that former convergent and divergent plate boundaries frequently are sites of ore deposits. What evidence would you look for to determine whether the area you're exploring might be an ancient convergent or divergent plate boundary? Is there anything you can do before visiting the area that might help you to determine the geology of the area?

eastern Mediterranean region to Pakistan. In addition, the majority of the world's gold is associated with deposits located at ancient convergent plate boundaries in such areas as Canada, Alaska, California, Venezuela, Brazil, Russia, southern India, and western Australia.

The copper deposits of western North and South America are an excellent example of the relationship between convergent plate boundaries and the distribution, concentration, and exploitation of valuable metallic ores (Figure 2.27a). The world's largest copper deposits are found along this belt. The majority of the copper deposits in the Andes and the southwestern United States were formed less than 60 million years ago when oceanic plates were subducted under the North and South American plates. The rising magma and associated hydrothermal fluids carried minute amounts of copper, which were originally widely disseminated but eventually became concentrated in the cracks and fractures of the surrounding andesites. These low-grade copper deposits contain from 0.2 to 2% copper and are extracted from large open-pit mines (Figure 2.27b).

Divergent plate boundaries also yield valuable ore resources. The island of Cyprus in the Mediterranean is rich in copper and has been supplying all or part of the world's needs for the past 3000 years. The concentration of copper on Cyprus formed as a result of precipitation adjacent to hydrothermal vents along a divergent plate boundary. This deposit was brought to the surface when the copper-rich seafloor collided with the European plate, warping the seafloor and forming Cyprus.

Studies indicate that minerals of such metals as copper, gold, iron, lead, silver, and zinc are currently forming in the Red Sea. The Red Sea is opening as a result of plate divergence and represents the earliest stage in the growth of an ocean basin (Figures 2.16c and 2.17b).

PLATE TECTONICS AND THE DISTRIBUTION OF LIFE

Plate tectonic theory is as revolutionary and far-reaching in its implications for geology as the theory of evolution was for biology when it was proposed. Interestingly, it was the fossil evidence that convinced Wegener, Suess, and du Toit, as well as many other geologists, of the correctness of continental drift. Together, the theories of plate tectonics and evolution have changed the way we view our planet, and we should

Figure 2.27 Copper Deposits and Convergent Plate Boundaries

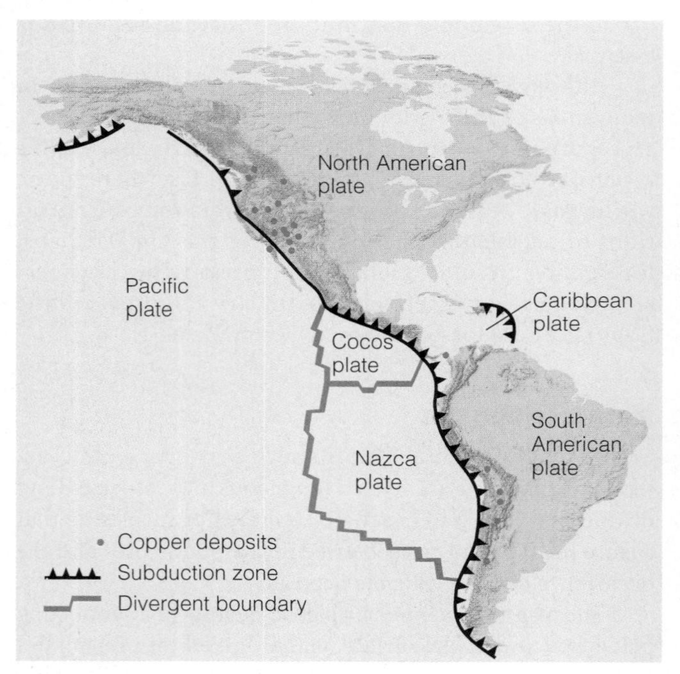

a Valuable copper deposits are located along the west coasts of North and South America in association with convergent plate boundaries. The rising magma and associated hydrothermal activity resulting from subduction carried small amounts of copper, which became trapped and concentrated in the surrounding rocks through time.

b Bingham Copper Mine, near Salt Lake City, Utah, is a huge open-pit copper mine with reserves estimated at 1.7 billion tons. More than 400,000 tons of rock are removed for processing each day. Note the small specks towards the middle of the photograph that are the 12-feet high dump trucks!

Geo-Focus Oil, Plate Tectonics, and Politics

It is certainly not surprising that oil and politics are closely linked. The Iran–Iraq War of 1980–1989 and the Gulf War of 1990–1991 were both fought over oil (Figure 1). Indeed, many of the conflicts in the Middle East have shared as their underlying cause the desire to control the vast deposits of petroleum in the region. Most people, however, are not aware of *why* there is so much oil in this part of the world.

Although significant concentrations of petroleum occur in many areas of the world, more than 50% of all proven reserves are in the Persian Gulf region. It is interesting, however, that this region did not become a significant petroleum-producing area until the economic recovery following World War II (1939–1945). After the war, Western Europe and Japan in particular became dependent on Persian Gulf oil, and they still rely heavily on this region for most of their supply. The United States is also dependent on imports from the Persian Gulf, but receives significant quantities of petroleum from other sources, such as Mexico and Venezuela.

Why is there so much oil in the Persian Gulf region? The answer lies in the ancient geography and plate movement of this region during the Mesozoic and Cenozoic eras. During the Mesozoic Era, and particularly the Cretaceous Period when most of the petroleum formed, the Persian Gulf area was a broad marine shelf extending eastward from Africa.

Reuters NewMedia Inc./Corbis

Figure 1 The Kuwaiti night skies were illuminated by 700 blazing oil wells set on fire by Iraqi troops during the 1991 Gulf War. The fires continued for nine months.

This continental margin lay near the equator where countless microorganisms lived in the surface waters. The remains of these organisms accumulated with the bottom sediments and were buried, beginning the complex process of petroleum generation and the formation of source beds in which petroleum forms.

As a consequence of rifting in the Red Sea and the Gulf of Aden during the Cenozoic Era, the Arabian plate is moving northeast away from Africa and subducting beneath Iran (Figure 2.17a). During the early stages of collision between Arabia and Iran, as the sediments of the passive continental margin were initially subducted, heating broke down the organic molecules and led to the formation of petroleum.

The tilting of the Arabian block to the northeast allowed the newly formed petroleum to migrate upward into the interior of the Arabian plate. The continued subduction and collision with Iran folded the rocks, creating traps for petroleum to accumulate, such that the vast area south of the collision zone (known as the Zagros suture) is a major oil-producing region.

not be surprised at the intimate association between them. Although the relationship between plate tectonic processes and the evolution of life is incredibly complex, paleontological data provide convincing evidence of the influence of plate movement on the distribution of organisms.

The present distribution of plants and animals is not random, but is controlled mostly by climate and geographic barriers. The world's biota occupy *biotic provinces*, which are regions characterized by a distinctive assemblage of plants and animals. Organisms within a province have similar ecological requirements, and the boundaries separating provinces are therefore natural ecological breaks. Climatic or geographic barriers are the most common province boundaries, and these are mostly controlled by plate movement.

The complex interaction between wind and ocean currents has a strong influence on the world's climates. Wind and ocean currents are thus strongly influenced by the number, distribution, topography, and orientation of continents. For example, the southern Andes Mountains act as an effective barrier to moist, easterly blowing Pacific winds, resulting in a desert east of the southern Andes that is virtually uninhabitable.

The distribution of continents and ocean basins not only influences wind and ocean currents, but also affects provinciality by creating physical barriers to, or pathways for, the migration of organisms. Intraplate volcanoes, island arcs, mid-oceanic ridges, mountain ranges, and subduction zones all result from the interaction of plates, and their orientation and distribution strongly influence the number of provinces and hence total global diversity. Thus, provinciality and diversity will be highest when there are numerous small continents spread across many zones of latitude.

When a geographic barrier separates a once-uniform fauna, species may undergo divergence. If conditions on opposite sides of the barrier are sufficiently different, then species must adapt to the new conditions, migrate, or become extinct. Adaptation to the new environment by various species may involve enough change that new species eventually evolve.

The marine invertebrates found on opposite sides of the Isthmus of Panama provide an excellent example of divergence caused by the formation of a geographic barrier. Prior to the rise of this land connection between North and South America, a homogeneous population of bottom-dwelling invertebrates inhabited the shallow seas of the area. After the rise of the Isthmus of Panama by subduction of the Pacific plate approximately 5 million years ago, the original population was divided. In response to the changing environment, new species evolved on opposite sides of the isthmus (Figure 2.28).

The formation of the Isthmus of Panama also influenced the evolution of the North and South American mammalian faunas. During most of the Cenozoic Era, South America was an island continent, and its mammalian fauna evolved in isolation from the rest of the world's faunas. When North and South America were connected by the Isthmus of Panama, most of the indigenous South American mammals were replaced by migrants from North America. Surprisingly, only a few South American mammal groups migrated northward.

Figure 2.28 Plate Tectonics and the Distribution of Organisms

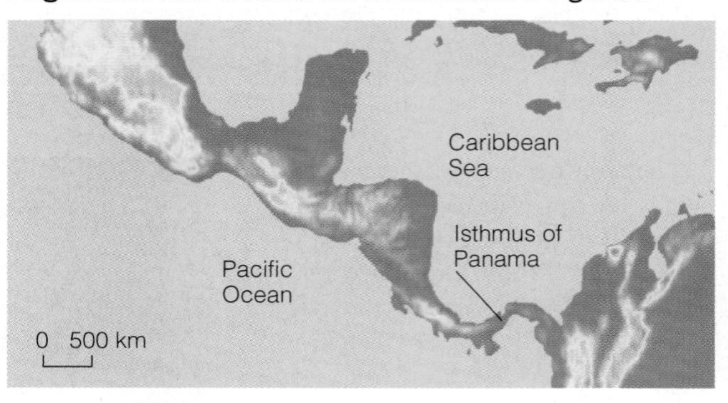

a The Isthmus of Panama forms a barrier that divides a once-uniform fauna of molluscs.

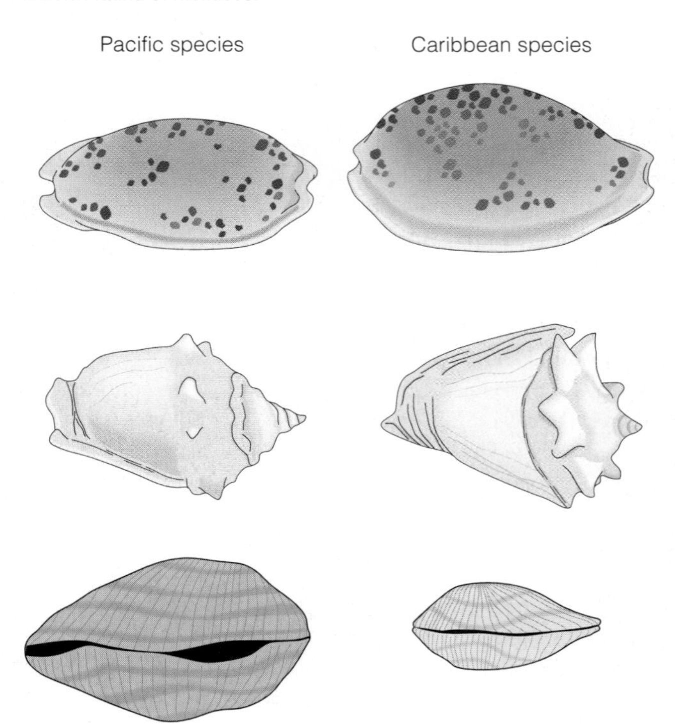

b Divergence of gastropod and bivalve species after the formation of the Isthmus of Panama. Each pair belongs to the same genus but is a different species.

Geo-Recap

Chapter Summary

- The concept of continental movement is not new. The earliest maps showing the similarity between the east coast of South America and the west coast of Africa provided the first evidence that continents may once have been united and subsequently separated from each other.
- Alfred Wegener is generally credited with developing the hypothesis of continental drift. He provided abundant geologic and paleontologic evidence to show that the continents were once united in one supercontinent, which he named Pangaea. Unfortunately, Wegener could not explain how the continents moved, and most geologists ignored his ideas.
- The hypothesis of continental drift was revived during the 1950s when paleomagnetic studies of rocks indicated the presence of multiple magnetic north poles instead of just one as there is today. This paradox was resolved by constructing a map in which the continents could be moved into different positions such that the paleomagnetic data would then be consistent with a single magnetic north pole.
- Seafloor spreading was confirmed by the discovery of magnetic anomalies in the ocean crust that were both parallel to and symmetric around the ocean ridges. The pattern of oceanic magnetic anomalies matched the pattern of magnetic reversals already known from continental lava flows.
- Plate tectonic theory became widely accepted by the 1970s because the evidence overwhelmingly supports it and because it provides geologists with a powerful theory for explaining such phenomena as volcanism, earthquake activity, mountain building, global climatic changes, the distribution of the world's biota, and the distribution of mineral resources.
- Three types of plate boundaries are recognized: divergent boundaries, where plates move away from each other; convergent boundaries, where two plates collide; and transform boundaries, where two plates slide past each other.
- Ancient plate boundaries can be recognized by their associated rock assemblages and geologic structures. For divergent boundaries, these may include rift valleys with thick sedimentary sequences and numerous dikes and sills. For convergent boundaries, ophiolites and andesitic rocks are two characteristic features. Transform faults generally do not leave any characteristic or diagnostic features in the geologic record.
- The average rate of movement and relative motion of the plates can be calculated in several ways. The results of these different methods all agree and indicate that the plates move at different average velocities.
- The absolute motion of plates can be determined by the movement of plates over mantle plumes. A mantle plume is an apparently stationary column of magma that rises to the surface where it becomes a hot spot and forms a volcano.
- Although a comprehensive theory of plate movement has yet to be developed, geologists think that some type of convective heat system is the major driving force.
- The supercontinent cycle indicates that all or most of Earth's landmasses form, break up, and form again in a cycle spanning approximately 500 million years.
- A close relationship exists between the formation of some mineral deposits and petroleum, and plate boundaries. Furthermore, the formation and distribution of some natural resources are related to plate movement.
- The relationship between plate tectonic processes and the evolution of life is complex. The distribution of plants and animals is not random, but is controlled mostly by climate and geographic barriers, which are controlled, to a great extent, by the movement of plates.

Important Terms

continental–continental plate boundary (p. 46)
continental drift (p. 31)
convergent plate boundary (p. 44)
Curie point (p. 36)
divergent plate boundary (p. 40)
Glossopteris flora (p. 30)
Gondwana (p. 31)
hot spot (p. 48)
Laurasia (p. 31)
magnetic anomaly (p. 38)
magnetic field (p. 35)
magnetic reversal (p. 37)
magnetism (p. 35)
oceanic–continental plate boundary (p. 44)
oceanic–oceanic plate boundary (p. 44)
paleomagnetism (p. 36)
Pangaea (p. 31)
plate tectonic theory (p. 39)
seafloor spreading (p. 37)
thermal convection cell (p. 37)
transform fault (p. 48)
transform plate boundary (p. 48)

Review Questions

1. The man credited with developing the continental drift hypothesis is
   a. ______ Wilson;
   b. ______ Wegener;
   c. ______ Hess;
   d. ______ du Toit;
   e. ______ Vine.
2. The southern part of Pangaea, consisting of South America, Africa, India, Australia, and Antarctica, is called
   a. ______ Laurasia;
   b. ______ Gondwana;
   c. ______ Panthalassa;
   d. ______ Laurentia;
   e. ______ Pacifica.
3. Hot spots and aseismic ridges can be used to determine the
   a. ______ location of divergent plate boundaries;
   b. ______ absolute motion of plates;
   c. ______ location of magnetic anomalies in oceanic crust;
   d. ______ relative motion of plates;
   e. ______ location of convergent plate boundaries.
4. Along what type of boundary does subduction occur?
   a. ______ divergent;
   b. ______ transform;
   c. ______ convergent;
   d. ______ answers a and b;
   e. ______ answers a and c.
5. The Himalayas are a good example of what type of plate boundary?
   a. ______ continental–continental;
   b. ______ oceanic–oceanic;
   c. ______ oceanic–continental;
   d. ______ divergent;
   e. ______ transform.
6. The most common biotic province boundaries are
   a. ______ geographic barriers;
   b. ______ biologic barriers;
   c. ______ climatic barriers;
   d. ______ answers a and b;
   e. ______ answers a and c.
7. Magnetic surveys of the ocean basins indicate that
   a. ______ the oceanic crust is youngest adjacent to mid-oceanic ridges;
   b. ______ the oceanic crust is oldest adjacent to mid-oceanic ridges;
   c. ______ the oceanic crust is youngest adjacent to the continents;
   d. ______ the oceanic crust is the same age everywhere;
   e. ______ answers b and c.
8. Convergent plate boundaries are areas where
   a. ______ new continental lithosphere is forming;
   b. ______ new oceanic lithosphere is forming;
   c. ______ two plates come together;
   d. ______ two plates slide past each other;
   e. ______ two plates move away from each other.
9. Iron-bearing minerals in magma gain their magnetism and align themselves with the magnetic field when they cool through the
   a. ______ Curie point;
   b. ______ magnetic anomaly point;
   c. ______ thermal convection point;
   d. ______ hot spot point;
   e. ______ isostatic point.
10. The San Andreas fault is an example of what type of plate boundary?
    a. ______ divergent;
    b. ______ convergent;
    c. ______ transform;
    d. ______ oceanic–continental;
    e. ______ continental–continental.
11. Using the age for each of the Hawaiian Islands in Figure 2.22 and an atlas in which you can measure the distance between islands, calculate the average rate of movement per year for the Pacific plate since each island formed. Is the average rate of movement the same for each island? Would you expect it to be? Explain why it may not be.
12. What evidence convinced Wegener and others that continents must have moved in the past and at one time formed a supercontinent?
13. Estimate the age of the seafloor crust and the age and thickness of the oldest sediment off the East Coast of the United States (e.g., Virginia). In so doing, refer to Figure 2.13 for the ages and to the deep-sea sediment accumulation rate stated in this chapter.
14. In addition to the volcanic eruptions and earthquakes associated with convergent and divergent plate boundaries, why are these boundaries also associated with the formation and accumulation of various metallic ore deposits?
15. If the movement along the San Andreas fault, which separates the Pacific plate from the North American plate, averages 5.5 cm per year, how long will it take before Los Angeles is opposite San Francisco?
16. Plate tectonic theory builds on the continental drift hypothesis and the theory of seafloor spreading. As such, it is a unifying theory of geology. Explain why it is a unifying theory.
17. Why is some type of thermal convection system thought to be the major force driving plate movement? How have slab-pull and ridge-push, both mainly

gravity driven, modified a purely thermal convection model for plate movement?

18. Based on your knowledge of biology and the distribution of organisms throughout the world, how do you think plate tectonics has affected this distribution both on land and in the oceans?
19. What is the supercontinent cycle? Who proposed this concept, and what elements of continental drift and seafloor spreading are embodied in the cycle?
20. Explain why global diversity increases with an increase in the number of biotic provinces. How does plate movement affect the number of biotic provinces?

CHAPTER 3

Minerals—The Building Blocks of Rocks

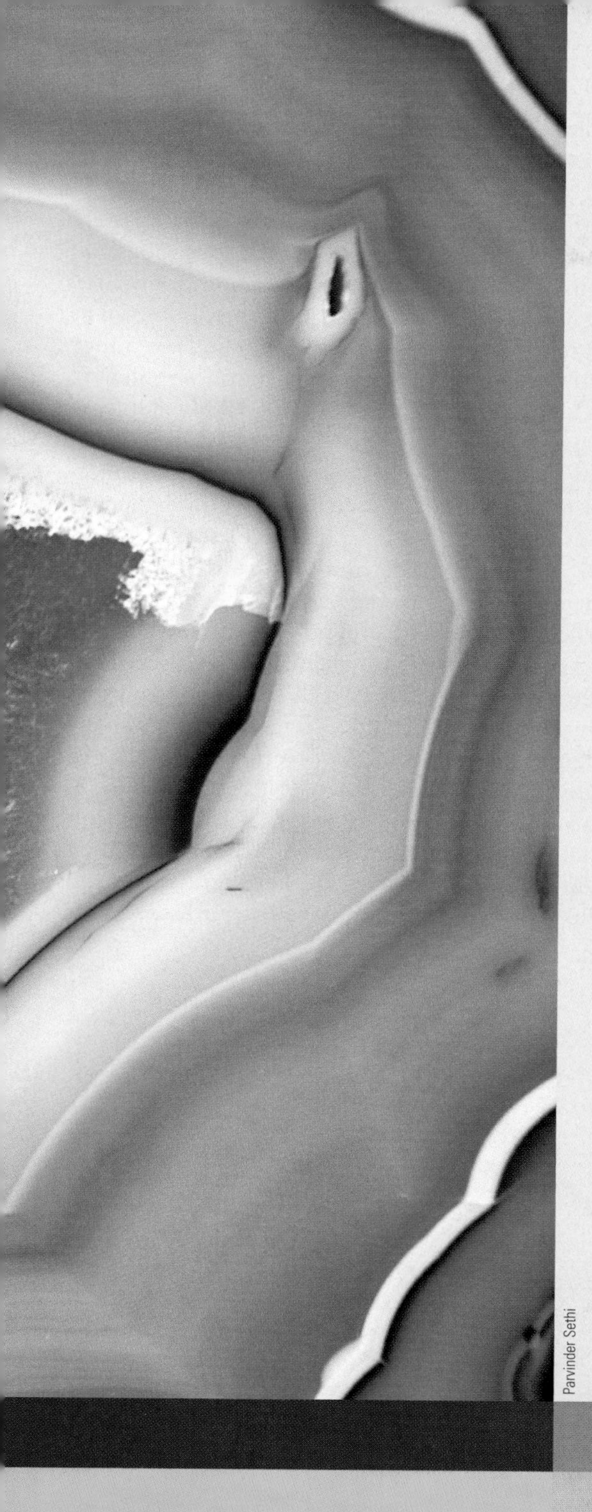

OUTLINE

OBJECTIVES

At the end of this chapter, you will have learned that

- All matter, including minerals, is made up of atoms that bond to form elements and compounds.
- Geologists have a specific definition for the term *mineral.*
- You can distinguish minerals from other naturally occurring and manufactured substances.
- Minerals are incredibly varied, yet only a few are particularly common.
- Geologists use physical properties such as color, hardness, and density to identify minerals.
- Minerals originate in various ways and under varied conditions.
- Some minerals, designated rock-forming minerals, are particularly common in rocks, whereas others are found in small amounts.
- Some minerals and rocks are important natural resources that are essential to industrialized societies.

In this specimen of polished agate, multi-colored rings surround a central mass of tiny quartz crystals. Agate forms by filling a hollow inside a host rock, and as a result it is often found with concentric bands that resemble tree rings.

INTRODUCTION

The term *mineral* brings to mind substances that we need for good nutrition, such as iron, calcium, and magnesium; however, these are actually chemical elements, not minerals, at least in the geological sense. Ice, on the other hand, is a mineral because it meets all criteria in the definition of a mineral—that is, it is an inorganic, naturally occurring, *crystalline solid*, meaning that its atoms are arranged in a specific three-dimensional pattern, as opposed to glass, which has no such orderly arrangement of atoms. In addition, ice has a specific chemical composition (H_2O) and characteristic physical properties such as hardness, density, and color. In sum, a **mineral** is an inorganic, naturally occurring, crystalline solid, with a narrowly defined chemical composition and characteristic physical properties.

We cannot overstate the importance of minerals in many human endeavors. Ores of iron, copper, manganese, and chromium, as well as minerals and rocks for fertilizers and animal feed supplements, are essential for our economic well-being. Much of the ink used in the brilliantly colored medieval manuscripts came from minerals, and the luster of lipstick, glitter, eye shadow, and paints for appliances and automobiles comes from the mineral muscovite. The United States and Canada owe much of their economic success to abundant minerals and energy resources (natural resources), although both nations must also import some essential commodities. Indeed, the distribution of natural resources is one important consideration in foreign policy decisions.

One important reason to study minerals is that they are building blocks of rocks; so rocks, with few exceptions, are combinations of one or more minerals. Granite, for instance, is made up of specified percentages of minerals known as quartz and feldspars along with other minerals in minor quantities. In several of the following chapters, we will have more to say about the importance of minerals in identifying and classifying rocks.

Some minerals are attractive and eagerly sought by collectors and for museum displays (Figure 3.1). Other minerals are known as *gemstones*—that is, precious or semi-precious minerals or rocks used for decorative purposes, especially jewelry. The precious gemstones, such as diamond (Figure 3.1a), ruby, sapphire, and emerald, are most desirable and most expensive. Amber and pearl are included among the semiprecious gemstones, but are they really minerals? Amber is hardened resin (sap) from coniferous trees and thus an organic substance and not a mineral; nevertheless, it is prized as a decorative "stone" (Figure 3.1b). It is best known from the Baltic Sea region of Europe where sun-worshiping cultures, noting its golden translucence resembling the Sun's rays, thought that it possessed mystical powers. Pearls form when molluscs, such as clams or oysters, deposit successive

Figure 3.1 Precious and Semiprecious Gemstones

Parvinder Sethi

a The diamond pendant in this necklace in the Smithsonian Institution is the 68-carat Victoria Transvaal diamond from South Africa.

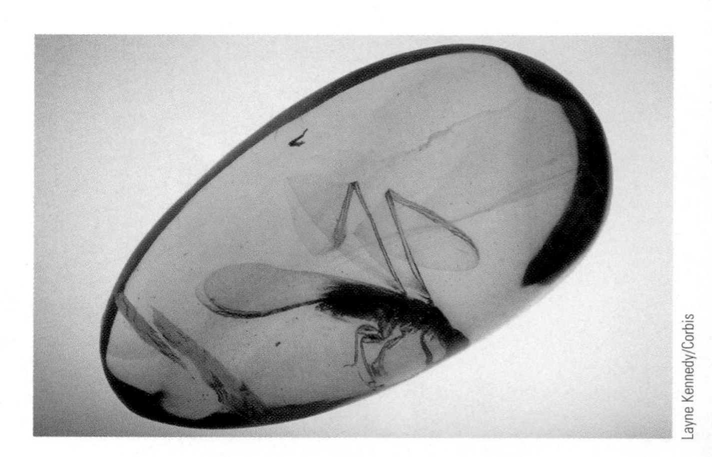

Layne Kennedy/Corbis

b Even though amber is an organic substance, it is nevertheless valued as a semiprecious gemstone.

Parvinder Sethi

c Strands of variously-colored pearls at the world-famous Pearl Shopping Center in Shanghai, China.

layers of tiny mineral crystals around some irritant, perhaps a sand grain. Most pearls are lustrous white, but some are silver gray, green, or black (Figure 3.1c).

From our discussion so far, we have a formal definition of the term *mineral* and we know that minerals are the basic constituents of rocks. Now let's delve deeper into what minerals are made of by considering matter, atoms, elements, and bonding.

MATTER—WHAT IS IT?

Matter is anything that has mass and occupies space, and accordingly includes water, plants, animals, the atmosphere, and minerals and rocks. Physicists generally recognize three states or phases of matter: *liquids*, *gases*, and *solids*.* Liquids, such as surface water and groundwater, as well as atmospheric gases, are important in our considerations of several surface processes, such as running water and wind, but here our main concern is with solids because, by definition, minerals are solids.

Atoms and Elements

Matter is made up of chemical **elements**, which, in turn, are composed of **atoms**, the smallest units of matter that retain the characteristics of a particular element (Figure 3.2). That is, elements cannot be changed into different substances except through radioactive decay (discussed in Chapter 17). Thus, an element is made up of atoms, all of which have the same properties. Scientists have discovered 92 naturally occurring elements, some of which are listed in Figure 3.2, and several others have been made in laboratories. All naturally occurring elements and most artificial ones have a name and a symbol—for example, oxygen (O), aluminum (Al), and potassium (K) (Figure 3.3).

At the center of an atom is a tiny **nucleus** made up of one or more particles known as **protons**, which have a positive electrical charge, and **neutrons**, which are electrically neutral. The nucleus is only about 1/100,000 of the diameter of an atom, yet it contains virtually all of the atom's mass. **Electrons**, particles with a negative electrical charge, orbit rapidly around the nucleus at specific distances in one or more **electron shells**. The electrons determine how an atom interacts with other atoms, but the nucleus determines how many electrons an atom has, because the positively charged protons attract and hold the negatively charged electrons in their orbits.

The number of protons in its nucleus determines an atom's identity and its **atomic number**. Hydrogen (H), for instance, has one proton in its nucleus and thus has an atomic number of 1. The nuclei of helium (He) atoms possess 2 protons, whereas those of carbon (C) have 6, and uranium (U) 92, so their atomic numbers are 2, 6, and 92, respectively. Atoms also have an **atomic mass number**, which is the sum of protons and neutrons in the nucleus (electrons contribute negligible mass to atoms). However, atoms of the same chemical element might have different atomic mass numbers because the number of neutrons can vary. All carbon (C) atoms have six protons—otherwise they would not be carbon—but the number of neutrons can be 6, 7, or 8. Thus we recognize three types of carbon, or what are known as *isotopes* (Figure 3.4), each with a different atomic mass number.

The isotopes of carbon, or those of any other element, behave the same chemically; carbon 12 and carbon 14 are both present in carbon dioxide (CO_2), for example. However, some isotopes are radioactive, meaning that they

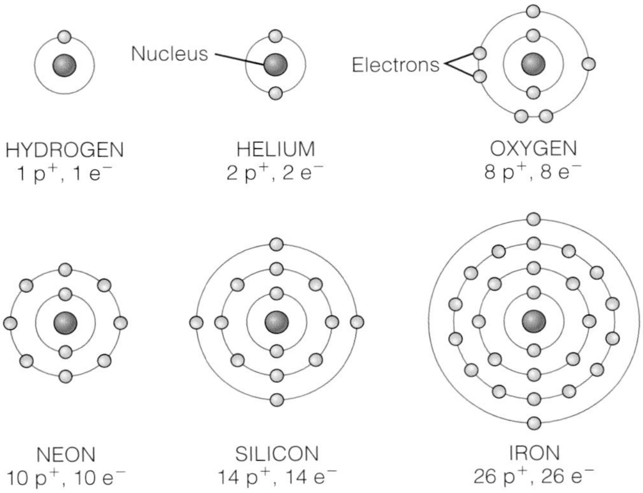

Nucleus
Electrons
HYDROGEN 1 p^+, 1 e^-
HELIUM 2 p^+, 2 e^-
OXYGEN 8 p^+, 8 e^-
NEON 10 p^+, 10 e^-
SILICON 14 p^+, 14 e^-
IRON 26 p^+, 26 e^-

Element	Symbol	Atomic Number	Distribution of Electrons			
			First Shell	Second Shell	Third Shell	Fourth Shell
Hydrogen	H	1	1	—	—	—
Helium	He	2	2	—	—	—
Carbon	C	6	2	4	—	—
Oxygen	O	8	2	6	—	—
Neon	Ne	10	2	8	—	—
Sodium	Na	11	2	8	1	—
Magnesium	Mg	12	2	8	2	—
Aluminum	Al	13	2	8	3	—
Silicon	Si	14	2	8	4	—
Phosphorus	P	15	2	8	5	—
Sulfur	S	16	2	8	6	—
Chlorine	Cl	17	2	8	7	—
Potassium	K	19	2	8	8	1
Calcium	Ca	20	2	8	8	2
Iron	Fe	26	2	8	14	2

Figure 3.2 Shell Models for Common Atoms The shell model for several atoms and their electron configurations. A blue circle represents the nucleus of each atom, but remember that atomic nuclei are made up of protons and neutrons, as shown in Figure 3.4.

*Actually, scientists also recognize a fourth state of matter known as *plasma*, an ionized gas as in fluorescent and neon lights and matter in the Sun and stars.

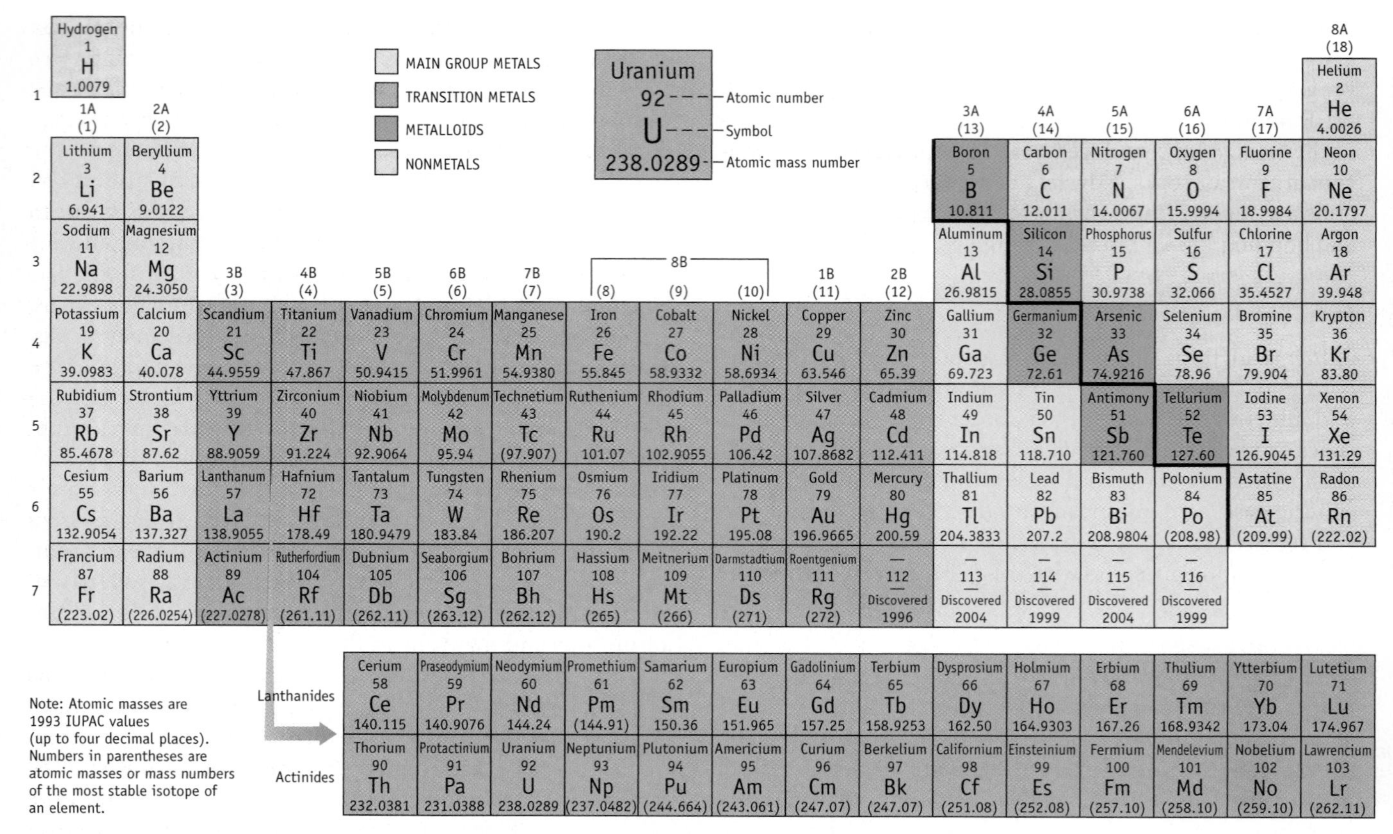

Period	1A (1)	2A (2)	3B (3)	4B (4)	5B (5)	6B (6)	7B (7)	8B (8)	8B (9)	8B (10)	1B (11)	2B (12)	3A (13)	4A (14)	5A (15)	6A (16)	7A (17)	8A (18)
1	Hydrogen 1 H 1.0079																	Helium 2 He 4.0026
2	Lithium 3 Li 6.941	Beryllium 4 Be 9.0122											Boron 5 B 10.811	Carbon 6 C 12.011	Nitrogen 7 N 14.0067	Oxygen 8 O 15.9994	Fluorine 9 F 18.9984	Neon 10 Ne 20.1797
3	Sodium 11 Na 22.9898	Magnesium 12 Mg 24.3050											Aluminum 13 Al 26.9815	Silicon 14 Si 28.0855	Phosphorus 15 P 30.9738	Sulfur 16 S 32.066	Chlorine 17 Cl 35.4527	Argon 18 Ar 39.948
4	Potassium 19 K 39.0983	Calcium 20 Ca 40.078	Scandium 21 Sc 44.9559	Titanium 22 Ti 47.867	Vanadium 23 V 50.9415	Chromium 24 Cr 51.9961	Manganese 25 Mn 54.9380	Iron 26 Fe 55.845	Cobalt 27 Co 58.9332	Nickel 28 Ni 58.6934	Copper 29 Cu 63.546	Zinc 30 Zn 65.39	Gallium 31 Ga 69.723	Germanium 32 Ge 72.61	Arsenic 33 As 74.9216	Selenium 34 Se 78.96	Bromine 35 Br 79.904	Krypton 36 Kr 83.80
5	Rubidium 37 Rb 85.4678	Strontium 38 Sr 87.62	Yttrium 39 Y 88.9059	Zirconium 40 Zr 91.224	Niobium 41 Nb 92.9064	Molybdenum 42 Mo 95.94	Technetium 43 Tc (97.907)	Ruthenium 44 Ru 101.07	Rhodium 45 Rh 102.9055	Palladium 46 Pd 106.42	Silver 47 Ag 107.8682	Cadmium 48 Cd 112.411	Indium 49 In 114.818	Tin 50 Sn 118.710	Antimony 51 Sb 121.760	Tellurium 52 Te 127.60	Iodine 53 I 126.9045	Xenon 54 Xe 131.29
6	Cesium 55 Cs 132.9054	Barium 56 Ba 137.327	Lanthanum 57 La 138.9055	Hafnium 72 Hf 178.49	Tantalum 73 Ta 180.9479	Tungsten 74 W 183.84	Rhenium 75 Re 186.207	Osmium 76 Os 190.2	Iridium 77 Ir 192.22	Platinum 78 Pt 195.08	Gold 79 Au 196.9665	Mercury 80 Hg 200.59	Thallium 81 Tl 204.3833	Lead 82 Pb 207.2	Bismuth 83 Bi 208.9804	Polonium 84 Po (208.98)	Astatine 85 At (209.99)	Radon 86 Rn (222.02)
7	Francium 87 Fr (223.02)	Radium 88 Ra (226.0254)	Actinium 89 Ac (227.0278)	Rutherfordium 104 Rf (261.11)	Dubnium 105 Db (262.11)	Seaborgium 106 Sg (263.12)	Bohrium 107 Bh (262.12)	Hassium 108 Hs (265)	Meitnerium 109 Mt (266)	Darmstadtium 110 Ds (271)	Roentgenium 111 Rg (272)	— 112 — Discovered 1996	— 113 — Discovered 2004	— 114 — Discovered 1999	— 115 — Discovered 2004	— 116 — Discovered 1999		

Lanthanides	Cerium 58 Ce 140.115	Praseodymium 59 Pr 140.9076	Neodymium 60 Nd 144.24	Promethium 61 Pm (144.91)	Samarium 62 Sm 150.36	Europium 63 Eu 151.965	Gadolinium 64 Gd 157.25	Terbium 65 Tb 158.9253	Dysprosium 66 Dy 162.50	Holmium 67 Ho 164.9303	Erbium 68 Er 167.26	Thulium 69 Tm 168.9342	Ytterbium 70 Yb 173.04	Lutetium 71 Lu 174.967
Actinides	Thorium 90 Th 232.0381	Protactinium 91 Pa 231.0388	Uranium 92 U 238.0289	Neptunium 93 Np (237.0482)	Plutonium 94 Pu (244.664)	Americium 95 Am (243.061)	Curium 96 Cm (247.07)	Berkelium 97 Bk (247.07)	Californium 98 Cf (251.08)	Einsteinium 99 Es (252.08)	Fermium 100 Fm (257.10)	Mendelevium 101 Md (258.10)	Nobelium 102 No (259.10)	Lawrencium 103 Lr (262.11)

Figure 3.3 The Periodic Table of Elements Only about a dozen elements are common in minerals and rocks, but many uncommon ones are important sources of natural resources. For example, lead (Pb) is not found in many minerals, but it is present in the mineral galena, the main ore of lead. Silicon (Si) and oxygen (O), in contrast, are important elements in most of the minerals in Earth's crust.

spontaneously decay or change to other elements. Carbon 14 is radioactive, whereas both carbon 12 and carbon 13 are not. Radioactive isotopes are important for determining the absolute ages of rocks (see Chapter 17).

Bonding and Compounds

Interactions among electrons around atoms can result in two or more atoms joining together, a process known as **bonding**. If atoms of two or more elements bond, the resulting substance is a **compound**. Gaseous oxygen consists of only oxygen atoms and is thus an element, whereas the mineral quartz, consisting of silicon and oxygen atoms, is a compound. Most minerals are compounds, although gold, platinum, and several others are important exceptions.

To understand bonding, it is necessary to delve deeper into the structure of atoms. Recall that negatively charged electrons orbit the nuclei of atoms in electron shells. With the exception of hydrogen, which has only one proton and one electron, the innermost electron shell of an atom contains only two electrons. The other shells contain various numbers of electrons, but the outermost shell never has more than eight (Figure 3.2). The electrons in the outermost shell are those that are usually involved in chemical bonding.

Two types of chemical bonds, *ionic* and *covalent*, are particularly important in minerals, and many minerals contain both types of bonds. Two other types of chemical bonds, *metallic* and *van der Waals*, are much less common, but are

Nucleus
6 p
6 n
^{12}C (Carbon 12)

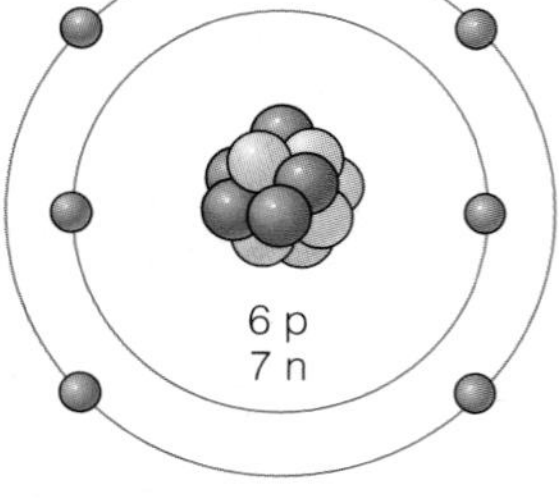

^{13}C (Carbon 13)

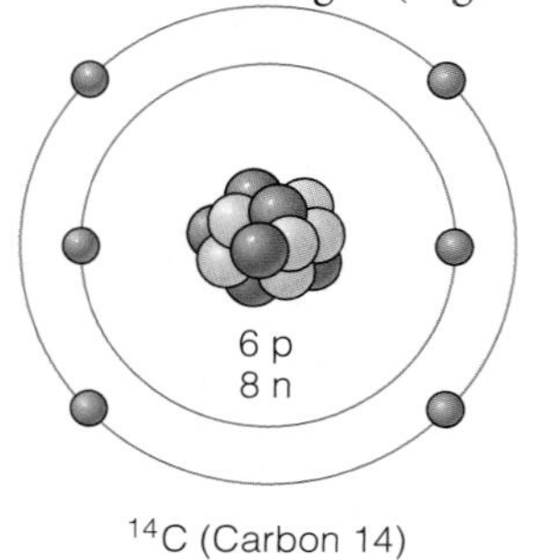

^{14}C (Carbon 14)

Figure 3.4 Carbon Isotopes Schematic representation of the isotopes of carbon. Carbon has an atomic number of 6 and an atomic mass number of 12, 13, or 14, depending on the number of neutrons in its nucleus.

extremely important in determining the properties of some useful minerals.

Ionic Bonding Notice in Figure 3.2 that most atoms have fewer than eight electrons in their outermost electron shell. However, some elements, including neon and argon, have complete outer shells with eight electrons; because of this electron configuration, these elements, known as the *noble gases*, do not react readily with other elements to form compounds. Interactions among atoms tend to produce electron configurations similar to those of the noble gases. That is, atoms interact so that their outermost electron shell is filled with eight electrons, unless the first shell (with two electrons) is also the outermost electron shell, as in helium.

One way that the noble gas configuration is attained is by the transfer of one or more electrons from one atom to another. Common salt is composed of the elements sodium (Na) and chlorine (Cl); each element is poisonous, but when combined chemically they form the compound sodium chloride (NaCl), better known as the mineral halite. Notice in Figure 3.5a that sodium has 11 protons and 11 electrons; thus the positive electrical charges of the protons are exactly balanced by the negative charges of the electrons, and the atom is electrically neutral. Likewise, chlorine with 17 protons and 17 electrons is electrically neutral (Figure 3.5a). However, neither sodium nor chlorine has eight electrons in its outermost electron shell; sodium has only one, whereas chlorine has seven. To attain a stable configuration, sodium loses the electron in its outermost electron shell, leaving its next shell with eight electrons as the outermost one (Figure 3.5a). Sodium now has one fewer electron (negative charge) than it has protons (positive charge), so it is an electrically charged **ion** and is symbolized Na^+.

The electron lost by sodium is transferred to the outermost electron shell of chlorine, which had seven electrons to begin with. The addition of one more electron gives chlorine an outermost electron shell of eight electrons, the configuration of a noble gas. But its total number of electrons is now 18, which exceeds by 1 the number of protons. Accordingly, chlorine also becomes an ion, but it is negatively charged (Cl^-). An **ionic bond** forms between sodium and chlorine because of the attractive force between the positively charged sodium ion and the negatively charged chlorine ion (Figure 3.5a).

In ionic compounds, such as sodium chloride (the mineral halite), the ions are arranged in a three-dimensional framework that results in overall electrical neutrality. In halite, sodium ions are bonded to chlorine ions on all sides, and chlorine ions are surrounded by sodium ions (Figure 3.5b).

Covalent Bonding **Covalent bonds** form between atoms when their electron shells overlap and they share electrons. For example, atoms of the same element, such as carbon, cannot bond by transferring electrons from one atom to another. Carbon (C), which forms the minerals graphite and diamond, has four electrons in its outermost electron shell (Figure 3.6a). If these four electrons were transferred

Figure 3.5 Ionic Bond to Form the Mineral Halite (NaCl)

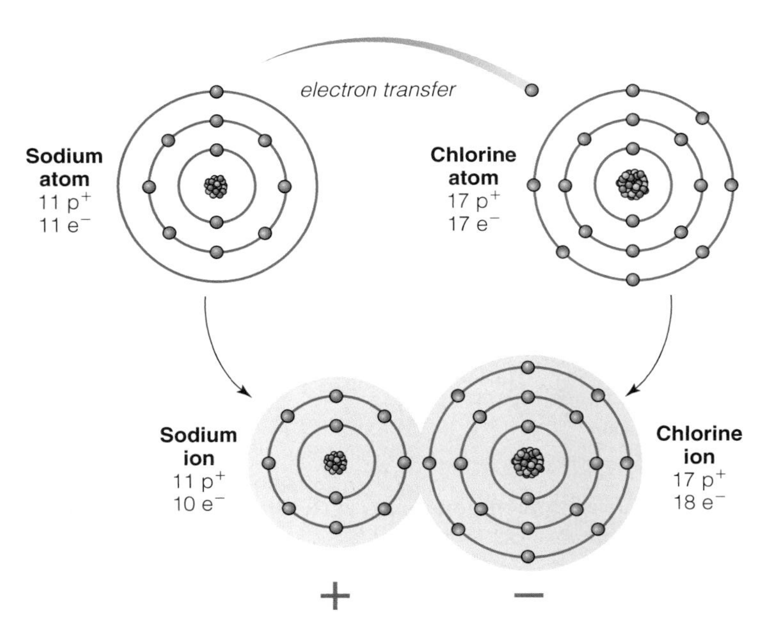

a Transfer of the electron in the outermost shell of sodium to the outermost shell of chlorine. After electron transfer, the sodium and chlorine atoms are positively and negatively charged ions, respectively.

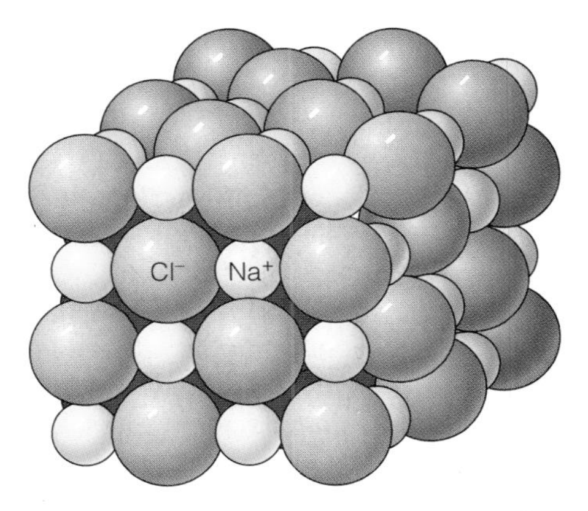

b This diagram shows the relative sizes of the sodium and chlorine atoms and their locations in a crystal of halite.

c Tiny crystals of halite.

Figure 3.6 Covalent Bonds

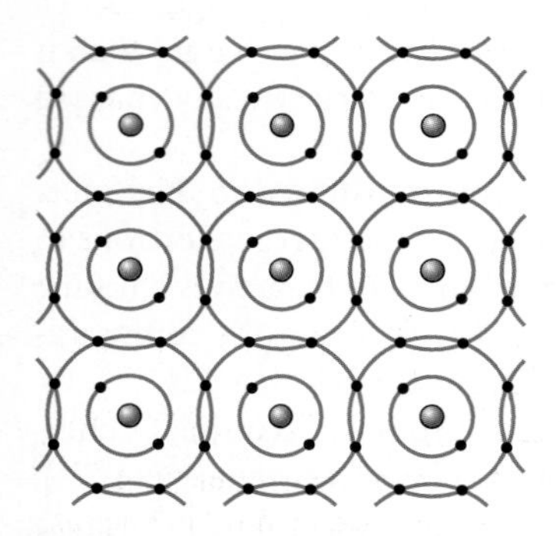

a The orbits in the outermost electron shell overlap, so electrons are shared in diamond.

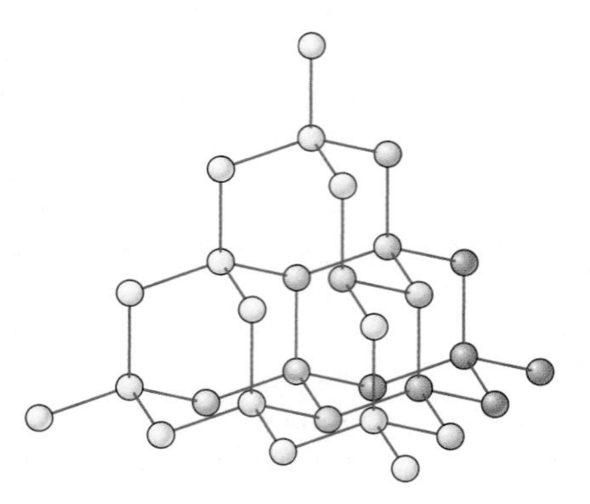

b Covalent bonding of carbon atoms in diamond forms a three-dimensional framework.

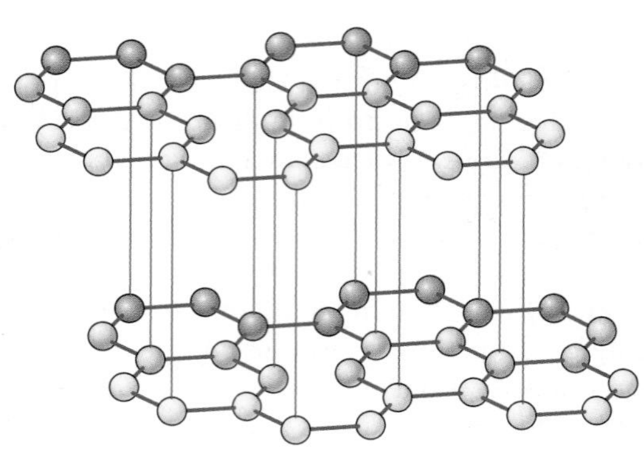

c Covalent bonds in graphite form strong sheets, but the van der Waals bonds between sheets are weak.

to another carbon atom, the atom receiving the electrons would have the noble gas configuration of eight electrons in its outermost electron shell, but the atom contributing the electrons would not.

In such situations, adjacent atoms share electrons by overlapping their electron shells. A carbon atom in diamond, for instance, shares all four of its outermost electrons with a neighbor to produce a stable noble gas configuration (Figure 3.6a).

Covalent bonds are not restricted to substances composed of atoms of a single kind. Among the most common minerals, the silicates (discussed later in this chapter), the element silicon forms partly covalent and partly ionic bonds with oxygen.

Metallic and van der Waals Bonds *Metallic bonding* results from an extreme type of electron sharing. The electrons of the outermost electron shell of metals such as gold, silver, and copper readily move about from one atom to another. This electron mobility accounts for the fact that metals have a metallic luster (their appearance in reflected light), provide good electrical and thermal conductivity, and can be easily reshaped. Only a few minerals possess metallic bonds, but those that do are very useful; copper, for example, is used for electrical wiring because of its high electrical conductivity.

Some electrically neutral atoms and molecules* have no electrons available for ionic, covalent, or metallic bonding. They nevertheless have a weak attractive force between them, called a *van der Waals* or *residual bond*, when in proximity. The carbon atoms in the mineral graphite are covalently bonded to form sheets, but the sheets are weakly held together by van der Waals bonds (Figure 3.6c). This type of bonding makes graphite useful for pencil lead—when a pencil point is moved across a piece of paper, small pieces of graphite flake off along the planes held together by van der Waals bonds and adhere to the paper.

*A molecule is the smallest unit of a substance that has the properties of that substance. A water molecule (H_2O), for example, possesses two hydrogen atoms and one oxygen atom.

EXPLORE THE WORLD OF MINERALS

We defined a *mineral* as an inorganic, naturally occurring, crystalline solid with a narrowly defined chemical composition and characteristic physical properties. Furthermore, we know from the preceding section that most minerals are compounds of two or more chemically bonded elements as in quartz (SiO_2). In the following sections, we will examine each part of the formal definition of the term mineral.

Naturally Occurring Inorganic Substances

The criterion *naturally occurring* excludes from minerals all substances manufactured by humans, such as synthetic diamonds and rubies. This criterion is particularly important to those who buy and sell gemstones, most of which are minerals, because some human-made substances are very difficult to distinguish from natural gem minerals.

Some geologists think the term *inorganic* in the mineral definition is unnecessary, but it does remind us that animal matter and vegetable matter are not minerals. Nevertheless, some organisms, including corals, clams, and a number of other animals and plants, construct their shells of the compound calcium carbonate ($CaCO_3$), which is either the mineral aragonite or calcite, or their shells are made of silicon dioxide (SiO_2) as in quartz.

Mineral Crystals

By definition, minerals are **crystalline solids**, in which the constituent atoms are arranged in a regular, three-dimensional framework (Figure 3.5b). Under ideal conditions, such as in a cavity, these crystalline solids can grow and form perfect **crystals** that possess planar surfaces (crystal faces), sharp corners, and straight edges (Figure 3.7). In other words, the regular geometric shape of a well-formed mineral crystal is

Figure 3.7 A Variety of Mineral Crystal Shapes

a Cubic crystals are typical of the minerals halite and galena.

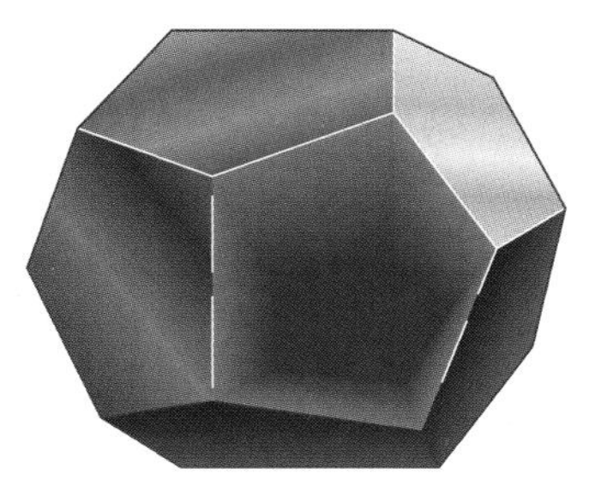

b Pyritohedron crystals such as those of pyrite have 12 sides.

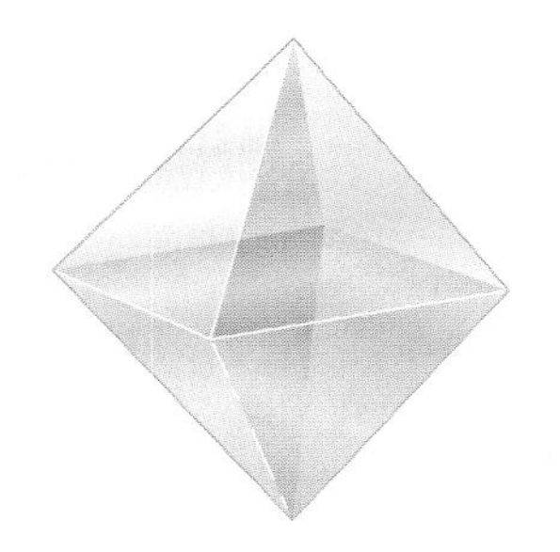

c Diamond has octahedral, or eight-sided, crystals.

d A prism terminated by a pyramid is found in quartz.

the exterior manifestation of an ordered internal atomic arrangement. Not all rigid substances are crystalline solids; natural and manufactured glass lack the ordered arrangement of atoms and is said to be *amorphous*, meaning "without form." Minerals are, by definition, crystalline solids; however, crystalline solids do not always yield well-formed crystals. The reason is that when crystals form, they may grow in proximity and form an interlocking mosaic in which individual crystals are not apparent or easily discerned (Figure 3.8).

So how do we know that the mineral in Figure 3.8b is actually crystalline? X-ray beams and light transmitted through mineral crystals or crystalline solids behave in a predictable

Figure 3.8 Quartz and the Constancy of Interfacial Angles

Sue Monroe

a Well-shaped crystal of smoky quartz.

Albert J. Copley/Visuals Unlimited

b Specimen of rose quartz in which no obvious crystals can be discerned.

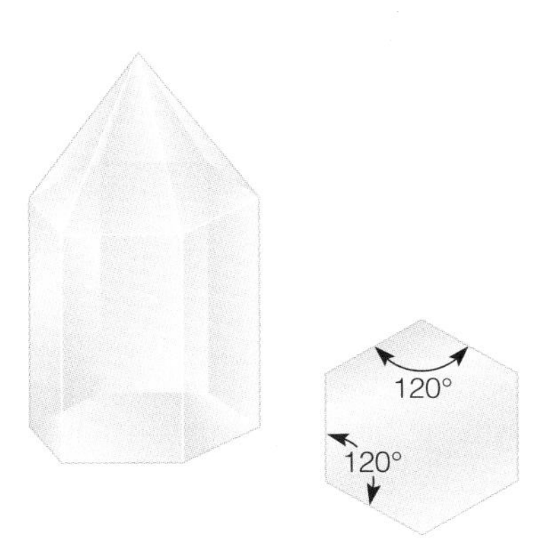

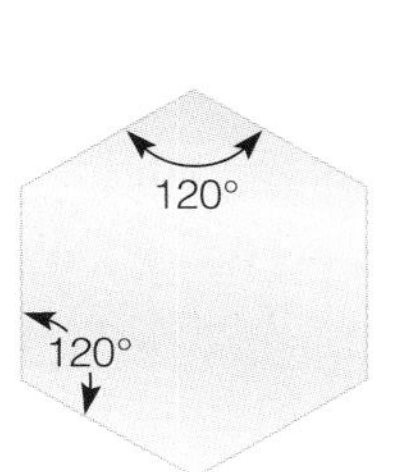

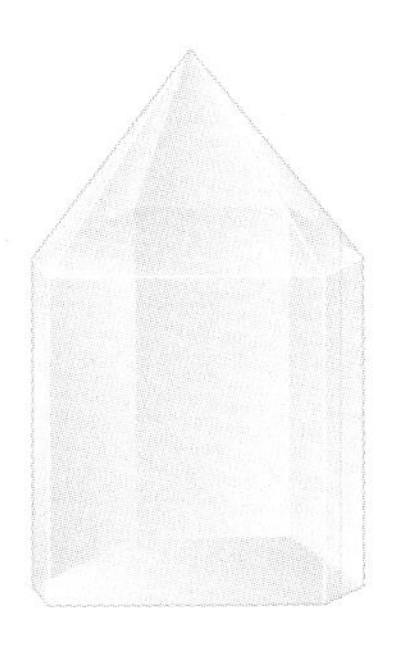

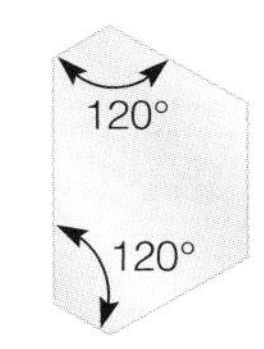

c Side views and cross sections of quartz crystals showing the constancy of interfacial angles. A well-shaped crystal (left), a larger well-shaped crystal (middle), and a poorly shaped crystal (right). The angles formed between equivalent crystal faces on different specimens of the same mineral are the same regardless of size, shape, age, or geographic occurrence of the specimens.

manner, which provides compelling evidence for an internal orderly structure. Another way we can determine that minerals with no obvious crystals are actually crystalline is by their *cleavage*, the property of breaking or splitting repeatedly along smooth, closely spaced planes. Not all minerals have cleavage planes, but many do, and such regularity certainly indicates that splitting is controlled by internal structure.

As early as 1669, the Danish scientist Nicholas Steno determined that the angles of intersection of equivalent crystal faces on different specimens of quartz are identical. Since then, this *constancy of interfacial angles* has been demonstrated for many other minerals, regardless of their size, shape, age, or geographic occurrence (Figure 3.8c). Steno postulated that mineral crystals are made up of very small, identical building blocks, and that the arrangement of these building blocks determines the external form of mineral crystals, a proposal that has since been verified.

Chemical Composition of Minerals

Mineral composition is shown by a chemical formula, which is a shorthand way of indicating the numbers of atoms of different elements that make up a mineral. The mineral quartz consists of one silicon (Si) atom for every two oxygen (O) atoms and thus has the formula SiO_2; the subscript number indicates the number of atoms. Orthoclase is composed of one potassium, one aluminum, three silicon, and eight oxygen atoms, so its formula is $KAlSi_3O_8$. Some minerals known as *native elements* consist of a single element and include silver (Ag), platinum (Pt), gold (Au), and graphite and diamond, both of which are composed of carbon (C).

The definition of a mineral contains the phrase *a narrowly defined chemical composition* because some minerals actually have a range of compositions. For many minerals, the chemical composition does not vary. Quartz is composed of only silicon and oxygen (SiO_2), and halite contains only sodium and chlorine (NaCl). Other minerals have a range of compositions because one element can substitute for another if the atoms of two or more elements are nearly the same size and the same charge. Notice in Figure 3.9 that iron and magnesium atoms are about the same size; therefore, they can substitute for each other. The chemical formula for the mineral olivine is $(Mg,Fe)_2SiO_4$, meaning that, in addition to silicon and oxygen, it may contain only magnesium, only iron, or a combination of both. A number of other minerals also have ranges of compositions, so these are actually mineral groups with several members.

Physical Properties of Minerals

The last criterion in our definition of a mineral, *characteristic physical properties*, refers to such properties as hardness, color, and crystal form. These properties are controlled by composition and structure. We will have more to say about the physical properties of minerals later in this chapter.

MINERAL GROUPS RECOGNIZED BY GEOLOGISTS

Geologists have identified and described more than 3500 minerals, but only a few—perhaps two dozen—are common. One might think that an extremely large number of minerals

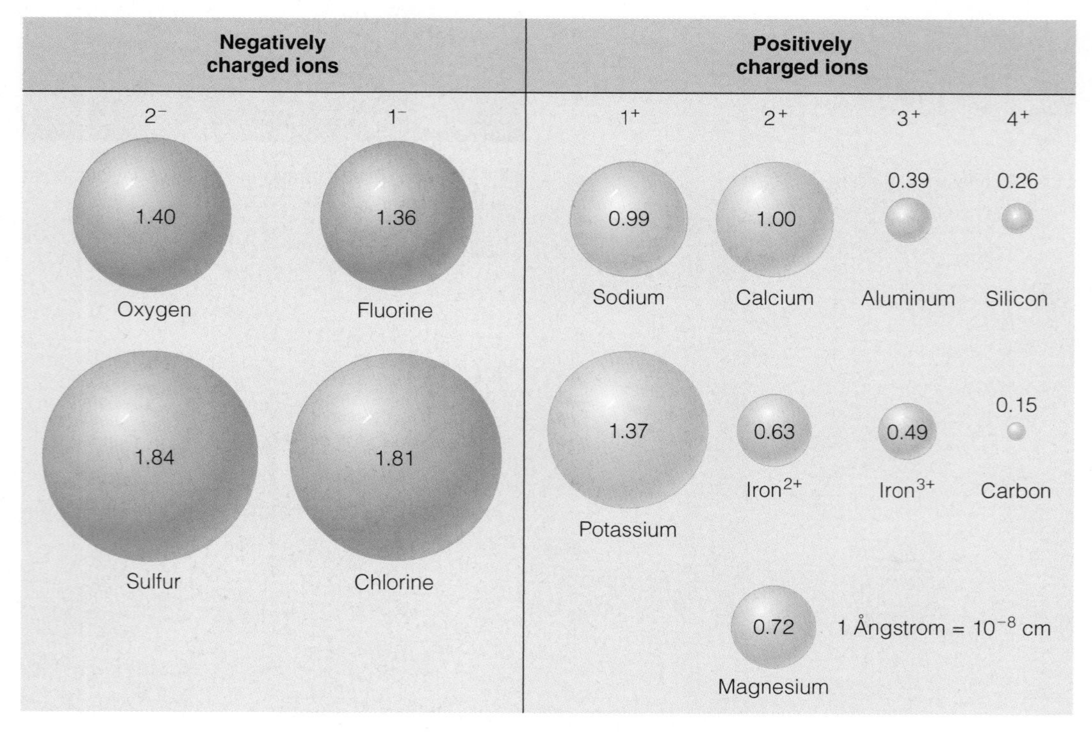

Figure 3.9 Sizes and Charges of Ions Electrical charges and relative sizes of ions common in minerals. The numbers within the ions are the radii shown in Angstrom units.

could form from 92 naturally occurring elements; however, several factors limit the number possible. For one thing, many combinations of elements simply do not occur; no compounds are composed of only potassium and sodium or of silicon and iron, for example. Another important factor is that the bulk of Earth's crust is made up of only eight chemical elements, and even among these eight, silicon and oxygen are by far the most common. In fact, most common minerals in Earth's crust consist of silicon, oxygen, and one or more of the elements in Figure 3.10.

Geologists recognize mineral classes or groups, each with members that share the same negatively charged ion or ion group (Table 3.1). We have mentioned that ions are atoms that have either a positive or negative electrical charge resulting from the loss or gain of electrons in their outermost shell. In addition to ions, some minerals contain tightly bonded, complex groups of different atoms known as *radicals* that act as single units. A good example is the carbonate radical, consisting of a carbon atom bonded to three oxygen atoms and thus having the formula CO_3 and a –2 electrical charge. Other common radicals and their charges are sulfate (SO_4, –2), hydroxyl (OH, –1), and silicate (SiO_4, –4) (Figure 3.11).

Silicate Minerals

Because silicon and oxygen are the two most abundant elements in Earth's crust, it is not surprising that many

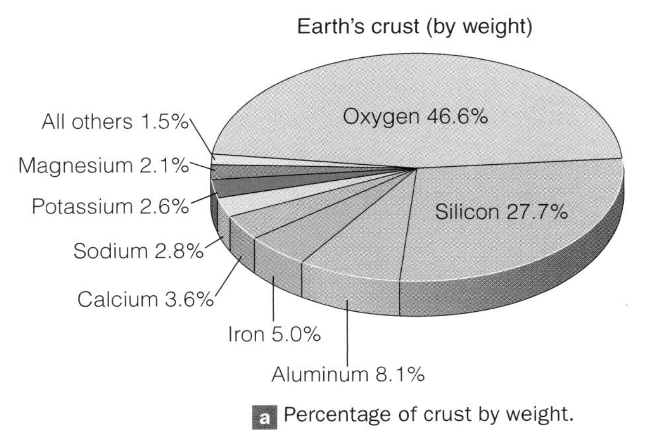

a Percentage of crust by weight.

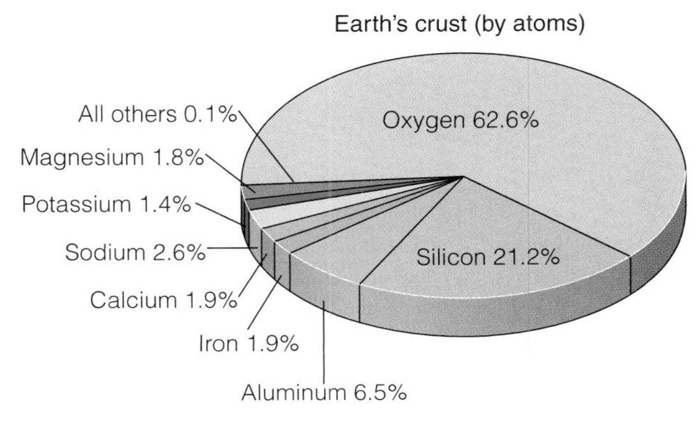

b Percentage of crust by atoms.

Figure 3.10 Common Elements in Earth's Crust a Source: From Miller, G. T., 1996. *Living in the Environment: Principles, Concepts, and Solutions*. Wadsworth Publishing. Figure 8.3.

TABLE 3.1 Mineral Groups Recognized by Geologists

Mineral Group	Negatively Charged Ion or Radical	Examples	Composition
Carbonate	$(CO_3)^{-2}$	Calcite	$CaCO_3$
		Dolomite	$CaMg(CO_3)_2$
Halide	Cl^{-1}, F^{-1}	Halite	NaCl
		Fluorite	CaF_2
Hydroxide	$(OH)^{-1}$	Brucite	$Mg(OH)_2$
Native element	—	Gold	Au
		Silver	Ag*
		Diamond	C
Phosphate	$(PO_4)^{-3}$	Apatite	$Ca_5(PO_4)_3(F,Cl)$
Oxide	O^{-2}	Hematite	Fe_2O_3
		Magnetite	Fe_3O_4
Silicate	$(SiO_4)^{-4}$	Quartz	SiO_2
		Potassium feldspar	$KAlSi_3O_8$
		Olivine	$(Mg,Fe)_2SiO_4$
Sulfate	$(SO_4)^{-2}$	Anhydrite	$CaSO_4$
		Gypsum	$CaSO_4 \cdot 2H_2O$
Sulfide	S^{-2}	Galena	PbS
		Pyrite	FeS_2
		Argentite	Ag^2S*

*Note that silver is found as both a native element and a sulfide mineral.

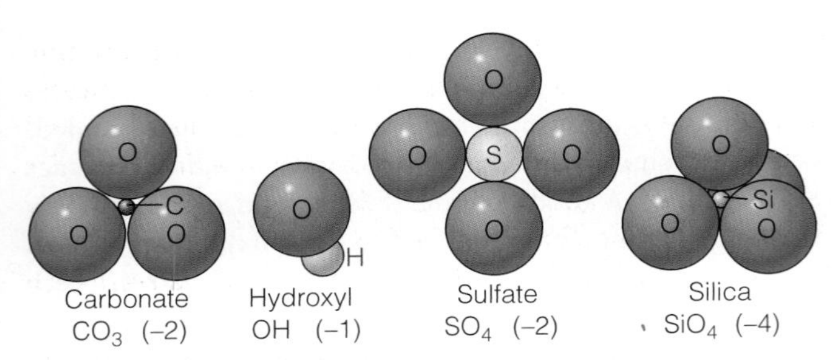

Figure 3.11 Radicals Many minerals contain radicals, which are complex groups of atoms tightly bonded together. The silica and carbonate radicals are particularly common in many minerals, such as quartz (SiO_2) and calcite ($CaCO_3$).

minerals contain these elements. A combination of silicon and oxygen is known as **silica**, and minerals that contain silica are **silicates**. Quartz (SiO_2) is pure silica because it is composed entirely of silicon and oxygen. But most silicates have one or more additional elements, as in orthoclase ($KAlSi_3O_8$) and olivine [$(Mg,Fe)_2SiO_4$]. Silicate minerals include about one-third of all known minerals, but their abundance is even more impressive when one considers that they make up perhaps 95% of Earth's crust.

The basic building block of all silicate minerals is the **silica tetrahedron**, consisting of one silicon atom and four oxygen atoms (Figure 3.12a). These atoms are arranged so

Figure 3.12 The Silica Tetrahedron and Silicate Materials

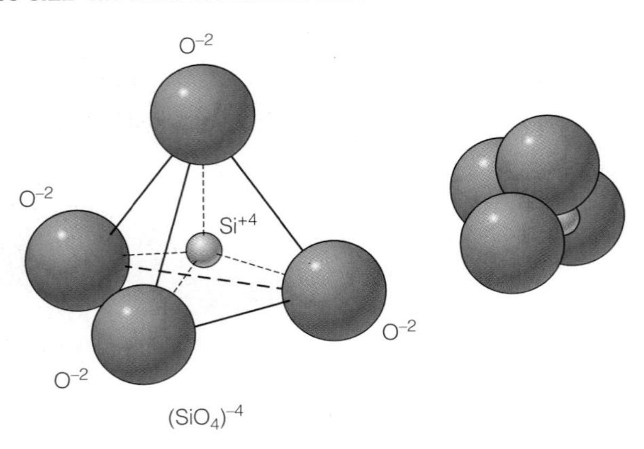

a Expanded view of the silica tetrahedron (left) and how it actually exists with its oxygen atoms touching.

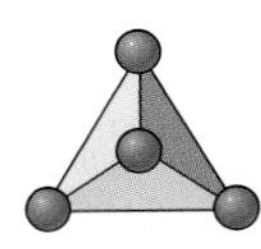

b View of the silica tetrahedron from above. Only the oxygen atoms are visible.

			Formula of negatively charged ion group		Example
c	Isolated tetrahedra		$(SiO_4)^{-4}$	No oxygen atoms shared	Olivine
d	Continuous chains of tetrahedra	Single chain	$(SiO_3)^{-2}$	Each tetrahedra shares two oxygen atoms with adjacent tetrahedra	Pyroxene group (augite)
		Double chain	$(Si_4O_{11})^{-6}$	Single chains linked by sharing oxygen atoms	Amphibole group (hornblende)
e	Continuous sheets		$(Si_4O_{10})^{-4}$	Three oxygen atoms shared with adjacent tetrahedra	Micas (muscovite)
f	Three-dimensional networks	SiO_2	$(SiO_2)^0$	All four oxygen atoms in tetrahedra shared	Quartz Potassium feldspars Plagioclase feldspars

(**c**–**f**) Structures of the common silicate minerals shown by various arrangements of the silica tetrahedra.

that the four oxygen atoms surround a silicon atom, which occupies the space between the oxygen atoms, thus forming a four-faced pyramidal structure (Figure 3.12b). The silicon atom has a positive charge of 4, and each of the four oxygen atoms has a negative charge of 2, resulting in a radical with a total negative charge of 4 $(SiO_4)^{-4}$.

Because the silica tetrahedron has a negative charge, it does not exist in nature as an isolated ion group; rather, it combines with positively charged ions or shares its oxygen atoms with other silica tetrahedra. In the simplest silicate minerals, the silica tetrahedra exist as single units bonded to positively charged ions. In minerals that contain isolated tetrahedra, the silicon-to-oxygen ratio is 1:4, and the negative charge of the silica ion is balanced by positive ions (Figure 3.12c). Olivine $[(Mg,Fe)_2SiO_4]$, for example, has either two magnesium (Mg^{+2}) ions, two iron (Fe^{+2}) ions, or one of each to offset the –4 charge of the silica ion.

Silica tetrahedra may also join together to form chains of indefinite length (Figure 3.12d). Single chains, as in the pyroxene minerals, form when each tetrahedron shares two of its oxygens with an adjacent tetrahedron, resulting in a silicon-to-oxygen ratio of 1:3. Enstatite, a pyroxene-group mineral, reflects this ratio in its chemical formula $MgSiO_3$. Individual chains, however, possess a net –2 electrical charge, so they are balanced by positive ions, such as Mg^{+2}, that link parallel chains together (Figure 3.12d).

The amphibole group of minerals is characterized by a double-chain structure in which alternate tetrahedra in two parallel rows are cross-linked (Figure 3.12d). The formation of double chains results in a silicon-to-oxygen ratio of 4:11, so each double chain possesses a –6 electrical charge. Mg^{+2}, Fe^{+2}, and Al^{+2} are usually involved in linking the double chains together.

In sheet-structure silicates, three oxygens of each tetrahedron are shared by adjacent tetrahedra (Figure 3.12e). Such structures result in continuous sheets of silica tetrahedra with silicon-to-oxygen ratios of 2:5. Continuous sheets also possess a negative electrical charge satisfied by positive ions located between the sheets. This particular structure accounts for the characteristic sheet structure of the *micas*, such as biotite and muscovite, and the *clay minerals.*

Three-dimensional networks of silica tetrahedra form when all four oxygens of the silica tetrahedra are shared by adjacent tetrahedra (Figure 3.12f). Such sharing of oxygen atoms results in a silicon-to-oxygen ratio of 1:2, which is electrically neutral. Quartz is a common framework silicate.

Geologists recognize two subgroups of silicates: ferromagnesian and nonferromagnesian silicates. The **ferromagnesian silicates** are those that contain iron (Fe), magnesium (Mg), or both. These minerals are commonly dark and more dense than nonferromagnesian silicates. Some of the common ferromagnesian silicate minerals are olivine, the pyroxenes, the amphiboles, and biotite (Figure 3.13a).

The **nonferromagnesian silicates** lack iron and magnesium, are generally light colored, and are less dense than ferromagnesian silicates (Figure 3.13b). The most common minerals in Earth's crust are nonferromagnesian silicates known as *feldspars.* Feldspar is a general name, however, and two distinct groups are recognized, each of which includes several species. The *potassium feldspars* are represented by microcline and orthoclase $(KAlSi_3O_8)$. The second group of feldspars, the *plagioclase feldspars,* range from calcium-rich $(CaAl_2Si_2O_8)$ to sodium-rich $(NaAlSi_3O_8)$ varieties.

Quartz (SiO_2) is another common nonferromagnesian silicate. It is a framework silicate that can usually be recognized by its glassy appearance and hardness. Another fairly common nonferromagnesian silicate is muscovite, which is a mica (Figure 3.13b and see Geo-Focus on page 76 and 77).

Carbonate Minerals

Carbonate minerals, those that contain the negatively charged carbonate radical $(CO_3)^{-2}$, include calcium carbonate $(CaCO_3)$ as the minerals *aragonite* or *calcite* (Figure 3.14a). Aragonite is unstable and commonly changes to calcite, the main constituent of the sedimentary rock *limestone.* A number of other carbonate minerals are known, but only one of these need concern us: *Dolomite* $[CaMg(CO_3)_2]$ forms by the chemical alteration of calcite by the addition of magnesium. Sedimentary rock composed of the mineral dolomite is *dolostone* (see Chapter 7).

Other Mineral Groups

In addition to silicates and carbonates, geologists recognize several other mineral groups (Table 3.1). Even though minerals from these groups are less common than silicates and carbonates, many are found in rocks in small quantities and others are important resources. In the oxides, an element combines with oxygen as in hematite (Fe_2O_3) and magnetite (Fe_3O_4). Rocks with high concentrations of these minerals in the Lake Superior region of Canada and the United States are sources of iron ores for the manufacture of steel. The related hydroxides form mostly by the chemical alteration of other minerals.

We have noted that the *native elements* are minerals composed of a single element, such as diamond and graphite (C) and the precious metals gold (Au), silver (Ag), and platinum (Pt) (see Geo-inSight on pages 72 and 73). Some elements, such as silver and copper, are found both as native elements and as compounds and are thus also included in other mineral groups; argentite (Ag_2S), a silver sulfide, is an example.

Several minerals and rocks that contain the phosphate radical $(PO_4)^{-3}$ are important sources of phosphorus for fertilizers. The sulfides, such as galena (PbS), the ore of lead, have a positively charged ion combined with sulfur (S^{-2}) (Figure 3.14b), whereas the sulfates have an element combined with the complex radical $(SO_4)^{-2}$, as in gypsum $(CaSO_4{\cdot}2H_2O)$ (Figure 3.14c). The halides contain the halogen elements, fluorine (F^{-1}) and chlorine (Cl^{-1}); examples are halite (NaCl) (Figure 3.14d) and fluorite (CaF_2).

Geo-inSight The Precious Metals

The discovery of gold by James Marshall at Sutter's Mill near Coloma in 1848 sparked the California gold rush (1849–1853) during which $200 million in gold was recovered.

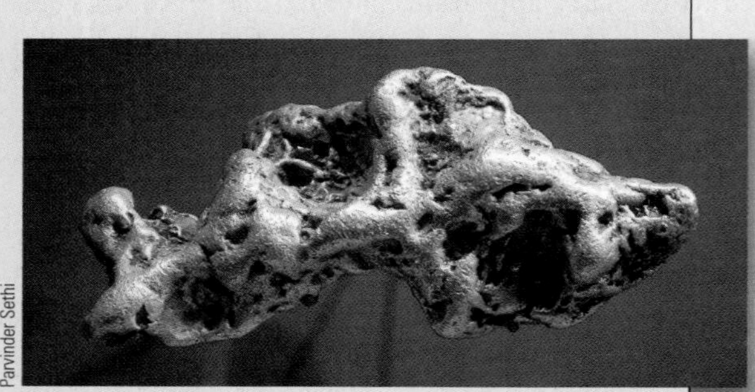

Parvinder Sethi

▲ **1.** Specimen of gold from Grass Valley, California. Gold is too heavy and too soft for tools and weapons, so it has been prized for jewelry and as a symbol of wealth, but it is also used in glass making, electrical circuitry, gold plating, the chemical industry, and dentistry.

Bettmann/Corbis

▲ **3.** Gold miners on the American River near Sacramento, California. Most of the gold came from placer deposits in which running water separated and concentrated minerals and rock fragments by their density.

▼ **2.** A miner pans for gold (foreground) by swirling water, sand, and gravel in a broad, shallow pan. The heavier gold sinks to the bottom. At the far left, a miner washes sediment in a cradle. As in panning, the cradle separates the heavier gold from other materials.

Bettmann/Corbis

Bettmann/Corbis

◀ **4.** Hydraulic mining in California in which strong jets of water washed gold-bearing sand and gravel into sluices. In this image taken in 1905 at Junction City, California, water is directed through a monitor onto a hillside. Hydraulic mining was efficient from the mining point of view, but caused considerable environmental damage.

J.C.H. Grabill/Corbis

▶ **5.** Reports in 1876 of gold in the Black Hills of South Dakota resulted in a flood of miners that led to hostilities with the Sioux Indians, and the annihilation of Lt. Col. George Armstrong Custer and 260 of his men at the Battle of the Little Big Horn in Montana. This view shows the headworks (upper right) of the Homestake Mine at Lead, South Dakota, in 1900. The headworks is the cluster of buildings near the opening to a mine.

Ken Lucas/Visuals Unlimited

◀ **6.** Like gold, silver is found as a native element as in this specimen, but it also occurs as a compound in the sulfide mineral argentite (Ag_2S). Silver is used in North America for silver halide film, jewelry, flatware, surgical instruments, and backing for mirrors.

James S. Monroe

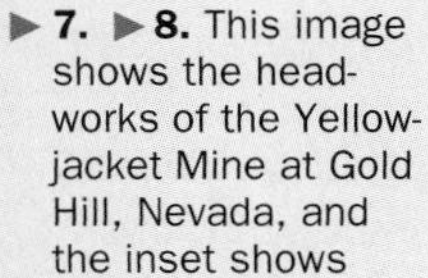

▶ **7.** ▶ **8.** This image shows the headworks of the Yellowjacket Mine at Gold Hill, Nevada, and the inset shows silver-bearing quartz (white) in volcanic rock. This largest silver discovery in North America, called the Comstock Lode, was responsible for bringing Nevada into the Union in 1864 during the Civil War, even though it had too few people to qualify for statehood. The Comstock Lode was mined for silver and gold from 1859 until 1898.

Sue. Monroe

Figure 3.13 Common Rock-Forming Silicate Minerals

a The ferromagnesian silicates.

b The nonferromagnesian silicates.

PHYSICAL PROPERTIES OF MINERALS

Many physical properties of minerals are remarkably constant for a given mineral species, but some, especially color, may vary. Although professional geologists use sophisticated techniques to study and identify minerals, most common minerals can be identified by using the physical properties described next.

Luster and Color

Luster (not to be confused with *color*) is the quality and intensity of light reflected from a mineral's surface. Geologists define two basic types of luster: *metallic*, having the appearance of a metal, and *nonmetallic.* Notice that of the four minerals shown in Figure 3.14, only galena has a metallic luster. Among the several types of nonmetallic luster are glassy or vitreous (as in quartz), dull or earthy, waxy, greasy, and brilliant (as in diamond) (Figure 3.1a).

Beginning geology students are distressed by the fact that the color of some minerals varies considerably, making the most obvious physical property of little use for mineral identification. In any case, we can make some helpful generalizations about color. Ferromagnesian silicates are typically black, brown, or dark green, although olivine is olive green (Figure 3.13a). Nonferromagnesian silicates, on the other hand, vary considerably in color, but are rarely very dark. White, cream, colorless, and shades of pink and pale green are more typical (Figure 3.13b).

Another helpful generalization is that the color of minerals with a metallic luster is more consistent than is the color of nonmetallic minerals. For example, galena is always lead-gray (Figure 3.14b) and pyrite is invariably brassy yellow. In contrast, quartz, a nonmetallic mineral, may be colorless, smoky brown to almost black, rose, yellow-brown, milky white, blue, or violet to purple (Figure 3.8a, b).

Crystal Form

As we noted, many mineral specimens do not show the perfect crystal form typical of that mineral species. Nevertheless, some minerals do typically occur as crystals (Figure 3.15a–d). For example, 12-sided crystals of garnet are common, as are 6- and 12-sided crystals of pyrite. Minerals that grow in cavities or are precipitated from circulating hot water (hydrothermal solutions) in cracks and crevices in rocks also commonly occur as crystals and, under ideal conditions, some minerals grow to incredible sizes (Figure 3.15d).

Crystal form can be a useful characteristic for mineral identification, but a number of minerals have the same crystal form. Pyrite (FeS_2), galena (PbS), and halite ($NaCl$) all occur as cubic crystals, but they can be easily identified by other properties such as color, luster, hardness, and density.

Cleavage and Fracture

Not all minerals possess **cleavage**, but those that do break, or split, along a smooth plane or planes of weakness determined by the strength of their chemical bonds. Cleavage is characterized in terms of quality (perfect, good, poor),

Figure 3.14 Representative Specimens from Four Mineral Groups

a Calcite ($CaCO_3$) is the most common carbonate mineral.

b The sulfide mineral galena (PbS) is the ore of lead.

c Gypsum ($CaSO_4 \cdot 2H_2O$) is a common sulfate mineral.

d Halite (NaCl) is a good example of a halide mineral.

direction, and angles of intersection of cleavage planes. Biotite, a common ferromagnesian silicate has perfect cleavage in one direction (Figure 3.16a). Biotite is a sheet silicate with the sheets of silica tetrahedra weakly bonded to one another by iron and magnesium ions.

Feldspars possess two directions of cleavage that intersect at right angles (Figure 3.16b), and the mineral halite has three directions of cleavage, all of which intersect at right angles (Figure 3.16c). Calcite also possesses three directions of cleavage, but none of the intersection angles is a right angle, so cleavage fragments of calcite are rhombohedrons (Figure 3.16d). Minerals with four directions of cleavage include fluorite and diamond (Figure 3.16e). Ironically, diamond, the hardest mineral, can be cleaved easily. A few minerals, such as sphalerite, an ore of zinc, have six directions of cleavage (Figure 3.16f).

Cleavage is an important diagnostic property of minerals, and recognizing it is essential in distinguishing between some minerals. The pyroxene mineral augite and the amphibole mineral hornblende, for example, look much alike: Both are dark green to black, have the same hardness, and possess two directions of cleavage. But the cleavage planes of augite intersect at about 90 degrees, whereas the cleavage planes of hornblende intersect at angles of 56 degrees and 124 degrees (Figure 3.17).

In contrast to cleavage, *fracture* is mineral breakage along irregular surfaces. Any mineral can be fractured if enough force is applied, but the fracture surfaces are commonly uneven or conchoidal (curved) rather than smooth.

Hardness

An Austrian geologist, Friedrich Mohs, devised a relative hardness scale for 10 minerals. He arbitrarily assigned a hardness value of 10 to diamond, the hardest mineral known, and lesser values to the other minerals. Relative hardness is easily determined by the use of Mohs hardness scale (Table 3.2). Quartz will scratch fluorite but cannot be scratched by fluorite, gypsum can be scratched by a fingernail, and so on. So **hardness** is defined as a mineral's resistance to abrasion and is controlled mostly by internal structure. For example, both graphite and diamond are composed of carbon, but the former has a hardness of 1 to 2, whereas the latter has a hardness of 10.

Specific Gravity (Density)

Specific gravity and density are two separate concepts, but here we will use them more or less as synonyms. A mineral's **specific gravity** is the ratio of its weight to the weight of an equal volume of pure water at 4°C. Thus, a mineral with a specific gravity of 3.0 is three times as heavy as water. **Density**, in contrast, is a mineral's mass (weight) per unit of volume expressed in grams per cubic centimeter. So the specific gravity of galena (Figure 3.14b) is 7.58 and its density is 7.58 g/cm^3. In most instances, we will refer to a mineral's density, and in some of the following chapters, we will mention the density of various rocks.

Structure and composition control a mineral's specific gravity and density. Because ferromagnesian silicates contain iron, magnesium, or both, they tend to be denser than nonferromagnesian silicates. In general, the metallic minerals, such as galena and hematite, are denser than nonmetals. Pure gold with a density of 19.3 g/cm^3 is about 2.5 times as dense as lead. Diamond and graphite, both of which are composed of carbon (C), illustrate how structure controls specific gravity or density. The specific gravity of diamond is 3.5, whereas that of graphite varies from 2.09 to 2.33.

Other Useful Mineral Properties

Talc has a distinctive soapy feel, graphite writes on paper, halite tastes salty, and magnetite is magnetic (Figure 3.18). Calcite possesses the property of *double refraction*, meaning that an object when viewed through a transparent piece of calcite will have a double image. Some sheet silicates are plastic and, when bent into a new shape, will retain that shape; others are flexible and, if bent, will return to their original position when the forces that bent them are removed.

Geo-Focus Welcome to the Wonderful World of Micas

What makes the paint on cars and appliances so lustrous? Why are lipstick, eyeliner, and glitter so attractive? Do you enjoy the amber glow seen through the isinglass window of a wood stove? Some of you may remember the line in the song "The Surrey With the Fringe on Top" from the musical *Oklahoma*: "isinglass curtains that can roll right down in case there's a change in the weather." What is this remarkable substance in paint, lipstick, and isinglass? The answer—micas.

Mica is the name given to a group of 37 sheet silicate minerals that have similar physical properties, particularly the way they split or cleave. Remember that mineral cleavage is breakage along a plane or planes of weakness determined by atomic structure. Some minerals have no cleavage planes, others have two, three, four, or six; but micas have only one. Nevertheless, their cleavage is perfect, meaning that when cleaved they yield very smooth planes. Indeed, the micas split into thin, flexible sheets. The name *mica* probably comes from the Latin term *micare*, "to shine," a reference to their shiny luster.

Even though there are more than three dozen varieties of mica, only a few are common in rocks. One of the most common, *biotite* (black mica) (Figure 3.13a), has no commercial uses, although geologists use it in potassium–argon dating (see Chapter 17). *Muscovite* (colorless, white, or pale red or green) mica is also common (Figure 1); it was named for *Moskva* (Moscow), where much of Europe's mica was mined. Isinglass, mentioned above, consists of thin, transparent sheets of muscovite. Muscovite and another mica called *phlogophite* (from Greek *phologopos*, "fiery") have commercial value.

Micas used for products are called scrap and flake micas, which either occur naturally or are ground into small pieces, and sheet mica, which is cut into various shapes and sizes for use in the electronics and electrical industries. Scrap and flake micas are produced in many countries and U.S. states, but about half the U.S. production comes from North Carolina. Micas are found in many types of rocks, but our main concern here is its uses.

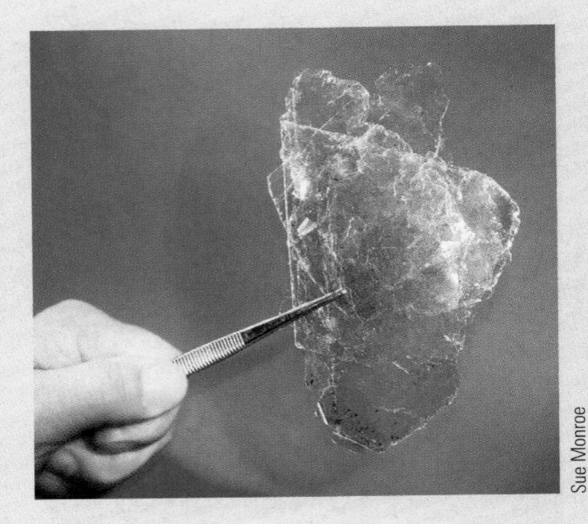

Sue Monroe

Figure 1 Muscovite mica is colorless, white, pale red, or green. It is a sheet silicate that has many industrial uses in paint, wallboard compound, eyeliner, lipstick, and nail polish.

When mica is ground up dry, it loses much of its luster, but retains its platy nature and is ideal for wallboard

A simple chemical test to identify the minerals calcite and dolomite involves applying a drop of dilute hydrochloric acid to the mineral specimen. If the mineral is calcite, it will react vigorously with the acid and release carbon dioxide, which causes the acid to bubble or effervesce. Dolomite, in contrast, will not react with hydrochloric acid unless it is powdered.

ROCK-FORMING MINERALS

Geologists use the term **rock** for a solid aggregate of one or more minerals, but the term also refers to masses of mineral-like matter, as in the natural glass obsidian (see Chapter 4), and masses of solid organic matter, as in coal (see Chapter 6). And even though some rocks may contain many minerals, only a few, designated **rock-forming minerals**, are sufficiently common for rock identification and classification (Table 3.3 and Figure 3.19). Others, known as *accessory minerals*, are present in such small quantities that they can be disregarded.

Given that silicate minerals are by far the most common ones in Earth's crust, it follows that most rocks are composed of these minerals. Indeed, feldspar minerals (plagioclase feldspars and potassium feldspars) and quartz make up more than 60% of Earth's crust. So, even though there are hundreds of silicates, only a few are particularly common in rocks.

The most common nonsilicate rock-forming minerals are the carbonates calcite ($CaCO_3$) and dolomite [$CaMg(CO_3)_2$], the main constituents of the sedimentary rocks limestone and dolostone, respectively (see Chapter 7). Among the sulfates and halides, gypsum ($CaSO_4 \cdot 2H_2O$) in rock gypsum and halite ($NaCl$) in rock salt (see Chapter 7) are common enough to qualify as rock-forming minerals. Even though these minerals and their corresponding rocks might be common in some areas, however, their overall abundance is limited compared to the silicate and carbonate rock-forming minerals.

HOW DO MINERALS FORM?

Thus far, we have discussed the composition, structure, and physical properties of minerals, but have not addressed how they originate. One phenomenon that accounts for the origin of minerals is the cooling of molten rock material known as *magma* (magma that reaches the surface is called *lava*).

Laureen Middley/Getty Images

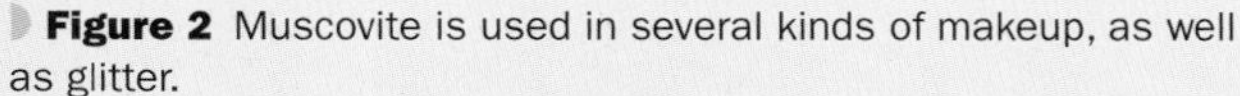

Figure 2 Muscovite is used in several kinds of makeup, as well as glitter.

Parvinder Sethi

Figure 3 Muscovite in paint on automobiles and appliances gives them their lustrous sheen.

joint compound and as an additive to paint. It is an essential component of joint compound because it makes the compound smoother and easier to work with and it prevents cracking. In fact, wallboard compounds and paints account for about 80% of all mica used. In addition, ground dry mica is used in plastics, roofing, rubber, and welding rods.

When mica is ground up wet, however, it retains its sparkling shine and is used in many cosmetics (Figure 2). Mica body powder is brushed onto the skin for an overall sheen. Fortunately, it is chemically inert and poses no risk when applied to the skin. Eye shadow, eyeliner, lipstick, blush, and nail polish have mica added to give them a resinous sheen, or mica-based powder may be added to lipstick. The brilliant sheen of some paints applied to automobiles (Figure 3) and the changing color depending on viewing angle come from micas. Perhaps the statement that *mica is the most amazing stuff on Earth* is an exaggeration, but mica certainly enhances the visual appeal of many products.

As magma or lava cools, minerals crystallize and grow, thereby determining the mineral composition of various igneous rocks such as basalt (dominated by ferromagnesian silicates) and granite (dominated by nonferromagnesian silicates) (see Chapter 4). Hot-water solutions derived from magma commonly invade cracks and crevasses in adjacent rocks, and from these solutions, a variety of minerals crystallize, some of economic importance. Minerals also originate when water in hot springs cools (see Chapter 13), and when hot, mineral-rich water discharges onto the seafloor at hot springs known as hydrothermal vents (see Chapter 9).

Dissolved materials in seawater, more rarely in lake water, combine to form minerals such as halite (NaCl), gypsum ($CaSO_4 \cdot 2H_2O$), and several others when the water evaporates. Aragonite and/or calcite, both varieties of calcium carbonate ($CaCO_3$), might also form from evaporating water, but most originate when organisms such as clams, oysters, corals, and floating microorganisms use this compound to construct their shells. A few plants and animals use silicon dioxide (SiO_2) for their skeletons, which accumulate as mineral matter on the seafloor when the organisms die (see Chapter 9).

Some clay minerals form when chemical processes compositionally and structurally alter other minerals (see Chapter 6), and others originate when rocks are changed during metamorphism (see Chapter 7). In fact, the agents that cause metamorphism—heat, pressure, and chemically active fluids—are responsible for the origin of many minerals. A few minerals even originate when gases such as hydrogen sulfide (H_2S) and sulfur dioxide (SO_2) react at volcanic vents to produce sulfur.

NATURAL RESOURCES AND RESERVES

Geologists at the U.S. Geological Survey define a **resource** as a concentration of naturally occurring solid, liquid, or gaseous material in or on Earth's crust in such form and amount that economic extraction of a commodity from the concentration is currently or potentially feasible.

Natural resources are mostly concentrations of minerals, rocks, or both, but liquid petroleum and natural gas are also included. In fact, some of the resources we refer to are *metallic resources* (copper, tin, iron ore, etc.), *nonmetallic resources*

What Would You Do?

The distinction between minerals and rocks is not easy for beginning students to understand. As a teacher, you know that minerals are made up of chemical elements and that rocks consist of one or more minerals, but despite your best efforts to clearly define them, your students commonly mistake one for the other. Can you think of analogies that might help students understand the difference between minerals and rocks?

(sand and gravel, crushed stone, salt, sulfur, etc.), and *energy resources* (petroleum, natural gas, coal, and uranium). All of these are indeed resources, but we must make a distinction between a resource, the total amount of a commodity, whether discovered or undiscovered, and a **reserve**, which is only that part of the resource base that is known and can be economically recovered.

The distinction between a resource and a reserve is simple enough in principle, but in practice it depends on several factors, not all of which remain constant. Geographic location may be important. For instance, a resource in a remote region might not be mined because transportation costs are

Figure 3.15 Mineral Crystals Mineral crystals are found in a variety of shapes, but different minerals may have the same kinds of crystals as shown by (a–c) below. Even so, differentiating one from the other is easy.

Sue Monroe

a Pyrite (FeS_2) is typically brassy yellow (the silvery color results from the way light is reflected), and it is much denser and harder than fluorite or halite.

Sue Monroe

b Fluorite (CaF_2) can be identified by its cleavage.

Parvinder Sethi

c Halite (NaCl) can be identified by its salty taste.

Richard D. Fisher

d Some of these gypsum ($CaSO_4 \cdot 2H_2O$) crystals in a cavern in Chihuahua, Mexico, measure up to 15.2 m long and may be the world's largest crystals. They were discovered in April 2000.

Figure 3.16 Several Types of Mineral Cleavage

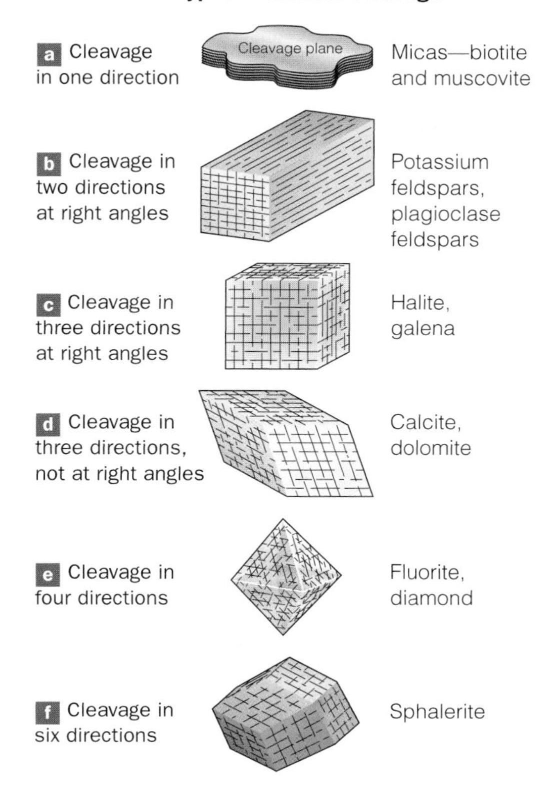

Figure 3.17 Cleavage in Augite and Hornblende

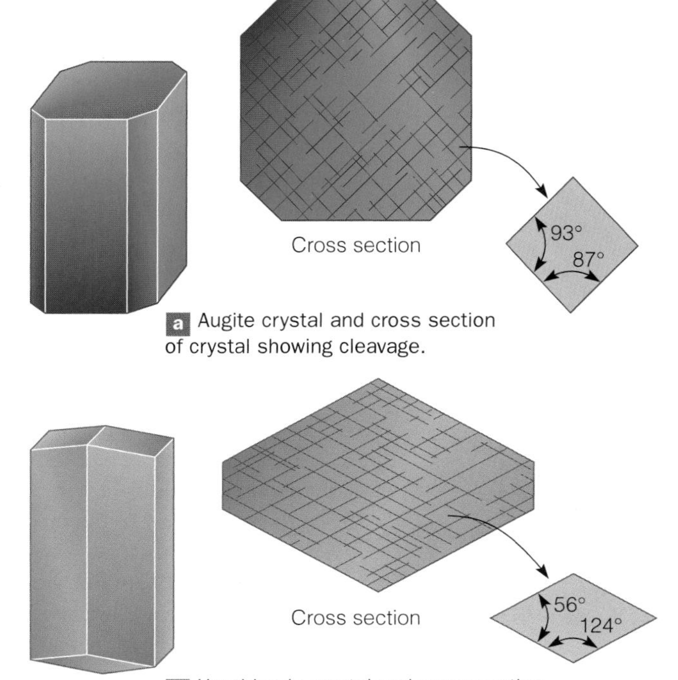

a Augite crystal and cross section of crystal showing cleavage.

b Hornblende crystal and cross section of crystal showing cleavage.

TABLE 3.2 Mohs Hardness Scale

Hardness	Mineral	Hardness of Some Common Objects
10	Diamond	
9	Corundum	
8	Topaz	
7	Quartz	
		Steel file (6½)
6	Orthoclase	
		Glass (5½–6)
5	Apatite	
4	Fluorite	
3	Calcite	Copper penny (3)
		Fingernail (2½)
2	Gypsum	
1	Talc	

Figure 3.18 Graphite and Magnetite

a Graphite is the mineral used to make pencil "lead" write on paper.

b Magnetite is magnetic. Magnetite is an ore of iron and is used in toys like Etch A Sketch®.

TABLE 3.3 Important Rock-Forming Minerals

Mineral	Primary Occurrence
Ferromagnesian silicates	
Olivine	Igneous and metamorphic rocks
Pyroxene group	
Augite most common	Igneous and metamorphic rocks
Amphibole group	
Hornblende most common	Igneous and metamorphic rocks
Biotite	All rock types
Nonferromagnesian silicates	
Quartz	All rock types
Potassium feldspar group	
Orthoclase, microcline	All rock types
Plagioclase feldspar group	All rock types
Muscovite	All rock types
Clay mineral group	Soils, sedimentary rocks, and some metamorphic rocks
Carbonates	
Calcite	Sedimentary rocks
Dolomite	Sedimentary rocks
Sulfates	
Anhydrite	Sedimentary rocks
Gypsum	Sedimentary rocks
Halides	
Halite	Sedimentary rocks

too high, and what might be deemed a resource rather than a reserve in the United States and Canada may be mined in a developing country where labor costs are low. The commodity in question is also important. Gold or diamonds in sufficient quantity can be mined profitably just about anywhere, whereas sand and gravel deposits must be close to their market areas.

Obviously the market price is important in evaluating any resource. From 1935 until 1968, the U.S. government maintained the price of gold at $35 per troy ounce (1 troy ounce = 31.1 g). When this restriction was removed, demand determined the market price and gold prices rose, reaching an all-time high of $843 per troy ounce in 1980. As a result, many marginal deposits became reserves and a number of abandoned mines were reopened.

The status of a resource is also affected by changes in technology. By the time of World War II (1939–1945), the richest iron ore deposits of the Great Lakes region in the United States and Canada had been mostly depleted. But the development of a method for separating the iron from unusable rock and shaping it into pellets ideal for use in blast furnaces made it profitable to mine rocks that contained less iron.

Most people know that industrialized societies depend on a variety of natural resources, but they know little about their occurrence, methods of recovery, and economics. Geologists are, of course, essential in finding and evaluating deposits; however, extraction involves engineers, chemists, miners, and many people in support industries that supply mining equipment. Ultimately, though, the decision about whether a deposit should be mined or not is made by people trained in business and economics. In short, extraction must yield a profit.

In addition to resources such as petroleum, gold, and ores of iron and copper, some quite common minerals are also essential. For example, pure quartz sand is used to manufacture glass and optical instruments, as well as sandpaper and steel alloys. Clay minerals are needed to make ceramics and paper; feldspars are used for porcelain, ceramics, enamel, and glass; and phosphate-bearing rock is used for fertilizers. Micas are used in a variety of products, including lipstick, glitter, and eye shadow, as well as the lustrous paints on appliances and automobiles (see Geo-Focus on pages 76 and 77).

Access to resources is essential for industrialization and the high standard of living enjoyed in many countries. The United States and Canada are resource-rich nations; however, many resources are *nonrenewable*, which means that there is a limited supply and they cannot be replenished by natu-

Figure 3.19 Minerals in Granite The igneous rock granite (see Chapter 4) is made up of mostly three minerals—quartz, potassium feldspar, and plagioclase feldspar—but it may also contain small amounts of biotite, muscovite, and hornblende.

ral processes as fast as they are depleted. Accordingly, once a resource is depleted, suitable substitutes, if available, must be found. For some essential resources, the United States is totally dependent on imports; for example, no cobalt was mined in this country during 2003. Yet the United States, the world's largest consumer of cobalt, uses this essential metal in gas-turbine aircraft engines and magnets, and for corrosion– and wear-resistant alloys. Obviously, all cobalt is imported, as is all manganese, an element essential for making steel.

The United States also imports all of the aluminum ore it uses, as well as all or some of many other resources (Figure 3.20). Canada, in contrast, is more self-reliant, meeting most of its domestic mineral and energy needs. Nevertheless, it must import phosphate, chromium, manganese, and aluminum ore. Canada also produces more crude oil and natural gas than it uses, and it is among the world leaders in producing and exporting uranium.

To ensure continued supplies of essential minerals and energy resources, geologists and other scientists, government agencies, and leaders in business and industry continually assess the status of resources in view of changing economic and political conditions and changes in science and technology. The U.S. Geological Survey, for instance, keeps detailed statistical records of mine production, imports, and exports, and regularly publishes reports on the status of numerous commodities. Similar reports appear regularly in the *Canadian Minerals Yearbook.* In several of the following chapters, we will discuss the geologic occurrence of resources.

What Would You Do?

Some reputable businesspeople tell you of opportunities to invest in natural resources. Two ventures look promising: a gold mine and a sand-and-gravel pit. Given that gold sells for about $800 per ounce, whereas sand and gravel are worth $4 or $5 per ton, would it be more prudent to invest in the gold mine? Explain not only how market price would influence your decision, but also what other factors you might need to consider.

Commodity	Percentage imported (0–100)	Major Import Sources	Uses
Bauxite	~100	Australia, Jamaica, Guinea, Suriname	Ore of aluminum
Columbium	~100	Brazil, Canada, Estonia, Germany	Carbon steel, superalloys
Graphite	~100	China, Mexico, Canada, Brazil	Brake linings, lubricants
Manganese	~100	South Africa, Gabon, Australia, Mexico	Steel production, dry cell batteries
Vanadium	~100	South Africa, Czech Republic, Canada, China	Steel alloys
Platinum	~93	South Africa, United Kingdom, Germany, Canada	Catalytic converters, jewelry
Tin	~79	Peru, China, Bolivia, Brazil, Indonesia	Tin cans and containers
Cobalt	~75	Finland, Norway, Russia, Canada	Superalloys
Tungsten	~70	China, Russia, Canada	Carbide parts for cutting tools
Chromium	~63	South Africa, Kazakhstan, Zimbabwe, Turkey, Russia	Stainless and heat-resistant steel
Silver	~61	Mexico, Canada, Peru, United Kingdom	Silver halide film, jewelry
Zinc	~60	Canada, Mexico, Peru	Galvanized metal, zinc alloys
Gold	~50	Canada, Brazil, Peru, Australia	Jewelry and arts, electrical industry
Nickel	~43	Canada, Norway, Russia, Australia	Stainless steel, electroplating
Copper	~37	Canada, Peru, Chile, Mexico	Copper and copper alloys, wiring
Lead	~18	Canada, China, Mexico, Australia	Lead for batteries, protective coatings
Iron ore	~11	Canada, Brazil, Australia, Venezuela	Steel, cast iron

Sources: USGS Minerals Information: http://minerals.usgs.gov/minerals/
USGS Mineral Commodity Summaries 2004: http://usgs.gov/minerals/pubs/mcs/2004.pdf

Figure 3.20 Mineral Commodities The dependence of the United States on imports of various mineral commodities is apparent from this chart. The lengths of the green bars correspond to the amounts of the resources imported.

Geo-Recap

Chapter Summary

- Matter is composed of chemical elements, each of which consists of atoms. Protons and neutrons are present in an atom's nucleus, and electrons orbit around the nucleus in electron shells.
- The number of protons in an atom's nucleus determines its atomic number. The atomic mass number is the number of protons plus neutrons in the nucleus.
- Bonding results when atoms join with other atoms; different elements bond to form a compound. With few exceptions, minerals are compounds.
- Ionic and covalent bonds are most common in minerals, but metallic and van der Waals bonds are found in some.
- Minerals are crystalline solids, which means that they possess an ordered internal arrangement of atoms.
- Mineral composition is indicated by a chemical formula, such as SiO_2 for quartz.
- Some minerals have a range of compositions because different elements substitute for one another if their atoms are about the same size and have the same electrical charge.
- More than 3500 minerals are known, and most of them are silicates. The two types of silicates are ferromagnesian and nonferromagnesian.
- In addition to silicates, geologists recognize carbonates, native elements, hydroxides, oxides, phosphates, halides, sulfates, and sulfides.
- Structure and composition control the physical properties of minerals, such as luster, crystal form, hardness, color, cleavage, fracture, and specific gravity.
- Several processes account for the origin of minerals, including cooling magma, weathering, evaporation of seawater, metamorphism, and organisms using dissolved substances in seawater to build their shells.
- A few minerals, designated rock-forming minerals, are common enough in rocks to be essential in their identification and classification. Most rock-forming minerals are silicates, but some carbonates are also common.
- Many resources are concentrations of minerals or rocks of economic importance. They are further characterized as metallic resources, nonmetallic resources, and energy resources.
- Reserves are that part of the resource base that can be extracted profitably. Distinguishing a resource from a reserve depends on market price, labor costs, geographic location, and developments in science and technology.
- The United States must import many resources to maintain its industrial capacity. Canada is more self-reliant, but it too must import some commodities.

Important Terms

atom (p. 63)
atomic mass number (p. 63)
atomic number (p. 63)
bonding (p. 64)
carbonate mineral (p. 71)
cleavage (p. 74)
compound (p. 64)
covalent bond (p. 65)
crystal (p. 66)
crystalline solid (p. 66)
density (p. 75)
electron (p. 63)
electron shell (p. 63)
element (p. 63)
ferromagnesian silicate (p. 71)
hardness (p. 75)
ion (p. 65)
ionic bond (p. 65)
luster (p. 74)
mineral (p. 62)
neutron (p. 63)
nonferromagnesian silicate (p. 71)
nucleus (p. 63)
proton (p. 63)
reserve (p. 78)
resource (p. 77)
rock (p. 76)
rock-forming mineral (p. 76)
silica (p. 70)
silica tetrahedron (p. 70)
silicate (p. 70)
specific gravity (p. 75)

Review Questions

1. A pyramid-shaped molecule made up of four oxygen atoms and one silicon atom is a/an
   a. ______ calcium carbonate cube;
   b. ______ silica tetrahedron;
   c. ______ oxide triangle;
   d. ______ sulfide polygram;
   e. ______ ionic bond.
2. An atom with six protons and eight neutrons in its nucleus has an atomic mass number of
   a. ______ 6;
   b. ______ 8;
   c. ______ 14;
   d. ______ 48;
   e. ______ 2.
3. The two most abundant elements in Earth's crust are
   a. ______ iron and magnesium;
   b. ______ potassium and chlorine;
   c. ______ aluminum and calcium;
   d. ______ oxygen and silicon;
   e. ______ carbon and phosphorous.
4. The quality and intensity of light reflected from a mineral is called its
   a. ______ density;
   b. ______ specific gravity;
   c. ______ hardness;
   d. ______ atomic number;
   e. ______ luster.
5. Which one of the following is a common carbonate rock-forming mineral?
   a. ______ calcite;
   b. ______ quartz;
   c. ______ biotite;
   d. ______ muscovite;
   e. ______ galena.
6. Minerals that possess the property of cleavage
   a. ______ are composed of the noble gases;
   b. ______ are noncrystalline solids;
   c. ______ break along planes of weakness;
   d. ______ are denser than minerals without cleavage;
   e. ______ exhibit double refraction.
7. The chemical formula for olivine is $(Fe,Mg)_2SiO_4$, which means that
   a. ______ silicon and oxygen may or may not be present;
   b. ______ ferromagnesian silicates are denser than nonferromagnesian silicates;
   c. ______ iron and magnesium can substitute for one another;
   d. ______ olivine is a nonferromagnesian silicate;
   e. ______ olivine contains iron or magnesium but never both.
8. Minerals composed of only one chemical element are known as
   a. ______ sulfides;
   b. ______ native elements;
   c. ______ phosphates;
   d. ______ hydroxides;
   e. ______ carbonates.
9. The atomic number of an element is determined by
   a. ______ the number of protons in its nucleus;
   b. ______ how many electrons it loses during bonding;
   c. ______ its color and density;
   d. ______ the amount of oxygen present;
   e. ______ whether or not it is crystalline.
10. In what type of chemical bonding are electrons shared by adjacent atoms?
    a. ______ octahedral;
    b. ______ polymorphic;
    c. ______ covalent;
    d. ______ ionic;
    e. ______ silicate.
11. How are the atomic number and atomic mass number of an element determined?
12. What is a silica tetrahedron and how may they bond to form silicate minerals?
13. How do rock-forming minerals differ from any other mineral? Give examples from three categories of rock-forming minerals.
14. How do resources and reserves differ? Also, why is the United States so dependent on imports of some resources?
15. Give three examples of mineral cleavage and explain how it occurs. Why do some minerals lack cleavage?
16. What is the distinction between the terms crystalline and crystal? Where might you find well-formed crystals?
17. What accounts for the fact that some minerals, such as olivine, have a range of chemical compositions rather than one specific composition?
18. How would the color and density of a rock made up mostly of ferromagnesian silicates differ from one composed primarily of nonferromagnesian silicates?
19. What are rock-forming minerals and accessory minerals?
20. Briefly discuss the ways in which minerals form.

CHAPTER 4

Igneous Rocks and Intrusive Igneous Activity

Sue Monroe

OUTLINE

OBJECTIVES

At the end of this chapter, you will have learned that

- With few exceptions, magma is composed of silicon and oxygen with lesser amounts of several other chemical elements.
- Temperature and composition are the most important controls on the mobility of magma and lava.
- Most magma originates within Earth's upper mantle or lower crust at or near divergent and convergent plate boundaries.
- Several processes bring about chemical changes in magma, so magma may evolve from one kind into another.
- All igneous rocks form when magma or lava cools and crystallizes, or by the consolidation of pyroclastic materials ejected during explosive eruptions.
- Geologists use texture and composition to classify igneous rocks.
- Intrusive igneous bodies called plutons form when magma cools below Earth's surface. The origin of the largest plutons is not fully understood.

View of the Sierra Nevada taken west of Lone Pine, California. The rocks in this view are part of the Sierra Nevada batholith, a huge mass of granite and related rocks made up of many intrusive bodies. The high peak toward the right is Mount Whitney, which at 4421 m is the highest peak in the continental United States.

INTRODUCTION

You already know that the term *rock* applies to solid aggregates of minerals, mineral-like matter as in obsidian, and masses of altered organic matter such as coal. In Chapter 1, we also briefly defined the three main families of rocks—igneous, sedimentary, and metamorphic (see Figure 1.15). Our concern in this chapter is with igneous rocks, that is, rocks that formed from molten rock matter (magma) that cooled and crystallized to form minerals, and rocks made up of particulate matter ejected during explosive volcanic eruptions. We are most familiar with eruptions of lava and particulate matter from volcanoes because they can be observed. And yet, most magma never reaches Earth's surface, but rather cools and crystallizes underground, thus forming several types of igneous rock bodies that geologists call *plutons*, named for Pluto, the Roman god of the underworld.

Granite, composed mostly of feldspars and quartz, and similar-appearing rocks are the most common in the large plutons, such as the one in the Sierra Nevada in California (see this chapter's opening photograph) and in Acadia National Park in Maine (Figure 4.1a). The images of presidents Washington, Jefferson, T. Roosevelt, and Lincoln at Mount Rushmore National Memorial in South Dakota (Figure 4.1b), as well as the nearby Crazy Horse Memorial (under construction), are in the 1.7-billion-year-old Harney Peak Granite. All of these exposures of granite formed far below the surface and were revealed in their present form only after uplift and deep erosion.

Granite and related rocks are attractive, especially when sawed and polished. They are used for tombstones, mantelpieces, kitchen counters, facing stones on buildings, monuments, pedestals for statues, and for statuary itself. More importantly, though, is the fact that when granite and allied rocks form as plutons, fluids emanating from them follow cracks and crevasses in adjacent rocks where ore minerals or metals such as copper form.

Although we discuss intrusive igneous activity and the origin of plutons in this chapter, whereas volcanism is considered in Chapter 5, both processes are related. In fact, the same kinds of magmas are involved; however, magma varies in its mobility, so only some of it reaches the surface. Furthermore, plutons that lie beneath areas of volcanism are the source of the overlying lava flows and particulate matter erupted by volcanoes.

One important reason to study igneous rocks and intrusive igneous processes is that these rocks constitute one of the three families of rocks. Furthermore, igneous rocks make up large parts of the continents and all of the oceanic crust, which forms continuously at divergent plate boundaries. And, as already noted, some igneous rocks and processes are responsible for valuable resources. Another reason to study this topic is that most plutons and volcanoes are found at or near divergent and convergent plate boundaries, so the presence of igneous rocks is one criterion for recognizing ancient plate boundaries.

Figure 4.1 Granitic Rocks

Richard Thom/Visuals Unlimited

a Light-colored granitic rocks exposed along the shoreline in Acadia National Park in Maine. The dark rock is basalt that formed when magma intruded along a fracture in the granitic rock.

National Park Service

b The presidents' images at Mount Rushmore National Memorial in South Dakota are in the Harney Peak Granite.

THE PROPERTIES AND BEHAVIOR OF MAGMA AND LAVA

In Chapter 3, we mentioned that one process that accounts for the origin of minerals is cooling and crystallization of magma and lava. Geologists define **magma** as any mass of molten rock material below Earth's surface, whereas **lava** is simply magma that reaches the surface. Thus, it is the same material in both locations, although when it reaches the surface, it tends to lose some of its gas content because of decreased pressure. Any magma is less dense than the rock from which it formed, so it tends to rise toward the surface, but much of it solidifies deep underground, thereby accounting for the origin of plutons. The magma that does reach the surface issues forth as streams of lava, or **lava flows**, and some of it is forcefully ejected into the atmosphere as particles known as **pyroclastic materials** (from the Greek *pyro*, "fire," and *klastos*, "broken").

All **igneous rocks** derive ultimately from magma; however, two separate processes account for them. They form when (1) magma or lava cools and crystallizes to form aggregates of minerals, or (2) pyroclastic materials are consolidated. Those igneous rocks derived from lava flows and pyroclastic materials, both of which are extruded onto the surface, are known as **volcanic rocks** or **extrusive igneous rocks. Plutonic rocks** or **intrusive igneous rocks**, on the other hand, are the ones that formed when magma cooled below the surface, that is, from magma intruded into the crust.

Composition of Magma

By far the most abundant minerals in Earth's crust are silicates such as quartz, feldspars, and several ferromagnesian silicates, all made up of silicon and oxygen, and other elements shown in Figure 3.10. As a result, melting of the crust yields mostly silica-rich magmas that also contain considerable aluminum, calcium, sodium, iron, magnesium, and potassium, and several other elements in lesser quantities. Another source of magma is Earth's upper mantle, which is composed of rocks that contain mostly ferromagnesian silicates. Thus, magma from this source contains comparatively less silicon and oxygen (silica) and more iron and magnesium.

Although there are a few exceptions, the primary constituent of magma is silica, which varies enough to distinguish magmas classified as felsic, intermediate, and mafic.* **Felsic magma**, with more than 65% silica, is silica rich and contains considerable sodium, potassium, and aluminum, but little calcium, iron, and magnesium. In contrast, **mafic magma**, with less than 52% silica, is silica poor and contains proportionately more calcium, iron, and magnesium. And as you would expect, **intermediate magma** has a composition between felsic and mafic magma (Table 4.1).

How Hot Are Magma and Lava?

Everyone knows that lava is very hot, but how hot is hot? Erupting lava generally has a temperature in the range of 700° to 1200°C, although a temperature of 1350°C was recorded above a lava lake in Hawaii where volcanic gases reacted with the atmosphere. Magma must be even hotter than lava; however, no direct measurements of magma temperatures have ever been made.

*Lava from some volcanoes in Africa cools to form carbonitite, an igneous rock with at least 50% carbonate minerals, mostly calcite and dolomite.

TABLE 4.1 The Most Common Types of Magmas and Their Characteristics

Type of Magma	Silica Content (%)	Sodium, Potassium, and Aluminum	Calcium, Iron, and Magnesium
Ultramafic	<45	↓	Increase ↑
Mafic	45–52	↓	↑
Intermediate	53–65	↓	↑
Felsic	>65	Increase	↑

Most lava temperatures are taken at volcanoes that show little or no explosive activity, so our best information comes from mafic lava flows such as those in Hawaii (Figure 4.2). In contrast, eruptions of felsic lava are not as common, and the volcanoes that these flows issue from tend to be explosive and thus cannot be approached safely. Nevertheless, the temperatures of some bulbous masses of felsic lava in lava domes have been measured at a distance with an optical pyrometer. The surfaces of these lava domes are as hot as 900°C, but their interiors must surely be even hotter.

When Mount St. Helens erupted in 1980 in Washington State, it ejected felsic magma as particulate matter in pyroclastic flows. Two weeks later, these flows had temperatures between 300° and 420°C, and a steam explosion took

Parvinder Sethi

Figure 4.2 How Hot Is Lava? A geologist uses a remotely-operated, hand-held device for measuring temperature of a lava flow visible through an opening in the rocks in Hawaiian Volcanoes National Park, Hawaii.

place more than a year later when water encountered some of the still-hot pyroclastic materials. The reason that lava and magma retain heat so well is that rock conducts heat so poorly. Accordingly, the interiors of thick lava flows and pyroclastic flow deposits may remain hot for months or years, whereas plutons, depending on their size and depth, may not cool completely for thousands to millions of years.

Viscosity—Resistance to Flow

All liquids have the property of **viscosity**, or resistance to flow. Water's viscosity is very low so it is highly fluid and flows readily. For other liquids, viscosity is so high that they flow much more slowly. Good examples are cold motor oil and syrup, both of which are quite viscous and thus flow only with difficulty. But when these liquids are heated, their viscosity is much lower and they flow more easily; that is, they become more fluid with increasing temperature. Accordingly, you might suspect that temperature controls the viscosity of magma and lava, and this inference is partly correct. We can generalize and say that hot magma or lava moves more readily than cooler magma or lava, but we must qualify this statement by noting that temperature is not the only control of viscosity.

Silica content strongly controls magma and lava viscosity. With increasing silica content, numerous networks of silica tetrahedra form and retard flow because for flow to take place, the strong bonds of the networks must be ruptured. Mafic magma and lava with 45–52% silica have fewer silica tetrahedra networks and, as a result, are more mobile than felsic magma and lava flows (Figure 4.3). One mafic flow in 1783 in Iceland flowed nearly 80 km, and geologists traced ancient flows in Washington State for more than 500 km. Felsic magma, in contrast, because of its higher viscosity, does not reach the surface as often as mafic magma. And when felsic lava flows do occur, they tend to be slow moving and thick, and move only short distances. A thick, pasty lava flow that erupted in 1915 from Lassen Peak in California flowed only about 300 m before it ceased moving.

Temperature and silica content are important controls on the viscosity of magma and lava, but other factors include gases, mostly CO_2, as well as the presence of mineral crystals and friction from the surface over which lava flows. Lava with a high content of dissolved gases flows more readily than one with a lesser amount of gases, whereas lava with many crystals or that flows over a rough surface tends to be more viscous.

HOW DOES MAGMA ORIGINATE AND CHANGE?

Most of us have not witnessed a volcanic eruption, but we have seen news reports or documentaries showing magma issuing forth as lava flows or pyroclastic materials. In any case, we are familiar with some aspects of igneous activity, but most people are unaware of how and where magma originates, how it rises, and how it might change. Indeed, many believe the misconception that lava comes from a continuous layer of molten rock beneath the crust or that it comes from Earth's molten outer core.

Figure 4.3 Viscosity of Magma and Lava Temperature is an important control on viscosity, but so is composition. Mafic lava tends to be fluid, whereas felsic lava is much more viscous.

J.D. Griggs

a A mafic lava flow in 1984 on Mauna Loa Volcano in Hawaii. These flows move rapidly and form thin layers.

USGS

b The Novarupta lava dome in Katmai National Park in Alaska. The lava is felsic and viscous, so it was extruded as a bulbous mass. This image was taken in 1987.

First, let us address how magma originates. We know that the atoms in a solid are in constant motion and that, when a solid is heated, the energy of motion exceeds the binding forces and the solid melts. We are all familiar with this phenomenon, and we are also aware that not all solids melt at the same temperature. Once magma forms, it tends to rise because it is less dense than the rock that melted, and some actually makes it to the surface.

Magma may come from 100 to 300 km deep, but most forms at much shallower depths in the upper mantle or lower crust and accumulates in reservoirs known as **magma chambers**. Beneath spreading ridges, where the crust is thin, magma chambers exist at a depth of only a few kilometers; however, along convergent plate boundaries, magma chambers are commonly a few tens of kilometers deep. The volume of a magma chamber ranges from a few to many hundreds of cubic kilometers of molten rock within the otherwise solid lithosphere. Some simply cools and crystallizes within Earth's crust, thus accounting for the origin of plutons, whereas some rises to the surface and is erupted as lava flows or pyroclastic materials.

What Would You Do?

You are a high school science teacher interested in developing experiments to show your students that (1) composition and temperature affect the viscosity of a lava flow, and (2) when magma or lava cools, some minerals crystallize before others. Describe the experiments you might devise to illustrate these points.

Bowen's Reaction Series

During the early part of the last century, N. L. Bowen knew that minerals do not all crystallize simultaneously from cooling magma, but rather crystallize in a predictable sequence. Based on his observations and laboratory experiments, Bowen proposed a mechanism, now called **Bowen's reaction series**, to account for the derivation of intermediate and felsic magmas from mafic magma. Bowen's reaction series consists of two branches: a *discontinuous branch* and a *continuous branch* (Figure 4.4). As the temperature of magma decreases, minerals crystallize along both branches simultaneously, but for convenience, we will discuss them separately.

In the discontinuous branch, which contains only ferromagnesian silicates, one mineral changes to another over specific temperature ranges (Figure 4.4). As the temperature decreases, a temperature range is reached in which a given mineral begins to crystallize. A previously formed mineral reacts with the remaining liquid magma (the melt) so that it forms the next mineral in the sequence. For instance, olivine [$(Mg,Fe)_2SiO_4$] is the first ferromagnesian silicate to crystallize. As the magma continues to cool, it reaches the temperature range at which pyroxene is stable; a reaction occurs between the olivine and the remaining melt, and pyroxene forms.

With continued cooling, a reaction takes place between pyroxene and the melt, and the pyroxene structure is rearranged to form amphibole. Further cooling causes a reaction between the amphibole and the melt, and its structure is rearranged so that the sheet structure of biotite mica forms. Although the reactions just described tend to convert one mineral to the next in the series, the reactions are not always complete. Olivine, for example, might have a rim of pyroxene, indicating an incomplete reaction. If magma cools rapidly enough, the early-formed minerals do not have time to react with the melt, and thus all the ferromagnesian silicates in the discontinuous branch can be in one rock. In any case, by the time biotite has crystallized, essentially all the magnesium and iron present in the original magma have been used up.

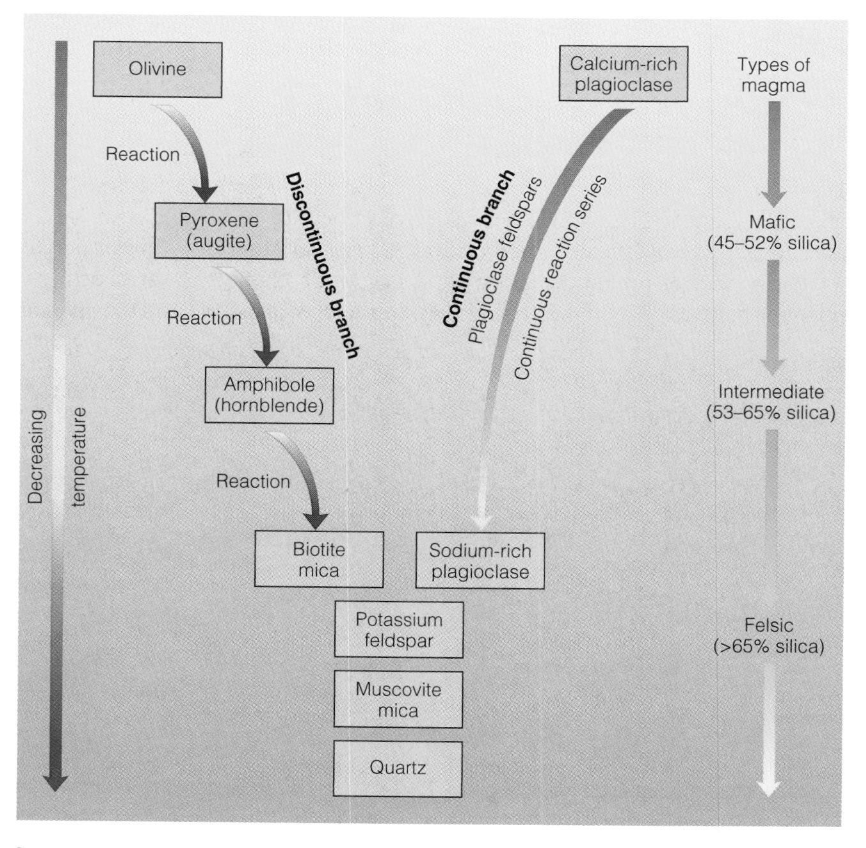

Figure 4.4 Bowen's Reaction Series Bowen's reaction series consists of a discontinuous branch along which a succession of ferromagnesian silicates crystallize as the magma's temperature decreases, and a continuous branch along which plagioclase feldspars with increasing amounts of sodium crystallize. Notice also that the composition of the initial mafic magma changes as crystallization takes place along the two branches.

Plagioclase feldspars, which are nonferromagnesian silicates, are the only minerals in the continuous branch of Bowen's reaction series (Figure 4.4). Calcium-rich plagioclase crystallizes first. As the magma continues to cool, calcium-rich plagioclase reacts with the melt, and plagioclase containing proportionately more sodium crystallizes until all of the calcium and sodium are used up. In many cases, cooling is too rapid for a complete transformation from calcium-rich to sodium-rich plagioclase to take place. Plagioclase forming under these conditions is *zoned*, meaning that it has a calcium-rich core surrounded by zones progressively richer in sodium.

As minerals crystallize simultaneously along the two branches of Bowen's reaction series, iron and magnesium are depleted because they are used in ferromagnesian silicates, whereas calcium and sodium are used up in plagioclase feldspars. At this point, any leftover magma is enriched in potassium, aluminum, and silicon, which combine to form orthoclase ($KAlSi_3O_8$), a potassium feldspar, and if water pressure is high, the sheet silicate muscovite forms. Any remaining magma is enriched in silicon and oxygen (silica) and forms the mineral quartz (SiO_2). The crystallization of orthoclase and quartz is not a true reaction series because they form independently rather than by a reaction of orthoclase with the melt.

The Origin of Magma at Spreading Ridges

One fundamental observation regarding the origin of magma is that Earth's temperature increases with depth. Known as the *geothermal gradient*, this temperature increase averages about 25°C/km. Accordingly, rocks at depth are hot but remain solid because their melting temperature rises with increasing pressure. However, beneath spreading ridges, the temperature locally exceeds the melting temperature, at least in part because pressure decreases. That is, plate separation at ridges probably causes a decrease in pressure on the already hot rocks at depth, thus initiating melting (Figure 4.5). In addition, the presence of water decreases the melting temperature beneath spreading ridges because water aids thermal energy in breaking the chemical bonds in minerals.

Magma formed beneath spreading ridges is invariably mafic (45–52% silica). However, the upper mantle rocks from which this magma is derived are characterized as ultramafic (<45% silica), consisting mostly of ferromagnesian silicates and lesser amounts of nonferromagnesian silicates. To explain how mafic magma originates from ultramafic rock, geologists propose that the magma forms from source rock that only partially melts. This phenomenon of partial melting takes place because not all of the minerals in rocks melt at the same temperature.

Recall the sequence of minerals in Bowen's reaction series (Figure 4.4). The order in which these minerals melt is the opposite of their order of crystallization. Accordingly, rocks made up of quartz, potassium feldspar, and sodium-rich plagioclase begin melting at lower temperatures than those composed of ferromagnesian silicates and the calcic varieties of plagioclase. So when ultramafic rock starts to melt, the minerals richest in silica melt first, followed by those containing less silica. Therefore, if melting is not complete, mafic magma containing proportionately more silica than the source rock results.

Subduction Zones and the Origin of Magma

Another fundamental observation regarding magma is that, where an oceanic plate is subducted beneath either a continental plate or another oceanic plate, a belt of volcanoes and plutons is found near the leading edge of the overriding plate (Figure 4.5). It would seem, then, that subduction and the origin of magma must be related in some way, and indeed they are. Furthermore, magma at these convergent plate boundaries is mostly intermediate (53–65% silica) or felsic (>65% silica).

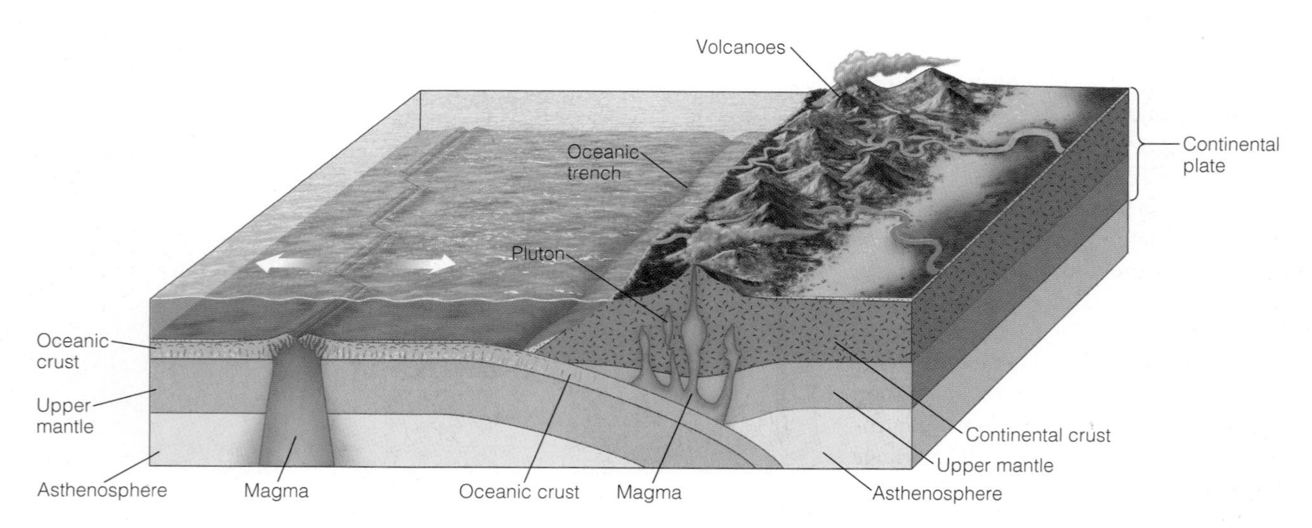

Figure 4.5 The Origin of Magma Magma forms beneath spreading ridges, because as plates separate, pressure is reduced on the hot rocks and partial melting of the upper mantle begins. Invariably, the magma formed is mafic. Magma also forms at subduction zones where water from the subducted plate causes partial melting of the upper mantle. This magma is also mafic, but as it rises, melting of the lower crust makes it more felsic.

Once again, geologists invoke the phenomenon of partial melting to explain the origin and composition of magma at subduction zones. As a subducted plate descends toward the asthenosphere, it eventually reaches the depth where the temperature is high enough to initiate partial melting. In addition, the oceanic crust descends to a depth at which dewatering of hydrous minerals takes place, and as the water rises into the overlying mantle, it enhances melting and magma forms (Figure 4.5).

Recall that partial melting of ultramafic rock at spreading ridges yields mafic magma. Similarly, partial melting of mafic rocks of the oceanic crust yields intermediate (53–65% silica) and felsic (>65% silica) magmas, both of which are richer in silica than the source rock. Moreover, some of the silica-rich sediments and sedimentary rocks of continental margins are probably carried downward with the subducted plate and contribute their silica to the magma. Also, mafic magma rising through the lower continental crust must be contaminated with silica-rich materials, which changes its composition.

Hot Spot and the Origin of Magma

Most volcanism occurs at divergent and convergent plate boundaries; however, there are some chains of volcanic outpourings in the ocean basins and on continents that are not near either of these boundaries. The Emperor Seamount–Hawaiian Islands, for instance, form a chain of volcanic islands 6000 km long, and the volcanic rocks become progressively older toward the northwest (see Figure 2.22). In 1963, Canadian geologist J. Tuzo Wilson proposed that the Hawaiian Islands and other areas showing similar trends lay above a **hot spot** over which a plate moves, thereby yielding a succession of volcanoes (Figure 4.6a).

Many geologists now think that hot-spot volcanism results from a rising **mantle plume**, a cylindrical plume of hot mantle rock that rises from perhaps near the core-mantle boundary. As it rises toward the surface, the pressure decreases on the hot rock and melting begins, thus yielding magma. Hot-spot volcanism may also account for vast flat-lying areas of overlapping lava flows or what geologists call *flood basalts* (Figure 4.6b). Figure 2.15 shows the locations of many hot spots, but of particular interest to us is the Yellowstone hot-spot in Wyoming. We will have more to say about the Yellowstone region in Chapter 23.

Not all geologists agree with the mantle plume theory. They cite evidence based on studies of earthquake waves that do not seem to be consistent with the theory. Nevertheless, mantle plumes and hot-spot volcanism are the most widely accepted explanation for chains of volcanoes in the oceans, as well as linear associations of volcanic rocks on land.

Processes That Bring About Compositional Changes in Magma

Once magma forms, its composition may change by **crystal settling**, which involves the physical separation of minerals by crystallization and gravitational settling (Figure 4.7). Olivine, the first ferromagnesian silicate to form in the discontinuous branch of Bowen's reaction series, has a density greater than the remaining magma and tends to sink. Accordingly, the remaining magma becomes richer in silica, sodium, and potassium because much of the iron and magnesium were removed as olivine and perhaps pyroxene minerals crystallized.

Although crystal settling does take place, it does not do so on a scale that would yield very much felsic magma from

Figure 4.6 Mantle Plume and Hot Spot

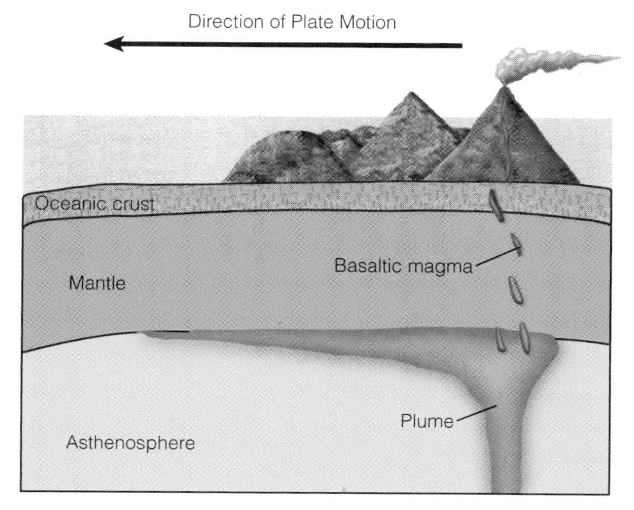

a A mantle plume beneath oceanic crust with a hot spot. Rising magma forms a series of volcanoes that become younger in the direction of plate movement.

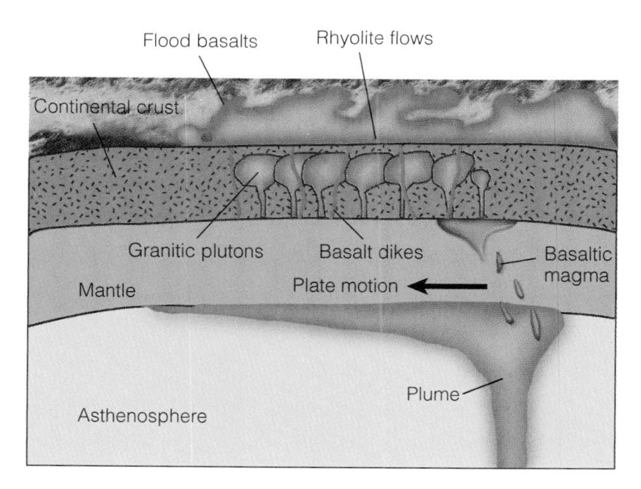

b A mantle plume with an overlying hot spot yields flood basalts and some of the continental crust melts to form felsic magma.

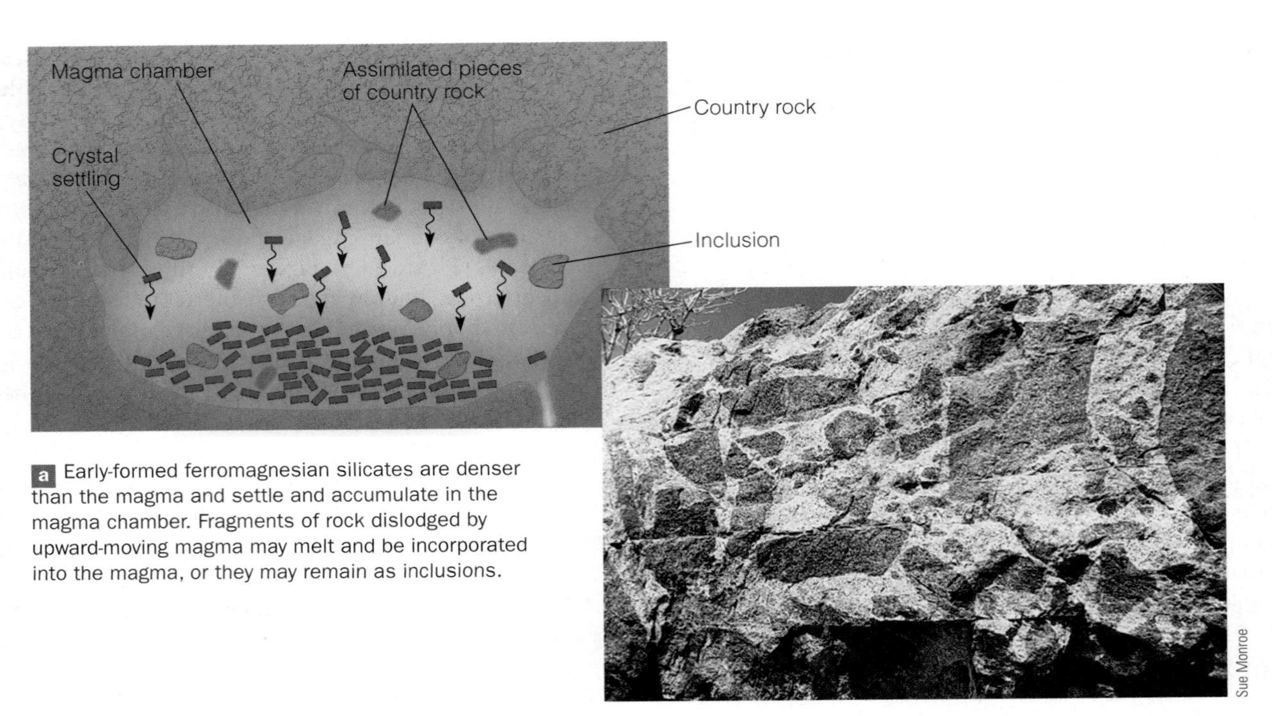

a Early-formed ferromagnesian silicates are denser than the magma and settle and accumulate in the magma chamber. Fragments of rock dislodged by upward-moving magma may melt and be incorporated into the magma, or they may remain as inclusions.

b Dark inclusions in granitic rock.

Figure 4.7 Crystal Settling and Assimilation

mafic magma. In some thick, sheetlike plutons called *sills*, the first-formed ferromagnesian silicates are indeed concentrated in their lower parts, thus making their upper parts less mafic. But even in these plutons, crystal settling has yielded very little felsic magma.

To yield a particular volume of granite (a felsic igneous rock), approximately 10 times as much mafic magma would have to be present initially for crystal settling to yield the volume of granite in question. If this were so, then mafic intrusive igneous rocks would be much more common than felsic ones. However, just the opposite is the case, so it appears that mechanisms other than crystal settling must account for the large volume of felsic magma. Partial melting of mafic oceanic crust and silica-rich sediments of continental margins during subduction yields magma richer in silica than the source rock. Furthermore, magma rising through the continental crust absorbs some felsic materials and becomes more enriched in silica.

The composition of magma also changes by **assimilation**, a process by which magma reacts with preexisting rock, called **country rock**, with which it comes in contact (Figure 4.7). The walls of a volcanic conduit or magma chamber are, of course, heated by the adjacent magma, which may reach temperatures of 1300°C. Some of these rocks partially or completely melt, provided that their melting temperature is lower than that of the magma. Because the assimilated rocks seldom have the same composition as the magma, the composition of the magma changes.

The fact that assimilation occurs is indicated by *inclusions*, incompletely melted pieces of rock that are fairly common in igneous rocks. Many inclusions were simply wedged loose from the country rock as magma forced its way into preexisting fractures (Figure 4.7). No one doubts that assimilation takes place, but its effect on the bulk composition of magma must be slight. The reason is that the heat for melting comes from the magma itself, and this has the effect of cooling the magma. Only a limited amount of rock can be assimilated by magma, and that amount is insufficient to bring about a major compositional change.

Neither crystal settling nor assimilation can produce a significant amount of felsic magma from a mafic one. But both processes, if operating concurrently, can bring about greater changes than either process acting alone. Some geologists think that this is one way that intermediate magma forms where oceanic lithosphere is subducted beneath continental lithosphere.

A single volcano can erupt lavas of different composition, indicating that magmas of differing composition are present. It seems likely that some of these magmas would come into contact and mix with one another. If this is the case, we would expect that the composition of the magma resulting from **magma mixing** would be a modified version of the parent magmas. Suppose rising mafic magma mixes with felsic magma of about the same volume (Figure 4.8). The resulting "new" magma would have a more intermediate composition.

IGNEOUS ROCKS—THEIR CHARACTERISTICS AND CLASSIFICATION

We have already defined *plutonic* or *intrusive igneous rocks* and *volcanic* or *extrusive igneous rocks.* Here we will have considerably more to say about the texture, composition, and classification of these rocks, which constitute one of the three major rock families depicted in the rock cycle (see Figure 1.14).

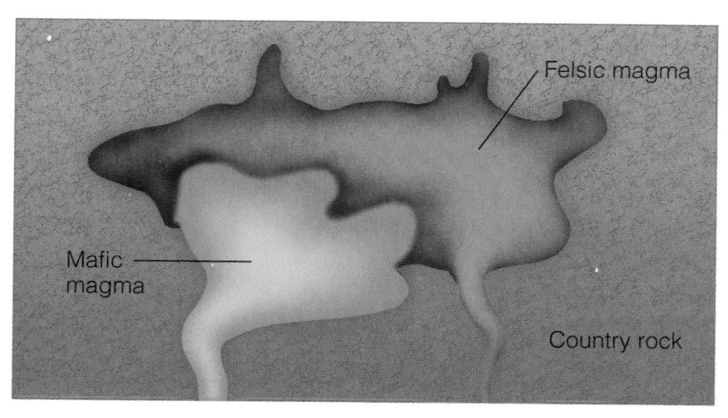

Figure 4.8 Magma Mixing Two magmas mix and produce magma with a composition different from either of the parent magmas. In this case, the resulting magma would have an intermediate composition.

Igneous Rock Textures

The term *texture* refers to the size, shape, and arrangement of the minerals that make up igneous rocks. Size is the most important because mineral crystal size is related to the cooling history of magma or lava and generally indicates whether an igneous rock is volcanic or plutonic. The atoms in magma and lava are in constant motion, but when cooling begins, some atoms bond to form small nuclei. As other atoms in the liquid chemically bond to these nuclei, they do so in an orderly geometric arrangement and the nuclei grow into crystalline *mineral grains*, the individual particles that make up igneous rocks.

During rapid cooling, as takes place in lava flows, the rate at which mineral nuclei form exceeds the rate of growth and an aggregate of many small mineral grains is formed. The result is a fine-grained or **aphanitic texture**, in which individual minerals are too small to be seen without magnification (Figure 4.9a). With slow cooling, the rate of growth exceeds the rate of nuclei formation, and large mineral grains form, thus yielding a coarse-grained or **phaneritic texture**, in which minerals are clearly visible (Figure 4.9b). Aphanitic textures usually indicate an extrusive origin, whereas rocks with phaneritic textures are usually intrusive. However, shallow plutons might have an aphanitic texture, and the rocks that form in the interiors of thick lava flows might be phaneritic.

Another common texture in igneous rocks is one termed **porphyritic**, in which minerals of markedly different size are present in the same rock. The larger minerals are *phenocrysts* and the smaller ones collectively make up the *groundmass*, which is simply the grains between phenocrysts (Figure 4.9c).

Figure 4.9 Textures of Igneous Rocks a Rapid cooling as in lava flows results in many small minerals and an aphanitic (fine-grained) texture. b Slower cooling in plutons yields a phaneritic texture. c These porphyritic textures indicate a complex cooling history. d Obsidian has a glassy texture because magma cooled too quickly for mineral crystals to form. e Gases expand in lava to yield a vesicular texture. f Microscopic view of a rock with a fragmental texture. The colorless, angular particles of volcanic glass measure up to 2 mm.

The groundmass can be either aphanitic or phaneritic; the only requirement for a porphyritic texture is that the phenocrysts be considerably larger than the minerals in the groundmass. Igneous rocks with porphyritic textures are designated *porphyry*, as in basalt porphyry. These rocks have more complex cooling histories than those with aphanitic or phaneritic textures and might involve, for example, magma partly cooling beneath the surface followed by its eruption and rapid cooling at the surface.

Lava may cool so rapidly that its constituent atoms do not have time to become arranged in the ordered, three-dimensional frameworks of minerals. As a consequence, *natural glass*, such as *obsidian*, forms (Figure 4.9d). Even though obsidian with its glassy texture is not composed of minerals, geologists nevertheless classify it as an igneous rock.

Some magmas contain large amounts of water vapor and other gases. These gases may be trapped in cooling lava where they form numerous small holes or cavities known as **vesicles**; rocks with many vesicles are termed *vesicular*, as in vesicular basalt (Figure 4.9e).

A **pyroclastic** or **fragmental texture** characterizes igneous rocks formed by explosive volcanic activity (Figure 4.9f). For example, ash discharged high into the atmosphere eventually settles to the surface where it accumulates; if consolidated, it forms pyroclastic igneous rock.

Composition of Igneous Rocks

Most igneous rocks, like the magma from which they originate, are mafic (45–52% silica), intermediate (53–65% silica), or felsic (>65% silica). A few are called *ultramafic* (<45% silica); however, these are probably derived from mafic magma by a process to be discussed later. The parent magma plays an important role in determining the mineral composition of igneous rocks, yet it is possible for the same magma to yield a variety of igneous rocks because its composition can change as a result of the sequence in which minerals crystallize, or by crystal settling, assimilation, and magma mixing (Figures 4.4, 4.7, and 4.8).

Classifying Igneous Rocks

Geologists use texture and composition to classify most igneous rocks. Notice in Figure 4.10 that all rocks except peridotite are in pairs; the members of a pair have the same composition but different textures. Basalt and gabbro, andesite and diorite, and rhyolite and granite are compositional (mineralogical) pairs, but basalt, andesite, and rhyolite are aphanitic and most commonly extrusive (volcanic), whereas gabbro, diorite, and granite are phaneritic and mostly intrusive (plutonic). The extrusive and intrusive members of each pair can usually be distinguished by texture, but remember that rocks in some shallow plutons may be aphanitic and rocks that formed in thick lava flows may be phaneritic. In other words, all of these rocks exist in a textural continuum.

The igneous rocks in Figure 4.10 are also differentiated by composition—that is, by their mineral content. Reading across the chart from rhyolite to andesite to basalt, for example, we see that the proportions of nonferromagnesian and ferromagnesian silicates change. The differences in composition, however, are gradual along a compositional continuum. In other words, there are rocks with compositions that correspond to the lines between granite and diorite, basalt and andesite, and so on. For example, Lassen Peak in California, which erupted from 1914 through 1917, is made up mostly of *dacite*, a rock with a composition between andesite and rhyolite.

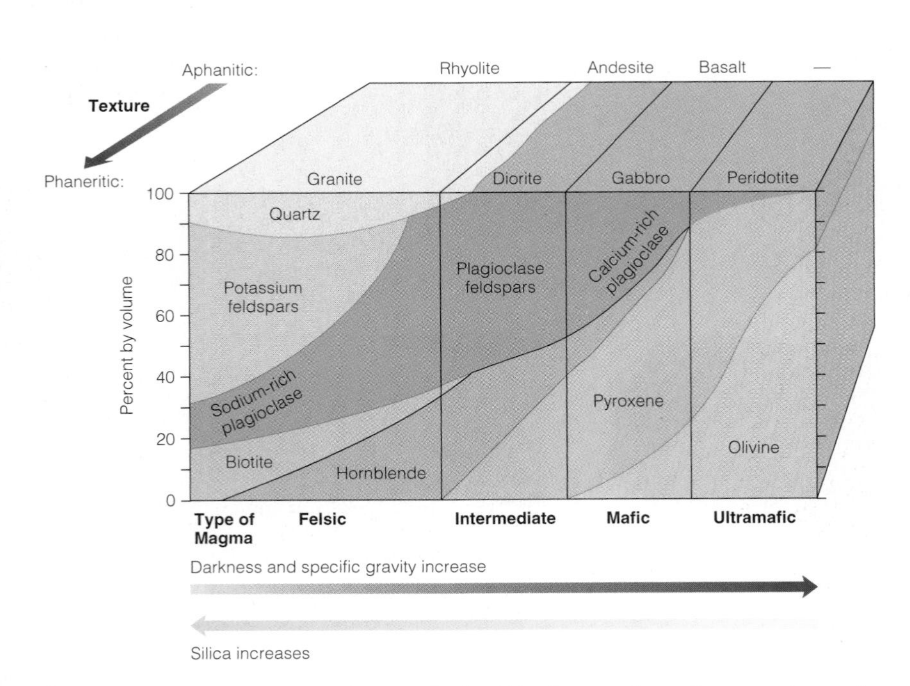

Figure 4.10 Classification of Igneous Rocks This diagram shows the percentages of minerals, as well as the textures of common igneous rocks. For example, an aphanitic (fine-grained) rock of mostly calcium-rich plagioclase and pyroxene is basalt.

Ultramafic Rocks Ultramafic rocks (<45% silica) are composed mostly of ferromagnesian silicates. *Peridotite* contains mostly olivine, lesser amounts of pyroxene, and usually a little plagioclase feldspar (Figure 4.10 and Figure 4.11), and pyroxenite is composed predominately of pyroxene. Because these minerals are dark, the rocks are black or green. Peridotite is a likely candidate for the rock that makes up the upper mantle (see Chapter 8). Ultramafic rocks in Earth's crust probably originate by concentration of the early-formed ferromagnesian minerals that separated from mafic magmas.

Ultramafic lava flows, called *komatiites,* are known in rocks older than 2.5 billion years, but are rare or absent in younger ones. The reason is that to erupt, ultramafic lava must have a near-surface temperature of approximately 1600°C; the surface temperatures of present-day mafic lava flows are rarely more than 1200°C. During early Earth history, though, more radioactive decay heated the mantle to as much as 300°C

Sue Monroe

Figure 4.11 Peridotite This specimen of the ultramafic rock peridotite is made up mostly of olivine. Notice in Figure 4.10 that peridotite is the only phaneritic rock that does not have an aphanitic counterpart. Peridotite is rare at Earth's surface, but is very likely the rock that makes up the mantle.

hotter than now and ultramafic lavas could erupt onto the surface. Because the amount of heat has decreased over time, Earth has cooled, and eruptions of ultramafic lava flows have ceased.

Basalt-Gabbro *Basalt* and *gabbro* are the aphanitic and phaneritic rocks that crystallize from mafic magma (45–52% silica) (Figure 4.12). Thus, both have the same composition—mostly calcium-rich plagioclase and pyroxene, with smaller amounts of olivine and amphibole (Figure 4.10). Because they contain a large proportion of ferromagnesian silicates, basalt and gabbro are dark; those that are porphyritic typically contain calcium plagioclase or olivine phenocrysts.

Extensive basalt lava flows cover vast areas in Washington, Oregon, Idaho, and northern California (see Chapter 5). Oceanic islands such as Iceland, the Galápagos, the Azores, and the Hawaiian Islands are composed mostly of basalt, and basalt makes up the upper part of the oceanic crust.

Gabbro is much less common than basalt, at least in the continental crust or where it can be easily observed. Small intrusive bodies of gabbro are present in the continental crust, but intermediate to felsic intrusive rocks are much more common. However, the lower part of the oceanic crust is composed of gabbro.

Andesite-Diorite Intermediate-composition magma (53–65% silica) crystallizes to form *andesite* and *diorite*, which are compositionally equivalent fine- and coarse-grained igneous rocks (Figure 4.13). Andesite and diorite are composed predominately of plagioclase feldspar, with the typical ferromagnesian component being amphibole or biotite (Figure 4.10). Andesite is generally medium to dark gray, but diorite has a salt-and-pepper appearance because of its white to light gray plagioclase and dark ferromagnesian silicates (Figure 4.13).

Andesite is a common extrusive igneous rock formed from lava erupted in volcanic chains at convergent plate boundaries. The volcanoes of the Andes Mountains of South America and the Cascade Range in western North America are composed, in part, of andesite. Intrusive bodies of diorite are fairly common in the continental crust.

Rhyolite-Granite *Rhyolite* and *granite* crystallize from felsic magma (>65% silica) and are therefore silica-rich rocks (Figure 4.14). They consist mostly of potassium feldspar, sodium-rich plagioclase, and quartz, with perhaps some biotite and rarely amphibole (Figure 4.10). Because nonferromagnesian silicates predominate, rhyolite and granite are typically light colored. Rhyolite is fine grained, although most often it contains phenocrysts of potassium feldspar or quartz, and granite is coarse grained. Granite porphyry is also fairly common.

Rhyolite lava flows are much less common than andesite and basalt flows. Recall that one control of magma viscosity is silica content. Thus, if felsic magma rises to the surface, it begins to cool, the pressure on it decreases, and gases are released explosively, usually yielding rhyolitic pyroclastic materials.

Parvinder Sethi

a Basalt is aphanitic.

Parvinder Sethi

b Gabbro is phaneritic. Notice the light reflected from the crystal faces.

Figure 4.12 Mafic Igneous Rocks

Parvinder Sethi

a This specimen of andesite has hornblende phenocrysts that are so numerous the rock may be classified as an andesite hornblende porphyry.

Parvinder Sethi

b Diorite has a salt-and-pepper appearance because it contains light-colored nonferromagnesian silicates and dark-colored ferromagnesian silicates.

Figure 4.13 Intermediate Igneous Rocks

The rhyolitic lava flows that do occur are thick and highly viscous and move only short distances.

Granite is a coarsely crystalline igneous rock with a composition corresponding to that of the field shown in Figure 4.10. Strictly speaking, not all rocks in this field are granites. For example, a rock with a composition close to the line separating granite and diorite is called *granodiorite.* To avoid the confusion that might result from introducing more rock names, we will follow the practice of referring to rocks to the left of the granite-diorite line in Figure 4.10 as *granitic.*

Granitic rocks are by far the most common intrusive igneous rocks, although they are restricted to the continents. Most granitic rocks were intruded at or near convergent plate margins during mountain-building episodes. When these mountainous regions are uplifted and eroded, the vast bodies of granitic rocks forming their cores are exposed. The granitic rocks of the Sierra Nevada of California form a composite body measuring approximately 640 km long and 110 km wide (see the chapter opening photograph), and the granitic rocks of the Coast Ranges of British Columbia, Canada, are even more voluminous.

Pegmatite The term *pegmatite* refers to a particular texture rather than a specific composition, but most pegmatites are composed mostly of quartz, potassium feldspar, and sodium-rich plagioclase, thus corresponding closely to granite. A few pegmatites are mafic or intermediate in composition and are appropriately called *gabbro* and *diorite pegmatites.* The most remarkable feature of pegmatites is the size of their minerals, which measure at least 1 cm across, and in some pegmatites they measure tens of centimeters or meters (Figure 4.15). Many pegmatites are adjacent to large granite plutons and are composed of minerals that formed from the water-rich magma that remained after most of the granite crystallized.

When felsic magma cools and forms granite, the remaining water-rich magma has properties that differ from the magma from which it separated. It has a lower density and viscosity and commonly invades cracks in the nearby

Parvinder Sethi

a Rhyolite with small phenocrysts of orthoclase (K) feldspar.

Parvinder Sethi

b Granite.

Figure 4.14 Felsic Igneous Rocks These rocks are typically light colored because they contain mostly nonferromagnesian silicate minerals. The dark spots in the granite specimen are biotite mica. The white and pinkish minerals are feldspars, whereas the glassy-appearing minerals are quartz.

Figure 4.15 Pegmatite

Steve Stahl

a This pegmatite, the light colored rock, is exposed in the Black Hills of South Dakota.

Parvinder Sethi

b Close-up view of a specimen from a pegmatite with minerals measuring 8 to 10 cm across.

rocks where minerals crystallize. This water-rich magma also contains elements that rarely enter into the common minerals that form granite. Pegmatites that are essentially very coarsely crystalline granite are simple pegmatites, whereas those with minerals containing elements such as lithium, beryllium, cesium, boron, and several others are complex pegmatites. Some complex pegmatites contain 300 different mineral species, a few of which are important economically. In addition, several gem minerals such as emerald and aquamarine, both of which are varieties of the silicate mineral beryl, and tourmaline are found in some pegmatites.

The formation and growth of mineral-crystal nuclei in pegmatites are similar to those processes in other magmas, but with one critical difference: The water-rich magma from which pegmatites crystallize inhibits the formation of nuclei. However, some nuclei do form, and because the appropriate ions in the liquid can move easily and attach themselves to a growing crystal, individual minerals have the opportunity to grow very large.

Other Igneous Rocks Geologists classify the igneous rocks in Figure 4.10 by texture and composition, but a few others are identified primarily by their textures (Figure 4.16). Much of the fragmental material erupted by volcanoes is *ash*, a designation for pyroclastic materials measuring less than 2.0 mm, most of which consists of pieces of minerals or shards of volcanic glass (Figure 4.9f). The consolidation of ash forms the pyroclastic rock *tuff* (Figure 4.17a). Most tuff is silica rich and light colored and is appropriately called *rhyolite tuff.* Some ash flows are so hot that as they come to rest, the ash particles fuse together and form a *welded tuff.* Consolidated deposits of larger pyroclastic materials, such as cinders, blocks, and bombs, are *volcanic breccia* (Figure 4.16).

Both *obsidian* and *pumice* are varieties of volcanic glass (Figure 4.17b, c). Obsidian may be black, dark gray, red, or brown, depending on the presence of iron. Obsidian breaks with the conchoidal (smoothly curved) fracture typical of glass. Analyses of many samples indicate that most obsidian has a high silica content and is compositionally similar to rhyolite.

Pumice is a variety of volcanic glass containing numerous vesicles that develop when gas escapes through lava and forms a froth (Figure 4.17c). If pumice falls into water, it can be carried great distances because it is so porous and light that it floats. Another vesicular rock is *scoria.* It is more crystalline and denser than pumice, but it has more vesicles than solid rock (Figure 4.17d).

INTRUSIVE IGNEOUS BODIES—PLUTONS

Unlike volcanism and the origin of volcanic rocks, we can study intrusive igneous bodies, collectively called **plutons**, only indirectly because intrusive rocks form when magma

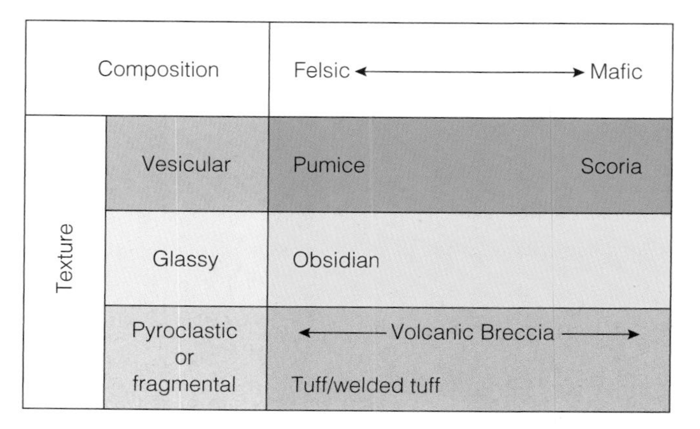

	Composition	Felsic ← → Mafic	
Texture	Vesicular	Pumice	Scoria
	Glassy	Obsidian	
	Pyroclastic or fragmental	← Volcanic Breccia → Tuff/welded tuff	

Figure 4.16 Texture Classification Classification of igneous rocks for which texture is not the main consideration. Composition is shown, but it is not essential for naming these rocks.

What Would You Do?

As the only member of your community with any geology background, you are considered the local expert on minerals and rocks. Suppose one of your friends brings you a rock specimen with the following features—composition: mostly potassium feldspar and plagioclase feldspar with about 10% quartz and minor amounts of biotite; texture: minerals average 3 mm across, but several potassium feldspars are up to 3 cm. Give the specimen a rock name and tell your friend as much as you can about the rock's history. Why are the minerals so large?

Figure 4.17 Igneous Rocks Classified Primarily by Their Texture

James S. Monroe

a Tuff is made up of pyroclastic materials such as those shown in Figure 4.9g.

Parvinder Sethi

b The natural glass obsidian.

Parvinder Sethi

c Pumice is glassy and extremely vesicular.

Parvinder Sethi

d Scoria is also vesicular, but it is darker, denser, and more crystalline than pumice.

cools and crystallizes within Earth's crust (see Geo-inSight on pages 100 and 101). We can observe these rock bodies only when uplift and deep erosion have taken place, thereby exposing them at the surface. Furthermore, geologists cannot duplicate the conditions under which intrusive rocks form except in small laboratory experiments.

Geologists recognize several types of plutons based on their geometry (three-dimensional shape) and relationships to the country rocks. In terms of their geometry, plutons are tabular, cylindrical, or irregular (massive). Furthermore, they may be **concordant**, meaning they have boundaries that parallel the layering in the country rock, or **discordant**, with boundaries that cut across the country rock's layering (see Geo-inSight on pages 100 and 101).

Dikes, Sills, and Laccoliths

Dikes and **sills** are tabular or sheetlike igneous bodies that differ only in that dikes are discordant and sills are concordant. Dikes are quite common and range from a few centimeters to more than 100 m thick. Invariably, they are intruded into preexisting fractures or where fluid pressure is great enough for them to form their own fractures as they move upward into county rock. Erosion of the Hawaiian volcanoes exposes dikes in fracture zones, and the Columbia River basalts in Washington State (see Chapter 5) issued from long fissures in which magma solidified to form dikes. Dikes underlie the Laki fissure and Eldgja fissure in Iceland, both of which have been the sites of voluminous eruptions during the past 1000 years.

Sills are tabular just as dikes are, but they are concordant. Many sills are a meter or less thick, although some are much thicker. A well-known sill in the United States is the Palisades sill that forms the Palisades along the west side of the Hudson River in New York and New Jersey (see Figure 22.7d). This 300 m-thick sill is exposed for 60 km along the river. Most sills were intruded into sedimentary rocks, but eroded volcanoes also reveal that sills are injected into piles of volcanic rocks. In fact, some of the inflation of a volcano that precedes an eruption may be caused by the injection of sills (see Chapter 5). In contrast to dikes, which follow zones of weakness, sills are intruded between layers in country rock when the fluid pressure is great enough for the magma to actually lift the overlying rocks.

Under some circumstances, a sill inflates and causes the overlying rocks to bow upward, forming an igneous body called a **laccolith**. A laccolith has a flat floor and is domed up in its central part, giving it a mushroom-like geometry (see Geo-inSight on pages 100 and 101). Like sills, laccoliths are rather shallow intrusions that lift the overlying rocks. Well-known laccoliths in the United States are in the Henry Mountains of southeastern Utah, and several buttes in Montana are eroded laccoliths.

Volcanic Pipes and Necks

Volcanoes have a cylindrical conduit known as a **volcanic pipe** that connects to an underlying magma chamber. Magma rises through this structure; however, when a volcano ceases to erupt, its slopes are attacked by weathering and erosion, but the magma in the pipe is commonly more resistant to erosion and is left as a remnant called a **volcanic neck**. Several volcanic necks are found in the southwestern United States, especially in Arizona and New Mexico, and others are recognized elsewhere (see Geo-inSight on pages 100 and 101 and Geo-Focus on pages 102 and 103).

Batholiths and Stocks

By definition, a **batholith**, the largest of all plutons, must have at least 100 km^2 of surface area, and most are far larger. A **stock**, in contrast, is similar but smaller. Some stocks are simply parts of large plutons that once exposed by erosion are batholiths (see Geo-inSight on pages 100 and 101). Both batholiths and stocks are mostly discordant, although locally they may be concordant, and batholiths, especially, consist of multiple intrusions. In other words, a batholith is a large composite body produced by repeated, voluminous intrusions of magma in the same region. The coastal batholith of Peru, for instance, was emplaced during a period of 60 to 70 million years and is made up of as many as 800 individual plutons.

The igneous rocks that make up batholiths are mostly granitic, although diorite may also be present. Batholiths and stocks are emplaced mostly near convergent plate boundaries during episodes of mountain building. One example is the Sierra Nevada batholith of California (see the chapter opening photograph), which formed over millions of years. Other large batholiths in North America include the Idaho batholith, the Boulder batholith in Montana, and the Coast Range batholith in British Columbia, Canada.

Mineral resources are found in rocks of batholiths and stocks, and in the adjacent country rocks. The copper deposits at Butte, Montana, are in rocks near the margins of the granitic rocks of the Boulder batholith. Near Salt Lake City, Utah, copper is mined from the mineralized rocks of the Bingham stock, a composite pluton composed of granite and granite porphyry. Granitic rocks also are the primary source of gold, which forms from mineral-rich solutions moving through cracks and fractures of the igneous body.

HOW ARE BATHOLITHS INTRUDED INTO EARTH'S CRUST?

Geologists realized long ago that the origin of batholiths posed a space problem. What happened to the rock that was once in the space now occupied by a batholith?

One solution to the space problem is that these large igneous bodies melted their way into the crust. In other words, they simply assimilated the country rock as they moved upward (Figure 4.7). The presence of inclusions, especially near the tops of some plutons, indicates that assimilation does occur. Nevertheless, as we noted, assimilation is a limited process because magma cools as country rock is assimilated. Calculations indicate that far too little heat is available in magma to assimilate the huge quantities of country rock necessary to make room for a batholith.

Geologists now generally agree that batholiths were emplaced by *forceful injection* as magma moved upward. Recall that granite is derived from viscous felsic magma and therefore rises slowly. It appears that the magma deforms and shoulders aside the country rock, and as it rises farther, some of the country rock fills the space beneath the magma.

Some batholiths do indeed show evidence of having been emplaced forcefully by shouldering aside and deforming the country rock. This mechanism probably occurs in the deeper parts of the crust where temperature and pressure are high and the country rocks are easily deformed in the manner described. At shallower depths, the crust is more rigid and tends to deform by fracturing. In this environment, batholiths may move upward by **stoping**, a process in which rising magma detaches and engulfs pieces of country rock (Figure 4.18). According to this concept, magma moves up along fractures and the planes separating layers of country rock. Eventually, pieces of country rock detach and settle into the magma. No new room is created during stoping; the magma simply fills the space formerly occupied by country rock (Figure 4.18).

Figure 4.18 Emplacement of a Batholith by Forceful Injection and Stoping

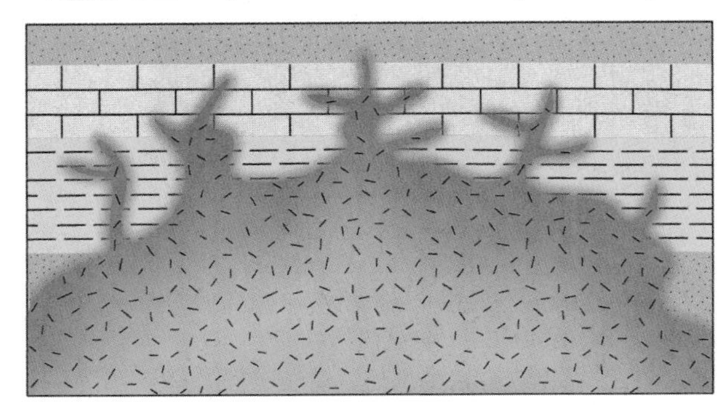

a Magma rises and forces its way into fractures and planes between layers in the country rock.

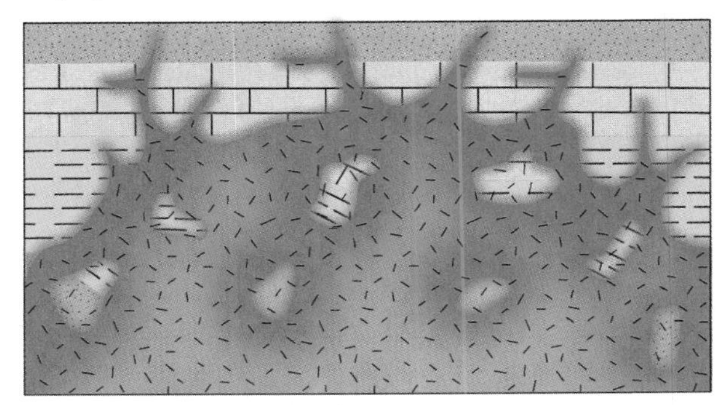

b Blocks of country rock are detached and engulfed in the magma, thereby making room for the magma to rise farther. Some of the engulfed blocks might be assimilated, and some may remain as inclusions (Figure 4.7b).

Intrusive bodies called plutons are common, but we see them at the surface only after deep erosion. Notice that they vary in geometry and their relationships to the country rock.

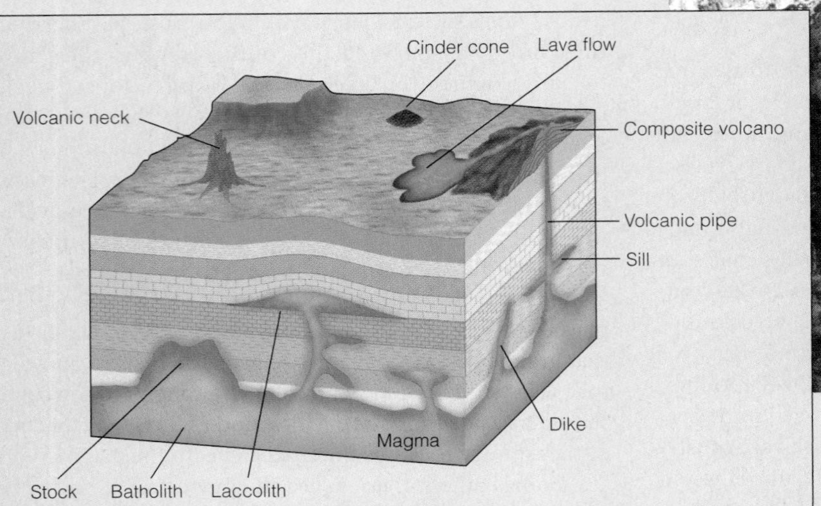

▲ **1.** Block diagram showing various plutons. Some plutons cut across the layering in country rock and are discordant, whereas others parallel the layering and are concordant.

James S. Monroe

▲ **2.** Part of the Sierra Nevada batholith in Yosemite National Park, California. The batholith, consisting of multiple intrusions of granitic rock, is more than 600 km long and up to 110 km wide. To appreciate the scale in this image, the waterfall has a descent of 435 m.

James S. Monroe

◀ **3.** A volcanic neck in Monument Valley Tribal Park, Arizona. This landform is 457 m high. Most of the original volcano was eroded, leaving only this remnant.

▶ **4.** Granitic rocks of a small stock at Castle Crags State Park, California.

Sue Monroe

► **5.** The dark materials in this image are igneous rocks, whereas the light layers are sedimentary. Notice that the sill parallels the layering, so it is concordant. The dike, though, clearly cuts across the layering and is discordant. Sills and dikes have sheetlike geometry but in this view we can see them in only two dimensions.

Martin G. Miller/Visuals Unlimited

http://formontana.net/cb.html

▲ **6.** Crown Butte in Montana is an eroded laccolith standing about 300 m above the surrounding plain. The magma making up this small pluton was intruded about 50 million years ago.

▲ ► **7.** Diagrams showing the evolution of an eroded laccolith.

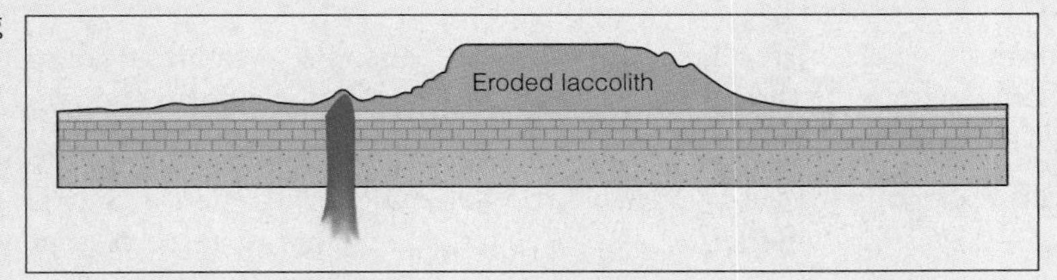

Geo-Focus Some Remarkable Volcanic Necks

We mentioned that as an extinct volcano weathers and erodes, a remnant of the original mountain may persist as a volcanic neck. The origin of volcanic necks is well known, but these isolated monoliths rising above otherwise rather flat land are scenic, awe-inspiring, and the subject of legends. They are found in many areas of recently active volcanism. A small volcanic neck rising only 79 m above the surface in the town of Le Puy, France, is the site of the 11th-century chapel of Saint Michel d'Aiguilhe (Figure 1). It is so steep that materials and tools used in its construction had to be hauled up in baskets.

Perhaps the most famous volcanic neck in the United States is Shiprock, New Mexico, which rises nearly 550 m above the surrounding plain and is visible from 160 km away. Radiating outward from this conical structure are three vertical dikes that stand like walls above the adjacent countryside (Figure 2). According to one legend, Shiprock, or *Tsae-bidahi*, meaning "winged rock," represents a giant bird that brought the Navajo people from the north. The same legend holds that the dikes are snakes that turned to stone.

Richard List/Corbis

Figure 1 This volcanic neck in Le Puy, France, rises 79 m above the surface of the town. Workers on the Chapel of Saint Michel d'Aiguilhe had to haul building materials and tools up in baskets.

Frank Hanna

Figure 2 Shiprock, a volcanic neck in northwestern New Mexico, rises nearly 550 m above the surrounding plain. One of the dikes radiating from Shiprock is in the foreground.

Geo-Recap

Chapter Summary

- Magma is the term for molten rock below Earth's surface, whereas the same material at the surface is called lava.
- Silica content distinguishes among mafic (45–52% silica), intermediate (53–65% silica), and felsic (>65% silica) magmas.
- Magma and lava viscosity depends mostly on temperature and composition: The more silica, the greater the viscosity.
- Minerals crystallize from magma and lava when small crystal nuclei form and grow.
- Rapid cooling accounts for the aphanitic textures of volcanic rocks, whereas comparatively slow cooling yields the phaneritic textures of plutonic rocks. Igneous rocks with markedly different-sized minerals are porphyritic.
- Igneous rock composition is determined mostly by the composition of the parent magma, but magma

An absolute age determined for one of the dikes indicates that Shiprock is about 27 million years old. When the original volcano formed, apparently during explosive eruptions, rising magma penetrated various rocks, including the Mancos Shale, the rock unit now exposed at the surface adjacent to Shiprock. The rock that makes up Shiprock itself is tuff breccia, consisting of fragmented volcanic debris, as well as pieces of metamorphic, sedimentary, and igneous rocks.

Geologists agree that Devil's Tower in northeastern Wyoming cooled from a small body of magma and that erosion has exposed it in its present form (Figure 3). However, opinion is divided on whether it is a volcanic neck or an eroded laccolith. In either case, the rock that makes up Devil's Tower is 45 to 50 million years old, and President Theodore Roosevelt designated this impressive landform as our first national monument in 1906. At 260 m high, Devil's Tower is visible from 48 km away and served as a landmark for early travelers in this area. It achieved further distinction in 1977 when it was featured in the film *Close Encounters of the Third Kind.*

James S. Monroe

Figure 3 Devil's Tower in northeastern Wyoming rises approximately 260 m above its base. It may be a volcanic neck or an eroded laccolith. The vertical lines result from intersections of fractures called columnar joints. According to Cheyenne legend, a gigantic grizzly bear made the deep scratches.

The Cheyenne and Sioux nations call Devil's Tower *Mateo Tepee*, meaning "Grizzly Bear Lodge." It was also called the "Bad God's Tower," and reportedly "Devil's Tower" is a translation taken from this phrase. The tower's most conspicuous features are the near-vertical lines that, according to Cheyenne legends, are scratch marks made by a gigantic grizzly bear. One legend holds that the bear made the scratches while pursuing a group of children. Another tells of six brothers and a woman also pursued by a grizzly bear. One brother carried a rock, and when he sang a song, it grew into Devil's Tower, safely carrying the brothers and woman out of the bear's reach.

Although not nearly as interesting as the Cheyenne legends, the origin of the "scratch marks" is well understood. These lines actually formed at the intersections of *columnar joints*, fractures that form in response to the cooling and contraction that occur in some plutons and lava flows (see Chapter 5). The columns outlined by these fractures are up to 2.5 m across, and the pile of rubble at the tower's base is simply an accumulation of collapsed columns.

composition can change so that the same magma may yield more than one kind of igneous rock.

- According to Bowen's reaction series, cooling mafic magma yields a sequence of minerals, each of which is stable within specific temperature ranges. Only ferromagnesian silicates are found in the discontinuous branch of Bowen's reaction series. The continuous branch of the reaction series yields only plagioclase feldspars that become increasingly enriched with sodium as cooling occurs.
- A chemical change in magma may take place as early ferromagnesian silicates form and, because of their density, settle in the magma.
- Compositional changes also take place in magma when it assimilates country rock or one magma mixes with another.
- Geologists recognize two broad categories of igneous rocks: volcanic or extrusive and plutonic or intrusive.
- Texture and composition are the criteria used to classify igneous rocks, although a few are defined only by texture.
- Crystallization from water-rich magma results in very large minerals that form rocks known as pegmatite. Most pegmatite has an overall composition similar to granite.
- Intrusive igneous bodies known as plutons vary in their geometry and their relationship to country rock: Some are concordant, whereas others are discordant.
- The largest plutons, known as batholiths, consist of multiple intrusions of magma during long periods of time.
- Most plutons, including batholiths, are found at or near divergent and convergent plate boundaries.

Important Terms

aphanitic texture (p. 93)
assimilation (p. 92)
batholith (p. 99)
Bowen's reaction series (p. 89)
concordant pluton (p. 98)
country rock (p. 92)
crystal settling (p. 91)
dike (p. 98)
discordant pluton (p. 98)
felsic magma (p. 87)
hot spot (p. 91)
igneous rock (p. 87)
intermediate magma (p. 87)
laccolith (p. 98)
lava (p. 86)
lava flow (p. 87)
mafic magma (p. 87)
magma (p. 86)
magma chamber (p. 89)
magma mixing (p. 92)
mantle plume (p. 91)
phaneritic texture (p. 93)
pluton (p. 97)
plutonic (intrusive igneous) rock (p. 87)
porphyritic texture (p. 93)
pyroclastic (fragmental) texture (p. 94)
pyroclastic materials (p. 87)
sill (p. 98)
stock (p. 99)
stoping (p. 99)
vesicle (p. 94)
viscosity (p. 88)
volcanic neck (p. 99)
volcanic pipe (p. 99)
volcanic (extrusive igneous) rock (p. 87)

Review Questions

1. An igneous rock such as granite with minerals large enough to see without magnification has a __________ texture.
   a. _____ assimilated;
   b. _____ phaneritic;
   c. _____ volcanic;
   d. _____ pyroclastic;
   e. _____ clastic.
2. Which one of the following is a tabular discordant pluton?
   a. _____ batholith;
   b. _____ laccolith;
   c. _____ sill;
   d. _____ dike;
   e. _____ stock.
3. Magma characterized as intermediate
   a. _____ flows more readily than mafic magma;
   b. _____ has between 53% and 65% silica;
   c. _____ crystallizes to form rhyolite;
   d. _____ is one from which ultramafic rocks crystallize;
   e. _____ makes up most of the mantle.
4. Which pair of the following igneous rocks has the same composition?
   a. _____ diorite-andesite;
   b. _____ peridotite-pegmatite;
   c. _____ gabbro-rhyolite;
   d. _____ basalt-granite;
   e. _____ obsidian-scoria.
5. One process that brings about chemical changes in magma is
   a. _____ magma mixing;
   b. _____ crystal inversion;
   c. _____ discordant separation;
   d. _____ mafic differentiation;
   e. _____ ultramafic accumulation.
6. Islands such as the Hawaiian chain form at hot spots that overlie a
   a. _____ volcanic neck;
   b. _____ convergent plate boundary;
   c. _____ divergent plate boundary;
   d. _____ granite batholith;
   e. _____ mantle plume.
7. Which one the following statements about batholiths is correct?
   a. _____ They are mostly concordant;
   b. _____ They are made up of basalt and gabbro;
   c. _____ They form mostly at convergent plate boundaries;
   d. _____ They formed from a single huge mass of magma;
   e. _____ They have a mushroom-like geometry and are concordant.
8. The igneous rock tuff is composed of
   a. _____ pyroclastic materials;
   b. _____ calcium-rich plagioclase and pyroxene;
   c. _____ obsidian and pumice;
   d. _____ granite and vesicles;
   e. _____ phenocrysts.
9. The phenomenon by which pieces of country rock are detached and engulfed by rising magma is
   a. _____ assimilation;
   b. _____ crystal settling;
   c. _____ stoping;
   d. _____ Bowen's reaction series;
   e. _____ viscosity.
10. A fluid's resistance to flow is called its
    a. _____ plutonic inertia;
    b. _____ composition;
    c. _____ texture;
    d. _____ melting point;
    e. _____ viscosity.
11. What factors control the viscosity of magma and lava?

12. What are the similarities and differences among dikes, sills, and laccoliths?
13. How does pegmatite form and why are its minerals so big?
14. Explain or diagram how a volcanic neck forms.
15. Two phaneritic igneous rocks have the following compositions: Specimen 1–5% biotite; 15% sodium-rich plagioclase; 60% potassium feldspar; and 10% quartz. Specimen 2–10% olivine; 55% pyroxene; 5% hornblende; and 30% calcium-rich plagioclase. Use Figure 4.10 to classify these rocks. Which one would be darkest and densest? Explain how you know.
16. How do assimilation and stoping contribute to the emplacement of a batholith in the crust?
17. Explain what a porphyritic texture is and how one might develop.
18. Use Bowen's reaction series to explain the sequence in which minerals crystallize from mafic magma.
19. What generalizations can you make about the size of minerals in igneous rocks and whether they are volcanic or plutonic? Are there exceptions to your generalizations?
20. What three processes account for the origin of magma?

CHAPTER 5

Volcanoes and Volcanism

OUTLINE

Introduction

Volcanism and Volcanoes

What Are the Types of Volcanoes?

Other Volcanic Landforms

GEO-INSIGHT: Types of Volcanoes

Distribution of Volcanoes

North America's Active Volcanoes

Plate Tectonics, Volcanoes, and Plutons

GEO-FOCUS: The Bronze Age Eruption of Santorini

Volcanic Hazards, Volcano Monitoring, and Forecasting Eruptions

Geo-Recap

OBJECTIVES

At the end of this chapter, you will have learned that

- In addition to lava flows, erupting volcanoes eject pyroclastic materials, especially ash, and gases.
- Geologists identify the basic types of volcanoes by their eruptive style, composition, and shape.
- Although all volcanoes are unique, most are identified as shield volcanoes, cinder cones, or composite volcanoes.
- Volcanoes characterized as lava domes tend to erupt explosively and thus are dangerous.
- Active volcanoes in the United States are found in Hawaii, Alaska, and the Cascade Range of the Pacific Northwest.
- Eruptions in Hawaii and Alaska are commonplace, but only two eruptions have occurred in the continental United States during the 1900s and one in 2004. Canada has had no eruptions during historic time.
- Some eruptions yield vast sheets of lava or pyroclastic materials rather than volcanoes.

Lava flowing into the sea forms new land and Hawaii becomes larger.

- Geologists have devised the volcanic explosivity index as a measure of an eruption's size.
- Most volcanoes are located in belts at or near divergent and convergent plate boundaries.
- Some volcanoes are carefully monitored to help geologists anticipate eruptions.

INTRODUCTION

No other geologic phenomenon has captured the public imagination more than erupting volcanoes, especially lava flows issuing forth as fiery streams or particulate matter blasted into the atmosphere in sensational pyrotechnic displays. What better subject for a disaster movie? Several such movies of varying quality and scientific accuracy have been released in recent years, but one of the best was *Dante's Peak* in 1997. Certainly the writers and the director exaggerated some aspects of volcanism, but the movie rather accurately depicted the phenomenal power of an explosive eruption. Incidentally, the volcano called Dante's Peak was a 10-m-high model built of wood and steel.

Incandescent streams of molten rock are often portrayed in movies as a great danger to humans, and, in fact, on a few occasions lava flows have caused fatalities. In 2002 lava flows in Goma, Zarie (Democratic Republic of the Congo), caused gasoline storage tanks to explode and kill 147 people. Furthermore, lava flows destroy buildings and cover otherwise productive land, but actually they are the least dangerous manifestation of volcanic eruptions. Explosive eruptions, in contrast, are quite dangerous, especially if they occur near populated areas.

One of the best-known volcanic catastrophes ever recorded was the A.D. 79 eruption of Mount Vesuvius that destroyed the thriving Roman communities of Pompeii, Herculaneum, and Stabiae in what is now Italy (Figure 5.1). Fortunately for us, Pliny the Younger recorded the event in detail; his uncle, Pliny the Elder, died while trying to investigate the eruption.

Pompeii, a city of approximately 20,000 people only 9 km downwind from the volcano, was buried in nearly 3 m of pyroclastic materials that covered all but the tallest buildings (Figure 5.1). About 2000 victims have been discovered in the city, but certainly far more were killed. Pompeii was covered by volcanic debris rather gradually, but surges of incandescent volcanic materials in glowing avalanches swept through Herculaneum, quickly burying the town to a depth of about 20 m. Since A.D. 79, Mount Vesuvius has erupted 80 times, most violently in 1631 and 1906; it last erupted in 1944. Ongoing volcanic and seismic activity in this area poses a continuing threat to the many cities and towns along the shores of the Bay of Naples (Figure 5.1).

Ironically, when considered in the context of Earth history, volcanism is actually a constructive process. The atmosphere and surface waters most likely resulted from the emission of gases during Earth's early history, and oceanic crust is continuously produced by volcanism at spreading ridges. Oceanic islands such as the Hawaiian Islands, Iceland, and the Azores owe their existence to volcanism, and weathering of lava flows, pyroclastic materials, and volcanic mudflows in tropical areas such as Indonesia converts them to productive soils.

People who live in Hawaii, southern Alaska, the Philippines, Japan, and Iceland are well aware of volcanic eruptions; however, eruptions in the continental United States have occurred only three times since 1914, all in the Cascade Range, which stretches from northern California through Oregon and Washington and into southern British Columbia, Canada. Canada has had no eruptions during historic time. Ancient and ongoing volcanism in the western United States has yielded interesting features, several of which are featured in this chapter.

One very good reason to study volcanic eruptions is that they illustrate the complex interactions among Earth's systems. Volcanism, especially the emission of gases and pyroclastic materials, has an immediate and profound impact on the atmosphere, hydrosphere, and biosphere, at least in the vicinity of an eruption. And in some cases, the effects are worldwide, as they were following the eruptions of Tambora in 1815, Krakatau in 1883, and Pinatubo in 1991. Furthermore, the fact that lava flows and explosive eruptions cause property damage, injuries, fatalities (Table 5.1), and at least short-term atmospheric changes indicates that volcanic eruptions are catastrophic events, at least from the human perspective.

VOLCANISM AND VOLCANOES

What do we mean by the terms *volcanism* and *volcano*? The latter is a landform—that is, a feature on Earth's surface—whereas **volcanism** is the process by which magma rises through Earth's crust and issues forth at the surface as lava flows and/or pyroclastic materials and gases. Volcanism is responsible for all extrusive igneous rocks, such as basalt, tuff, and obsidian.

Volcanic eruptions are common; about 550 volcanoes are *active*, that is, they have erupted during historic time, but only about a dozen are erupting at any one time. Most eruptions are minor and go unreported in the popular press unless they occur near populated areas or have tragic consequences. However, large eruptions that cause extensive property damage, injuries, and fatalities are not uncommon, so a great amount of effort is devoted to understanding and more effectively anticipating large eruptions.

In addition to active volcanoes, Earth has numerous *dormant* volcanoes that have not erupted during historic time but may do so in the future. Prior to its eruption in A.D. 79, Mount Vesuvius had not been active in human memory. The largest volcanic outburst in the last 50 years was when Mount Pinatubo in the Philippines erupted in 1991 after lying dormant for 600 years. Some volcanoes have not erupted in historic time and show no signs of doing so again; thousands of these *extinct* or *inactive* volcanoes are known.

All terrestrial planets and Earth's Moon were volcanically active during their early histories, but now volcanoes are known only on Earth and on one or two other bodies in

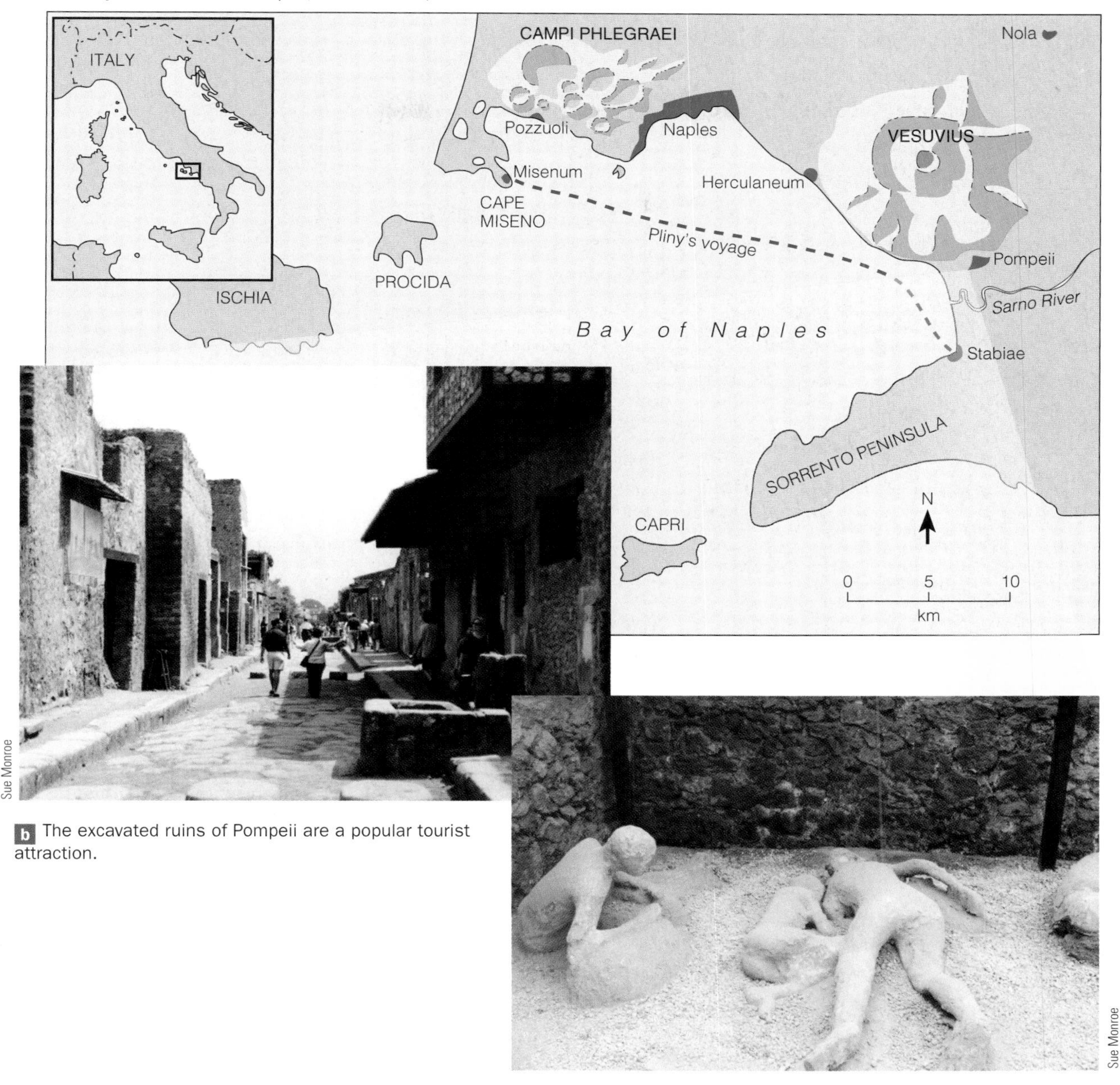

a The Mount Vesuvius region on the shore of the Bay of Naples in Italy. Vesuvius erupted in A.D. 79 and destroyed the cities of Pompeii, Herculaneum, and Stabiae.

b The excavated ruins of Pompeii are a popular tourist attraction.

c Body casts of some of the volcano's victims in Pompeii.

Figure 5.1 The 79 A.D. Eruption of Mount Vesuvius

the solar system. Triton, one of Neptune's moons, probably has active volcanoes, and Jupiter's moon Io is by far the most volcanically active body in the solar system. Many of its hundred or so volcanoes are erupting at any given time.

Volcanic Gases

Volcanic gases from present-day volcanoes are 50% to 80% water vapor, with lesser amounts of carbon dioxide, nitrogen, sulfur gases, especially sulfur dioxide and hydrogen sulfide, and very small amounts of carbon monoxide, hydrogen, and chlorine. In areas of recent volcanism, such as Lassen Volcanic National Park in California, emission of gases continues, and one cannot help noticing the rotten-egg odor of hydrogen sulfide gas (Figure 5.2).

When magma rises toward the surface, the pressure is reduced and the contained gases begin to expand. In highly viscous, felsic magma, expansion is inhibited and gas pressure increases. Eventually, the pressure may become great enough to cause an explosion and produce pyroclastic materials such as volcanic ash. In contrast, low-viscosity mafic magma allows gases to expand and escape easily. Accordingly, mafic magma generally erupts rather quietly.

Most volcanic gases quickly dissipate in the atmosphere and pose little danger to humans, but on several occasions, they have caused fatalities. In 1783, toxic gases, probably

TABLE 5.1 Some Notable Volcanic Eruptions

Date	Volcano	Deaths
Apr 10, 1815	Tambora, Indonesia	117,000 killed, including deaths from eruption, famine, and disease.
Oct 8, 1822	Galunggung, Java	Pyroclastic flows and mudflows killed 4011.
Mar 2, 1856	Awu, Indonesia	2806 died in pyroclastic flows.
Aug 27, 1883	Krakatau, Indonesia	More than 36,000 died; most killed by tsunami.
June 7, 1892	Awu, Indonesia	1532 died in pyroclastic flows.
May 8, 1902	Mount Pelée, Martinique	Nuée ardente engulfed St. Pierre and killed 28,000.
Oct 24, 1902	Santa Maria, Guatemala	5000 died during eruption.
May 19, 1919	Kelut, Java	Mudflows devastated 104 villages and killed 5110.
Jan 21, 1951	Lamington, New Guinea	Pyroclastic flows killed 2942.
Mar 17, 1963	Agung, Indonesia	1148 perished during eruption.
May 18, 1980	Mount St. Helens, Washington	63 killed, 600 km^2 of forest devastated.
Mar 28, 1982	El Chichon. Mexico	Pyroclastic flows killed 1877.
Nov 13, 1985	Nevado del Ruiz, Colombia	Minor eruption triggered mudflows that killed 23,000.
Aug 21, 1986	Oku volcanic field, Cameroon	Cloud of CO_2 released from Lake Nyos killed 1746.
June 15, 1991	Mount Pinatubo, Philippines	281 killed during eruption, 83 died in later mudflows; 358 died of illness.
July 1999	Soufriére Hills, Montserrat	19 killed; 12,000 evacuated.
Jan 17, 2002	Nyiragongo, Zaire	Lava flow killed 147 in Goma.
Aug 16, 2006	Tungurahua, Ecuador	Explosive eruption killed 7; continuously active since 1999; periodic evacuations and several villages destroyed.

sulfur dioxide, erupting from Laki fissure in Iceland had devastating effects. About 75% of the nation's livestock died, and the haze resulting from the gas caused lower temperatures and crop failures; about 24% of Iceland's population died as a result of the ensuing Blue Haze Famine. The country suffered its coldest winter in 225 years in 1783–1784, with temperatures 4.8°C below the long-term average. The eruption also produced what Benjamin Franklin called a "dry fog" that was responsible for dimming the intensity of sunlight in Europe.

Sue Monroe

Figure 5.2 Fumeroles Gases emitted from vents (fumeroles) at the Sulfur Works in Lassen Volcanic National Park in California. Hot, acidic gases and fluids have altered the original igneous rocks to clay. Several other vents are also present in this area, but the two shown here opened up only a few years ago.

In 1986, in the African nation of Cameroon, 1746 people died when a cloud of carbon dioxide engulfed them. The gas accumulated in the waters of Lake Nyos, which occupies a volcanic caldera. Scientists disagree about what caused the gas to suddenly burst forth from the lake, but once it did, it flowed downhill along the surface because it was denser than air. In fact, the density and velocity of the gas cloud were great enough to flatten vegetation, including trees, a few kilometers from the lake. Unfortunately, thousands of animals and many people, some as far as 23 km from the lake, were asphyxiated.

Residents of the island of Hawaii have coined the term *vog* for volcanic smog. Kilauea volcano has been erupting continuously since 1983, releasing small amounts of lava, and copious quantities of carbon dioxide and sulfur dioxide. Carbon dioxide is no problem because it dissipates quickly in the atmosphere, but sulfur dioxide produces a haze and the unpleasant odor of sulfur. Vog probably poses little or no health risk for tourists, but a long-term threat exists for residents of the west side of the island where vog is most common.

Lava Flows

Although lava flows are portrayed in movies and on television as a great danger to humans, they only rarely cause fatalities. The reason is that most lava flows do not move very fast, and because they are fluid, they follow existing low areas. Thus, once a flow erupts from a volcano, determining the path it will take is fairly easy, and anyone in areas likely to be affected can be evacuated.

Even low-viscosity lava flows generally do not move very rapidly. Flows can move much faster, though, when their margins cool to form a channel, and especially when insulated on all sides as in a **lava tube**, where a speed of more than 50 km/hr has been recorded. A conduit known as a lava tube within a lava flow forms when the margins and upper surface of the flow solidify. Thus confined and insulated, the flow moves rapidly and over great distances. As an eruption ceases, the tube drains, leaving an empty tunnel-like structure (Figure 5.3a). Part of the roof of a lava tube may collapse to form a *skylight*

a Entrance to a lava tube on Medicine Lake Volcano, a huge shield volcano in northern California. It has a circumference of 240 km and it last erupted about 1000 years ago.

J.B. Judd/USGS

b An active lava tube in Hawaii. Part of the tube's roof has collapsed, forming a skylight.

Figure 5.3 Lava Tubes Lava tubes consisting of hollow spaces beneath the surfaces of lava flows are common in many areas.

through which an active flow can be observed (Figure 5.3b), or access can be gained to an inactive lava tube. In Hawaii, lava moves through lava tubes many kilometers long and, in some cases, discharges into the sea.

Geologists define two types of lava flows, both named for Hawaiian flows. A **pahoehoe** (pronounced *pah-hoy-hoy*) flow has a ropy surface much like taffy (▶ Figure 5.4a). The surface of an **aa** (pronounced *ah-ah*) flow is characterized by rough, jagged, angular blocks and fragments (Figure 5.4b). Pahoehoe flows are less viscous than aa flows; indeed, the latter are viscous enough to break up into blocks and move forward as a wall of rubble.

Pressure on the partly solidified crust of a still-moving lava flow causes the surface to buckle into *pressure ridges* (▶ Figure 5.5a). Gases escaping from a flow hurl globs of lava into the air, which fall back to the surface and adhere to one another, thus forming small, steep-sided *spatter cones*, or spatter ramparts if they are elongated (Figure 5.5b). Spatter cones a few meters high are common on lava flows in Hawaii, and you can see ancient ones in many areas.

Many lava flows, especially mafic ones, have a distinctive pattern of columns bounded by fractures, or what geologists call **columnar joints**. Once a lava flow stops moving, it contracts as it cools and produces forces that cause fractures called *joints* to open. On the surface of a lava flow, the joints are commonly polygonal (often six-sided) cracks that extend downward, thus forming parallel columns with their long axes perpendicular to the cooling surface (▶ Figure 5.6). Excellent examples of columnar jointing are found in many areas.

Much of the igneous rock in the upper part of the oceanic crust is a distinctive type consisting of bulbous masses of basalt that resemble pillows, hence the name **pillow lava**. It was long recognized that pillow lava forms when lava is rapidly chilled beneath water, but its formation was not observed until 1971. Divers near Hawaii saw pillows form when a blob of lava broke through the crust of an underwater lava flow and cooled almost instantly, forming a pillow-shaped structure with a glassy exterior. The remaining fluid inside then broke through the crust of the pillow, repeating the process and resulting in an accumulation of interconnected pillows (▶ Figure 5.7).

Parvinder Sethi

a An excellent example of the "taffy-like" appearance of pahoehoe.

Robert Tilling/USGS

b An aa lava flow advances over an older pahoehoe flow. Notice the rubbly nature of the aa flow.

▶ **Figure 5.4 Pahoehoe and aa Lava Flows** Pahoehoe and aa were named for lava flows in Hawaii, but the same kinds of flows are found in many other areas.

Pyroclastic Materials

In addition to lava flows, erupting volcanoes eject pyroclastic materials, especially **volcanic ash**, a designation for pyroclastic particles that measure less than 2.0 mm (▶ Figure 5.8). In some cases, ash is ejected into the atmosphere and settles to the surface as an *ash fall*. In 1947, ash that erupted from Mount Hekla in Iceland fell 3800 km away on Helsinki, Finland. In contrast to an ash fall, an *ash flow* is a cloud of ash and gas that flows along or close to the land surface. Ash flows can move faster than 100 km/hr, and some cover vast areas.

In populated areas adjacent to volcanoes, ash falls and ash flows pose serious problems, and volcanic ash in the atmosphere is a hazard to aviation. Since 1980, approximately

Figure 5.5 Pressure Ridge and Spatter Cones

Sue Monroe

a The buckled surface of this lava flow on Medicine Lake Volcano in California is a pressure ridge. It formed when the solidified surface of a lava flow was bent upward by a still-moving flow beneath.

Sue Monroe

b These two chimney-like pillars of rocks are spatter cones on the surface of a lava flow in the Coso volcanic field of California.

80 aircraft have been damaged when they encountered clouds of volcanic ash. The most serious incident took place in 1989 when ash from Redoubt volcano in Alaska caused all four jet engines to fail on KLM Flight 867. The plane, carrying 231 passengers, nearly crashed when it fell more than 3 km before the crew could restart the engines. The plane landed safely in Anchorage, Alaska, but it required $80 million in repairs.

In addition to volcanic ash, volcanoes erupt *lapilli*, consisting of pyroclastic materials that measure from 2 to 64 mm, and *blocks* and *bombs*, both larger than 64 mm (Figure 5.8). Bombs have a twisted, streamlined shape, which indicates that they were erupted as globs of magma that cooled and solidified during their flight through the air. Blocks, in contrast, are angular pieces of rock ripped from a volcanic conduit or pieces

Figure 5.6 Columnar Jointing Columnar jointing is seen mostly in mafic lava flows and related intrusive rocks.

b Columnar joints in a basalt lava flow at Devil's Postpile National Monument in California. The rubble in the foreground is collapsed columns.

c Surface view of the columns from (b). The straight lines and polish resulted from abrasion by a glacier that moved over this surface.

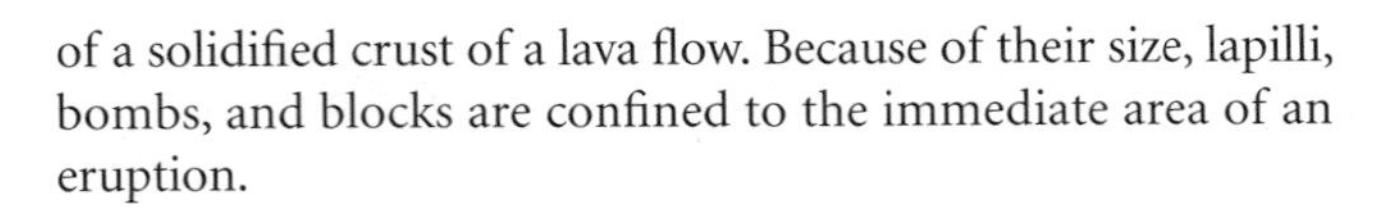

a As lava cools and contracts, three-pronged cracks form that grow and intersect to form four- to seven-sided columns, most of which are six-sided.

of a solidified crust of a lava flow. Because of their size, lapilli, bombs, and blocks are confined to the immediate area of an eruption.

WHAT ARE THE TYPES OF VOLCANOES?

Simply put, a **volcano** is a hill or mountain that forms around a vent where lava, pyroclastic materials, and gases erupt. Although volcanoes vary in size and shape, all have a conduit or conduits leading to a magma chamber beneath the surface. Vulcan, the Roman deity of fire, was the inspiration for calling these mountains volcanoes, and because of their danger and obvious connection to Earth's interior, they have been held in awe by many cultures.

In Hawaiian legends, the volcano goddess Pele resides in the crater of Kilauea on Hawaii. During one of her frequent rages, Pele causes earthquakes and lava flows, and she may hurl flaming boulders at those who offend her. Native Americans in the Pacific Northwest tell of a titanic battle between the volcano gods Skel and Llao to account for huge eruptions that took place about 7000 years ago in Oregon and California.

Most volcanoes have a circular depression known as a **crater** at their summit, or on their flanks, that forms by explosions or collapse. Craters are generally less than 1 km across, whereas much larger rimmed depressions on

Figure 5.7 Pillow Lava Much of the upper part of the oceanic crust is made up of pillow lava that formed when lava erupted underwater.

a Pillow lava on the seafloor in the Pacific Ocean about 150 miles west of Oregon that formed about five years before the photo was taken.

b Ancient pillow lava that formed on the seafloor is now on land in Marin County, California. The largest pillow measures about 0.6 m across.

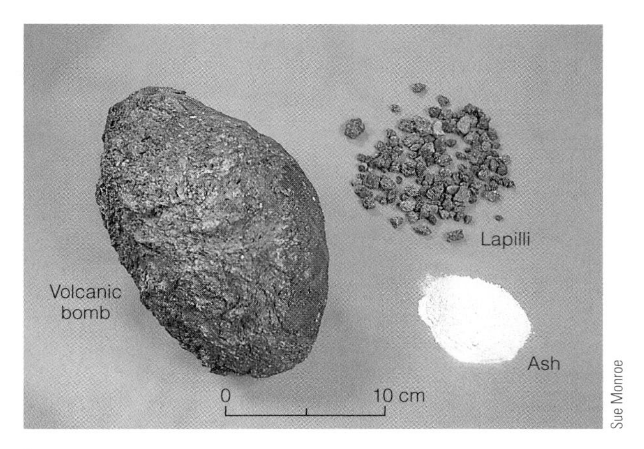

Figure 5.8 Pyroclastic Materials Pyroclastic materials are all particles ejected from volcanoes, especially during explosive eruptions. The volcanic bomb is elongate because it was molten when it descended through the air. The lapilli was collected at a small volcano in Oregon, whereas the ash came from the 1980 eruption of Mount St. Helens in Washington.

volcanoes are **calderas**. In fact, some volcanoes have a summit crater within a caldera. Calderas are huge structures that form following voluminous eruptions during which part of a magma chamber drains and the mountain's summit collapses into the vacated space below. An excellent example is misnamed Crater Lake in Oregon (Figure 5.9). Crater Lake is actually a steep-rimmed caldera that formed about 7700 years ago in the manner just described; it is more than 1200 m deep and measures 9.7 × 6.5 km. As impressive as Crater Lake is, it is not nearly as large as some other calderas, such as the Toba caldera in Sumatra, which is 100 km long and 30 km wide.

Geologists recognize several major types of volcanoes, but one must realize that each volcano is unique in its history of eruptions and development. For instance, the frequency of eruptions varies considerably; the Hawaiian volcanoes and Mount Etna on Sicily have erupted repeatedly, whereas Pinatubo in the Philippines erupted in 1991 for the first time in 600 years. Some volcanoes are complex mountains that have the characteristics of more than one type of volcano.

Shield Volcanoes

A **shield volcano** looks like the outer surface of a shield lying on the ground with the convex side up (see Geo-inSight on pages 120 and 121). They have low, rounded profiles with gentle slopes ranging from about 2 to 10 degrees; they are composed mostly of mafic flows that had low viscosity, so the flows spread out and formed thin, gently sloping layers. Eruptions from shield volcanoes, sometimes called *Hawaiian-type eruptions*, are quiet compared to those of volcanoes such as Mount St. Helens. Lava most commonly rises to the surface with little explosive activity, and poses little danger to humans. Lava fountains, some up to 400 m high, contribute some pyroclastic materials to shield volcanoes, but otherwise they are composed largely of basalt lava flows.

Although eruptions of shield volcanoes tend to be rather quiet, some of the Hawaiian volcanoes have, on occasion, produced sizable explosions when groundwater instantly vaporizes as it comes in contact with magma. One such explosion in 1790 killed about 80 warriors in a party headed by Chief Keoua, who was leading them across the summit of Kilauea volcano.

Kilauea volcano is impressive because it has been erupting continuously since January 3, 1983, making it the longest recorded eruption. During these 25 years, more than 2.5 km^3 of molten rock has flowed out at the surface, much of it reaching the sea and forming 2.2 km^2 of new property on the island of Hawaii. Unfortunately, lava flows from Kilauea have also destroyed about 200 homes and caused some $61 million in damages.

Shield volcanoes are most common in the ocean basins, but some are also present on the continents—in East Africa, for instance. The island of Hawaii is made up of five huge shield volcanoes, two of which, Kilauea and Mauna Loa, are active much of the time. Mauna Loa is nearly 100 km across

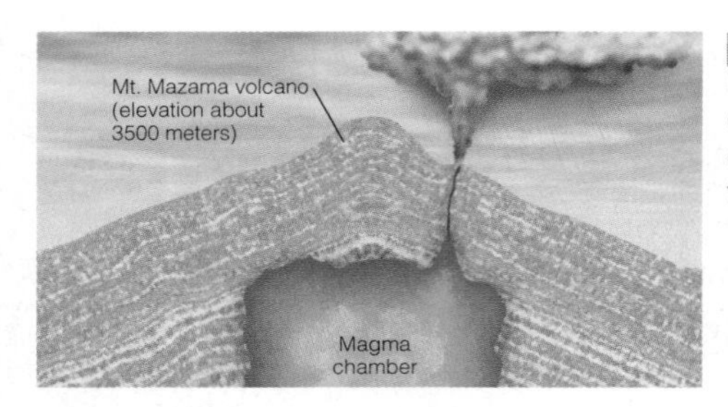

a Eruption begins as huge quantities of ash are ejected from the volcano.

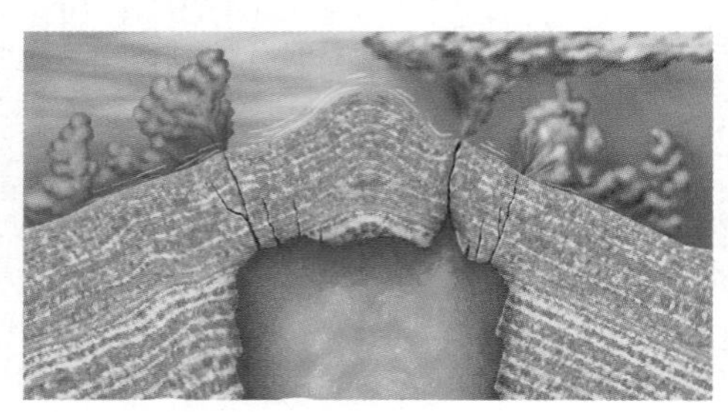

b The eruption continues as more ash and pumice are ejected into the air and pyroclastic flows move down the flanks of the mountain.

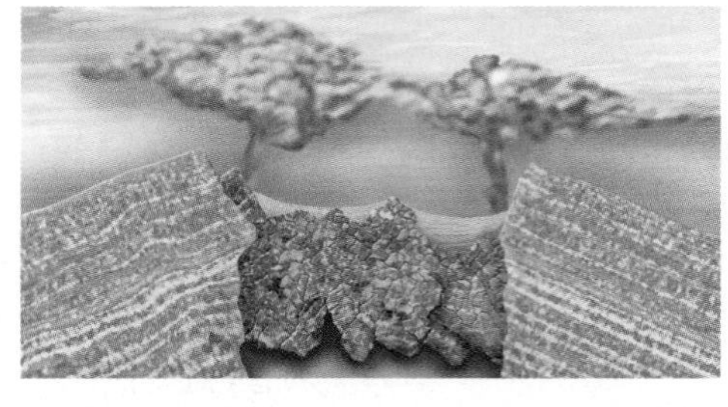

c The collapse of the summit into the partially drained magma chamber forms a huge caldera.

d Postcaldera eruptions partly cover the caldera floor, and the small cinder cone called Wizard Island forms.

e View from the rim of Crater Lake showing Wizard Island. The lake is 594 m deep, making it the second deepest in North America.

Figure 5.9 The Origin of Crater Lake, Oregon Remember, Crater Lake is actually a caldera that formed by partial draining of a magma chamber.

its base and stands more than 9.5 km above the surrounding seafloor; it has a volume estimated at 50,000 km^3, making it the world's largest volcano.

Cinder Cones

Small, steep-sided **cinder cones** made up of particles resembling cinders form when pyroclastic materials accumulate around a vent from which they erupted (see Geo-inSight on pages 120 and 121). Cinder cones are small, rarely exceeding 400 m high, with slope angles up to 33 degrees, depending on the angle that can be maintained by the angular pyroclastic materials. Many of these small volcanoes have a large, bowl-shaped crater, and if they issue any lava flows, they usually break through the base or lower flanks of the mountains. Although all cinder cones are conical, their symmetry varies from those that are almost perfectly symmetrical to those that formed when prevailing winds caused pyroclastic materials to build up higher on the downwind side of the vent.

Many cinder cones form on the flanks or within the calderas of larger volcanoes and represent the final stages of activity, particularly in areas of basaltic volcanism. Wizard Island in Crater Lake, Oregon, is a small cinder cone that formed after the summit of Mount Mazama collapsed to form a caldera (Figure 5.9). Cinder cones are common in the southern Rocky Mountain states, particularly New Mexico and Arizona, and many others are in California, Oregon, and Washington.

In 1973, on the Icelandic island of Heimaey, the town of Vestmannaeyjar was threatened by a new cinder cone. The initial eruption began on January 23, and within two days, a cinder cone, later named Eldfell, rose to about 100 m above the surrounding area (Figure 5.10). Pyroclastic materials from the volcano buried parts of the town, and by February, a massive aa lava flow was advancing toward the town. The flow's leading edge ranged from 10 to 20 m thick, and its central part was as much as 100 m thick. The residents of Vestmannaeyjar sprayed the leading edge of the flow with seawater in an effort to divert it from the town. The flow was in fact diverted, but how effective the efforts of the townspeople were is not clear; they may have been simply lucky.

Solarfilm/GeoScience Features

Figure 5.10 Eldfell, a Cinder Cone in Iceland Eldfell began erupting in 1973 and in two days grew to 100 m high. The steam visible on the left side of the image resulted from aa lava flowing into the sea. Another cinder cone, known as Helgafel, is also visible.

Composite Volcanoes (Stratovolcanoes)

Pyroclastic layers, as well as lava flows, both of intermediate composition, are found in **composite volcanoes**, which are also called *stratovolcanoes* (see Geo-inSight on pages 120 and 121). As the lava flows cool, they typically form andesite; recall that intermediate lava flows are more viscous than mafic ones. Geologists use the term **lahar** for volcanic mudflows, which are also common on composite volcanoes. A lahar may form when rain falls on unconsolidated pyroclastic materials and creates a muddy slurry that moves downslope (Figure 5.11). On November 13, 1985, a minor eruption of Nevado del Ruiz in Colombia melted snow and ice on the volcano, causing lahars that killed approximately 23,000 people (Table 5.1).

Composite volcanoes differ from shield volcanoes and cinder cones in composition, and their overall shape differs, too. Remember that shield volcanoes have very low slopes, whereas cinder cones are small, steep-sided, conical mountains. In contrast, composite volcanoes are steep-sided near their summits, perhaps as much as 30 degrees, but the slope decreases toward the base, where it may be no more than 5 degrees. Mayon volcano in the Philippines is one of the most nearly symmetrical composite volcanoes anywhere. It erupted in 1999 for the 13th time during the 1900s.

When most people think of volcanoes, they picture the graceful profiles of composite volcanoes, which are the typical large volcanoes found on the continents and island arcs. And some of these volcanoes are indeed large; Mount Shasta in northern California is made up of about 350 km^3 of material and measures 20 km across its base. Mount Pinatubo in the Philippines erupted violently on June 15, 1991. Huge quantities of gas and an estimated 3–5 km^3 of ash were discharged into the atmosphere, making this the world's largest eruption since 1912. Fortunately, warnings of an impending eruption were heeded, and 200,000 people were evacuated from around the volcano. Nevertheless, the eruption was responsible for 722 deaths (Table 5.1).

Figure 5.11 Lahars or volcanic mud flows are common on composite volcanoes.

U. S. Dept. of Interior/USGS/David Johnson

a Homes partly buried by a lahar on June 15, 1991, following the eruption of Mount Pinatubo in the Philippines.

Darrell G. Herd/USGS

b Aerial view of Armero, Colombia, where at least 23,000 people were killed by lahars that inundated the area in 1985.

Lava Domes

Most volcanoes can be classified as shield volcanoes, cinder cones, or composite volcanoes. Less common are **lava domes**, also known as *volcanic domes* and *plug domes*, which are steep-sided, bulbous mountains that form when viscous felsic magma, and occasionally intermediate magma, is forced toward the surface (see Geo-inSight on pages 120 and 121). Because felsic magma is so viscous, it moves upward very slowly and only when the pressure from below is great.

Beginning in 1980, a number of lava domes were emplaced in the crater of Mount St. Helens in Washington; most of these were destroyed during subsequent eruptions. Since 1983, Mount St. Helens has been characterized by sporadic dome growth, and renewed eruptions in 2004. In June 1991, a lava dome in Japan's Unzen volcano collapsed under its own weight, causing a flow of debris and hot ash that killed 43 people in a nearby town.

Lava dome eruptions are some of the most violent and destructive. In 1902, viscous magma accumulated beneath the summit of Mount Pelée on the island of Martinique. Eventually, the pressure increased until the side of the mountain blew out in a tremendous explosion, ejecting a mobile, dense cloud of pyroclastic materials and a glowing cloud of gases and dust called a **nuée ardente** (French for "glowing cloud"). The pyroclastic

Figure 5.12 Nuée Ardente

a St. Pierre, Martinique, after it was destroyed by a nuée ardente from Mount Pelée in 1902. Only 2 of the city's 28,000 inhabitants survived.

b An April 1986 pyroclastic flow rushing down Augustine volcano in Alaska. This flow is similar to the one that wiped out St. Pierre.

flow followed a valley to the sea, but the nuée ardente jumped a ridge and engulfed the city of St. Pierre (Figure 5.12).

A tremendous blast hit St. Pierre, leveling buildings; hurling boulders, trees, and pieces of masonry down the streets; and moving a 3-ton statue 16 m. Accompanying the blast was a swirling cloud of incandescent ash and gases with an internal temperature of 700°C that incinerated everything in its path. The nuée ardente passed through St. Pierre in two or three minutes, only to be followed by a firestorm as combustible materials burned and casks of rum exploded. But by then most of the 28,000 residents of the city were already dead. In fact, in the area covered by the nuée ardente, only two survived!* One survivor was on the outer edge of the nuée ardente, but even there, he was terribly burned and his family and neighbors were all killed. The other survivor, a stevedore incarcerated the night before for disorderly conduct, was in a windowless cell partly below ground level. He remained in his cell badly burned for four days after the eruption until rescue workers heard his cries for help. He later became an attraction in the Barnum and Bailey Circus where he was advertised as "The only living object that survived in the 'Silent City of Death' where 40,000 beings were suffocated, burned or buried by one belching blast of Mont Pelée's terrible volcanic eruption."**

Supervolcano Eruptions

Geologists have no formal definition for *supervolcano eruptions*, but we can take it to mean an explosive eruption of hundreds of cubic kilometers of pyroclastic materials and the origin of a huge caldera. No supervolcano eruptions have occurred in historic times, but geologists know of several that took place during the past two million years—Long Valley in eastern Califormia, Toba in Indonesia, and Taupo in New Zealand, for example.

On three occasions, supervolcano eruptions followed the accumulation of rhyolitic magma beneath Yellowstone National Park, which is mostly in Wyoming, each yielding a widespread blanket of volcanic ash and pumice and gigantic calderas. We can summarize Yellowstone's volcanic history by noting that supervolcano eruptions took place 2 million years ago, 1.3 million years ago, and 600,000 years ago. Then, between 150,000 and 75,000 years ago, an additional 1000 km^3 of pyroclastic materials were erupted within the Yellowstone caldera (Figure 5.13).

Geologists think that these huge eruptions were caused by a rising mantle plume, a cylindrical mass of magma probably of rhyolitic composition. Because this type of magma is viscous, it triggers explosive eruptions when it nears the surface. Personnel from the U.S. Geological Survey (USGS) and the University of Utah continue to monitor the Yellowstone area for any signs of renewed activity.

---

* Although reports commonly claim that only two people survived the eruption, at least 69 and possibly as many as 111 people survived beyond the extreme margins of the nuée ardente and on ships in the harbor. Many, however, were badly injured.

**Quoted from A. Scarth, Vulcan's Fury: Man Against the Volcano (New Haven, CT: Yale University Press, 1999), p. 177.

Figure 5.13 **The Yellowstone Tuff** The walls of the Grand Canyon of the Yellowstone River are made up of the hydrothermally altered Yellowstone Tuff that partly fills the Yellowstone caldera.

OTHER VOLCANIC LANDFORMS

In some areas of volcanism, volcanoes fail to develop at all. For instance, during *fissure eruptions*, fluid lava pours out and simply builds up rather flat-lying areas, whereas huge explosive eruptions might yield *pyroclastic sheet deposits*, which, as their name implies, have a sheetlike geometry.

Fissure Eruptions and Basalt Plateaus

Rather then erupting from central vents, the lava flows making up **basalt plateaus** issue from long cracks or fissures during **fissure eruptions**. The lava is so fluid (has such low viscosity) that is spreads out and covers vast areas. A good example is the Columbia River basalt in eastern Washington and parts of Oregon and Idaho. This huge accumulation of 17-to-6-million-year-old overlapping lava flows covers about 164,000 km^2 (Figure 5.14a and b), and has an aggregate thickness of more than 1000 m. And some individual flows are enormous—the Roza flow advanced along a front about 100 km wide and covered 40,000 km^2. Geologists have identified 300 huge flows here, one of which flowed 600 km from its source.

Similar accumulations of vast, overlapping lava flows are also found in the Snake River Plain in Idaho (Figure 5.14a and c). These flows are 5.0 to 1.6 million years old, and they represent a style of eruption between fissure eruptions and

USGS/NPS

a Relief map of the northwestern United States showing the location of the Columbia River basalt and the Snake River Plain.

Frank Kujawa/University of Central Florida/GeoPhoto

b About 20 lava flows of the Columbia River basalt are exposed in the canyon of the Grand Ronde River in Washington.

Sue Monroe

c Basalt lava flows of the Snake River Plain near Twin Falls,

Figure 5.14 **Basalt Plateaus** Basalt plateaus are vast areas of overlapping lava flows that issued from long fissures. Fissure eruptions take place today in Iceland; however, in the past, they formed basalt plateaus in various areas.

Geo-inSight Types of Volcanoes

All volcanoes are structures resulting from the eruption of lava and pyroclastic materials, but they are all unique in their history of eruptions and development. Nevertheless, most are conveniently classified as one of the types shown here: shield, cinder cone, composite, and lava dome. There are also places where eruptions of very fluid lava take place along fissures and volcanoes do not develop.

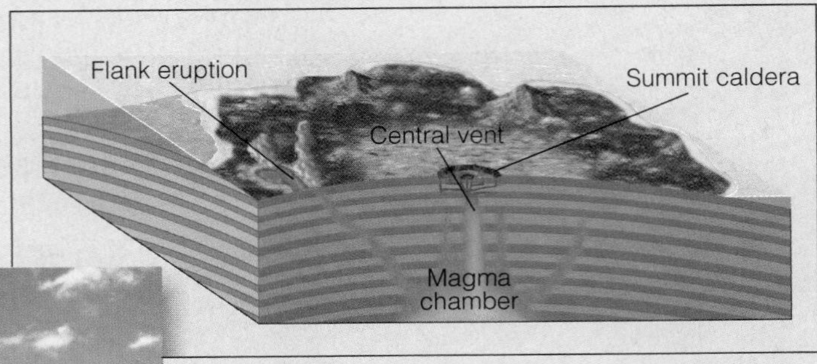

▲ **1.** Shield volcanoes consist of numerous thin basalt lava flows that build up mountains with slopes rarely exceeding 10 degrees.

James S. Monroe

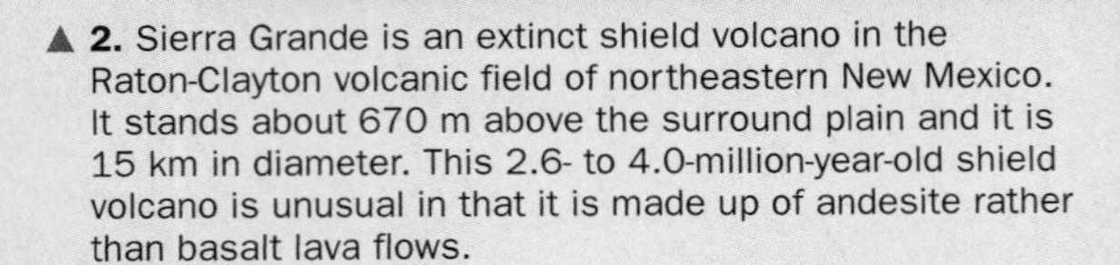

▲ **2.** Sierra Grande is an extinct shield volcano in the Raton-Clayton volcanic field of northeastern New Mexico. It stands about 670 m above the surround plain and it is 15 km in diameter. This 2.6- to 4.0-million-year-old shield volcano is unusual in that it is made up of andesite rather than basalt lava flows.

Robert Tilling/USGS

▲ **3.** View of Mauna Loa, an active shield volcano on Hawaii, with its upper 1.5 km covered by snow. Mauna Loa is the largest mountain on Earth; it measures about 100 km across its base, stands more than 9.5 km above the seafloor, and is made up of an estimated 50,000 km^3 of material.

James S. Monroe

▲ **4.** A cinder cone in the Big Pine volcanic field near Big Pine, California. Notice the large crater and the eroded cinder cone in the right foreground. The Sierra Nevada is in the background.

James S. Monroe

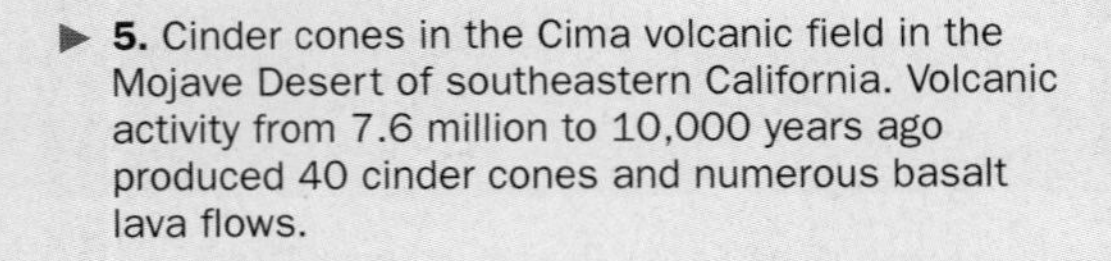

▶ **5.** Cinder cones in the Cima volcanic field in the Mojave Desert of southeastern California. Volcanic activity from 7.6 million to 10,000 years ago produced 40 cinder cones and numerous basalt lava flows.

► **6.** Composite volcanoes, or stratovolcanoes, are composed mostly of lava flows and pyroclastic materials of intermediate composition, although volcanic mudflow deposits are also common.

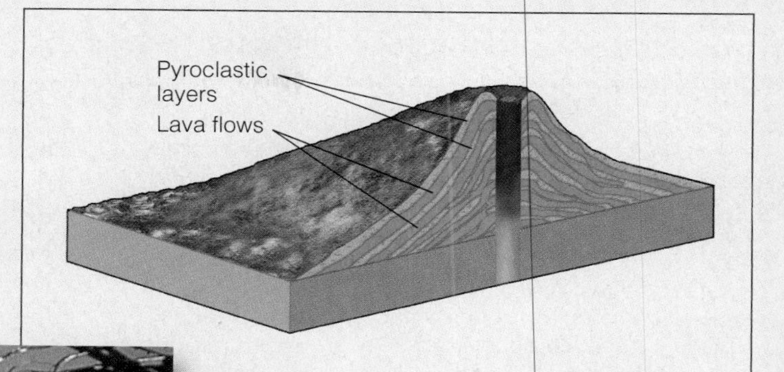

▼ **7.** Mayon volcano in the Philippines, a nearly symmetrical composite volcano that last erupted during 1999.

▼ **8.** Two views of Mount Shasta, a huge composite volcano in northern California. Mount Shasta is about 24 km across its base and rises more than 3400 m above its surroundings.

Sue Monroe

► **9.** This view of Mount Shasta from the north shows a cone known as Shastina on the flank of the larger mountain.

▼ **10.** Chaos Crags in the distance are made up of at least four lava domes that formed less than 1200 years ago in Lassen Volcanic National Park in California. The debris in the foreground, called Chaos Jumbles, formed when parts of the domes collapsed.

▲ **11.** This steep-sided lava dome lies atop Novarupta in Katmai National Park and Preserve in Alaska.

What Would You Do?

You are an enthusiast of natural history and would like to share your interests with your family. Accordingly, you plan a vacation to see some of the volcanic features in U.S. national parks and monuments. Let's assume your planned route will take you through Wyoming, Idaho, Washington, Oregon, and California. What specific areas might you visit, and what kinds of volcanic features would you see in these areas? What other parts of the United States might you visit in the future to see additional evidence for volcanism?

those of shield volcanoes. In fact, there are small, low shields, as well as fissure flows, in the Snake River Plain.

Currently, fissure eruptions occur only in Iceland. Iceland has a number of volcanoes, but the bulk of the island is composed of basalt lava flows that issued from fissures. In fact, about half of the lava erupted during historic time in Iceland came from two fissure eruptions, one in A.D. 930 and the other in 1783. The 1783 eruption from Laki fissure, which is more than 30 km long, accounted for lava that covered 560 km^2 and, in one place, filled a valley to a depth of about 200 m.

Pyroclastic Sheet Deposits

Geologists have long been aware of vast areas covered by felsic volcanic rocks a few meters to hundreds of meters thick. It seemed improbable that these could be vast lava flows, but it seemed equally unlikely that they were ash fall deposits. Based on observations of historic pyroclastic flows, such as the nuée ardente erupted by Mount Pelée in 1902, it seems that these ancient rocks originated as pyroclastic flows—hence the name **pyroclastic sheet deposits**.

They cover far greater areas than any observed during historic time, however, and apparently erupted from long fissures rather than from a central vent. The pyroclastic materials of many of these flows were so hot that they fused together to form *welded tuff.*

Geologists now think that major pyroclastic flows issue from fissures formed during the origin of calderas. For instance, pyroclastic flows erupted during the formation of a large caldera now occupied by Crater Lake in Oregon (Figure 5.9), and in the Yellowstone caldera in Wyoming.

Similarly, the Bishop Tuff of eastern California erupted shortly before the formation of the Long Valley caldera. Interestingly, earthquake activity in the Long Valley caldera and nearby areas beginning in 1978 may indicate that magma is moving upward beneath part of the caldera. Thus, the possibility of future eruptions in that area cannot be discounted.

DISTRIBUTION OF VOLCANOES

Most of the world's active volcanoes are in well-defined zones or belts rather than randomly distributed. The **circum-Pacific belt**, popularly called the Ring of Fire, has more than 60% of all active volcanoes. It includes volcanoes in the Andes of South America; the volcanoes of Central America, Mexico, and the Cascade Range of North America; as well as the Alaskan volcanoes and those in Japan, the Philippines, Indonesia, and New Zealand (Figure 5.15). Also in the

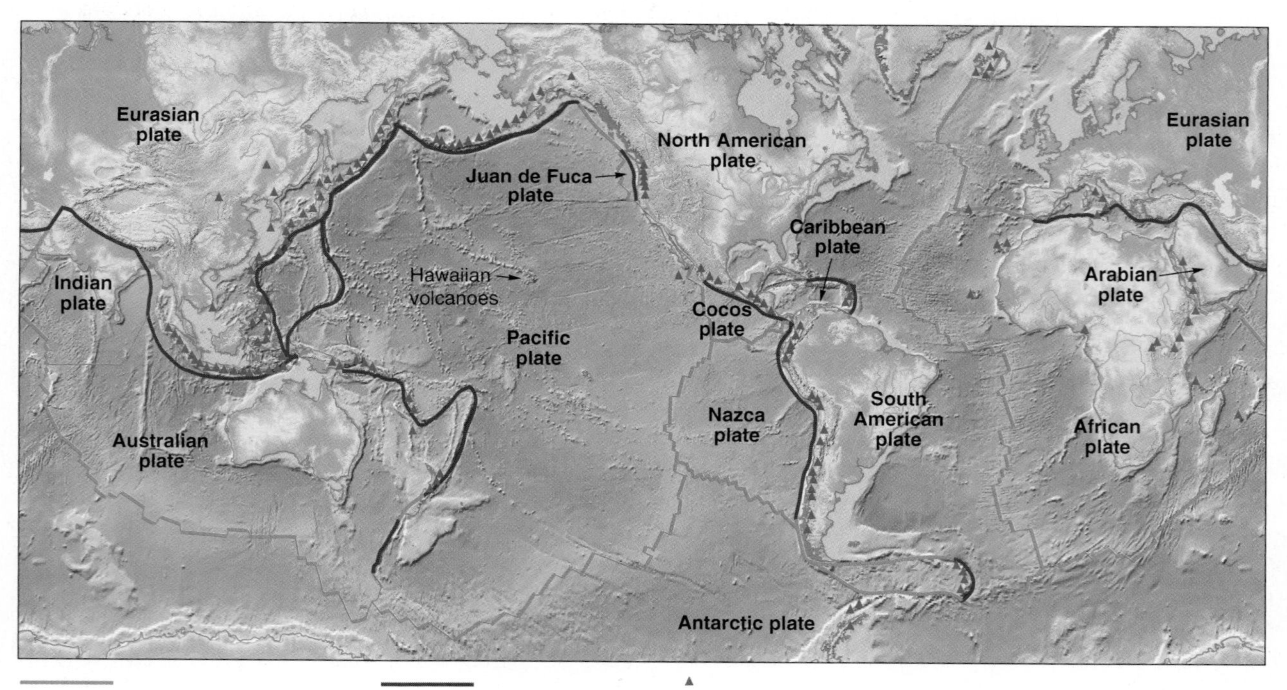

Figure 5.15 Volcanoes of the World Most volcanoes are at or near convergent and divergent plate boundaries. The two major volcano belts are the circum-Pacific belt, commonly known as the Ring of Fire, with about 60% of all active volcanoes, and the Mediterranean belt, with 20% of active volcanoes. Most of the rest lie near the mid-oceanic ridges.

circum-Pacific belt are the southernmost active volcanoes at Mount Erebus in Antarctica and a large caldera at Deception Island that erupted most recently during 1970.

The second area of active volcanism is the **Mediterranean belt** (Figure 5.15). About 20% of all active volcanism takes place in this belt, where the famous Italian volcanoes such as Mounts Etna and Vesuvius and the Greek volcano Santorini are found (see Geo-Focus on pages 124 and 125). Mount Etna has issued lava flows 190 times since 1500 B.C., when activity was first recorded. A particularly violent eruption of Santorini in 1390 B.C. might be the basis for the myth about the lost continent of Atlantis, and, in A.D. 79, an eruption of Mount Vesuvius destroyed Pompeii and other nearby cities (see the Introduction).

Nearly all the remaining active volcanoes are at or near mid-oceanic ridges or the extensions of these ridges onto land (Figure 5.15). These include the East Pacific Rise and the longest of all mid-oceanic ridges, the Mid-Atlantic Ridge. The latter is located near the center of the Atlantic Ocean basin, accounting for the volcanism in Iceland and elsewhere. It continues around the southern tip of Africa, where it connects with the Indian Ridge. Branches of the Indian Ridge extend into the Red Sea and East Africa, where such volcanoes as Kilamanjaro in Tanzania, Nyiragongo in Zaire, and Erta Ale in Ethiopia with its continuously active lava lake are found.

NORTH AMERICA'S ACTIVE VOLCANOES

Part of the circum-Pacific belt includes volcanoes in the Pacific Northwest, as well as those in Alaska. Both of these areas of volcanism are at convergent plate boundaries. Of the 80 or so potentially active volcanoes in Alaska, at least half have erupted since 1760. The other active North American volcanoes are in the Cascade Range in the Pacific Northwest where the Juan de Fuca plate is subducted beneath North America. Many of these volcanoes have been historically active, although since 1900, only Lassen Peak in California and Mount St. Helens in Washington have erupted.

Alaska's Volcanoes

Many of the volcanoes in mainland Alaska and in the Aleutian Islands (Figure 5.15) are composite volcanoes, some with huge calderas. Mount Spurr has erupted explosively at least 35 times during the last 5000 years, but its eruptions pale by comparison with that of Novarupta in 1912. Its defining event was the June 1912 eruption, the largest in the world since the late 1800s. At least 15 km^3 of mostly pyroclastic materials erupted during about 60 hours.

When the eruption was over, 120 km^2 of land was buried beneath pyroclastic deposits as deep as 213 m. In fact, the deposits filled the Valley of Ten Thousand Smokes—so named because of the hundreds of fumaroles where gases vented through the hot deposits for as long as 15 years following the eruption. Fortunately, the eruption took place in a remote area so there were no injures or fatalities, but enough ash, gases, and pumice were ejected that, for several days, the sky was darkened over much of the Northern Hemisphere.

By the time you read this chapter, several more volcanoes in Alaska will have erupted as the Pacific Plate moves relentlessly northward only to be subducted at the Aleutian Trench. The Alaska Volcanoes Observatory in Anchorage, Alaska, continues to monitor these volcanoes and issue warnings about potential eruptions.

The Cascade Range

The **Cascade Range** (◗ Figure 5.16) stretches from Lassen Peak in northern California north through Oregon and Washington to Meager Mountain in British Columbia, Canada, which erupted 2350 years ago. Most of the large volcanoes in the range are composite volcanoes, but Lassen Peak in California is the world's largest lava dome. Actually, it is a rather small volcano that developed 27,000 years ago on the flank of a much larger, deeply eroded composite volcano. It erupted from 1914 to 1917, but has since been quiet except for ongoing hydrothermal activity (Figure 5.16b).

Two large shield volcanoes lie just to the east of the main Cascade Range volcanoes—Medicine Lake Volcano in California and Newberry Volcano in Oregon. Distinctive features at Newberry Volcano are a 1600-year-old obsidian flow and casts of trees that formed when lava flowed around them and solidified. Cinder cones are common throughout the range (Figure 5.9d).

What was once a nearly symmetrical composite volcano changed markedly on May 6, 1980, when Mount St. Helens in Washington erupted explosively, killing 63 people and leveling some 600 km^2 of forest (Figure 5.16c). A huge lateral blast caused much of the damage and fatalities, but snow and ice on the volcano melted and pyroclastic materials displaced water in lakes and rivers, causing lahars and extensive flooding.

Mount St. Helens's renewed activity, beginning in late September 2004, has resulted in dome growth and small steam and ash explosions. Scientists at the Cascades Volcano Observatory in Vancouver, Washington, issued a low-level alert for an eruption and continue to monitor the volcano.

Several of the Cascade Range volcanoes will almost certainly erupt again, but the most dangerous is probably Mount Rainier in Washington. Rather than lava flows or even a colossal explosion, the greatest danger from Mount Rainier is volcanic mudflows or huge debris flows. Of the 60 large flows that have occurred during the last 100,000 years, the largest, consisting of 4 km^3 of debris, covered an area now occupied by more than 120,000 people. Indeed, in August 2001, a sizable debris flow took place on the south side of the mountain, but it caused no injuries or fatalities. No one knows when the next flow will take place, but at least one community has taken the threat seriously enough to formulate an emergency evacuation plan. Unfortunately, the residents would have only 1 or 2 hours to carry out the plan.

PLATE TECTONICS, VOLCANOES, AND PLUTONS

In Chapter 4, we discussed the origin and evolution of magma and concluded that (1) mafic magma is generated beneath spreading ridges, and (2) intermediate and felsic magma forms where an oceanic plate is subducted beneath another oceanic plate or a continental plate. Accordingly,

Geo-Focus The Bronze Age Eruption of Santorini

Crater Lake in Oregon, which formed approximately 7700 years ago, is the best-known caldera in the United States (Figure 5.9), but many others are equally impressive. One that formed recently, geologically speaking, resulted from a Bronze Age eruption of Santorini, an event that figured importantly in Mediterranean history (Figure 1a). Actually, Santorini consists of five islands in that part of the Mediterranean called the Aegean Sea. The islands have a total area of

b These 350-m-high cliffs are part of the caldera wall just west of Fira.

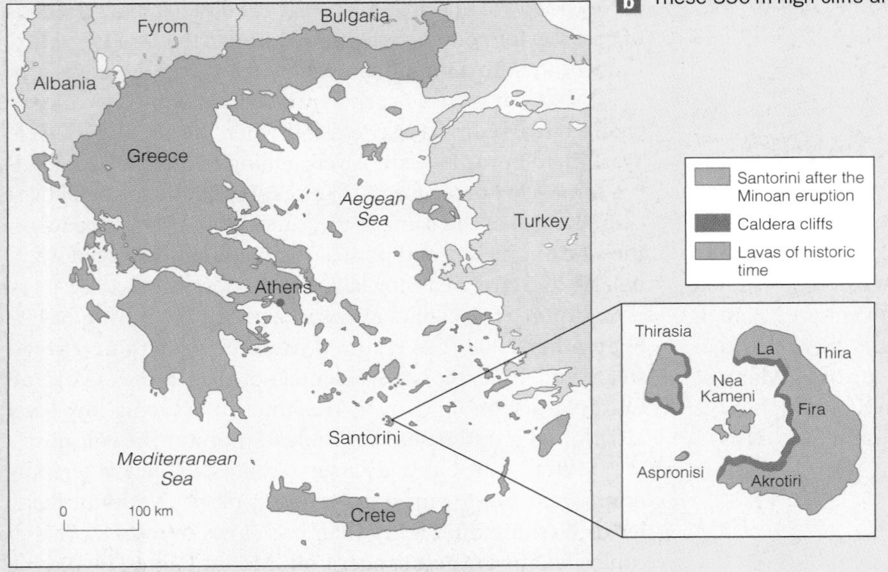

a Map showing Santorini and nearby areas in the Aegean Sea, which is part of the northwestern Mediterranean Sea.

Figure 1 The Minoan eruption Santorini that took place between 1650 and 1596 B.C. formed a large caldera. It may have contributed to the demise of the Minoan culture on Crete and may be the basis for Plato's account of the sinking of Atlantis.

most volcanism and emplacement of plutons take place at or near divergent and convergent plate boundaries.

Igneous Activity at Divergent Plate Boundaries

Much of the mafic magma that originates at spreading ridges is emplaced as vertical dikes and gabbro plutons, thus composing the lower part of the oceanic crust. However, some rises to the surface and issues forth as submarine lava flows and pillow lava (Figure 5.7), which constitutes the upper part of the oceanic crust. Much of this volcanism goes undetected, but researchers in submersibles have seen the results of recent eruptions.

Mafic lava is very fluid, allowing gases to escape easily, and at great depth in the oceans, the water pressure is so great that explosive volcanism is prevented. In short, pyroclastic materials are rare to absent unless, of course, a volcanic center builds up above sea level. Even if this occurs, however, the mafic magma is so fluid that it forms the gently sloping layers found on shield volcanoes.

Excellent examples of divergent plate boundary volcanism are found along the Mid-Atlantic Ridge, particularly where it rises above sea level as in Iceland (Figure 5.15). In November 1963, a new volcanic island, later named Surtsey, rose from the sea south of Iceland. The East Pacific Rise and the Indian Ridge are areas of similar volcanism. A divergent

76 km^2, all of which owe their present configuration to a colossal volcanic eruption that took place about 3600 years ago (estimates range from 1650 B.C. to 1596 B.C.). Indeed, the eruption was responsible for the present islands, the origin of the huge caldera, and it probably accounted for or at least contributed to the demise of the Minoan culture on Crete. Furthermore, some authorities think that the disappearance of much of the original island during this eruption was the basis for Plato's story about Atlantis (see Chapter 9).

As you approach Santorini from the sea, the first impression is snow-covered cliffs in the distance. On closer inspection, though, the "snow" is actually closely spaced white buildings that cover much of the higher parts of the largest island (Figure 1b). Perhaps the most impressive features of Santorini are the near vertical cliffs rising as much as 350 m from the sea. Actually, these cliffs are the walls of a caldera that measures about 6 by 12 km and is as much as 400 m deep. The caldera-forming eruption, "known as the 'Minoan eruption,' ejected into the air 30 cubic kilometers of magma in the form of pumice and volcanic ash. This material buried the island [as much as 50 m deep] and its civilization. . . ."[1] The two small islands within the caldera, where volcanic activity continues, appeared above sea level in 197 B.C. and since then have grown to their present size. The most recent activity occurred in 1950 on the larger of the two islands (Figure 1a).

Santorini volcano began forming two million years ago, and during the last 400,000 years it has erupted at least 100 times, each eruption adding new layers to the island, making it larger. Today, approximately 8000 people live on the islands, and we know from archaeological evidence that several tens of thousands of people resided there before the Minoan eruption when Santorini was larger. However, a year or so before the catastrophic eruption, a devastating earthquake occurred and many people left the island then; perhaps there were signs of an impending eruption by this time.

The fact that the island's residents escaped is indicated by the lack of human and animal skeletons in the ruins of the civilization, the only exception being one pig skeleton. In fact, archaeological excavations, many still in progress, show that the people had time to collect their valuables and tools before evacuating the island. Their destination, however, remains a mystery.

[1]Vougioukalakis, G., Santorini: The Volcano (Institute for the Study and Monitoring of the Santorini Volcano, 1995) p. 7.

plate boundary is also present in Africa as the East African Rift system, which is well known for its volcanoes (Figure 5.15).

Igneous Activity at Convergent Plate Boundaries

Nearly all of the large active volcanoes in both the circum-Pacific and Mediterranean belts are composite volcanoes near the leading edges of overriding plates at convergent plate boundaries (Figure 5.15). The overriding plate, with its chain of volcanoes, may be oceanic as in the case of the Aleutian Islands, or it may be continental as is, for instance, the South American plate with its chain of volcanoes along its western edge.

As we noted, these volcanoes at convergent plate boundaries consist mostly of lava flows and pyroclastic materials of intermediate to felsic composition. Remember that when mafic oceanic crust partially melts, some of the magma generated is emplaced near plate boundaries as plutons and some is erupted to build up composite volcanoes. More viscous magmas, usually of felsic composition, are emplaced as lava domes, thus accounting for the explosive eruptions that typically occur at convergent plate boundaries.

Good examples of volcanism at convergent plate boundaries are the explosive eruptions of Mount Pinatubo and Mayon volcano in the Philippines; both are near a plate boundary beneath which an oceanic plate is subducted.

Figure 5.16 The Cascade Range of the Pacific Northwest

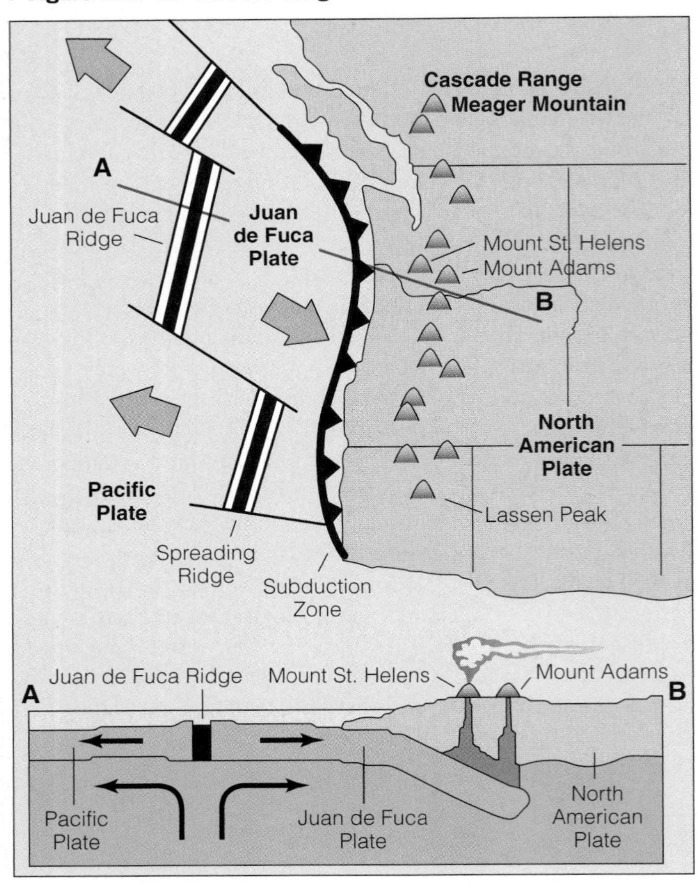

a Plate tectonic setting for the Pacific Northwest. Subduction of the Juan de Fuca plate accounts for ongoing volcanism in the region.

b Lassen Peak in California erupted from 1914 to 1917. This eruption took place in 1915.

Courtesy of Keith Ronnholm

c Mount St. Helens, Washington. The lateral blast on May 18, 1980, took place when a bulge on the volcano's north face collapsed, reducing the pressure on gas-charged magma.

Mount St. Helens, Washington, is similarly situated, but it is on a continental rather than an oceanic plate.

Intraplate Volcanism

Mauna Loa and Kilauea on the island of Hawaii and Loihi just 32 km to the south are within the interior of a rigid plate far from any divergent or convergent plate boundary (Figure 5.15). The magma is derived from the upper mantle, as it is at spreading ridges, and accordingly is mafic, so it builds up shield volcanoes. Loihi is particularly interesting because it represents an early stage in the origin of a new Hawaiian island. It is a submarine volcano that rises more than 3000 m above the adjacent seafloor, but its summit is still about 940 m below sea level.

Even though the Hawaiian volcanoes are not at or near a spreading ridge or a subduction zone, their evolution is nevertheless related to plate movements. Notice in Figure 2.22 that the ages of the rocks that make up the Hawaiian islands increase toward the northwest. Kauai formed 5.6 to 3.8 million years ago, whereas Hawaii began forming less than 1 million years ago, and Loihi began to form even more recently. The islands have formed in succession as the Pacific plate moves continuously over a hot spot now beneath Hawaii and just to the south at Loihi.

VOLCANIC HAZARDS, VOLCANO MONITORING, AND FORECASTING ERUPTIONS

Undoubtedly you suspect that living near an active volcano poses some risk, and of course this assessment is correct. But what exactly are volcanic hazards, is there any way to anticipate eruptions, and what can we do to minimize the dangers of eruptions? We have already mentioned that lava flows, with few exceptions, pose little threat to humans although they may destroy property. Lava flows, nuée ardentes, and volcanic gases are threats during an eruption; however, lahars and landslides may take place even when no eruption has taken place for a long time (Figure 5.17). Certainly, the most vulnerable areas in the United States are Alaska, Hawaii, California, Oregon, and Washington, but some other parts of the West might also experience renewed volcanism.

Figure 5.17 Volcanic Hazards A volcanic hazard is any manifestation of volcanism that poses a threat, including lava flows and, more importantly, volcanic gas, ash, and lahars.

AFP/Getty Images

a This 2002 lava flow in Goma, Democratic Republic of Congo, killed 147 people, mostly by causing gasoline storage tanks to explode.

Sue Monroe

b This sign at Mammoth Mountain volcano in California warns of the potential danger of CO_2 gas, which has killed 170 acres of trees.

Reuters/Corbis

c When Mount Pinatubo in the Philippines erupted on June 15, 1991, this huge cloud of ash and steam formed over the volcano.

What Would You Do?

No one doubts that some of the Cascade Range volcanoes will erupt again, but we don't know when or how large these eruptions will be. A job transfer takes you to a community in Oregon that has several nearby large volcanoes. You have some concerns about future eruptions. What kinds of information would you seek out before buying a home in this area? In addition, as a concerned citizen, can you make any suggestions about what should be done in the case of a large eruption?

How Large Is an Eruption, and How Long Do Eruptions Last?

The most widely used indication of the size of a volcanic eruption is the **volcanic explosivity index (VEI)** (Figure 5.18). Unlike the Richter Magnitude Scale for earthquakes (see Chapter 8), the VEI is only semiquantitative, being based partly on subjective criteria.

The VEI ranges from 0 (gentle) to 8 (cataclysmic) and is based on several aspects of an eruption, such as the volume of material explosively ejected and the height of the eruption plume. However, the volume of lava, fatalities, and property damage are not considered. For instance, the 1985 eruption of Nevado del Ruiz in Colombia killed 23,000 people, yet has a VEI value of only 3. In contrast, the huge eruption (VEI = 6) of Novarupta in Alaska in 1912 caused no fatalities or injuries. Since A.D. 1500, only the 1815 eruption of Tambora had a value of 7; it was both large and deadly (Table 5.1). Nearly 5700 eruptions during the last 10,000 years have been assigned VEI numbers, but none has exceeded 7, and most (62%) were assigned a value of 2.

The duration of eruptions varies considerably. Fully 42% of about 3300 historic eruptions lasted less than one month. About 33% erupted for one to six months, but some 16 volcanoes have been active more or less continuously for more than 20 years. Stromboli and Mount Etna in Italy and Erta Ale in Ethiopia are good examples. For some explosive volcanoes, the time from the onset of their eruptions to the climactic event is weeks or months. A case in point is the colossal explosive eruption of Mount St. Helens on May 18, 1980, that occurred two months after eruptive activity began. Unfortunately, many volcanoes give little or no warning of such large-scale events; of 252 explosive eruptions, 42% erupted most violently during their first day of activity. As one might imagine, predicting eruptions is complicated by those volcanoes that give so little warning of impending activity.

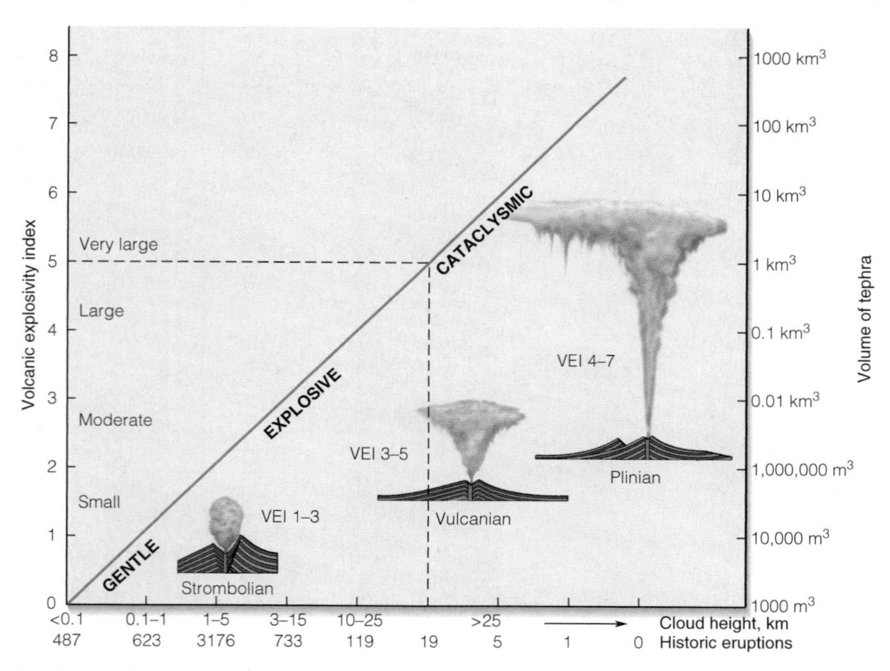

Figure 5.18 The Volcanic Explosivity Index In this example, an eruption with a VEI of 5 has an eruption cloud up to 25 km high and ejects at least 1 km³ of tephra, a collective term for all pyroclastic materials. Geologists characterize eruptions as Hawaiian (nonexplosive), Strombolian, Vulcanian, and Plinian.

Is It Possible to Forecast Eruptions?

Only a few of Earth's potentially dangerous volcanoes are monitored, including some in Japan, Italy, Russia, New Zealand, and the United States. Volcano monitoring involves recording and analyzing physical and chemical changes at volcanoes (Figure 5.19). Tiltmeters detect changes in the slopes of a volcano as it inflates when magma rises beneath it, and a geodimeter uses a laser beam to measure horizontal distances, which change as a volcano inflates. Geologists also monitor gas emissions, changes in groundwater level and temperature, hot springs activity, and changes in the local magnetic and electrical fields. Even the accumulating snow and ice, if any, are evaluated to anticipate hazards from floods should an eruption take place.

Of critical importance in volcano monitoring and warning of an imminent eruption is the detection of **volcanic tremor**, continuous ground motion that lasts for minutes to hours as opposed to the sudden, sharp jolts produced by most earthquakes. Volcanic tremor, also known as *harmonic tremor*, indicates that magma is moving beneath the surface.

To more fully anticipate the future activity of a volcano, its eruptive history must be known. Accordingly, geologists study the record of past eruptions preserved in rocks. Detailed studies before 1980 indicated that Mount St. Helens, Washington, had erupted explosively 14 or 15 times during the last 4500 years, so geologists concluded that it was one of the most likely Cascade Range volcanoes to erupt again. In fact, maps they prepared showing areas in which damage from an eruption could be expected were helpful in determining which areas should have restricted access and evacuations once an eruption did take place.

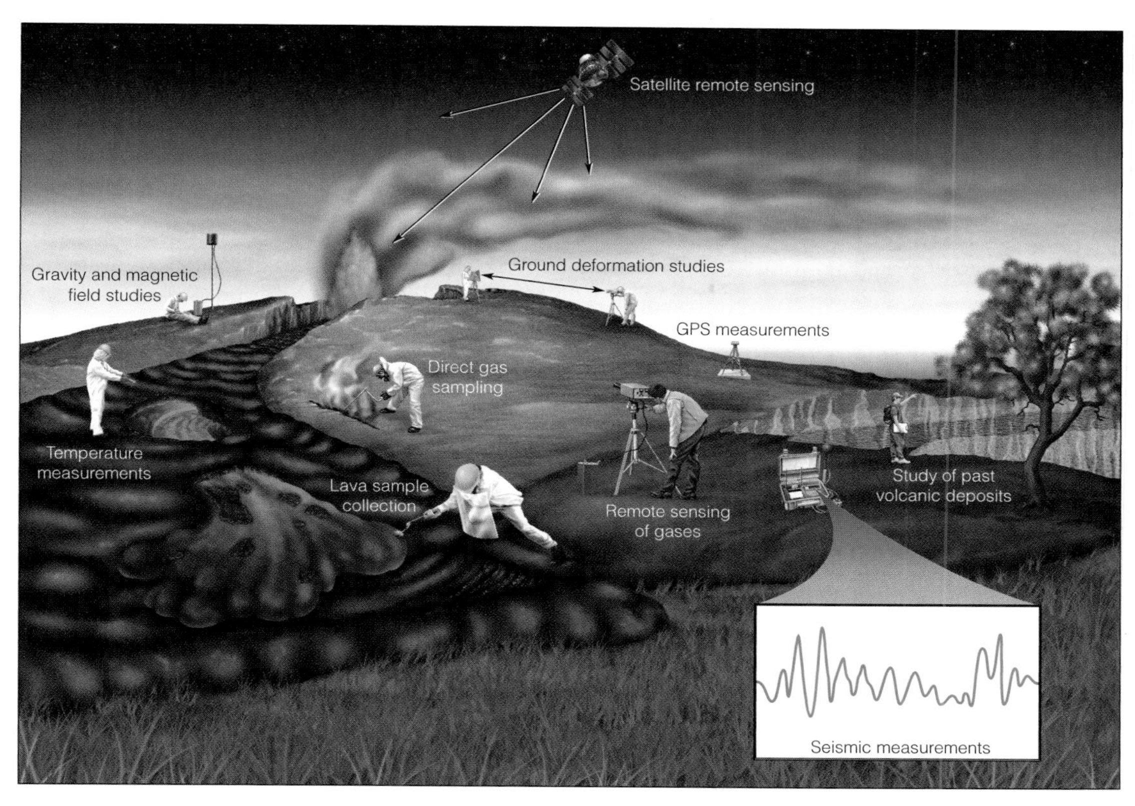

Figure 5.19 Volcanic Monitoring Some important techniques used to monitor volcanoes.

Geologists successfully gave timely warnings of impending eruptions of Mount St. Helens in Washington and Mount Pinatubo in the Philippines, but in both cases, the climactic eruptions were preceded by eruptive activity of lesser intensity. In some cases, however, the warning signs are much more subtle and difficult to interpret. Numerous small earthquakes and other warning signs indicated to USGS geologists that magma was moving beneath the surface of the Long Valley caldera in eastern California, so in 1987, they issued a low-level warning, and then nothing happened.

Volcanic activity in the Long Valley caldera occurred as recently as 250 years ago, and there is every reason to think that it will occur again. Unfortunately, the local populace was largely unaware of the geologic history of the region, the USGS did a poor job in communicating its concerns, and premature news releases caused more concern than was justified. In any case, local residents were outraged because the warnings caused a decrease in tourism (Mammoth Mountain on the margins of the caldera is the second largest ski area in the country) and property values plummeted. Monitoring continues in the Long Valley caldera, and the signs of renewed volcanism, including earthquake swarms, trees being killed by carbon dioxide gas apparently emanating from magma, and hot spring activity, cannot be ignored. In April 2006, three members of a ski patrol were killed by carbon dioxide gas that accumulated in a low area.

Geo-Recap

Chapter Summary

- Volcanism encompasses those processes by which magma rises to the surface as lava flows and pyroclastic materials and gases are released into the atmosphere.
- Gases make up only a few percent by weight of magma. Most is water vapor, but sulfur gases may have far-reaching climatic effects.
- Aa lava flows have surfaces of jagged, angular blocks, whereas the surfaces of pahoehoe flows are smoothly wrinkled.
- Several other features of lava flows are spatter cones, pressure ridges, lava tubes, and columnar joints. Lava erupted under water typically forms bulbous masses known as pillow lava.

- Volcanoes are found in various shapes and sizes, but all form where lava and pyroclastic materials are erupted from a vent.
- The summits of volcanoes have either a crater or a much larger caldera. Most calderas form following voluminous eruptions, and the volcanic peak collapses into a partially drained magma chamber.
- Shield volcanoes have low, rounded profiles and are composed mostly of mafic flows that cool and form basalt. Small, steep-sided cinder cones form around a vent where pyroclastic materials erupt and accumulate. Composite volcanoes are made up of lava flows and pyroclastic materials of intermediate composition and volcanic mudflows.
- Viscous bulbous masses of lava, generally of felsic composition, form lava domes, which are dangerous because they erupt explosively.
- Fluid mafic lava from fissure eruptions spreads over large areas to form a basalt plateau.
- Pyroclastic sheet deposits result from huge eruptions of ash and other pyroclastic materials, particularly when calderas form.
- Geologists have devised a volcanic explosivity index (VEI) to give a semiquantitative measure of the size of an eruption. Volume of material erupted and height of the eruption plume are criteria used to determine the VEI; fatalities and property damage are not considered.
- Approximately 80% of all volcanic eruptions take place in the circum-Pacific and the Mediterranean belts, mostly at convergent plate boundaries. Most of the rest of the eruptions occur along mid-oceanic ridges or their extensions onto land.
- The two active volcanoes on the island of Hawaii and one just to the south lie above a hot spot over which the Pacific plate moves.
- To effectively monitor volcanoes, geologists evaluate several physical and chemical aspects of volcanic regions. Of particular importance in monitoring volcanoes and forecasting eruptions is detecting volcanic tremor and determining the eruptive history of a volcano.

Important Terms

aa (p. 112)
basalt plateau (p. 119)
caldera (p. 115)
Cascade Range (p. 123)
cinder cone (p. 116)
circum-Pacific belt (p. 122)
columnar jointing (p. 112)
composite volcano (stratovolcano) (p. 117)
crater (p. 114)
fissure eruption (p. 119)
lahar (p. 117)
lava dome (p. 117)
lava tube (p. 111)
Mediterranean belt (p. 123)
nuée ardente (p. 117)
pahoehoe (p. 112)
pillow lava (p. 112)
pyroclastic sheet deposit (p. 122)
shield volcano (p. 115)
volcanic ash (p. 112)
volcanic explosivity index (VEI) (p. 128)
volcanic tremor (p. 128)
volcanism (p. 108)
volcano (p. 114)

Review Questions

1. Water-saturated flows of volcanic debris called _____ are common on _____ volcanoes.
   a. _____ nuée ardentes/shield;
   b. _____ lava flows/lava dome;
   c. _____ lahars/composite;
   d. _____ pillow lava/submarine;
   e. _____ fissure eruptions/cinder cones.
2. The shaking that takes place when magma moves beneath the surface is called
   a. _____ volcanic tremor;
   b. _____ columnar joints;
   c. _____ pyroclastic intrusion;
   d. _____ basalt plateau;
   e. _____ volcanic explosivity index.
3. A lava flow made up of angular blocks and fragments is a/an _____ flow.
   a. _____ lava tube;
   b. _____ aa;
   c. _____ pahoehoe;
   d. _____ lahar;
   e. _____ caldera.
4. The chain of volcanoes stretching from northern California into British Columbia, Canada, is known as the
   a. _____ circum-Pacific Mountains;
   b. _____ mid-oceanic ridge volcanic chain;
   c. _____ Rocky Mountains;
   d. _____ Cascade Range;
   e. _____ Columbia River basalt.

5. An incandescent cloud of gas and particles erupted from a volcano is a
   a. ______ spatter cone;
   b. ______ lapilli;
   c. ______ nuée ardente;
   d. ______ caldera;
   e. ______ volcanic tremor.
6. Pillow lava forms when
   a. ______ thick layers of pyroclastic layers accumulate;
   b. ______ lava erupts under water;
   c. ______ globs of lava stick together on a lava flow's surface;
   d. ______ pressure within a flow causes its surface to buckle;
   e. ______ a volcano's summit collapses.
7. Which one of the following is one of Earth's main volcanic belts?
   a. ______ Appalachian belt;
   b. ______ East African belt;
   c. ______ Indian Ridge belt;
   d. ______ East Pacific belt;
   e. ______ Mediterranean belt.
8. A small, steep-sided volcano made up of pyroclastic materials is a
   a. ______ cinder cone;
   b. ______ lava dome;
   c. ______ shield volcano;
   d. ______ caldera;
   e. ______ nuée ardente.
9. Volcanoes emit several gases, but the most common one is
   a. ______ carbon dioxide;
   b. ______ fluorine;
   c. ______ hydrogen sulfide;
   d. ______ water vapor;
   e. ______ methane.
10. The Hawaiian volcanoes are made up mostly of lava that cools to form
    a. ______ rhyolite;
    b. ______ granite;
    c. ______ basalt;
    d. ______ welded tuff;
    e. ______ andesite.
11. Why are eruptions of mafic magma rather quiet, whereas eruptions of felsic magma are commonly explosive?
12. How do columnar joints and spatter cones form? Where are good places to see each?
13. Why do shield volcanoes have such gentle slopes, whereas cinder cones have very steep slopes?
14. What does volcanic tremor indicate, and how does it differ from the shaking caused by most earthquakes?
15. Suppose that you find rocks on land that consist of layers of pillow lava overlain by deep-sea sedimentary rocks. Where and how did the pillow lava form, and what type of rock would you expect to find beneath the pillow lava?
16. What criteria are used to assign a volcanic explosivity index (VEI) value to an eruption? Why do you think that the number of fatalities and property damage are not considered when assigning a value?
17. Explain what types of volcanoes form at divergent and convergent plate boundaries. What accounts for the difference(s)?
18. What kinds of information do geologists evaluate when they monitor volcanoes and warn of imminent eruptions?
19. How does a caldera form? Where would you go to see one?
20. What is a lava dome and why are they so dangerous?

CHAPTER 6

Weathering, Soil, and Sedimentary Rocks

OUTLINE

OBJECTIVES

At the end of this chapter, you will have learned that

- Weathering yields the raw materials for both soils and sedimentary rocks.
- Some weathering processes bring about physical changes in Earth materials with no change in composition, whereas others result in compositional changes.
- A variety of factors are important in the origin and evolution of soils.
- Soil degradation involves any loss of soil productivity that results from erosion, chemical pollution, or compaction.
- Sediments are deposited as aggregates of loose solids that may become sedimentary rocks if they are compacted and/or cemented.
- Geologists use texture and composition to classify sedimentary rocks.
- A variety of features preserved in sedimentary rocks are good indicators of how the original sediment was deposited.

An arch in the Entrada Sandstone in Arches National Park in Utah. The park is famous for its pillars, spires, and arches, all of which formed by weathering and erosion.

- Most evidence of prehistoric life in the form of fossils is found in sedimentary rocks.
- Weathering is important in the origin and concentration of some resources, and sediments and sedimentary rocks are resources themselves or contain resources such as petroleum and natural gas.

INTRODUCTION

All rocks at or near Earth's surface, as well as rocklike substances such as pavement and concrete in sidewalks, bridges, and foundations, decay and crumble with age. In short, they experience **weathering**, defined as the physical breakdown and chemical alteration of Earth materials as they are exposed to the atmosphere, hydrosphere, and biosphere. Actually, weathering is a group of physical and chemical processes that alter Earth materials so that they are more nearly in equilibrium with a new set of environmental conditions. Many rocks form within the crust where little or no oxygen or water is present, but at or near the surface they are exposed to both, as well as to lower temperature and pressure and the activities of organisms.

During weathering, the rock acted on by weathering, or **parent material**, is disaggregated to form smaller pieces (Figure 6.1), and some of its constituent minerals are altered or dissolved. Some of this weathered material may accumulate and be further modified to form *soil.* Much of it, however, is removed by **erosion**, which is the wearing away of soil and rock by geologic agents such as running water. This eroded material is transported elsewhere by running water, wind, glaciers, and marine currents and is eventually deposited as *sediment*, the raw material for *sedimentary rocks.*

Earth's crust is composed mostly of *crystalline rock*, a term that refers loosely to metamorphic and igneous rocks, except those made up of pyroclastic materials. Nevertheless, sediment and sedimentary rocks, making up perhaps only 5% of the crust, are by far the most common materials in surface exposures and in the shallow subsurface. They cover approximately two-thirds of the continents and most of the seafloor, except spreading ridges. All rocks are important in deciphering Earth history, but sedimentary rocks have a special place in this endeavor because they preserve evidence of surface processes responsible for them, as well as most fossils, which are evidence of prehistoric life.

Weathering and erosion have yielded many areas of exceptional scenery, including Bryce Canyon and Arches National Parks, both in Utah (Figure 6.2 and the chapter opening photograph). In addition to interesting landscapes, weathering is also responsible for the origin of some natural resources, aluminum ore, for example, and it

Figure 6.1 Weathering of Granite

Sue Monroe

a This exposure of granite has been so thoroughly weathered that only a few spherical masses of the original rock are visible.

Sue Monroe

b Closeup view of the weathered material. Mechanical weathering has predominated, so the particles are mostly small pieces of granite and materials such as quartz and feldspars.

Parvinder Sethi

Figure 6.2 Differential Weathering The spectacular scenery at Bryce Canyon National Park in Utah resulted from differential weathering and erosion of the 40- to 50-million-year-old Wasatch Formation that was deposited in ancient lakes. Weathering and erosion that took place along closely spaced fractures yielded this panorama of spires, pillars, gullies, and ravines.

enriches other resources by removing soluble materials. In fact, some sediments and sedimentary rocks are resources in their own right, or they are the host rocks for petroleum and natural gas.

HOW ARE EARTH MATERIALS ALTERED?

Weathering is a surface or near-surface process, but the rocks it acts on are not structurally and compositionally homogeneous throughout, which accounts for **differential weathering**. That is, weathering takes place at different rates even in the same area, so it commonly results in uneven surfaces. Differential weathering and *differential erosion*—that is, variable rates of erosion—combine to yield some unusual and even bizarre features, such as hoodoos, spires, and arches (Figure 6.2) (see Geo-inSight on pages 136 and 137).

The two recognized types of weathering, *mechanical* and *chemical*, both proceed simultaneously on parent material, as well as on materials in transport and those deposited as sediment. In short, all surface or near-surface materials weather, although one type of weathering may predominate depending on such variables as climate and rock type.

Mechanical Weathering

Mechanical weathering takes place when physical forces break earth materials into smaller pieces that retain the composition of the parent material. Granite, for instance, might be mechanically weathered and yield smaller pieces of granite or individual grains of quartz, potassium feldspars, plagioclase feldspars, and biotite (Figure 6.1). Several physical processes are responsible for mechanical weathering.

Frost action involving water repeatedly freezing and thawing in cracks and pores in rocks is particularly effective where temperatures commonly fluctuate above and below freezing. The reason frost action is so effective is that water expands by about 9% when it freezes, thus exerting great force on the walls of a crack, widening and extending it by *frost wedging* (Figure 6.3a). Repeated freezing and thawing dislodge angular pieces of rock from the parent material that tumble downslope and accumulate as **talus** (Figure 6.3b). Frost action is most effective in high mountains, even during the summer months, but it has little or no effect where the temperature rarely drops below freezing, or where earth materials are permanently frozen.

Some rocks form at depth and are stable under tremendous pressure. Granite crystallizes far below the surface, so when it is uplifted and eroded, its contained energy is released by outward expansion, a phenomenon known as **pressure release**. The outward expansion results in the origin of fractures called *sheet joints* that more or less parallel the exposed rock surface. Sheet-joint-bounded slabs of rock slip or slide off the parent rock, leaving large, rounded masses known as **exfoliation domes** (Figure 6.4a).

That solid rock expands and produces fractures might be counterintuitive, but is nevertheless a well-known phenomenon. In deep mines, masses of rock detach from the sides of the excavation, often explosively. These *rock bursts* and less violent *popping* pose a danger to mine workers, and in South Africa they are responsible for approximately 20 deaths per year. In some quarries for building stone, excavations to only 7 or 8 m exposed rocks in which sheet joints formed (Figure 6.4b), in some cases with enough force to throw quarrying machines weighing more than a ton from their tracks.

During **thermal expansion and contraction,** the volume of rocks changes as they heat up and then cool down. The temperature may vary as much as 30°C a day in a desert, and rock, being a poor conductor of heat, heats and expands on its outside more than its inside. Even dark minerals absorb heat faster than light-colored ones, so differential expansion takes place between minerals. Surface expansion might generate enough stress to cause fracturing, but experiments in which rocks are heated and cooled repeatedly to simulate years of such activity indicate that thermal expansion and contraction are of minor importance in mechanical weathering.

The formation of salt crystals can exert enough force to widen cracks and dislodge particles in porous, granular rocks such as sandstone. And even in rocks with an interlocking mosaic of crystals, such as granite, **salt crystal growth** pries loose individual minerals. It takes place mostly in hot, arid regions, but also probably affects rocks in some coastal areas.

Animals, plants, lichens, and bacteria all participate in the mechanical and chemical alteration of rocks (Figure 6.5a). Burrowing animals, such as worms, reptiles, rodents, termites, and ants, constantly mix soil and sediment particles and bring material from depth to the surface where further weathering occurs. The roots of plants, especially large bushes and trees, wedge themselves into cracks in rocks and further widen them (Figure 6.5b).

Geo-inSight Arches National Park, Utah

The sandstone arches at Arches National Park near Moab, Utah, are in the Jurassic-age Entrada Sandstone. The park is noted for its arches, spires, balanced rocks, pinnacales, and other features that resulted from differential weathering and erosion.

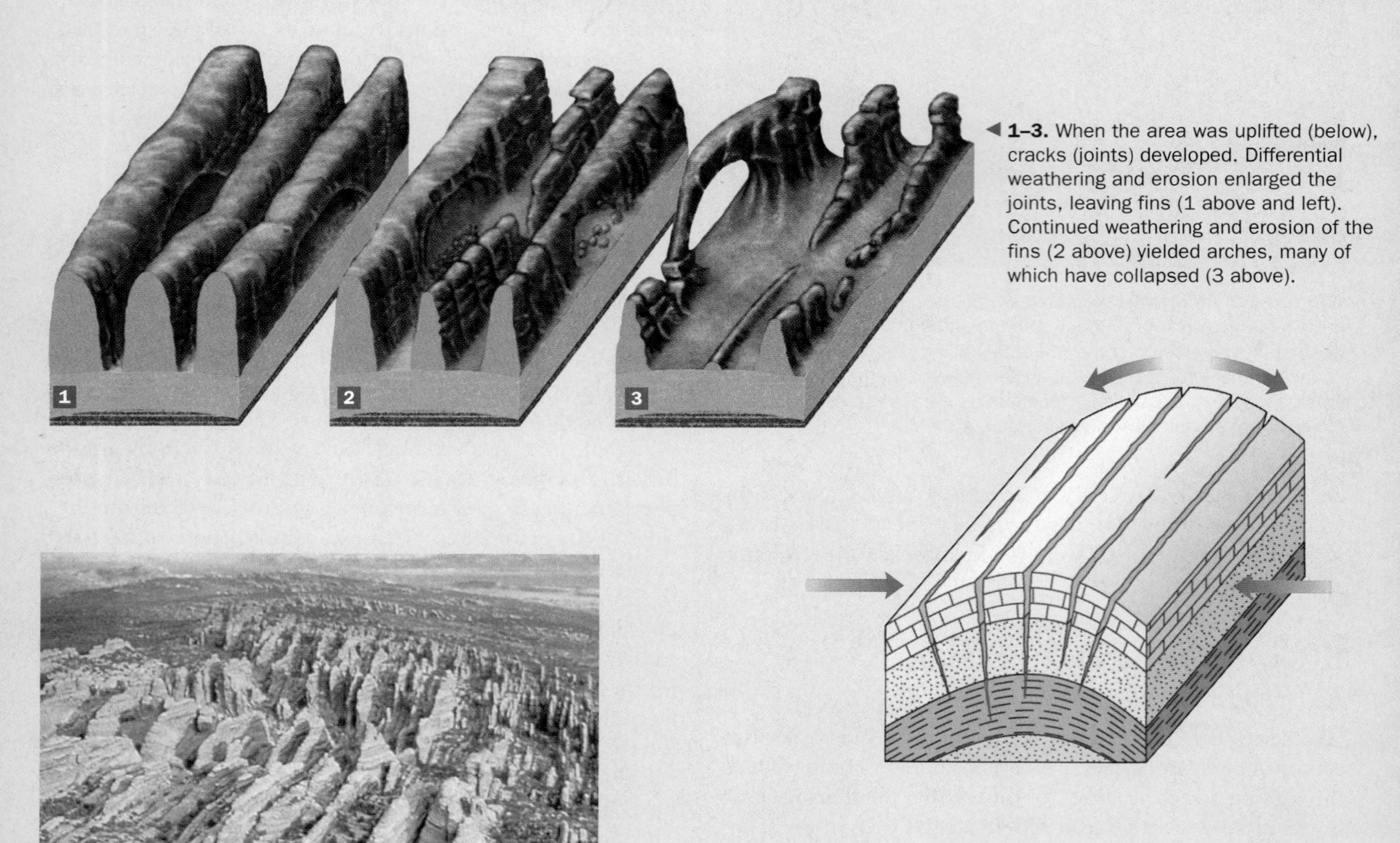

◀ **1–3.** When the area was uplifted (below), cracks (joints) developed. Differential weathering and erosion enlarged the joints, leaving fins (1 above and left). Continued weathering and erosion of the fins (2 above) yielded arches, many of which have collapsed (3 above).

Galen Rowell/Peter Arnold, Inc

▼ **4.** Delicate Arch shows an advanced stage in arch formation. It measures 9.7 m wide and 14 m high.

Sue Monroe

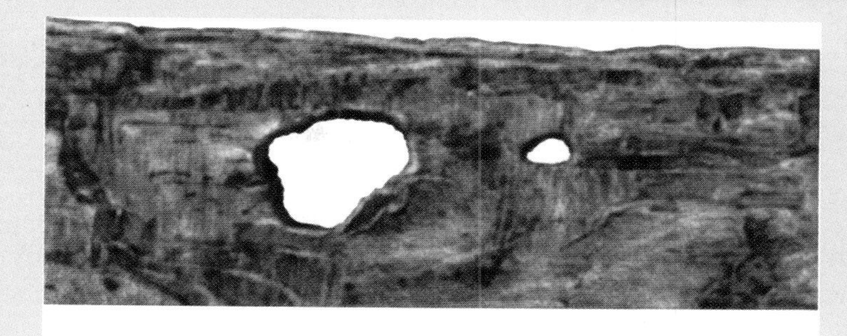

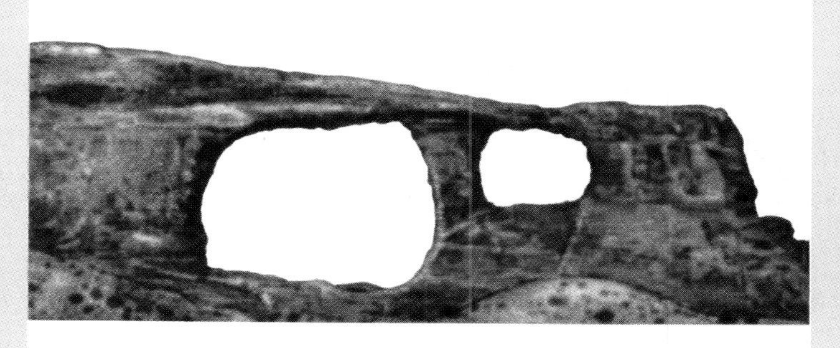

▶ **5–7.** Probable evolution of Sheep Rock and the fin with Baby Arch or Hole-in-the-Wall (right and above). Close-up view of Baby Arch (below). It measures 7.6 m wide and 4.5 m high.

Sue Monroe

Sue Monroe

7

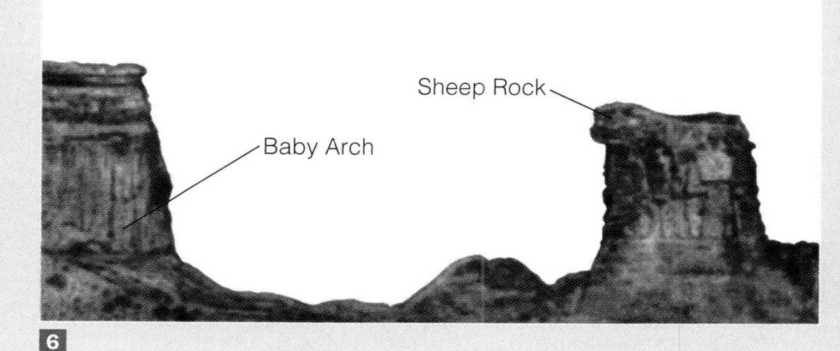

6

▼ **8.** The opening scene in the movie *Indiana Jones and the Last Crusade* (1989), showed a young Indiana Jones, played by River Phoenix, at Double Arch.

Sue Monroe

S. W. Lohman/USGS

◀ **9.** Balanced Rock rises 39 m above its base. Isolated columns and pillars like this one are called *hoodoos*, many of which are in the park.

Figure 6.3 Frost Wedging

a Frost wedging takes place when water seeps into cracks and expands as it freezes. Angular pieces of rock are pried loose by repeated freezing and thawing.

b Frost wedging and other mechanical weathering processes produced these talus accumulations in Banff National Park in the Canadian Rocky Mountains.

Figure 6.4 Sheet Joints and Exfoliation Domes

a Sheet joints in this body of granite in the Sierra Nevada of California parallel the surface of the exposed rock. The rounded mass formed by this process is an exfoliation dome.

b The sheet joint indicated by the hammer formed by expansion in the Mount Airy Granite in North Carolina. The hammer is about 30 cm long.

Chemical Weathering

Chemical weathering decomposes rocks and minerals by chemical alteration of the parent material. In contrast to mechanical weathering, chemical weathering changes the composition of weathered materials. For example, several clay minerals (sheet silicates) form by the chemical and structural alteration of other minerals, such as potassium feldspars and plagioclase feldspars, both of which are framework silicates. Other minerals are completely

Figure 6.5 Organisms and Weathering

a The orange and gray masses on this rock at Grimes Point Archaeological Site in Nevada are lichens, composite organisms made up of fungi and alage. Lichens derive their nutrients from the rock and thus contribute to chemical weathering.

b These trees in the Black Hills of South Dakota contribute to mechanical weathering as they grow in cracks in the rocks, thereby breaking the parent material into smaller pieces.

decomposed during chemical weathering, but some chemically stable minerals are simply liberated from the parent material.

Important agents of chemical weathering include atmospheric gases, especially oxygen, water, and acids. Organisms also play an important role. Rocks with lichens (composite organisms made up of fungi and algae) on their surfaces undergo more rapid chemical alteration than lichen-free rocks (Figure 6.5a). In addition, plants remove ions from soil water and reduce the chemical stability of soil minerals, and plant roots release organic acids.

When **solution** takes place, the ions of a substance separate in a liquid, and the solid substance dissolves. Water is a remarkable solvent because its molecules have an asymmetric shape, consisting of one oxygen atom with two hydrogen atoms arranged so that the angle between the two hydrogen atoms is about 104 degrees (Figure 6.6). Because of this asymmetry, the oxygen end of the molecule retains a slight negative electrical charge, whereas the hydrogen end retains a slight positive charge. When a soluble substance such as the mineral halite (NaCl) comes in contact with a water molecule, the positively charged sodium ions are attracted to the negative end of the water molecule, and the negatively charged chloride ions are attracted to the positively charged end of the molecule (Figure 6.6). Thus, ions are liberated from the crystal structure, and the solid goes into solution; in other words, it dissolves.

Most minerals are not very soluble in pure water because the attractive forces of water molecules are not sufficient to overcome the forces between particles in minerals. The mineral calcite ($CaCO_3$), the major constituent of the sedimentary rock limestone and the metamorphic rock marble, is practically insoluble in pure water, but it rapidly dissolves if a small amount of acid is present. One way to make water acidic is by dissociating the ions of carbonic acid as follows:

$$\underset{\text{water}}{H_2O} + \underset{\substack{\text{carbon}\\ \text{dioxide}}}{CO_2} \rightleftharpoons \underset{\substack{\text{carbonic}\\ \text{acid}}}{H_2CO_3} \rightleftharpoons \underset{\substack{\text{hydrogen}\\ \text{ion}}}{H^+} + \underset{\substack{\text{bicarbonate}\\ \text{ion}}}{HCO_3^-}$$

According to this chemical equation, water and carbon dioxide combine to form *carbonic acid*, a small amount of which dissociates to yield hydrogen and bicarbonate ions. The concentration of hydrogen ions determines the acidity of a solution; the more hydrogen ions present, the stronger the acid.

Carbon dioxide from several sources may combine with water and react to form acid solutions. The atmosphere is mostly nitrogen and oxygen, but approximately 0.03% is carbon dioxide, causing rain to be slightly acidic. Decaying organic matter and the respiration of organisms produce carbon dioxide in soils, so groundwater is also generally slightly acidic. Climate affects the acidity, however, with arid regions tending to have alkaline groundwater (that is, it has a low concentration of hydrogen ions).

Whatever the source of carbon dioxide, once an acidic solution is present, calcite rapidly dissolves according to the reaction.

Figure 6.6 The Solution of Halite

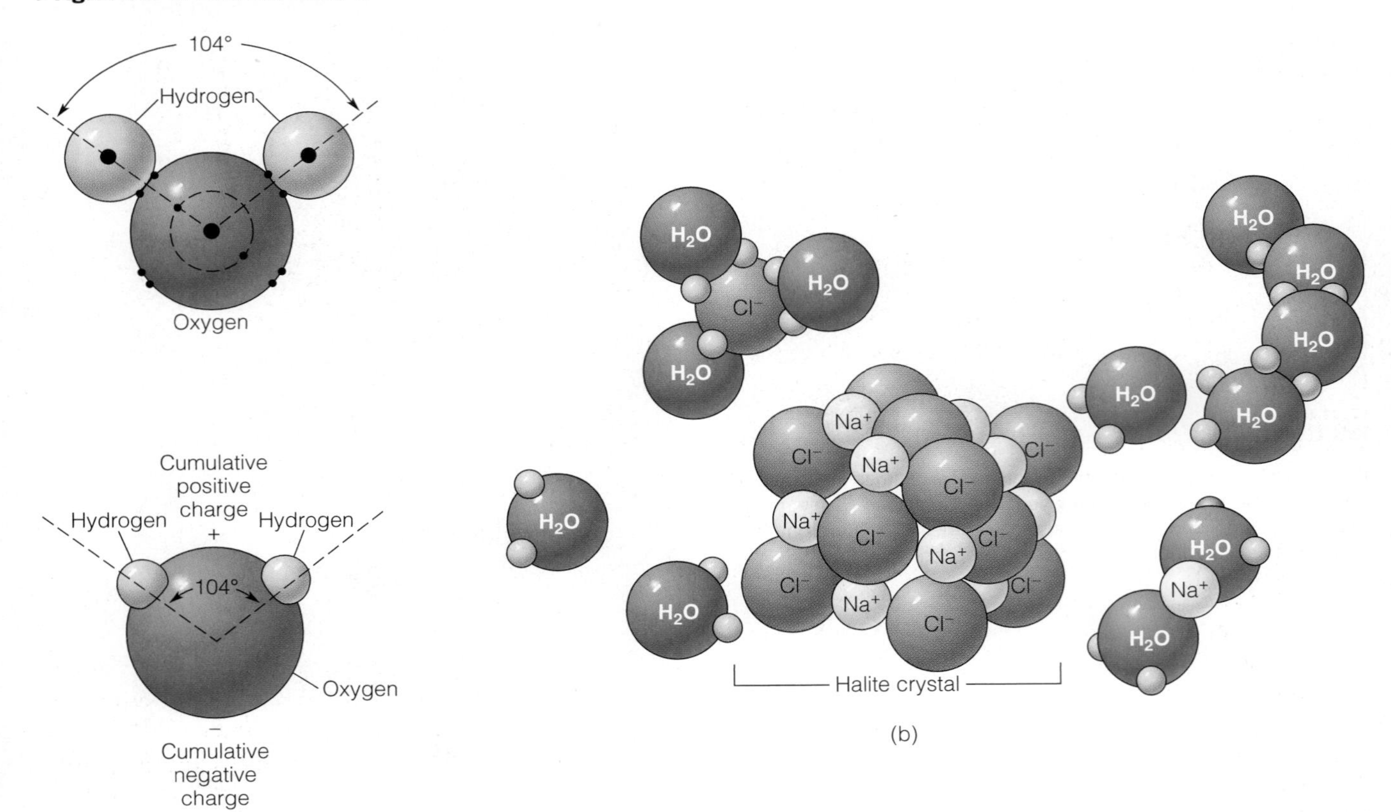

a The structure of a water molecule. The asymmetric arrangement of hydrogen atoms causes the molecule to have a slight positive electrical charge at its hydrogen end and a slight negative charge at its oxygen end.

b Solution of sodium chloride (NaCl), the mineral halite, in water. Note that the sodium atoms are attracted to the oxygen end of a water molecule, whereas chloride ions are attracted to the hydrogen end of the molecule.

$$\underset{\text{calcite}}{CaCO_3} + \underset{\text{water}}{H_2O} + \underset{\text{carbon dioxide}}{CO_2} \rightleftharpoons \underset{\text{calcium ion}}{Ca^{++}} + \underset{\text{bicarbonate ion}}{2HCO_3^-}$$

The term **oxidation** has a variety of meanings for chemists, but in chemical weathering, it refers to reactions with oxygen to form an oxide (one or more metallic elements combined with oxygen) or, if water is present, a hydroxide (a metallic element or radical combined with OH^-). For example, iron rusts when it combines with oxygen to form the iron oxide hematite:

$$\underset{\text{iron}}{4Fe} + \underset{\text{oxygen}}{3O_2} \rightarrow \underset{\text{iron oxide (hematite)}}{2Fe_2O_3}$$

Atmospheric oxygen is abundantly available for oxidation reactions, but oxidation is generally a slow process unless water is present. Thus, most oxidation is carried out by oxygen dissolved in water.

Oxidation is important in the alteration of ferromagnesian silicates such as olivine, pyroxenes, amphiboles, and biotite. Iron in these minerals combines with oxygen to form the reddish iron oxide hematite (Fe_2O_3) or the yellowish or brown hydroxide limonite [$FeO(OH)\cdot nH_2O$]. The yellow, brown, and red colors of many soils and sedimentary rocks are caused by the presence of small amounts of hematite or limonite.

The chemical reaction between the hydrogen (H^+) ions and hydroxyl (OH^-) ions of water and a mineral's ions is known as **hydrolysis.** In hydrolysis, hydrogen ions actually replace positive ions in minerals. Such replacement changes the composition of minerals and liberates iron that then may be oxidized.

The chemical alteration of the potassium feldspar orthoclase provides a good example of hydrolysis. All feldspars are framework silicates, but when altered, they yield soluble salts and clay minerals, such as kaolinite, which are sheet silicates. The chemical weathering of orthoclase by hydrolysis occurs as follows:

$$\underset{\text{orthoclase}}{2KAlSi_3O_8} + \underset{\text{hydrogen ion}}{2H^+} + \underset{\text{bicarbonate ion}}{2HCO_3^-} + \underset{\text{water}}{H_2O} \rightarrow$$

$$\underset{\text{clay (kaolinite)}}{Al_2Si_2O_5(OH)_4} + \underset{\text{potassium ion}}{2K^+} + \underset{\text{bicarbonate ion}}{2HCO_3^-} + \underset{\text{silica}}{4SiO_2}$$

In this reaction, hydrogen ions attack the ions in the orthoclase structure, and some liberated ions are incorporated in a developing clay mineral. The potassium and bicarbonate ions go into solution and combine to form a soluble salt. On the right side of the equation is excess silica that would not fit into the crystal structure of the clay mineral.

USGS

Figure 6.7 **Weathering Along Fractures** These granitic rocks in Joshua Tree National Park in California have been chemically weathered more intensely along fractures than in unfractured parts of the same rock outcrop.

Factors That Control the Rate of Chemical Weathering

Chemical weathering operates on the surfaces of particles, so it alters rocks and minerals from the outside inward. In fact, if you break open a weathered stone, you will see a rind of weathering at and near the surface, but the stone is completely unaltered inside. The rate at which chemical weathering proceeds depends on several factors. One is simply the presence or absence of fractures because fluids seep along fractures, and weathering is more intense along these surfaces (Figure 6.7). Other factors also control chemical weathering, including particle size, climate, and parent material.

Because chemical weathering affects particle surfaces, the greater the surface area, the more effective the weathering. It is important to realize that small particles have larger surface areas compared to their volume than do large particles. Notice in Figure 6.8 that a block measuring 1 m on a side has a total surface area of 6 m^2, but when the block is broken into particles measuring 0.5 m on a side, the total surface area increases to 12 m^2. And if these particles are all reduced to 0.25 m on a side, the total surface area increases to 24 m^2. Note that although the surface area in this example increases, the total volume remains the same at 1 m^3.

We can conclude that mechanical weathering contributes to chemical weathering by yielding smaller particles with greater surface area compared to their volume. Actually, your own experiences with particle size verify our contention about surface area and volume. Because of its very small particle size, powdered sugar gives an intense burst of sweetness as the tiny pieces dissolve rapidly, but otherwise it is the same as the granular sugar we use on our cereal or in our coffee.

It is not surprising that chemical weathering is more effective in the tropics than in arid and arctic regions because temperatures and rainfall are high and evaporation rates are low. In addition, vegetation and animal life are much more abundant. Consequently, the effects of weathering extend to depths of several tens of meters, but they extend only centimeters to a few meters deep in arid and arctic regions.

Some rocks are more resistant to chemical alteration than others, so parent material is another control on the rate of chemical weathering. The metamorphic rock quartzite is an extremely stable substance that alters slowly compared to most other rock types. In contrast, basalt, which contains large amounts of calcium-rich plagioclase and pyroxene minerals, decomposes rapidly because these minerals are chemically unstable. In fact, the stability of common minerals is just the opposite of their order of crystallization in Bowen's reaction series (Table 6.1, also see Figure 4.4): The minerals that form last in this series are more stable, whereas those that form early are easily altered because they are most out of equilibrium with their conditions of formation.

One manifestation of chemical weathering is **spheroidal weathering** (Figure 6.9). In spheroidal weathering, a stone, even one that is rectangular to begin with, weathers to form a more spherical shape because that is the most stable shape

Figure 6.8 **Particle Size and Chemical Weathering**

Surface area = 6 m^2

1 m

1 m

a As a rock is divided into smaller particles, its surface area increases, but its volume remains the same. The surface area is 6 m^2.

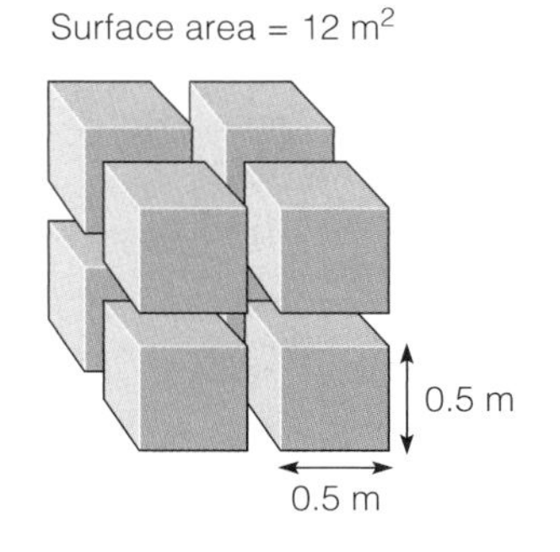

b The surface area is 12 m^2.

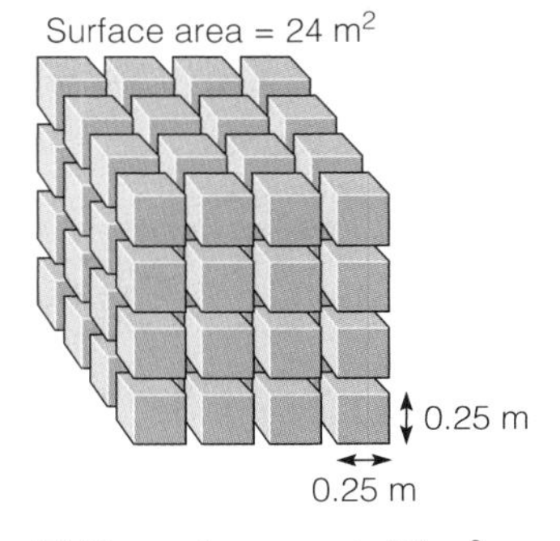

c The surface area is 24 m^2, but the volume remains the same at 1 m^3. Small particles have more surface area in relation to their volume than do large particles.

TABLE 6.1 Stability of Silicate Minerals

	Ferromagnesian Silicates	Nonferromagnesian Silicates
Increasing Stability ↓	Olivine	Calcium plagioclase
	Pyroxene	
	Amphibole	Sodium plagioclase
	Biotite	Potassium feldspar
		Muscovite
		Quartz

it can assume. The reason is that on a rectangular stone, the corners are attacked by weathering from three sides, and the edges are attacked from two sides, but the flat surfaces weather more or less uniformly (Figure 6.9). Consequently, the corners and edges are altered more rapidly, the material sloughs off, a more spherical shape develops, and all surfaces weather at the same rate.

HOW DOES SOIL FORM AND DETERIORATE?

Most of Earth's land surface is covered by a layer of **regolith**, a collective term for sediment, as well as layers of pyroclastic materials and the residue formed in place by weathering. Some regolith consisting of weathered materials, air, water, and organic matter supports vegetation and is called **soil** (Figure 6.10a). Almost all land-dwelling organisms depend directly or indirectly on soil for their existence. Plants grow in soil from which they derive their nutrients and most of their water, whereas many land-dwelling animals depend on plants for nutrients.

About 45% of good soil for farming and gardening is composed of weathered particles, with much of the remaining volume simply void spaces filled with air and/or water. In addition, a small but important amount of humus is usually present. *Humus* is carbon derived by bacterial decay of organic matter and is highly resistant to further decay. Even a fertile soil might have as little as 5% humus, but it is nevertheless important as a source of plant nutrients and it enhances a soil's capacity to retain moisture.

Some weathered materials in soils are sand- and silt-sized minerals, especially quartz, but other minerals may be present as well. These solids hold soil particles apart, allowing oxygen and water to circulate more freely. Clay minerals are also important in soils and aid in the retention of water, as well as supplying nutrients to plants. Soils with excess clay minerals, however, drain poorly and are sticky when wet and hard when dry.

Residual soils form when parent material weathers in place. For example, if a body of granite weathers, and the weathering

Figure 6.9 Spheroidal Weathering

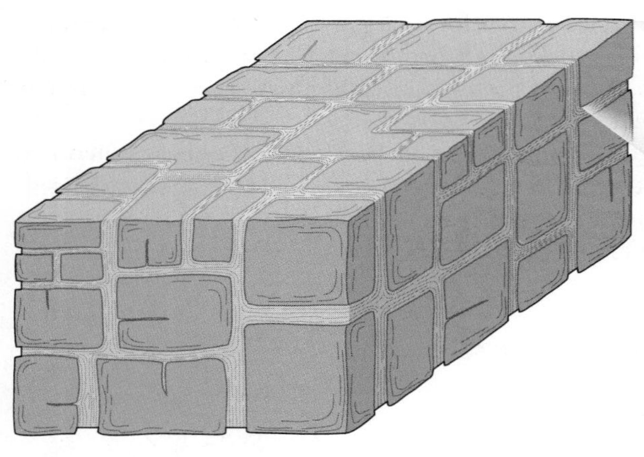

a The rectangular blocks outlined by fractures are attacked by chemical weathering processes.

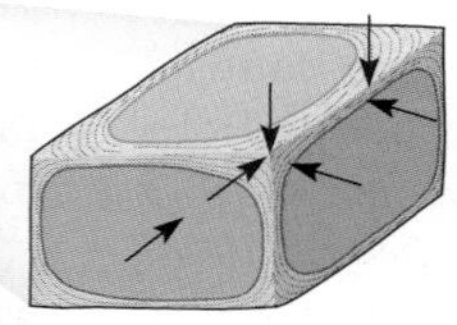

b Corners and edges weather most rapidly.

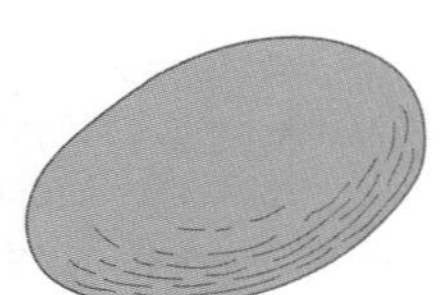

c When the blocks are weathered so that they are nearly spherical, their surfaces weather evenly and no further change in shape takes place.

d An exposure of granite showing spheroidal weathering in Joshua Tree National Park in California.

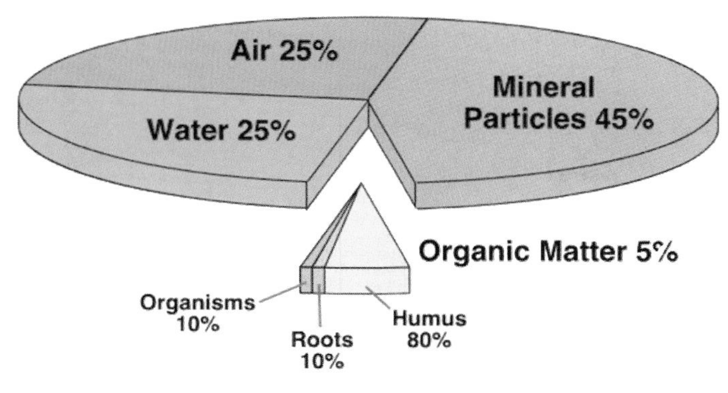

a Soils are made up mostly of minerals and rock fragments derived by weathering, air, water, and organic matter. Most of the organic matter is humus.

Soil horizons
O Humus
A Zone of leaching (topsoil)
B Zone of accumulation (subsoil)
C Weathered parent material (bedrock)
Gradational contact
Fresh parent material (bedrock)

b The soil horizons of a fully developed soil.

Figure 6.10 The Composition of Soil and Soil Horizons

residue accumulates over the granite and is converted to soil, the soil thus formed is residual. In contrast, *transported soils* develop on weathered material that was eroded and carried to a new location, where it is altered to soil.

The Soil Profile

Observed in vertical cross section, soil consists of distinct layers or **soil horizons** that differ from one another in texture, structure, composition, and color (Figure 6.10b). Starting from the top, the soil horizons are designated O, A, B, and C, but the boundaries between horizons are transitional. Because soil formation begins at the surface and works downward, horizon A is more altered from the parent material than the layers below.

Horizon O, which is only a few centimeters thick, consists of organic matter. The remains of plant materials are clearly recognizable in the upper part of horizon O, but its lower part consists of humus.

Horizon A, called *topsoil*, contains more organic matter than horizons B and C below. It is also characterized by intense biological activity because plant roots, bacteria, fungi, and animals such as worms are abundant. Threadlike soil bacteria give freshly plowed soil its earthy aroma. In soils developed over a long period of time, horizon A consists mostly of clays and chemically stable minerals such as quartz. Water percolating down through horizon A dissolves soluble minerals and carries them away or downward to lower levels in the soil, a process called *leaching*. Accordingly, horizon A is also referred to as the *zone of leaching*.

Horizon B, or *subsoil*, contains fewer organisms and less organic matter than horizon A (Figure 6.10b). Horizon B is also known as the *zone of accumulation* because soluble minerals leached from horizon A accumulate as irregular masses. If horizon A is eroded, leaving horizon B exposed, plants do not grow as well, and if it is clayey, it is harder when dry and stickier when wet than other soil horizons.

Horizon C has little organic matter and consists of partially altered parent material grading down into unaltered parent material (Figure 6.10b). In horizons A and B, the composition and texture of the parent material have been so thoroughly altered that it is no longer recognizable. In contrast, rock fragments and minerals of the parent material retain their identity in horizon C.

Factors That Control Soil Formation

Climate is the single most important factor in soil origins, but complex interactions among several factors account for soil type, thickness, and fertility (Figure 6.11). A very general classification recognizes three major soil types characteristic of different climatic settings. Soils that develop in humid regions such as the eastern United States and much of Canada are **pedalfers,** a name derived from the Greek word *pedon*, meaning "soil," and from the chemical symbols for aluminum (Al) and iron (Fe) (Figure 6.12a). Because these soils form where abundant moisture is present, most of the soluble minerals have been leached from horizon A. Although it may be gray, horizon A is commonly dark because of abundant organic matter, and aluminum-rich clays and iron oxides tend to accumulate in horizon B.

Soils found in much of the arid and semiarid western United States, especially the Southwest, are **pedocals.** Pedocal derives its name, in part, from the first three letters

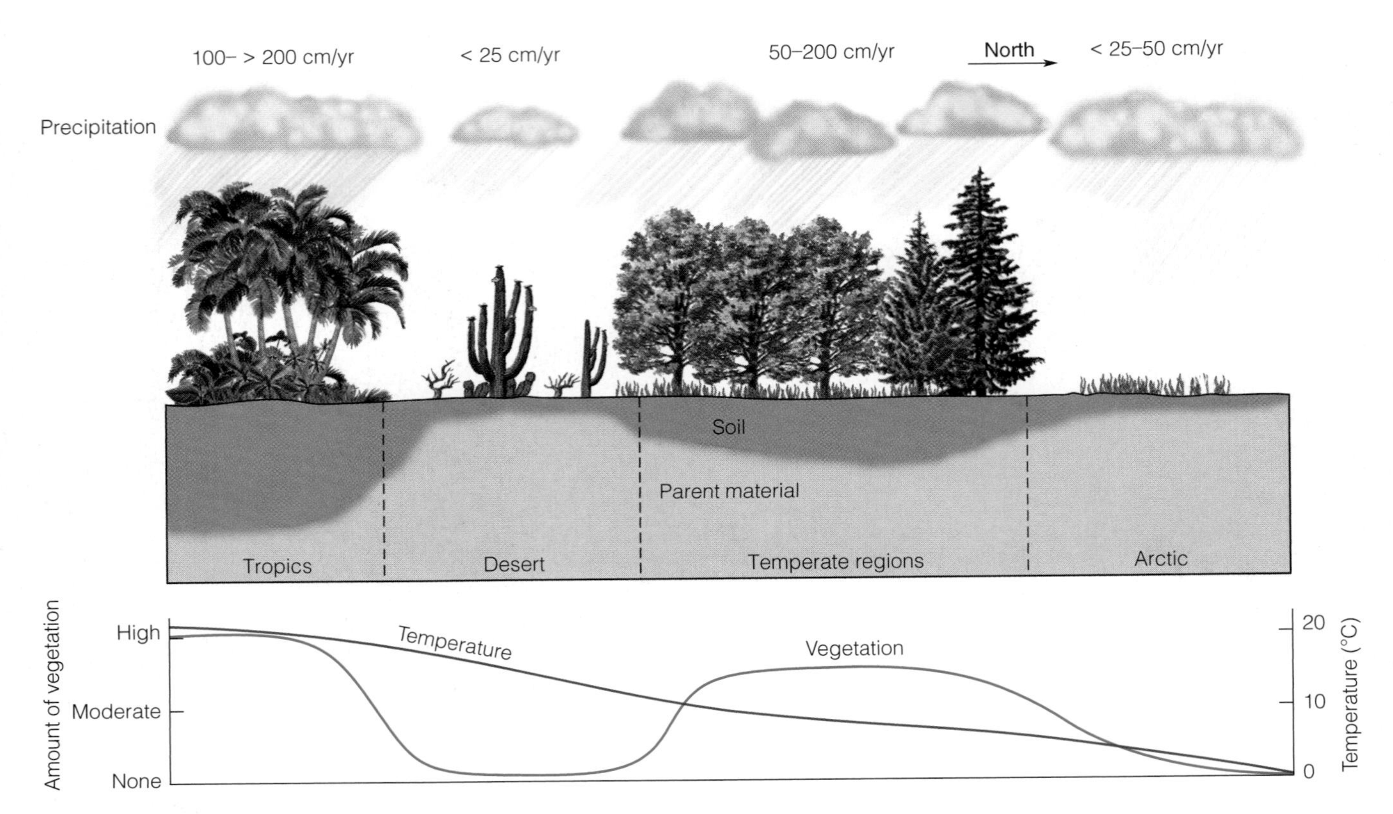

Figure 6.11 Climate and Soil Formation Generalized diagram showing soil formation as a function of the relationships between climate and vegetation, which alter parent material over time. Soil formation operates most vigorously where precipitation and temperatures are high, as in the tropics.

of "calcite" (Figure 6.12b). These soils contain less organic matter than pedalfers, so horizon A is lighter in color and contains more unstable minerals because of less intense chemical weathering. As soil water evaporates, calcium carbonate leached from above precipitates in horizon B where it forms irregular masses of *caliche*. Precipitation of sodium salts in some desert areas where soil water evaporation is intense yields *alkali soils* that are so alkaline that they cannot support plants.

Laterite forms in the tropics where chemical weathering is intense and leaching of soluble minerals is complete. These soils are red, extend to depths of several tens of meters, and are composed mostly of aluminum hydroxides, iron oxides, and clay minerals; even quartz, a chemically stable mineral, is leached out (Figure 6.12c).

Although laterite supports lush vegetation, it is not very fertile. The native vegetation is sustained by nutrients derived mostly from the surface layer of organic matter, but little humus is present in the soil because bacterial action destroys it. When laterite is cleared of its native vegetation, the surface accumulation of organic matter rapidly oxidizes, and there is little to replace it. Consequently, societies that practice slash-and-burn agriculture clear these soils and raise crops for only a few years at best. Then the soil is depleted of plant nutrients, the clay-rich laterite bakes brick hard in the tropical sun, and the farmers move on to another area where the process is repeated.

The same rock type can yield different soils in different climatic regimes and, in the same climatic regime, the same soils can develop on different rock types. Thus, climate is more important than parent material in determining the type of soil. Nevertheless, rock type does exert some control. For example, the metamorphic rock quartzite will have a thin soil over it because it is chemically stable, whereas an adjacent body of granite will have a much deeper soil.

Soils depend on organisms for their fertility, and in return they provide a suitable habitat for many organisms. Earthworms, ants, sowbugs, termites, centipedes, millipedes, and nematodes, along with fungi, algae, and single-celled organisms, make their homes in soil. All contribute to soil formation and provide humus when they die and decompose by bacterial action.

Much of the humus in soils comes from grasses or leaf litter that microorganisms decompose to obtain food. In so doing, they break down organic compounds in plants and release nutrients back into the soil. In addition, organic acids from decaying soil organisms are important in further weathering of parent materials and soil particles.

Burrowing animals constantly churn and mix soils, and their burrows provide avenues for gases and water. Soil organisms, especially some types of bacteria, are extremely important in changing atmospheric nitrogen into a form of soil nitrogen suitable for use by plants.

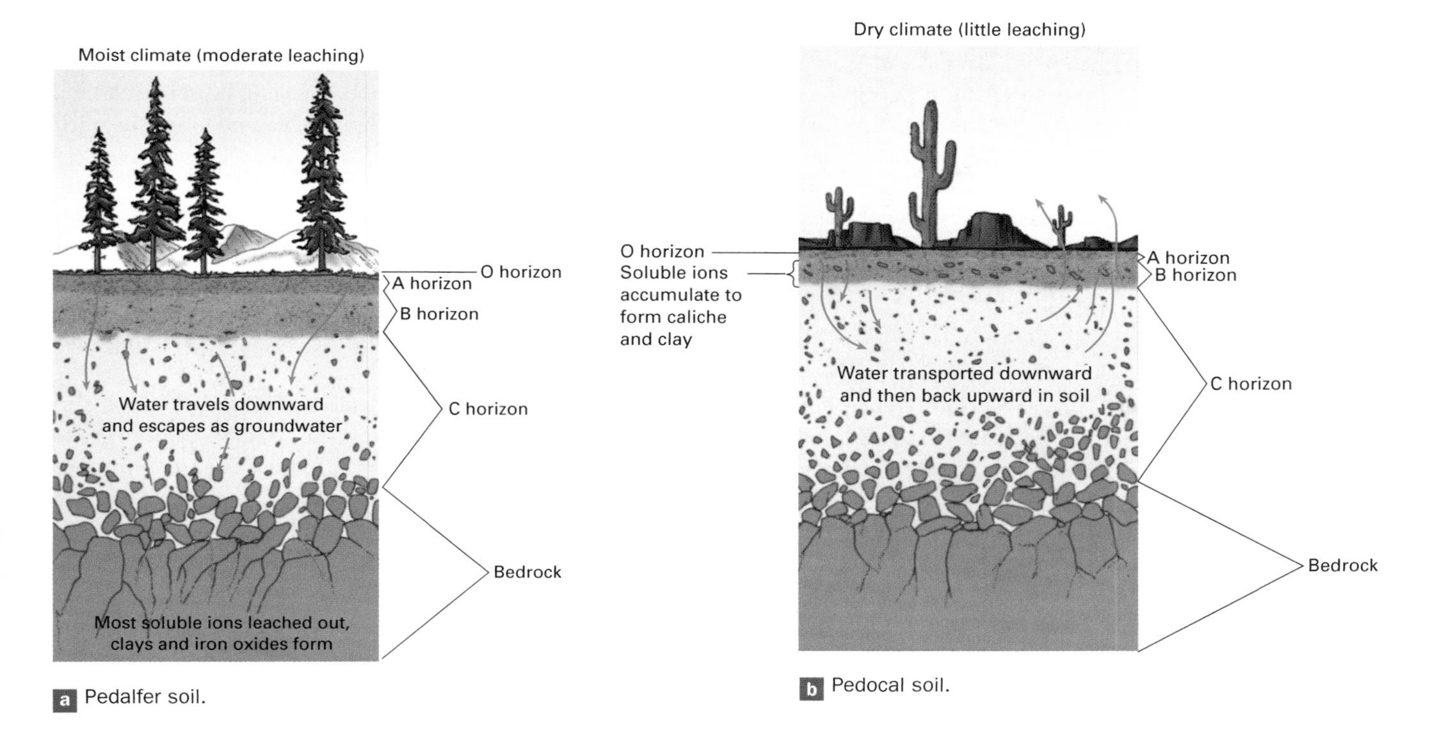

a Pedalfer soil.

b Pedocal soil.

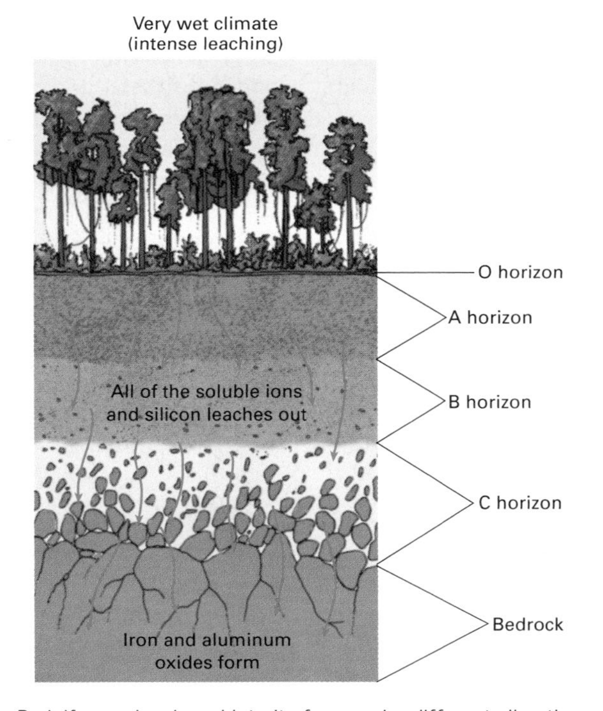

c Laterite soil.

Figure 6.12 Three Major Soil Types Pedalfer, pedocal, and laterite form under different climatic conditions.

The difference in elevation between high and low points in a region is called *relief.* And because climate is such an important factor in soil formation and climate changes with elevation, areas with considerable relief have different soils in mountains and adjacent lowlands. *Slope*, another important control, influences soil formation in two ways. One is simply *slope angle;* steep slopes have little or no soil because weathered materials are eroded faster than soil-forming processes can operate. The other factor is *slope direction.* In the Northern Hemisphere, north-facing slopes receive less sunlight than south-facing slopes and have cooler internal temperatures, support different vegetation, and if in a cold climate, remain snow covered or frozen longer.

How much time is needed to develop a centimeter of soil or a fully developed soil a meter or so deep? We can give no definite answer because weathering proceeds at vastly different rates depending on climate and parent material, but an overall average might be about 2.5 cm per century. Nevertheless, a lava flow a few centuries old in Hawaii may have a well-developed soil on it, whereas a flow the same age in Iceland will have considerably less soil. Given the same climatic conditions, soil develops faster on unconsolidated sediment than it does on bedrock.

Soil Degradation

In the context of geologic time, soils form rapidly; however, from the human perspective, the processes are so slow that soil is a nonrenewable resource. So any soil losses that exceed the rate of soil formation are viewed with alarm. Likewise, any reduction in soil fertility or production is cause for concern, especially in areas where soils provide only a marginal existence. Any process that removes soil or makes it less productive is defined as **soil degradation**, a serious problem in many parts of the world that includes erosion, chemical deterioration, and physical changes.

Erosion, an ongoing natural process, is usually slow enough for soil formation to keep pace, but unfortunately, some human practices add to the problem. Removing natural vegetation by plowing, overgrazing, overexploitation for fire wood, and deforestation all contribute to erosion by wind and running water. The Dust Bowl that developed in several Great Plains states during the 1930s is a poignant example of just how effective wind erosion is on soil pulverized and exposed by plowing (see Geo-Focus 6.1 on pages 148 and 149).

Wind has caused considerable soil erosion in some areas, but running water is much more powerful. Some soil is removed by *sheet erosion*, which involves the removal of thin layers of soil more or less evenly over a broad, sloping surface. *Rill erosion*, in contrast, takes place when running water scours small, troughlike channels. Channels shallow enough to be eliminated by plowing are *rills*, but those too deep (about 30 cm) to be plowed over are *gullies* (Figure 6.13). Where gullying is extensive, croplands can no longer be tilled and must be abandoned.

Soil undergoes chemical deterioration when its nutrients are depleted and its productivity decreases. Loss of soil nutrients is most notable in many of the populous developing nations where soils are overused to maintain high levels of agricultural productivity. Chemical deterioration is also caused by insufficient use of fertilizers and by clearing soils of their natural vegetation.

Other types of chemical deterioration are pollution and *salinization*, which occurs when the concentration of salts increases in a soil, making it unfit for agriculture. Improper disposal of domestic and industrial wastes, oil and chemical spills, and the concentration of insecticides and pesticides in soils all cause pollution. Soil pollution is a particularly serious problem in some parts of Eastern Europe.

Soil deteriorates physically when it is compacted by the weight of heavy machinery and livestock, especially cattle. Compacted soils are more costly to plow, and plants have a more difficult time emerging from them. Furthermore, water does not readily infiltrate, so more runoff occurs; this in turn accelerates the rate of water erosion.

In North America, the rich prairie soils of the midwestern United States and the Great Plains of the United States and Canada are suffering soil degradation. Nevertheless, this degradation is moderate and less serious than in many other parts of the world. Problems experienced in the past have stimulated the development of methods to minimize soil erosion on agricultural lands. Crop rotation, contour plowing (Figure 6.14), and the construction of terraces have all proved helpful. So has no-till planting, in which the residue from the harvested crop is left on the ground to protect the surface from the ravages of wind and water.

James S. Monroe

a Rill erosion in a field in Michigan during a rainstorm. The rill was later plowed over.

H.H. Walsron/USGS

b A large gully in the upper basin of the Rio Reventado in Costa Rica. Notice the man at the right for scale.

Figure 6.13 Soil Degradation Resulting from Erosion

WEATHERING AND RESOURCES

Soils are certainly one of our most precious natural resources. Indeed, if it were not for soils, food production on Earth would be vastly different and capable of supporting far fewer people. In addition, other aspects of soils are important economically. We discussed the origin of laterite in response to intense chemical weathering in the tropics, and we noted further that laterite is not very productive. If the parent material is rich in aluminum, however, the ore of aluminum called *bauxite* accumulates in horizon B. Some bauxite is found in Arkansas, Alabama, and Georgia, but at present, it is cheaper to import rather than mine these deposits, so both the United States and Canada depend on foreign sources of aluminum ore.

Bauxite and other accumulations of valuable minerals by the selective removal of soluble substances during chemical weathering are known as *residual concentrations*. Certainly, bauxite is a good example, but other deposits that formed in a similar manner are those rich in iron, manganese, clays, nickel, phosphate, tin, diamonds, and gold. Some of the sedimentary iron deposits in the Lake Superior region of the United States and Canada were enriched by chemical weathering when soluble parts of the deposits were carried away. Some kaolinite deposits in the southern United States formed when chemical weathering altered feldspars in pegmatites or as residual concentrations of clay-rich limestones and dolostones. Kaolinite is a clay mineral used in the manufacture of paper and ceramics.

A *gossan* is a yellow to red deposit made up mostly of hydrated iron oxides that formed by oxidation and leaching of sulfide minerals such as pyrite (FeS_2). The dissolution of pyrite and other sulfides forms sulfuric acid, which causes other metallic minerals to dissolve, and these tend to be carried down toward the groundwater table, where the descending solutions form minerals containing copper, lead, and zinc. Gossans have been mined for iron, but they are far more important as indicators of underlying ore deposits.

What Would You Do?

In the past few years, many gullies have appeared in farmers' fields in your area, and residents of your area are concerned because agriculture is the main source of jobs and tax revenue. Obviously, a decrease in agricultural production would be an economic disaster. You are appointed to a county board charged with making recommendations to prevent or at least minimize erosion on local croplands. How would you determine what caused the problem, and what specific recommendations would you make to reduce gullying?

SEDIMENT AND SEDIMENTARY ROCKS

Mechanical and chemical weathering, erosion, transport, and deposition are essential parts of the rock cycle (see Figure 1.14) because they are responsible for the origin and deposition of *sediment* that may become lithified, that is, transformed into *sedimentary rock* (Figure 6.15). The term **sediment** refers to all solid particles derived from preexisting rocks, which is *detrital sediment*. It also encompasses *chemical sediment*, which includes (1) minerals derived from solutions that contain minerals dissolved during chemical weathering, and (2) minerals extracted from water, mostly seawater, by organisms to build their shells. **Sedimentary rock** is simply rock made up of consolidated sediment.

One important criterion for classifying detrital sediment is particle size. Particles described as *gravel* measure more than 2 mm, whereas sand measures 1/16 –2 mm, and silt is any particle between 1/256 and 1/16 mm. None of these designations implies anything about composition; most gravel is made up of rock fragments—that is, small pieces of granite, basalt, or any other rock type—but sand and silt grains are usually single minerals, especially quartz. Particles smaller than 1/256 mm are termed clay, but clay has

Science VU/Visuals Unlimited

Figure 6.14 Soil Conservation Practices Contour plowing and strip cropping are two soil conservation practices used on this farm. Contour plowing involves plowing parallel to the contours of the land to inhibit runoff and soil erosion. In strip cropping, row crops such as corn alternate with other crops such as alfalfa.

Geo-Focus The Dust Bowl—An American Tragedy

The stock market crash of 1929 ushered in the Great Depression, a time when millions of people were unemployed and many had no means to acquire food and shelter. Urban areas were affected most severely by the depression, but rural areas suffered as well, especially during the great drought of the 1930s. Prior to the 1930s, farmers had enjoyed a degree of success unparalleled in U.S. history. During World War I (1914–1918), the price of wheat soared, and after the war, when Europe was recovering, the government subsidized wheat prices. High prices and mechanized farming resulted in more and more land being tilled. Even the weather cooperated, and land in the western United States that would otherwise have been marginally productive was plowed. Deep-rooted prairie grasses that held the soil in place were replaced by shallow-rooted wheat.

Beginning about 1930, drought prevailed throughout the country. Drought conditions varied from moderate to severe, but the consequences were particularly severe in the southern Great Plains. And because the land, even marginal land, had been tilled, the native vegetation was no longer present to keep the topsoil from blowing away. And blow away it did—in huge quantities.

A large region in the southern Great Plains that was particularly hard hit by drought, dust storms, and soil erosion came to be known as the Dust Bowl. Although its boundaries were not well defined, it included parts of Kansas, Colorado, and New Mexico, as well as the panhandles of Oklahoma and Texas (Figure 1a); together the Dust Bowl and its less affected fringe area covered more than 400,000 km^2!

Dust storms were common during the 1930s, and some reached phenomenal sizes (Figure 1b). One of the largest storms occurred in 1934 and covered more than 3.5 million km^2. It lifted dust nearly 5 km into the air, obscured the sky over large parts of six states, and blew hundreds of millions of tons of soil eastward where it settled on several eastern cities, as well as on ships 480 km out in the Atlantic Ocean. The Soil Conservation Service reported dust storms of regional extent on 140 occasions during 1936 and 1937. Dust was everywhere. It seeped into houses, suffocated wild animals and livestock, and adversely affected human health.

The dust was, of course, the topsoil from the tilled lands. Blowing dust was not the only problem; sand piled up along fences, drifted against houses and farm machinery, and covered what otherwise might have been productive soils. Agricultural production fell precipitously, farmers could not meet their mortgage payments, and by 1935, tens of thousands were homeless, on relief, or leaving (Figure 1c). Many went west to California and became the migrant farm workers immortalized in John Steinbeck's novel *The Grapes of Wrath.*

The Dust Bowl was an economic disaster of great magnitude. Droughts had stricken the southern Great Plains before and have done so since, but the drought of the 1930s was especially severe. Political and economic factors contributed to the disaster. Due in part to the artificially inflated wheat prices, many farmers were deeply in debt—mostly because they had purchased farm machinery in order to produce more and benefit from the high prices. Feeling economic pressure because of their huge debts, they tilled marginal land and employed few, if any, soil conservation measures.

If the Dust Bowl has a bright side, it is that the government, farmers, and the public in general no longer take soil for granted or regard it as a substance that needs no nurturing. In addition, a number of soil conservation methods developed then have now become standard practices.

two meanings. One is a size designation, but the term also refers to certain types of sheet silicates known as *clay minerals.* However, most clay minerals are also clay sized.

Sediment Transport and Deposition

Weathering is fundamental to the origin of sediment and sedimentary rocks, and so are erosion and *deposition*—that is, the movement of sediment by natural processes and its accumulation in some area. Because glaciers are moving solids, they can carry sediment of any size, whereas wind transports only sand and smaller sediment. Waves and marine currents transport sediment along shorelines, but running water is by far the most common way to transport sediment from its source to other locations.

During transport, *abrasion* reduces the size of particles, and the sharp corners and edges are worn smooth, a process known as *rounding*, as pieces of sand and gravel collide with one another (Figure 6.16a, b). Transport and processes that operate where sediment accumulates also result in *sorting*, which refers to the particle-size distribution in a sedimentary deposit. Sediment is characterized as well sorted if all particles are about the same size, and poorly sorted if a wide range of particle sizes is present (Figure 6.16c). Both rounding and sorting have important implications for other aspects of sediment and sedimentary rocks, such as how readily fluids move through them, and they also help geologists decipher the history of a deposit.

Regardless of how sediment is transported, it is eventually deposited in some geographic area known as a **depositional environment.** Deposition might take place on a

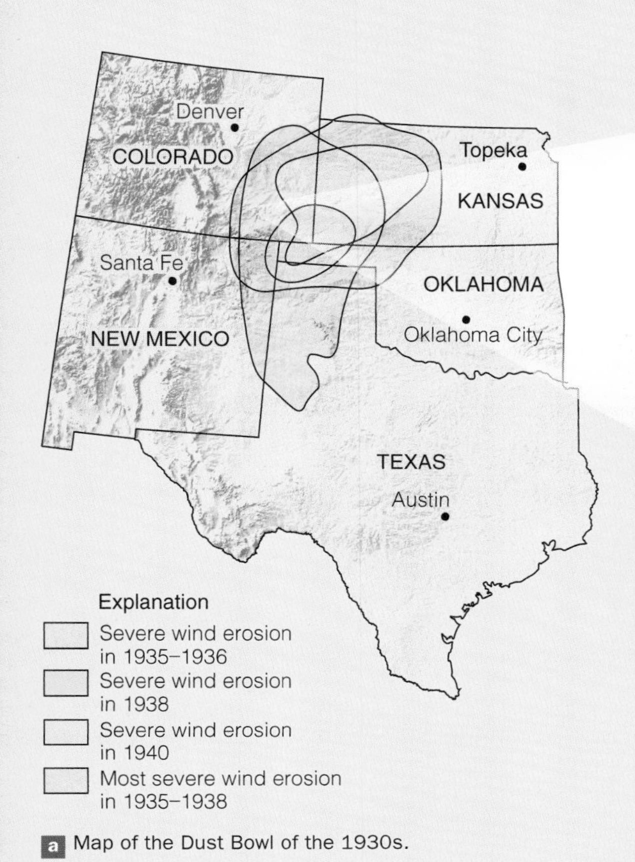

a Map of the Dust Bowl of the 1930s.

b This huge dust storm was photographed at Lamar, Colorado, in 1934.

Figure 1 The dust bowl of the 1930s was a time of drought, dust storms resulting from wind erosion, and economic hardship.

c By the mid-1930s, tens of thousands of people were on relief, homeless, or leaving the Dust Bowl. In 1939, Dorthea Lange photographed this homeless family of seven in Pittsburg County, Oklahoma.

Source: a "Map: Extent of Area Subject to Severe Wind Erosion," p. 30, from *Dust Bowl: The Southern Plains in the 1930s*, by Donald Worster. Copyright © 1979, 1982 by Oxford University Press, Inc. Used by permission of Oxford University Press, Inc.

floodplain, in a stream channel, on a beach, or on the seafloor, where physical, chemical, and biological processes impart various characteristics to the accumulating sediment. Geologists recognize three major depositional settings: continental (on the land), transitional (on or near seashores), and marine, each with several specific depositional environments (Figure 6.17).

How Does Sediment Become Sedimentary Rock?

A deposit of detrital sediment consists of a loose aggregate of particles: Mud in lakes and sand and gravel in stream channels or on beaches are good examples. To convert these aggregates of particles into sedimentary rocks requires **lithification** by compaction, cementation, or both (Figure 6.18).

To illustrate the relative importance of compaction and cementation, consider a detrital deposit made up of mud and another composed of sand. In both cases, the sediment consists of solid particles and *pore spaces*, the voids between particles. These deposits are subjected to **compaction** from their own weight and the weight of any additional sediment deposited on top of them, thereby reducing the amount of pore space and the volume of the deposit. Our hypothetical mud deposit may have 80% water-filled pore space, but after compaction, its volume is reduced by as much as 40% (Figure 6.18). The sand deposit, with as much as 50% pore space, is also compacted, but far less than the mud deposit, so that the grains fit more tightly together (Figure 6.18).

Compaction alone is sufficient for lithification of mud, but for sand and gravel, **cementation** involving the precipitation

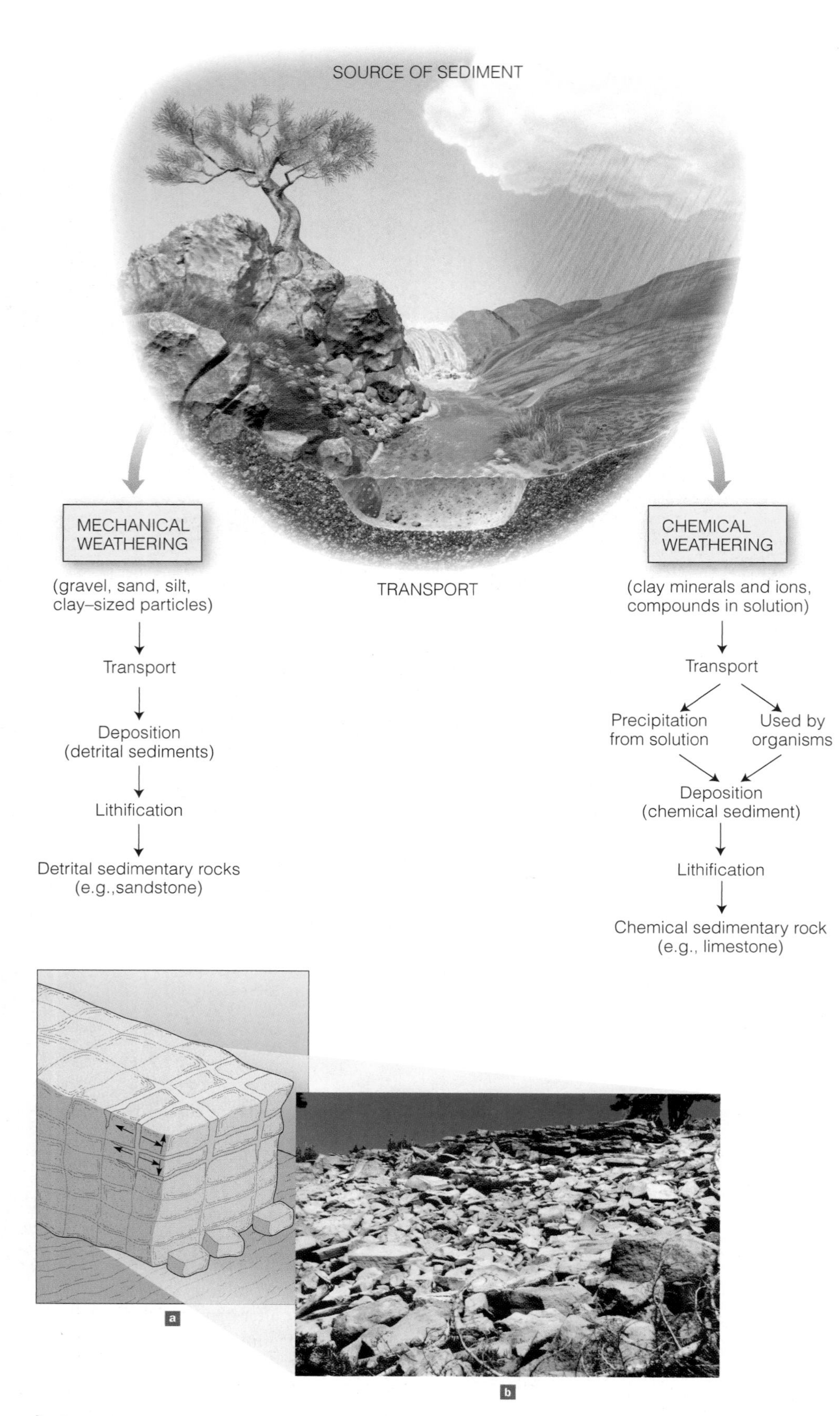

Figure 6.15 The Origin of Sedimentary Rocks Notice that several steps are involved in the origin of sedimentary rocks, including weathering, transport, deposition, and lithification. This illustration simply shows part of the rock cycle in more detail (see Figure 1.14).

of minerals in pore spaces is also necessary. The two most common chemical cements are calcium carbonate ($CaCO_3$) and silicon dioxide (SiO_2), but iron oxide and hydroxide cement, such as hematite (Fe_2O_3) and limonite [$FeO(OH)\cdot nH_2O$], are found in some sedimentary rocks. Recall that calcium carbonate readily dissolves in water that contains a small amount of carbonic acid, and chemical weathering of feldspars and other minerals yields silica in solution. Cementation takes place when minerals precipitate in the pore spaces of sediment from circulating water, thereby binding the loose particles together. Iron oxide and hydroxide cements account for the red, yellow, and brown sedimentary rocks found in many areas (see the chapter opening photograph).

We have explained lithification of detrital sediments, but we have not yet considered this process in chemical sediments. By far the most common chemical sediments are calcium carbonate mud and sand- and gravel-sized accumulations of calcium carbonate grains, such as shells and shell fragments. Compaction and cementation also take place in these sediments, converting them into various types of limestone, but compaction is generally less effective because cementation takes place soon after deposition. In any case, the cement is calcium carbonate derived by partial solution of some of the particles in the deposit.

TYPES OF SEDIMENTARY ROCKS

Thus far, we have considered the origin of sediment, its

Figure 6.16 Rounding and Sorting in Sediments

a Beginning students mistake rounding to mean ball-shaped or spherical. These three stones are all rounded, but only the one at the upper left is spherical.

b Deposit of moderately sorted gravel. The largest particle is about 5 cm across.

c Angular, poorly sorted gravel. Note the quarter for scale.

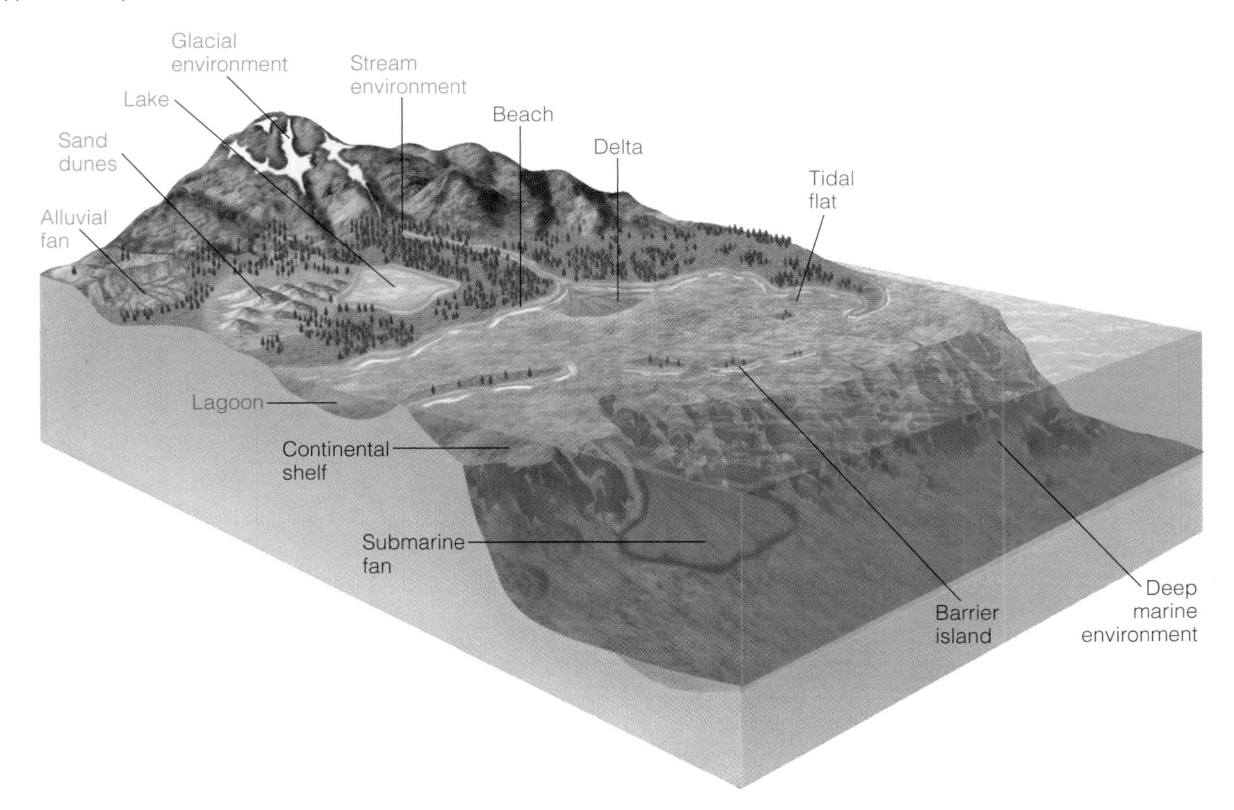

Figure 6.17 Depositional Environment Continental environments are shown in red type. The environments along the shoreline, shown in blue type, are transitional between continental and marine. The others, shown in black type, are marine environments.

transport, deposition, and lithification. We now turn to the types of sedimentary rocks and how they are classified. The two broad classes or types of sedimentary rocks are *detrital* and *chemical*, although the latter has a subcategory known as *biochemical* (Table 6.2).

Detrital Sedimentary Rocks

Detrital sedimentary rocks are made up of solid particles such as sand and gravel derived from parent material. All detrital sedimentary rocks have a *clastic texture*, meaning they are composed of particles or fragments known as *clasts*. The several varieties of detrital rocks are classified by the size of their constituent particles, although composition is used to modify some rock names.

Both *conglomerate* and *sedimentary breccia* are composed of gravel-sized particles (Figure 6.18 and Figure 6.19a, b); but conglomerate has rounded gravel, whereas sedimentary breccia has angular gravel. Conglomerate is common, but sedimentary breccia is rare because gravel becomes rounded very quickly during transport. Thus, if you encounter sedimentary breccia, you can conclude that its angular

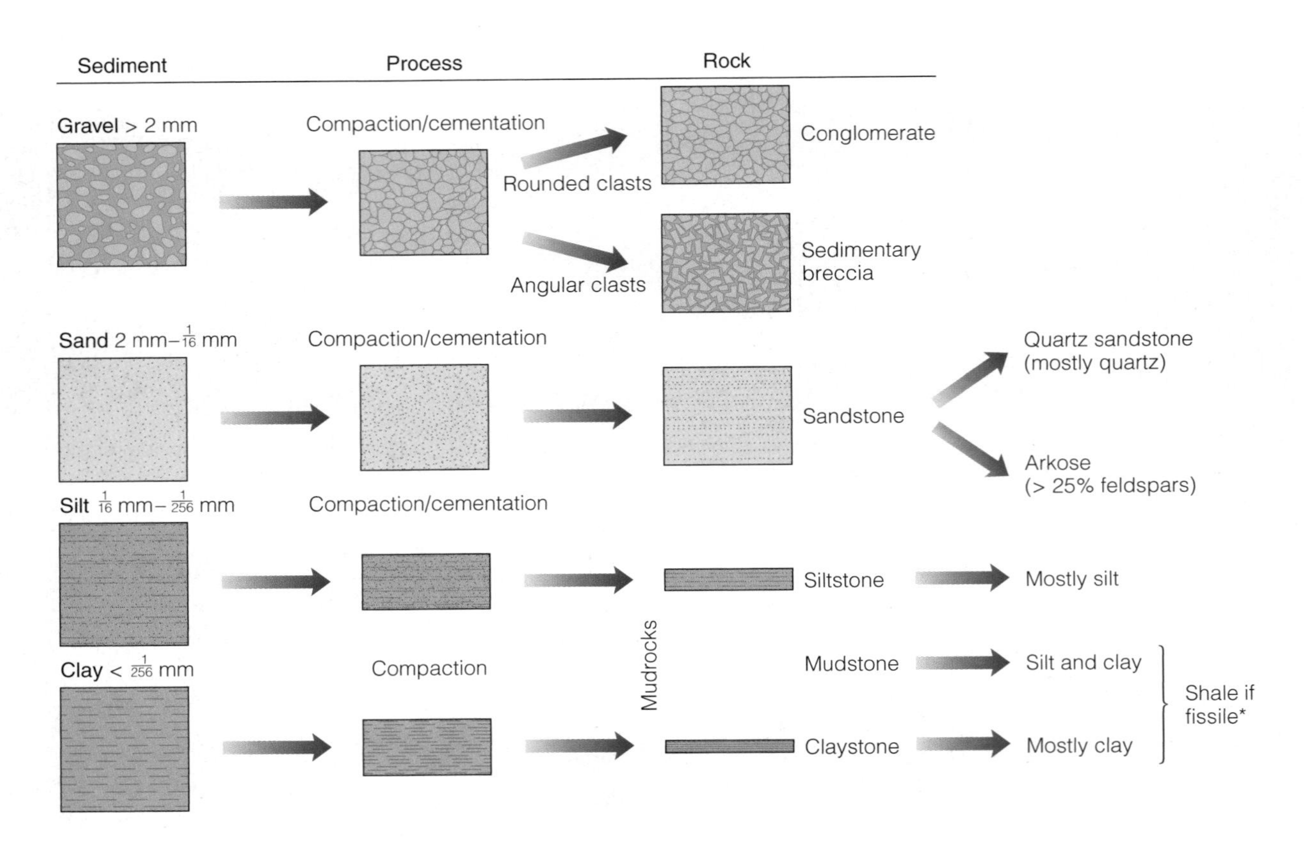

Figure 6.18 Lithification and Classification of Detrital Sedimentary Rocks Notice that little compaction takes place in gravel and sand.

TABLE 6.2 Classification of Chemical and Biochemical Sedimentary Rocks

Texture	Composition	Rock Name	
Chemical Sedimentary Rocks			
Varies	Calcite ($CaCO_3$)	Limestone	Carbonate rocks
Varies	Dolomite [$CaMg(CO_3)_2$]	Dolostone	Carbonate rocks
Crystalline	Gypsum ($CaSO_4 \cdot 2H_2O$)	Rock gypsum	Evaporites
Crystalline	Halite (NaCl)	Rock salt	Evaporites
Biochemical Sedimentary Rocks			
Clastic	Calcite ($CaCO_3$) shells	Limestone (various types, such as chalk and coquina)	
Usually crystalline	Altered microscopic shells of SiO_2	Chert (various color varieties)	
	Carbon from altered land plants	Coal (lignite, bituminous, anthracite)	

gravel has experienced little transport, probably less than a kilometer. Considerable energy is needed to transport gravel, so conglomerate is usually found in environments such as stream channels and beaches.

Sand is a size designation for particles between 1/16 and 2 mm, so any mineral or rock fragment can be in *sandstone*. Geologists recognize varieties of sandstone based on mineral content (Figure 6.18). *Quartz sandstone* is the most common and, as the name implies, is made up mostly of quartz sand. Another variety of sandstone called *arkose* contains at least 25% feldspar minerals. Sandstone is found in many depositional environments, including stream channels, sand dunes, beaches, barrier islands, deltas, and the continental shelf.

Mudrock is a general term that encompasses all detrital sedimentary rocks composed of silt- and clay-sized particles

Figure 6.19 Detrital and Sedimentary Rocks

a A layer of conglomerate overlying sandstone. The largest clast measures about 30 cm across.

b Sedimentary breccia in Death Valley, California. Notice the angular gravel-sized particles. The largest clast is about 12 cm across.

c Exposure of shale in Tennessee.

(Figure 6.18). Varieties include *siltstone* (mostly silt-sized particles), *mudstone* (a mixture of silt and clay), and *claystone* (primarily clay-sized particles). Some mudstones and claystones are designated *shale* if they are fissile, meaning that they break along closely spaced parallel planes (Figure 6.19c). Even weak currents transport silt- and clay-sized particles, and deposition takes place only where currents and fluid turbulence are minimal, as in the quiet offshore waters of lakes or in lagoons.

Chemical and Biochemical Sedimentary Rocks

Several compounds and ions taken into solution during chemical weathering are the raw materials for **chemical sedimentary rocks**. Some of these rocks have a *crystalline texture*, meaning they are composed of a mosaic of interlocking mineral crystals as in rock salt. Others, though, have a clastic texture; some limestones, for instance, are composed of fragmented seashells. Organisms play an important role in the origin of chemical sedimentary rocks designated **biochemical sedimentary rocks**.

Limestone and dolostone, the most abundant chemical sedimentary rocks, are known as **carbonate rocks** because each is made up of minerals that contain the carbonate radical (CO_3^{-2}). Limestone consists of calcite ($CaCO_3$), and dolostone is made up of dolomite [$CaMg(CO_3)_2$] (see Chapter 3). Recall that calcite rapidly dissolves in acidic water, but the chemical reaction leading to dissolution is reversible, so calcite can precipitate from solution in some circumstances. Thus, some limestone, though probably not very much, forms by inorganic chemical precipitation. Most limestone is biochemical because organisms are so important in its origin—the rock in coral reefs and limestone composed of seashells, for instance (Figure 6.20a, b).

A type of limestone composed almost entirely of fragmented seashells is known as *coquina* (Figure 6.20c), and

Figure 6.20 Varieties of Limestone

Sue Monroe

a Limestone with numerous fossil shells.

R.V. Dietrich

b The rock in these sea cliffs in Denmark is chalk, a type of limestone consisting of microscopic shells.

Sue Monroe

c Coquina is limestone composed of broken shells.

Sue Monroe

Stan Celestian/Glendale Community College

d This limestone is made up partly of ooids (see inset), which are rather spherical grains of calcium carbonate.

chalk is a soft variety of limestone made up mostly of microscopic shells (Figure 6.20b). One distinctive variety of limestone contains small spherical grains called *ooids* that have a small nucleus around which concentric layers of calcite precipitated (Figure 20d). Lithified deposits of ooids form *oolitic limestones.*

Dolostone is similar to limestone, but most or all of it formed secondarily by the alteration of limestone. The consensus among geologists is that dolostone originates when magnesium replaces some of the calcium in calcite, thereby converting calcite to dolomite.

Some of the dissolved substances derived by chemical weathering precipitate from evaporating water and thus form chemical sedimentary rocks known as **evaporites** (Table 6.2). *Rock salt*, composed of halite (NaCl), and *rock gypsum* ($CaSO_4 \cdot 2H_2O$) are the most common (Figure 6.21a, b), although several others are known and some are important resources. Compared with mudrocks, sandstone, and limestone, evaporites are not very common, but nevertheless are significant deposits in areas such as Michigan, Ohio, New York, the Gulf Coast region, and Saskatchewan, Canada.

Chert is a hard rock composed of microscopic crystals of quartz (Table 6.2 and Figure 6.21c). Some of the color varieties of chert are *flint*, which is black because of inclusions of organic matter, and *jasper*, which is colored red or brown by iron oxides. Because chert is hard and lacks cleavage, it can be shaped to form sharp cutting edges, so it has been used to manufacture tools, spear points, and arrowheads. Chert is found as irregular masses or *nodules* in other rocks, especially limestone, and as distinct layers of *bedded chert* made up of tiny shells of silica-secreting organisms.

Coal consists of compressed, altered remains of land plants, but it is nevertheless a biochemical sedimentary rock (Figure 6.21d). It forms in swamps and bogs where the water is oxygen deficient or where organic matter accumulates faster than it decomposes. In oxygen-deficient swamps and bogs, the bacteria that decompose vegetation can live without oxygen, but their wastes must be oxidized, and because little or no oxygen is present, wastes accumulate and kill the

Figure 6.21 Evaporites, Chert, and Coal

a This cylindrical core of rock salt was taken from an oil well in Michigan.

b Rock gypsum. When deeply buried, gypsum ($CaSO_4 \cdot 2H_2O$) loses its water and is converted to anhydrite ($CaSO_4$).

c Bedded chert exposed in Marin County, California. Most of the layers are about 5 cm thick.

d Bituminous coal is the most common type of coal used for fuel.

bacteria. Bacterial decay ceases, and the vegetation is not completely decomposed and forms organic muck. When buried and compressed, the muck becomes *peat*, which looks somewhat like coarse pipe tobacco. Where peat is abundant, as in Ireland and Scotland, it is used for fuel.

Peat represents the first step in forming coal. If peat is more deeply buried and compressed, and especially if it is heated, too, it is converted to dull black coal called *lignite*. During this change, the easily vaporized or volatile elements are driven off, enriching the residue in carbon; lignite has about 70% carbon, whereas only about 50% is present in peat. *Bituminous coal*, with about 80% carbon, is dense, black, and so thoroughly altered that plant remains are rarely seen. It burns more efficiently than lignite, but the highest-grade coal is *anthracite*, a metamorphic type of coal (see Chapter 7), with up to 98% carbon.

SEDIMENTARY FACIES

Long ago, geologists realized that if they traced a layer of sediment or sedimentary rock laterally, it generally changed in composition, texture, or both. These changes resulted from the simultaneous operation of different processes in adjacent depositional environments. For example, sand may be deposited in a high-energy nearshore marine environment, whereas mud and carbonate sediments accumulate simultaneously in the laterally adjacent low-energy offshore environments (Figure 6.22). Deposition in each environment produces **sedimentary facies**, bodies of sediment each possessing distinctive physical, chemical, and biological attributes. Figure 6.22 illustrates three sedimentary facies: a sand facies, a mud facies, and a carbonate facies. If these sediments become lithified, they are sandstone, mudstone (or shale), and limestone facies, respectively.

Many sedimentary rocks in the interiors of continents show clear evidence of deposition in marine environments. The rock layers in Figure 6.22 (left), for example, consist of a sandstone facies that was deposited in a nearshore marine environment overlain by shale and limestone facies deposited in offshore environments. Geologists explain this vertical sequence of facies by deposition occurring during a time when sea level rose with respect to the continents. As sea level rises, the shoreline moves inland, giving rise to a **marine transgression** (Figure 6.22), and the depositional environments parallel to the shoreline migrate landward. As a result, offshore facies are superimposed over nearshore facies, thus accounting for the vertical succession of sedimentary facies. Even though the nearshore environment is long and narrow at any particular time, deposition takes place continuously as the environment migrates landward. The sand deposit may be tens to hundreds of meters thick, but have

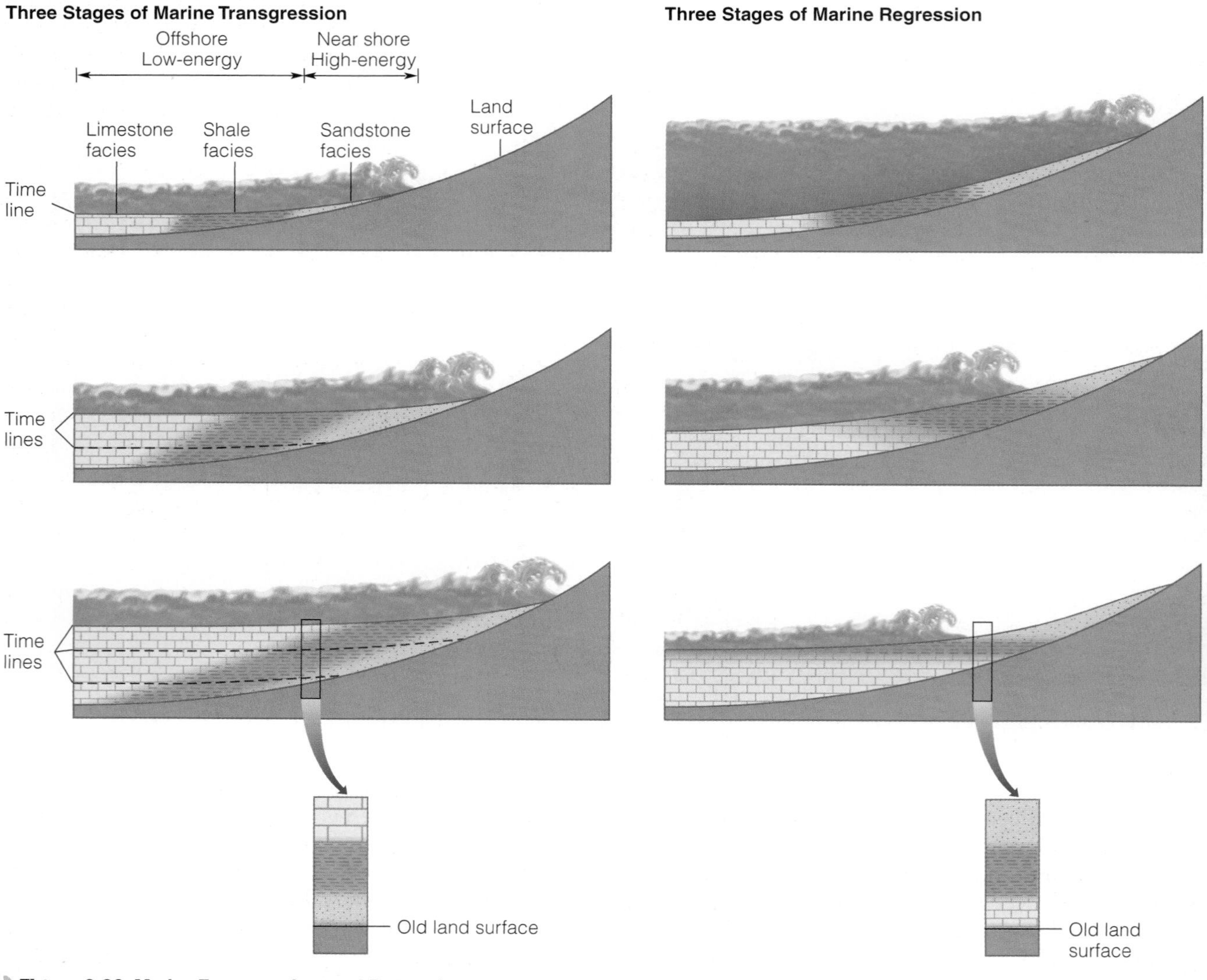

Figure 6.22 Marine Transgressions and Regressions

horizontal dimensions of length and width measured in hundreds of kilometers.

The opposite of a marine transgression is a **marine regression** (Figure 6.22). If sea level falls with respect to a continent, the shoreline and environments that parallel the shoreline move seaward. The vertical sequence produced by a marine regression has facies of the nearshore environment superposed over facies of offshore environments.

READING THE STORY IN SEDIMENTARY ROCKS

No one was present when ancient sediments were deposited, so geologists must evaluate those aspects of sedimentary rocks that allow them to make inferences about the original depositional environment. And making such determinations is of more than academic interest. For instance, barrier island sand deposits make good reservoirs for hydrocarbons, so knowing the environment of deposition and the geometry of these deposits is helpful in exploration for resources.

Sedimentary textures such as sorting and rounding can give clues to depositional processes. Windblown dune sands tend to be well sorted and well rounded, but poor sorting is typical of glacial deposits. The geometry or three-dimensional shape is another important aspect of sedimentary rock bodies. Marine transgressions and regressions yield sediment bodies with a blanket or sheetlike geometry, but sand deposits in stream channels are long and narrow and are described as having a shoestring geometry. Sedimentary textures and geometry alone are usually insufficient to determine depositional environment, but when considered with other sedimentary rock properties, especially *sedimentary structures* and *fossils*, they enable geologists to reliably determine the history of a deposit.

Sedimentary Structures

Physical and biological processes operating in depositional environments are responsible for a variety of features known as **sedimentary structures.** One of the most common is distinct layers known as **strata** or **beds** (Figure 6.23a), with individual layers of less than a millimeter up to many meters thick. These strata or beds are separated from one another by surfaces above and below in which the rocks differ in composition, texture, color, or a combination of features. Layering of some kind is present in almost all sedimentary rocks; however, a few, such as limestone that formed in coral reefs, lack this feature.

Many sedimentary rocks are characterized by **cross-bedding**, in which layers are arranged at an angle to the surface on which they are deposited (Figure 6.23b, c). Cross-beds are found in many depositional environments such as sand dunes in deserts and along shorelines, as well as in stream-channel deposits and shallow marine sediments. Invariably, cross-beds result from transport and deposition by wind or water currents, and the cross-beds are inclined downward in the same direction that the current flowed. Thus, ancient deposits with cross-beds inclined down toward the south, for example, indicate that the currents responsible for them flowed from north to south.

Some individual sedimentary rock layers show an upward decrease in grain size, termed **graded bedding**, mostly formed by turbidity current deposition. A *turbidity current* is an underwater flow of sediment and water with a greater density than sediment-free water. Because of its greater density, a turbidity current flows downslope until it reaches the relatively flat seafloor, or lakefloor, where it slows and begins depositing large particles followed by progressively smaller ones (Figure 6.24).

The surfaces that separate layers in sand deposits commonly have **ripple marks**, small ridges with intervening troughs, giving them a somewhat corrugated appearance. Some ripple marks are asymmetrical in cross section, with a gentle slope on one side and a steep slope on the other. Currents that flow in one direction, as in stream channels, generate these so-called *current ripple marks* (Figure 6.25a, b). And because the steep slope of these ripples is on the downstream side, they are good indications of ancient current directions. In contrast, *wave-formed ripple marks* tend to be symmetrical in cross section and, as their name implies, are generated by the to-and-fro motion of waves (Figure 6.25c, d).

When clay-rich sediment dries, it shrinks and develops intersecting fractures called **mud cracks** (Figure 6.26). Mud cracks in ancient sedimentary rocks indicate that the sediment was deposited in an environment where periodic drying took place, such as on a river floodplain, near a lakeshore, or where muddy deposits are exposed along seacoasts at low tide.

Fossils—Remains and Traces of Ancient Life

Fossils, the remains or traces of ancient organisms, are interesting as evidence of prehistoric life (Figure 6.27), and are also important for determining depositional environments. Most people are familiar with fossils of dinosaurs and some other land-dwelling animals, but are unaware that fossils of invertebrates, animals lacking a segmented vertebral column, such as corals, clams, oysters, and a variety of microorganisms, are much more useful because they are so common.

What Would You Do?

You live in the continental interior where flat-lying sedimentary rock layers are well exposed. Some local residents tell you of a location nearby where sandstone and mudstone with dinosaur fossils are overlain first by a seashell-bearing sandstone, followed upward by shale and finally limestone containing the remains of clams, oysters, and corals. How would you explain the presence of fossils, especially marine fossils so far from the sea, and how this vertical sequence of rocks came to be deposited?

Figure 6.23 Bedding (Stratification) and Cross Bedding

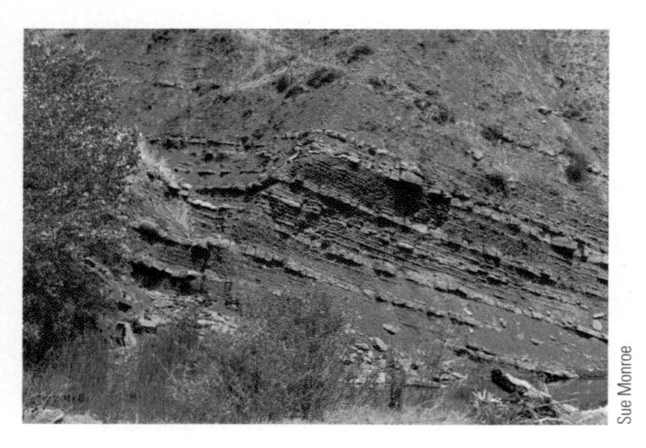

a Bedding or stratification is obvious in these alternating layers of sandstone and mudrock (shale).

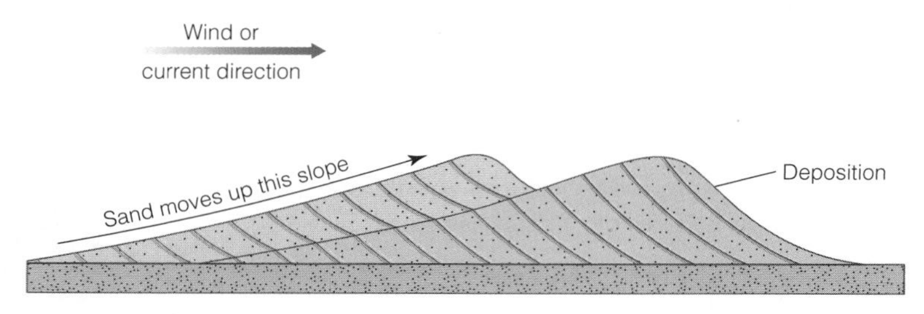

b Origin of cross-bedding by deposition on the sloping surface of a desert dune. Cross-bedding is also common in dunelike structures in stream and river channels.

c Cross-bedding in sandstone at Natural Bridges National Monument in Utah. The current moved from left to right.

Figure 6.24 Turbidity Currents and the Origin of Graded Bedding

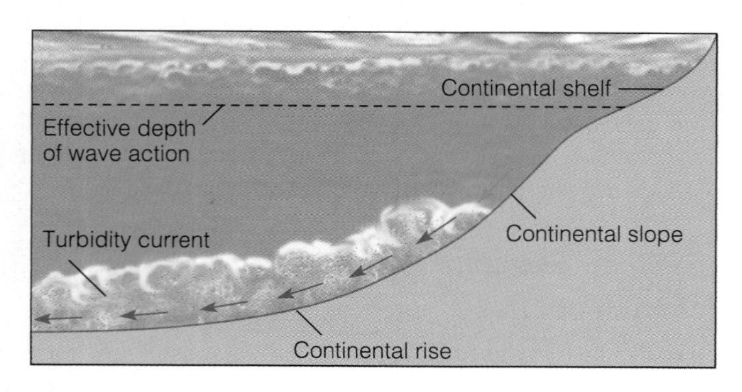

a A turbidity current flows downslope along the seafloor (or a lake bottom) because it is denser than sediment-free water.

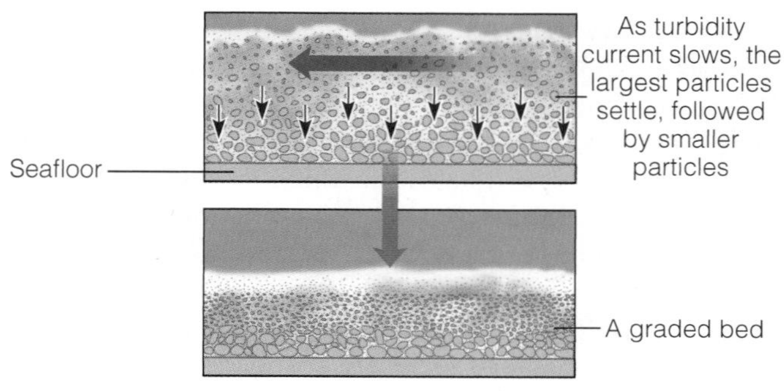

b The flow slows and deposits progressively smaller particles, thus forming a graded bed.

It is true that the remains of land-dwelling creatures and plants can be washed into marine environments, but most are preserved in rocks deposited on land or perhaps transitional environments such as deltas. In contrast, fossils of corals tell us that the rocks in which they are preserved were deposited in the ocean.

Figure 6.25 Current and Wave-Formed Ripple Marks

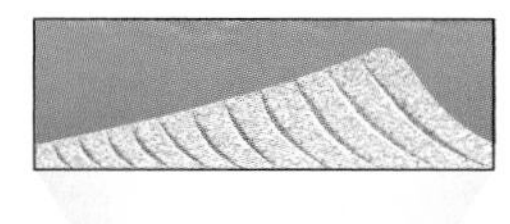

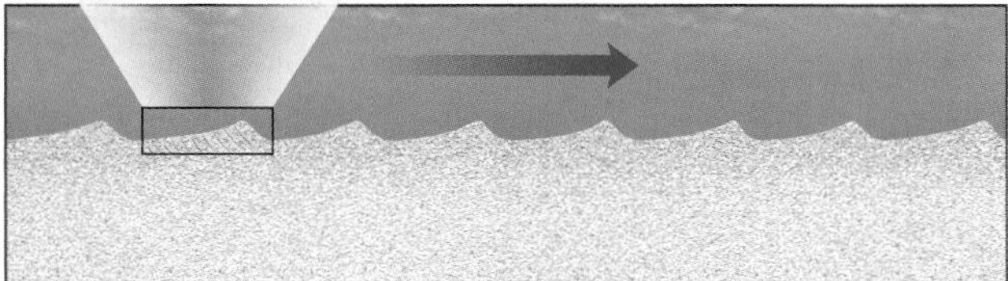

a Current ripple marks form in response to currents that flow in one direction, as in a stream. The enlargement shows the cross-beds in an individual ripple.

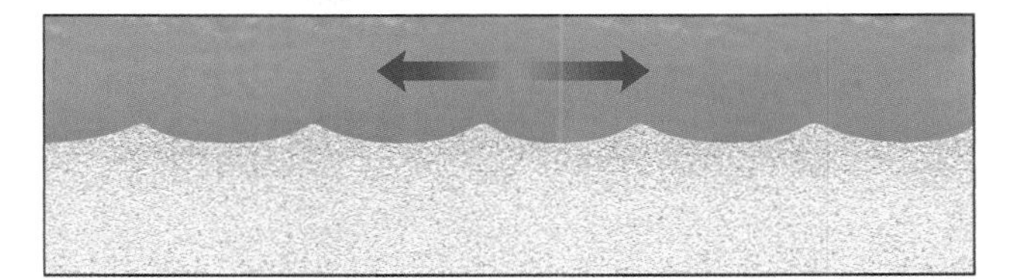

c The to-and-fro motion of waves in shallow water yield wave-formed ripple marks.

b Current ripple marks that formed in a stream channel. Flow was from right to left.

d Wave-formed ripple marks in sand in shallow seawater.

Figure 6.26 Mud Cracks Mud cracks form in clay-rich sediments when they dry and contract.

a Mud cracks in a present-day environment.

b Ancient mud cracks in Glacier National Park in Montana. Note that the cracks have been filled in with sediment.

Clams with heavily constructed shells typically live in shallow turbulent seawater, whereas organisms living in low-energy environments commonly have thin, fragile shells. Marine organisms that carry on photosynthesis are restricted to the zone of sunlight penetration, which is usually less than 200 m. The amount of sediment is also a limiting factor; for example, many corals live in shallow, clear seawater because suspended sediment clogs their respiratory and food-gathering organs, and some have photosynthesizing algae living in their tissues.

Figure 6.27 Fossils

Sue Monroe

a Skull of the dinosaur *Allosaurus* on display at the Natural History Museum in Vienna, Austria.

Sue Monroe

b Shells of extinct ocean-dwelling animals known as horn corals.

Microfossils are particularly useful for environmental studies because hundreds or even thousands can be recovered from small rock samples. In oil-drilling operations, small rock chips known as *well cuttings* are brought to the surface. These samples may contain numerous microfossils, but rarely have entire fossils of larger organisms. These fossils are routinely used to determine depositional environments and to match up rocks of the same relative age (see Chapter 17).

Determining the Environment of Deposition

Geologists rely on textures, sedimentary structures, and fossils to interpret how a particular sedimentary rock body was deposited. Furthermore, they compare the features seen in ancient rocks with those in deposits forming today. But are we justified in using present-day processes and environments to make inferences about what happened when no human observers were present? Perhaps some examples will help answer this question.

The Navajo Sandstone of the southwestern United States is an ancient desert dune deposit that formed when the prevailing winds blew from the northeast. What evidence justifies this conclusion? This 300-m-thick sandstone is made up of well-sorted, well-rounded sand grains measuring 0.2–0.5 mm in diameter. Furthermore, it has cross-beds up to 30 m high (Figure 6.28a) and current ripple marks, both typical of desert dunes. Some of the sand layers have preserved dinosaur tracks and tracks of other land-dwelling animals, ruling out the possibility of a marine origin. In short, the Navajo Sandstone possesses features that point to a desert dune depositional environment. Finally, the cross-beds are inclined downward toward the southwest, indicating that the prevailing winds were from the northeast.

In the Grand Canyon of Arizona, several formations are well exposed; a *formation* is a widespread unit of rock, especially sedimentary rock, that is recognizably different from the rocks above and below. A vertical sequence consisting of the Tapeats Sandstone, Bright Angel Shale, and Muav Limestone is present in the lower part of the canyon (Figure 6.28b), all of which contain features, including fossils, clearly indicating that they were deposited in transitional and marine environments. As a matter of fact, all three were forming simultaneously in different adjacent environments, and during a marine transgression they were deposited in the vertical sequence now seen. They conform closely to the sequence shown in Figure 6.22.

In some of the later chapters on Earth history, we refer to river deposits, ancient deltas, carbonate shelf deposits, and transgressive-regressive sequences. We cannot include all of the supporting evidence for these interpretations, but we can say that they are based on the kinds of criteria discussed in this chapter.

IMPORTANT RESOURCES IN SEDIMENTARY ROCKS

Sand and gravel are essential to the construction industry; pure clay deposits are used for ceramics, and limestone is used in the manufacture of cement and in blast furnaces where iron ore is refined to make steel. Evaporites are the source of table salt, as well as chemical compounds, and rock gypsum is used to manufacture wallboard. Phosphate-bearing sedimentary rock is used in fertilizers and animal feed supplements.

Some valuable sedimentary deposits are found in streams and on beaches where minerals were concentrated during transport and deposition. These *placer deposits* are surface accumulations resulting from the separation and concentration of materials of greater density from those of lesser density. Much of the gold recovered during the initial stages of the California gold rush (1849–1853) was mined from placer deposits, and placers of other minerals such as diamonds and tin are important.

Figure 6.28 Ancient Sedimentary Rocks and Their Interpretation

James S. Monroe

a The Jurassic-aged Navajo Sandstone in Zion National Park in Utah is a wind-blown dune deposit. Vertical fractures intersect cross-beds, hence the name Checkerboard Mesa for this rock exposure.

Alan Mayo/GeoPhoto

b View of three formations in the Grand Canyon of Arizona. These rocks were deposited during a marine transgression. Compare with the vertical sequence of rocks in Figure 6.22.

Historically, most coal mined in the United States has been bituminous coal from the Appalachian region that formed in coastal swamps during the Pennsylvanian Period (299–318 million years ago) (see Chapter 20). Huge lignite and subbituminous coal deposits in the western United States are becoming increasingly important. During 2005, more than a billion tons of coal were mined in this country, more than 60% of it from mines in Wyoming, West Virginia, and Kentucky.

Anthracite coal (see Chapter 7) is especially desirable because it burns more efficiently than other types of coal. Unfortunately, it is the least common variety, so most coal used for heating buildings and generating electricity is bituminous (Figure 6.21d). *Coke,* a hard, gray substance consisting of the fused ash of bituminous coal, is used in blast furnaces where steel is produced. Synthetic oil and gas and a number of other products are also made from bituminous coal and lignite.

Petroleum and Natural Gas

Petroleum and natural gas are *hydrocarbons,* meaning that they are composed of hydrogen and carbon. The remains of microscopic organisms settle to the seafloor, or lakefloor in some cases, where little oxygen is present to decompose them. If buried beneath layers of sediment, they are heated and transformed into petroleum and natural gas. The rock in which hydrocarbons form is known as *source rock,* but for them to accumulate in economic quantities, they must migrate from the source rock into some kind of *reservoir rock.* And finally, the reservoir rock must have an overlying, nearly impervious *cap rock;* otherwise, the hydrocarbons would eventually reach the surface and escape (Figure 6.29a, b). Effective reservoir rocks must have appreciable pore space and good permeability, the capacity to transmit fluids; otherwise, hydrocarbons cannot be extracted from them in reasonable quantities.

Many hydrocarbon reservoirs consist of nearshore marine sandstones with nearby fine-grained, organic-rich source rocks. Such oil and gas traps are called *stratigraphic traps* because they owe their existence to variations in the strata (Figure 6.29a). Indeed, some of the oil in the Persian Gulf region and Michigan is trapped in ancient reefs that are also good stratigraphic traps. *Structural traps* result when rocks are deformed by folding, fracturing, or both. In sedimentary rocks that have been deformed into a series of folds, hydrocarbons migrate to the high parts of these structures (Figure 6.29b). Displacement of rocks along faults (fractures along which movement has occurred) also yields traps for hydrocarbons (Figure 6.29b).

Other sources of petroleum that will probably become increasingly important in the future include *oil shales* and *tar sands.* The United States has about two-thirds of all known oil shales, although large deposits are known in South America, and all continents have some oil shale. The richest deposits in the United States are in the Green River Formation of Colorado, Utah, and Wyoming. When the appropriate extraction processes are used, liquid oil and combustible gases can be produced from an organic substance called *kerogen,* which is found in oil shale (Figure 6.29c). Oil shale in the Green River Formation yields between 10 and 140 gallons

Figure 6.29 Oil and Natural Gas The arrows in a and b indicate the direction of migration of hydrocarbons.

a Two examples of stratigraphic traps: one in sand within shale and the other in a buried reef.

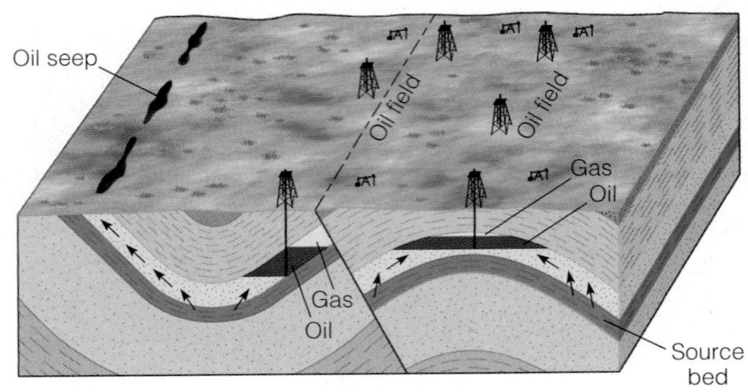

b Two examples of structural traps: one formed by folding and the other by faulting.

c The sedimentary rock oil shale (left) and oil extracted from it. The United States has vast oil shale deposits.

of oil per ton of rock processed, and the total amount of oil recoverable with present processes is estimated at 80 billion barrels. Currently, no oil is produced from oil shale in the United States because conventional drilling and pumping are less expensive.

Tar sand is a type of sandstone in which viscous, asphalt-like hydrocarbons fill the pore spaces. This substance is the sticky residue of once-liquid petroleum from which the volatile constituents have been lost. Liquid petroleum can be recovered from tar sand, but for this to happen, large quantities of rock must be mined and processed. Because the United States has few tar-sand deposits, it cannot look to this source as a significant future energy resource. The Athabaska tar sands in Alberta, Canada, however, are one of the largest deposits of this type. These deposits are currently being mined, and it is estimated that they contain several hundred billion barrels of recoverable petroleum.

Uranium

Most of the uranium used in nuclear reactors in North America comes from the complex potassium-, uranium-, vanadium-bearing mineral *carnotite*, which is found in some sedimentary rocks. Some uranium is also derived from *uraninite* (UO_2), a uranium oxide in granitic rocks and hydrothermal veins. Uraninite is easily oxidized and dissolved in groundwater, transported elsewhere, and chemically reduced and precipitated in the presence of organic matter.

The richest uranium ores in the United States are widespread in the Colorado Plateau area of Colorado and adjoining parts of Wyoming, Utah, Arizona, and New Mexico. These ores, consisting of fairly pure masses and encrustations of carnotite, are associated with plant remains in sandstones that formed in ancient stream channels. Although most of these ores are associated with fragmentary plant remains, some petrified trees also contain large quantities of uranium (Figure 6.30a).

Large reserves of low-grade uranium ore also are found in the Chattanooga Shale. The uranium is finely disseminated in this black, organic-rich mudrock that underlies large parts of several states, including Illinois, Indiana, Ohio, Kentucky, and Tennessee. Canada is the world's largest producer and exporter of uranium.

Figure 6.30 Carnotite and Banded Iron Formation

Wesley Pymm

a The yellowish mineral on this piece of petrified wood from Colorado is carnotite, which is an ore of uranium.

James S. Monroe

b This banded iron formation at Ishpeming, Michigan, consists of alternating layers of red chert and silver-colored iron minerals.

Banded Iron Formation

The chemical sedimentary rock known as *banded iron formation* consists of alternating thin layers of chert and iron minerals, mostly the iron oxides hematite and magnetite (Figure 6.30b). Banded iron formations are present on all continents and account for most of the iron ore mined in the world today. Vast banded iron formations are present in the Lake Superior region of the United States and Canada and in the Labrador Trough of eastern Canada. We will consider the origin of banded iron formations in Chapter 19.

Geo-Recap

Chapter Summary

- Mechanical and chemical weathering disintegrate and decompose parent material so that it is more nearly in equilibrium with new physical and chemical conditions. The products of weathering include solid particles and substances in solution.
- Mechanical weathering includes such processes as frost action, pressure release, salt crystal growth, thermal expansion and contraction, and the activities of organisms. Particles liberated by mechanical weathering retain the chemical composition of the parent material.
- The chemical weathering processes of solution, oxidation, and hydrolysis bring about chemical changes of the parent material. Clay minerals and substances in solution form during chemical weathering.
- Mechanical weathering aids chemical weathering by breaking parent material into smaller pieces, thereby exposing more surface area.
- Mechanical and chemical weathering produce regolith, some of which is soil if it consists of solids, air, water, and humus and supports plant growth.
- Soils are characterized by horizons that are designated, in descending order, as O, A, B, and C. Soil horizons differ from one another in texture, structure, composition, and color.
- Soils called pedalfers develop in humid regions, whereas arid and semiarid region soils are pedocals. Laterite is a soil that results from intense chemical weathering in the tropics.
- Soil erosion, caused mostly by sheet and rill erosion, is a problem in some areas. Human practices such as construction, agriculture, and deforestation can accelerate losses of soil to erosion.
- Sedimentary particles are designated in order of decreasing size as gravel, sand, silt, and clay.
- Sedimentary particles are rounded and sorted during transport, although the degree of rounding and sorting depends on particle size, transport distance, and depositional process.
- Any area in which sediment is deposited is a depositional environment. Major depositional settings are continental, transitional, and marine, each of which includes several specific depositional environments.

- Lithification involves compaction and cementation, which convert sediment into sedimentary rock. Silica and calcium carbonate are the most common chemical cements, but iron oxide and iron hydroxide cements are important in some rocks.
- Detrital sedimentary rocks consist of solid particles derived from preexisting rocks. Chemical sedimentary rocks are derived from substances in solution by inorganic chemical processes or the biochemical activities of organisms. Geologists also recognize a subcategory called biochemical sedimentary rocks.
- Sedimentary facies are bodies of sediment or sedimentary rock that are recognizably different from adjacent sediments or rocks.
- Some sedimentary facies are geographically widespread because they were deposited during marine transgressions or marine regressions.
- Sedimentary structures such as bedding, cross-bedding, and ripple marks commonly form in sediments when, or shortly after, they are deposited.
- Geologists determine the depositional environments of ancient sedimentary rocks by studying sedimentary textures and structures, examining fossils, and making comparisons with present-day depositional processes.
- Intense chemical weathering is responsible for the origin of residual concentrations, many of which contain valuable minerals such as iron, lead, copper, and clay.
- Many sediments and sedimentary rocks, including sand, gravel, evaporites, coal, and banded iron formations, are important resources. Most oil and natural gas are found in sedimentary rocks.

Important Terms

bed (p. 157)
biochemical sedimentary rock (p. 153)
carbonate rock (p. 153)
cementation (p. 149)
chemical sedimentary rock (p. 153)
chemical weathering (p. 138)
compaction (p. 149)
cross-bedding (p. 157)
depositional environment (p. 148)
detrital sedimentary rock (p. 151)
differential weathering (p. 135)
erosion (p. 134)
evaporite (p. 154)
exfoliation dome (p. 135)
fossil (p. 157)
frost action (p. 135)
graded bedding (p. 157)
hydrolysis (p. 140)
laterite (p. 144)
lithification (p. 149)
marine regression (p. 157)
marine transgression (p. 156)
mechanical weathering (p. 135)
mud crack (p. 157)
oxidation (p. 140)
parent material (p. 134)
pedalfer (p. 143)
pedocal (p. 143)
pressure release (p. 137)
regolith (p. 142)
ripple mark (p. 157)
salt crystal growth (p. 135)
sediment (p. 147)
sedimentary facies (p. 156)
sedimentary rock (p. 147)
sedimentary structure (p. 157)
soil (p. 142)
soil degradation (p. 146)
soil horizon (p. 143)
solution (p. 139)
spheroidal weathering (p. 141)
strata (p. 157)
talus (p. 135)
thermal expansion and contraction (p. 135)
weathering (p. 134)

Review Questions

1. Which one of the following is a chemical weathering process?
   a. _____ lithification;
   b. _____ oxidation;
   c. _____ frost wedging;
   d. _____ pressure release;
   e. _____ compaction.
2. The type of soil found in the semiarid to arid Southwest United States is
   a. _____ breccia;
   b. _____ laterite;
   c. _____ pedocal;
   d. _____ regolith;
   e. _____ detrital.
3. Which one of the following sedimentary structures would allow you to determine ancient current directions?
   a. _____ cross-beds;
   b. _____ mud cracks;
   c. _____ evaporites;
   d. _____ wave-formed ripples;
   e. _____ caliche.
4. Spheroidal weathering takes place because
   a. _____ thermal expansion and contraction are so effective;
   b. _____ aluminum oxides are nearly insoluble;
   c. _____ oxidation changes limestone to dolostone;
   d. _____ rounding takes place rapidly during transport;
   e. _____ corners and edges of stones weather faster than flat surfaces.

5. Exfoliation domes form by a mechanical weathering process called
   a. ______ pressure release;
   b. ______ salt crystal growth;
   c. ______ hydrolysis;
   d. ______ regression;
   e. ______ granitization.
6. Lithification involves compaction and
   a. ______ oxidation;
   b. ______ carbonization;
   c. ______ cementation;
   d. ______ graded bedding;
   e. ______ degradation.
7. A deposit of sand in which all of the particles are nearly the same size is characterized as
   a. ______ well sorted;
   b. ______ carbonate;
   c. ______ angular;
   d. ______ spheroidal;
   e. ______ moderately lithified.
8. A marine transgression yields a vertical sequence
   a. ______ of rounded domes of rock with sheet joints;
   b. ______ that includes interlayered igneous and metamorphic rocks;
   c. ______ with offshore facies overlying nearshore facies;
   d. ______ of evaporates, granite, and parent material;
   e. ______ of delta deposits overlain by turbidites.
9. Dolostone forms from limestone when
   a. ______ horizon A is eroded leaving horizon B exposed;
   b. ______ magnesium replaces some of the calcium in limestone;
   c. ______ organic matter accumulates in a swamp;
   d. ______ volcanic rocks are adjacent to sedimentary rocks;
   e. ______ seawater evaporates and calcium sulfate is deposited.
10. The unconsolidated material covering much of Earth's land surface, as well as most of the seafloor, is called
   a. ______ regolith;
   b. ______ conglomerate;
   c. ______ limestone;
   d. ______ rock gypsum;
   e. ______ parent material.
11. How do climate, particle size, and parent material control the rate of chemical weathering?
12. Describe two sedimentary structures that can be used to determine ancient current directions.
13. How is it possible to determine how sedimentary rocks were deposited given that no human observers were present to witness their deposition?
14. Explain how deposits of mud and sand are lithified.
15. What are residual concentrations and how do they form? Are any of them valuable?
16. How can you account for the fact that some of the rock layers in Figure 6.28b have steep slopes, whereas others have gentle slopes? Also, given that limestone is soluble in the presence of acidic groundwater, why does the Muav Limestone have such a bold exposure at this location?
17. Why do people in tropical areas practice slash-and-burn agriculture?
18. How do the following pairs of detrital sedimentary rocks compare? Conglomerate and sedimentary breccia; quartz sandstone and arkose.
19. What are the similarities between the mechanical weathering processes frost wedging and salt crystal growth?
20. Suppose you encounter the sedimentary rocks shown in the illustration below. Interpret the geologic history of this area as thoroughly as possible. (An answer is provided at the end of the book.)

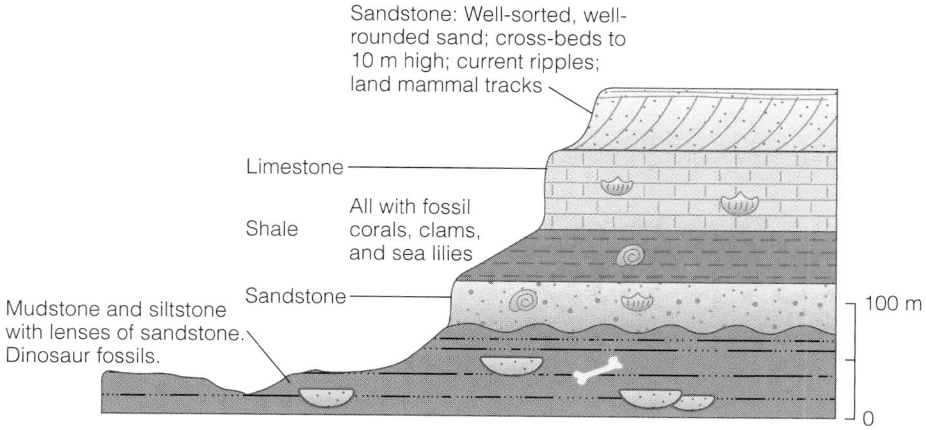

Sandstone: Well-sorted, well-rounded sand; cross-beds to 10 m high; current ripples; land mammal tracks
Limestone
Shale
All with fossil corals, clams, and sea lilies
Sandstone
Mudstone and siltstone with lenses of sandstone. Dinosaur fossils.
100 m
0

CHAPTER 7

Metamorphism and Metamorphic Rocks

OUTLINE

OBJECTIVES

At the end of this chapter, you will have learned that

- Metamorphic rocks result from the transformation of other rocks by various processes occurring beneath Earth's surface.
- Heat, pressure, and fluid activity are the three agents of metamorphism.
- Contact, dynamic, and regional metamorphism are the three types of metamorphism.
- Metamorphic rocks are typically divided into two groups, foliated and nonfoliated, primarily on the basis of texture.
- Metamorphic rocks with a foliated texture include slate, phyllite, schist, gneiss, amphibolite, and migmatite.
- Metamorphic rocks with a nonfoliated texture include marble, quartzite, greenstone, hornfels, and anthracite.
- Metamorphic rocks can be grouped into metamorphic zones based on the presence of index minerals that form under specific temperature and pressure conditions.
- The successive appearance of particular metamorphic minerals indicates increasing or decreasing metamorphic intensity.

Lining the Belvedere of Infinity balcony at the Hotel Villa Cimbrone in Ravello, Italy, are these beautifully sculpted white marble Roman busts. White marble is the result of metamorphism of very pure limestone. Because of its softness, relative resistance to shattering, and waxy, life-like glow, it has been the marble of choice for sculptors of the human body since classical times. In the background is the Tyrrhenian Sea and Amalfi coast.

- Metamorphism is associated with all three types of plate boundaries but is most widespread along convergent plate boundaries.
- Many metamorphic minerals and rocks are valuable metallic ores, building materials, and gemstones.

INTRODUCTION

Its homogeneity, softness, and various textures have made marble, a metamorphic rock formed from limestone or dolostone, a favorite rock of sculptors throughout history. As the value of authentic marble sculptures has increased over the years, the number of forgeries has also increased. With the price of some marble sculptures in the millions of dollars, private collectors and museums need some means of ensuring the authenticity of the work that they are buying. Aside from the monetary considerations, it is important that forgeries not become part of the historical and artistic legacy of human endeavor.

Experts have traditionally relied on artistic style and weathering characteristics to determine whether a marble sculpture is authentic or a forgery. Because marble is not very resistant to weathering, however, forgers have been able to produce the weathered appearance of an authentic work.

Using newly developed techniques, geologists can now distinguish a naturally weathered marble surface from one that has been artificially altered. Yet, there are examples in which expert opinion is still divided on whether a sculpture is authentic. One of the best examples is the Greek *kouros* (a sculptured figure of a Greek youth) that the J. Paul Getty Museum in Malibu, California, purchased for a reputed price of $7 million in 1984. Because some of its stylistic features caused some experts to question its authenticity, the museum had a variety of geochemical and mineralogical tests performed in an effort to determine the authenticity of the sculpture.

Although numerous scientific tests have not unequivocally proved authenticity, they have shown that the weathered surface layer of the kouros bears more similarities to naturally occurring weathered surfaces of dolomitic marble than to known artificially produced surfaces. Furthermore, no evidence indicates that the surface alteration of the kouros is of modern origin.

Unfortunately, despite intensive study by scientists, archaeologists, and art historians, opinion is still divided as to the authenticity of the Getty kouros. Most scientists accept that the kouros was carved sometime around 530 B.C. Pointing to inconsistencies in its style of sculpture for that period, other art historians think that it is a modern forgery.

Regardless of whether the Getty kouros is proven to be authentic or a forgery, geologic testing to authenticate marble sculptures is now an important part of many museums' curatorial functions. To help geologists in the authentication of marble sculptures, a large body of data about the characteristics and origin of marble is being amassed as more sculptures and their quarries are analyzed.

Having examined igneous and sedimentary rocks, we now turn our attention to the third major rock group, the **metamorphic rocks** (from the Greek *meta*, "change," and *morpho*, "shape"), which result from the transformation of other rocks by processes that typically occur beneath Earth's surface (see Figure 1.14). During **metamorphism**, rocks are subjected to sufficient heat, pressure, and fluid activity to change their mineral composition, texture, or both. These changes result in metamorphic rocks that usually do not look anything like the original rock before it was metamorphosed. These transformations take place below the melting temperature of the rock; otherwise, an igneous rock would result.

A useful analogy for metamorphism is baking a cake. Just like a metamorphic rock, the resulting cake depends on the ingredients, their proportions, how they are mixed together, how much water or milk is added, and the temperature and length of time used for baking the cake.

Except for marble and slate, most people are not familiar with metamorphic rocks. We are frequently asked by students why it is important to study metamorphic rocks and processes. Our answer is always, "Just look around you."

A large portion of Earth's continental crust is composed of metamorphic and igneous rocks. Together, they form the crystalline basement rocks underlying the sedimentary rocks of a continent's surface. These basement rocks are widely exposed in regions of the continents known as *shields*, which have been very stable during the past 600 million years (Figure 7.1). Some of the oldest known rocks, dated at 3.96 billion years from the Canadian Shield, are metamorphic, which means that they formed from even older rocks!

Metamorphic rocks such as marble and slate are used as building materials, and some metamorphic minerals, such as garnets (used as gemstones or abrasives) and talc (used in cosmetics, in manufacturing paint, and as a lubricant), are economically important. Asbestos, another valuable metamorphic mineral, is used for insulation and fireproofing and has been the subject of much debate over the danger it poses to the public's health (see Geo-Focus on pages 172 and 173).

THE AGENTS OF METAMORPHISM

The three principal agents of change that cause metamorphism are *heat*, *pressure*, and *fluid activity*. Time is also important to the metamorphic process because chemical reactions taking place in rocks undergoing metamorphism proceed at different rates and thus require different amounts of time to complete. Reactions involving silicate compounds are particularly slow, and because most metamorphic rocks are composed of silicate minerals, it is thought that metamorphism is a very slow geologic process.

During metamorphism, the original rock, which was in equilibrium with its environment, meaning that it was chemically and physically stable under those conditions, undergoes changes to achieve equilibrium with its new environment. These changes may result in the formation of new minerals, a change in the texture of the rock, or both. In some instances, the change is minor, and features of the original rock can still be recognized. In other cases, the rock changes so much that the identity of the original rock can be determined only with great difficulty, if at all.

Heat

Heat is an important agent of metamorphism because it increases the rate of chemical reactions that may produce minerals different from those in the original rock. Heat may come from extrusive lava, intrusive magma, or as a result of deep burial in the crust due to subduction along a convergent plate boundary.

When rocks are intruded by bodies of magma, they are subjected to intense heat that affects the surrounding rock; the most intense heating usually occurs adjacent to the magma body and gradually decreases with distance from the intrusion. The zone of metamorphosed rocks that forms in the country rock adjacent to an intrusive igneous body is usually distinct and easy to recognize.

Recall that temperature increases with depth and that Earth's geothermal gradient averages about 25°C/km. Rocks that form at the surface may be transported to great depths by subduction along a convergent plate boundary and thus subjected to increasing temperature and pressure. During subduction, some minerals may be transformed into other minerals that are more stable under the higher temperature and pressure conditions.

Pressure

During burial, rocks are subjected to increasingly greater pressure, just as you feel greater pressure the deeper you dive in a body of water. Whereas the pressure you feel is known as *hydrostatic pressure*, because it comes from the water surrounding you, rocks undergo **lithostatic pressure**. This means that the stress (force per unit area) on a rock in Earth's crust is the same in all directions (Figure 7.2a). A similar situation occurs when an object is immersed in water. For example, the deeper a cup composed of Styrofoam™ is submerged in the ocean, the smaller it gets because pressure increases with depth and is exerted on the cup equally in all directions, thereby compressing the cup composed of Styrofoam™ (Figure 7.2b).

Along with lithostatic pressure resulting from burial, rocks may also experience **differential pressure** (Figure 7.3). In this case, the stresses are not equal in all directions, but are stronger from some directions than from others. Differential pressures typically occur when two plates collide. In this case, the horizontal stress from collision is greater than the vertical stress from burial, resulting in distinctive metamorphic textures and features in the affected rocks.

What Would You Do?

As the director of a major museum, you have the opportunity to purchase, for a considerable sum of money, a newly discovered marble bust by a famous ancient sculptor. You want to be sure it is not a forgery. What would you do to ensure that the bust is authentic and not a clever forgery? After all, you are spending a large sum of the museum's money. As a nonscientist, how would you go about making sure that the proper tests are performed to authenticate the bust?

Fluid Activity

In almost every region of metamorphism, water and carbon dioxide (CO_2) are present in varying amounts along mineral grain boundaries or in the pore spaces of rocks. These fluids, which may contain ions in solution, enhance metamorphism by increasing the rate of chemical reactions. Under dry conditions, most minerals react very slowly, but when even small amounts of fluid are introduced, reaction rates increase, mainly because ions can move readily through the fluid and thus enhance chemical reactions and the formation of new minerals.

The following reaction provides a good example of how new minerals can be formed by **fluid activity**. Seawater moving through hot basaltic rock in the oceanic crust transforms olivine into the metamorphic mineral serpentine:

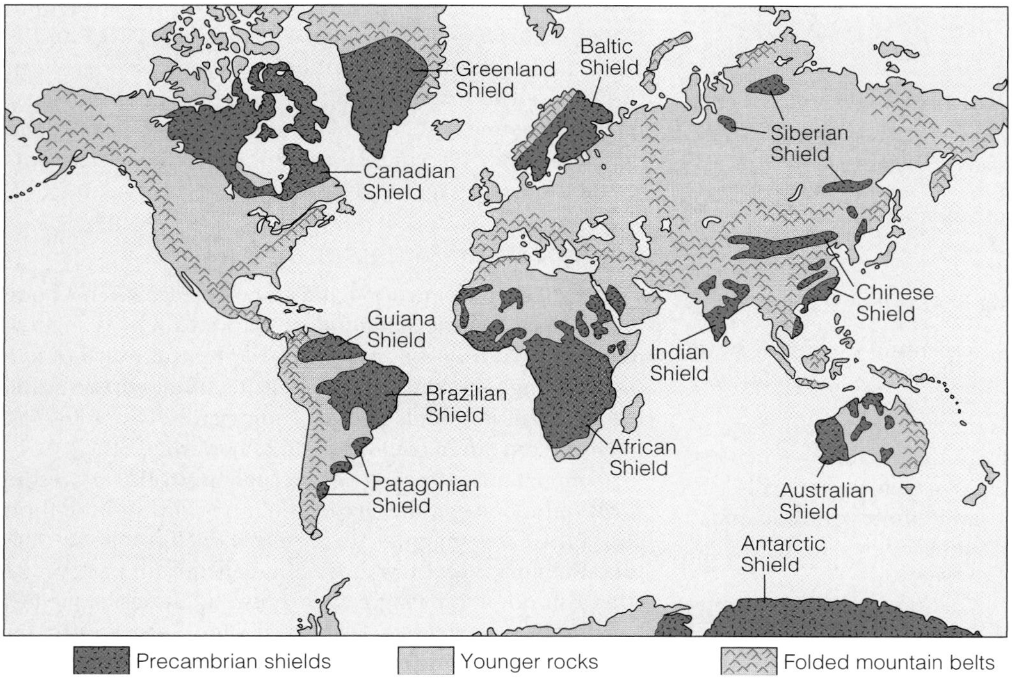

Figure 7.1 Metamorphic Rock Occurrences Shields are the exposed portions of the crystalline basement rocks underlying each continent. These areas have been very stable during the past 600 million years. Metamorphic rocks also constitute the crystalline core of major mountain belts.

Figure 7.2 Lithostatic Pressure

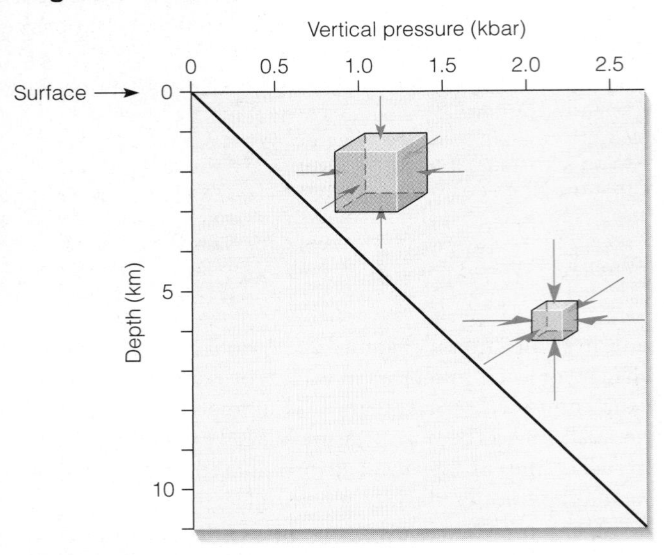

a Lithostatic pressure is applied equally in all directions in Earth's crust due to the weight of overlying rocks. Thus, pressure increases with depth, as indicated by the sloping black line.

David J. and Jane M. Matty

b A similar situation occurs when 200-ml cups composed of Styrofoam™ are lowered to ocean depths of approximately 750 m and 1500 m. Increased water pressure is exerted equally in all directions on the cups, and they consequently decrease in volume while maintaining their general shape.

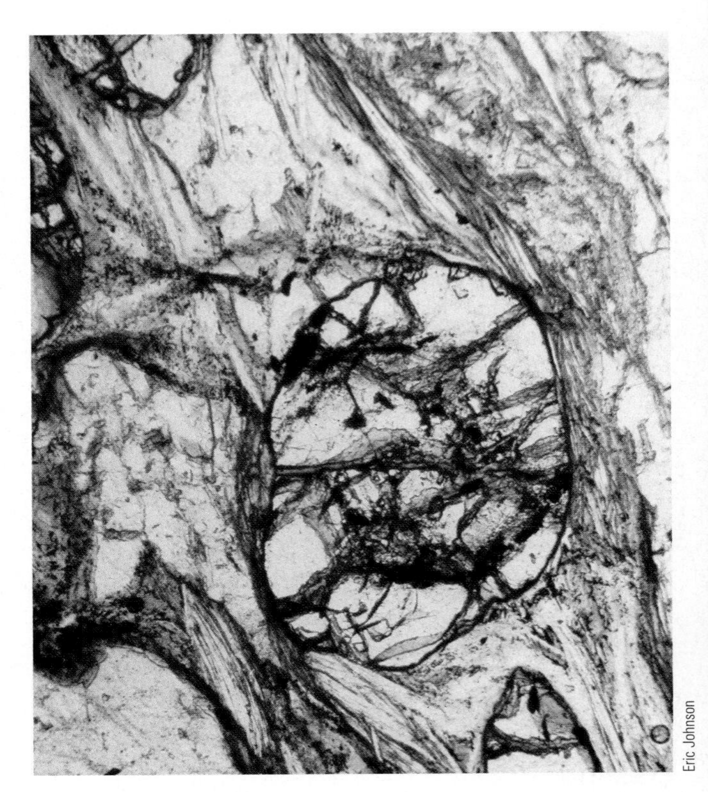

Eric Johnson

Figure 7.3 Differential Pressure Differential pressure results from stress that is unequally applied to an object. Rotated garnets are a good example of the effects of differential pressure applied to a rock during metamorphism. In this example from a schist in northeast Sardinia, stress was applied in opposite directions on the left and right side of the garnet (center), causing it to rotate.

$$2Mg_2SiO_4 + 2H_2O \longrightarrow Mg_3Si_2O_5(OH)_4 + MgO$$

olivine — water — serpentine — carried away in solution

The chemically active fluids important in the metamorphic process come primarily from three sources. The first is water trapped in the pore spaces of sedimentary rocks as they form. The second is the volatile fluid within magma. The third source is the dehydration of water-bearing minerals such as gypsum ($CaSO_4 \cdot 2H_2O$) and some clays.

THE THREE TYPES OF METAMORPHISM

Geologists recognize three major types of metamorphism: *contact (thermal) metamorphism*, in which magmatic heat and fluids act to produce change; *dynamic metamorphism*, which is principally the result of high differential pressures associated with intense deformation; and *regional metamorphism*, which occurs within a large area and is associated with major mountain-building episodes. Even though we will discuss each type of metamorphism separately, the boundary between them is not always distinct and depends largely on which of the three metamorphic agents was dominant (Figure 7.4).

Contact Metamorphism

Contact (thermal) metamorphism takes place when a body of magma alters the surrounding country rock. At shallow depths, intruding magma raises the temperature of the surrounding rock, causing thermal alteration. Furthermore, the release of hot fluids into the country rock by the cooling intrusion can aid in the formation of new minerals.

Important factors in contact metamorphism are the initial temperature, the size of the intrusion, and the fluid content of the magma, the country rock, or both. The initial temperature of an intrusion depends, in part, on its composition; mafic magmas are hotter than felsic magmas (see Chapter 4) and hence have a greater thermal effect on the rocks surrounding them. The size of the intrusion is also important. In the case of small intrusions, such as dikes and sills, usually only those rocks in immediate contact with the intrusion are affected. Because large intrusions, such as batholiths, take a long time to cool, the increased temperature in the surrounding rock may last long enough for a larger area to be affected.

An **aureole** (Figure 7.5) is the area of metamorphism surrounding an intrusion, and the boundary between an intrusion and its aureole may be either sharp or transitional. Metamorphic aureoles vary in width depending on the size, temperature, and composition of the intruding magma, as well as the mineralogy of the surrounding country rock.

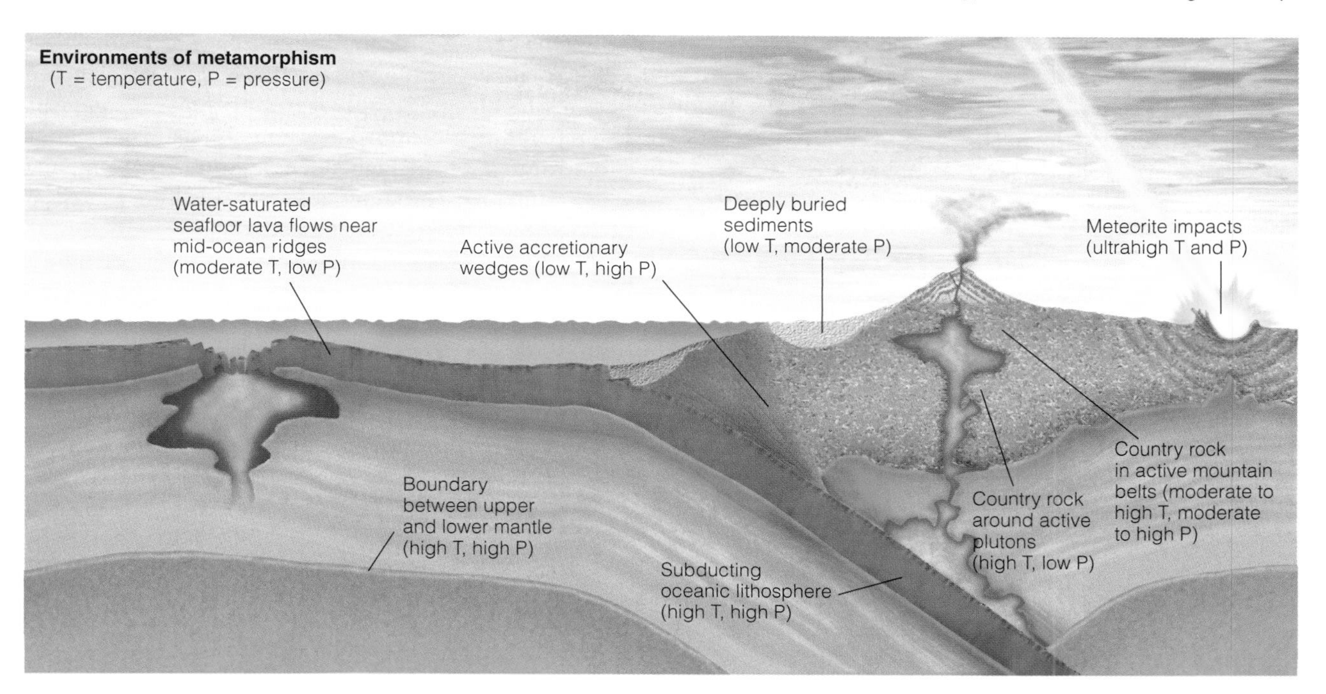

Figure 7.4 Environments of Metamorphism The type of metamorphism that results depends largely on which of the three metamorphic agents was dominant. Illustrated here are some of the common metamorphic environments associated with plate movement, and whether the temperature and pressure in this environment is considered low, moderate, or high. The third agent of metamorphism, fluid activity, although playing an important role in metamorphism, isn't shown here, but is discussed in the text.

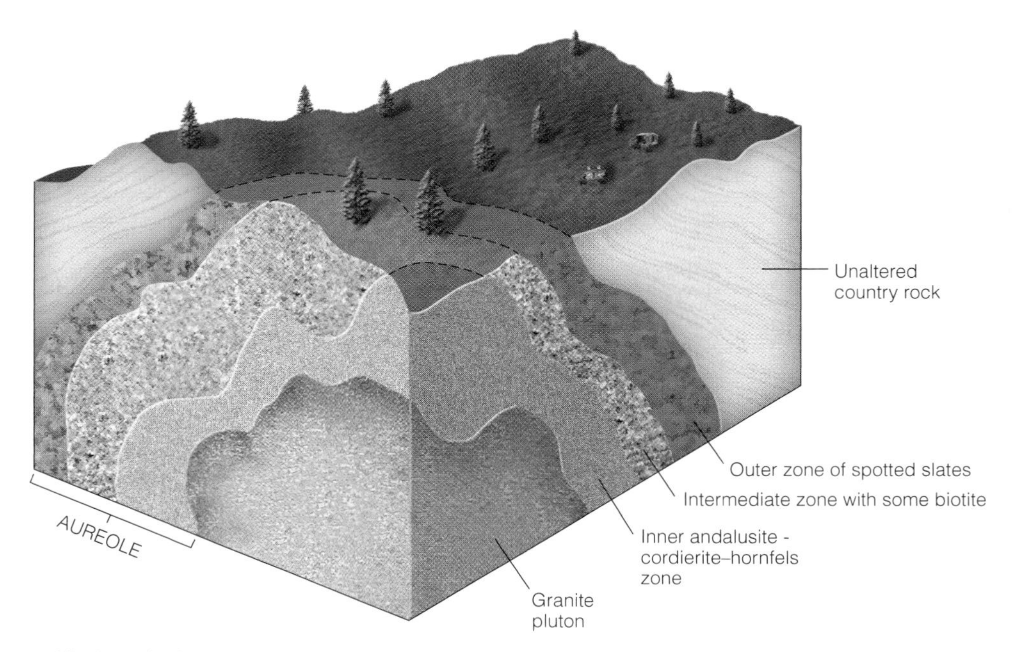

Figure 7.5 Metamorphic Aureole A metamorphic aureole, the area surrounding an intrusion, consists of zones that reflect the degree of metamorphism. The metamorphic aureole associated with this idealized granite pluton contains three zones of mineral assemblages reflecting the decrease in temperature with distance from the intrusion. An inner andalusite–cordierite hornfels zone forms adjacent to the pluton and is reflective of the high temperatures near the intrusion. This is followed by an intermediate zone of extensive recrystallization in which some biotite develops, and farthest from the intrusion is an outer zone characterized by spotted slates.

What Would You Do?

The problem of removing asbestos from public buildings is an important national health and political issue. The current policy of the U.S. Environmental Protection Agency (EPA) mandates that all forms of asbestos are treated as identical hazards. Yet studies indicate that only one form of asbestos is a known health hazard. Because the cost of asbestos removal has been estimated to be as high as $100 billion, many people are questioning whether it is cost effective to remove asbestos from all public buildings where it has been installed.

As a leading researcher on the health hazards of asbestos, you have been asked to testify before a congressional committee on whether it is worthwhile to spend so much money for asbestos removal. How would you address this issue in terms of formulating a policy that balances the risks and benefits of removing asbestos from public buildings? What role would geologists play in formulating this policy?

Aureoles range from a few centimeters wide bordering small dikes and sills, to several hundred meters or even several kilometers wide around large plutons.

The degree of metamorphic change within an aureole generally decreases with distance from the intrusion, reflecting the decrease in temperature from the original heat source. The region or zone closest to the intrusion, and hence subject to the highest temperatures, commonly contains high-temperature metamorphic minerals (that is, minerals in equilibrium with the higher-temperature environment) such as sillimanite. The outer zones, that is, those farthest from the intrusion, are typically characterized by lower-temperature metamorphic minerals such as chlorite, talc, and epidote.

Contact metamorphism can result not only from igneous intrusions, but also from lava flows, either along mid-ocean ridges (Figure 7.4) or from lava flowing over land and thermally altering the underlying rocks (Figure 7.6). Whereas recognizing a recent lava flow and the resulting contact

Geo-Focus: Asbestos: Good or Bad?

Asbestos (from the Latin, meaning "unquenchable") is a general term applied to any silicate mineral that easily separates into flexible fibers. The combination of such features as fire resistance and flexibility makes asbestos an important industrial material of considerable value. In fact, asbestos has more than 3000 known uses, including brake linings, fireproof fabrics, and heat insulators.

Asbestos is divided into two broad groups: *serpentine asbestos* and *amphibole asbestos*. *Chrysotile* is the fibrous form of serpentine asbestos (Figure 1); it is the most valuable type and constitutes the bulk of all commercial asbestos. Its strong, silky fibers are easily spun and can withstand temperatures as high as 2750°C.

The vast majority of chrysotile asbestos is in serpentine, a type of rock formed by the alteration of ultra-mafic igneous rocks such as peridotite under low- and intermediate-grade metamorphic conditions. Other chrysotile results when the metamorphism of magnesium limestone or dolostone produces discontinuous serpentine bands within the carbonate beds.

Among the varieties of amphibole asbestos, *crocidolite* is the most common. Also known as blue asbestos, crocidolite is a long, coarse, spinning fiber that is stronger but more brittle than chrysotile and also less resistant to heat. Crocidolite is found in such metamorphic rocks as slates and schists and is thought to form by the solid-state alteration of other minerals as a result of deep burial.

Parvinder Sethi

Figure 1 Specimen of chrysotile from Arizona. Chrysotile is the fibrous form of serpentine asbestos and the most commonly used in buildings and other structures.

Despite the widespread use of asbestos, the U.S. Environmental Protection Agency (EPA) instituted a gradual ban on all new asbestos products. The ban was imposed because some forms of asbestos can cause lung cancer and scarring of the lungs if fibers are inhaled.

Because the EPA apparently paid little attention to the issue of risks versus benefits when it enacted this rule, the U.S. Fifth Circuit Court of Appeals overturned the EPA ban on asbestos in 1991.

The threat of lung cancer has also resulted in legislation mandating the removal of asbestos already in place metamorphism of the rocks below is easy, less obvious is whether an igneous body is intrusive or extrusive in a rock outcrop where sedimentary rocks occur above and below the igneous body. Recognizing which sedimentary rock units have been metamorphosed enables geologists to determine whether the igneous body is intrusive (such as a sill or dike) or extrusive (lava flow). Such a determination is critical in reconstructing the geologic history of an area (see Chapter 17) and may have important economic implications as well.

Fluids also play an important role in contact metamorphism. Magma is usually wet and contains hot, chemically active fluids that may emanate into the surrounding rock. These fluids can react with the rock and aid in the formation of new minerals. In addition, the country rock may contain pore fluids that, when heated by magma, also increase reaction rates.

The formation of new minerals by contact metamorphism depends not only on proximity to the intrusion, but also on the composition of the country rock. Shales, mudstones, impure limestones, and impure dolostones are particularly susceptible to the formation of new minerals by contact metamorphism, whereas pure sandstones or pure limestones typically are not.

Because heat and fluids are the primary agents of contact metamorphism, two types of contact metamorphic rocks are generally recognized: those resulting from baking of country rock and those altered by hot solutions. Many of the rocks that result from contact metamorphism have the texture of porcelain; that is, they are hard and fine grained. This is particularly true for rocks with a high clay content, such as shale. Such texture results because the clay minerals in the rock are baked, just as a clay pot is baked when fired in a kiln.

During the final stages of cooling when an intruding magma begins to crystallize, large amounts of hot, watery solutions are often released. These solutions may react with the country rock and produce new metamorphic minerals.

in all public buildings, including all public and private schools. However, important questions have been raised concerning the threat posed by asbestos and the additional potential hazards that may arise from its improper removal.

Current EPA policy mandates that all forms of asbestos are to be treated as identical hazards. Yet studies indicate that only the amphibole forms constitute a known health hazard. Chrysotile, whose fibers tend to be curly, does not become lodged in the lungs. Furthermore, its fibers are generally soluble and disappear in tissue. In contrast, crocidolite has long, straight, thin fibers that penetrate the lungs and stay there. These fibers irritate the lung tissue and over a long period of time can lead to lung cancer (Figure 2). Thus crocidolite, and not chrysotile, is overwhelmingly responsible for asbestos-related lung cancer. Because approximately 95% of the asbestos in place in the United States is chrysotile, many people question whether the dangers from asbestos are exaggerated.

Removing asbestos from buildings where it has been installed could cost as much as $100 billion. Unless the material containing the asbestos is disturbed, asbestos does not shed fibers and thus does not contribute to airborne asbestos that can be inhaled. Furthermore, improper removal of asbestos can lead to contamination. In most cases of improper removal, the concentration of airborne asbestos fibers is far higher than if the asbestos had been left in place.

The problem of asbestos contamination is a good example of how geology affects our lives and why we should have a basic knowledge of science before making decisions that could have broad economic and societal impacts.

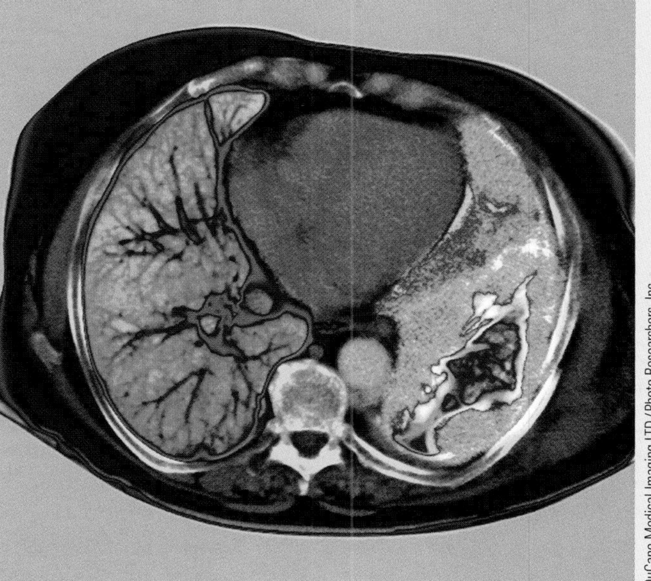
DuCane Medical Imaging LTD / Photo Researchers, Inc.

Figure 2 Lung cancer. Colored tomography (CT) scan of an axial section through the chest of a patient with a mesothelioma cancer (light red). It is surrounding and constricting the lung at right (pink). The other lung (dark blue) has a healthy pleura (dark red). The spine (lower center, light blue), the descending aorta (green), and the heart (dark green between lungs) are also seen. Mesothelioma is a malignancy of the pleura, the membrane lining the chest cavity and lungs. It is usually caused by asbestos exposure. It often reaches a large size, as here, before diagnosis, and the prognosis is then poor.

This process, which usually occurs near Earth's surface, is called *hydrothermal alteration* (from the Greek *hydro*, "water," and *therme*, "heat") and may result in valuable mineral deposits.

Geologists think that many of the world's ore deposits result from the migration of metallic ions in hydrothermal solutions. Examples are copper, gold, iron ores, tin, and zinc in various localities including Australia, Canada, China, Cyprus, Finland, Russia, and the western United States.

Figure 7.6 Contact Metamorphism from a Lava Flow A highly weathered basaltic lava flow near Susanville, California, has altered an underlying rhyolitic volcanic ash by contact metamorphism. The red zone below the lava flow has been baked by the heat of the lava when it flowed over the ash layer. The lava flow displays spheroidal weathering, a type of weathering common in fractured rocks (see Chapter 6).

Dynamic Metamorphism

Most **dynamic metamorphism** is associated with fault (fractures along which movement has occurred) zones in the shallow crust where rocks are subjected to concentrated high levels of differential pressure. The metamorphic rocks that result from pure dynamic metamorphism are called *mylonites* and typically they are restricted to narrow zones adjacent to faults. Mylonites are hard, dense, fine-grained rocks, many of which are characterized by thin laminations (Figure 7.7). Tectonic settings where mylonites occur include the Moine Thrust Zone in northwest Scotland and portions of the San Andreas fault in California (see Chapter 2), and parts of the northeastern United States.

Figure 7.7 Mylonite An outcrop of mylonite from the Adirondack Highlands, New York. Mylonites result from dynamic metamorphism, where rocks are subjected to high levels of differential pressure. Note the thin laminations (closely spaced layers), which are characteristic of many mylonites.

Regional Metamorphism

Most metamorphic rocks result from **regional metamorphism**, which occurs over a large area and is usually caused by tremendous temperatures, pressures, and deformation all occurring together within the deeper portions of the crust. Regional metamorphism is most obvious along convergent plate boundaries where rocks are intensely deformed and recrystallized during convergence and subduction (Figure 7.4). These metamorphic rocks usually reveal a gradation of metamorphic intensity from areas that were subjected to the most intense pressures and/or highest temperatures to areas of lower pressures and temperatures. Such a gradation in metamorphism can be recognized by the metamorphic minerals that are present.

Regional metamorphism is not, however, confined only to convergent margins. It can also occur in areas where plates diverge, although usually at much shallower depths because of the high geothermal gradient associated with these areas (Figure 7.4).

Index Minerals and Metamorphic Grade

From field studies and laboratory experiments, certain minerals are known to form only within specific temperature and pressure ranges. Such minerals are known as **index minerals** because their presence allows geologists to recognize low-, intermediate-, and high-grade metamorphism (**Figure 7.8**).

Metamorphic grade is a term that generally characterizes the degree to which a rock has undergone metamorphic change (Figure 7.8). Although the boundaries between the

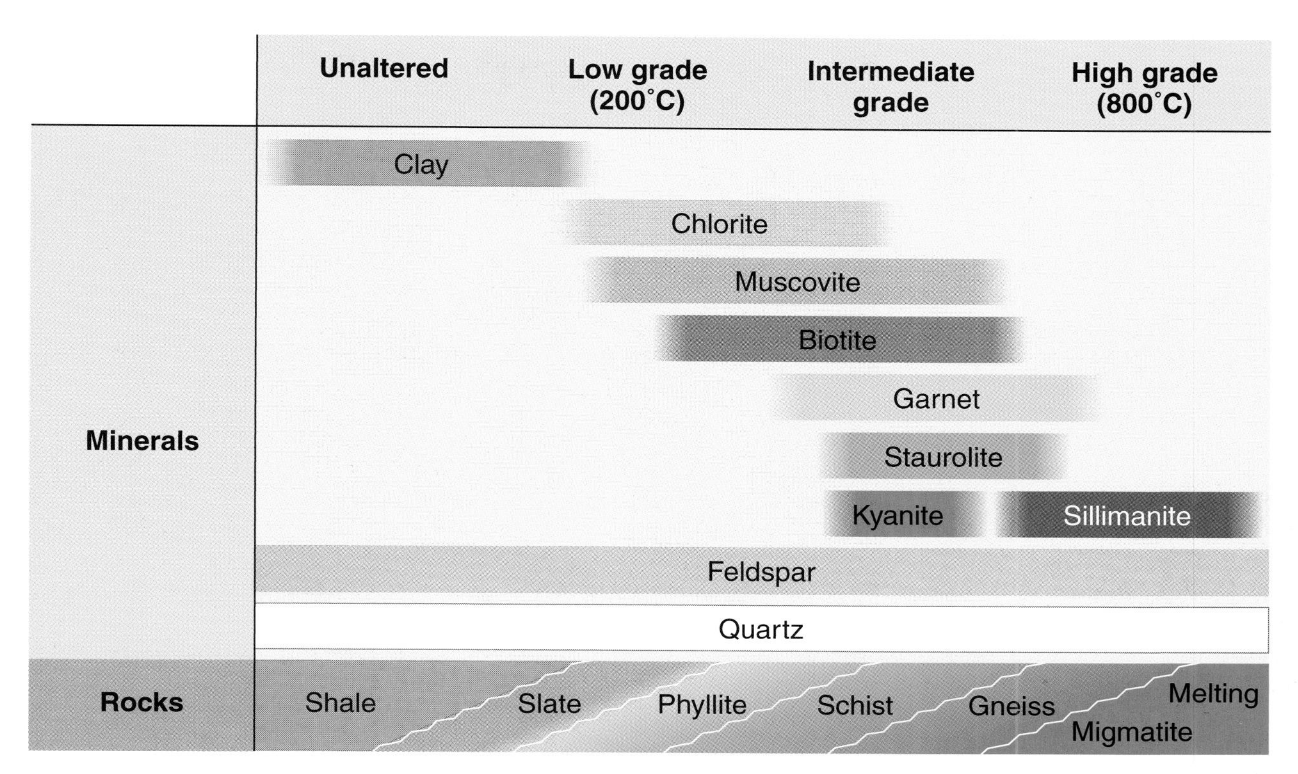

Figure 7.8 Metamorphic Grade Change in mineral assemblage and rock type with increasing metamorphism of shale. When a clay-rich rock such as shale is subjected to increasing metamorphism, new minerals form, as shown by the colored bars. The progressive appearance of certain minerals, known as index minerals, allows geologists to recognize low-, intermediate-, and high-grade metamorphism.

different metamorphic grades are not sharp, the distinction is nonetheless useful for communicating in a general way the degree to which rocks have been metamorphosed. The presence of index minerals thus helps determine metamorphic grade. For example, when a clay-rich rock such as shale undergoes regional metamorphism, the mineral chlorite first begins to crystallize under relatively low temperatures of about 200°C. Its presence in these rocks thus indicates low-grade metamorphism. If temperatures and pressures continue to increase, new minerals crystallize and replace chlorite because they are more stable under those changing conditions. Thus, there is a progression in the appearance of new minerals from chlorite, whose presence indicates low-grade metamorphism, to biotite and garnet, which are good index minerals for intermediate grade metamorphism, to sillimanite, whose presence indicates high-grade metamorphism and temperatures exceeding 500°C (Figure 7.8).

Different rock compositions develop different sets of index minerals. For example, clay-rich rocks such as shale will develop the index minerals shown in Figure 7.8. On the other hand, a sandy dolomite will produce a different set of index minerals as metamorphism progresses because it has a different mineral composition than shale. Thus, a particular set of index minerals will form based on the original composition of the parent rock undergoing metamorphism.

Although such common minerals as mica, quartz, and feldspar can occur in both igneous and metamorphic rocks, other minerals such as andalusite, sillimanite, and kyanite generally occur only in metamorphic rocks derived from clay-rich rocks such as shale. Whereas these three minerals all have the same chemical formula (Al_2SiO_5), they differ in crystal structure and other physical properties because each forms under a different range of pressures and temperatures, and is thus reflective of a different metamorphic grade. For this reason, they are useful index minerals for metamorphic rocks that formed from clay-rich rocks.

HOW ARE METAMORPHIC ROCKS CLASSIFIED?

For purposes of classification, metamorphic rocks are commonly divided into two groups: those exhibiting a *foliated texture* (from the Latin *folium*, "leaf") and those with a *nonfoliated texture* (Table 7.1).

Foliated Metamorphic Rocks

Rocks subjected to heat and differential pressure during metamorphism typically have minerals arranged in a parallel fashion, giving them a **foliated texture** (Figure 7.9). Low-grade

TABLE 7.1 Classification of Common Metamophic Rocks

Texture	Metamorphic Rock	Typical Mineral	Metamorphic Grade	Characteristics of Rocks	Parent Rock
Foliated	Slate	Clays, micas, chlorite	Low	Fine-grained, splits easily into flat pieces	Mudrocks, volcanic ash
	Phyllite	Fine-grained quartz, micas, chlorite	Low to medium	Fine-grained, glossy or lustrous sheen	Mudrocks
	Schist	Micas, chlorite, quartz, talc, hornblende, garnet, staurolite, graphite	Low to high	Distinct foliation, minerals visible	Mudrocks, carbonates, mafic igneous rocks
	Gneiss	Quartz, feldspars, hornblende, micas	High	Segregated light and dark bands visible	Mudrocks, sandstones, felsic igneous rocks
	Amphibolite	Hornblende, plagioclase	Medium to high	Dark, weakly foliated	Mafic igneous rocks
	Migmatite	Quartz, feldspars, hornblende, micas	High	Streaks or lenses of granite intermixed with gneiss	Felsic igneous rocks mixed with sedimentary rocks
Nonfoliated	Marble	Calcite, dolomite	Low to high	Interlocking grains of calcite or dolomite, reacts with HCl	Limestone or dolostone
	Quartzite	Quartz	Medium to high	Interlocking quartz grains, hard, dense	Quartz sandstone
	Greenstone	Chlorite, epidote, hornblende	Low to high	Fine-grained, green	Mafic igneous rocks
	Hornfels	Micas, garnets, andalusite, cordierite, quartz	Low to medium	Fine-grained, equidimensional grains, hard, dense	Mudrocks
	Anthracite	Carbon	High	Black, lustrous, subconcoidal fracture	Coal

metamorphic rocks, such as slate, have a finely foliated texture in which the mineral grains are so small that they cannot be distinguished without the aid of magnification. High-grade foliated rocks, such as gneiss, are coarse-grained, such that the individual grains can easily be seen with the unaided eye. Foliated metamorphic rocks can be arranged in order of increasingly coarse grain size and perfection of foliation.

Slate is the lowest-grade foliated metamorphic rock and commonly exhibits *slaty cleavage* (Figure 7.10). It results from regional metamorphism of shale or, more rarely, volcanic ash. Because it can easily be split along cleavage planes into flat pieces, slate is an excellent rock for roofing and floor tiles, billiard and pool table tops, and blackboards. The different colors of slate are caused by minute amounts of graphite (black), iron oxide (red and purple), and chlorite (green).

Phyllite is similar in composition to slate but is coarser grained. The minerals, however, are still too small to be identified without magnification. Phyllite can be distinguished from slate by its glossy or lustrous sheen (Figure 7.11), and represents an intermediate grain size between slate and schist.

Further coarsening in grain size produces *schist*—a metamorphic rock that is most commonly produced by regional metamorphism. The type of schist formed depends on the intensity of metamorphism and the character of the original rock (Figure 7.12). Metamorphism of many rock types can yield schist, but most schist appears to have formed from clay-rich sedimentary rocks (Table 7.1).

All schists contain more than 50% platy and elongated minerals, all of which are large enough to be clearly visible. Their mineral composition imparts a *schistosity* or *schistose foliation* to the rock that usually produces a wavy type of parting when split. Schistosity is common in low- to high-grade metamorphic environments, and each type of schist is known by its most conspicuous mineral or minerals, such as mica schist, chlorite schist, or garnet-mica schist (Figure 7.12).

Gneiss is a high-grade metamorphic rock that is streaked or has segregated bands of light and dark minerals. Gneisses consist of mostly granular minerals such as quartz, feldspar, or both, with lesser percentages of platy or elongated minerals such as micas or amphiboles (Figure 7.13). Quartz and feldspar characteristically make up the light-colored bands of minerals, whereas biotite and hornblende compose the dark mineral bands. Gneiss typically breaks in an irregular manner, much like coarsely crystalline nonfoliated rocks.

Most gneiss probably results from recrystallization of clay-rich sedimentary rocks during regional metamorphism (Table 7.1). Gneiss also can form from igneous rocks such as granite or older metamorphic rocks.

Figure 7.9 Foliated Texture

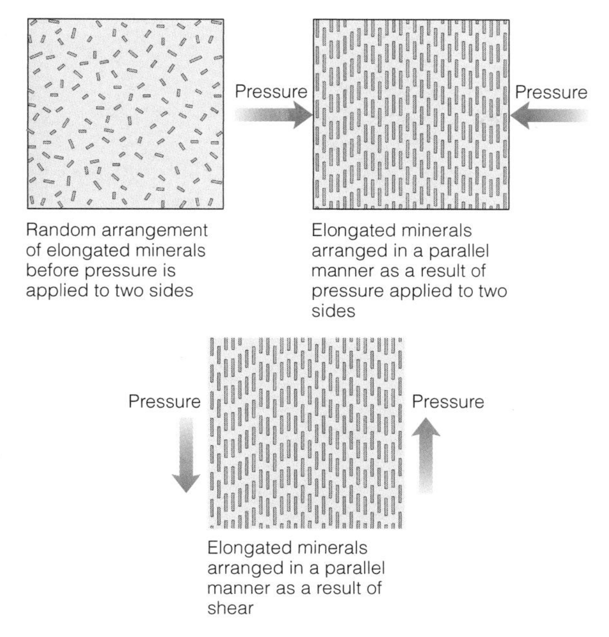

a When rocks are subjected to differential pressure, the mineral grains are typically arranged in a parallel manner, producing a foliated texture.

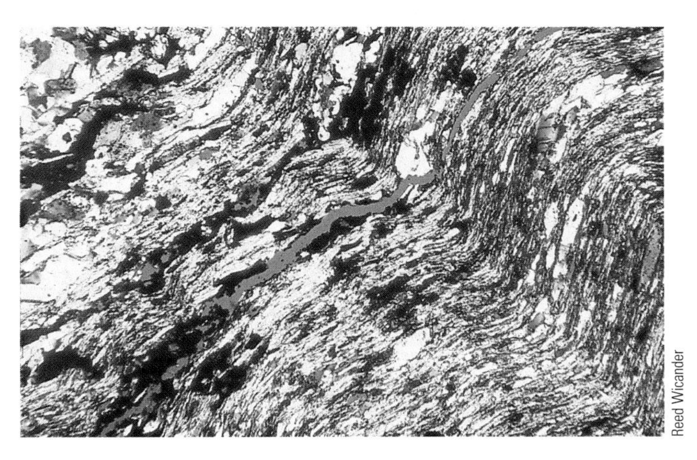

b Photomicrograph of a metamorphic rock with a foliated texture showing the parallel arrangement of mineral grains.

Another fairly common foliated metamorphic rock is *amphibolite*. A dark rock, it is composed mainly of hornblende and plagioclase. The alignment of the hornblende crystals produces a slightly foliated texture. Many amphibolites result from intermediate- to high-grade metamorphism of basalt and ferromagnesian-rich mafic rocks.

In some areas of regional metamorphism, exposures of "mixed rocks" called *migmatites*, having both igneous and high-grade metamorphic characteristics, are present (Figure 7.14). Migmatites are thought to result from the extremely high temperatures produced during metamorphism. However, part of the problem in determining the origin of migmatites is explaining how the granitic component formed. According to one model, the granitic magma formed in place

Figure 7.10 Slate

a Hand specimen of slate.

b This panel of Arvonia Slate from the Albermarle Slate Quarry, Virginia, shows bedding (upper right to lower left) at an angle to the slaty cleavage.

c Slate roof of Chalet Enzian, Switzerland.

Reed Wicander

Figure 7.11 Phyllite Hand specimen of phyllite. Note the lustrous sheen, as well as the bedding (upper left to lower right) at an angle to the cleavage of the specimen.

by the partial melting of rock during intense metamorphism. Such an origin is possible provided that the host rocks contained quartz and feldspars and that water was present.

Not all geologists agree that migmatites necessarily form from partial melting. Some argue that the characteristic layering or wavy appearance arises by the redistribution of minerals during recrystallization in the solid state—that is, through purely metamorphic processes.

Nonfoliated Metamorphic Rocks

In some metamorphic rocks, the mineral grains do not show a discernable preferred orientation. Instead, these rocks consist of a mosaic of roughly equidimensional minerals and are characterized as having a **nonfoliated texture** (Figure 7.15). Most nonfoliated metamorphic rocks result from contact or regional metamorphism of rocks with no platy or elongate minerals. Frequently, the only indication that a granular rock has been metamorphosed is the large grain size resulting from recrystallization. Such rocks may also have a sugary or shiny luster from the reflection of light from the many large crystal faces.

Nonfoliated metamorphic rocks are generally of two types: those composed of mainly one mineral—for example, marble or quartzite—and those in which the different mineral grains are too small to be seen without magnification, such as greenstone and hornfels.

Marble is a well-known metamorphic rock composed predominantly of calcite or dolomite; its grain size ranges from fine to coarsely granular. Marble results from either contact or regional metamorphism of limestones or dolostones (Figure 7.16 and Table 7.1). Pure marble is snowy white or bluish; however, many color varieties exist because of the presence of mineral impurities in the original sedimentary rock. The softness of marble, its uniform texture, and its varying colors have made it the favorite rock of builders and sculptors throughout history (see the Introduction and Geo-inSight on pages 180 and 181).

Quartzite is a hard, compact rock typically formed from quartz sandstone under intermediate- to high-grade metamorphic conditions during contact or regional metamorphism (Figure 7.17). Because recrystallization is so complete, metamorphic quartzite is of uniform strength and therefore usually breaks across the component quartz

Figure 7.12 Schist

Parvinder Sethi

a Almandine garnet crystals in a mica schist.

Sue Monroe

b Hornblende–mica–garnet schist.

Reed Wicander

Figure 7.13 Gneiss Gneiss is characterized by segregated bands of light and dark minerals. This folded gneiss is exposed at Wawa, Ontario, Canada.

grains rather than around them when it is struck. Pure quartzite is white; however, iron and other impurities commonly impart a red or other color to it. Quartzite is commonly used as foundation material for road and railway beds.

The name *greenstone* is applied to any compact, dark-green, altered, mafic igneous rock that formed under low- to high-grade metamorphic conditions. The green color results from the presence of chlorite, epidote, and hornblende.

Hornfels is a common, fine-grained, nonfoliated metamorphic rock resulting from contact metamorphism consisting of various equidimensional mineral grains. The composition of hornfels depends directly on the composition of the original rock, and many compositional varieties are known. The majority of hornfels, however, are apparently derived from contact metamorphism of clay-rich sedimentary rocks or impure dolostones.

Parvinder Sethi

Figure 7.14 Migmatites A migmatite boulder in the Rocky Mountain National Park, near Estes Park, Colorado. Migmatities consist of high-grade metamorphic rock intermixed with streaks or lenses of granite.

Reed Wicander

Figure 7.15 Nonfoliated Texture Nonfoliated textures are characterized by a mosaic of roughly equidimensional minerals, as in this photomicrograph of marble.

Anthracite is a black, lustrous, hard coal that contains a high percentage of fixed carbon and a low percentage of volatile matter. It is highly valued by people who burn coal for heating and power. Anthracite usually forms from the metamorphism of lower-grade coals by heat and pressure, and many geologists consider it to be a metamorphic rock.

METAMORPHIC ZONES AND FACIES

While mapping the 440- to 400-million-year-old Dalradian schists of Scotland in the late 1800s, George Barrow and other British geologists made the first systematic study of metamorphic zones. Here, clay-rich sedimentary rocks have been subjected to regional metamorphism, and the resulting metamorphic rocks can be divided into different

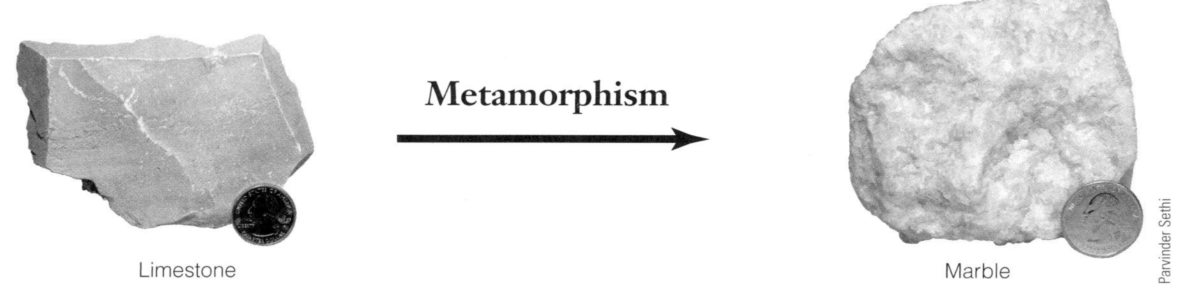

Parvinder Sethi

Figure 7.16 Marble Results from the Metamorphism of the Sedimentary Rock Limestone or Dolostone

Geo-inSight The Many Uses of Marble

Marble is a remarkable stone that has a variety of uses. Formed from limestone or dolostone by the metamorphic processes of heat and pressure, marble comes in a variety of colors and textures.

Marble has been used by sculptors and architects for many centuries in statuary, monuments, as a facing and main stone in buildings and structures, as well as for floor tiling and other ornamental and structural uses. Ground marble can also be found in toothpaste and as a source of lime in agricultural fertilizers.

Reed Wicander

▶ **1.** *Aphrodite of Melos,* also known as *Venus de Milo,* is one of the most recognizable works of art in the world. Dated at around 150 B.C., *Venus de Milo* was created by an unknown artist during the Hellenistic period and carved from the world-famous Parian marble from Paros in the Cyclades. Today *Venus de Milo* attracts thousands of visitors per year to the Louvre Museum in Paris, where she can be viewed and appreciated.

Photodisc/ Getty Images

◀ **2.** Marble has been used extensively as a building stone through the ages and throughout the world. For example, the Greek Parthenon was constructed of white Pentelic marble from Mt. Pentelicus in Attica.

Photodisc/ Getty Images

Parvindes Sethi

▶ **3.** The Taj Mahal in India is constructed mostly of Makrana marble quarried from hills just southwest of Jaipur in Rajasthan. In addition to its main use as a building material, marble was used throughout the structure in art works and intricately carved marble flowers (right). All in all, it took more than 20,000 workers 17 years to build the Taj Mahal from A.D. 1631 to 1648.

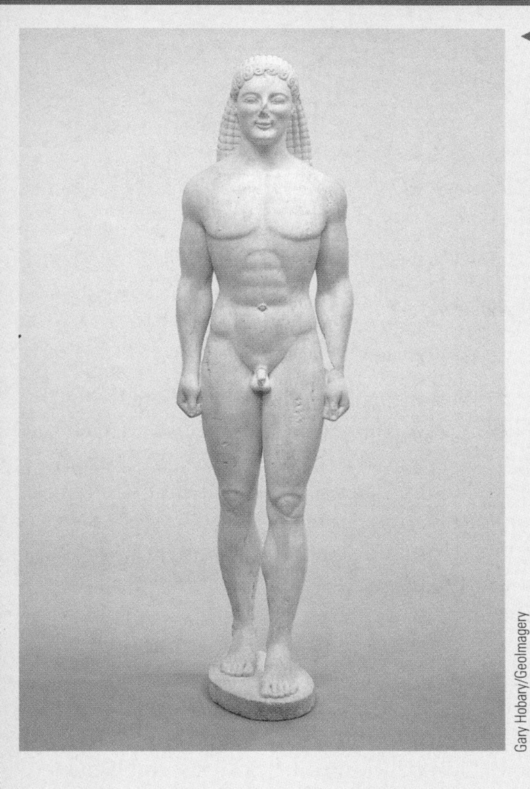

Gary Hobary/GeoImagery

◀ **4.** This Greek Kouros, which stands 206 cm tall, has been the object of an intensive authentication study by the Getty Museum. Using a variety of geologic tests, scientists have determined that the kouros was carved from dolomitic marble and probably came from the Cape Vathy quarries on the island of Thasos.

▶ **5.** The Peace Monument at Pennsylvania Avenue on the west side of the Capitol in Washington D.C., is constructed from white marble from Carrara, Italy, a locality famous for its marble.

Photodisc/ Getty Images

▼ **6.** A marble quarry in northcentral Vermont. Vermont is known for producing some of the finest marble in the United States.

R. V. Dietrich

Rick Hadell

▲ **7.** A public bath in the ancient Roman town of Herculaneum, preserved following the A.D. 79 eruption of Mt. Vesuvius in southern Italy. Marble heads ornament the upper wall. A thin layer of plaster still covers other parts of the wall, and includes a decorative scene made of blue mosaic tile near the floor.

"Plaster" is an artificial metamorphic material made from the baking of marble and limestone to drive off volatiles and convert calcite to lime, or pure calcium oxide. The calcium oxide is mixed with grains of sand, commonly including non-reactive quartz and feldspar. We make cement, which is a plaster-like substance, in much the same way.

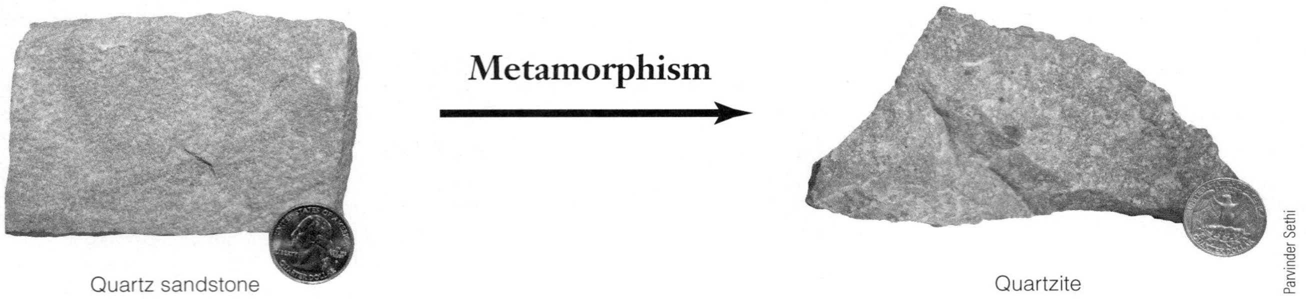

Metamorphism

Quartz sandstone

Quartzite

Parvinder Sethi

Figure 7.17 Quartzite Results from the Metamorphism of the Sedimentary Rock Quartz Sandstone

zones based on the presence of distinctive silicate mineral assemblages. These mineral assemblages, each recognized by the presence of one or more telltale index minerals, indicate different degrees of metamorphism. The index minerals that Barrow and his associates chose to represent increasing metamorphic intensity were chlorite, biotite, garnet, staurolite, kyanite, and sillimanite (Figure 7.8), which we now know all result from the recrystallization of clay-rich sedimentary rocks. As we already mentioned, other mineral assemblages and index minerals are produced from rocks with different original compositions.

The successive appearance of metamorphic index minerals indicates gradually increasing or decreasing intensity of metamorphism. Going from lower- to higher-grade metamorphic zones, the first appearance of a particular index mineral indicates the location of the minimum temperature and pressure conditions needed for the formation of that mineral. When the locations of the first appearances of that index mineral are connected on a map, the result is a line of equal metamorphic intensity or an *isograd*. The region between two adjacent isograds makes up a single **metamorphic zone**—a belt of rocks showing roughly the same general degree of metamorphism. Through mapping a series of adjoining metamorphic zones, geologists can reconstruct metamorphic conditions across entire regions (Figure 7.18).

Not long after Barrow and his coworkers completed their work, the geologists V. M. Goldschmidt and Pentii Eskola, working in Norway and Finland, developed another way of mapping metamorphism that was even more useful than the metamorphic zone approach. Because of the great variety of rock types that Goldschmidt and Eskola encountered in Scandanavia, Eskola defined a **metamorphic facies** as a group of metamorphic rocks characterized by its own distinctive mineral assemblages formed under broadly similar temperature and pressure conditions (Figure 7.19). He named each facies he encountered after its most characteristic rock or mineral. For example, the green metamorphic mineral chlorite, which forms under relatively low temperatures and pressures, yields rocks belonging to the *greenschist facies*. Under increasingly higher temperatures and pressures, mineral assemblages indicative of the *amphibolite* and *granulite facies* develop.

Students in their initial introduction to metamorphic rocks frequently confuse the concepts of metamorphic zones and facies. Remember that metamorphic zones are identified by the appearance of a single index mineral within rocks of the same general composition that occur throughout an area. On the other hand, rocks of greatly different composition within an area can belong to the same metamorphic facies, because each facies has its own characteristic assemblage of minerals whose presence indicates metamorphism with the same broad temperature–pressure range unique to that facies.

Under some conditions, neither metamorphic zones nor facies can be identified. For example, in areas where the

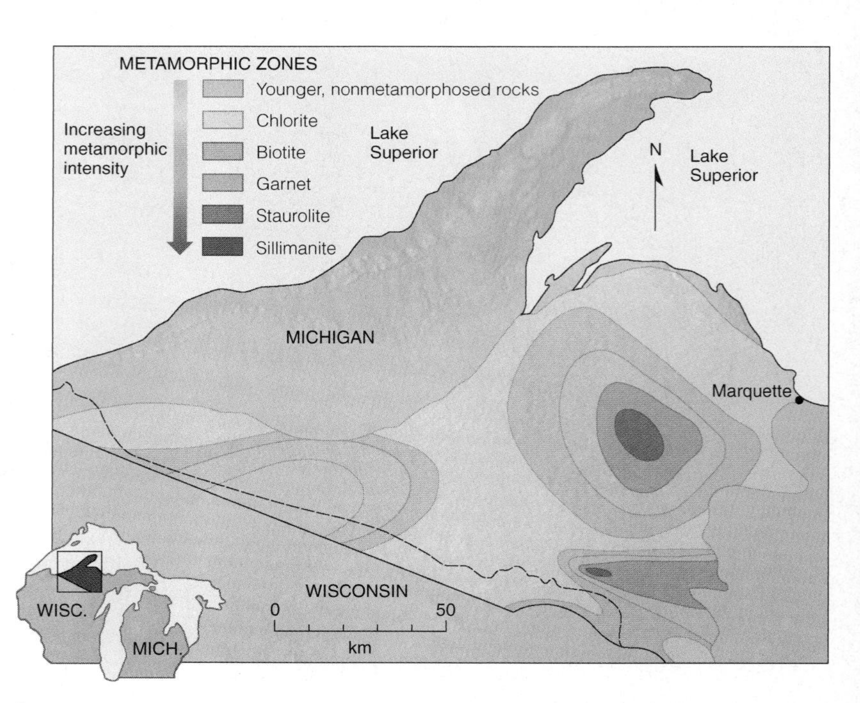

Figure 7.18 Metamorphic Zones in the Upper Peninsula of Michigan The zones in this region are based on the presence of distinctive silicate mineral assemblages resulting from the metamorphism of sedimentary rocks during an interval of mountain building and minor granitic intrusion during the Proterozoic Eon, approximately 1.5 billion years ago. The lines separating the different metamorphic zones are isograds.

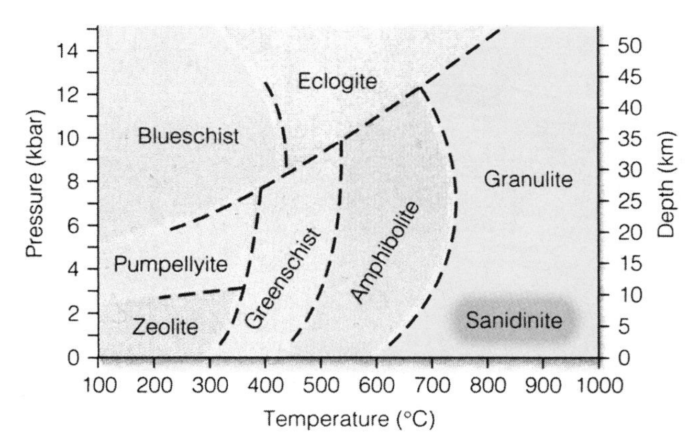

Figure 7.19 Metamorphic Facies and Their Associated Temperature–Pressure Conditions A temperature–pressure diagram showing under what conditions various metamorphic facies occur. A metamorphic facies is characterized by a particular mineral assemblage that formed under the same broad temperature–pressure conditions. Each facies is named after its most characteristic rock or mineral.

original rocks were pure quartz sandstone or pure limestone or dolostone, metamorphism, regardless of the imposed temperature and pressure conditions, will yield only quartzite and marble, respectively. In such cases, all one can say is that "metamorphism has happened."

PLATE TECTONICS AND METAMORPHISM

Although metamorphism is associated with all three types of plate boundaries (see Figure 1.13), it is most common along convergent plate margins. Metamorphic rocks form at convergent plate boundaries because temperature and pressure increase as a result of plate collisions.

Figure 7.20 illustrates the various metamorphic facies conditions present at a typical oceanic–continental convergent plate boundary. When an oceanic plate collides with a continental plate, tremendous pressure is generated as the oceanic plate is subducted. Because rock is a poor heat conductor, the cold descending oceanic plate heats slowly, and metamorphism is caused mostly by increasing pressure with depth. Metamorphism in such an environment produces rocks typical of the *blueschist facies* (low temperature, high pressure). Geologists use the presence of blueschist facies rocks as evidence of ancient subduction zones.

As subduction along the oceanic–continental convergent plate boundary continues, both temperature and pressure increase with depth and yield high-grade metamorphic rocks.

Eventually, the descending plate begins to melt and generate magma that moves upward. This rising magma may alter the surrounding rock by contact metamorphism, producing migmatites in the deeper portions of the crust and hornfels at shallower depths. High temperatures and low to medium pressures characterize such an environment.

Although metamorphism is most common along convergent plate margins, many divergent plate boundaries are characterized by contact metamorphism. At mid-oceanic ridges, infiltration of seawater into hot crust and mantle leads to patchy low- to intermediate-grade metamorphism (Figure 7.4). In addition, fluids emanating from the rising magma—and its reaction with seawater—very commonly produce metal-bearing hydrothermal solutions that may

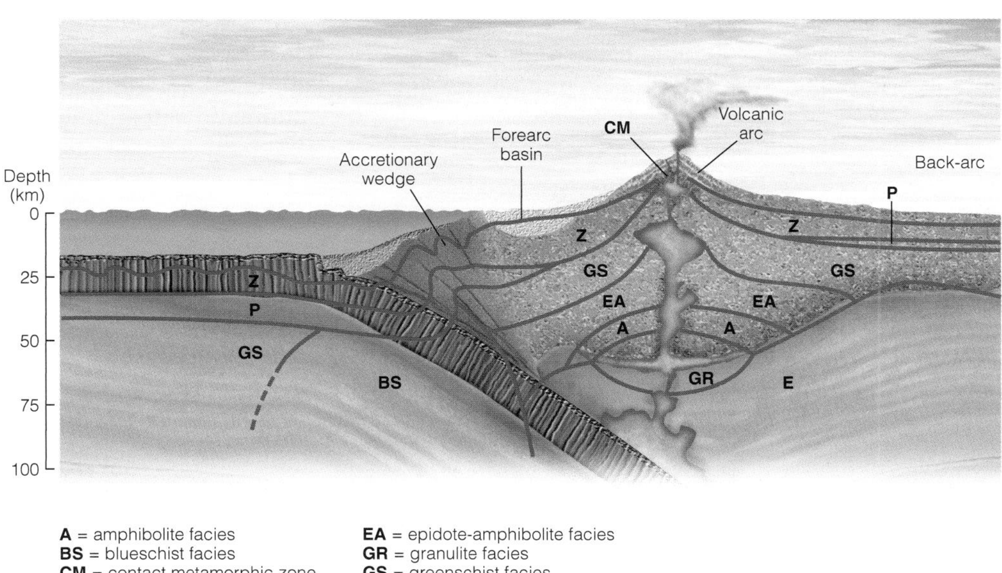

Figure 7.20 Relationship of Facies to Major Tectonic Features at an Oceanic–Continental Convergent Plate Boundary

precipitate minerals of economic value. These deposits may eventually be brought to Earth's surface by later tectonic activity. The copper ores of Cyprus are a good example of such hydrothermal activity (see Chapter 2).

METAMORPHISM AND NATURAL RESOURCES

Many metamorphic rocks and minerals are valuable natural resources. Although these resources include various types of ore deposits, the two most familiar and widely used metamorphic rocks are marble and slate, which, as we discussed earlier, have been used for centuries in a variety of ways (◗ Figure 7.21).

Many ore deposits result from contact metamorphism during which hot, ion-rich fluids migrate from igneous intrusions into the surrounding rock, thereby producing rich ore deposits. The most common sulfide ore minerals associated with contact metamorphism are bornite, chalcopyrite, galena (lead), pyrite, and sphalerite (zinc); two common iron oxide ore minerals are hematite and magnetite. Tin and tungsten are also important ores associated with contact metamorphism (Table 7.2).

Other economically important metamorphic minerals include talc for talcum powder, graphite for pencils and dry lubricants, and garnets and corundum, which are used as abrasives or gemstones, depending on their quality. In addition, andalusite, kyanite, and sillimanite, which, as we mentioned

Peter Hulme/Ecoscene/Corbis

◗ **Figure 7.21 Slate Quarry in Wales** Slate, which has a variety of uses, is the result of low-grade regional metamorphism of shale. These high-quality slates were formed by a mountain-building episode that took place approximately 400 to 440 million years ago in the present-day countries of Ireland, Scotland, Wales, and Norway.

TABLE 7.2 The Main Ore Deposits Resulting from Contact Metamorphism

Ore Deposit	Major Mineral	Formula	Use
Copper	Bornite Chalcopyrite	Cu_5FeS_4 $CuFeS_2$	Important sources of copper, which is used in manufacturing, transporation, communications, and construction
Iron	Hematite Magnetite	Fe_2O_3 Fe_3O_4	Major sources of iron for manufacture of steel, which is used in nearly every form of construction, manufacturing, transportation, and communications
Lead	Galena	PbS	Chief source of lead, which is used in batteries, pipes, solder, and elsewhere where resistance to corrosion is required
Tin	Cassiterite	SnO_2	Principal source of tin, which is used for tin plating, solder, alloys, and chemicals
Tungsten	Scheelite Wolframite	$CaWO_4$ $(Fe,Mn)WO_4$	Chief sources of tungsten, which is used in hardening metals and manufacturing carbides
Zinc	Sphalerite	(Zn, Fe)S	Major source of zinc, which is used in batteries and in galvanizing iron and making brass

earlier, all have the same chemical composition but differ in crystal structure, are used in manufacturing high-temperature porcelains and temperature-resistant minerals for products such as sparkplugs and furnace linings.

Asbestos is a metamorphic mineral that is widely used for insulation and fireproofing and is widespread in buildings and building materials. It is also generally regarded as an environmental hazard because the concentration of amphibole asbestos fibers in your lungs could have fatal consequences (see Geo-Focus on pages 172 and 173).

Geo-Recap

Chapter Summary

- Metamorphic rocks result from the transformation of other rocks, usually beneath Earth's surface, as a consequence of one or a combination of three agents: heat, pressure, and fluid activity.
- Heat for metamorphism comes from intrusive magmas, extrusive lava flows, or deep burial. Pressure is either lithostatic (uniformly applied stress) or differential (stress unequally applied from different directions). Fluids trapped in sedimentary rocks or emanating from intruding magmas can enhance chemical changes and the formation of new minerals.
- The three major types of metamorphism are contact, dynamic, and regional.
- Contact metamorphism takes place when a magma or lava alters the surrounding country rock.
- Dynamic metamorphism is associated with fault zones where rocks are subjected to high differential pressure.
- Regional metamorphism occurs over a large area and is usually caused by tremendous temperatures, pressures, and deformation within the deeper portions of the crust.
- Metamorphic grade generally characterizes the degree to which a rock has undergone metamorphic change.
- Index minerals—minerals that form only within specific temperature and pressure ranges—allow geologists to recognize low-, intermediate-, and high-grade metamorphism.
- Metamorphic rocks are primarily classified according to their texture. In a foliated texture, platy and elongate minerals have a preferred orientation. A nonfoliated texture does not exhibit any discernable preferred orientation of the mineral grains.
- Foliated metamorphic rocks can be arranged in order of increasing grain size, perfection of their foliation, or both. Slate is fine grained, followed by (in increasingly larger grain size) phyllite and schist; gneiss displays segregated bands of minerals. Amphibolite is another fairly common foliated metamorphic rock. Migmatities have both igneous and high-grade metamorphic characteristics.
- Marble, quartzite, greenstone, hornfels, and anthracite are common nonfoliated metamorphic rocks.
- Metamorphic zones are based on index minerals and are areas of rock that all have similar grades of metamorphism, that is, they have all experienced the same intensity of metamorphism.
- A metamorphic facies is a group of metamorphic rocks whose minerals all formed under a particular range of temperatures and pressures. Each facies is named after its most characteristic rock or mineral.
- Metamorphism occurs along all three types of plate boundaries, but is most common at convergent plate margins.
- Metamorphic rocks formed near Earth's surface along an oceanic–continental convergent plate boundary result from low-temperature, high-pressure conditions. As a subducted oceanic plate descends, it is subjected to increasingly higher temperatures and pressures that result in higher-grade metamorphism.
- Many metamorphic rocks and minerals, such as marble, slate, graphite, talc, and asbestos, are valuable natural resources. In addition, many ore deposits are the result of metamorphism and include copper, tin, tungsten, lead, iron, and zinc.

Important Terms

aureole (p. 171)
contact (thermal) metamorphism (p. 170)
differential pressure (p. 169)
dynamic metamorphism (p. 174)
fluid activity (p. 169)
foliated texture (p. 176)
heat (p. 169)
index mineral (p. 174)
lithostatic pressure (p. 169)
metamorphic facies (p. 182)
metamorphic grade (p. 174)
metamorphic rock (p. 168)
metamorphic zone (p. 182)
metamorphism (p. 168)
nonfoliated texture (p. 178)
regional metamorphism (p. 174)

Review Questions

1. To which metamorphic facies do metamorphic rocks formed under low-temperature, low-pressure conditions belong?
   a. ______ granulite;
   b. ______ zeolite;
   c. ______ amphibolite;
   d. ______ blueschist;
   e. ______ eclogite.
2. Which of the following metamorphic rocks displays a foliated texture?
   a. ______ marble;
   b. ______ quartzite;
   c. ______ greenstone;
   d. ______ schist;
   e. ______ hornfels.
3. Concentric zones surrounding an igneous intrusion and characterized by distinctive mineral assemblages are
   a. ______ thermodynamic rings;
   b. ______ hydrothermal regions;
   c. ______ metamorphic layers;
   d. ______ regional facies;
   e. ______ aureoles.
4. Which is the correct metamorphic sequence of increasingly coarser grain size?
   a. ______ gneiss —> schist —> phyllite —> slate;
   b. ______ phyllite —> slate —> schist —> gneiss;
   c. ______ schist —> slate —> gneiss —> phyllite;
   d. ______ slate —> phyllite —> schist —> gneiss;
   e. ______ slate —> schist —> phyllite —> gneiss.
5. The metamorphic rock formed from limestone or dolostone is called
   a. ______ quartzite;
   b. ______ marble;
   c. ______ hornfels;
   d. ______ greenstone;
   e. ______ schist.
6. In what type of metamorphism are magmatic heat and fluid activity the primary agents of change?
   a. ______ dynamic;
   b. ______ lithostatic;
   c. ______ contact;
   d. ______ regional;
   e. ______ thermodynamic.
7. From which of the following rock groups can metamorphic rocks form?
   a. ______ plutonic;
   b. ______ sedimentary;
   c. ______ metamorphic;
   d. ______ volcanic;
   e. ______ all of these.
8. Along what type of plate boundary is metamorphism most common?
   a. ______ divergent;
   b. ______ transform;
   c. ______ aseismic;
   d. ______ convergent;
   e. ______ lithospheric.
9. Pressure resulting from deep burial and applied equally in all directions on a rock is
   a. ______ directional;
   b. ______ differential;
   c. ______ lithostatic;
   d. ______ shear;
   e. ______ unilateral.
10. Which of the following is not an agent of metamorphism?
    a. ______ pressure;
    b. ______ heat;
    c. ______ fluid activity;
    d. ______ gravity;
    e. ______ none of these.
11. Where does contact metamorphism occur, and what types of changes does it produce?

12. Why is metamorphism more widespread along convergent plate boundaries than along any other type of plate boundary?
13. Using Figure 7.19, go to a point that is represented by 200°C and 2 kbar of pressure. What metamorphic facies is represented by those conditions? If the pressure is raised to 12 kbar, what facies is represented by the new conditions? What change in depth of burial is required to effect the pressure change from 2 to 12 kbar?
14. How do metamorphic rocks record the influence of differential pressure in their structures and mineral textures?
15. What specific features about foliated metamorphic rocks would make them unsuitable as foundations for dams? Are there any metamorphic rocks that would make good foundations? Explain your answer.
16. How can aureoles be used to determine the effects of metamorphism?
17. Describe the two types of metamorphic texture, and discuss how they are produced.
18. If plate tectonic movement did not exist, could there be metamorphism? Do you think metamorphic rocks exist on other planets in our solar system? Why?
19. Why should the average citizen know about metamorphic rocks and how they form?
20. Discuss the role each of the three major agents of metamorphism plays in transforming any rock into a metamorphic rock.

CHAPTER 8

Earthquakes and Earth's Interior

OUTLINE

OBJECTIVES

At the end of this chapter, you will have learned that

- Energy is stored in rocks and is released when they fracture, thus producing various types of waves that travel outward in all directions from their source.
- Most earthquakes take place in well-defined zones at transform, divergent, and convergent plate boundaries.
- An earthquake's epicenter is found by analyzing earthquake waves at no fewer than three seismic stations.

The San Andreas fault, as seen in this aerial view where it crosses the Carrizo Plain in central California, is famous for generating large and destructive earthquakes. Movement along this fault caused the famous 1906 San Francisco earthquake. Along with associated faults, the San Andreas has also been responsible for many other devastating earthquakes.

- Intensity is a qualitative assessment of the damage done by an earthquake.
- The Richter Magnitude Scale and Moment Magnitude Scale are used to express the amount of energy released during an earthquake.
- Great hazards are associated with earthquakes, such as ground shaking, fire, tsunami, and ground failure.
- Efforts by scientists to make accurate, short-term earthquake predictions have thus far met with only limited success.
- Geologists use seismic waves to determine Earth's internal structure.
- Earth has a central core overlain by a thick mantle and a thin outer layer of crust.
- Earth possesses considerable internal heat that continuously escapes at the surface.

INTRODUCTION

At 5:54 a.m., on May 27, 2006, a 6.3-magnitude earthquake struck Java, Indonesia. When the earthquake was over, more than 6200 people were dead, at least 38,600 were injured, more than 127,000 houses were destroyed, and approximately 451,000 structures were damaged. The amount of economic destruction inflicted by this earthquake is estimated to be a staggering $3.1 billion. All in all, this was a disaster of epic proportions. Yet it was not the first, nor will it be the last major devastating earthquake in this region or other parts of the world.

Earthquakes, along with volcanic eruptions, are manifestations of Earth's dynamic and active makeup. As one of nature's most frightening and destructive phenomena, earthquakes have always aroused feelings of fear and have been the subject of myths and legends. What makes an earthquake so frightening is that when it begins, there is no way to tell how long it will last or how violent it will be. Approximately 13 million people have died in earthquakes during the past 4000 years, with about 2.7 million of these deaths occurring during the last century (Table 8.1). This increase in fatalities shows that the rapid rise in numbers of humans living in hazardous conditions has trumped our improved understanding of how to build and live safely in earthquake-prone areas.

Geologists define an **earthquake** as the shaking or trembling of the ground caused by the sudden release of energy, usually as a result of faulting, which involves the displacement of rocks along fractures (we discuss the different types of faults in Chapter 10). After an earthquake, continuing adjustments along a fault may generate a series of earthquakes known as *aftershocks*. Most aftershocks are smaller than the main shock, but they can still cause considerable damage to already weakened structures.

Although the geologic definition of an earthquake is accurate, it is not nearly as imaginative or colorful as the explanations many people held in the past. Many cultures attributed the cause of earthquakes to movements of some kind of animal on which Earth rested. In Japan, it was a giant catfish; in Mongolia, a giant frog; in China, an ox; in South America, a whale; and to the Algonquin of North America, an immense tortoise. A legend from Mexico holds that earthquakes occur when the devil, El Diablo, rips open the crust so that he and his friends can reach the surface.

If earthquakes are not the result of animal movement or the devil ripping open the crust, what does cause earthquakes? Geologists know that most earthquakes result from energy released along plate boundaries, and as such, earthquakes are a manifestation of Earth's dynamic nature and the fact that Earth is an internally active planet.

Why should you study earthquakes? The obvious reason is that they are destructive and cause many deaths and injuries to the people living in earthquake-prone areas. Earthquakes also affect the economies of many countries in terms of cleanup costs, lost jobs, and lost business revenues. From a purely personal standpoint, you someday may be caught in an earthquake. Even if you don't live in an area subject to earthquakes, you probably will sooner or later travel where there is the threat of earthquakes, and you should know what to do if you experience one. Such knowledge may help you avoid serious injury or even death.

ELASTIC REBOUND THEORY

Based on studies conducted after the 1906 San Francisco earthquake, H. F. Reid of The Johns Hopkins University proposed the **elastic rebound theory** to explain how energy is released during earthquakes. Reid studied three sets of measurements taken across a portion of the San Andreas fault that had broken during the 1906 earthquake. The measurements revealed that points on opposite sides of the fault had moved 3.2 m during the 50-year period prior to breakage in 1906, with the west side moving northward (Figure 8.1).

According to Reid, rocks on opposite sides of the San Andreas fault had been storing energy and bending slightly for at least 50 years before the 1906 earthquake. Any straight line, such as a fence or road that crossed the San Andreas fault, was gradually bent because rocks on one side of the fault moved relative to rocks on the other side (Figure 8.1). Eventually, the strength of the rocks was exceeded, the rocks on opposite sides of the fault rebounded or "snapped back" to their former undeformed shape, and the energy stored was released as earthquake waves radiating outward from the break (Figure 8.1). Additional field and laboratory studies conducted by Reid and others have confirmed that elastic rebound is the mechanism by which energy is released during earthquakes.

The energy stored in rocks undergoing deformation is analogous to the energy stored in a tightly wound watch spring. The tighter the spring is wound, the more energy is stored, thus making more energy available for release. If the spring is wound so tightly that it breaks, then the stored energy is released as the spring rapidly unwinds and partially regains its original shape.

Perhaps an even more meaningful analogy is simply bending a long, straight stick over your knee. As the stick

TABLE 8.1 Some Significant Earthquakes

Year	Location	Magnitude (estimated before 1935)	Deaths (estimated)
1556	China (Shanxi Province)	8.0	1,000,000
1755	Portugal (Lisbon)	8.6	70,000
1906	USA (San Francisco, California)	8.3	3000
1923	Japan (Tokyo)	8.3	143,000
1960	Chile	9.5	5700
1964	USA (Anchorage, Alaska)	8.6	131
1976	China (Tangshan)	8.0	242,000
1985	Mexico (Mexico City)	8.1	9500
1988	Armenia	6.9	25,000
1990	Iran	7.3	50,000
1993	India	6.4	30,000
1995	Japan (Kobe)	7.2	6000+
1998	Afghanistan	6.9	5000+
1999	Turkey	7.4	17,000
2001	India	7.9	14,000+
2003	Iran	6.6	43,000
2004	Indonesia	9.0	>220,000
2005	Pakistan	7.6	>86,000
2006	Indonesia	6.3	>6200

Figure 8.1 The Elastic Rebound Theory

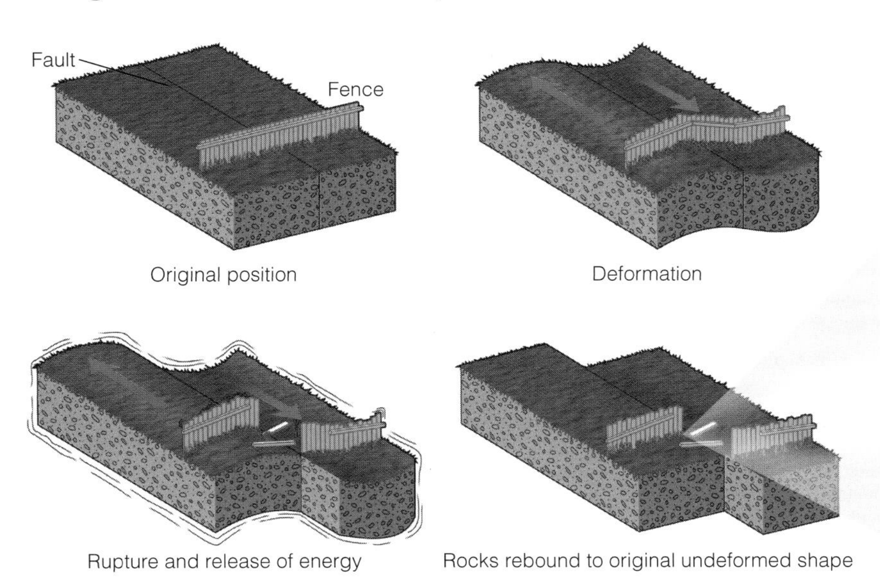

a According to the elastic rebound theory, rocks experiencing defomation store energy and bend. When the initial strength of the rocks is exceeded, they rupture, releasing their accumulated energy, and "snap back" or rebound to their former undeformed shape. This sudden release of energy is what causes an earthquake.

b During the 1906 San Francisco earthquake, this fence in Marin County was displaced by 2.5 m. Whereas many people would see a broken fence, a geologist sees that the fence has moved or been displaced and would look for evidence of a fault. A geologist would also notice that the ground has been displaced toward the right side, relative to his or her view. Regardless of what side of the fence you stand on, you must look to the right to see the other part of the fence. Try it!

bends, it deforms and eventually reaches the point at which it breaks. When this happens, the two pieces of the original stick snap back into their original straight position. Likewise, rocks subjected to intense forces bend until they break and then return to their original position, releasing energy in the process.

SEISMOLOGY

Seismology, the study of earthquakes, emerged as a true science during the 1880s with the development of **seismographs,** instruments that detect, record, and measure the vibrations produced by an earthquake (Figure 8.2). The record made by a seismograph is called a *seismogram*. Modern seismographs have electronic sensors and record moments precisely using computers rather than simply relying on the drum strip charts commonly used on older seismographs.

When an earthquake occurs, energy in the form of *seismic waves* radiates out from the point of release (Figure 8.3). These waves are somewhat analogous to the ripples that move out concentrically from the point where a stone is thrown into a pond. Unlike waves on a pond, however, seismic waves move outward in all directions from their source.

Earthquakes take place because rocks are capable of storing energy; however, their strength is limited, so if enough force is present, they rupture and thus release their stored energy. In other words, most earthquakes result when movement occurs along fractures (faults), most of which are related, at least indirectly, to plate movements. Once a fracture begins, it moves along the fault at several kilometers per second for as long as conditions for failure exist. The longer the fracture along which movement occurs, the more time it takes for the stored energy to be released, and therefore the longer the

Figure 8.2 Seismographs

a Seismographs record ground motion during an earthquake. The record produced is a seismogram. This seismograph records earthquakes on a strip of paper attached to a rotating drum.

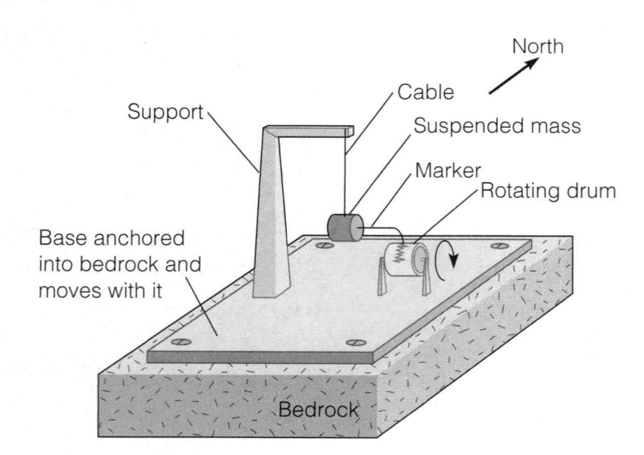

b A horizontal-motion seismograph. Because of its inertia, the heavy mass that contains the marker remains stationary while the rest of the structure moves along with the ground during an earthquake. As long as the length of the arm is not parallel to the direction of ground movement, the marker will record the earthquake waves on the rotating drum. This seismograph would record waves from west or east, but to record waves from the north or south, another seismograph at right angles to this one is needed.

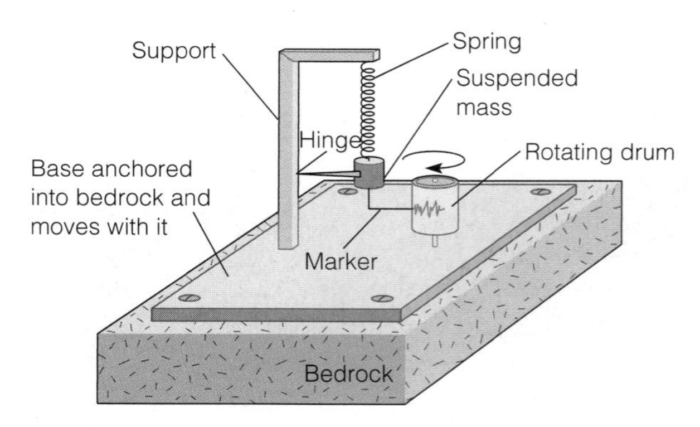

c A vertical-motion seismograph. This seismograph operates on the same principle as the horizontal-motion instrument and records vertical ground movement.

Figure 8.3 The Focus and Epicenter of an Earthquake

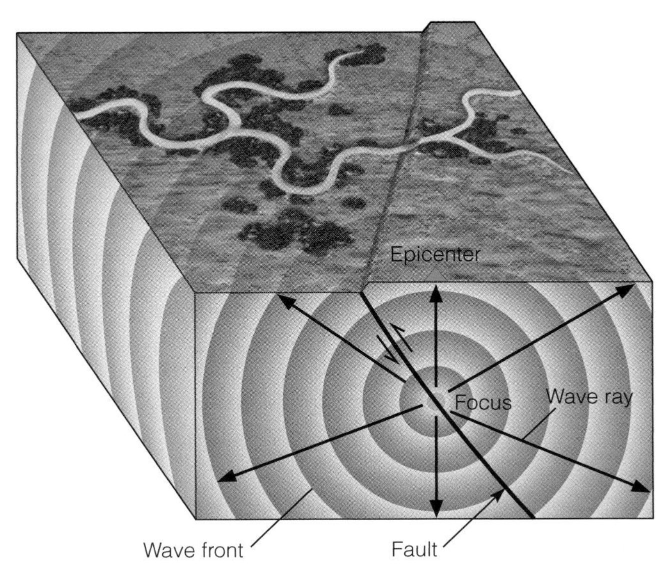

a The focus of an earthquake is the location where the rupture begins and energy is released. The place on the surface vertically above the focus is the epicenter. Seismic wave fronts move out in all directions from their source, the focus of an earthquake.

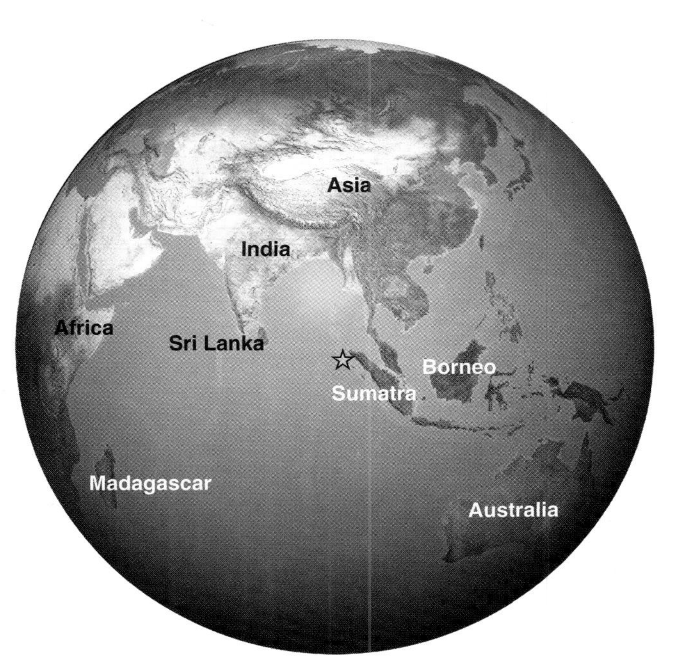

b The epicenter of the December 26, 2004 earthquake that caused the devastating tsunami in the Indian Ocean. The epicenter was located 160 km off the west coast of northern Sumatra and had a focal depth of 30 km.

ground will shake. During some very large earthquakes, the ground might shake for 3 minutes, a seemingly brief time, but interminable if you are experiencing the earthquake firsthand.

The Focus and Epicenter of an Earthquake

The location within Earth's lithosphere where fracturing begins—that is, the point at which energy is first released—is an earthquake's **focus**, or *hypocenter*. What we usually hear in news reports, however, is the location of the **epicenter**, the point on Earth's surface directly above the focus (Figure 8.3a). For example, according to the U.S. Geological Survey, the December 26, 2004, earthquake that triggered the devastating tsunami in the Indian Ocean had an epicenter 160 km off the west coast of northern Sumatra (3°18' N and 95°52' E) and a focal depth of 30 km (Figure 8.3b).

Seismologists recognize three categories of earthquakes based on focal depth. *Shallow-focus* earthquakes have focal depths of less than 70 km from the surface, whereas those with foci between 70 and 300 km are *intermediate-focus*, and the foci of those characterized as *deep-focus* are more than 300 km deep. However, earthquakes are not evenly distributed among these three categories. Approximately 90% of all earthquake foci are at depths of less than 100 km, whereas only about 3% of all earthquakes are deep. Shallow-focus earthquakes are, with few exceptions, the most destructive, because the energy they release has little time to dissipate before reaching the surface.

A definite relationship exists between earthquake foci and plate boundaries. Earthquakes generated along divergent or transform plate boundaries are invariably shallow focus, whereas many shallow-focus earthquakes and nearly all intermediate- and deep-focus earthquakes occur along convergent margins (Figure 8.4). Furthermore, a pattern emerges when the focal depths of earthquakes near island arcs and their adjacent ocean trenches are plotted. Notice in Figure 8.5 that the focal depth increases beneath the Tonga Trench in a narrow, well-defined zone that dips approximately 45 degrees. Dipping seismic zones, called *Benioff* or *Benioff–Wadati zones*, are common to convergent plate boundaries where one plate is subducted beneath another. Such dipping seismic zones indicate the angle of plate descent along a convergent plate boundary.

WHERE DO EARTHQUAKES OCCUR, AND HOW OFTEN?

No place on Earth is immune to earthquakes, but almost 95% take place in seismic belts corresponding to plate boundaries where plates converge, diverge, and slide past each other.

Earthquake activity distant from plate margins is minimal, but can be devastating when it occurs. The relationship between plate margins and the distribution of earthquakes is readily apparent when the locations of earthquake epicenters are superimposed on a map showing the boundaries of Earth's plates.

The majority of all earthquakes (approximately 80%) occur in the *circum-Pacific belt*, a zone of seismic activity nearly encircling the Pacific Ocean basin. Most of these earthquakes result from convergence along plate margins, as in the case of the 1995 Kobe, Japan, earthquake (Figure 8.6a). The earthquakes along the North American Pacific Coast, especially in California, are also in this belt, but here plates

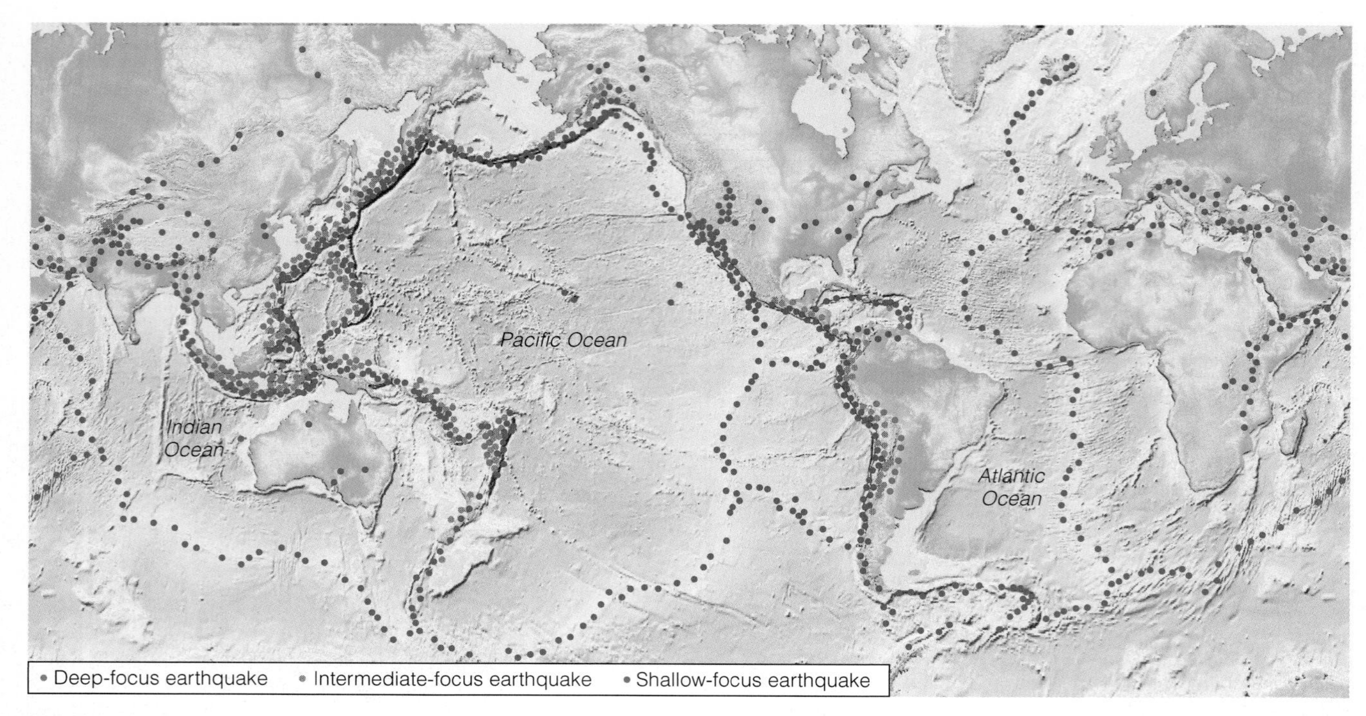

Figure 8.4 Earthquake Epicenters and Plate Boundaries This map of earthquake epicenters shows that most earthquakes occur within seismic zones that correspond closely to plate boundaries. Approximately 80% of earthquakes occur within the circum-Pacific belt, 15% within the Mediterranean–Asiatic belt, and the remaining 5% within plate interiors and along oceanic spreading ridges. The dots represent earthquake epicenters and are divided into shallow-, intermediate-, and deep-focus earthquakes. Along with many shallow-focus earthquakes, nearly all intermediate- and deep-focus earthquakes occur along convergent plate boundaries.

slide past one another rather than converge. The October 17, 1989, Loma Prieta earthquake in the San Francisco area (Figure 8.6b) and the January 17, 1994, Northridge earthquake (Figure 8.6c) occurred along this plate boundary.

The second major seismic belt, accounting for 15% of all earthquakes, is the *Mediterranean–Asiatic belt.* This belt extends westward from Indonesia through the Himalayas, across Iran and Turkey, and westward through the Mediterranean region of Europe. The devastating 1990 and 2003 earthquakes in Iran that killed 40,000 and 43,000 people, respectively; the 1999 Turkey earthquake that killed approximately 17,000; the 2001 India earthquake that killed more than 20,000 people;

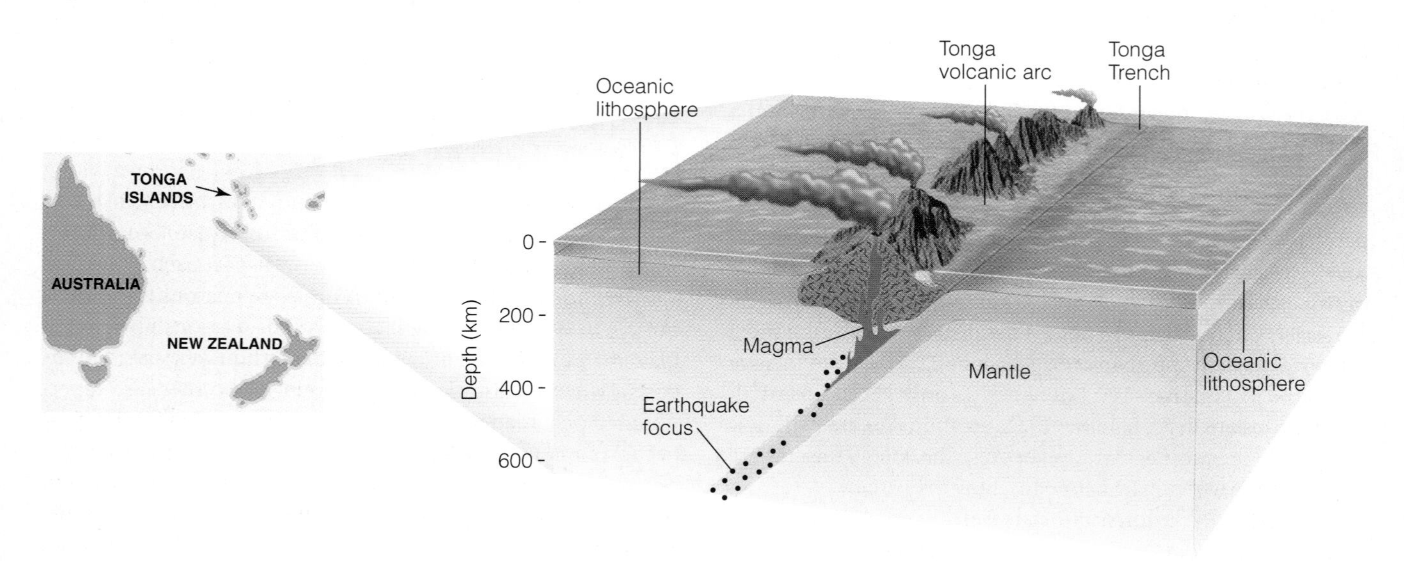

Figure 8.5 Benioff Zones Focal depth increases in a well-defined zone that dips approximately 45 degrees beneath the Tonga volcanic arc in the South Pacific. Dipping seismic zones are called *Benioff* or *Benioff–Wadati* zones.

Figure 8.6 Earthquake Damage in the Circum-Pacific Belt

Eric Draper/AP/Wide World Photos

a Some of the damage in Kobe, Japan, caused by the January 1995 earthquake in which more than 5000 people died.

Dennis Fox

b Damage in Oakland, California, resulting from the October 1989 Loma Prieta earthquake. The columns supporting the upper deck of Interstate 880 failed, causing the upper deck to collapse onto the lower one.

Roger Ressmeyer/Corbis

c View of the severe exterior damage to the Northridge Meadows apartments in which 16 people were killed as a result of the January 1994 Northridge, California earthquake.

and the 2005 earthquake in Pakistan that killed more than 86,000 people are recent examples of the destructive earthquakes that strike this region (Table 8.1).

The remaining 5% of earthquakes occur mostly in the interiors of plates and along oceanic spreading-ridge systems. Most of these earthquakes are not strong, although several major intraplate earthquakes are worthy of mention. For example, the 1811 and 1812 earthquakes near New Madrid, Missouri, killed approximately 20 people and nearly destroyed the town. So strong were these earthquakes that they were felt from the Rocky Mountains to the Atlantic Ocean and from the Canadian border to the Gulf of Mexico. Within the immediate area, numerous buildings were destroyed and forests were flattened. The land sank several meters in some areas, causing flooding, and reportedly the Mississippi River reversed its flow during the shaking and changed its course slightly.

Another major intraplate earthquake struck Charleston, South Carolina, on August 31, 1886, killing 60 people and causing $23 million in property damage. In December 1988, a large intraplate earthquake struck near Tennant Creek in Australia's Northern Territory.

The cause of intraplate earthquakes is not well understood, but geologists think that they arise from localized stresses caused by the compression that most plates experience along their margins. A useful analogy is moving a house. Regardless of how careful the movers are, moving something so large without its internal parts shifting slightly is impossible. Similarly, plates are not likely to move without some internal stresses that occasionally cause earthquakes. It is interesting that many intraplate earthquakes are associated with very ancient and presumed inactive faults that are reactivated at various intervals.

More than 900,000 earthquakes are recorded annually by the worldwide network of seismograph stations. Many of these are too small to be felt, but are nonetheless recorded. These small earthquakes result from the energy released as continual adjustments take place between the various plates. However, on average, more than 31,000 earthquakes per year are strong enough to be felt, and can cause various amounts of damage, depending on how strong they are and where they occur.

SEISMIC WAVES

Many people have experienced an earthquake, but most are probably unaware that the shaking they feel and the damage to structures are caused by the arrival of *seismic waves*, a general term encompassing all waves generated by an earthquake. When movement on a fault takes place, energy is released in the form of two kinds of seismic waves that radiate outward in all directions from an earthquake's focus. *Body waves*, so called because they travel through the solid body of Earth, are somewhat like sound waves, and *surface waves*, which travel along the ground surface, are analogous to undulations or waves on water surfaces.

Body Waves

An earthquake generates two types of body waves: P-waves and S-waves (Figure 8.7). **P-waves** or *primary waves* are the fastest seismic waves and can travel through solids, liquids, and gases. P-waves are compressional, or push-pull, waves and are similar to sound waves in that they move material forward and backward along a line in the same direction that the waves themselves are moving (Figure 8.7b). Thus, the material through which P-waves travel is expanded and compressed as the waves move through it and returns to its original size and shape after the wave passes by.

S-waves or *secondary waves* are somewhat slower than P-waves and can travel only through solids. S-waves are *shear waves* because they move the material perpendicular to the direction of travel, thereby producing shear stresses in the material they move through (Figure 8.7c). Because liquids (as well as gases) are not rigid, they have no shear strength and S-waves cannot be transmitted through them.

The velocities of P- and S-waves are determined by the density and elasticity of the materials through which they travel. For example, seismic waves travel more slowly through rocks of greater density, but more rapidly through rocks with greater elasticity. *Elasticity* is a property of solids, such as rocks, and means that once they have been deformed by an applied force, they return to their original shape when the force is no longer present. Because P-wave velocity is greater than S-wave velocity in all materials, P-waves always arrive at seismic stations first.

Surface Waves

Surface waves travel along the surface of the ground, or just below it, and are slower than body waves. Unlike the sharp jolting and shaking that body waves cause, surface waves generally produce a rolling or swaying motion, much like the experience of being on a boat.

Several types of surface waves are recognized. The two most important are Rayleigh waves and Love waves, named after the British scientists who discovered them, Lord Rayleigh and A. E. H. Love. **Rayleigh waves (R-waves)** are generally the slower of the two and behave like water waves in that they move forward while the individual particles of material move in an elliptical path within a vertical plane oriented in the direction of wave movement (Figure 8.8b).

Figure 8.7 Primary and Secondary Seismic Body Waves Body waves travel through Earth.

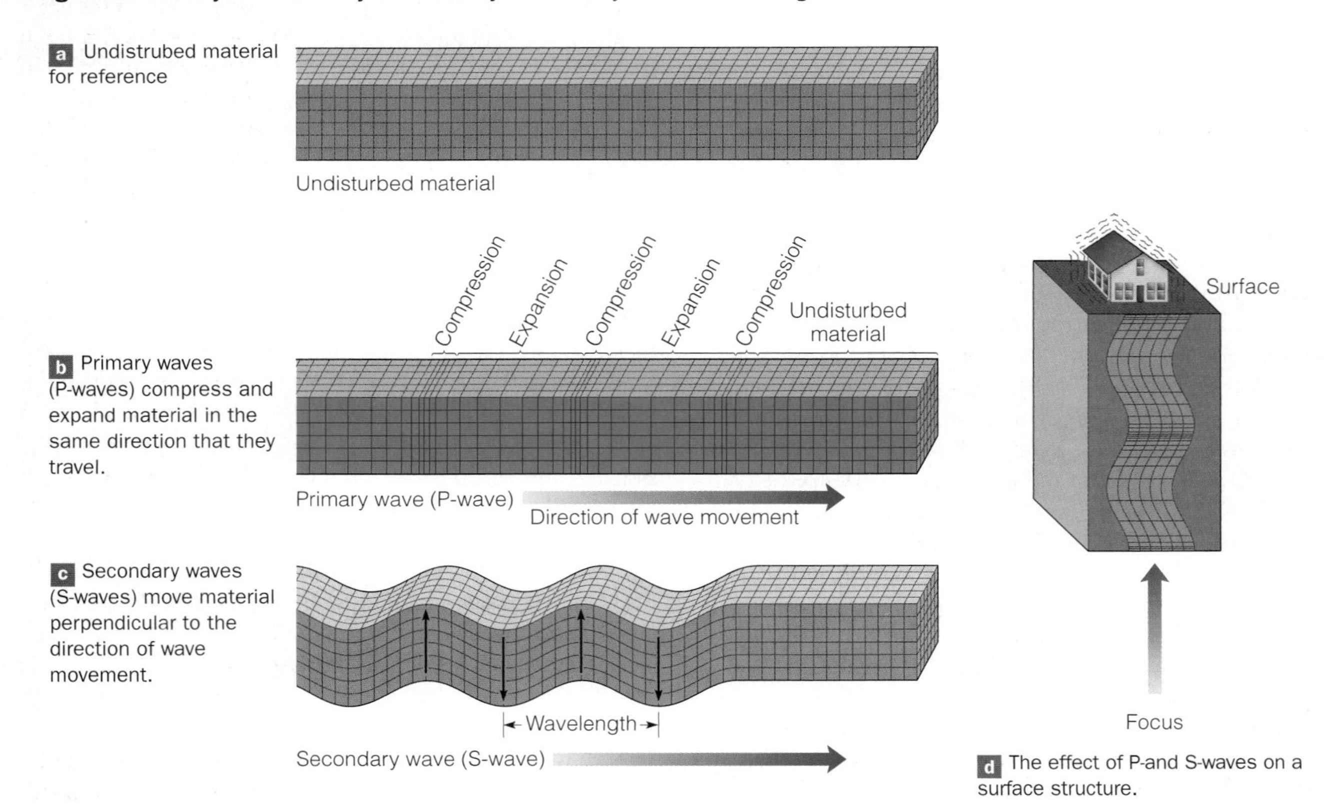

Figure 8.8 Rayleigh and Love Seismic Surface Waves Surface waves travel along Earth's surface or just below it.

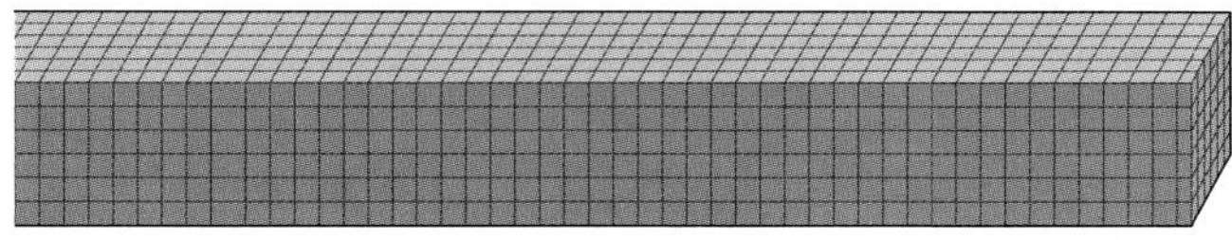

a Undisturbed material for reference.

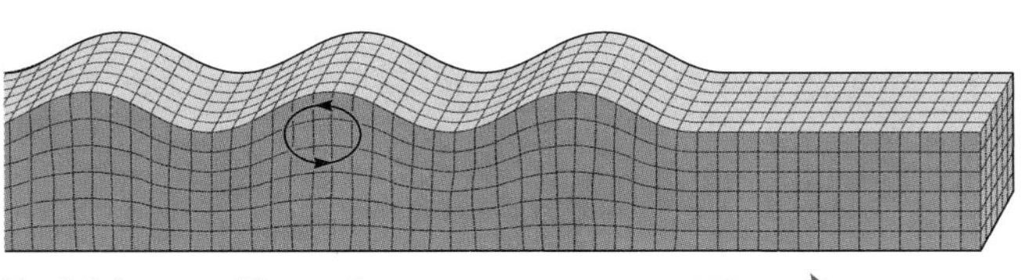

b Rayleigh waves (R-waves) move material in an elliptical path in a plane oriented parallel to the direction of wave movement.

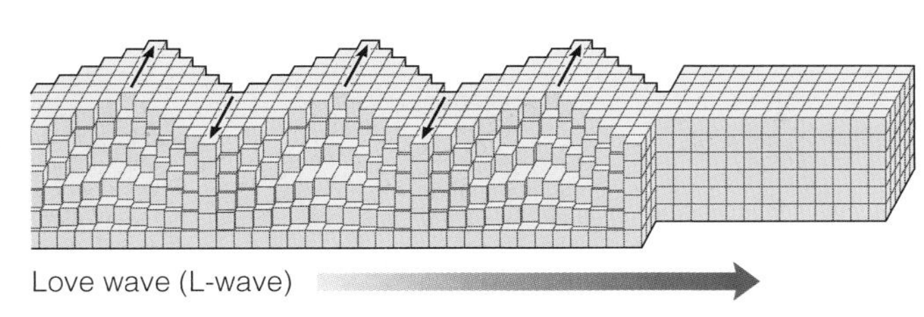

c Love waves (L-waves) move material back and forth in a horizontal plane perpendicular to the direction of wave movement.

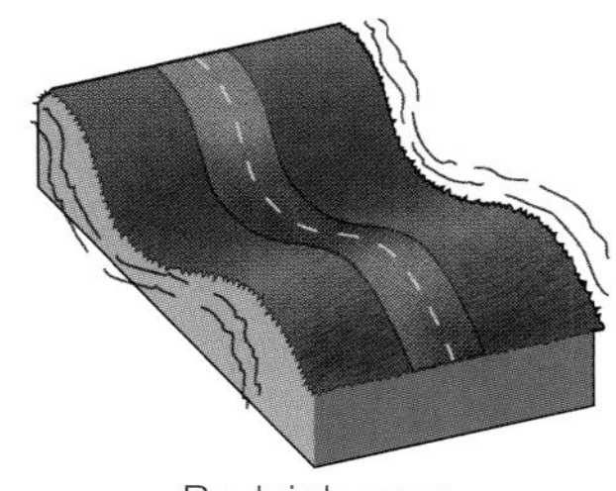

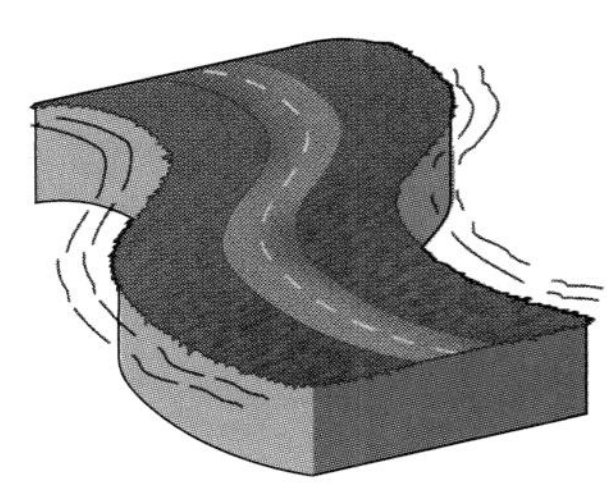

d The arrival of R- and L-waves causes the surface to undulate and shake from side to side.

The motion of a **Love wave (L-wave)** is similar to that of an S-wave, but the individual particles of the material move only back and forth in a horizontal plane perpendicular to the direction of wave travel (Figure 8.8c).

LOCATING AN EARTHQUAKE

We mentioned that news articles commonly report an earthquake's epicenter, but just how is the location of an epicenter determined? Once again, geologists rely on the study of seismic waves. We know that P-waves travel faster than S-waves, nearly twice as fast in all substances, so P-waves arrive at a seismograph station first, followed some time later by S-waves. Both P- and S-waves travel directly from the focus to the seismograph station through Earth's interior, but L- and R-waves arrive last because they are the slowest, and they also travel the longest route along the surface (Figure 8.9a). However, only the P- and S-waves need concern us here because they are the ones important in finding an epicenter.

Seismologists, geologists who study seismology, have accumulated a tremendous amount of data over the years and now know the average speeds of P- and S-waves for any specific distance from their source. These P- and S-wave travel times are published in *time–distance graphs* that illustrate the difference between the arrival times of the two waves as a function of the distance between a seismograph and an earthquake's focus (Figure 8.9b). That is, the farther the waves travel, the greater the *P–S time interval*, which is simply the time difference between the arrivals of P- and S-waves (Figure 8.9a, b).

If the P–S time intervals are known from at least three seismograph stations, then the epicenter of any earthquake can be determined (Figure 8.10). Here is how it works. Subtracting the arrival time of the first P-wave from the arrival time of the first S-wave gives the P–S time interval for each seismic station. Each of these time intervals is plotted on a time–distance graph, and a line is drawn straight down to the distance axis of the graph, thus giving the distance from the focus to each seismic station (Figure 8.9b). Next, a circle whose radius equals the distance shown on the time–distance graph from each of the seismic stations is drawn on a map (Figure 8.10). The intersection of the three circles is the location of the earthquake's epicenter. It should be obvious from Figure 8.10 that P–S time intervals from at least three seismic stations are needed. If only one were used, the epicenter could be at any location on the circle drawn around that station, and using two stations would give two possible locations for the epicenter.

Determining the focal depth of an earthquake is much more difficult and considerably less precise than finding its epicenter. The focal depth is usually found by making computations based on several assumptions, comparing the results with those obtained at other seismic stations, and then recalculating and approximating the depth as closely as possible. Even so, the results are not highly accurate, but they do tell us that most earthquakes, probably 75%, have foci no deeper than 10 to 15 km and that a few are as deep as 680 km.

MEASURING THE STRENGTH OF AN EARTHQUAKE

Following any earthquake that causes extensive damage, fatalities, and injuries, graphic reports of the quake's

Figure 8.9 Determining the Distance from an Earthquake

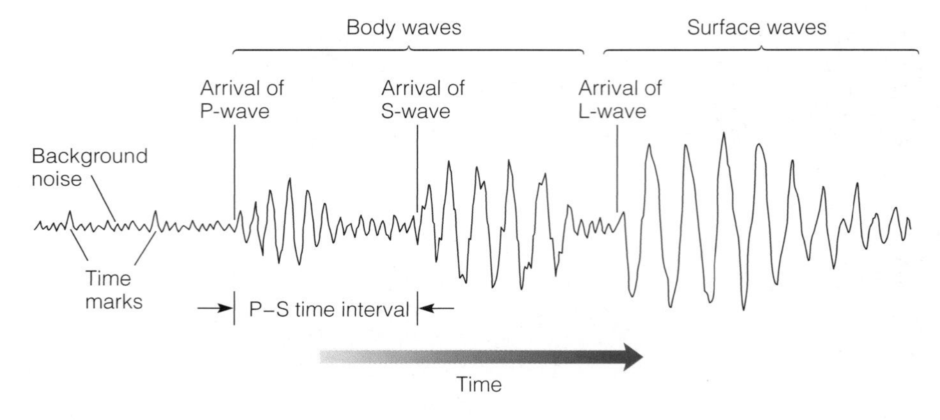

a A schematic seismogram showing the arrival order and pattern produced by P-, S-, and L-waves. When an earthquake occurs, body and surface waves radiate out from the focus at the same time. Because P-waves are the fastest, they arrive at a seismograph first, followed by S-waves and then by surface waves, which are the slowest waves. The difference between the arrival times of the P- and S-waves is the P–S time interval; it is a function of the distance the seismograph station is from the focus.

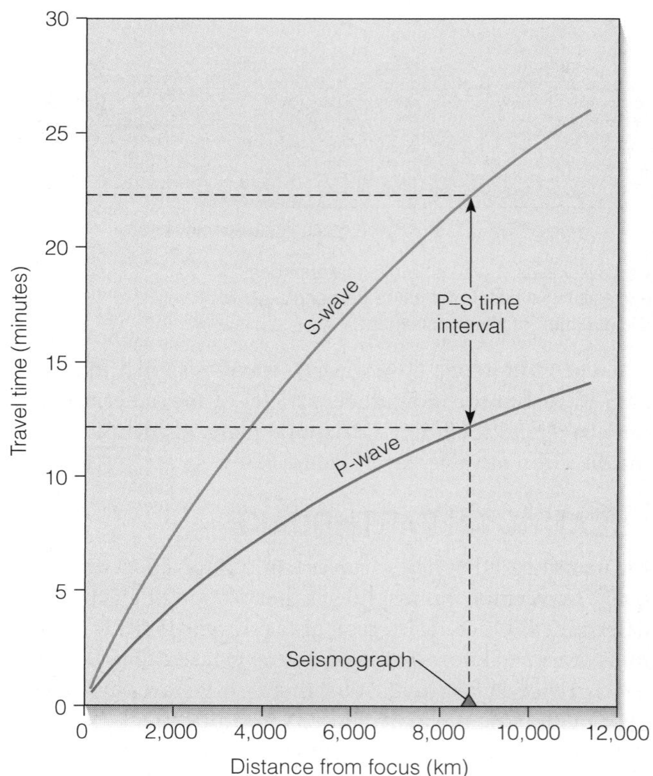

b A time–distance graph showing the average travel times for P- and S-waves. The farther away a seismograph station is from the focus of an earthquake, the longer the interval between the arrival of the P- and S-waves, and hence the greater the distance between the P- and S-wave curves on the time–distance graph as indicated by the P–S time interval. For example, let's assume the difference in arrival times between the P- and S-waves in **a** is 10 minutes (P–S time interval). Using the travel time (minutes) scale, measure how long 10 minutes is (P–S time interval), and move that distance between the S-wave curve and the P-wave curve until the line touches both curves as shown. Then draw a line straight down to the Distance from focus (km) scale. That number is the distance the seismograph is from the earthquake's focus. In this example, the distance is almost 9000 km.

violence and human suffering are common. Headlines tell us that thousands died, many more were injured or left homeless, and property damage is in the millions and possibly billions of dollars. Few other natural processes account for such tragic consequences. Although descriptions of fatalities and damage give some indication of the size of an earthquake, geologists are interested in more reliable methods for determining an earthquake's size.

Two measures of an earthquake's strength are commonly used. One is *intensity*, a qualitative assessment of the kinds of damage done by an earthquake. The other, *magnitude*, is a quantitative measure of the amount of energy released by an earthquake. Each method provides important information that can be used to prepare for future earthquakes.

Intensity

Intensity is a subjective or qualitative measure of the kind of damage done by an earthquake as well as people's reaction to it. Since the mid-19th century, geologists have used intensity as a rough approximation of the size and strength of an earthquake. The most common intensity scale used in the United States is the **Modified Mercalli Intensity Scale**, which has values ranging from I to XII (Table 8.2).

Intensity maps can be constructed for regions hit by earthquakes by dividing the affected region into various intensity zones. The intensity value given for each zone is the maximum intensity that the earthquake produced for that zone. Even though intensity maps are not precise because of the subjective nature of the measurements, they do provide geologists

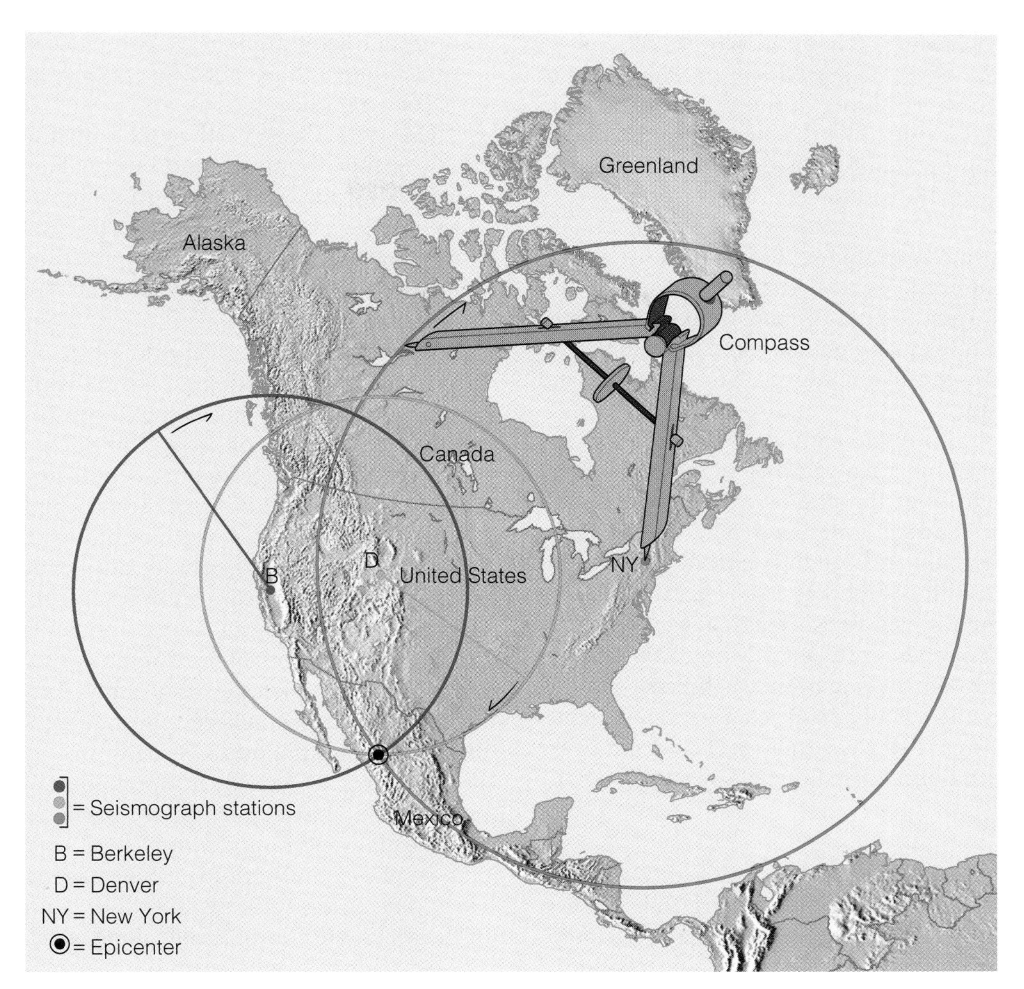

Figure 8.10 Determining the Epicenter of an Earthquake Three seismograph stations are needed to locate the epicenter of an earthquake. The P–S time interval is plotted on a time–distance graph for each seismograph station to determine the distance that station is from the epicenter. A circle with that radius is drawn from each station, and the intersection of the three circles is the epicenter of the earthquake.

TABLE 8.2 Modified Mercalli Intensity Scale

I Not felt except by a very few under especially favorable circumstances.

II Felt by only a few people at rest, especially on upper floors of buildings.

III Felt quite noticeably indoors, especially on upper floors of buildings, but many people do not recognize it as an earthquake. Standing automobiles may rock slightly.

IV During the day felt indoors by many, outdoors by few. At night some awakened. Sensation like heavy truck striking building, standing automobiles rocked noticeably.

V Felt by nearly everyone, many awakened. Some dishes, windows, etc. broken, a few instances of cracked plaster. Disturbance of trees, poles, and other tall objects sometimes noticed.

VI Felt by all, many frightened and run outdoors. Some heavy furniture moved. A few instances of fallen plaster or damaged chimneys. Damage slight.

VII Everybody runs outdoors. Damage negligible in buildings of good design and construction; slight to moderate in well-built ordinary structures; considerable in poorly built or badly designed structures; some chimneys broken. Noticed by people driving automobiles.

VIII Damage slight in specially designed structures; considerable in normally constructed buildings with possible partial collapse; great in poorly built structures. Fall of chimneys, monuments, walls. Heavy furniture overturned. Sand and mud ejected in small amounts.

IX Damage considerable in specially designed structures. Buildings shifted off foundations. Ground noticeably cracked. Underground pipes broken.

X Some well-built wooden structures destroyed; most masonry and frame structures with foundations destroyed; ground badly cracked. Rails bent. Landsides considerable from river banks and steep slopes. Water splashed over river banks.

XI Few, if any (masonry) structures remain standing. Bridges destroyed. Broad fissures in ground. Underground pipelines completely out of service.

XII Damage total. Waves seen on ground surface. Objects thrown upward into the air.

Source: U.S. Geological Survey.

with a rough approximation of the location of the earthquake, the kind and extent of the damage done, and the effects of local geology on different types of building construction (Figure 8.11). Because intensity is a measure of the kind of damage done by an earthquake, insurance companies still classify earthquakes on the basis of intensity.

Generally, a large earthquake will produce higher intensity values than a small earthquake, but many other factors besides the amount of energy released by an earthquake also affect its intensity. These include distance from the epicenter, focal depth of the earthquake, population density and geology of the area, type of building construction employed, and duration of shaking.

A comparison of the intensity map for the 1906 San Francisco earthquake and a geologic map of the area shows a strong correlation between the amount of damage done and the underlying rock and soil conditions (Figure 8.11). Damage was greatest in those areas underlain by poorly consolidated material or artificial fill because the effects of shaking are amplified in these materials, whereas damage was less in areas of solid bedrock. The correlation between the geology and the amount of damge done by an earthquake was further reinforced by the 1989 Loma Prieta earthquake, when many of the same areas that were extensively damaged in the 1906 earthquake were once again heavily damaged.

Magnitude

If earthquakes are to be compared quantitatively, we must use a scale that measures the amount of energy released and is independent of intensity. Charles F. Richter, a seismologist at the California Institute of Technology, developed such a scale in 1935. The **Richter Magnitude Scale** measures earthquake **magnitude**, which is the total amount of energy released by an earthquake at its source. It is an open-ended scale with values beginning at zero. The largest magnitude recorded was a magnitude-9.5 earthquake in Chile on May 22, 1960 (Table 8.1).

Scientists determine the magnitude of an earthquake by measuring the amplitude of the largest seismic wave as recorded on a seismogram (Figure 8.12). To avoid large numbers, Richter used a conventional base-10 logarithmic scale to convert the amplitude of the largest recorded seismic wave to a numerical magnitude value. Therefore, each whole-number increase in magnitude represents a 10-fold increase in wave amplitude. For example, the amplitude of the largest seismic wave for an earthquake of magnitude 6 is 10 times that produced by an earthquake of magnitude 5, 100 times as large as a magnitude-4 earthquake, and 1000 times that of an earthquake of magnitude 3 ($10 \times 10 \times 10 = 1000$).

A common misconception about the size of earthquakes is that an increase of one unit on the Richter Magnitude Scale—a 7 versus a 6, for instance—means a 10-fold increase in size. It is true that each whole-number increase in magnitude represents a 10-fold increase in the wave amplitude, but each magnitude increase of one unit corresponds to a roughly 30-fold increase in the amount of energy released (actually, it is 31.5, but 30 is close enough for our purposes). This means that it would take approximately 30 earthquakes of magnitude 6 to equal the energy released in one earthquake of magnitude 7.

The 1964 Alaska earthquake with a magnitude of 8.6 released nearly 900 times more energy than the 1994 Northridge, California, earthquake of magnitude 6.7! And the Alaska earthquake released more than 27,000 times as much energy as an earthquake with a magnitude of 5.6 would have.

We mentioned that more than 900,000 earthquakes are recorded around the world each year. This figure can be placed in better perspective by reference to Table 8.3, which shows that the vast majority of earthquakes have a Richter magnitude of less than 2.5 and that great earthquakes (those with a magnitude greater than 8.0) occur, on average, only once every five years.

The Richter Magnitude Scale was devised to measure earthquake waves on a particular seismograph and at a specific distance from an earthquake. One of its limitations is that it underestimates the energy of very large earthquakes because it measures the highest peak on a seismogram, which represents only an instant during an earthquake. For large earthquakes, though, the energy might be released over several minutes and along hundreds of kilometers of a fault. For example, during the 1857 Fort Tejon, California, earthquake, the ground shook for longer than 2 minutes and energy was released for 360 km along the fault. Despite their shortcomings, however, Richter magnitudes still usually appear in news releases.

Seismologists now commonly use a somewhat different scale to measure magnitude. Known as the *seismic-moment magnitude scale*, this scale takes into account the strength of the rocks, the area of a fault along which rupture occurs, and the amount of movement of rocks adjacent to the fault. Because larger earthquakes rupture more rocks than smaller earthquakes and rupture usually occurs along a longer segment of a fault and therefore for a longer duration, these very large earthquakes release more energy. For example, the December 26, 2004, Sumatra, Indonesia, earthquake that generated the devastating tsunami created the longest fault rupture and had the longest duration ever recorded.

Thus, magnitude is now frequently given in terms of both Richter magnitude and seismic-moment magnitude. For example, the 1964 Alaska earthquake is given a Richter magnitude of 8.6 and a seismic-moment magnitude of 9.2. Because the Richter Magnitude Scale is most commonly used in the news, we will use that scale here.

WHAT ARE THE DESTRUCTIVE EFFECTS OF EARTHQUAKES?

Certainly, earthquakes are one of nature's most destructive phenomena. Little or no warning precedes earthquakes, and once they begin, little or nothing can be done to minimize their destructive effects, although planning before an earthquake can help.

The number of deaths and injuries, as well as the amount of property damage, depends on several factors. Generally speaking, earthquakes that occur during working hours and school hours in densely populated urban areas are the most

Figure 8.11 Relationship between Intensity and Geology for the 1906 San Francisco Earthquake A close correlation exists between the geology of the San Francisco area and intensity during the 1906 earthquake. Areas underlain by bedrock correspond to the lowest-intensity values, followed by areas underlain by thin alluvium (sediment) and thick alluvium. Bay mud, artificial fill, or both lie beneath the areas shaken most violently.

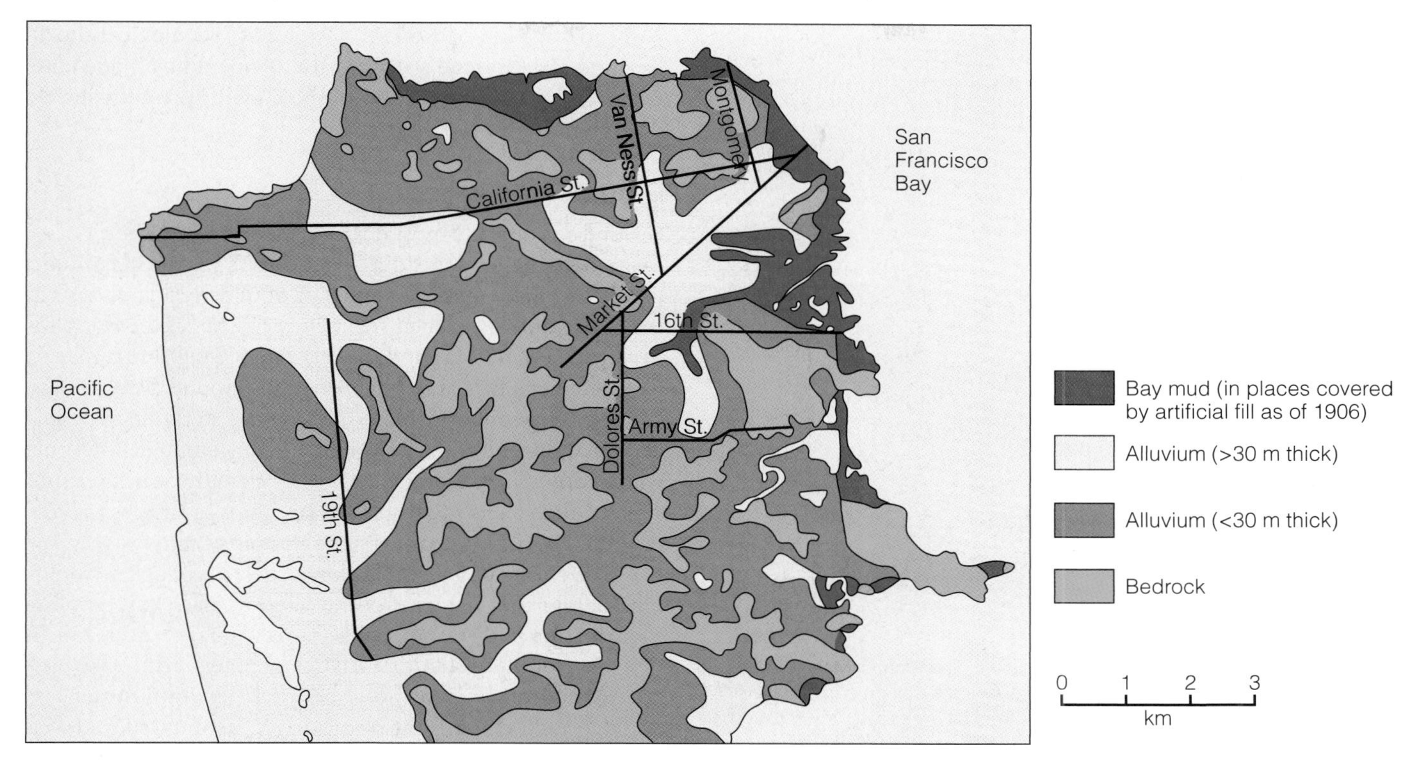

a Geology of the San Francisco area.

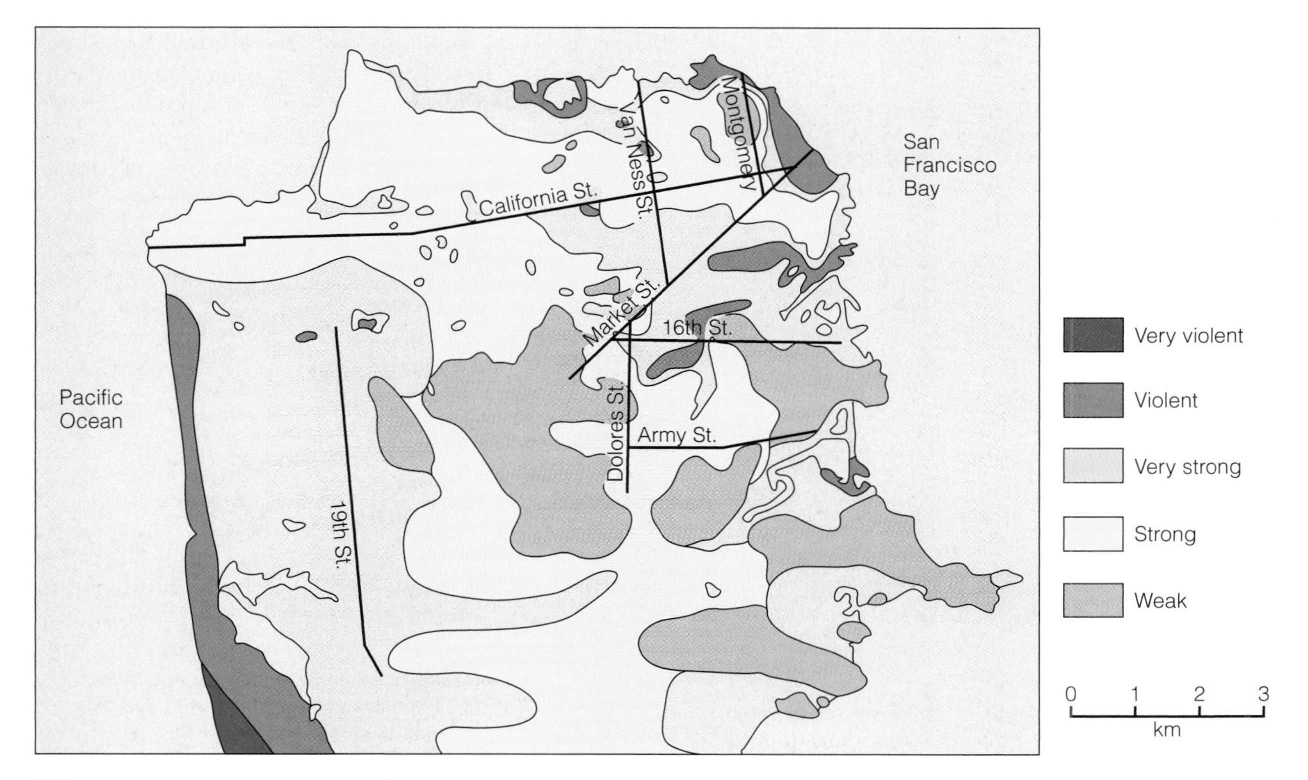

b Intensity of ground shaking in the San Francisco area during the 1906 earthquake.

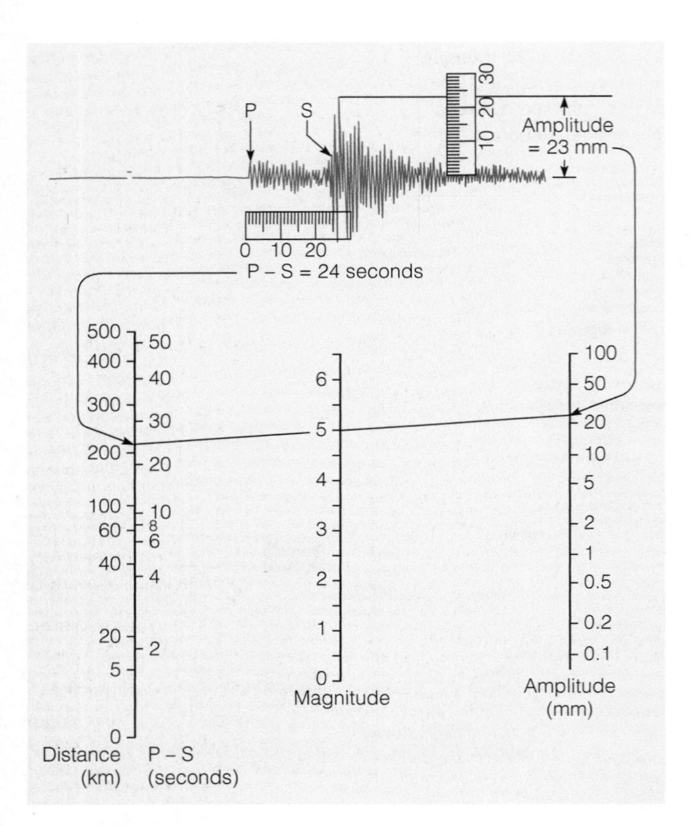

Figure 8.12 Richter Magnitude Scale The Richter Magnitude Scale measures the total amount of energy released by an earthquake at its source. The magnitude is determined by measuring the maximum amplitude of the largest seismic wave and marking it on the right-hand scale. The difference between the arrival times of the P- and S-waves (recorded in seconds) is marked on the left-hand scale. When a line is drawn between the two points, the magnitude of the earthquake is the point at which the line crosses the center scale.

TABLE 8.3 Average Number of Earthquakes of Various Magnitudes per Year Worldwide

Magnitude	Effects	Average Number per Year
<2.5	Typically not felt but recorded	900,000
2.5–6.0	Usually felt; minor to moderate damage to structures	31,000
6.1–6.9	Potentially destructive, especially in populated areas	100
7.0–7.9	Major earthquakes; serious damage results	20
>8.0	Great earthquakes; usually result in total destruction	1 every 5 years

Source: Modified from *Earthquake Information Bulletin*, and B. Gutenberg and C. F. Richter, *Seismicity of the Earth and Associated Phenomena* (Princeton, NJ: Princeton University Press, 1949).

destructive and cause the most fatalities and injuries. However, magnitude, duration of shaking, distance from the epicenter, geology of the affected region, and type of structures are also important considerations. Given these variables, it should not be surprising that a comparatively small earthquake can have disastrous effects, whereas a much larger one might go largely unnoticed, except perhaps by seismologists.

The destructive effects of earthquakes include ground shaking, fire, seismic sea waves, and landslides, as well as panic, disruption of vital services, and psychological shock. In some cases, rescue attempts are hampered by inadequate resources or planning, conditions of civil unrest, or simply the magnitude of the disaster.

Ground Shaking

Ground shaking, the most obvious and immediate effect of an earthquake, varies depending on the earthquake's magnitude, distance from the epicenter, and type of underlying materials in the area—unconsolidated sediment or fill versus bedrock, for instance. Certainly, ground shaking is terrifying, and it may be violent enough for fissures to open in the ground. Nevertheless, contrary to popular myth, fissures do not swallow up people and buildings and then close on them. And although California will no doubt have big earthquakes in the future, rocks cannot store enough energy to displace a landmass as large as California into the Pacific Ocean, as some alarmists claim.

The effects of ground shaking, such as collapsing buildings, falling building facades and window glass, and toppling monuments and statues, cause more damage and result in more loss of life and injuries than any other earthquake hazard. Structures built on solid bedrock generally suffer less damage than those built on poorly consolidated material such as water-saturated sediments or artificial fill.

Structures built on poorly consolidated or water-saturated material are subjected to ground shaking of longer duration and greater S-wave amplitude than structures built on bedrock (Figure 8.13). In addition, fill and water-saturated sediments tend to liquefy, or behave as a fluid, a process known as *liquefaction*. When shaken, the individual grains lose cohesion and the ground flows. Two dramatic examples of damage

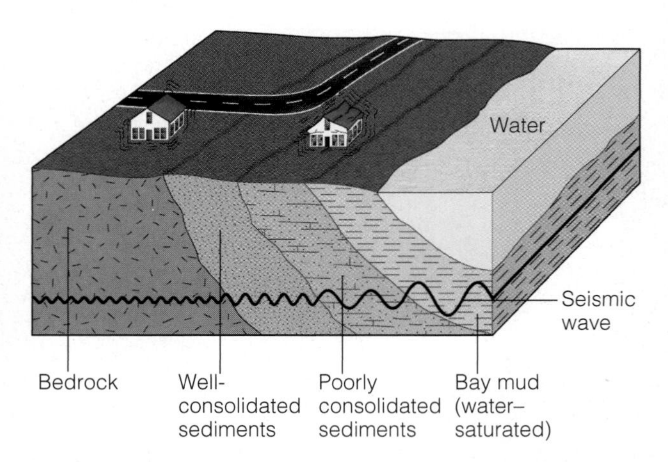

Figure 8.13 Relationship between Seismic Wave Amplitude and Underlying Geology The amplitude and duration of seismic waves generally increase as the waves pass from bedrock to poorly consolidated or water-saturated material. Thus, structures built on weaker material typically suffer greater damage than similar structures built on bedrock because the shaking lasts longer.

resulting from liquefaction are Niigata, Japan, and Turnagain Heights, Alaska. In Niigata, Japan, large apartment buildings were tipped to their sides after the water-saturated soil of the hillside collapsed (Figure 8.14). In Turnagain Heights, Alaska, many homes were destroyed when the Bootlegger Cover Clay lost all of its strength when it was shaken by the 1964 earthquake (see Figure 11.18).

Besides the magnitude of an earthquake and the underlying geology, the material used and the type of construction also affect the amount of damage done. Adobe and mud-walled structures are the weakest and almost always collapse during an earthquake. Unreinforced brick structures and poorly built concrete structures are also particularly susceptible to collapse, such as was the case in the 1999 Turkey earthquake in which an estimated 17,000 people died (Figure 8.15). The 1976 earthquake in

National Geophysical Data Center

Figure 8.14 Liquefaction The effects of ground shaking on water-saturated soil are dramatically illustrated by the collapse of these buildings in Niigata, Japan, during a 1964 earthquake. The buildings, designed to be earthquake resistant, fell over on their sides intact when the ground below them underwent liquefaction.

Yann Arhus-Bertrand/Peter Arnold, Inc.

Figure 8.15 Ground Shaking Most of the buildings collapsed or were severely damaged as a result of ground shaking during the August 17, 1999, Turkey earthquake, which killed more than 17,000 people.

Geo-inSight The San Andreas Fault

The circum-Pacific belt is well known for its volcanic activity and earthquakes. Indeed, about 60% of all volcanic eruptions and 80% of all earthquakes take place in this belt that nearly encircles the Pacific Ocean basin (Figure 8.4).

One well-known and well-studied segment of the circum-Pacific belt is the 1300-km-long San Andreas fault extending from the Gulf of California north through coastal California until it terminates at the Mendocino fracture zone off California's north coast. In plate tectonic terminology, it marks a transform plate boundary between the North American and Pacific plates (see Chapter 2).

Earthquakes along the San Andreas and related faults will continue to occur. But other segments of the circum-Pacific belt, as well as the Mediterranean–Asiatic belt, are also quite active and will continue to experience earthquakes.

USGS

▲ **1.** Aerial view of the San Andreas fault.

James S. Monroe

▲ **2.** View across the San Andreas fault at Tomales Bay, north of San Francisco. The low area occupied by the bay is underlain by shattered rocks of the San Andreas fault zone. Rocks underlying the hills in the distance are on the North American plate, whereas those at the point where this photograph was taken are on the Pacific plate.

James S. Monroe

▲ **3.** This shop in Olema, California, is rather whimsically called The Epicenter, alluding to the fact that it is in the San Andreas fault zone.

Steinbrugge Collection, Earthquake Engineering Research Center, University of California Berkeley

San Francisco following the 1906 earthquake. This view along Sacramento Street shows damaged buildings and the approaching fire.

◄ **4.** Rocks on opposite sides of the San Andreas fault periodically lurch past one another, generating large earthquakes. The most famous one destroyed San Francisco on April 18, 1906. It resulted when 465 km of the fault ruptured, causing about 6 m of horizontal displacment in some areas (see Figure 8.1b). It is estimated that 3000 people died. The shaking lasted nearly 1 minute and caused property damage estimated at $ 400 million in 1906 dollars! Approximately 28,000 buildings were destroyed, many of them by the three-day fire that raged out of control and devastated about 12 km^2 of the city.

► **5.** Since 1906, the San Andreas fault and its subsidiary faults have spawned many more earthquakes; one of the most tragic was centered at Northridge, California, a small community north of Los Angeles. During the early morning hours of January 17, 1994, Northridge and surrounding areas were shaken for 40 seconds. When it was over, 61 people were dead and thousands injured; an oil main and at least 250 gas lines had ruptured, igniting numerous fires; nine freeways were destroyed; and thousands of homes and other buildings were damaged or destroyed by ground shaking. The nearly total destruction of their apartment complex resulted in 16 deaths.

Tim Clary/AFP/Getty Images

Jonathan Novrok/Photoedit

▲ **6.** A portion of Interstate 5 Golden State Freeway collapsed during the 1994 Northridge earthquake. Fortunately, no one was killed at this location.

Ted Soqui/Sygma/Corbis

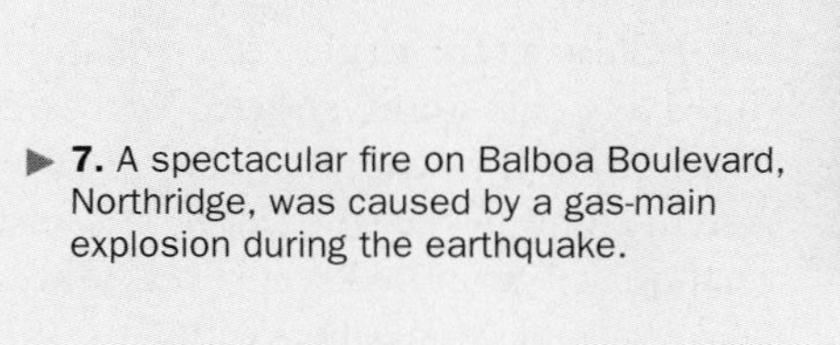

► **7.** A spectacular fire on Balboa Boulevard, Northridge, was caused by a gas-main explosion during the earthquake.

Tangshan, China, completely leveled the city because hardly any structures were built to resist seismic forces. In fact, most had unreinforced brick walls, which have no flexibility, and consequently they collapsed during the shaking.

The magnitude-6.4 earthquake that struck India in 1993 killed approximately 30,000 people, whereas the magnitude-6.7 Northridge, California, earthquake one year later resulted in only 61 deaths. What is the reason for such a difference in the death toll? Both earthquakes occurred in densely populated regions, but in India, the brick and stone buildings could not withstand ground shaking; most collapsed and entombed their occupants.

Fire

In many earthquakes, particularly in urban areas, fire is a major hazard. Nearly 90% of the damage done in the 1906 San Francisco earthquake was caused by fire. The shaking severed many of the electrical and gas lines, which touched off flames and started fires throughout the city. Because the earthquake ruptured water mains, there was no effective way to fight the fires that raged out of control for three days, destroying much of the city.

Eighty-three years later, during the 1989 Loma Prieta earthquake, a fire broke out in the Marina district of San Francisco. This time, however, the fire was contained within a small area because San Francisco had a system of valves throughout its water and gas pipeline system so that lines could be isolated from breaks (see Geo-inSight on pages 204 and 205).

During the September 1, 1923, earthquake in Japan, fires destroyed 71% of the houses in Tokyo and practically all of the houses in Yokohama. In all, 576,262 houses were destroyed by fire, and 143,000 people died, many as a result of fire. A horrible example occurred in Tokyo where thousands of people gathered along the banks of the Sumida River to escape the raging fires. Suddenly, a firestorm swept over the area, killing more than 38,000 people. The fires from this earthquake were particularly devastating because most of the buildings were constructed of wood, and were fanned by 20-km/hr winds.

Tsunami: Killer Waves

On December 26, 2004, a magnitude-9.0 earthquake struck 160 km off the west coast of northern Sumatra, Indonesia, generating the deadliest tsunami in history (▶ Figure 8.16a). Within hours, walls of water as high as 10.5 m pounded the coasts of Indonesia, Sri Lanka, India, Thailand, Somalia, Myanmar, Malaysia, and the Maldives, killing more than 220,000 people and causing billions of dollars in damage.

This earthquake generated what is popularly called a "tidal wave," but is more correctly termed a *seismic sea wave* or **tsunami**, a Japanese term meaning "harbor wave." The term *tidal wave* nevertheless persists in popular literature and some news accounts, but these waves are not caused by or related to tides. Indeed, tsunami are destructive sea waves generated when the sea floor undergoes sudden, vertical movements. Many result from submarine earthquakes, but volcanoes at sea or submarine landslides can also cause them. For example, the 1883 eruption of Krakatau between Java and Sumatra generated a large sea wave that killed 36,000 on nearby islands.

Once a tsunami is generated, it can travel across an entire ocean and cause devastation far from its source. In the open sea, tsunami travel at several hundred kilometers per hour and commonly go unnoticed as they pass beneath ships because they are usually less than 1 m high and the distance between wave crests is typically hundreds of kilometers. When they enter shallow water, however, the wave slows down and water piles up to heights anywhere from a meter or two to many meters high. The 1946 tsunami that struck Hilo, Hawaii, was 16.5 m high! In any case, the tremendous energy possessed by a tsunami is concentrated on a shoreline when it hits either as a large breaking wave or, in some cases, as what appears to be a very rapidly rising tide.

A common popular belief is that a tsunami is a single large wave that crashes onto a shoreline. Any tsunami consists of a series of waves that pour onshore for as long as 30 minutes followed by an equal time during which water rushes back to sea. Furthermore, after the first wave hits, more waves follow at 20- to 60-minute intervals. Approximately 80 minutes after the 1755 Lisbon, Portugal, earthquake, the first of three tsunami, the largest more than 12 m high, destroyed the waterfront area and killed thousands of people. Following the arrival of a 2-m-high tsunami in Crescent City, California, in 1964, curious people went to the waterfront to inspect the damage. Unfortunately, 10 were killed by a following 4-m-high wave!

One of nature's warning signs of an approaching tsunami is a sudden withdrawal of the sea from a coastal region. In fact, the sea might withdraw so far that it cannot be seen and the seafloor is laid bare over a huge area. On more than one occasion, people have rushed out to inspect exposed reefs or to collect fish and shells only to be swept away when the tsunami arrived.

During the December 2004 tsunami, however, a 10-year-old British girl saved numerous lives because she recognized the warning signs that she had learned in a school lesson on tsunami only two weeks before. While vacationing with her mother on the island of Phuket, Thailand, a popular resort area, the girl noticed the water quickly receding from the beach. She immediately told her mother that she thought a tsunami was coming, and her mother, along with the resort staff, quickly warned everyone standing around watching the water recede to clear the beach area. Their quick action resulted in many lives being saved.

Following the tragic 1946 tsunami that hit Hilo, Hawaii, the U.S. Coast and Geodetic Survey established a Pacific Tsunami Early Warning System in Ewa Beach, Hawaii. This system combines seismographs and instruments that detect earthquake-generated sea waves. Whenever a strong earthquake takes place anywhere within the Pacific basin, its location

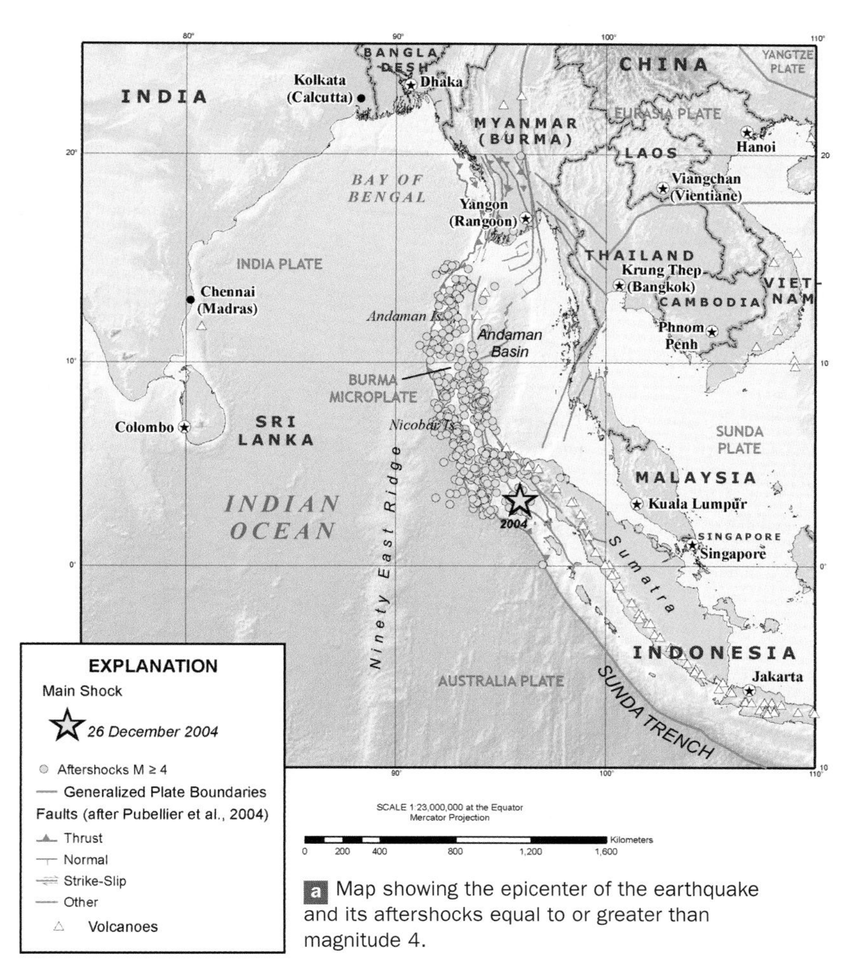

a Map showing the epicenter of the earthquake and its aftershocks equal to or greater than magnitude 4.

Figure 8.16 2004 Indian Ocean Tsunami The magnitude-9.0 earthquake off the coast of northwest Sumatra in December 2004 generated a devastating tsunami throughout the Indian Ocean.

Courtesy of DigitalGlobe

b Satellite image of the north shore of Banda Aceh, the capital city of Aceh Provence, Sumatra, Indonesia, taken on June 23, 2004.

Courtesy of DigitalGlobe

c A similar satellite image of the same area on December 28, 2004, two days after the tsunami struck. Notice the total destruction of all the buildings and one bridge.

is determined, and instruments are checked to see whether a tsunami has been generated. If it has, a warning is sent out to evacuate people from low-lying areas that may be affected.

Nevertheless, tsunami remain a threat to people in coastal areas, especially around the Pacific Ocean (Table 8.4). Unfortunately, no such warning system exists for the Indian Ocean. If one had been in place, it is possible that the death toll from the December 26, 2004 tsunami would have been significantly lower.

Ground Failure

Earthquake-triggered landslides are particularly dangerous in mountainous regions and have been responsible for tremendous amounts of damage and many deaths. The 1959 earthquake in Madison Canyon, Montana, for example, caused a huge rock slide (Figure 8.17), and the 1970 Peru earthquake caused an avalanche that destroyed the town of Yungay and killed an estimated 66,000 people. Most of the

TABLE 8.4 Tsunami Fatalities Since 1990

Date	Location	Maximum Wave Height	Fatalities
September 2, 1992	Nicaragua	10 m	170
December 12, 1992	Flores Island	26 m	>1000
July 12, 1993	Okushiri, Japan	31 m	239
June 2, 1994	East Java	14 m	238
November 14, 1994	Mindoro Island	7 m	49
October 9, 1995	Jalisco, Mexico	11 m	1
January 1, 1996	Sulawesi Island	3.4 m	9
February 17, 1996	Irian Jaya	7.7 m	161
February 21, 1996	North coast of Peru	5 m	12
July 17, 1998	Papua New Guinea	15 m	>2200
December 26, 2004	Sumatra, Indonesia	10.5 m	>220,000

Source: F. I. Gonzales, Tsunami! *Scientific American* 280, no. 5 (1999): 59. and United States Geologic Survey.

Figure 8.17 Ground Failure On August 17, 1959, an earthquake with a Richter magnitude of 7.3 shook southwestern Montana and a large area in adjacent states.

a The fault scarp in this image was produced when the block in the background moved up several meters compared to the one in the foreground.

b The earthquake triggered a landslide (visible in the distance) that blocked the Madison River in Montana and created Earthquake Lake (foreground). The slide entombed approximately 26 people in a campground at the very bottom.

100,000 deaths from the 1920 earthquake in Gansu, China, resulted when cliffs composed of loess (wind-deposited silt) collapsed. More than 20,000 people were killed when two-thirds of the town of Port Royal, Jamaica, slid into the sea following an earthquake on June 7, 1692.

EARTHQUAKE PREDICTION

A successful prediction must include a time frame for the occurrence of an earthquake, its location, and its strength. Despite the tremendous amount of information geologists have gathered about the cause of earthquakes, successful predictions are still rare. Nevertheless, if reliable predictions can be made, they can greatly reduce the number of deaths and injuries.

From an analysis of historic records and the distribution of known faults, geologists construct *seismic risk maps* that indicate the likelihood and potential severity of future earthquakes based on the intensity of past earthquakes. An international effort by scientists from several countries resulted in the publication of the first Global Seismic Hazard Assessment Map in December 1999 (▶ Figure 8.18). Although such maps cannot be used to predict when an earthquake will take place in any particular area, they are useful in anticipating future earthquakes and helping people plan and prepare for them (see Geo-Focus on page 213).

Earthquake Precursors

Studies conducted during the past several decades indicate that most earthquakes are preceded by both short-term and long-term changes within Earth. Such changes are called *precursors* and may be useful in earthquake prediction.

One long-range prediction technique used in seismically active areas involves plotting the location of major earthquakes and their aftershocks to detect areas that have had major earthquakes in the past, but are currently inactive. Such regions are said to be locked and not releasing energy. Nevertheless, pressure is continuing to accumulate in these regions because of plate motions, making these *seismic gaps* prime locations for future earthquakes. Several seismic gaps along the San Andreas fault have the potential for future major earthquakes (▶ Figure 8.19). A major earthquake that damaged Mexico City in 1985 occurred along a seismic gap in the convergence zone along the west coast of Mexico.

Earthquake precursors that may be useful in making short-term predictions include slight changes in elevation and tilting of the land surface, fluctuations in the water level in wells, changes in Earth's magnetic field, and the electrical resistance of the ground.

Earthquake Prediction Programs

Currently, only four nations—the United States, Japan, Russia, and China—have government-sponsored earthquake prediction programs. These programs include laboratory and field studies of rock behavior before, during, and after large earthquakes, as well as monitoring activity along major active faults. Most earthquake prediction work in the United States is done by the U.S. Geological Survey (USGS) and involves research into all aspects of earthquake-related phenomena.

The Chinese have perhaps the most ambitious earthquake prediction program in the world, which is understandable considering their long history of destructive earthquakes.

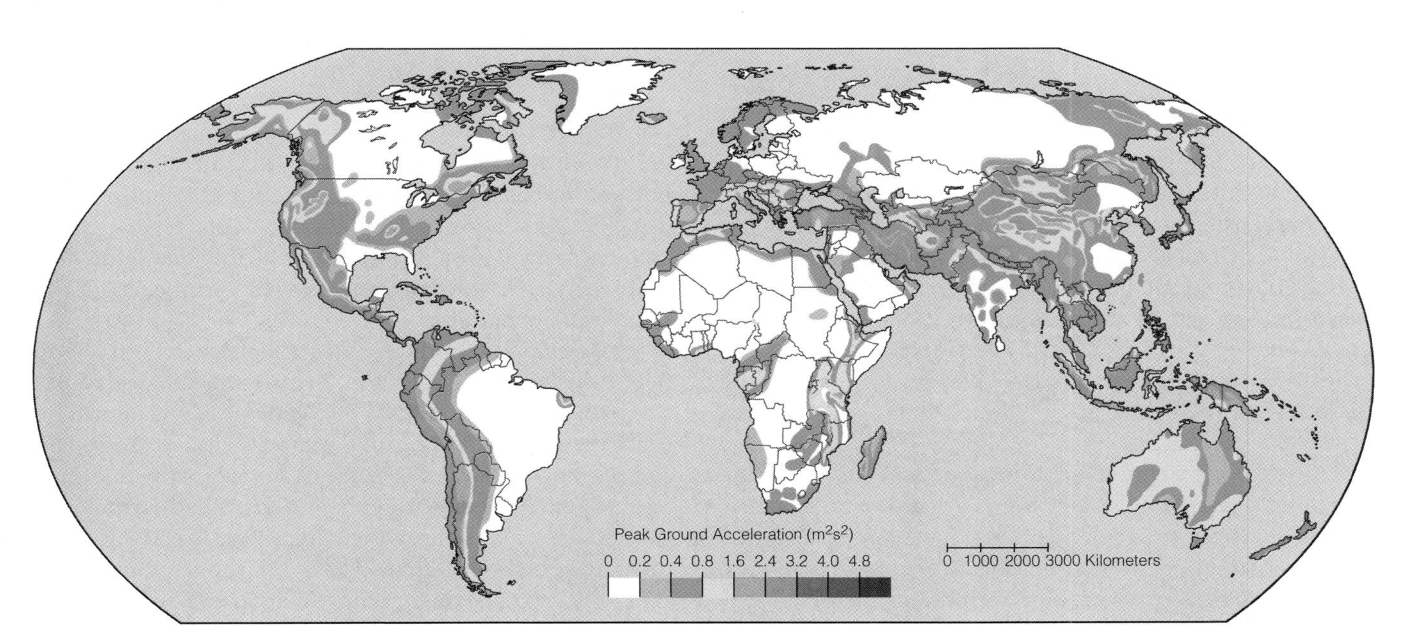

▶ **Figure 8.18 Global Seismic Hazard Assessment Map** The Global Seismic Hazard Assessment Program published this seismic hazard map showing peak ground accelerations. The values are based on a 90% probability that the indicated horizontal ground acceleration during an earthquake is not likely to be exceeded in 50 years. The higher the number, the greater the hazard. As expected, the greatest seismic risks are in the circum-Pacific belt and the Mediterranean–Asiatic belt.

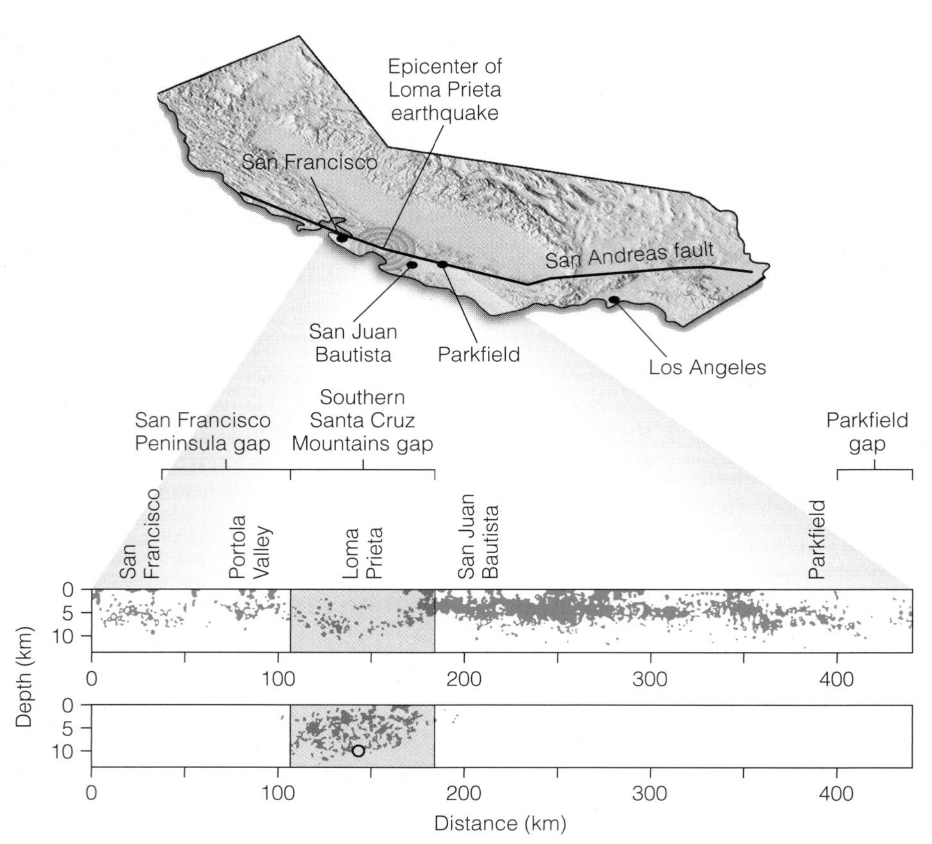

Figure 8.19 Earthquake Precursors Seismic gaps are one type of earthquake precursor that can indicate a potential earthquake in the future. Seismic gaps are regions along a fault that are locked; that is, they are not moving and releasing energy. Three seismic gaps are evident in this cross section along the San Andreas fault from north of San Francisco to south of Parkfield. The first is between San Francisco and Portola Valley, the second near Loma Prieta Mountain, and the third is southeast of Parkfield. The top section shows the epicenters of earthquakes between January 1969 and July 1989. The bottom section shows the southern Santa Cruz Mountains gap after it was filled by the October 17, 1989, Loma Prieta earthquake (open circle) and its aftershocks.

Their earthquake prediction program was initiated soon after two large earthquakes occurred at Xingtai (300 km southwest of Beijing) in 1966. This program includes extensive study and monitoring of all possible earthquake precursors. In addition, the Chinese emphasize changes in phenomena that can be observed and heard without the use of sophisticated instruments, such as observing changes in animal behavior or changes in well water levels. They successfully predicted the 1975 Haicheng earthquake, but failed to predict the devastating 1976 Tangshan earthquake that killed at least 242,000 people.

Progress is being made toward dependable, accurate earthquake predictions, and studies are underway to assess public reactions to long-, medium-, and short-term earthquake warnings. However, unless short-term warnings are actually followed by an earthquake, most people will probably ignore the warnings as they frequently do now for hurricanes, tornadoes, and tsunami. Perhaps the best we can hope for is that people in seismically active areas will take measures to minimize their risk from the next major earthquake (Table 8.5).

EARTHQUAKE CONTROL

Reliable earthquake prediction is still in the future, but can anything be done to control or at least partly control earthquakes? Because of the tremendous energy involved, it seems unlikely that humans will ever be able to prevent earthquakes. However, it may be possible to gradually release the energy stored in rocks, thus decreasing the probability of a large earthquake and extensive damage.

During the early- to mid-1960s, Denver, Colorado, experienced numerous small earthquakes. This was surprising because Denver had not been prone to earthquakes in the past. In 1962, geologist David M. Evans suggested that Denver's earthquakes were directly related to the injection of contaminated wastewater into a disposal well 3674 m deep at the Rocky Mountain Arsenal, northeast of Denver (Figure 8.20a). The U.S. Army initially denied that a connection existed, but a USGS study concluded that the pumping of waste fluids into fractured rocks beneath the disposal well decreased the friction on opposite sides of fractures and, in effect, lubricated them so that movement occurred, causing the earthquakes that Denver experienced.

Figure 8.20b shows the relationship between the average number of earthquakes in Denver per month and the average amount of contaminated fluids injected into the disposal well per month. Obviously, a high degree of correlation between the two exists, and the correlation is particularly convincing considering that during the time when no waste fluids were injected, earthquake activity decreased dramatically.

Experiments conducted in 1969 at an abandoned oil field near Rangely, Colorado, confirmed the arsenal hypothesis. Water was pumped into and out of abandoned oil wells, the pore-water pressure in these wells was measured, and seismographs were installed in the area to measure any seismic activity. Monitoring showed that small earthquakes were occurring in the area when fluids were injected and that earthquake activity declined when fluids were pumped out. What the geologists were doing was starting and stopping earthquakes at will, and the relationship between pore-water pressures and earthquakes was established.

Based on these results, some geologists have proposed that fluids be pumped into the locked segments or seismic gaps of active faults to cause small- to moderate-sized earthquakes. They think that this would relieve the pressure on the fault and prevent a major earthquake from occurring.

Although this plan is intriguing, it also has many potential problems. For instance, there is no guarantee that only a small earthquake might result. Instead, a major earthquake might occur, causing tremendous property damage and loss of life. Who would be responsible? Certainly, a great deal more research is needed before such an experiment is performed, even in an area of low population density.

What Would You Do?

Some geologists think that by pumping liquids into locked segments of active faults, they can generate small- to moderate-sized earthquakes. These earthquakes would relieve the buildup of pressure along a fault and thus prevent very large earthquakes from taking place. What do you think of this proposal? What kind of social, political, and economic consequences would there be? Do you think such an effort will ever actually reduce the threat of earthquakes?

WHAT IS EARTH'S INTERIOR LIKE?

During most of historic time, Earth's interior was perceived as an underground world of vast caverns, heat, and sulfur gases, populated by demons. By the 1860s, scientists knew what the average density of Earth was and that pressure and temperature increase with depth. And even though Earth's interior is hidden from direct observation, scientists now have a reasonably good idea of its internal structure and composition.

Earth is generally depicted as consisting of concentric layers that differ in composition and density separated from adjacent layers by rather distinct boundaries (Figure 8.21). Recall that the outermost layer, or *crust*, is Earth's thin skin. Below the crust and extending about halfway to Earth's center is the *mantle*, which comprises more than 80% of Earth's volume. The central part of Earth consists of a *core*, which is divided into a solid inner portion and a liquid outer part (Figure 8.21).

The behavior and travel times of P- and S-waves provide geologists with information about Earth's internal structure. Seismic waves travel outward as wave fronts from their source areas, although it is most convenient to depict them as *wave rays*, which are lines showing the direction of movement of small parts of wave fronts (Figure 8.3).

TABLE 8.5 What You Can Do to Prepare for an Earthquake

Anyone who lives in an area that is subject to earthquakes or who will be visiting or moving to such an area can take certain precautions to reduce the risks and losses resulting from an earthquake.

Before an earthquake:

1. Become familiar with the geologic hazards of the area where you live and work.
2. Make sure your house is securely attached to the foundation by anchor bolts and that the walls, floors, and roof are all firmly connected together.
3. Heavy furniture such as bookcases should be bolted to the walls; semiflexible natural gas lines should be used so that they can give without breaking; water heaters and furnaces should be strapped and the straps bolted to wall studs to prevent gas-line rupture and fire. Brick chimneys should have a bracket or brace that can be anchored to the roof.
4. Maintain a several-day supply of freshwater and canned foods, and keep a fresh supply of flashlight and radio batteries, as well as a fire extinguisher.
5. Maintain a basic first-aid kit and have a working knowledge of first-aid procedures.
6. Learn how to turn off the various utilities at your house.
7. Above all, have a planned course of action for when an earthquake strikes.

During an earthquake:

1. Remain calm and avoid panic.
2. If you are indoors, get under a desk or table if possible, or stand in an interior doorway or room corner as these are the structurally strongest parts of a room; avoid windows and falling debris.
3. In a tall building, do not rush for the stairwells or elevators.
4. In an unreinforced or other hazardous building, it may be better to get out of the building rather than to stay in it. Be on the alert for fallen power lines and the possibility of falling debris.
5. If you are outside, get to an open area away from buildings if possible.
6. If you are in an automobile, stay in the car, and avoid tall buildings, overpasses, and bridges if possible.

After an earthquake:

1. If you are uninjured, remain calm and assess the situation.
2. Help anyone who is injured.
3. Make sure there are no fires or fire hazards.
4. Check for damage to utilities and turn off gas valves if you smell gas.
5. Use your telephone only for emergencies.
6. Do not go sightseeing or move around the streets unnecessarily.
7. Avoid landslide and beach areas.
8. Be prepared for aftershocks.

Figure 8.20 Controlling Earthquakes

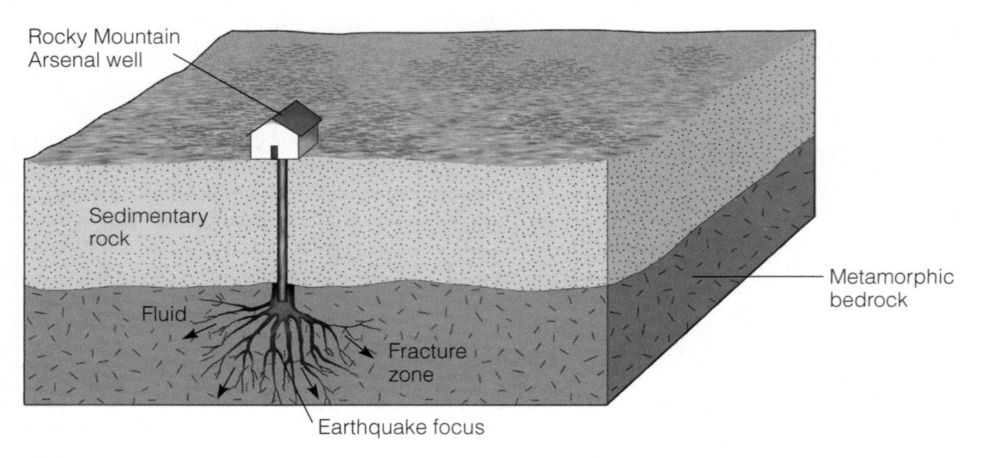

a A block diagram of the Rocky Mountain Arsenal well and the underlying geology.

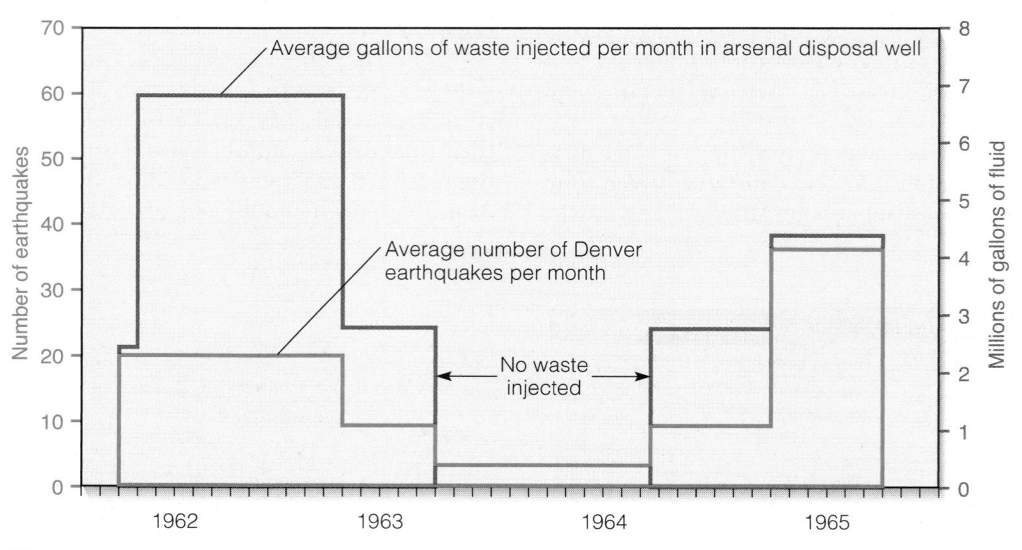

b A graph showing the relationship between the amount of wastewater injected into the well per month and the average number of Denver earthquakes per month. There have been no significant earthquakes in Denver since injection of wastewater into the disposal well ceased in 1965.

Any disturbance, such as a passing train or construction equipment, can cause seismic waves, but only those generated by large earthquakes, explosive volcanism, asteroid impacts, and nuclear explosions can travel completely through Earth.

As we noted earlier, P- and S-wave velocity is determined by the density and elasticity of the materials they travel through, both of which increase with depth. Wave velocity is slowed by increasing density, but increases in materials with greater elasticity. Because elasticity increases with depth faster than density, a general increase in seismic wave velocity takes place as the waves penetrate to greater depths. P-waves travel faster than S-waves under all circumstances, but unlike P-waves, S-waves are not transmitted through liquids because liquids have no shear strength (rigidity); liquids simply flow in response to shear stress.

If Earth were a homogeneous body, P- and S-waves would travel in straight paths as shown in Figure 8.22a. However, as a seismic wave travels from one material into another of different density and elasticity, its velocity and direction of travel change. That is, the wave is bent, a phenomenon known as **refraction**, in much the same way as light waves are refracted as they pass from air into more dense water. Because seismic waves pass through materials of differing density and elasticity, they are continually refracted so that their paths are curved; wave rays travel in a straight line only when their direction of travel is perpendicular to a boundary (Figure 8.22b, c).

In addition to refraction, seismic rays are **reflected**, much as light is reflected from a mirror. When seismic rays encounter a boundary separating materials of different density or elasticity, some of a wave's energy is reflected back to

Geo-Focus Paleoseismology

Paleoseismology is the study of prehistoric earthquakes. As more people move into seismically active areas, it is important to know how frequently earthquakes in the area have occurred in the past, and how strong those earthquakes were. In this way, prudent decisions can be made about what precautions need to be taken in developing an area and how stringent the building codes for a region need to be.

A typical technique in paleoseismology is to excavate trenches across active faults in an area to be studied and date the sediments disturbed by prehistoric earthquakes (Figure 1). By exposing the upper few meters of material along an active fault, geologists can find evidence of previous earthquakes in the ancient soil layers. Furthermore, by dating the paleosoils by carbon-14 or other dating techniques, geologists can determine the frequency of past earthquakes and when the last earthquake occurred, and thus have a basis for estimating the probability of future earthquakes.

John Karachewski

Figure 1 Geologists examine a trench across an active fault in California to determine possible seismic hazards. Excavating trenches is a common method used by geologists to gather information about ancient earthquakes in a region and to help assess the potential for future earthquakes and the damage that they might cause.

Paleoseismic studies are currently underway in many areas of North America, particularly along the San Andreas fault in California and in the coastal regions of Washington. An interesting case in point concerns an ancient earthquake in what is now Seattle, Washington.

Data from a variety of sources have convinced many geologists that a shallow-focus earthquake of at least magnitude 7 occurred beneath Seattle, Washington, less than 1100 years ago. In a point not lost on officials, they noted the catastrophic effects that a similar-sized earthquake would have if it occurred in the same area today.

The first link in the chain of evidence for a paleoearthquake came from the discovery of a marine terrace that had been uplifted some 7 m at Restoration Point, 5 km west of Seattle. Carbon-14 analysis of peat within sediments of the terrace indicates that uplift occurred between 500 and 1700 years ago. Carbon-14 dating of other sites to the north, south, and east also indicates a sudden uplift in the area within the same time period. The amount of uplift suggests to geologists a magnitude-7 or greater earthquake.

Because many earthquakes in and around the Pacific Ocean basin can cause tsunami, geologists looked for evidence that a tsunami occurred, and they found it in the form of unusual sand layers in nearby tidal marsh deposits. Carbon-14 dating of organic matter associated with the sands yielded an age of 850 to 1250 years ago, well within the time period during which the terraces were uplifted.

Geologists also found evidence of rock avalanches in the Olympic Mountains that dammed streams, thereby forming lakes. Drowned trees in the lakes were dated as having died between 1000 and 1300 years ago, again fitting in nicely with the date of the ancient earthquake.

One of the final pieces of evidence is the deposits found on the bottom of Lake Washington. The earthquake apparently caused the bottom sediment of the lake to be resuspended and to move downslope as a turbidity current. Dating of these sediments indicates that they were deposited between 940 and 1280 years ago, consistent with their being caused by an earthquake.

All evidence points to a large (magnitude 7 or greater) shallow-focus earthquake occurring in the Seattle area approximately 1000 years ago. If history and the events of the geologic past are any guide, it is very likely that another large earthquake will hit the Seattle area in the future. When this will occur can't yet be predicted, but it would be wise to plan for such an eventuality. After all, metropolitan Seattle has a population of more than 2.5 million people, and its entire port area is built on fill that would probably be hard hit by an earthquake. Furthermore, most of Seattle's schools, hospitals, utilities, and fire and police stations are not built to withstand a strong earthquake. What the future holds for Seattle has yet to be determined. Geologists have provided a window on what has happened seismically in the past, and it is up to today's government and its various agencies to decide how they want to use this information.

Earth's Composition and Density		
	Composition	**Density (g/cm^3)**
Continental crust	Average composition of granodiorite	~2.7
Oceanic crust	Upper part basalt, lower part gabbro	~3.0
Mantle	Peridotite (made up of ferromagnesian silicates)	3.3–5.7
Outer core	Iron with perhaps 12% sulfur silicon, oxygen, nickel, and potassium	9.9–12.2
Inner core	Iron with 10–20% nickel	12.6–13.0

Figure 8.21 Earth's Internal Structure The inset shows Earth's outer part in more detail. The asthenosphere is solid, but behaves plastically and flows.

Figure 8.22 What If Earth Were Homogeneous?

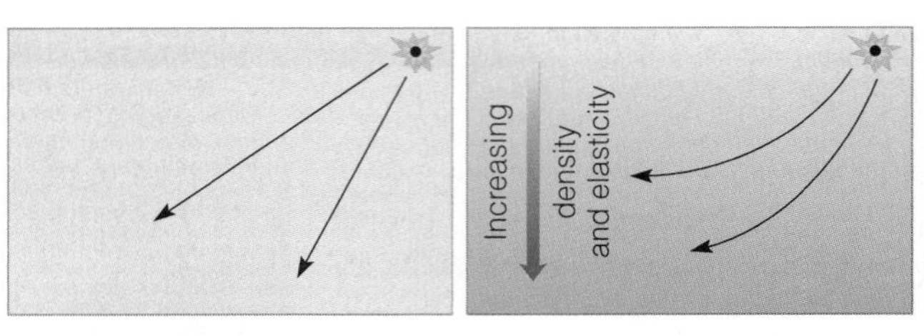

a If Earth were homogeneous throughout, seismic wave rays would follow straight paths.

b Because density and elasticity increase with depth, wave rays are continuously refracted so that their paths are curved.

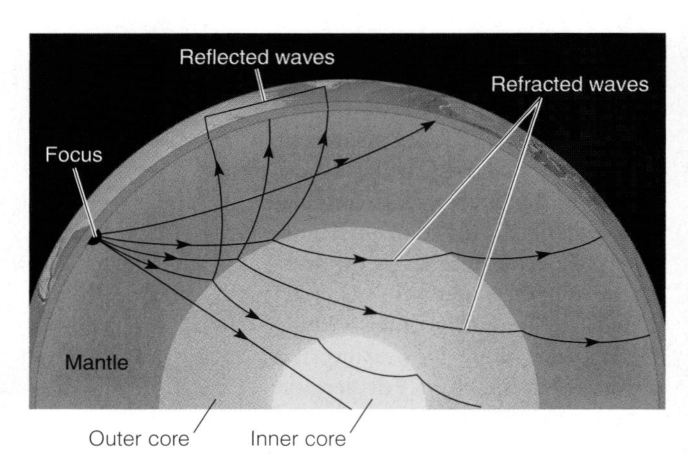

c Refraction and reflection of P-waves as they encounter boundaries separating materials of different density or elasticity. Notice that the only wave ray not refracted is the one perpendicular to boundaries.

the surface (Figure 8.22c). If we know the wave velocity and the time required for the wave to travel from its source to the boundary and back to the surface, we can calculate the depth of the reflecting boundary. Such information is useful in determining not only Earth's internal structure, but also the depths of sedimentary rocks that may contain petroleum. Seismic reflection is a common tool used in petroleum exploration.

Although changes in seismic wave velocity occur continuously with depth, P-wave velocity increases suddenly at the base of the crust and decreases abruptly at a depth of approximately 2900 km (Figure 8.23). These marked changes in seismic wave velocity indicate a boundary called a **discontinuity** across which a significant change in Earth materials or their properties occurs. Discontinuities are the basis for subdividing Earth's interior into concentric layers.

THE CORE

In 1906, R. D. Oldham of the Geological Survey of India realized that seismic waves arrived later than expected at seismic stations more than 130 degrees from an earthquake focus. He postulated that Earth has a core that transmits seismic waves more slowly than shallower Earth materials. We now know that P-wave velocity decreases markedly at a depth of 2900 km, which indicates an important discontinuity now recognized as the core–mantle boundary (Figure 8.23).

Because of the sudden decrease in P-wave velocity at the core–mantle boundary, P-waves are refracted in the core so that little P-wave energy reaches the surface in the area between 103 degrees and 143 degrees from an earthquake focus (Figure 8.24a). This **P-wave shadow zone,** as it is called, is an area in which little P-wave energy is recorded by seismographs.

The P-wave shadow zone is not a perfect shadow zone because some weak P-wave energy is recorded within it. Scientists proposed several hypotheses to account for this observation, but all were rejected by the Danish seismologist Inge Lehmann, who in 1936 postulated that the core is not entirely liquid as previously thought. She proposed that seismic wave reflection from a solid inner core accounts for the arrival of weak P-wave energy in the P-wave shadow zone, a proposal that was quickly accepted by seismologists.

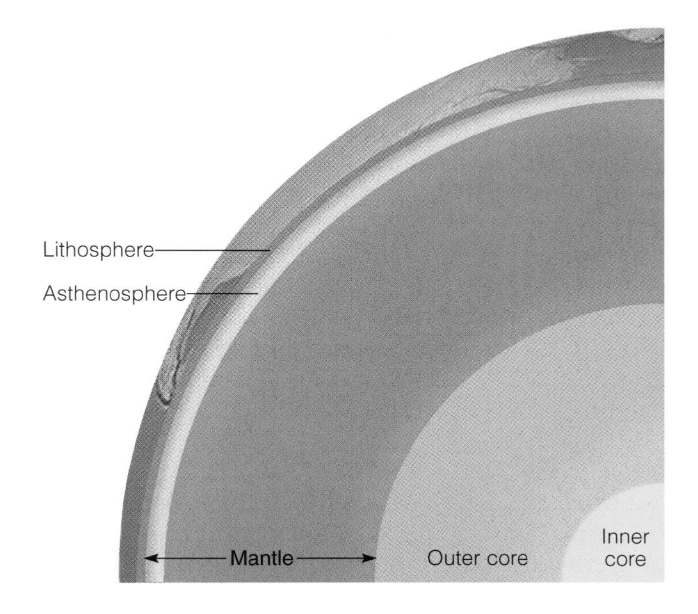

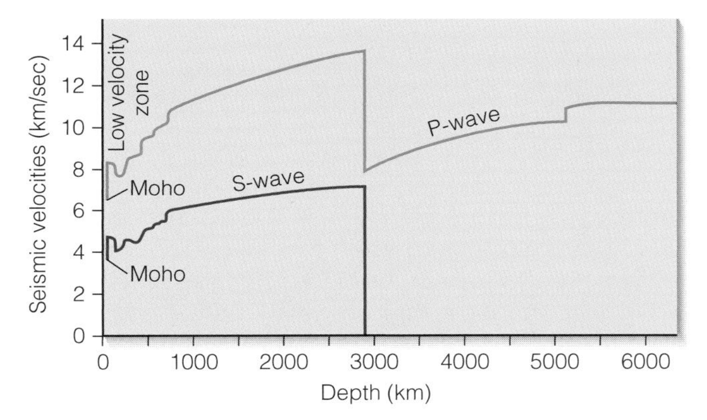

Figure 8.23 Seismic Wave Velocities Profiles showing seismic wave velocities versus depth. Several discontinuities are shown, across which seismic wave velocities change rapidly.

In 1926, the British physicist Harold Jeffreys realized that S-waves were not simply slowed by the core, but were completely blocked by it. So, besides a P-wave shadow zone, a much larger and more complete **S-wave shadow zone** also exists (Figure 8.24b). At locations greater than 103 degrees from an earthquake focus, no S-waves are recorded, which indicates that S-waves cannot be transmitted through the core. S-waves will not pass through a liquid, so it seems that the outer core must be liquid or behave as a liquid. The inner core, however, is thought to be solid because P-wave velocity increases at the base of the outer core.

Density and Composition of the Core

The core constitutes 16.4% of Earth's volume and nearly one-third of its mass. Geologists can estimate the core's density and composition by using seismic evidence and laboratory experiments. For instance, geologists use a diamond-anvil pressure cell in which small samples are studied while subjected to pressures and temperatures similar to those in the core. Furthermore, meteorites, which are thought to represent remnants of the material from which the solar system formed, are used to make estimates of density and composition. For example, meteorites composed of iron and nickel alloys may represent the differentiated interiors of large asteroids and approximate the density and composition of Earth's core. The density of the

What Would You Do?

Of course, novels such as *Journey to the Center of the Earth* are fiction, but it is surprising how many people think that vast caverns and cavities exist deep within the planet. How would you explain that even though we have no direct observations at great depth, we can still be sure that these proposed openings do not exist?

Figure 8.24 P-Wave and S-Wave Shadow Zones

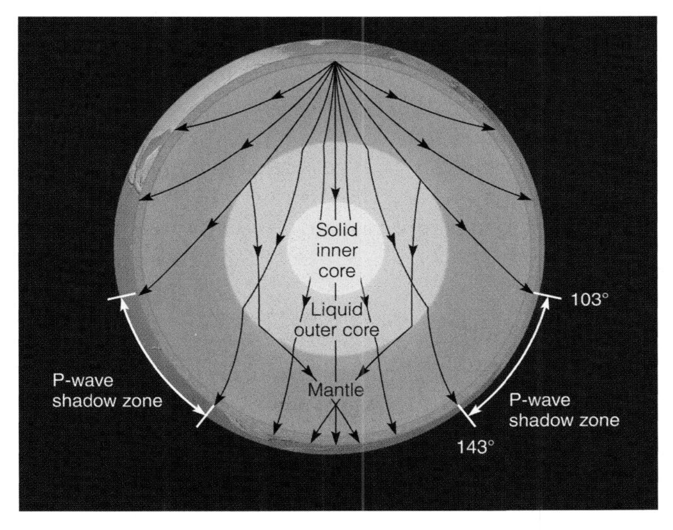

a P-waves are refracted so that no direct P-wave energy reaches the surface in the P-wave shadow zone.

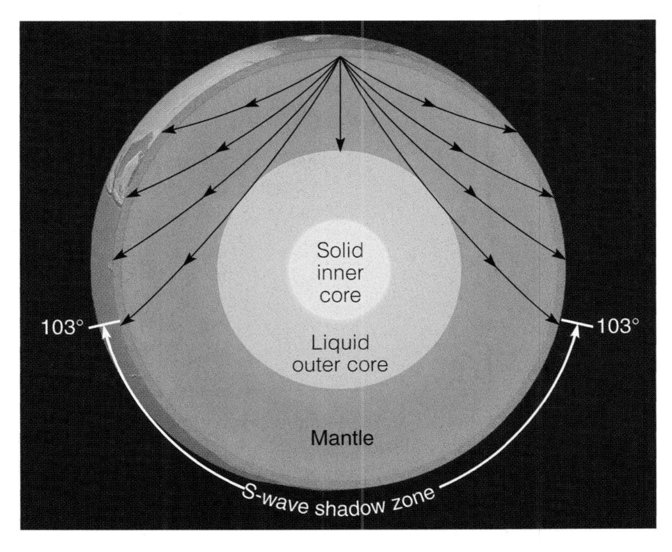

b The presence of an S-wave shadow zone indicates that S-waves are being blocked within Earth.

outer core varies from 9.9 to 12.2 g/cm^3. At Earth's center, the pressure is equivalent to approximately 3.5 million times normal atmospheric pressure.

The core cannot be composed of minerals common at the surface because, even under the tremendous pressures at great depth, they would still not be dense enough to yield an average density of 5.5 g/cm^3 for Earth. Both the outer and inner cores are thought to be composed largely of iron, but pure iron is too dense to be the sole constituent of the outer core. It must be "diluted" with elements of lesser density. Laboratory experiments and comparisons with iron meteorites indicate that perhaps 12% of the outer core consists of sulfur and possibly some silicon, oxygen, nickel, and potassium (Figure 8.21).

In contrast, pure iron is not dense enough to account for the estimated density of the inner core, so perhaps 10–20% of the inner core consists of nickel. These metals form an iron–nickel alloy thought to be sufficiently dense under the pressure at that depth to account for the density of the inner core.

When the core formed during early Earth history, it was probably entirely molten and has since cooled so that its interior has crystallized. Indeed, the inner core continues to grow as Earth slowly cools, and the liquid of the outer core crystallizes as iron. Recent evidence also indicates that at present the inner core rotates faster than the outer core, moving approximately 20 km/yr relative to the outer core.

EARTH'S MANTLE

Another significant discovery about Earth's interior was made in 1909 when the Yugoslavian seismologist Andrija Mohorovičić detected a seismic discontinuity at a depth of about 30 km. While studying the arrival times of seismic waves from Balkan (part of southeastern Europe) earthquakes, Mohorovičić noticed that seismic stations a few hundred kilometers from an earthquake's epicenter were recording two distinct sets of P- and S-waves.

From his observations, Mohorovičić concluded that a sharp boundary separates rocks with different properties at a depth of about 30 km. He postulated that P-waves below this boundary travel at 8 km/sec, whereas those above the boundary travel at 6.75 km/sec. When an earthquake occurs, some waves travel directly from the focus to a seismic station, whereas others travel through the deeper layer and some of their energy is refracted back to the surface (◗ Figure 8.25). The waves traveling through the deeper layer (the mantle) travel farther to a seismic station, but they do so more rapidly and arrive before those that travel more slowly in the shallower layer.

The boundary identified by Mohorovičić separates the crust from the mantle and is now called the **Mohorovičić discontinuity**, or simply the **Moho**. It is present everywhere except beneath spreading ridges. However, its depth varies: Beneath continents, it ranges from 20 to 90 km, with an average of 35 km; beneath the sea floor, it is 5 to 10 km deep.

The Mantle's Structure, Density, and Composition

Although seismic wave velocity in the mantle increases with depth, several discontinuities exist. Between depths of 100 and 250 km, both P- and S-wave velocities decrease markedly (◗ Figure 8.26). This 100- to 250-km-deep layer is the *low-velocity zone*, which corresponds closely to the *asthenosphere*, a layer in which the rocks are close to their melting point and are less elastic, accounting for the observed decrease in seismic wave velocity. The asthenosphere is an important zone because it is where most magma is generated, especially under the ocean basins. Furthermore, it lacks strength, flows plastically, and is thought to be the layer over which the plates of the outer, rigid *lithosphere* move.

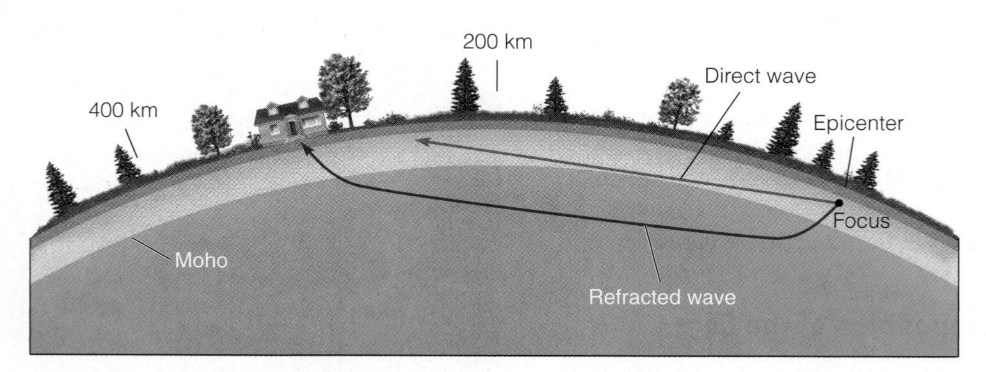

◗ **Figure 8.25 Seismic Discontinuity** Andrija Mohorovičić studied seismic waves and detected a seismic discontinuity at a depth of about 30 km. The deeper, faster seismic waves arrive at seismic stations first, even though they travel farther. This discontinuity, now known as the Moho, is between the crust and mantle.

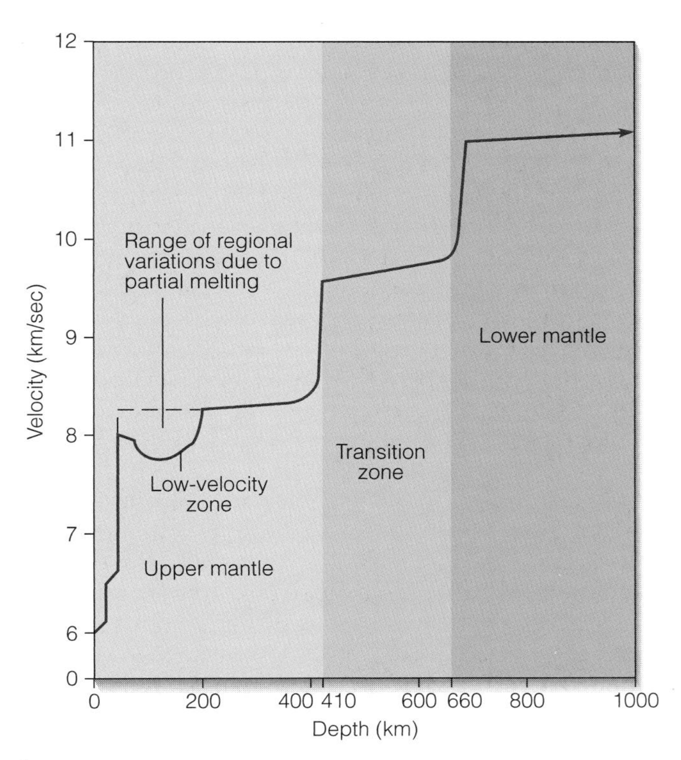

Figure 8.26 Variation in P-Wave Velocity Variations in P-wave velocity in the upper mantle and transition zone.

Other discontinuities are also present at deeper levels within the mantle. But unlike those between the crust and mantle or between the mantle and core, these probably represent structural changes in minerals rather than compositional changes. In other words, geologists think that the mantle is composed of the same material throughout, but the structural states of minerals such as olivine change with depth.

At a depth of 410 km, seismic wave velocity increases slightly as a consequence of such changes in mineral structure (Figure 8.26). Another velocity increase occurs at about 660 km, where the minerals break down into metal oxides, such as FeO (iron oxide) and MgO (magnesium oxide), and silicon dioxide (SiO_2). These two discontinuities define the top and base of a *transition zone* separating the upper mantle from the lower mantle (Figure 8.26).

Although the mantle's density, which varies from 3.3 to 5.7 g/cm^3, can be inferred rather accurately from seismic waves, its composition is less certain. The igneous rock *peridotite*, containing mostly ferromagnesian silicates, is considered the most likely component of the upper mantle. Laboratory experiments indicate that it possesses physical properties that account for the mantle's density and observed rates of seismic wave transmissions. Peridotite also forms the lower parts of igneous rock sequences thought to be fragments of the oceanic crust and upper mantle emplaced on land. In addition, peridotite occurs as inclusions in volcanic rock bodies such as *kimberlite pipes* that are known to have come from depths of 100 to 300 km.

SEISMIC TOMOGRAPHY

The model of Earth's interior consisting of a core and a mantle is probably accurate, but not very precise. In recent years, geophysicists have developed a technique called *seismic tomography* that allows them to develop more accurate models of Earth's interior. In seismic tomography, numerous crossing seismic waves are analyzed much as CAT (computerized axial tomography) scans are analyzed. In CAT scans, X-rays penetrate the body and a two-dimensional image of its interior is formed. Repeated CAT scans from slightly different angles are stacked to produce a three-dimensional image.

In a similar manner, geophysicists use seismic waves to probe Earth's interior. In seismic tomography, the average velocities of numerous crossing seismic waves are analyzed so that "slow" and "fast" areas of wave travel are detected. Remember that seismic wave velocity depends partly on elasticity; cold rocks have greater elasticity and therefore transmit seismic waves faster than hotter rocks.

As a result of studies in seismic tomography, a much clearer picture of Earth's interior is emerging. It has already given us a better understanding of complex convection within the mantle and a clearer picture of the nature of the core–mantle boundary.

EARTH'S INTERNAL HEAT

During the 19th century, scientists realized that the temperature in deep mines increases with depth. More recently, the same trend has been observed in deep drill holes. This temperature increase with depth, or **geothermal gradient**, is approximately 25°C/km near the surface. In areas of active or recently active volcanism, the geothermal gradient is greater than in adjacent nonvolcanic areas, and temperature rises faster beneath spreading ridges than elsewhere beneath the seafloor.

Much of Earth's internal heat is generated by radioactive decay, especially the decay of isotopes of uranium and thorium and, to a lesser degree, potassium-40. When these isotopes decay, they emit energetic particles and gamma rays that heat surrounding rocks. Because rock is such a poor conductor of heat, it takes little radioactive decay to build up considerable heat, given enough time.

Unfortunately, the geothermal gradient is not useful for estimating temperatures at great depth. If we were simply to extrapolate from the surface downward, the temperature at 100 km would be so high that, despite the great pressure, all known rocks would melt. Yet except for pockets of magma, it appears that the mantle is solid rather than liquid because it transmits S-waves. Accordingly, the geothermal gradient must decrease markedly.

Current estimates of the temperature at the base of the crust are 800 to 1200°C. The latter figure seems to be an upper limit; if it were any higher, melting would be expected.

Furthermore, fragments of mantle rock in kimberlite pipes, thought to have come from depths of 100–300 km, appear to have reached equilibrium at these depths at a temperature of approximately 1200°C. At the core–mantle boundary, the temperature is probably between 2500 and 5000°C; the wide range of values indicates the uncertainties of such estimates. If these figures are reasonably accurate, the geothermal gradient in the mantle is only about 1°C/km.

Because the core is so remote and its composition uncertain, only very general estimates of its temperature are possible. Based on various experiments, the maximum temperature at the center of the core is thought to be 6500°C, very close to the estimated temperature for the surface of the Sun!

EARTH'S CRUST

Our main concern in the latter part of this chapter is Earth's interior; however, to be complete, we must briefly discuss the crust, which along with the upper mantle constitutes the lithosphere.

Continental crust is complex, consisting of all rock types, but it is usually described as "granitic," meaning that its overall composition is similar to that of granitic rocks. With the exception of metal-rich rocks such as iron ore deposits, most rocks of the continental crust have densities between 2.5 and 3.0 g/cm^3, with the average density of the crust being about 2.7 g/cm^3. P-wave velocity in continental crust is about 6.75 km/sec, but at the base of the crust, P-wave velocity abruptly increases to about 8 km/sec. Continental crust averages 35 km thick, but its thickness varies from 20 to 90 km. Beneath mountain ranges such as the Rocky Mountains, the Alps in Europe, and the Himalayas in Asia, continental crust is much thicker than it is in adjacent areas. In contrast, continental crust is much thinner than average beneath the Rift Valleys of East Africa and in a large area called the Basin and Range Province in the western United States and northern Mexico. The crust in these areas has been stretched and thinned in what appear to be the initial stages of rifting (see Chapter 2).

In contrast to continental crust, oceanic crust is simpler, consisting of gabbro in its lower part and overlain by basalt. It is thinnest, about 5 km, at spreading ridges, and nowhere is it thicker than 10 km. Its average density of 3.0 g/cm^3 accounts for the fact that it transmits P-waves at about 7 km/sec. In fact, this P-wave velocity is what one would expect if oceanic crust is composed of basalt and gabbro. We present a more detailed description of the oceanic crust's composition and structure in Chapter 9.

Geo-Recap

Chapter Summary

- Earthquakes are vibrations caused by the sudden release of energy, usually along a fault.
- The elastic rebound theory is an explanation for how energy is released during earthquakes. As rocks on opposite sides of a fault are subjected to force, they accumulate energy and slowly deform until their internal strength is exceeded. At that time, a sudden movement occurs along the fault, releasing the accumulated energy, and the rocks snap back to their original undeformed shape.
- Seismology is the study of earthquakes. Earthquakes are recorded on seismographs, and the record of an earthquake is a seismogram.
- An earthquake's focus is the location where rupture within Earth's lithosphere occurs and energy is released. The epicenter is the point on Earth's surface directly above the focus.
- Approximately 80% of all earthquakes occur in the circum-Pacific belt, 15% within the Mediterranean–Asiatic belt, and the remaining 5% mostly in the interior of the plates and along oceanic spreading ridges.
- The two types of body waves are P-waves (primary waves) and S-waves (secondary waves). P-waves are the fastest seismic waves and travel through all materials, whereas S-waves are somewhat slower and can travel only through solids. P-waves are compressional (expanding and compressing the material through which they travel), whereas S-waves are shear (moving material perpendicular to the direction of travel).
- Rayleigh (R-waves) and Love waves (L-waves) move along or just below Earth's surface.
- An earthquake's epicenter is determined using a time–distance graph of the P- and S-waves to calculate how

far away a seismic station is from an earthquake. The greater the difference in arrival times between the two waves, the farther away the seismic station is from the earthquake. Three seismographs are needed to locate the epicenter.

- Intensity is a subjective, or qualitative, measure of the kind of damage done by an earthquake. It is expressed in values from I to XII in the Modified Mercalli Intensity Scale.
- The Richter Magnitude Scale measures an earthquake's magnitude, which is the total amount of energy released by an earthquake at its source. It is an open-ended scale with values beginning at 1. Each increase in magnitude number represents about a 30-fold increase in energy released.
- The seismic-moment magnitude scale more accurately measures the total energy released by very large earthquakes.
- The destructive effects of earthquakes include ground shaking, fire, tsunami, landslides, and disruption of vital services.
- Seismic risk maps help geologists in determining the likelihood and potential severity of future earthquakes based on the intensity of past earthquakes.
- Earthquake precursors are changes preceding an earthquake and include seismic gaps, changes in surface elevations, and fluctuations of water levels in wells.
- A variety of earthquake research programs are underway in the United States, Japan, Russia, and China. Studies indicate that most people would probably not heed a short-term earthquake warning.
- Because of the tremendous energy involved, it seems unlikely that humans will ever be able to prevent earthquakes. However, it might be possible to release small amounts of the energy stored in rocks and thus avoid a large earthquake and the extensive damage that typically results.
- Earth has an outer layer of oceanic and continental crust below which lies a rocky mantle and an iron-rich core with a solid inner part and a liquid outer part.
- Studies of P- and S-waves, laboratory experiments, comparisons with meteorites, and studies of inclusions in volcanic rocks provide evidence for Earth's internal structure and composition.
- Density and elasticity of Earth materials determine the velocity of seismic waves. Seismic waves are refracted when their direction of travel changes. Wave reflection occurs at boundaries across which the properties of rocks change.
- Geologists use the behavior of P- and S-waves and the presence of the P- and S-wave shadow zones to estimate the density and composition of Earth's interior, as well as to estimate the size and depth of the core and mantle.
- Earth's inner core is probably made up of iron and nickel, whereas the outer core is mostly iron with 10–20% other substances.
- Peridotite, an igneous rock composed mostly of ferromagnesian silicates, is the most likely rock making up Earth's mantle.
- Oceanic crust is composed of basalt and gabbro, whereas continental crust has an overall composition similar to granite. The Moho is the boundary between the crust and the mantle.
- The geothermal gradient of 25°C/km cannot continue to great depths; within the mantle and core, it is probably about 1°C/km. The temperature at Earth's center is estimated to be 6500°C.

Important Terms

discontinuity (p. 214)
earthquake (p. 190)
elastic rebound theory (p. 190)
epicenter (p. 193)
focus (p. 193)
geothermal gradient (p. 217)
intensity (p. 198)
Love wave (L-wave) (p. 197)
magnitude (p. 200)
Modified Mercalli Intensity Scale (p. 198)
Mohorovičić discontinuity (Moho) (p. 216)
P-wave (p. 196)
P-wave shadow zone (p. 214)
Rayleigh wave (R-wave) (p. 196)
reflection (p. 212)
refraction (p. 212)
Richter Magnitude Scale (p. 200)
seismograph (p. 192)
seismology (p. 192)
S-wave (p. 196)
S-wave shadow zone (p. 215)
tsunami (p. 206)

Review Questions

1. The minimum number of seismographs needed to determine an earthquake's epicenter is
   a. _____ 1;
   b. _____ 2;
   c. _____ 3;
   d. _____ 4;
   e. _____ 5.
2. Most of Earth's internal heat is generated by
   a. _____ moving plates;
   b. _____ volcanism;
   c. _____ earthquakes;
   d. _____ radioactive decay;
   e. _____ meteorite impacts.
3. The majority of all earthquakes take place in the
   a. _____ spreading-ridge zone;
   b. _____ Mediterranean–Asiatic belt;
   c. _____ rifts in continental interiors;
   d. _____ circum-Pacific belt;
   e. _____ Appalachian fault zone.
4. A tsunami is a(n)
   a. _____ part of a fault with a seismic gap;
   b. _____ precursor to an earthquake;
   c. _____ seismic sea wave;
   d. _____ particularly large and destructive earthquake;
   e. _____ earthquake with a focal depth exceeding 300 km.
5. The seismic discontinuity at the base of the crust is known as the
   a. _____ transition zone;
   b. _____ magnetic reflection point;
   c. _____ low-velocity zone;
   d. _____ Moho;
   e. _____ high-velocity zone.
6. The total amount of energy released by an earthquake at its source is its
   a. _____ intensity;
   b. _____ dilatancy;
   c. _____ seismicity;
   d. _____ magnitude;
   e. _____ liquefaction.
7. An epicenter is
   a. _____ the location where rupture begins;
   b. _____ the point on the Earth's surface vertically above the focus;
   c. _____ the same as the hypocenter;
   d. _____ the location where energy is released;
   e. _____ none of these.
8. How much more energy is released by a magnitude-7 earthquake than a magnitude-3 earthquake?
   a. _____ 4;
   b. _____ 810,000;
   c. _____ 27,000;
   d. _____ 90;
   e. _____ 2,500,000.
9. The geothermal gradient is Earth's
   a. _____ capacity to reflect and refract seismic waves;
   b. _____ most destructive aspect of earthquakes;
   c. _____ temperature increase with depth;
   d. _____ average rate of seismic wave velocity in the mantle;
   e. _____ elastic rebound potential.
10. A P-wave is one in which
    a. _____ material moves perpendicular to the direction of wave travel;
    b. _____ Earth's surface moves as a series of waves;
    c. _____ material is expanded and compressed as the wave moves through it;
    d. _____ large waves crash onto a shoreline following a submarine earthquake;
    e. _____ none of these.
11. How does the lithosphere differ from the asthenosphere?
12. Refer to the graph in Figure 8.9b. A seismograph in Berkeley, California, records the arrival time of an earthquake's P-waves as 6:59:54 P.M. and the S-waves as 7:00:02 P.M. The maximum amplitude of the S-waves as recorded on the seismogram was 75 mm. What was the magnitude of the earthquake, and how far away from Berkeley did it occur?
13. What is the difference between an earthquake's focus and its epicenter? Why is an earthquake's epicenter the location that is usually reported in the news?
14. How does the elastic rebound theory account for the energy released during an earthquake?
15. Why do insurance companies use the qualitative Modified Mercalli Intensity Scale instead of the quantitative Richter Magnitude Scale in classifying earthquakes?
16. What is the likely explanation for changes with depth in the mantle?
17. Describe the various ways that earthquakes are destructive.

18. If Earth were completely solid and homogeneous throughout, how would P- and S-waves behave as they traveled through the planet? How do they actually behave?
19. Why do scientists think that the inner core is solid and the outer core is liquid? What is the core composed of?
20. From the arrival times of P- and S-waves shown in the accompanying chart and the graph in Figure 8.9b, calculate how far away from each seismograph station the earthquake occurred. How would you determine the epicenter of this earthquake?

Arrival Time of P-Wave	Arrival Time of S-Wave
Station A: 2:59:03 P.M.	3:04:03 P.M.
Station B: 2:51:16 P.M.	3:01:16 P.M.
Station C: 2:48:25 P.M.	2:55:55 P.M.

CHAPTER 9

The Seafloor

USGS/NASA

OUTLINE

OBJECTIVES

At the end of this chapter, you will have learned that

- Scientists use echo sounding, seismic profiling, sampling, and observations from submersibles to study the largely hidden seafloor.
- Oceanic crust is thinner and compositionally less complex than continental crust.
- The margins of continents consist of a continental shelf and slope and, in some cases, a continental rise with adjacent abyssal plains. The elements that make up a continental margin depend on the geologic activity that takes place in these marginal areas.
- Although the seafloor is flat and featureless in some places, it also has ridges, trenches, seamounts, and other features.
- Geologic activities at or near divergent and convergent plate boundaries account for distinctive seafloor features such as submarine volcanoes and deep-sea trenches.
- Most seafloor sediment comes from the weathering and erosion of continents and oceanic islands, and from the shells of tiny marine organisms.
- Organisms in warm, shallow seas build wave-resistant structures known as reefs.

Heart Reef in Australia formed naturally into the shape of a heart. It is part of the 2000-km-long Great Barrier Reef in Australia, the largest coral reef in the world.

- Several important resources, such as common salt, come from seawater, and hydrocarbons are found in some seafloor sediments.

INTRODUCTION

According to two dialogues written in about 350 B.C. by the Greek philosopher Plato, there was a huge continent called Atlantis in the Atlantic Ocean west of the Pillars of Hercules, or what we now call the Strait of Gibraltar (Figure 9.1) According to Plato's account, Atlantis controlled a large area extending as far east as Egypt. Yet despite its vast wealth, advanced technology, and large army and navy, Atlantis was defeated in war by Athens. Following the conquest of Atlantis,

> there were violent earthquakes and floods and one terrible day and night came when . . . Atlantis . . . disappeared beneath the sea. And for this reason even now the sea there has become unnavigable and unsearchable, blocked as it is by the mud shallows which the island produced as it sank.*

No "mud shallows" exist in the Atlantic, as Plato asserted. In fact, no geologic evidence indicates that Atlantis ever existed, so why has the legend persisted for so long?

One reason is that sensational stories of lost civilizations are popular, but another is that until recently no one had much knowledge of what lies beneath the oceans. Much of the seafloor is a hidden domain, so myths and legends were widely accepted. The most basic observation we can make about Earth is that it has vast water-covered areas and continents, which, at first glance, might seem to be nothing more than parts of the planet not covered by water. Nevertheless, the continents and the ocean basins are very different.

Oceanic crust is thinner and denser than continental crust, and it is made up of basalt and gabbro, whereas continental crust has an overall composition similar to that of granite. In addition, oceanic crust is produced continually at spreading ridges and consumed at subduction zones, so none of it is older than about 180 million years. Continental crust, on the other hand, varies in age from recent to 3.96 billion years old.

One important reason to study the seafloor is that it makes up the largest part of Earth's surface (Figure 9.2). Despite the commonly held misconception that the seafloor is flat and featureless, its topography is as varied as that of the continents. Furthermore, many seafloor features, as well as several aspects of the oceanic crust, provide important evidence for plate tectonic theory (see Chapter 2). And finally, natural resources are found on the marginal parts of continents, in seawater, and on the seafloor.

Our discussion of the seafloor focuses on (1) the physical attributes and composition of the oceanic crust, (2) the composition and distribution of seafloor sediments, (3) seafloor topography, and (4) the origin and evolution of the continental margins. *Oceanographers* study these topics, too, but they also study the chemistry and physics of seawater, as well as oceanic circulation and marine biology.

EXPLORING THE OCEANS

An interconnected body of saltwater that we call oceans and seas covers 71% of Earth's surface. Nevertheless, this world ocean has areas distinct enough for us to recognize the Pacific, Atlantic, Indian, and Arctic Oceans. The term *ocean* refers to these large areas of saltwater, whereas *sea* designates a smaller body of water, usually a marginal part of an ocean (Figure 9.2). We should point out that whereas the oceans and their marginal seas are largely underlain by oceanic crust, the same is not true of the Dead Sea, Salton Sea, and Caspian Sea; these are actually saline lakes on the continents.

During most of historic time, people knew little of the oceans, and until recently they thought the seafloor was a vast, featureless plain. In fact, through most of this time, the seafloor, in one sense, was more remote than the Moon's surface because it could not be observed.

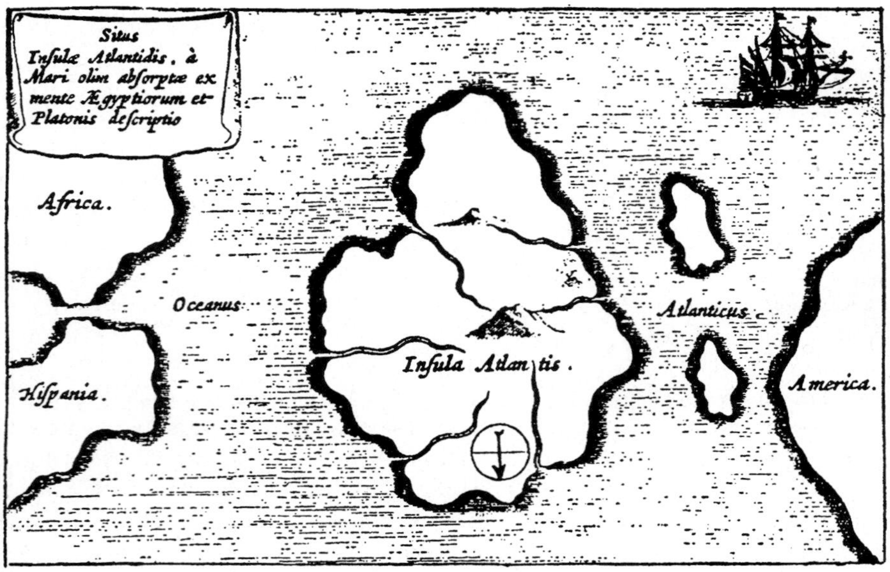

Figure 9.1 Atlantis According to Plato, Atlantis was a continent west of the Pillars of Hercules, now called the Strait of Gibraltar. In this map from Anthanasium Kircher's *Mundus Subterraneus* (1664), north is toward the bottom of the map. The Strait of Gibraltar is the narrow area between Hispania (Spain) and Africa.

Early Exploration

The ancient Greeks had determined Earth's size and shape rather accurately, but western Europeans were not aware of the vastness of the oceans until the 1400s and 1500s, when explorers sought trade routes to the Indies. Even when Christopher Columbus set sail on August 3, 1492, in an effort to find a route to the Indies,

*From the Timaeus, quoted in E. W. Ramage, Ed., Atlantis: Fact or Fiction? (Bloomington: Indiana University Press, 1978), p. 13.

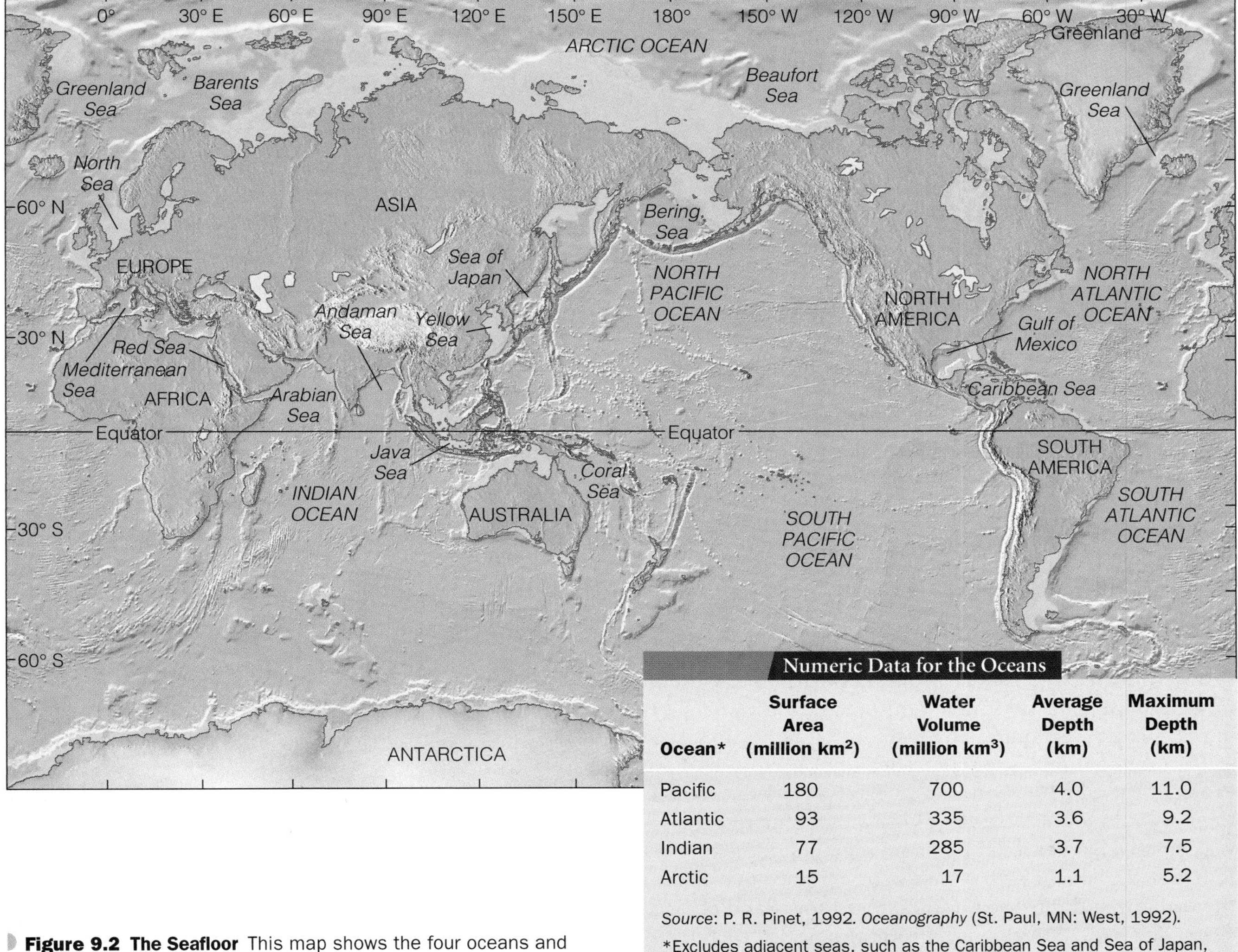

Numeric Data for the Oceans

Ocean*	Surface Area (million km²)	Water Volume (million km³)	Average Depth (km)	Maximum Depth (km)
Pacific	180	700	4.0	11.0
Atlantic	93	335	3.6	9.2
Indian	77	285	3.7	7.5
Arctic	15	17	1.1	5.2

Source: P. R. Pinet, 1992. *Oceanography* (St. Paul, MN: West, 1992).

*Excludes adjacent seas, such as the Caribbean Sea and Sea of Japan, which are marginal parts of oceans.

Figure 9.2 The Seafloor This map shows the four oceans and many of the seas, which are marginal parts of oceans. The seafloor constitutes the largest part of Earth's surface.

he greatly underestimated the width of the Atlantic Ocean. Contrary to popular belief, he was not attempting to demonstrate Earth's spherical shape; its shape was well accepted by then. The controversy was over Earth's circumference and the shortest route to China; on these points, Columbus's critics were correct.

These and similar voyages added considerably to the growing body of knowledge about the oceans, but truly scientific investigations did not begin until the late 1700s. At that time, Great Britain was the dominant maritime power, and to maintain that dominance, the British sought to increase their knowledge of the oceans. So, scientific voyages led by Captain James Cook were launched in 1768, 1772, and 1777. From 1831 until 1836, the HMS *Beagle* sailed the seas. Aboard was Charles Darwin, who is well known for his views of organic evolution, but who also proposed a theory on the evolution of coral reefs. In 1872, the converted British warship HMS *Challenger* began a four-year voyage to sample seawater, determine oceanic depths, collect samples of seafloor sediment and rock, and name and classify thousands of species of marine organisms.

During these voyages, many oceanic islands previously unknown to Europeans were visited. And even though exploration of the oceans was limited, it was becoming increasingly apparent that the seafloor was not flat and featureless as formerly believed. Indeed, scientists discovered that the seafloor has varied topography just as continents do, and they recognized such features as oceanic trenches, submarine ridges, broad plateaus, hills, and vast plains.

How Are Oceans Explored Today?

Measuring the length of a weighted line lowered to the seafloor was the first method for determining ocean depths. Now scientists use an instrument called an *echo sounder*, which detects sound waves that travel from a ship to the seafloor and back (Figure 9.3). Depth is calculated by knowing the velocity of sound in water and the time required for the waves to reach the seafloor and return to the ship, thus yielding a continuous profile of seafloor depths along the ship's route. **Seismic profiling** is similar to echo sounding, but is even more useful. Strong waves from an energy source reflect from the seafloor, and some of the waves penetrate seafloor layers and reflect from various horizons back to the

Figure 9.3 Echo Sounding and Seismic Profiling Diagram showing how echo sounding and seismic profiling are used to study the seafloor. With seismic profiling, the energy generated at the energy source is reflected from various horizons back to the surface, where it is detected by hydrophones.

surface (Figure 9.3). Seismic profiling is particularly useful for determining the structure of the oceanic crust where it is buried beneath seafloor sediments.

An international program called the Deep Sea Drilling Project began in 1968 with its research vessel the *Glomar Challenger*, but after 15 years of oceanographic research, it was replaced by the Ocean Drilling Program's research vessel JOIDES* *Resolution.* Both vessels were equipped to drill into the seafloor and retrieve samples of sediment and oceanic crust. Then, in 2003, the responsibilities passed to an even larger research vessel, the R/V *Chikyu* ("Earth"), a Japanese ship that can drill as much as 11 kilometers into the seafloor (Figure 9.4a).

In addition to surface vessels, submersibles are also used extensively in oceanographic research. One of the most famous of these is *Alvin*, which has carried scientists to the seafloor in many areas to make observations and collect samples (Figure 9.4b). Other submersibles are remotely controlled and towed by surface ships. In 1985, the *Argo*, equipped with sonar and radar systems, provided the first views of the British ocean liner HMS *Titanic* since it sank after hitting an iceberg in 1912 (Figure 9.5). Another remotely operated submersible, the *Kaiko*, operated by the Japanese, has descended to the greatest oceanic depths.

Scientific investigations have yielded important information about the oceans for more than 200 years, but much of our current knowledge has been acquired since World War II (1939–1945). This is particularly true of the seafloor because only in recent decades has instrumentation been available to study this largely hidden domain.

OCEANIC CRUST—ITS STRUCTURE AND COMPOSITION

We have mentioned that oceanic crust is composed of basalt and gabbro and is generated continuously at spreading ridges. Of course, drilling into the oceanic crust provides some details about its composition and structure, but it has never been completely penetrated and sampled. So how do we know what it is composed of and how it varies with depth? Actually, even before it was sampled and observed, these details were known.

Remember that oceanic crust is consumed at subduction zones and thus most of it is recycled, but a small amount is found in mountain ranges on land where it was emplaced by

JAMSTEC

a The R/V *Chikyu* ("Earth"), a research ship in the Integrated Ocean Drilling Program.

WHOI-R. Chatarach Visuals Unlimited

b The submersible *Alvin* is used for observing and sampling the deep seafloor.

Figure 9.4 Oceanographic Research Vessels

*JOIDES is an acronym for Joint Oceanographic Institutions for Deep Earth Sampling.

Figure 9.5 Bow of the HMS *Titanic* This view of the British ocean liner that sank in 1912 after hitting an iceberg was taken in 1986. Scientists in submersibles made several visits to the ship.

moving along large fractures called thrust faults (faults are discussed more fully in Chapter 10). These preserved slivers of oceanic crust, along with part of the underlying upper mantle, are known as **ophiolites**. Detailed studies reveal that an ideal ophiolite consists of rocks of the upper oceanic crust, especially pillow lava and sheet lava flows (Figure 9.6) underlain by a sheeted dike complex consisting of vertical basaltic dikes, and then massive gabbro and layered gabbro that probably formed in the upper part of a magma chamber. And finally, the lowermost unit is peridotite from the upper mantle; this is sometimes altered by metamorphism to a greenish rock known as serpentinite. Thus, a complete ophiolite consists of deep-sea sedimentary rocks underlain by rocks of the oceanic crust and upper mantle (Figure 9.6).

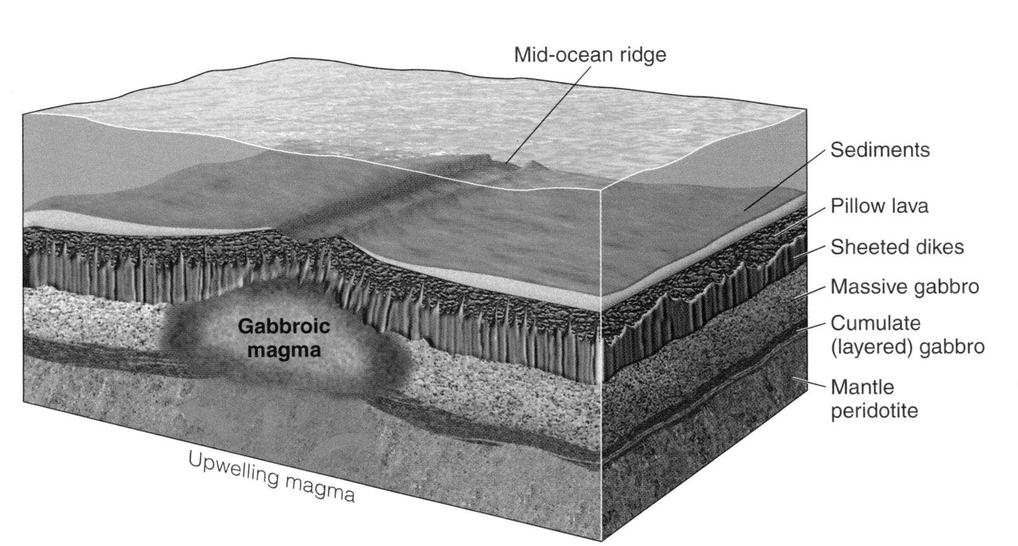

Figure 9.6 Composition of Oceanic Crust New oceanic crust made up of the layers shown here forms as magma rises beneath oceanic ridges. The composition of oceanic crust was known from ophiolites, sequences of rock on land consisting of deep-sea sediments, oceanic crust, and upper mantle, before scientists observed it when they descended in submersibles to seafloor fractures.

What Would You Do?

As the only person in your community with any geologic training, you are often called on to explain local geologic features and identify fossils. Several school children on a natural history field trip picked up some rocks that you recognize as peridotite. When you visit the site where the rocks were collected, you also notice some pillow lava in the area and what appear to be dikes composed of basalt. What other rock types might you expect to find here? How would you explain (1) the association of these rocks with one another, and (2) how they came to be on land?

Sampling and drilling at oceanic ridges reveal that oceanic crust is indeed made up of pillow lava and sheet lava flows underlain by a sheeted dike complex, just as predicted from studies of ophiolites. But it was not until 1989 that a submersible carrying scientists descended to the walls of a seafloor fracture in the North Atlantic and verified what lay below the sheeted dike complex. Just as expected, the lower oceanic crust consists of gabbro and the upper mantle is made up of peridotite.

THE CONTINENTAL MARGINS

Most people think of continents as land areas outlined by the oceans; however, the true geologic margin of a continent—where granitic continental crust changes to basalt and gabbro oceanic crust—is below sea level. A **continental margin** is made up of a gently sloping continental shelf, a more steeply inclined continental slope, and, in some cases, a deeper, gently sloping continental rise (Figure 9.7). Seaward of the continental margin lies the deep ocean basin. Thus, the continental margins extend to increasingly greater depths until they merge with the deep seafloor. Continental crust changes to oceanic crust somewhere beneath the continental rise, so part of the continental slope and the continental rise actually rest on oceanic crust.

The Continental Shelf

As one proceeds seaward from the shoreline across the continental margin, the first area encountered is the gently sloping **continental shelf** lying between the shore and the more steeply dipping continental slope (Figure 9.7). The width of the continental shelf varies considerably, ranging from a few tens of meters to more than 1000 km; the shelf

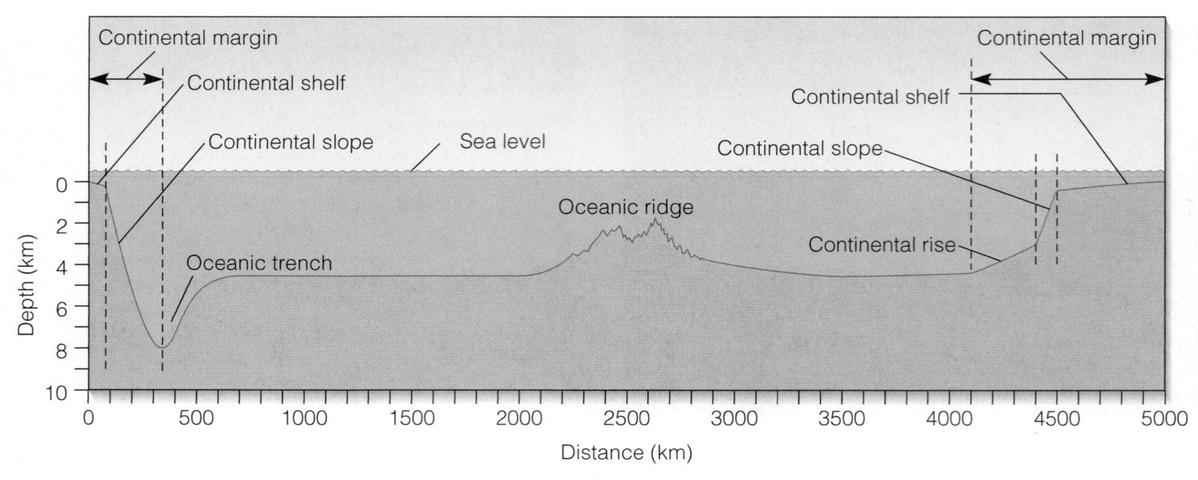

Figure 9.7 Features of Continental Margins A generalized profile showing features of the continental margins. The vertical dimensions of the features in this profile are greatly exaggerated because the vertical and horizontal scales differ.

terminates where the inclination of the seafloor increases abruptly from 1 degree or less to several degrees. The outer margin of the continental shelf, or simply the *shelf–slope break*, is at an average depth of 135 m, so by oceanic standards, the continental shelves are covered by shallow water.

At times during the Pleistocene Epoch (1.8 million to 10,000 years ago), sea level was as much as 130 m lower than it is now. As a result, the continental shelves were above sea level where deposition in stream channels and floodplains took place. In addition, in many parts of northern Europe and North America, glaciers extended well out onto the continental shelves and deposited gravel, sand, and mud. Since the Pleistocene ended, sea level has risen, submerging these deposits, which are now being reworked by marine processes. Evidence that these sediments were in fact deposited on land includes remains of human settlements and fossils of land-dwelling animals (see Chapter 23).

The Continental Slope and Rise

The seaward margin of the continental shelf is marked by the *shelf–slope break* (at an average depth of 135 m), where the more steeply inclined **continental slope** begins (Figure 9.7). In most areas around the margins of the Atlantic, the continental slope merges with a more gently sloping **continental rise**. This rise is absent around the margins of the Pacific, where continental slopes descend directly into an oceanic trench (Figure 9.7).

The shelf–slope break is an important feature in terms of sediment transport and deposition. Landward of the break—that is, on the shelf—sediments are affected by waves and tidal currents, but these processes have no effect on sediments seaward of the break, where gravity is responsible for their transport and deposition on the slope and rise. In fact, much of the land-derived sediment crosses the shelves and is eventually deposited on the continental slopes and rises, where more than 70% of all sediments in the oceans are found. Much of this sediment is transported through submarine canyons by turbidity currents.

Submarine Canyons, Turbidity Currents, and Submarine Fans

In Chapter 6, we discussed the origin of graded bedding, most of which results from **turbidity currents**, underwater flows of sediment–water mixtures with densities greater than that of sediment-free water. As a turbidity current flows onto the relatively flat seafloor, it slows and begins depositing sediment, the largest particles first, followed by progressively smaller particles, thus forming a layer with graded bedding (see Figure 6.24). Deposition by turbidity currents yields a series of overlapping **submarine fans**, which constitute a large part of the continental rise (Figure 9.8). Submarine fans are distinctive features, but their outer margins are difficult to discern because they grade into deposits of the deep-ocean basin.

No one has ever observed a turbidity current in progress in the oceans, so for many years, some doubted their existence; however, seafloor samples from many areas show a succession of layers with graded bedding and the remains of shallow-water organisms that were displaced into deeper water by turbidity currents (Figure 9.8).

Perhaps the most compelling evidence for turbidity currents is the pattern of trans-Atlantic cable breaks that took place in the North Atlantic near Newfoundland on November 18, 1929. Initially, an earthquake was assumed to have ruptured telephone and telegraph cables. However, the breaks on the continental shelf near the epicenter occurred when the earthquake struck, but cables farther seaward were broken later and in succession (Figure 9.8b). The last cable to break was 720 km from the source of the earthquake, and it did not snap until 13 hours after the first break. In 1949, geologists realized that an earthquake-generated turbidity current had moved downslope, breaking the cables in succession. The precise time at which each cable broke was known, so calculating the velocity of the turbidity current was simple. It moved at about 80 km/hr on the continental slope, but slowed to about 27 km/hr when it reached the continental rise.

Figure 9.8 Submarine Fans and Graded Bedding

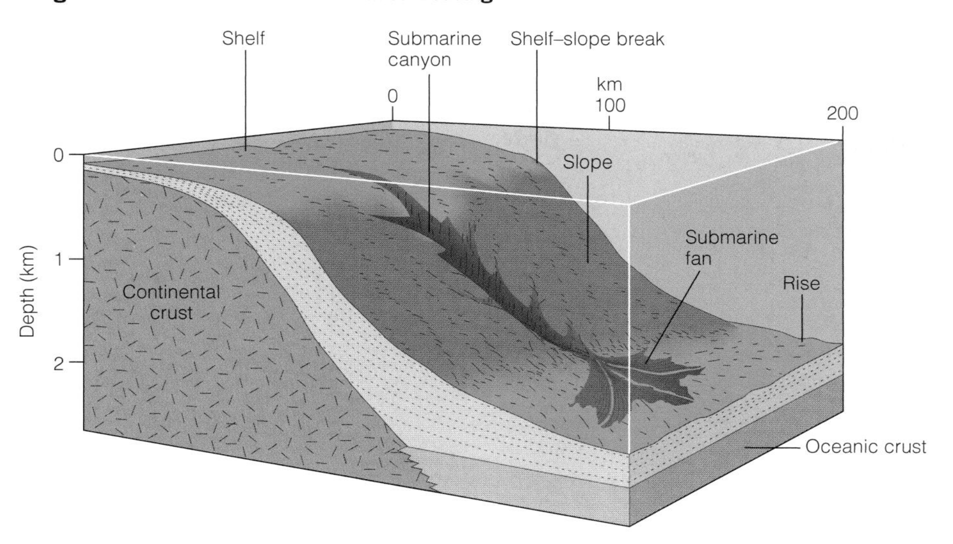

a Much of the continental rise is made up of submarine fans that form by deposition of sediments carried down submarine canyons by turbidity currents.

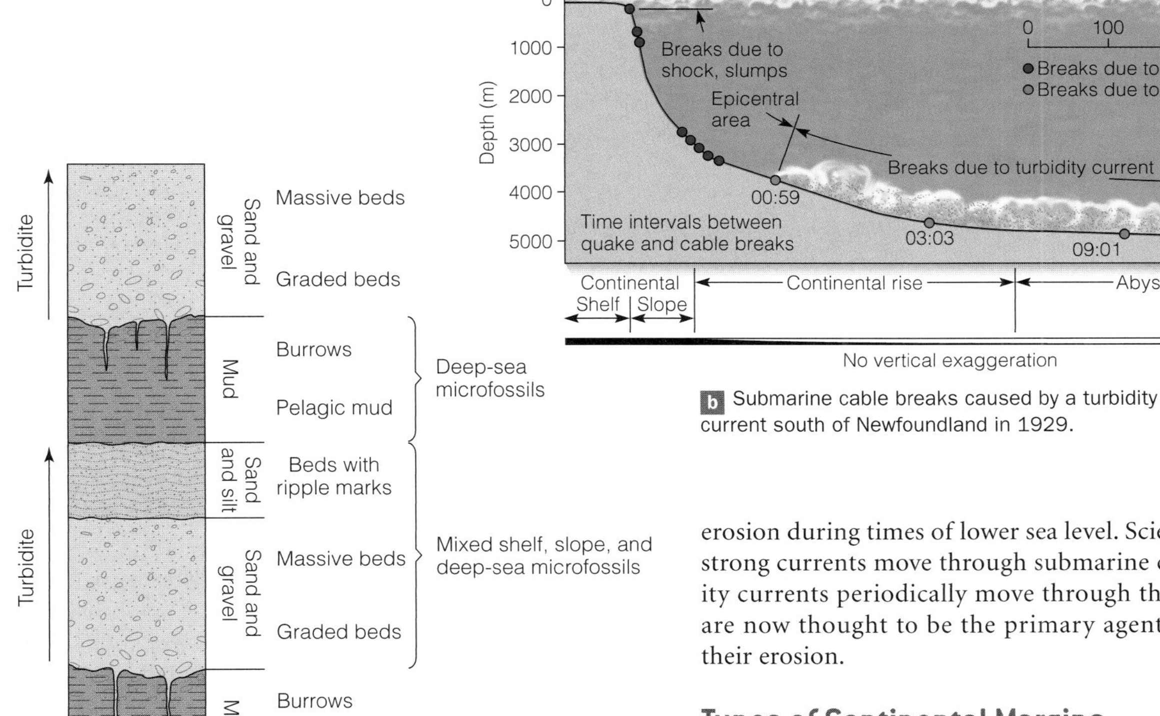

b Submarine cable breaks caused by a turbidity current south of Newfoundland in 1929.

c A vertical section from a submarine fan shows numerous turbidity current deposits.

Deep, steep-sided **submarine canyons** are present on continental shelves, but they are best developed on continental slopes (Figure 9.8a). Some submarine canyons extend across the shelf to rivers on land and apparently formed as river valleys when sea level was lower during the Pleistocene. However, many have no such association, and some extend far deeper than can be accounted for by river erosion during times of lower sea level. Scientists know that strong currents move through submarine canyons. Turbidity currents periodically move through these canyons and are now thought to be the primary agent responsible for their erosion.

Types of Continental Margins

Continental margins are *active* or *passive*, depending on their relationship to plate boundaries. An **active continental margin** develops at the leading edge of a continental plate where oceanic lithosphere is subducted. The western margin of South America is a good example of where an oceanic plate is subducted beneath the continent, resulting in seismic activity, a geologically young mountain range, and active volcanism (Figure 9.9). In addition, the continental shelf is narrow, and the continental slope descends directly into the Peru–Chile Trench, so sediment is dumped into the trench and no continental rise develops. The western margin of North America is also considered an active continental margin, although much of it is now

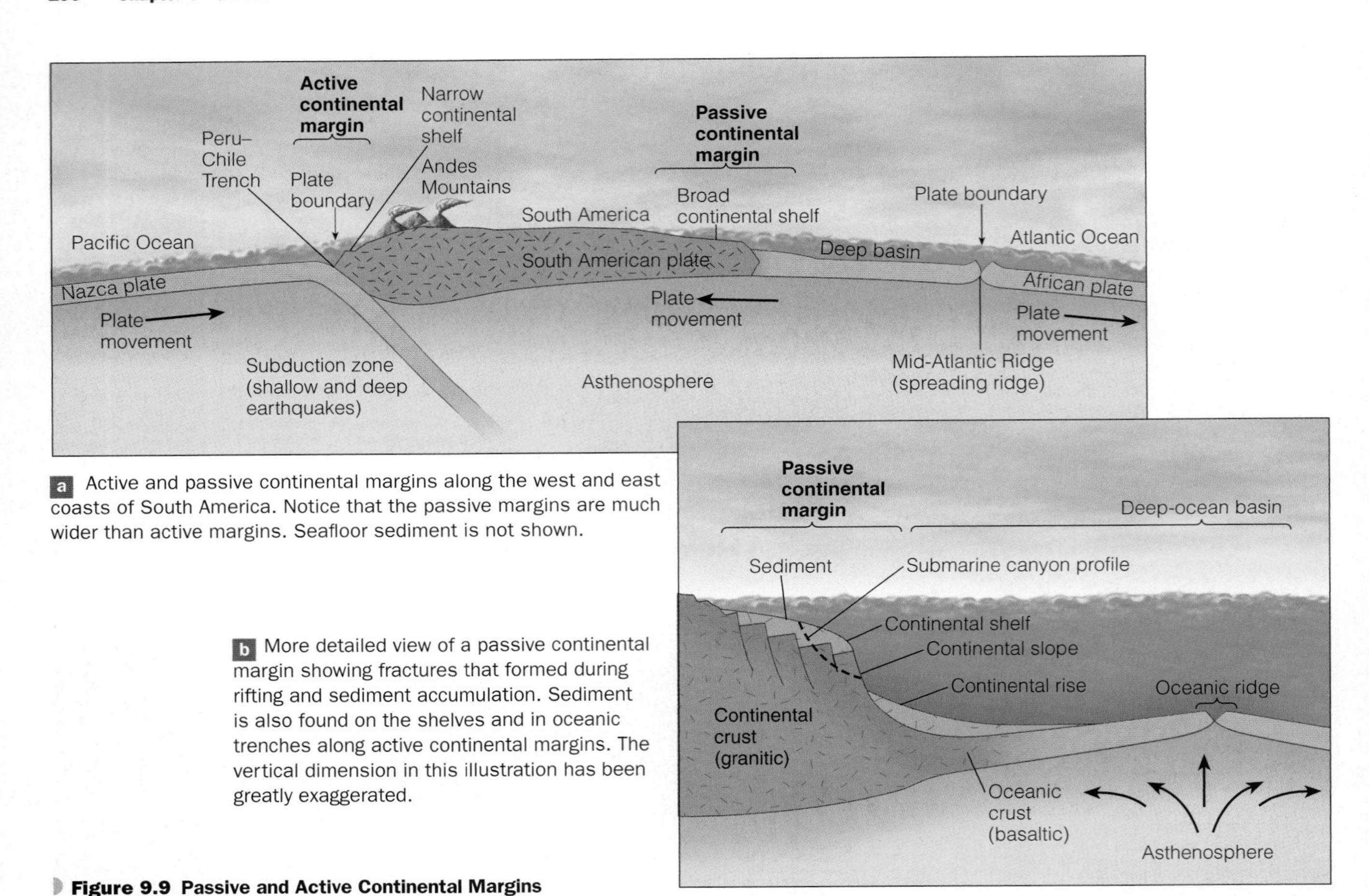

a Active and passive continental margins along the west and east coasts of South America. Notice that the passive margins are much wider than active margins. Seafloor sediment is not shown.

b More detailed view of a passive continental margin showing fractures that formed during rifting and sediment accumulation. Sediment is also found on the shelves and in oceanic trenches along active continental margins. The vertical dimension in this illustration has been greatly exaggerated.

Figure 9.9 Passive and Active Continental Margins

bounded by transform faults rather than a subduction zone. However, plate convergence and subduction still take place in the Pacific Northwest along the continental margins of northern California, Oregon, and Washington.

The continental margins of eastern North America and South America differ considerably from their western margins. For one thing, they possess broad continental shelves, as well as a continental slope and rise; also, vast, flat *abyssal plains* are present adjacent to the rises (Figure 9.9). Furthermore, these **passive continental margins** are within a plate rather than at a plate boundary, and they lack the volcanic and seismic activity found at active continental margins. Nevertheless, earthquakes do take place occasionally at these margins.

Active and passive continental margins share some features, but they are notably different in the widths of their continental shelves, and active margins have an oceanic trench but no continental rise. Why the differences? At both types of continental margins, turbidity currents transport sediment into deeper water. At passive margins, the sediment forms a series of overlapping submarine fans and thus develops a continental rise, whereas at an active margin, sediment is simply dumped into the trench and no rise forms. The proximity of a trench to a continent also explains why the continental shelves of active margins are so narrow. In contrast, land-derived sedimentary deposits at passive margins have built a broad platform extending far out into the ocean.

WHAT FEATURES ARE FOUND IN THE DEEP-OCEAN BASINS?

The seafloor has an average depth of 3.8 km, so most of it lies far below the depth of sunlight penetration, which is generally less than 100 m. Accordingly, the deep seafloor is completely dark, no plant life exists, the temperature is just above freezing, and the pressure varies from 200 to more than 1000 atmospheres depending on depth. In fact, biologic productivity is low on the deep seafloor with the exception of hydrothermal vent communities (discussed later).

Scientists have descended to the greatest oceanic depths, submarine ridges, and elsewhere in submersibles, so they have observed some of the seafloor. Nevertheless, much of the seafloor has been studied only by echo sounding, seismic profiling, sampling of seafloor sediments and oceanic crust, and remotely controlled submersibles.

Abyssal Plains

Beyond the continental rises of passive continental margins are **abyssal plains,** flat surfaces covering vast areas of the seafloor. In some areas, they are interrupted by peaks rising more than 1 km, but abyssal plains are nevertheless the flattest, most featureless areas on Earth (Figure 9.10). Their flatness is a result of sediment deposition covering the rugged topography of the seafloor (Figure 9.11).

Abyssal plains are invariably found adjacent to continental rises, which are composed mostly of overlapping submarine fans. Along active continental margins, sediments derived from the shelf and slope are trapped in an oceanic trench, and abyssal plains fail to develop. Accordingly, abyssal plains are common in the Atlantic Ocean basin, but rare in the Pacific Ocean basin (Figure 9.10).

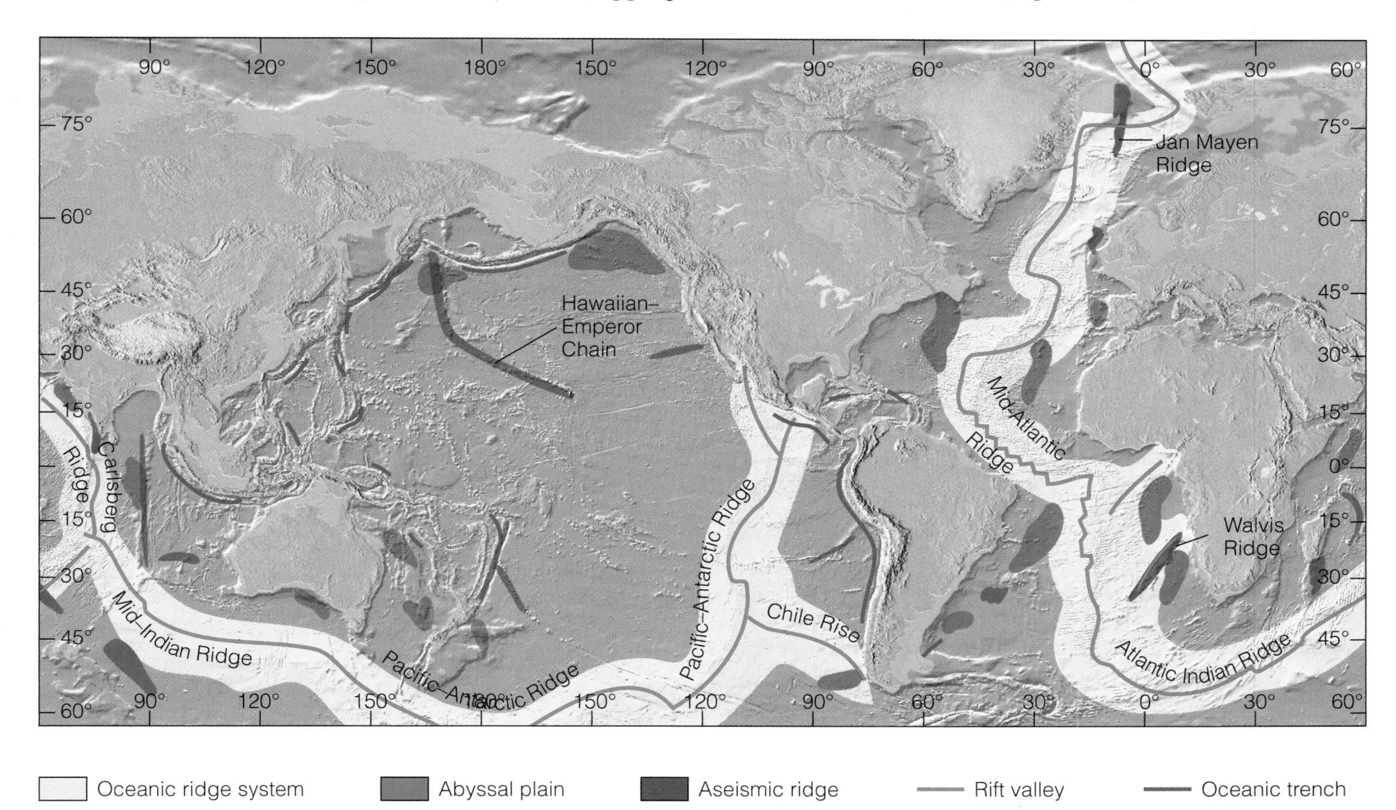

Figure 9.10 Deep Seafloor Fractures Features found on the deep seafloor include oceanic trenches (brown), abyssal plains (green), the oceanic ridge system (yellow), rift valleys (red), and some aseismic ridges (blue). Other features such as seamounts and guyots are shown in Figure 9.14. Source: From Alyn and Alison Duxbury, *An Introduction to the World's Oceans*. Copyright 1984 McGraw-Hill.

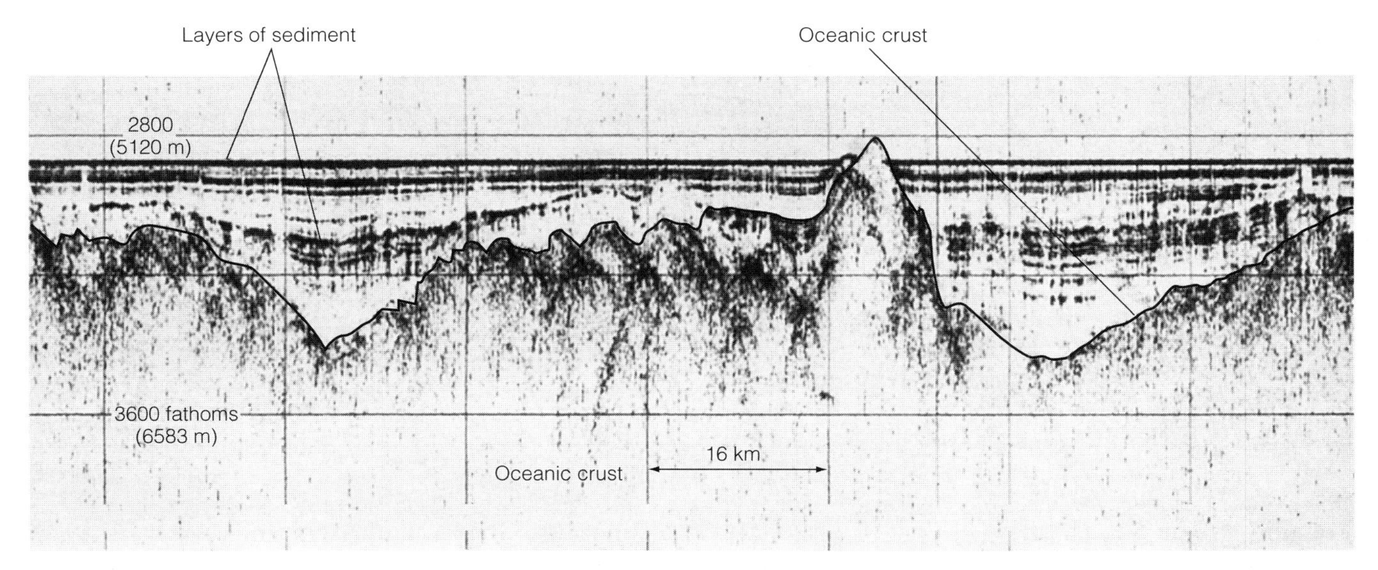

Figure 9.11 Rugged Seafloor Topography Seismic profile showing rugged seafloor topography covered by sediments of the Northern Madeira Abyssal Plain in the Atlantic Ocean. Source: From Bruce E. Heezen and Charles D. Hollister, *The Face of the Deep* (New York: Oxford University Press, 1971), Figure 8.38, page 329.

Oceanic Trenches

Long, steep-sided depressions on the seafloor near convergent plate boundaries, called **oceanic trenches**, constitute no more than 2% of the seafloor, but they are important features because it is here that oceanic lithosphere is consumed by subduction (see Chapter 2). Because oceanic trenches are found along active continental margins, they are common in the Pacific Ocean basin, but largely lacking in the Atlantic, those in the Caribbean being notable exceptions (Figure 9.10). On the landward sides of oceanic trenches, the continental slope descends into them at up to 25 degrees, and many have thick accumulations of sediments. The greatest oceanic depths are found in trenches; the Challenger Deep of the Marianas Trench in the Pacific is more than 11,000 m deep!

Sensitive instruments can detect the amount of heat escaping from Earth's interior by the phenomenon of *heat flow.* As one might expect, heat flow is greatest in areas of active or recently active volcanism. For instance, higher-than-average heat flow takes place at spreading ridges, but at subduction zones heat flow values are less than the average for Earth as a whole. Oceanic crust at oceanic trenches is cooler and slightly denser than elsewhere.

Seismic activity also takes place at or near oceanic trenches along planes dipping at about 45 degrees. In Chapter 8, we discussed these inclined seismic zones called Benioff zones (see Figure 8.5), where most of Earth's intermediate and deep earthquakes occur. Volcanism does not take place in trenches, but because these are zones where oceanic lithosphere is subducted beneath either oceanic or continental lithosphere, an arcuate chain of volcanoes is found on the overriding plate (Figure 9.9). The Aleutian Islands and the volcanoes along the western margin of South America are good examples of such chains.

Oceanic Ridges

When the first submarine cable was laid between North America and Europe during the late 1800s, a feature called the Telegraph Plateau was discovered in the North Atlantic. Using this data and data from a 1925–27 voyage of the German research vessel *Meteor*, scientists proposed that the plateau was actually a continuous ridge extending the length of the Atlantic Ocean basin. Subsequent investigations revealed that this conjecture was correct, and we now call this feature the Mid-Atlantic Ridge (Figure 9.10).

The Mid-Atlantic Ridge is more than 2000 km wide and rises 2 to 2.5 km above the adjacent seafloor. Furthermore, it is part of a much larger **oceanic ridge** system of mostly submarine mountainous topography. This system runs from the Arctic Ocean through the middle of the Atlantic and curves around South Africa, where the Indian Ridge continues into the Indian Ocean; the Atlantic–Pacific Ridge extends eastward and a branch of this, the East Pacific Rise, trends northeast until it reaches the Gulf of California (Figure 9.10). The entire system is at least 65,000 km long, far exceeding the length of any mountain system on land. Oceanic ridges are composed almost entirely of basalt and gabbro and possess features produced by tensional forces. Mountain ranges on land, in contrast, consist of all rock types, and they formed when rocks were folded and fractured by compressive forces (see Chapter 10).

Oceanic ridges are mostly below sea level, but they rise above the sea in Iceland, the Azores, Easter Island, and several other places. Of course, oceanic ridges are the sites where new oceanic crust is generated and plates diverge. The rate of plate divergence is important because it determines the cross-sectional profile of a ridge. For example, the Mid-Atlantic Ridge has a comparatively steep profile because divergence is slow, allowing the new oceanic crust to cool, shrink, and subside closer to the ridge crest than it does in areas of faster divergence such as the East Pacific Rise. A ridge may also have a rift along its crest that opens in response to tension (Figure 9.12a). A rift is particularly obvious along the Mid-Atlantic Ridge, but it is absent along parts of the East Pacific Rise. These rifts are commonly 1 to 2 km deep and several kilometers wide. They open as seafloor spreading takes place (discussed in Chapter 2) and are characterized by shallow-focus earthquakes, basaltic volcanism, and high heat flow.

Scientists have been making direct observations of oceanic ridges and their rifts since 1974. As part of Project FAMOUS (French-American Mid-Ocean Undersea Study), submersibles have descended to the ridges and into their rifts in several areas. Researchers have not seen any active volcanism, but they did see pillow lavas (see Figure 5.7), lava tubes, and sheet lava flows, some of which formed very recently. In fact, on return visits to a site, they have seen the effects of volcanism that occurred since their previous visit. And on January 25, 1998, a submarine volcano began erupting along the Juan de Fuca Ridge west of Oregon. Researchers aboard submersibles have also observed hot water being discharged from the seafloor at or near ridges in submarine hydrothermal vents.

Submarine Hydrothermal Vents

Scientists first saw **submarine hydrothermal vents** on the seafloor in 1979 when they descended about 2500 m to the Galapagos Rift in the eastern Pacific Ocean. Since 1979, they have seen similar vents in several other areas in the Pacific (Figure 9.12b), Atlantic, the Indian Ocean, and the Sea of Japan. The vents are at or near spreading ridges where cold seawater seeps through oceanic crust, is heated by the hot rocks at depth, and then rises and discharges into the seawater as plumes of hot water with temperatures as high as 400°C. Many of the plumes are black because dissolved minerals giving them the appearance of black smoke—hence the name **black smoker** (Figure 9.12c).

Submarine hydrothermal vents are interesting from the biologic, geologic, and economic points of view. Near the vents live communities of organisms, such as bacteria, crabs, mussels, starfish, and tube worms, many of which had never been seen before. No sunlight is available, so these

Figure 9.12 Central Rift and Submarine Hydrothermal Vents

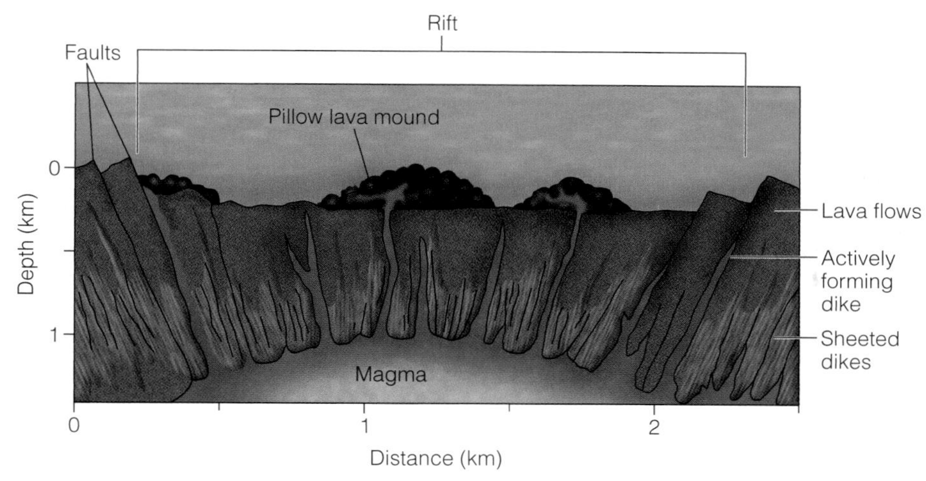

a Cross section of the Mid-Atlantic Ridge showing its central rift where recent moundlike accumulations of volcanic rocks, mostly pillow lava, are found.

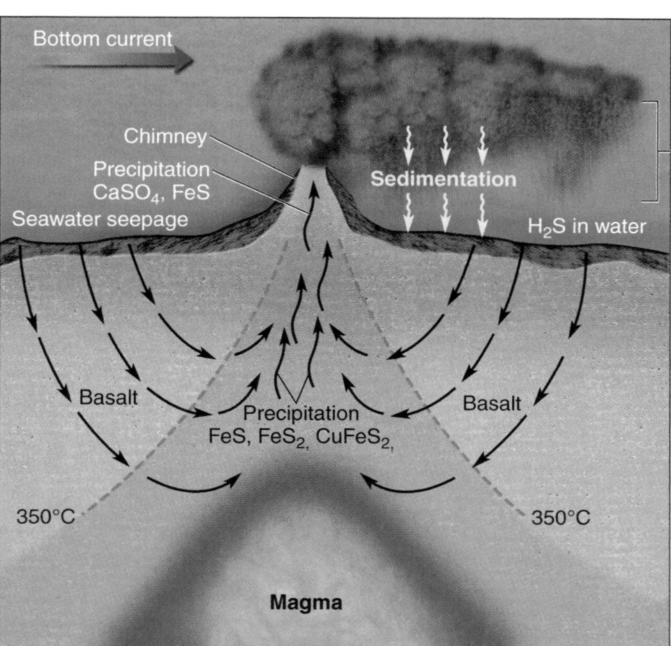

b Cross section showing the origin of a submarine hydrothermal vent called a black smoker.

c This black smoker on the East Pacific Rise is at a depth of 2800 m. The plume of "black smoke" is heated seawater with dissolved minerals.

organisms depend on bacteria that oxidize sulfur compounds for their ultimate source of nutrients. The vents are also interesting because of their economic potential. The heated seawater reacts with oceanic crust, transforming it into a metal-rich solution that discharges into seawater and cools, precipitating iron, copper, and zinc sulfides and other minerals. A chimney-like vent forms that eventually collapses and forms a mound of sediments rich in the elements mentioned above.

Apparently, the chimneys through which black smokers discharge grow rapidly. A 10-m-high chimney accidentally knocked over in 1991 by the submersible *Alvin* grew to 6 m in just three months. Also in 1991, scientists aboard *Alvin* saw the results of a submarine eruption on the East Pacific Rise, which they missed by less than two weeks. Fresh lava and ash covered the area, as well as the remains of tube worms killed

What Would You Do?

Hydrothermal vents on the seafloor are known sites of several metals of great importance to industrialized societies. Furthermore, it appears that these metals are being deposited even now, so if we mine one area, more of the same resources form elsewhere. Given these conditions, it would seem that our problems of diminishing resources are solved. So why not simply mine the seafloor? Also, many chemical elements are present in seawater. The technology exists to extract elements such as gold, uranium, and others, so why not do so?

during the eruption. And in a nearby area, a new fissure opened in the seafloor; by December 1993, a new hydrothermal vent community had become well established.

In 2001, scientists announced another kind of seafloor vent in the North Atlantic responsible for massive pillars and spires as tall as 60 m. Unlike the black smokers, though, these vents are 14–15 km from spreading ridges, and they consist of light-colored minerals that were derived by chemical reaction between seawater and minerals in the oceanic crust.

Seafloor Fractures

Oceanic ridges are not continuous features winding without interruption around the globe. They abruptly terminate where they are offset along fractures oriented more or less at right angles to ridge axes (Figure 9.13). These large-scale fractures are hundreds of kilometers long, although they are difficult to trace where they are buried beneath seafloor sediments. Many geologists are convinced that some geologic features on the continents are best explained by the extension of these fractures into continents.

Shallow-focus earthquakes take place along these fractures, but only between the displaced ridge segments. Furthermore, because ridges are higher than the adjacent seafloor, the offset segments yield nearly vertical escarpments 2 or 3 km high (Figure 9.13). The reason oceanic ridges have so many fractures is that plate divergence takes place irregularly on a sphere, resulting in stresses that cause fracturing. We discussed these fractures between offset ridge segments more fully in Chapter 2, where they are termed *transform faults.*

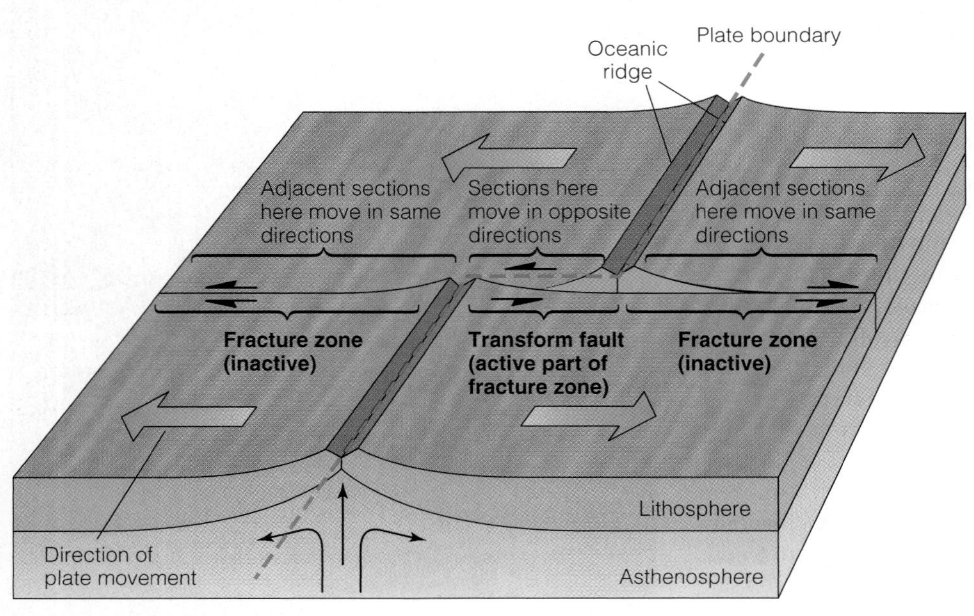

Figure 9.13 Seafloor Fractures Diagrammatic view of an oceanic ridge offset along a fracture. That part of the fracture between displaced segments of the ridge is a transform fault. Recall from Chapter 2 that transform faults are one type of plate boundary.

Seamounts, Guyots, and Aseismic Ridges

As noted, the seafloor is not a flat, featureless plain except for the abyssal plains, and even these are underlain by rugged topography (Figure 9.11). In fact, a large number of volcanic hills, seamounts, and guyots rise above the seafloor in all ocean basins, but they are particularly abundant in the Pacific. All are of volcanic origin and differ mostly in size. **Seamounts** rise more than 1 km above the seafloor, and if flat topped, they are called **guyots** (Figure 9.14). Guyots are volcanoes that originally extended above sea level. However, as the plate upon which they were located continued to move, they were carried away from a spreading ridge, and as the oceanic crust cooled, it descended to greater depths. Thus, what was once an island slowly sank beneath the sea, and as it did, wave erosion produced the typical flat-topped appearance (Figure 9.14). Many other volcanic features smaller than seamounts exist on the seafloor, but they probably originated in the same way. These so-called *abyssal hills* average only about 250 m high.

Other common features in the ocean basins are long, narrow ridges and broad plateau-like features rising as much as 2 to 3 km above the surrounding seafloor. These **aseismic ridges** are so called because they lack seismic activity. A few of these ridges are probably small fragments separated from continents during rifting and are referred to as *microcontinents.* The Jan Mayen Ridge in the North Atlantic is probably a microcontinent (Figure 9.10).

Most aseismic ridges form as a linear succession of hot-spot volcanoes. These may develop at or near an oceanic ridge, but each volcano so formed is carried laterally with the plate upon which it originated. The net result is a line of seamounts/guyots extending from an oceanic ridge (Figure 9.14); the Walvis

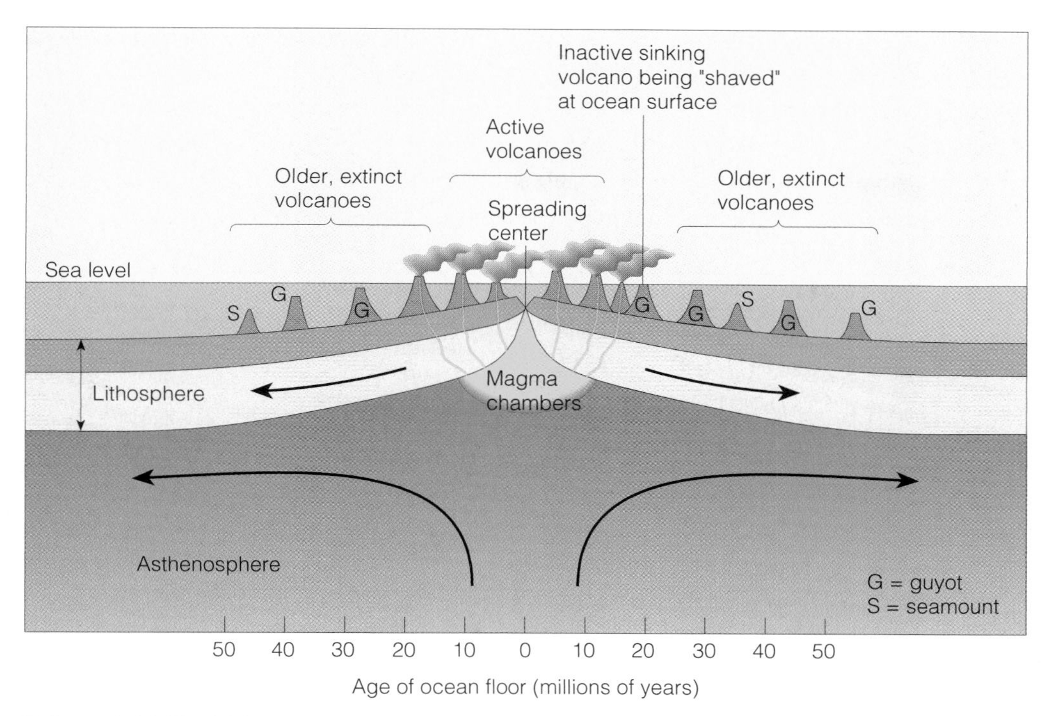

Figure 9.14 The Origin of Seamounts and Guyots As the plate on which a volcano rests moves into greater water depths, the submerged volcanic island is called a seamount. Those that are flat-topped are called guyots.

Ridge in the South Atlantic is a good example (Figure 9.10). Aseismic ridges also form over hot spots unrelated to ridges—the Hawaiian–Emperor chain in the Pacific, for example (Figure 9.10).

SEDIMENTATION AND SEDIMENTS ON THE DEEP SEAFLOOR

Sediments on the deep seafloor are mostly fine grained, consisting of silt- and clay-sized particles, because few processes transport sand and gravel very far from land. Certainly, icebergs carry sand and gravel, and, in fact, a broad band of glacial-marine sediment is adjacent to Antarctica and Greenland. Floating vegetation might also carry large particles far out to sea, but it contributes very little sediment to the deep seafloor.

Most of the fine-grained sediment on the deep seafloor is derived from (1) windblown dust and volcanic ash from the continents and volcanic islands, and (2) the shells of microscopic plants and animals that live in the near-surface waters. Minor sources are chemical reactions in seawater that yield manganese nodules found in all ocean basins (Figure 9.15) and cosmic dust. Researchers think that as many as 40,000 metric tons of cosmic dust fall to Earth each year, but this is a trivial quantity compared to the volume of sediment derived from the two primary sources.

Institute of Oceanographic Sciences/Nerc/SPL/Photo Researchers, Inc.

Figure 9.15 Manganese Nodules Chemical reactions in seawater yield manganese nodules on the seafloor.

Most sediment on the deep seafloor is *pelagic*, meaning that it settled from suspension far from land (Figure 9.16). Pelagic sediment is further characterized as pelagic clay and ooze. **Pelagic clay** is brown or red and, as its name implies, is composed of clay-sized particles from the continents or oceanic islands. **Ooze**, in contrast, is made up mostly of tiny shells of marine organisms. *Calcareous ooze* consists primarily of calcium carbonate ($CaCO_3$) skeletons of marine organisms such as foraminifera, and *siliceous ooze* is composed of the silica (SiO_2) skeletons of single-celled organisms such as radiolarians (animals) and diatoms (plants) (Figure 9.16).

REEFS

The term **reef** has a variety of meanings, such as shallowly submerged rocks that pose a hazard to navigation; however, here we restrict it to mean a moundlike, wave-resistant structure composed of the skeletons of marine organisms (see Geo-inSight on pages 236 and 237). Although commonly

Geo-inSight Reefs: Rocks Made by Organisms

Reefs are wave-resistant structures composed of the skeletons of corals, mollusks, sponges, and encrusting algae. Most reefs are characterized as fringing, barrier, or atolls, all of which actively grow in shallow, warm seawater where there is little or no influx of detrital sediment, especially mud. Reef rock is a type of limestone that forms directly as a solid, rather than from sediment that is later lithified. Ancient reefs are important reservoirs for hydrocarbons in some areas.

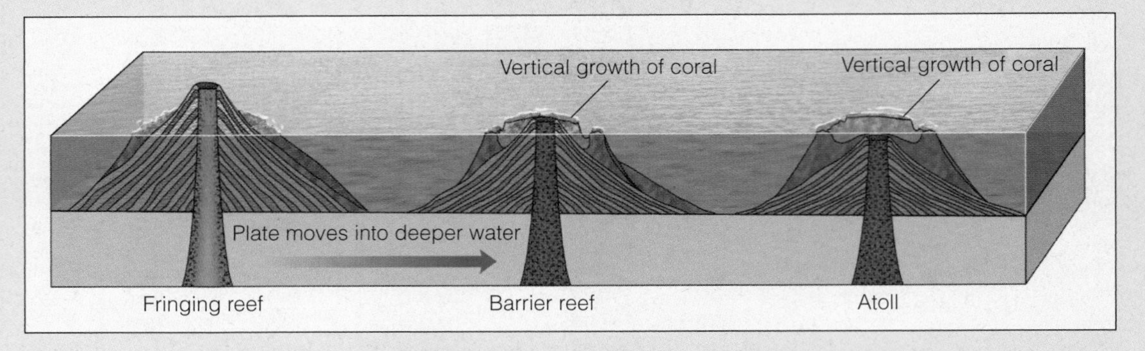

▲ **1.** Three stages in the evolution of a reef. A fringing reef forms around a volcanic island, but as the island is carried into deeper water on a moving plate, the reef is separated from the island by a lagoon and becomes a barrier reef. Continued plate movement carries the island into even deeper water. The island disappears below sea level but the reef grows upward, forming an atoll.

Courtesy of Carl Roessler

Courtesy of Jeff and Dianna Monroe

◄ **2.** Underwater views of reefs in the Red Sea (left) and in Hawaii.

▼ **3.** The white line of breaking waves marks the site of a barrier reef around Rarotonga in the Cook Islands in the Pacific Ocean. The island is only about 12 km long.

Patrick Ward/Corbis

Sue Monroe

▲ **4.** This oval reef with a central lagoon in the Pacific Ocean is an atoll. How does it differ from the reef shown at the bottom of the previous page?

◀ **5.** An ancient reef in Australia. You can see the reef talus on the left side of the image sloping away from the reef core, which has no layering. To the right of the reef core are back-reef deposits, which show horizontal bedding.

▶ **6.** Block diagram showing the various environments in a reef complex

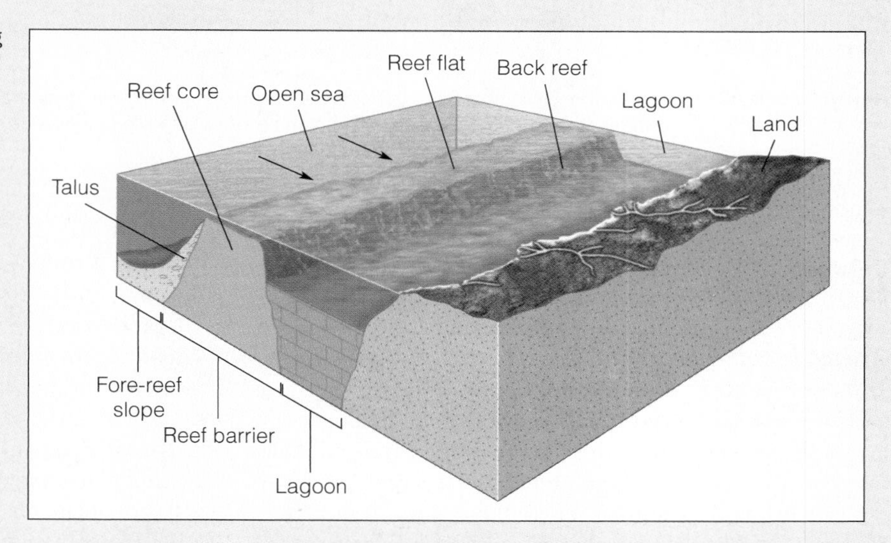

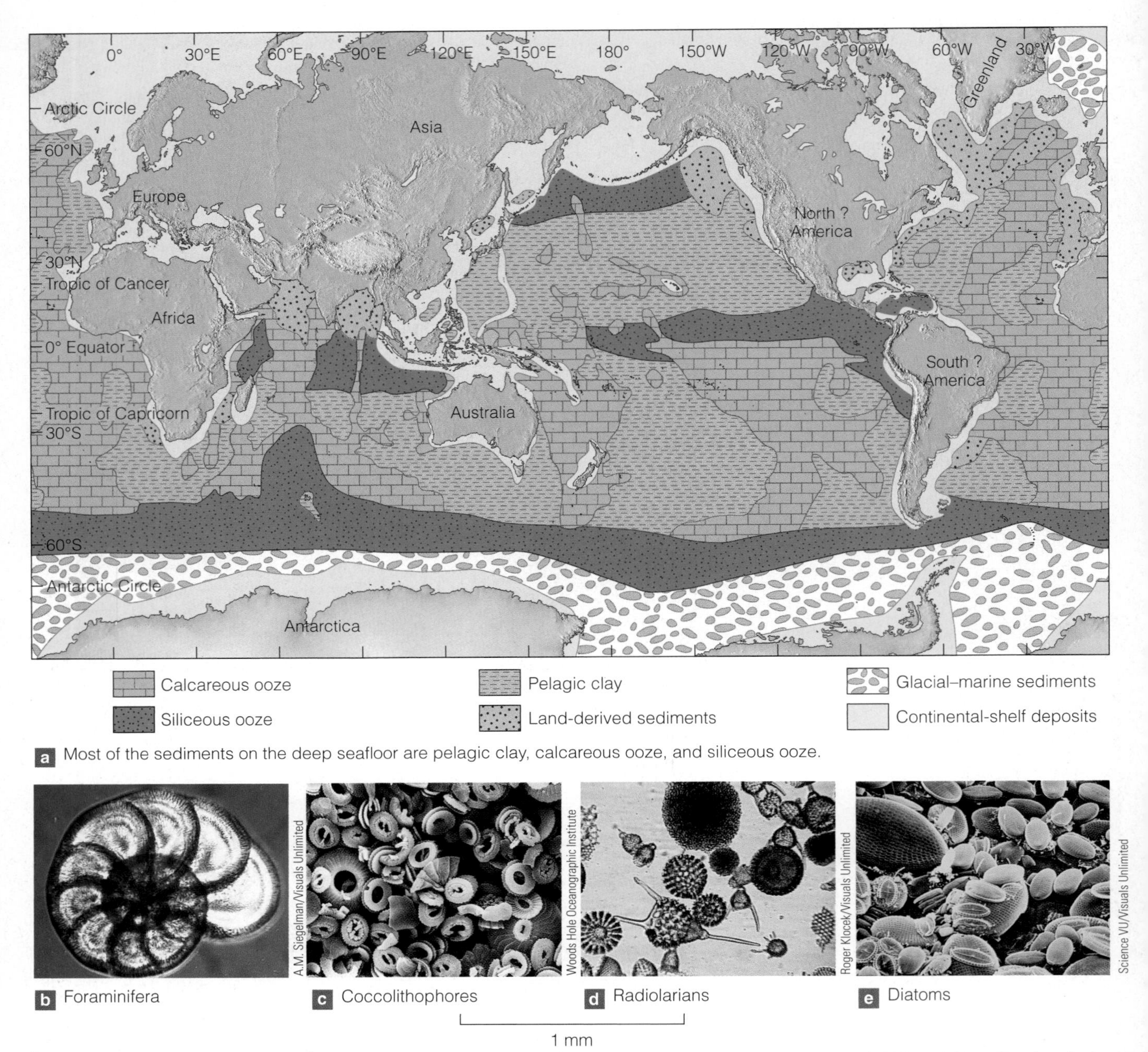

a Most of the sediments on the deep seafloor are pelagic clay, calcareous ooze, and siliceous ooze.

b Foraminifera **c** Coccolithophores **d** Radiolarians **e** Diatoms

Figure 9.16 Sediments on the Deep Seafloor The particles making up the calcareous ooze are skeletons of **b** foraminifera (floating single-celled animals) and **c** coccolithophores (floating single-celled plants), whereas siliceous ooze is composed of skeletons of **d** radiolarians (single-celled floating animals) and **e** diatoms (single-celled floating plants).

called coral reefs, they actually have a solid framework composed of skeletons of corals and mollusks, such as clams, and encrusting organisms, including sponges and algae. Reefs are restricted to shallow, tropical seas where the water is clear and its temperature does not fall below about 20°C. The depth to which reefs grow, rarely more than 50 m, depends on sunlight penetration because many of the corals rely on symbiotic algae that must have sunlight for energy.

Reefs of many shapes are known, but most are one of three basic varieties: fringing, barrier, and atoll. *Fringing reefs* are solidly attached to the margins of an island or continent. They have a rough, tablelike surface, are as much as 1 km wide, and, on their seaward side, slope steeply down to the seafloor. *Barrier reefs* are similar to fringing reefs, except that a lagoon separates them from the mainland. The best-known barrier reef in the world is the 2000-km-long Great Barrier Reef of Australia.

Circular to oval reefs surrounding a lagoon are *atolls.* They form around volcanic islands that subside below sea level as the plate on which they rest is carried progressively farther from an oceanic ridge. As subsidence proceeds, the reef organisms construct the reef upward so that the living

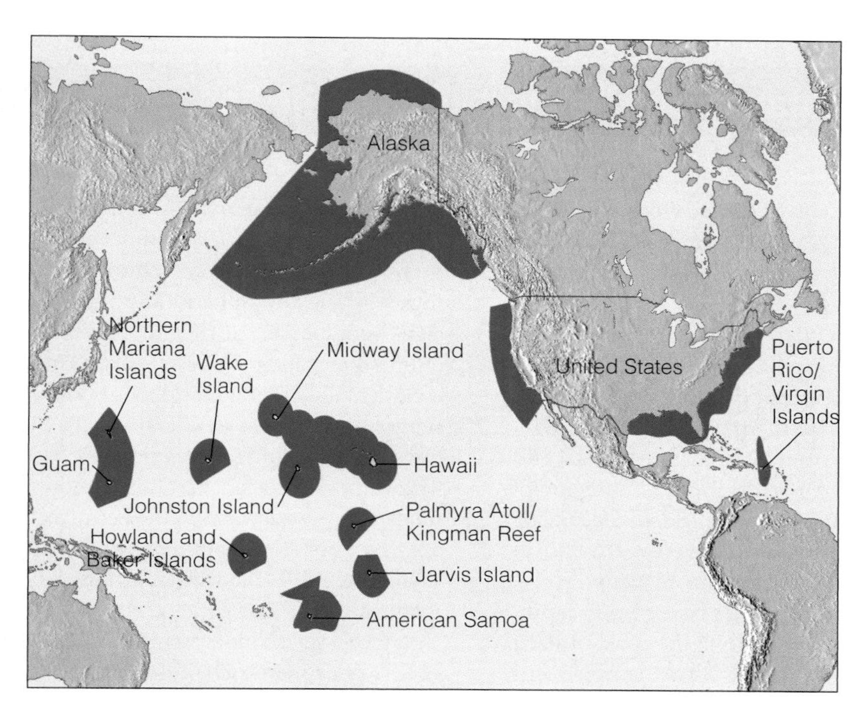

Figure 9.17 The Exclusive Economic Zone (EEZ) Shown in dark blue, the EEZ includes a vast area adjacent to the United States and its territories.

part of the reef remains in shallow water. However, the island eventually subsides below sea level, leaving a circular lagoon surrounded by a more or less continuous reef. Atolls are particularly common in the western Pacific Ocean basin. Many of them began as fringing reefs, but as the plate they were on was carried into deeper water, they evolved first to barrier reefs and finally to atolls (see Geo-inSight on pages 236 and 237).

RESOURCES FROM THE OCEANS

Seawater contains many elements in solution, some of which are extracted for industrial and domestic uses. Sodium chloride (table salt) is produced by the evaporation of seawater, and a large proportion of the world's magnesium comes from seawater. Numerous other elements and compounds can be extracted from seawater, but for many, such as gold, the cost is prohibitive.

In addition to substances in seawater, deposits on the seafloor or within seafloor sediments are becoming increasingly important. Many of these potential resources lie well beyond continental margins, so their ownership is a political and legal problem that has not yet been resolved.

Most nations bordering the ocean claim those resources within their adjacent continental margin. The United States, by a presidential proclamation issued on March 10, 1983, claims sovereign rights over an area designated the **Exclusive Economic Zone (EEZ)** (Figure 9.17). The EEZ extends seaward 200 nautical miles (371 km) from the coast and includes areas adjacent to U.S. territories such as Guam, American Samoa, Wake Island, and Puerto Rico. In short, the United States claims rights to all resources within an area about 1.7 times larger than its land area. Many other nations make similar claims.

Numerous resources are found within the EEZ, some of which have been exploited for many years. Sand and gravel for construction are mined from the continental shelf in several areas, and approximately 34% of U.S. oil production comes from wells on the continental shelf (Figure 9.18). Ancient shelf deposits in the Persian Gulf region contain the world's largest reserves of oil.

A potential resource within the EEZ is methane hydrate, consisting of single methane molecules bound up in networks formed by frozen water. These methane hydrates are stable at water depths of more than 500 m and near-freezing temperatures. According to one estimate, the carbon in these deposits is double that in all coal, oil, and natural gas reserves. However, no one knows yet whether methane hydrates can be effectively recovered and

Figure 9.18 Oil and Natural Gas Wells on the Continental Shelf This is Shell Oil Company's tension-leg platform *Ursa* in the Gulf of Mexico about 265 km south of New Orleans, Louisiana. It is anchored in about 1160 m of water by steel cables that hold the partly submerged platform in place. It is designed to withstand waves up to 22 m high and hurricane-force winds of 225 km/hr. Another tension-leg platform deployed in January 2004 in the Gulf of Mexico is in water more than 1300 m deep.

Geo-Focus Oceanic Circulation and Resources from the Sea

Earth's oceans are in constant motion. Huge quantities of water circulate in surface and deep currents as water is transferred from one part of an ocean basin to another. The Gulf Stream and South Equatorial Current carry great quantities of water toward the poles and have an important modifying effect on climate. In addition to surface and deep currents that carry water horizontally, vertical circulation takes place when *upwelling* slowly transfers cold water from depth to the surface and *downwelling* transfers warm surface water to depth.

Upwelling is of more than academic interest. It not only transfers water from depth to the surface, but also carries nutrients, especially nitrates and phosphate, into the zone of sunlight penetration. Here, these nutrients sustain huge concentrations of floating organisms, which in turn support other organisms. Other than the continental shelves and areas adjacent to hydrothermal vents on the seafloor, areas of upwelling are the only parts of the oceans where biological productivity is very high. In fact, they are so productive that even though constituting less than 1% of the ocean surface, they support more than 50% (by weight) of all fishes.

Scientists recognize three types of upwelling, but only *coastal upwelling* need concern us here. Most coastal upwelling takes place along the west coasts of Africa, North America, and South America, although one notable exception is in the Indian Ocean.

Coastal upwelling involves movement of water offshore, which is replaced by water rising from depth (Figure 1). Along the coast of Peru, for example, the winds, coupled with the Coriolis effect,* transport surface water seaward, and cold, nutrient-rich water rises to replace it. This area is a major fishery, and changes in the surface-water circulation every three to seven years adjacent to South America are associated with the onset of El Niño, a weather phenomenon with far-reaching consequences.

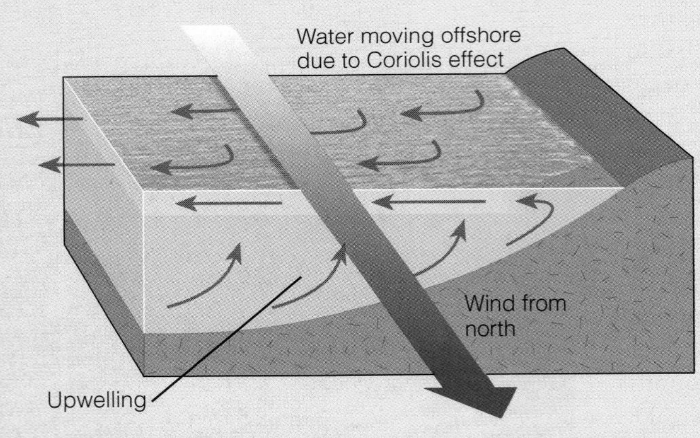

Figure 1 Wind from the north along the west coast of a continent, coupled with the Coriolis effect, causes surface water to move offshore, resulting in upwelling of cold, nutrient-rich deep water.

Among the nutrients in upwelling oceanic waters is considerable phosphorus, an essential element for animal and plant nutrition. Although present in minute quantities in many sedimentary rocks, most commercial phosphorus is derived from *phosphorite*, a sedimentary rock with such phosphate-rich minerals as fluorapatite [$Ca_5(PO_4)_3F$]. Areas of upwelling along the outer margins of continental shelves are the depositional sites of most of the so-called bedded phosphorites, which are interlayered with carbonate rocks, chert, shale, and sandstone. Vast deposits in the Permian-age Phosphoria Formation of Montana, Wyoming, and Idaho formed in this manner.

Upwelling accounts for most of Earth's phosphate-rich sedimentary rocks, but some forms by other processes. In *phosphatization*, carbonate grains such as animal skeletons and ooids are replaced by phosphate, and *guano* is made up of calcium phosphate from bird or bat excrement. Another type of phosphate deposit is essentially a placer deposit where the skeletons of vertebrate animals are found in large numbers [vertebrate skeletons are made up mostly of hydroxyapatite, [$Ca_5(PO_4)_3OH$]. The 3- to 15-million-year-old Bone Valley Formation in Florida is a good example.

The United States is the world leader in production and consumption of phosphate rock, most of it coming from deposits in Florida and North Carolina, but some is also mined in Idaho and Utah. More than 90% of all phosphate rock mined in this country is used to make chemical fertilizers and animal feed supplements. It also has several other uses in metallurgy, preserved foods, ceramics, and matches.

---

*The Coriolis effect is the apparent deflection of a moving object from its anticipated course resulting from Earth's rotation. Oceanic currents are deflected clockwise in the Northern Hemisphere and counterclockwise in the Southern Hemisphere.

used as an energy source. Additionally, their contribution to global warming must be assessed because a volume of methane 3000 times greater than in the atmosphere is present in seafloor deposits—and methane is 10 times more effective than carbon dioxide as a greenhouse gas.

The manganese nodules previously discussed are another potential seafloor resource (Figure 9.15). These spherical objects are composed mostly of manganese and iron oxides, but also contain copper, nickel, and cobalt. The United States, which must import most of the manganese and cobalt it uses, is particularly interested in these nodules as a potential resource.

Other seafloor resources of interest include massive sulfide deposits that form by submarine hydrothermal activity at spreading ridges. These deposits, containing iron, copper, zinc, and other metals, have been identified within the EEZ at the Gorda Ridge off the coasts of California and Oregon; similar deposits occur at the Juan de Fuca Ridge within the Canadian EEZ.

Within the EEZ, manganese nodules are found near Johnston Island in the Pacific Ocean and on the Blake Plateau off the east coast of South Carolina and Georgia. In addition, seamounts and seamount chains within the EEZ in the Pacific are known to have metalliferous oxide crusts several centimeters thick from which cobalt and manganese could be mined.

Another important resource found in shallow marine deposits is phosphate-rich sedimentary rock known as *phosphorite* (see Geo-Focus on page 240).

Geo-Recap

Chapter Summary

- Scientific investigations of the oceans began more than 200 years ago, but much of our knowledge comes from studies done during the last few decades.
- Present-day research vessels are equipped to investigate the seafloor by sampling and drilling, echo sounding, and seismic profiling. Scientists also use submersibles in their studies.
- Deep-sea drilling and observations on land and on the seafloor confirm that oceanic crust is made up, in descending order, of pillow lava/sheet lava flows, sheeted dikes, and gabbro.
- Continental margins consist of a gently sloping continental shelf, a more steeply inclined continental slope, and, in some cases, a continental rise.
- The width of continental shelves varies considerably. They slope seaward to the shelf–slope break at a depth averaging 135 m, where the seafloor slope increases abruptly.
- Submarine canyons, mostly on continental slopes, carry huge quantities of sediment by turbidity currents into deeper water, where it is deposited as overlapping submarine fans that make up a large part of the continental rise.
- Active continental margins at the leading edge of a tectonic plate have a narrow shelf and a slope that descends directly into an oceanic trench. Volcanism and seismic activity also characterize these margins.
- Passive continental margins lie within a tectonic plate and have wide continental shelves, and the slope merges with a continental rise that grades into an abyssal plain. These margins show little seismic activity and no volcanism.
- Long, narrow oceanic trenches are found where oceanic lithosphere is subducted beneath either oceanic lithosphere or continental lithosphere. The trenches are the sites of the greatest oceanic depths and low heat flow.
- Oceanic ridges are composed of volcanic rocks, and many have a central rift caused by tensional forces. Basaltic volcanism, hydrothermal vents, and shallow-focus earthquakes occur at ridges, which are offset by fractures that cut across them.
- Seamounts, guyots, and abyssal hills rising from the seafloor are common features that differ mostly in scale and shape. Many aseismic ridges on the seafloor consist of chains of seamounts, guyots, or both.
- Submarine hydrothermal vents known as black smokers found at or near spreading ridges support biologic communities and are potential sources of several resources.
- Moundlike, wave-resistant structures called reefs, consisting of animal skeletons, are found in a variety of shapes, but most are classified as fringing reefs, barrier reefs, or atolls.
- Sediments called pelagic clay and ooze cover vast areas of the seafloor.
- The United States claims rights to all resources within 200 nautical miles of its shorelines. Resources including sand and gravel, as well as metals, are found within this Exclusive Economic Zone (EEZ).

Important Terms

abyssal plain (p. 230)
active continental margin (p. 229)
aseismic ridge (p. 234)
black smoker (p. 232)
continental margin (p. 227)
continental rise (p. 228)
continental shelf (p. 227)
continental slope (p. 228)
Exclusive Economic Zone (EEZ) (p. 239)
guyot (p. 234)
oceanic ridge (p. 232)
oceanic trench (p. 232)
ooze (p. 235)
ophiolite (p. 227)
passive continental margin (p. 230)
pelagic clay (p. 235)
reef (p. 235)
seamount (p. 234)
seismic profiling (p. 225)
submarine canyon (p. 229)
submarine fan (p. 228)
submarine hydrothermal vent (p. 232)
turbidity current (p. 228)

Review Questions

1. A broad, flat area on the seafloor adjacent to a continental rise is a/an
   a. ______ seamount;
   b. ______ aseismic ridge;
   c. ______ abyssal plain;
   d. ______ active continental margin;
   e. ______ calcareous ooze.
2. Submarine fans make up a large part of
   a. ______ the continental rise;
   b. ______ hydrothermal vents;
   c. ______ ophiolite sequences;
   d. ______ the EEZ;
   e. ______ aseismic ridges.
3. The areas on the seafloor where oceanic lithosphere is subducted are called
   a. ______ submarine canyons;
   b. ______ oceanic trenches;
   c. ______ black smokers;
   d. ______ guyots;
   e. ______ pelagic clay.
4. Graded bedding is typically found in ________ deposits.
   a. ______ coral reef;
   b. ______ hydrothermal;
   c. ______ ophiolite;
   d. ______ pelagic;
   e. ______ turbidity current.
5. The gently sloping part of the continental margin adjacent to a continent is called the continental
   a. ______ ooze;
   b. ______ rise;
   c. ______ shelf;
   d. ______ reef;
   e. ______ plain.
6. A barrier reef is
   a. ______ found on the deep seafloor adjacent to a black smoker;
   b. ______ a deposit made up of copper, zinc, and iron;
   c. ______ a type of deposit made up of clay and seashells;
   d. ______ a reef separated from an island or continent by a lagoon;
   e. ______ made up of sediment that settles from suspension far from land.
7. Which one of the following statements is correct?
   a. ______ Volcanism takes place at passive continental margins;
   b. ______ A guyot is a reef that extends far above sea level;
   c. ______ Oceanic plates are subducted at oceanic trenches;
   d. ______ Abyssal plains are common around the margins of the Pacific;
   e. ______ Submarine fans are made up mostly of siliceous ooze.
8. Much of the sediment that ends up on submarine fans is transported through ________ by turbidity currents.
   a. ______ abyssal plains;
   b. ______ submarine canyons;
   c. ______ atolls;
   d. ______ continental depressions;
   e. ______ hydrothermal vents.
9. The greatest oceanic depths are found at
   a. ______ seamounts;
   b. ______ aseismic ridges;
   c. ______ oceanic trenches;
   d. ______ passive continental margins;
   e. ______ seafloor fractures.
10. Which one of the following is characteristic of an active continental margin?
    a. ______ wide continental shelf;
    b. ______ volcanic activity;
    c. ______ narrow abyssal plain;
    d. ______ submarine hydrothermal vents;
    e. ______ aseismic ridges.
11. During the Pleistocene Epoch (Ice Age), sea level was as much as 130 m lower than it is now. What effect did this

have on rivers? Is there any evidence from the continental shelves that might bear on this? If so, what?

12. The most distant part of a 30-million-year-old aseismic ridge is 1000 km from an oceanic ridge. How fast, on average, did the plate with this ridge move in centimeters per year?
13. How do mountains of mid-oceanic ridges differ from mountain ranges on land?
14. Although seafloor fractures may extend for hundreds of kilometers, earthquakes occur only on those parts of the fractures between offset ridge segments. Why?
15. How did geologists determine the nature of the upper mantle and oceanic crust even before they observed mantle and crust rocks in the ocean basins?
16. What kinds of evidence would you need to demonstrate that an atoll evolved from a fringing reef to a barrier reef and, finally, to an atoll?
17. What are calcareous ooze and pelagic clay, and where are they found?
18. Why are abyssal plains common around the margins of the Atlantic, but rare in the Pacific Ocean basin?
19. Identify the types of continental margins in Figure 9.7 and list the characteristics of each.
20. Explain how an aseismic ridge forms.

CHAPTER 10

Deformation, Mountain Building, and the Continents

OUTLINE

OBJECTIVES

At the end of this chapter, you will have learned that

- Rock deformation involves changes in the shape or volume or both of rocks in response to applied forces.
- Geologists use several criteria to differentiate among geologic structures such as folds, joints, and faults.
- Correctly interpreting geologic structures is important in human endeavors such as constructing highways and dams, choosing sites for power plants, and finding and extracting some resources.
- Deformation and the origin of geologic structures are important in the origin and evolution of mountains.
- Most of Earth's large mountain systems formed and, in some cases, continue to form, at or near the three types of convergent plate boundaries.
- Terranes have special significance in mountain building.
- Earth's continental crust, and especially mountains, stands higher than adjacent crust because of its composition and thickness.

The Teton Range in Wyoming is one of many mountain ranges in the Rocky Mountains. This area has a history of mountain building that goes back at least 90 million years, but the present range began forming less than 10 million years ago. The Teton Range formed when uplift of a block of Earth's crust took place along fractures (faults). Now the range stands 2100 m above the valley (Jackson Hole) to the east.

INTRODUCTION

"Solid as a rock" implies permanence and durability, but you know from earlier chapters that physical and chemical processes disaggregate and decompose rocks, and rocks behave very differently at great depth than they do at or near Earth's surface. Indeed, under the tremendous pressures and high temperature at several kilometers below the surface, rock layers actually crumple or fold, yet remain solid, and at shallower depths, they yield by fracturing or a combination of folding and fracturing. In either case, dynamic forces within Earth cause **deformation**, a general term encompassing all changes in the shape or volume of rocks (see the chapter opening photograph and Figure 10.1).

The action of dynamic forces within Earth is obvious from ongoing seismic activity, volcanism, plate movements, and the continuing evolution of mountains in South America, Asia, and elsewhere. In short, Earth is an active planet with a variety of processes driven by internal heat, particularly plate movements; most of Earth's seismic activity, volcanism, and rock deformation take place at divergent, convergent, and transform plate boundaries.

The origin of Earth's truly large mountain ranges on the continents involves tremendous deformation, usually accompanied by emplacement of plutons, volcanism, and metamorphism, at convergent plate boundaries. The Appalachians of North America, the Alps in Europe, the Himalayas of Asia, and the Andes in South America all owe their existence to deformation at convergent plate boundaries. And, in some cases, this activity continues even now. Thus, deformation and mountain building are closely related topics and accordingly we consider both in this chapter.

The past and continuing evolution of continents involves not only deformation at continental margins, but also additions of new material to existing continents, a phenomenon known as *continental accretion* (see Chapter 19). North America, for instance, has not always had its present shape and area. Indeed, it began evolving during the Archean Eon (4.6–2.5 billion years ago) as new material was added to the continent at deformation belts along its margins.

Much of this chapter is devoted to a review of *geologic structures*, such as folded and fractured rock layers resulting from deformation, their descriptive terminology, and the forces responsible for them. There are several practical reasons to study deformation and mountain building. For one thing, crumpled and fractured rock layers provide a record of the kinds and intensities of forces that operated during the past. Thus, interpretations of these structures allow us to satisfy our curiosity about Earth history, and, in addition, such studies are essential in engineering endeavors such as choosing sites for dams, bridges, and nuclear power plants, especially if they are in areas of ongoing deformation. Also, many aspects of mining and exploration for petroleum and natural gas rely on correctly identifying geologic structures.

ROCK DEFORMATION—HOW DOES IT OCCUR?

We defined *deformation* as a general term referring to changes in the shape or volume (or both) of rocks; that is, rocks may be crumpled into folds or fractured as a result of **stress**, which results from force applied to a given area of rock. If the intensity of the stress is greater than the rock's internal strength, the rock undergoes **strain**, which is simply deformation caused by stress. The terminology is a little confusing at first, but keep in mind that *deformation* and *strain* are synonyms, and stress is the force that causes deformation or strain. The following discussion and Figure 10.2 will help clarify the meaning of stress and the distinction between stress and strain.

Reed Wicander

Figure 10.1 Deformation Many rocks show the effect of deformation. These rocks have been deformed by folding and fracturing. Notice the light pole for scale. The nearly vertical fracture where light-colored rocks were displaced is a fault, a fracture along which rocks on opposite sides of the fracture have moved parallel with the fracture surface.

Stress and Strain

Remember that stress is the force applied to a given area of rock, usually expressed in kilograms per square centimeter (kg/cm^2). For example, the stress, or force, exerted by a person walking on an ice-covered pond is a function of the person's weight and the area beneath her or his feet.

Figure 10.2 Stress and Strain Exerted on an Ice-Covered Pond

a The woman weighs 65 kg (6500 g). Her weight is imparted to the ice through her feet, which have a contact area of 120 cm². The stress she exerts on the ice (6500/120 cm²) is 54 g/cm². This is sufficient stress to cause the ice to crack.

b To avoid plunging into the freezing water, the woman lays out flat, thereby decreasing the stress she exerts on the ice. Her weight remains the same, but her contact area with the ice is 3150 cm², so the stress is only about 2 g/cm² (6500 g/3150 cm²), which is well below the threshold needed to crack the ice.

The ice's internal strength resists the stress unless the stress is too great, in which case the ice may bend or crack as it is strained (deformed) (Figure 10.2). To avoid breaking through the ice, the person may lie down; this does not reduce the weight of the person on the ice, but it does distribute it over a larger area, thus reducing the stress per unit area.

Although stress is force per unit area, it comes in three varieties: *compression*, *tension*, and *shear*, depending on the direction of the applied forces. In **compression**, rocks or any other object are squeezed or compressed by forces directed toward one another along the same line, as when you squeeze a rubber ball in your hand. Rock layers in compression tend to be shortened in the direction of stress by either folding or fracturing (Figure 10.3a). **Tension** results from forces acting along the same line, but in opposite directions. Tension tends to lengthen rocks or pull them apart (Figure 10.3b). Incidentally, rocks are much stronger in compression than they are in tension. In **shear stress**, forces act parallel to one another, but in opposite directions, resulting in deformation by displacement along closely spaced planes (Figure 10.3c).

Types of Strain

Geologists characterize strain as **elastic strain** if deformed rocks return to their original shape when the deforming forces are relaxed. In Figure 10.2, the ice on the pond may bend under a person's weight, but return to its original shape once the person leaves. As you might expect, rocks are not very elastic, but Earth's crust behaves elastically when loaded by glacial ice and depressed into the mantle.

As stress is applied, rocks respond first by elastic strain, but when strained beyond their elastic limit, they undergo

Figure 10.3 Stress and Possible Types of Resulting Deformation

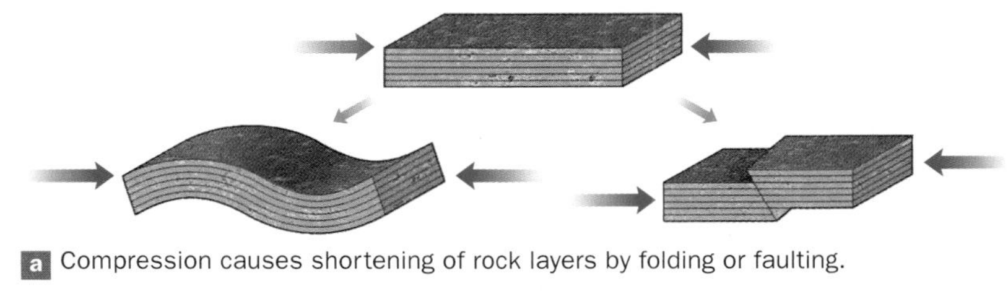

a Compression causes shortening of rock layers by folding or faulting.

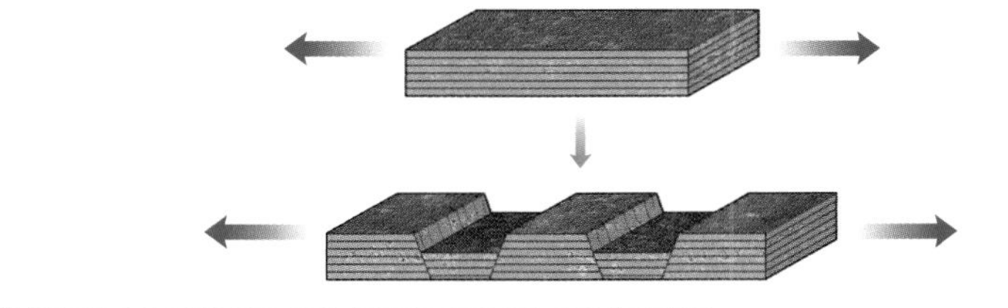

b Tension lengthens rock layers and causes faulting.

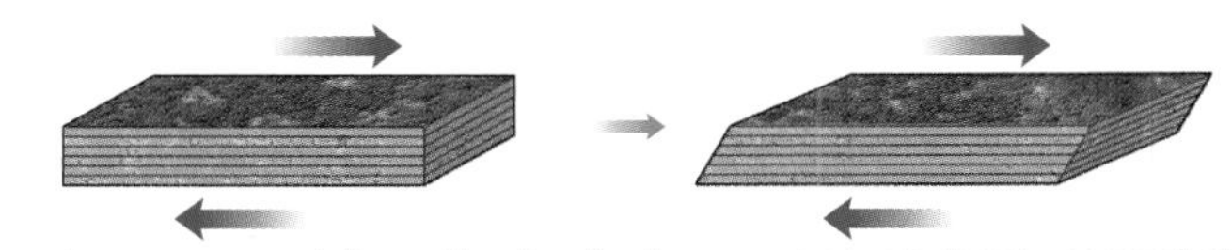

c Shear stress causes deformation by displacement along closely spaced planes.

What Would You Do?

The types of stresses, as well as elastic versus plastic strain, might seem rather esoteric, but perhaps understanding these concepts has some practical applications. What relevance do you think knowing about stress and strain has to some professions, other than geology, and what professions might these be? Can you think of stresses and strain that we contend with in our daily lives? As an example, what happens when a car smashes into a tree?

plastic strain as when they yield by folding, or they behave like brittle solids and **fracture** (Figure 10.4). In either folding or fracturing, the strain is permanent; that is, the rocks do not recover their original shape or volume even if the stress is removed.

Whether strain is elastic, plastic, or fracture depends on the kind of stress applied, pressure and temperature, rock type, and the length of time rocks are subjected to stress. A small stress applied over a long period, as on a mantelpiece supported only at its ends, will cause the rock to sag; that is, the rock deforms plastically (Figure 10.4). By contrast, a large stress applied rapidly to the same object, as when struck by a hammer, results in fracture. Rock type is important because not all rocks have the same internal strength and thus respond to stress differently. Some rocks are *ductile*, whereas others are *brittle*, depending on the amount of plastic strain they exhibit. Brittle rocks show little or no plastic strain before they fracture, but ductile rocks exhibit a great deal (Figure 10.4).

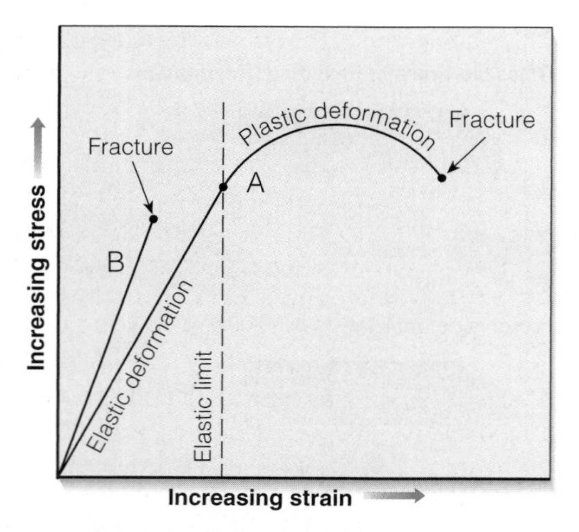

Figure 10.4 Rock Response to Stress Rocks initially respond to stress by elastic deformation and then return to their original shape when the stress is released. If the elastic limit is exceeded, as in curve A, rocks deform plastically, which is permanent deformation. The amount of plastic deformation rocks exhibit before fracturing depends on their ductility. If they are ductile, they show considerable plastic deformation (curve A), but if they are brittle, they show little or no plastic deformation before failing by fracture (curve B).

Many rocks show the effects of plastic deformation that must have taken place deep within the crust. At or near the surface, rocks commonly behave like brittle solids and fracture, but at depth, they more often yield by plastic deformation; they become more ductile with increasing pressure and temperature. Most earthquake foci are at depths of less than 30 km, indicating that deformation by fracturing becomes increasingly difficult with depth and no fracturing is known deeper than about 700 km.

STRIKE AND DIP—THE ORIENTATION OF DEFORMED ROCK LAYERS

During the 1660s, Nicholas Steno, a Danish anatomist, proposed several principles essential for deciphering Earth history from the record preserved in rocks. One is the *principle of original horizontality*, meaning that sediments accumulate in horizontal or nearly horizontal layers. Thus, if we observe steeply inclined sedimentary rocks, we are justified in inferring that they were deposited nearly horizontally, lithified, and then tilted into their present position (Figure 10.5a). Rock layers deformed by folding, faulting, or both are no longer in their original position, so geologists use *strike* and *dip* to describe their orientation with respect to a horizontal plane.

By definition, **strike** is the direction of a line formed by the intersection of a horizontal plane and an inclined plane. The surfaces of the rock layers in Figure 10.5b are good examples of inclined planes, whereas the water surface is a horizontal plane. The direction of the line formed at the intersection of these planes is the strike of the rock layers. The strike line's orientation is determined by using a compass to measure its angle with respect to north. **Dip** is a measure of an inclined plane's deviation from horizontal, so it must be measured at right angles to strike direction (Figure 10.5b).

Geologic maps showing the age, aerial distribution, and geologic structures of rocks in an area use a special symbol to indicate strike and dip. A long line oriented in the appropriate direction indicates strike, and a short line perpendicular to the strike line shows the direction of dip (Figure 10.5b). Adjacent to the strike and dip symbol is a number corresponding to the dip angle. The usefulness of strike and dip symbols will become apparent in the sections on folds and faults.

DEFORMATION AND GEOLOGIC STRUCTURES

Remember that deformation and its synonym strain refer to changes in the shape or volume of rocks. During deformation, rocks might be crumpled into folds, or they might be fractured, or perhaps folded and fractured. Any of these features resulting from deformation is referred to as a **geologic structure**. Geologic structures are present almost everywhere that rock exposures can be observed, and many are detected far below the surface by drilling and several geophysical techniques.

Figure 10.5 Strike and Dip of Deformed Rocks

Sue Monroe

a We can infer that these layers of sedimentary rocks in Death Valley National Park in California were deposited horizontally or nearly so, lithified, and then deformed. To describe their orientation, geologists use the terms strike and dip.

Water surface
Strike
Dip direction
50
Dip angle 50°

b Strike is the intersection of a horizontal plane (the water surface) with an inclined plane (the surface of the rock layer). Dip is the maximum angular deviation of the inclined layer from horizontal. Notice the strike and dip symbol.

Folded Rock Layers

Geologic structures known as **folds**, in which planar features are crumpled and bent, are quite common. Compression is responsible for most folding, as when you place your hands on a tablecloth and move them toward one another, thereby producing a series of up- and down-arches in the fabric. Rock layers in the crust respond similarly to compression, but unlike the table-cloth, folding in rock layers is permanent. That is, plastic strain has taken place: so once folded, the rocks stay folded. Most folding probably takes place deep in the crust where rocks are more ductile than they are at or near the surface. The configuration of folds and the intensity of folding vary considerably, but there are only three basic types of folds: *monoclines*, *anticlines*, and *synclines*.

Monoclines A simple bend or flexure in otherwise horizontal or uniformly dipping rock layers is a **monocline** (Figure 10.6a). The large monocline in Figure 10.6b formed when the Bighorn Mountains in Wyoming rose vertically along a fracture. The fracture did not penetrate to the surface, so as uplift of the mountains proceeded, the near-surface rocks were bent so that they now appear to be draped over the margin of the uplifted block. In a manner of speaking, a monocline is simply one-half of an anticline or syncline.

Anticlines and Synclines An **anticline** is an uparched or convex upward fold with the oldest rock layers in its core, whereas a **syncline** is a down-arched or concave downward fold in which the youngest rock layers are in its core (Figure 10.7). Anticlines and synclines have an axial plane connecting the points of maximum curvature of each folded layer (Figure 10.8); the axial plane divides folds into halves, each half being a *limb*. Because folds are most often found in a series of anticlines alternating with synclines, an anticline and adjacent syncline share a limb. It is important to remember that anticlines and synclines are simply folded rock layers and do not necessarily correspond to high and low areas at the surface (Figure 10.9).

Folds are commonly exposed to view in areas of deep erosion, but even where eroded, strike and dip and the relative ages of the folded rock layers easily distinguish anticlines from synclines. Notice in Figure 10.10 that in the surface

Figure 10.6 Monoclines

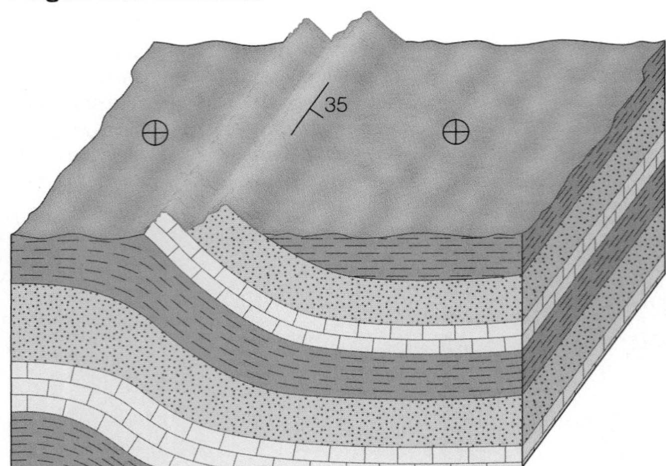

a A monocline. Notice the strike and dip symbol and the circled cross, which is the symbol for horizontal layers.

b A monocline in the Bighorn Mountains in Wyoming.

Sue Monroe

view of the anticline, each limb dips outward or away from the center of the fold, and the oldest exposed rocks are in the fold's core. In an eroded syncline, though, each limb dips inward toward the fold's center, where the youngest exposed rocks are found.

The folds described so far are *upright*, meaning that their axial planes are vertical and both fold limbs dip at the same angle (Figure 10.10). In many folds, the axial plane is not vertical, the limbs dip at different angles, and the folds are characterized as *inclined* (Figure 10.11a). If both limbs dip in the same direction, the fold is *overturned*. That is, one limb has been rotated more than 90 degrees from its original position so that it is now upside down (Figure 10.11b). In some areas, deformation has been so intense that axial planes of folds are now horizontal, giving rise to what geologists call *recumbent folds* (Figure 10.11c). Overturned and recumbent folds are particularly common in mountains resulting from compression at convergent plate boundaries (discussed later in this chapter).

For upright folds, the distinction between anticlines and synclines is straightforward, but interpreting complex folds that have been tipped on their sides or turned completely upside down is more difficult. Can you determine which of the two folds shown in Figure 10.11c is an anticline? Even if strike and dip symbols were shown, you would still not be able to resolve this question, but knowing the relative ages of the folded rock layers provides a solution. Remember that an anticline has the oldest rock layers in its core, so the fold nearest the surface is an anticline and the lower fold is a syncline.

Plunging Folds To complicate matters further, folds are also characterized as *nonplunging* or *plunging*. In some folds, the fold axis, a line formed by the intersection of the axial plane with the folded layers, is horizontal and the folds are nonplunging (Figure 10.11b). Much more commonly, though, fold axes are inclined so that they appear to plunge beneath adjacent rocks, and the folds are said to be plunging (Figure 10.12).

It might seem that with this additional complication, differentiating plunging anticlines from plunging synclines would be much more difficult; however, geologists use exactly the same criteria that they use for nonplunging folds. Therefore, all rock layers dip away from the fold axis in plunging

Figure 10.7 Anticlines and Synclines

a Folded rocks in the Calico Mountains of southeastern California. Compression was responsible for these folds, which are, from left to right, a syncline, an anticline, and a syncline.

b Anticlines and synclines are usually found in a series as in (a); however, the Barstow syncline shown here, also in southern California, is unusual because it is a single fold with no adjacent anticlines.

anticlines and toward the axis in plunging synclines. The oldest exposed rocks are in the core of an eroded plunging anticline, whereas the youngest exposed rock layers are found in the core of an eroded plunging syncline (Figure 10.12b).

In Chapter 6, we noted that anticlines form one type of structural trap in which petroleum and natural gas might accumulate (see Figure 6.29b). As a matter of fact, most of the world's petroleum production comes from anticlines, although other geologic structures and stratigraphic traps are important, too. Accordingly, geologists are particularly interested in correctly identifying geologic structures in areas of potential hydrocarbon production.

Domes and Basins Anticlines and synclines are elongate structures, meaning that their length greatly exceeds their width. In contrast, folds that are nearly equidimensional, that is, circular, are *domes* and *basins*. In a **dome**, all of the folded strata dip outward from a central point (as opposed to outward from a line as in an anticline), and the oldest exposed rocks are at the center of the fold (Figure 10.13a, b), so we characterize a dome as the circular equivalent of an anticline. In contrast, a **basin**, the circular counterpart of a syncline, has all strata dipping inward toward a central point and the youngest exposed rocks are at the fold's center (Figure 10.13c).

Geo-Focus Engineering and Geology

As you might expect, engineering geologists apply the principles and concepts of geology to engineering practices. Geologists with this specialized training may be involved in slope stability studies and in studies of acceptable areas for power plants, for highways in mountainous regions, for tunnels and canals, and for structures to protect riverbank and seashore communities.

A good example is the concern prior to building the Mackinac (pronounced "mack-in-aw") Bridge, a huge suspension bridge that connects the Upper and Lower Peninsulas of Michigan (Figure 1). Geologists and engineers were aware that some of the rock in the area, called Mackinac Breccia, was a collapse breccia, or rubble that formed when caverns collapsed. The concern was whether the breccia or any uncollapsed caverns beneath the area would support the weight of the huge piers and abutments for the bridge. Obviously, the project was completed successfully, but detailed studies were done before construction began.

Geologic engineers are invariably involved in planning for large-scale structures such as bridges, dams, power plants, and highways, especially in tectonically active regions. For example, the bridge in Figure 2 is only a short distance from the San Andreas fault, so engineers had to take into account the near certainty that it would be badly shaken during an earthquake. Many other sturctures on or near the San Andreas fault were constructed when codes were much less stringent, and now they are being retrofitted to make them safer during earthquakes.

Figure 1 Before the Mackinac Bridge that connects the Lower and Upper Peninsulas of Michigan could be built, geologists and engineers had to determine whether the Mackinac Breccia could support such a large structure.

Being aware of a problem and taking remedial action sometimes come too late. For instance, engineers were aware that the Santa Monica Freeway in the Los Angeles area would likely be damaged during an earthquake and retrofitting was scheduled for February 1994. Unfortunately, the Northridge earthquake struck on January 17, 1994, and part of the freeway collapsed. Of course, this and other similar events provide important

Many domes and basins are so large that they can be visualized only on geologic maps or aerial photographs. The Black Hills of South Dakota, for example, are a large oval dome (Figure 10.13b). One of the best-known basins in the United States is the Michigan Basin, most of which is buried beneath younger strata so it is not directly observable at the surface. Nevertheless, strike and dip of exposed strata near the basin margin and thousands of drill holes for oil and gas show that the strata are deformed into a large basin.

Unfortunately, the terms *dome* and *basin* are also used to distinguish high and low areas of Earth's surface, but as with anticlines and synclines, domes and basins resulting from deformation do not necessarily correspond with mountains or valleys. In some of the following discussions, we will have occasion to use these terms in other contexts, but we will try to be clear when we refer to surface elevations as opposed to geologic structures.

Joints

Besides folding, rocks are also permanently deformed by fracturing. **Joints** are fractures along which no movement has taken place parallel with the fracture surface (Figure 10.14), although they may open up; that is, they show movement perpendicular to the fracture. Coal miners used the term *joint* long ago for cracks in rocks that they thought were surfaces where the adjacent blocks were "joined" together.

Remember that rocks near the surface are brittle and therefore commonly fail by fracturing when subjected to stress. In fact, almost all near-surface rocks have joints that form in response to compression, tension, and shearing. They vary from minute fractures to those extending for many kilometers and are often arranged in two or perhaps three prominent sets. Regional mapping reveals that joints

James S. Monroe

Figure 2 This freeway crosses a small valley a short distance from the San Andreas fault in California. Engineers had to take into account the near certainty that this structure would be badly shaken during an earthquake.

information that can be incorporated into engineering practice to make freeways, buildings, and bridges safer during earthquakes.

Natural hazards include wildfires, swarms of insects, and severe weather phenomena, as well as various geologic hazards such as flooding, landslides, volcanic eruptions, earthquakes, tsunami, land subsidence, soil creep, and radon gas. Geologic hazards account for thousands of fatalities and billions of dollars in property damage every year, and whereas the incidence of hazards has not increased, fatalities and damages have grown because more and more people live in disaster-prone areas.

We cannot eliminate geologic hazards, but we can better understand these events, design structures to protect shoreline communities and withstand shaking during earthquakes, enact zoning and land-use regulations, and, at the very least, decrease the amount of damage and human suffering.

Unfortunately, geologic information that is readily available is often ignored or overlooked. A case in point is the Turnagain Heights subdivision in Anchorage, Alaska, that was so heavily damaged when the soil liquefied during the 1964 earthquake (see Figure 14.19). Not only were reports on soil stability ignored or overlooked before homes were built there, but since 1964, new homes were built on part of the same site!

In many mountainous or even hilly areas, highways are notoriously unstable and slump or slide from hillsides. When this happens, engineering geologists are consulted for their recommendations for stabilizing slopes. They may suggest building retaining walls or drainage systems to keep the slopes dry, planting vegetation, or, in some cases, simply rerouting the highway if it is too costly to maintain in its present position.

As you read the following chapters on surface processes, keep in mind that engineering geologists are involved in many aspects of stabilizing slopes, designing and constructing dams and flood-control projects, and building structures to protect seaside communities.

and sets of joints are usually related to other geologic structures such as large folds and faults.

We have discussed columnar joints that form when lava or magma in some shallow plutons cools and contracts (see Figure 5.6). A different type of jointing previously discussed is sheet jointing that forms in response to pressure release (see Figure 6.4).

Faults

Another type of fracture known as a **fault** is one along which blocks of rock on opposite sides of the fracture have moved parallel with the fracture surface, and the surface along which movement takes place is a **fault plane** (Figure 10.15a). Not all faults penetrate to the surface, but those that do might show a *fault scarp*, a bluff or cliff formed by vertical movement (Figure 10.15b). Fault scarps are usually quickly eroded and obscured. When movement takes place on a fault plane, the rocks on opposite sides may be scratched and polished or crushed and shattered into angular blocks, forming *fault breccia* (Figure 10.15c).

Notice the designations *hanging wall block* and *footwall block* in Figure 10.15a. The **hanging wall block** consists of the rock overlying the fault, whereas the **footwall block** lies beneath the fault plane. You can recognize these two blocks on any fault except a vertical one—that is, one that dips at 90 degrees. To identify some kinds of faults, you must not only correctly identify these two blocks, but also determine which one moved relatively up or down. We use the phrase *relative movement* because you cannot usually tell which block moved or if both moved. In Figure 10.15a, the footwall block may have moved up, the hanging wall block may have moved down, or both could have moved. Nevertheless, the hanging wall block appears to have moved down relative to the footwall block.

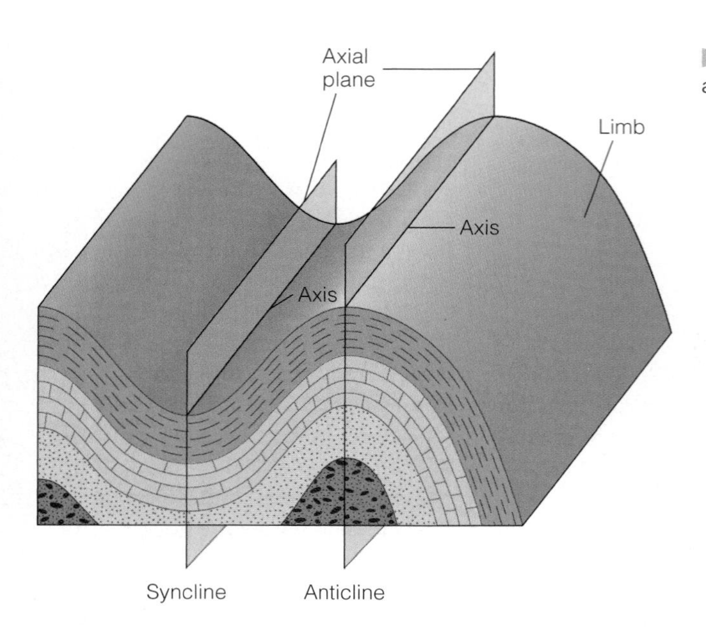

Figure 10.8 Syncline and Anticline Axial Planes Syncline and anticline showing the axial plane, axis, and fold limbs.

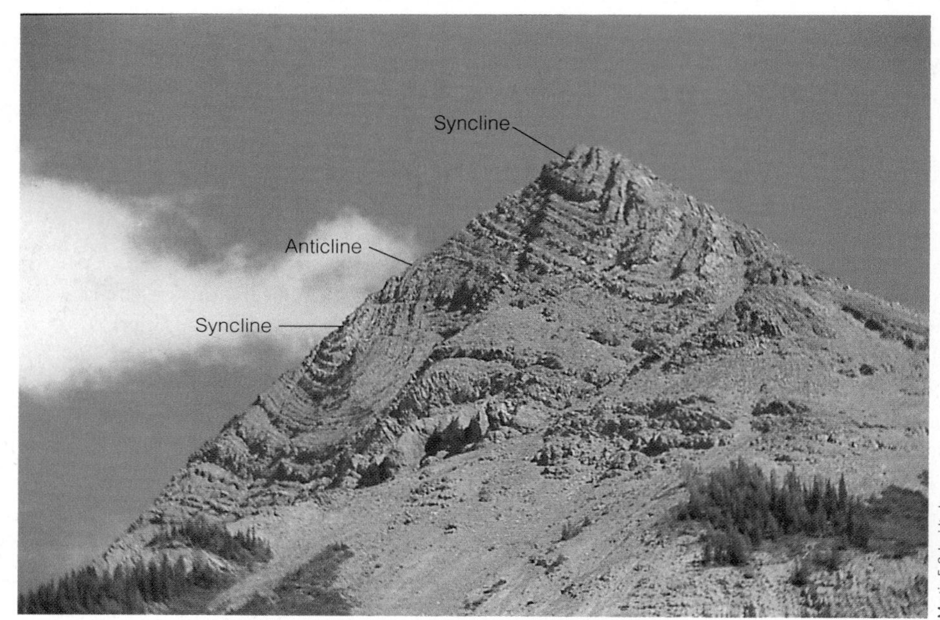

Figure 10.9 Folds and Their Relationship to Topography Anticlines and synclines do not necessarily correspond to high and low areas of the surface. A syncline is at the peak of this mountain in Kootenay National Park, British Columbia, Canada. Lower on the left flank of the mountain, an anticline and another syncline are also visible.

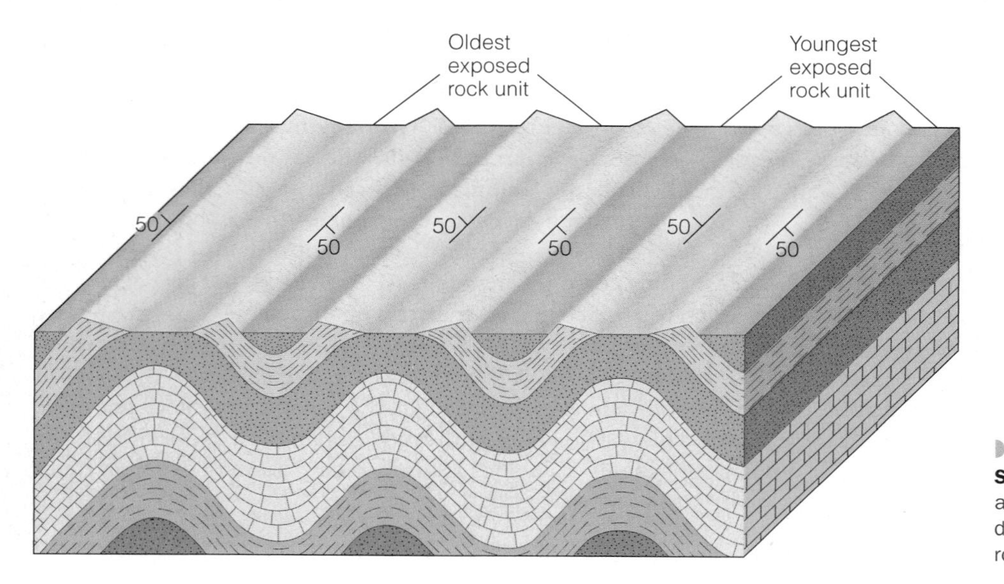

Figure 10.10 Eroded Anticlines and Synclines Geologists identify eroded anticlines and synclines by strike and dip and the relative ages of the folded rock layers.

Figure 10.11 Inclined, Overturned, and Recumbent Folds

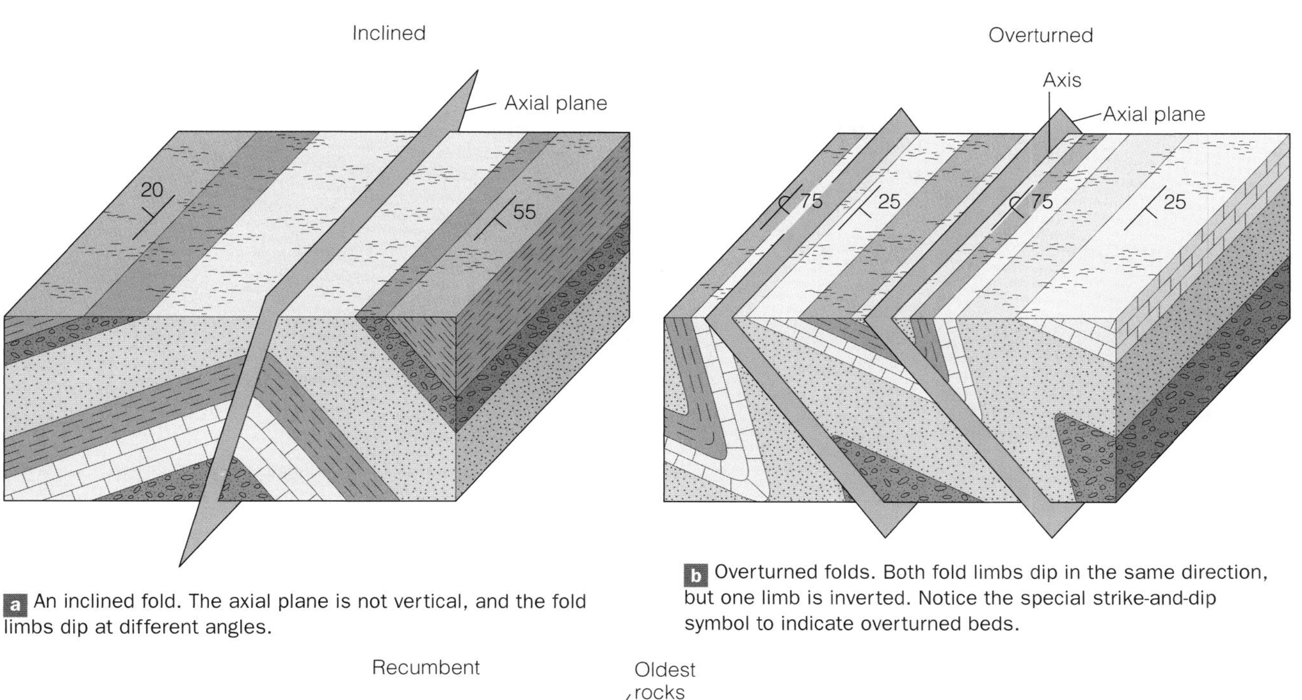

a An inclined fold. The axial plane is not vertical, and the fold limbs dip at different angles.

b Overturned folds. Both fold limbs dip in the same direction, but one limb is inverted. Notice the special strike-and-dip symbol to indicate overturned beds.

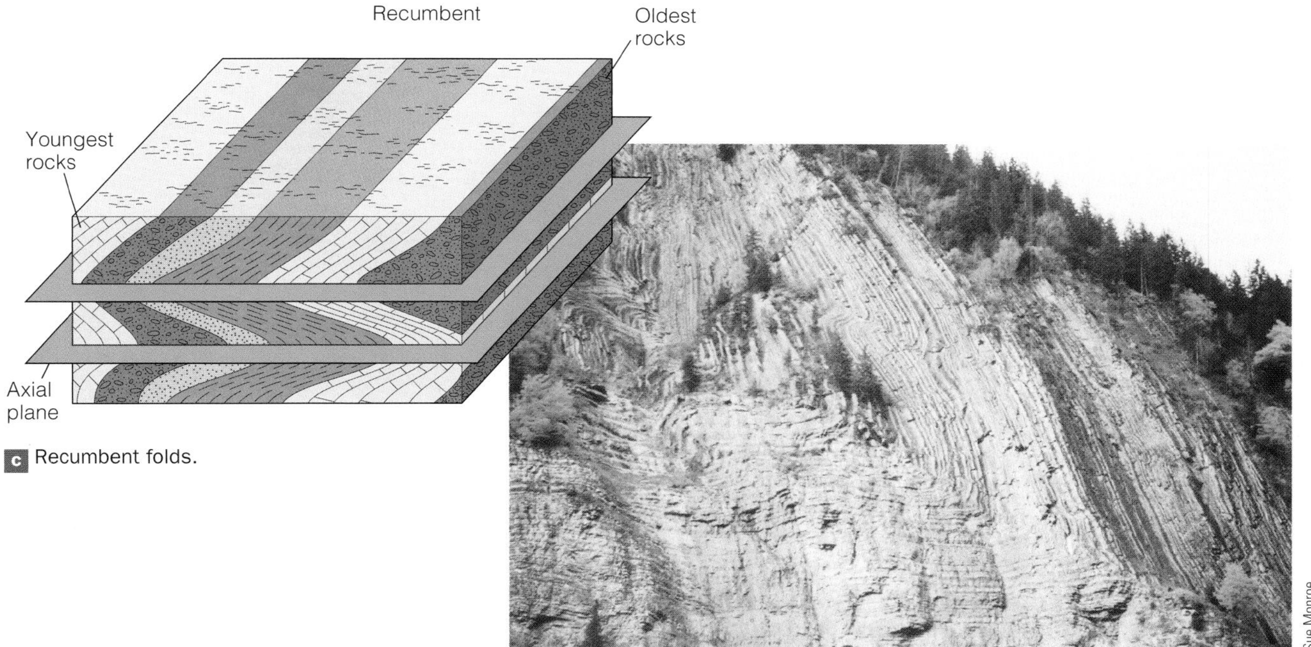

c Recumbent folds.

d An overturned fold in Switzerland.

Recall our discussion of strike and dip of rock layers. Fault planes are also inclined planes and they, too, are characterized by strike and dip (Figure 10.15a). In fact, the two basic varieties of faults are defined by whether the blocks on opposite sides of the fault plane moved parallel to the direction of dip (dip-slip faults) or along the direction of strike (strike-slip faults) (see Geo-inSight on pages 262 and 263).

Dip-Slip Faults All movement on **dip-slip faults** takes place parallel with the fault's dip; that is, movement is vertical, either up or down the fault plane. In Figure 10.16, for example, the hanging wall block moved down relative to the footwall block, giving rise to a **normal fault**. In contrast, in a **reverse fault**, the hanging wall block moves up relative to the footwall block (Figure 10.16b). In Figure 10.16c, the hanging wall block also moved up relative to the footwall block, but the fault has a dip of less than 45 degrees and is a special variety of reverse fault known as a **thrust fault.**

Figure 10.3b shows that normal faults result from tension. Numerous normal faults are present along one or both sides of mountain ranges in the Basin and Range Province of the western United States where the crust is being stretched and thinned. The Sierra Nevada at the

Figure 10.12 Plunging Folds

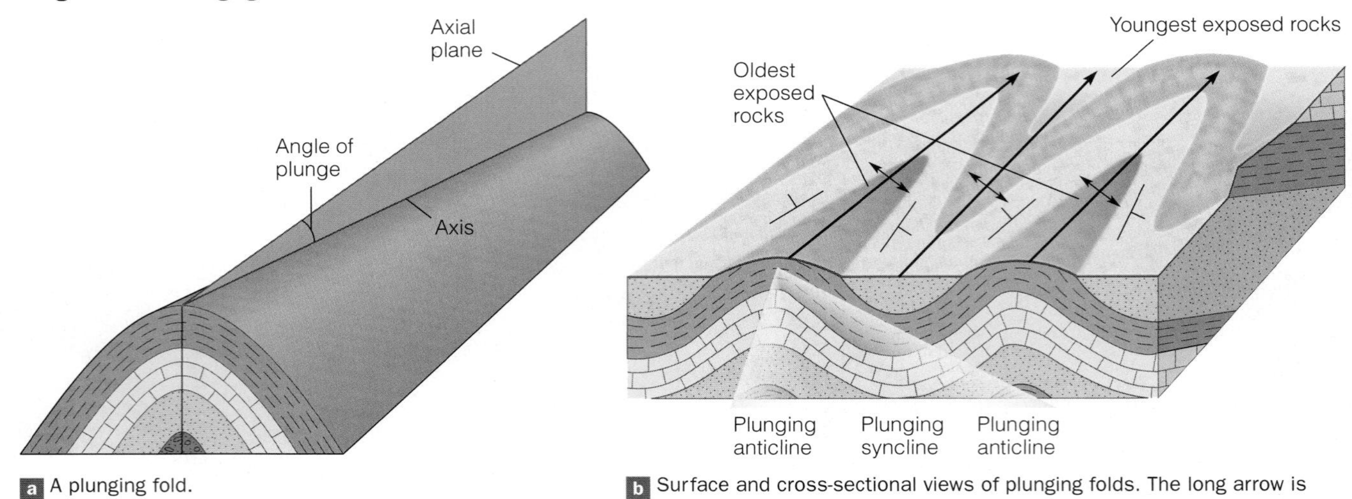

a A plunging fold.

b Surface and cross-sectional views of plunging folds. The long arrow is the geologic symbol for a plunging fold; it shows the direction of plunge.

c The Sheep Mountain anticline in Wyoming. You can tell from the strike and dip symbols that this is a plunging anticline because the rock layers dip outward or away from the fold axis.

John S. Shelton

western margin of the Basin and Range is bounded by normal faults, and the range has risen along these faults so that it now stands more than 3000 m above the lowlands to the east (see Chapter 23). Also, an active normal fault is found along the eastern margin of the Teton Range in Wyoming, accounting for the 2100-m elevation difference between the valley floor and the highest peaks in the mountains (Figure 10.17).

Reverse and thrust faults result from compression (Figure 10.3a; Figure 10.16b, c). Large-scale examples of both are found in mountain ranges that formed at convergent plate margins, where one would expect compression (discussed later in this chapter). A well-known thrust fault is the Lewis overthrust of Montana. (An overthrust is a low-angle thrust fault with movement measured in kilometers.) On this fault, a huge slab of Precambrian-age rocks moved at least 75 km eastward and now rests upon much younger Cretaceous-aged rocks (see Geo-inSight on pages 262 and 263).

Strike-Slip Faults **Strike-slip faults**, resulting from shear stresses, show horizontal movement with blocks on opposite sides of the fault sliding past one another (Figure 10.3c; Figure 10.16d). In other words, all movement is in the direction of the fault plane's strike—hence the name *strike-slip* fault. Several large strike-slip faults are known, but the best studied is the San Andreas fault, which cuts through coastal California. Recall from Chapter 2 that the San Andreas fault is called a *transform fault* in plate tectonics terminology.

Figure 10.13 Domes and Basins

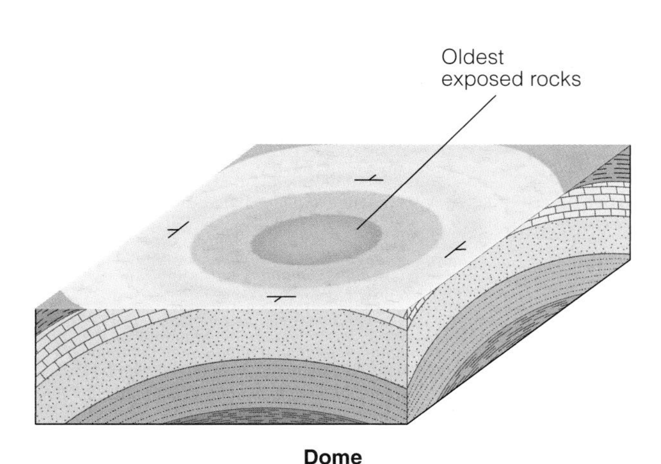

Dome

a Notice that in a dome, the oldest exposed rocks are in the center and all rocks dip outward from a central point.

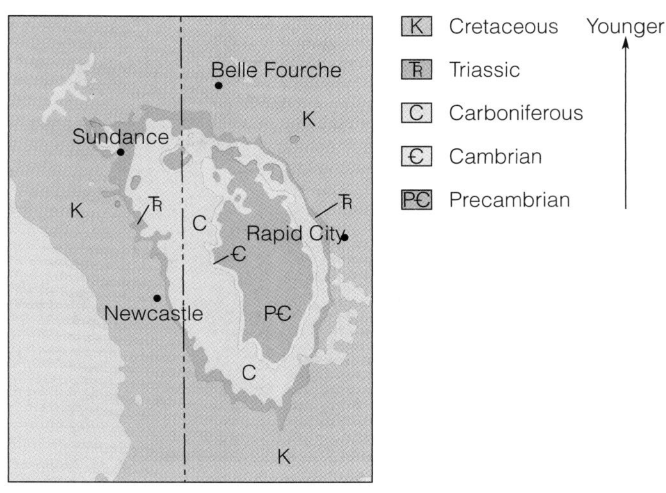

b This geologic map, a surface view only, uses colors and symbols to depict rocks and geologic structures in the Black Hills of South Dakota. You can tell this is a dome because the oldest rock layers are exposed at its center.

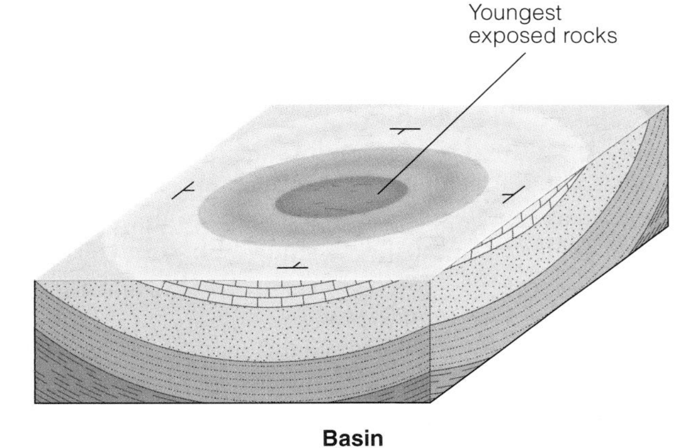

Basin

c In a basin, the youngest exposed rocks are in the center and all rocks dip inward toward a central point.

Strike-slip faults are characterized as right-lateral or left-lateral, depending on the apparent direction of offset. In Figure 10.16d, for example, observers looking at the block on the opposite side of the fault from their location notice that it appears to have moved to the left. Accordingly, this is a *left-lateral strike-slip fault.* If it had been a *right-lateral strike-slip fault*, the block across the fault from the observers would appear to have moved to the right. The San Andreas fault in California is a right-lateral strike-slip fault, whereas the Great Glen fault in Scotland is a left-lateral strike-slip fault (see Geo-inSight on pages 262 and 263).

Oblique-Slip Faults The movement on most faults is primarily dip-slip or strike-slip, but on **oblique-slip faults**, both types of movement take place. Strike-slip movement might be accompanied by a component of dip-slip, giving rise to a combined movement that includes left-lateral and reverse, or right-lateral and normal (Figure 10.16e and see Geo-inSight on pages 262 and 263).

As we mentioned in the introduction, the study of geologic structures is important in the exploration for minerals, oil, and natural gas. In addition, geologic engineers evaluate geologic structures in the planning stages for many projects, especially if they are in tectonically active areas (see Geo-Focus on pages 252 and 253).

DEFORMATION AND THE ORIGIN OF MOUNTAINS

Mountains form in several ways, but the truly large mountains on continents result mostly from compression-induced deformation at convergent plate boundaries. Before discussing mountain building, though, we should define what we mean by the term *mountain* and briefly discuss the types of mountains. *Mountain* is a designation for any area of land that stands significantly higher, at least 300 m, than the surrounding country and has a restricted summit area. Some mountains are single, isolated peaks, but more

Figure 10.14 Joints Fractures along which no movement has taken place parallel with the fracture surface are called joints.

C. G. Tillman

a Joints intersecting at right angles yield this rectangular pattern in Wales.

© Galen Rowell/Peter Arnold, Inc.

b Erosion along parallel joints in Arches National Park in Utah.

commonly they are parts of linear associations of peaks and ridges known as *mountain ranges* that are related in age and origin. A *mountain system*, a complex linear zone of deformation and crustal thickening, on the other hand, consists of several or many mountain ranges. The Teton Range in Wyoming is one of many ranges in the Rocky Mountains (Figure 10.17). The Appalachian Mountains of the eastern United States and Canada is another complex mountain system made up of many ranges, such as the Great Smoky Mountains of North Carolina and Tennessee, the Adirondack Mountains of New York, and the Green Mountains of Vermont.

Mountain Building

Mountains form in several ways, some involving little or no deformation. For example, differential weathering and erosion have yielded high areas with adjacent lowlands in the southwestern United States, but these erosional remnants are flat topped or pinnacle-shaped and go by the names *mesa* and *butte*, and most are less than 300 m high (see Chapter 15). *Block-faulting* is another way that mountains form, but this is caused by deformation of the crust. It involves movement on normal faults so that one or more blocks are elevated relative to adjacent blocks (Figure 10.18). A classic example is the Basin and Range Province, which is centered on Nevada, but extends into adjacent areas (see Chapter 23). Differential movement on faults has produced uplifted blocks called *horsts* and down-dropped blocks called *grabens*. Erosion of the horsts has yielded mountainous topography.

Volcanic outpourings form chains of volcanic mountains such as the Hawaiian Islands, where a plate moves over a hot spot (see Figure 2.22). Some mountains such as the Cascade Range of the Pacific Northwest are made up almost entirely of volcanic rocks (Figure 10.19), and the mid-ocean ridges are also mountains (see Figure 2.11). However, most mountains on land are composed of all rocks types and show clear evidence of deformation by compression.

Plate Tectonics and Mountain Building

Geologists define the term **orogeny** as an episode of mountain building during which intense deformation takes place, generally accompanied by metamorphism, the emplacement of plutons, especially batholiths, and thickening of Earth's crust. The processes responsible for an orogeny are still not fully understood, but it is known that mountain building is related to plate movements.

Any theory that accounts for mountain building must adequately explain the characteristics of mountain ranges, such as their geometry and location; they tend to be long and narrow and at or near plate margins. Mountains also show intense deformation, especially compression-induced overturned and recumbent folds, as well as reverse and thrust faults. Furthermore, granitic plutons and regional metamorphism characterize the interiors or cores of mountain ranges. Another feature is sedimentary rocks now far above

Figure 10.15 Faults Fractures along which movement has occurred parallel to the fracture surface are called faults.

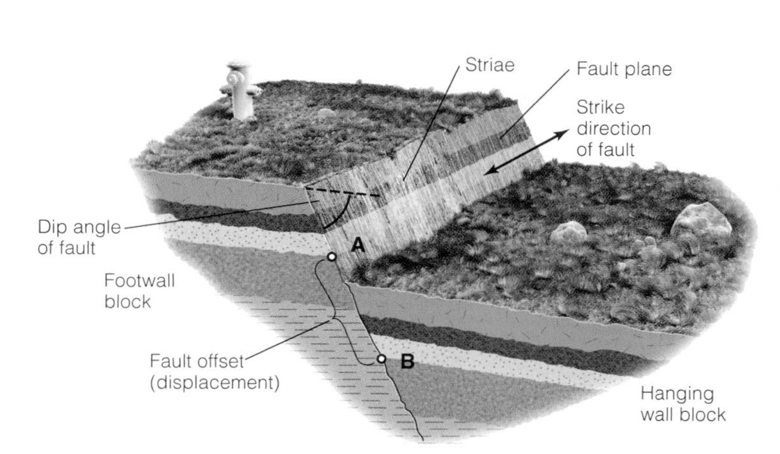

a Terms used to describe the orientation of a fault plane. Striae are scratch marks that form when one block slides past another. You can measure offset or displacement on a fault wherever the truncated end of one feature (point A) can be related to its equivalent across the fault (B).

b A polished, scratched fault plane and fault scarp near Klamath Falls, Oregon.

c Fault breccia, the zone of rubble along a fault in the Bighorn Mountains, Wyoming. The arrows show the direction of movement along the fault.

sea level that were deposited in shallow and deep marine environments.

Deformation and associated activities at convergent plate boundaries are certainly important processes in mountain building. They account for a mountain system's location and geometry, as well as complex geologic structures, plutons, and metamophism. Yet, the present-day topographic expression of mountains is also related to several surface processes, such as mass wasting (gravity-driven processes including landslides), glaciers, and running water. In other words, erosion also plays an important role in the evolution of mountains.

Most of Earth's geologically recent and present-day orogenies are found in two major zones or belts: the *Alpine–Himalayan orogenic belt* and the *circum-Pacific orogenic belt* (see Chapter 23). In fact, we can explain most of Earth's past and present orogenies in terms of the geologic activity at convergent plate boundaries.

Orogenies at Oceanic–Oceanic Plate Boundaries

Deformation, igneous activity, and the origin of a volcanic island arc characterize orogenies that take place where oceanic lithosphere is subducted beneath ocean lithosphere. Sediments derived from the island arc are deposited in an adjacent

Figure 10.16 Types of Faults

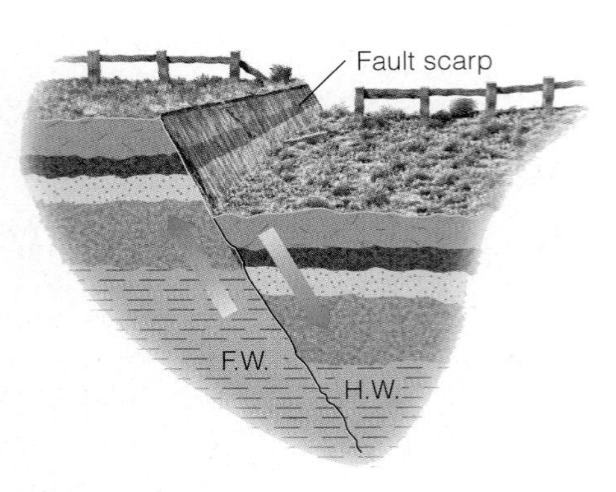

a Normal fault—hanging wall block (HW) moves down relative to the footwall block (FW).

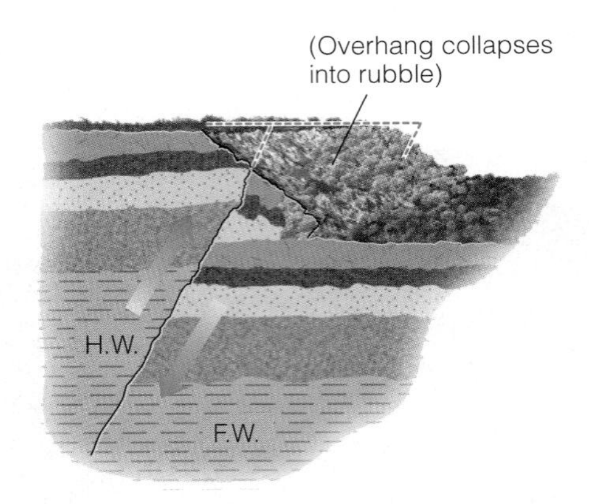

b Reverse fault—hanging wall block moves up relative to the footwall block.

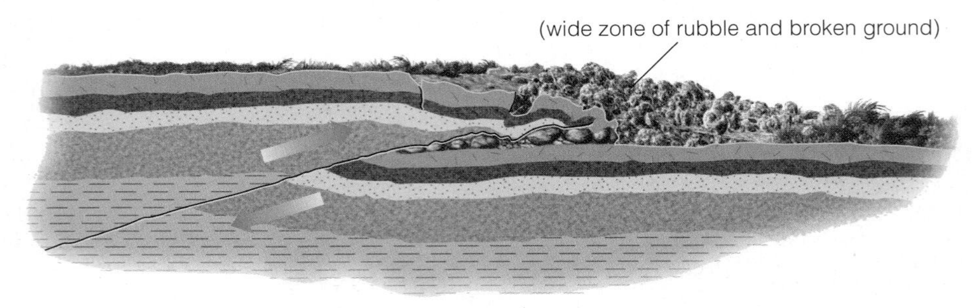

c A thrust is a type of reverse fault with a fault plane dipping at less than 45 degrees.

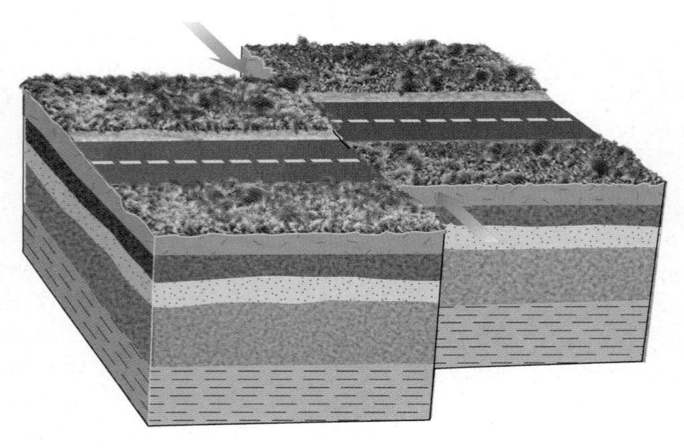

d Left-lateral strike-slip fault.

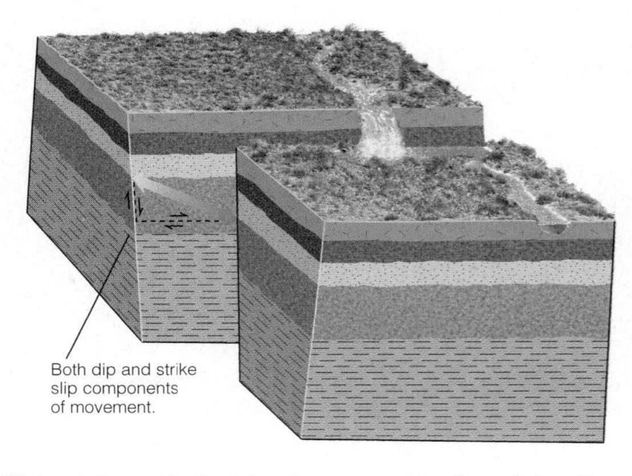

e An oblique-slip fault involves a combination of dip-slip and strike-slip movements.

oceanic trench, and then deformed and scraped off against the landward side of the trench (Figure 10.20). These deformed sediments are part of a subduction complex, or an *accretionary wedge*, of intricately folded rocks cut by numerous thrust faults, resulting from compression. In addition, orogenies in this setting are characterized by low-temperature, high-pressure metamorphism of the blueschist facies (see Figure 7.19).

Deformation caused largely by the emplacement of plutons also takes place in the island arc system where many rocks show evidence of high-temperature, low-pressure metamorphism. The overall effect of an island arc orogeny is the origin of two more or less parallel orogenic belts consisting of a landward volcanic island arc underlain by batholiths and a seaward belt of deformed trench rocks (Figure 10.20). The Japanese Islands are a good example.

In the area between an island arc and its nearby continent, the back-arc basin, volcanic rocks and sediments derived from the island arc and the adjacent continent are also deformed as the plates continue to converge. The sediments are intensely folded and displaced toward the continent along low-angle thrust faults. Eventually, the entire island arc complex is fused to the edge of the continent, and the back-arc basin sediments are thrust onto the continent, forming a thick stack of thrust sheets (Figure 10.20).

James S. Monroe

Figure 10.17 The Teton Range in Wyoming The Teton Range is one of many mountain ranges in the Rocky Mountain system. It began forming about 10 million years ago as uplift took place on normal faults that parallel the range front. As uplift of the Teton Range proceeded, it was eroded by running water, glaciers, and gravity-driven processes, giving it its rugged aspect.

Orogenies at Oceanic–Continental Plate Boundaries The Andes of South America are perhaps the best example of continuing orogeny at an oceanic–continental plate boundary. Among the ranges of the Andes are the highest mountain peaks in the Americas and many active volcanoes. Furthermore, the west coast of South America is an extremely active segment of the circum-Pacific earthquake belt, and one of Earth's great oceanic trench systems, the Peru–Chile Trench, lies just off the coast.

Prior to 200 million years ago, the western margin of South America was a passive continental margin where sediments accumulated much as they do now along the East Coast of North America. However, when Pangaea split apart along what is now the Mid-Atlantic Ridge, the South American plate moved westward. As a consequence, the oceanic lithosphere west of South America began subducting beneath the continent (Figure 10.21). Subduction resulted in partial melting of the descending plate, which produced the andesitic volcanic arc of composite volcanoes, and the west coast became an active continental margin. Felsic magmas, mostly of granitic composition, were emplaced as large plutons beneath the arc (Figure 10.21).

As a result of the events just described, the Andes Mountains consist of a central core of granitic rocks capped by andesitic volcanoes. To the west of this central core along the coast are the deformed rocks of the accretionary wedge. And to the east of the central core are intensely folded sedimentary rocks that were thrust eastward onto the continent (Figure 10.21). Present-day subduction, volcanism, and seismicity along South America's west coast indicate that the Andes Mountains are still forming.

Orogenies at Continental–Continental Plate Boundaries The best example of an orogeny along a continental–continental plate boundary is the Himalayas of Asia. The Himalayas began forming when India collided with Asia about 40 to 50 million years ago. Prior to that time, India was far south of Asia and separated from it by an ocean basin (Figure 10.22a). As the Indian plate moved northward, a subduction zone formed along the southern margin of Asia where oceanic lithosphere was consumed. Partial melting generated magma, which rose to form a volcanic arc, and large granite plutons were emplaced into what is now Tibet. At this stage, the activity along Asia's southern margin was similar to what is now occurring along the west coast of South America.

The ocean separating India from Asia continued to close and India eventually collided with Asia (Figure 10.22a). As a result, two continental plates became welded, or sutured, together. Thus, the Himalayas are now within a continent rather than along a continental margin. The exact time of India's collision with Asia is uncertain, but between 40 and 50 million years ago, India's rate of northward drift decreased abruptly from about 15–20 cm per year to about 5 cm per year. Because continental lithosphere is not dense enough to be subducted, this decrease seems to mark the time of collision and India's resistance to subduction. Consequently, the leading margin of India was thrust beneath Asia, causing crustal thickening, thrusting, and uplift. Sedimentary rocks that had been deposited in the sea south of Asia were thrust northward, and two major thrust faults carried rocks of Asian origin onto the Indian plate. Rocks deposited in the shallow seas along India's northern margin now form the higher parts of the Himalayas (Figure 10.22b). Since its collision with Asia, India has been thrust horizontally about 2000 km beneath Asia and now moves north at several centimeters per year.

Other mountain systems also formed as a result of collisions between two continental plates. The Urals in Russia and the Appalachians of North America formed by such collisions. In addition, the Arabian plate is now colliding with Asia along the Zagros Mountains of Iran.

Types of Faults

Faults are very common geologic structures. They are fractures along which movement takes place parallel to the fracture surface. A block of rock adjacent to a fault may move up or down a fault plane—that is, up or down the dip of the fault. These are thus called dip-slip faults. On the other hand, movement may take place along a fault's strike, giving rise to strike-slip faults.

Movement on faults and the release of stored energy are responsible for earthquakes (see Chapter 8). Most faults are found at the three major types of plate boundaries: convergent, divergent, and transform.

James S. Monroe

◀ **1.** Two small normal faults cutting through layers of volcanic ash in Oregon. ▼ **2.** Notice that the sandstone layers to the right of the hammer are cut by a reverse fault. Compare the sense of movement of the hanging wall blocks in these two images.

Martin Miller/Visuals Unlimited

David J. Matty

Sue Monroe

◀ **3.** ◀ **4.** Can you identify the type of faults shown in these two images?

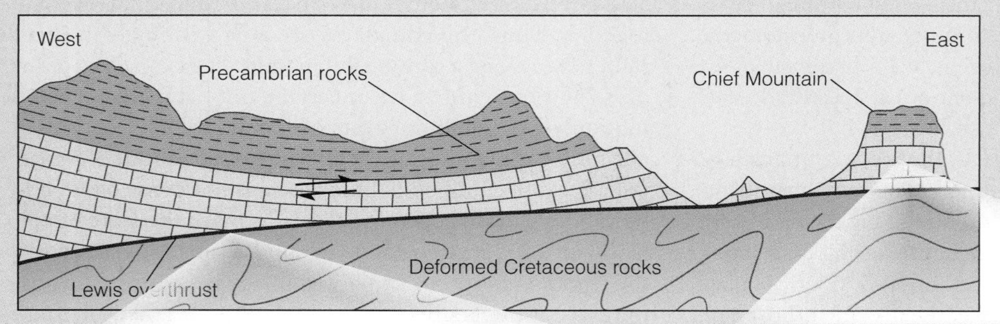

◀ **5.** Diagrammatic view of the Lewis overthrust fault (a low-angle thrust fault) in Glacier National Park, Montana. Ancient Precambrian-age rocks now rest on deformed Cretaceous-age sedimentary rocks.

James S. Monroe

▲ **6.** View from Marias Pass reveals the fault as a light-colored line on the mountainside.

James S. Monroe

▲ **7.** Erosion has isolated Chief Mountain from the rest of the slab of overthrust rock.

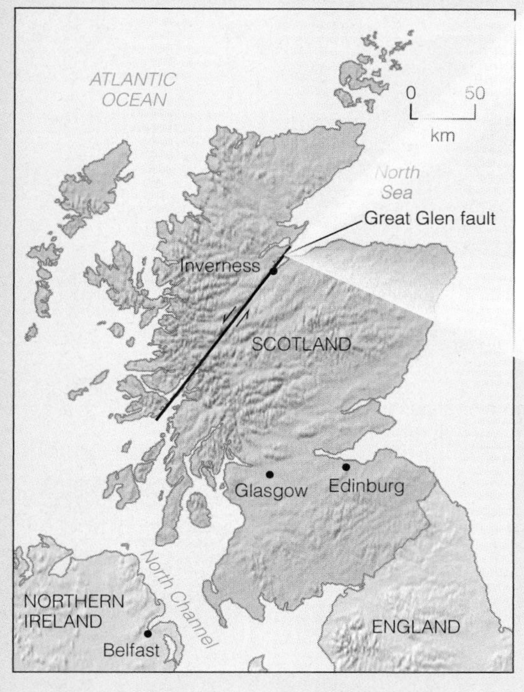

▲ **9.** View to the southwest along Loch Ness, Scotland, lying in the Great Glen fault zone, which, at this point, is more than 1.5 km wide.

◀ **8.** Map showing the location of the Great Glen fault, a left-lateral strike-slip fault that cuts across Scotland.

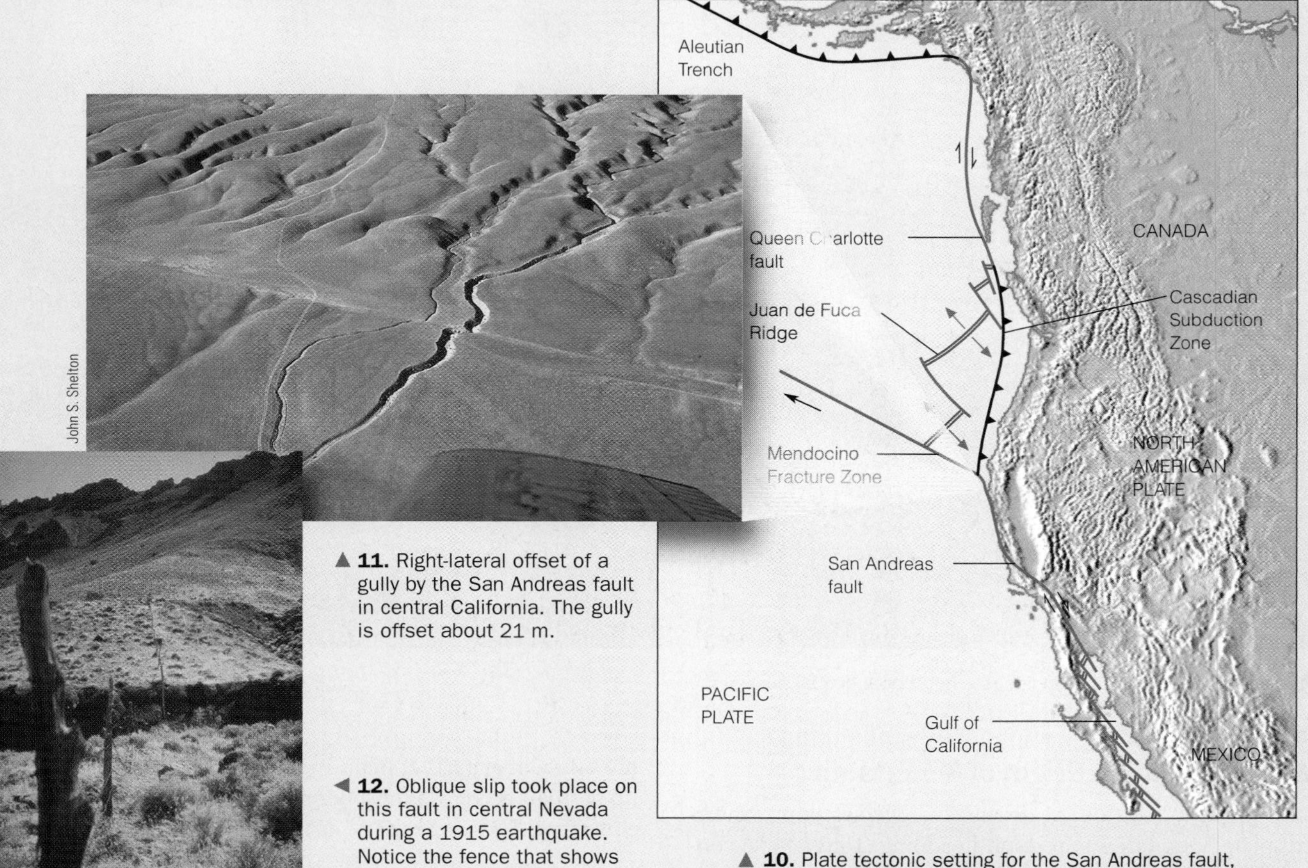

▲ **11.** Right-lateral offset of a gully by the San Andreas fault in central California. The gully is offset about 21 m.

◀ **12.** Oblique slip took place on this fault in central Nevada during a 1915 earthquake. Notice the fence that shows right-lateral displacement and dip-slip displacement.

▲ **10.** Plate tectonic setting for the San Andreas fault, a strike-slip fault. Remember that in plate tectonics terminology, this is a transform fault.

Figure 10.18 Origins of Horsts and Grabens

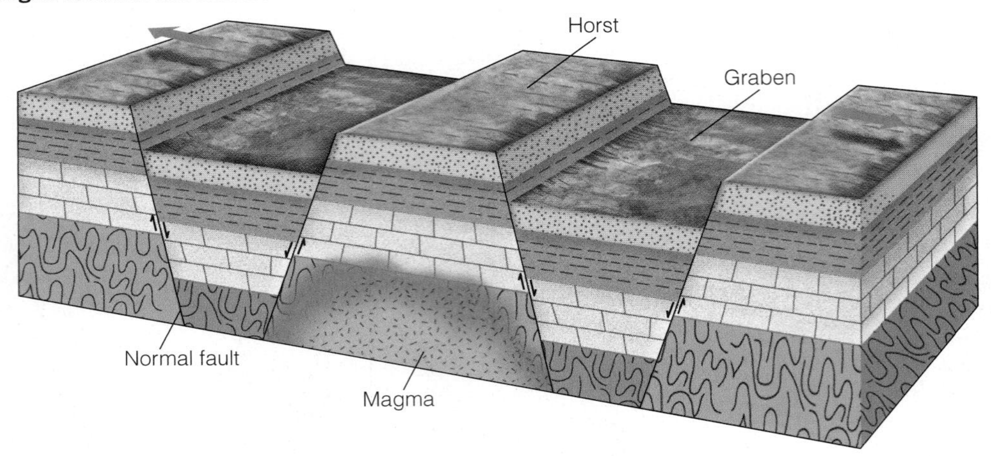

a Block-faulting and the origin of horsts and grabens. Many of the mountain ranges in the Basin and Range Province of the western United States and northern Mexico formed in this manner.

b The Stillwater Range in Nevada is a horst bounded by normal faults.

Terranes and the Origin of Mountains

In the preceding section, we discussed orogenies along convergent plate boundaries that result in adding material to a continent, a process termed **continental accretion**. Much of the material added to continental margins is eroded older continental crust, but some plutonic and volcanic rocks are new additions. During the 1970s and 1980s, however, geologists discovered that parts of many mountain systems are also made up of small, accreted lithospheric blocks that clearly originated elsewhere. These **terranes**,* as they are called, are fragments

*Some geologists prefer the terms *suspect terrane*, *exotic terrane*, or *displaced terrane*. Notice also the spelling of *terrane* as opposed to the more familiar *terrain*, the latter a geographic term indicating a particular area of land.

Figure 10.19 The Cascade Range The mountains of the Cascade Range are made up of volcanic rocks, many of which formed during the last few million years. Lassen Peak in California erupted from 1914 to 1917 and Mount St. Helens in Washington erupted in 1980 and 2004. This view in Oregon shows three of the volcanoes in the range, which stretches from northern California northward into British Columbia, Canada.

of seamounts, island arcs, and small pieces of continents that were carried on oceanic plates that collided with continental plates, thus adding them to the continental margins. We discuss this topic of terranes and their importance in mountain building more fully in Chapter 22.

EARTH'S CONTINENTAL CRUST

Continental crust stands higher than oceanic crust, but why should this be so? Also, why do mountains stand higher than surrounding areas? To answer these questions, we must examine continental crust in more detail. You already know that continental crust is granitic with an overall density of 2.7 g/cm^3, whereas oceanic crust is made up of basalt and gabbro and its density is 3.0 g/cm^3. In most places, continental crust is about 35 km thick except beneath mountain systems, where it is much thicker. The oceanic crust, in contrast, varies from only 5 to 10 km thick. So, these differences, as well as variations in crustal thickness, account for why mountains stand high and why continents stand higher than ocean basins (see the following section).

Floating Continents?

How is it possible for a solid (continental crust) to float in another solid (the mantle)? Floating brings to mind a ship at sea or a block of wood in water; however, continents do not behave in this manner. Or do they? Actually, they do float, in a manner of speaking, but a complete answer requires more discussion on the concept of gravity and on the principle of isostasy.

Isaac Newton formulated the law of universal gravitation in which the force of gravity (F) between two masses (m_1 and m_2) is directly proportional to the products of their masses and inversely proportional to the square of the distance between their centers of mass. This means that an attractive force exists between any two objects, and the magnitude of that force varies depending on the masses of the objects and the distance between their centers. We generally refer to the gravitational force between an object and Earth as its *weight.*

Gravitational attraction would be the same everywhere on the surface if Earth were perfectly spherical, homogeneous throughout, and not rotating. But because Earth varies in all of these aspects, the force of gravity varies from area to area. Geologists use a *gravimeter* to measure gravitational attraction and to detect **gravity anomalies**—that is, departures from the expected force of gravity (Figure 10.23). Gravity anomalies might be *positive*, meaning that an excess of mass is present at some location, or *negative*, when a mass deficiency exists. For instance, a buried iron ore deposit would yield a positive gravity anomaly because of the greater density of these rocks.

Principle of Isostasy

Geologists realized long ago that mountains are not simply piles of materials of Earth's surface, and in 1865 George Airy proposed that, in addition to projecting high above sea level, mountains also project far below the surface and thus have a low density root (Figure 10.23). In effect, he was saying that the thicker crust of mountains float on denser rock at depth, with their excess mass above sea level compensated for by low-density material at depth. Another explanation was proposed by J. H. Pratt, who thought that mountains were high because they were composed of rocks of lower density than those in adjacent regions.

Actually, both Airy and Pratt were correct, because there are places where density or thickness accounts for differences in the level of the crust. For example, Pratt's hyphothesis was confirmed because (1) continental crust is thicker and less dense than oceanic crust and thus stands high, and (2) the mid-oceanic ridges stand high because the crust there is hot and less dense than cooler crust elsewhere. Airy, on the other hand, was correct in his claim that the crust, continental or oceanic, "floats" on the mantle, which has a density of 3.3 g/cm^3 in its upper part. However, we have not yet explained what we mean by one solid floating in another solid.

This phenomenon of Earth's crust floating in the denser mantle is now known as the **principle of isostasy,** which is easy to understand by an analogy to an iceberg

Figure 10.20 Orogeny and the Origin of a Volcanic Island Arc at an Oceanic–Oceanic Plate Boundary

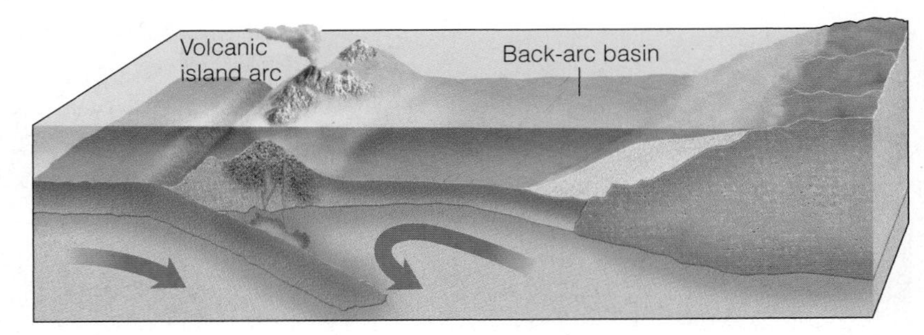

a Subduction of an oceanic plate and the origin of a volcanic island arc and a back-arc basin.

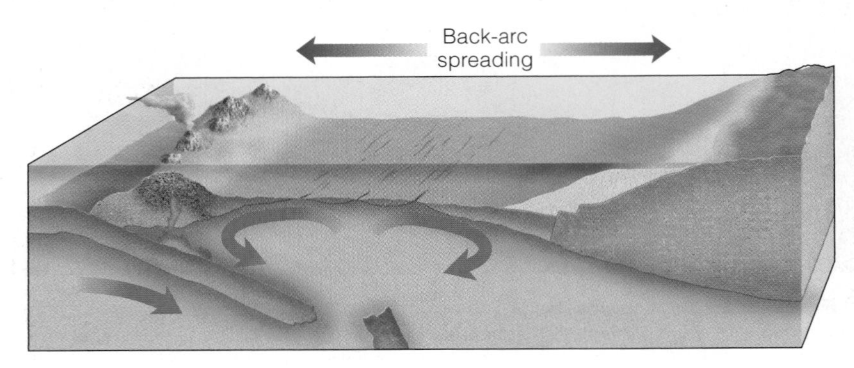

b Continued subduction and back-arc spreading.

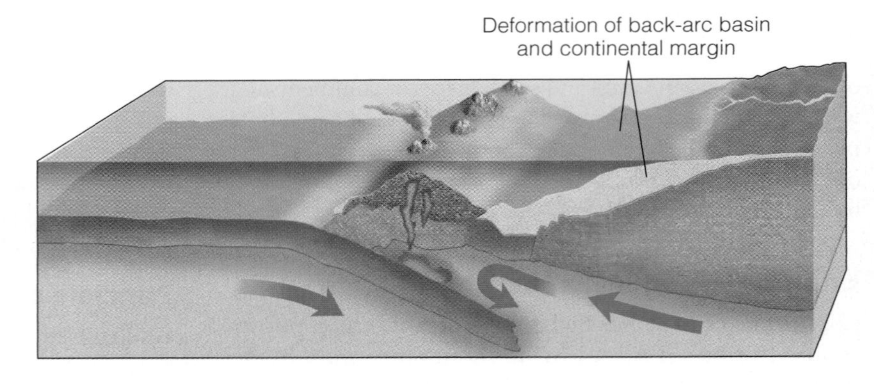

c Back-arc basin begins to close, resulting in deformation of back-arc basin and continental margin deposits.

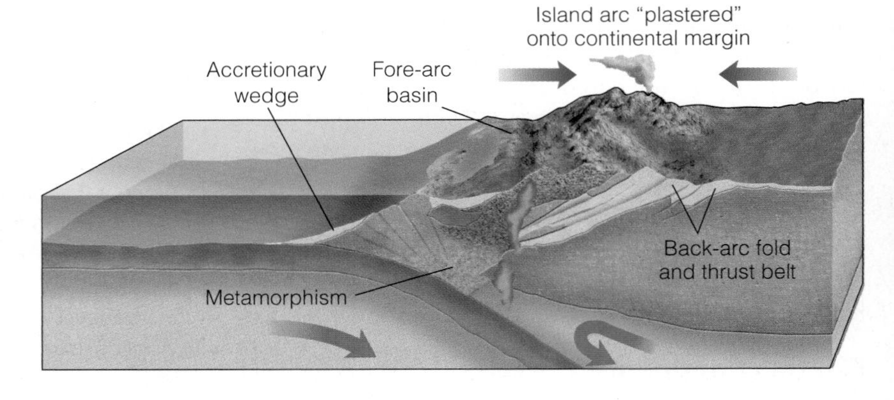

d Thrusting of back-arc sediments onto the adjacent continent and suturing of the island arc to the continent.

(Figure 10.24). Ice is slightly less dense than water, so it floats. According to Archimedes' principle of buoyancy, an iceberg sinks in water until it displaces a volume of water whose weight is equal to that of the ice. When the iceberg has sunk to an equilibrium position, only about 10% of its volume is above water level. If some of the ice above water level should melt, the iceberg rises to maintain equilibrium with the same proportion of ice above and below the water.

Earth's crust is similar to the iceberg in that it sinks into the mantle to its equilibrium level. Where the crust is thickest, as beneath mountains, it sinks farther down into the mantle and it also rises higher above the surface. And because continental crust is thicker and less dense than oceanic crust, it stands higher than the ocean basins. Remember, the mantle is hot, yet solid, and under tremendous pressure, so it behaves in a fluid-like manner.

Figure 10.21 The Andes Mountains in South America

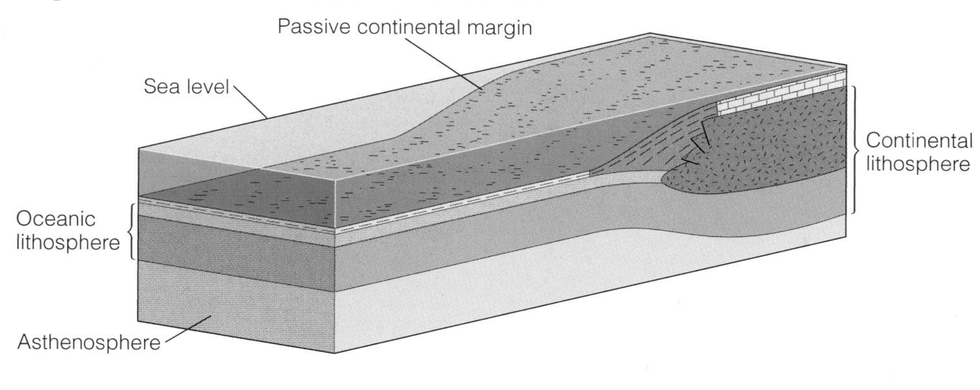

a Prior to 200 million years ago, the western margin of South America was a passive continental margin.

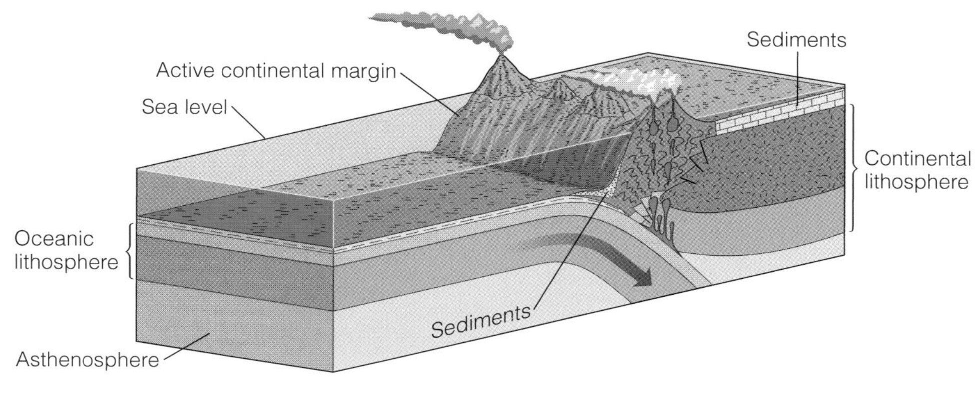

b Orogeny began when this area became an active continental margin as the South American plate moved to the west and collided with oceanic lithosphere.

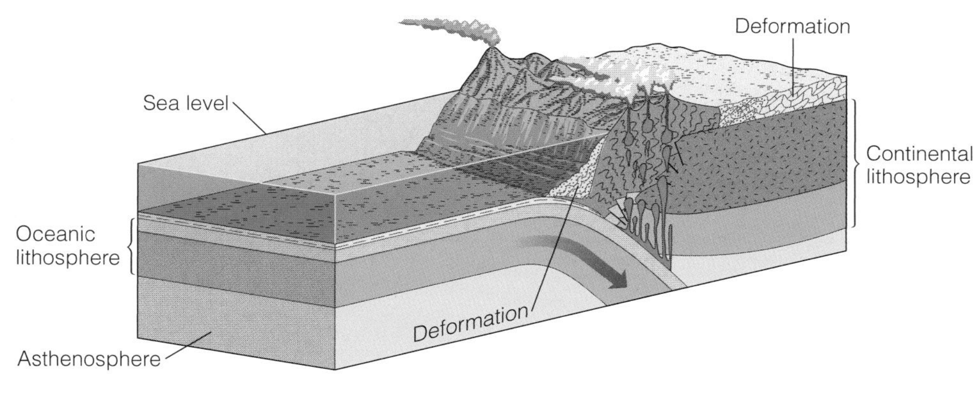

c Continued deformation, plutonism, and volcanism.

Tibor Bognar/Corbis

d View of the Andes in Chile.

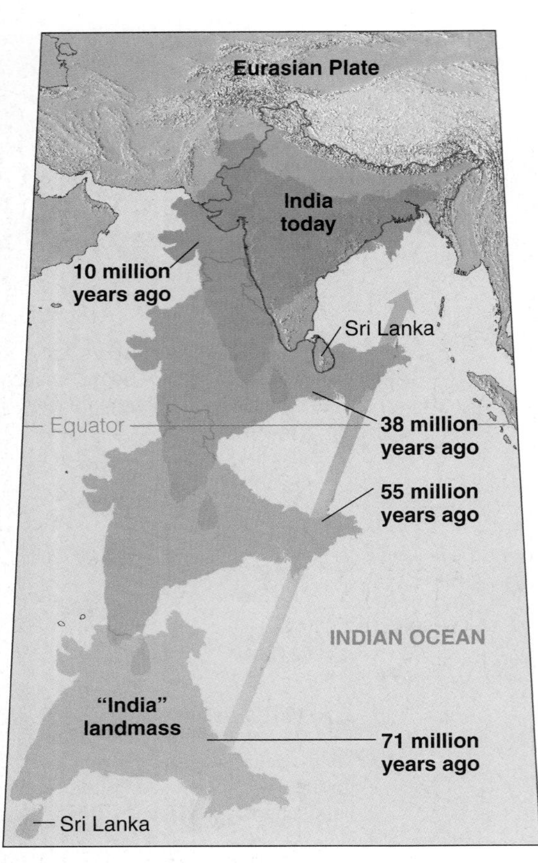

a During its long journey north, India moved 15 to 20 cm per year; however, beginning 40 to 50 million years ago, its rate of movement decreased markedly as it collided with the Eurasian plate.

b Aerial view of the Himalaya Mountains in Tibet.

Figure 10.22 Orogeny at a Continental–Continental Plate Boundary and the Origin of the Himalayas of Asia

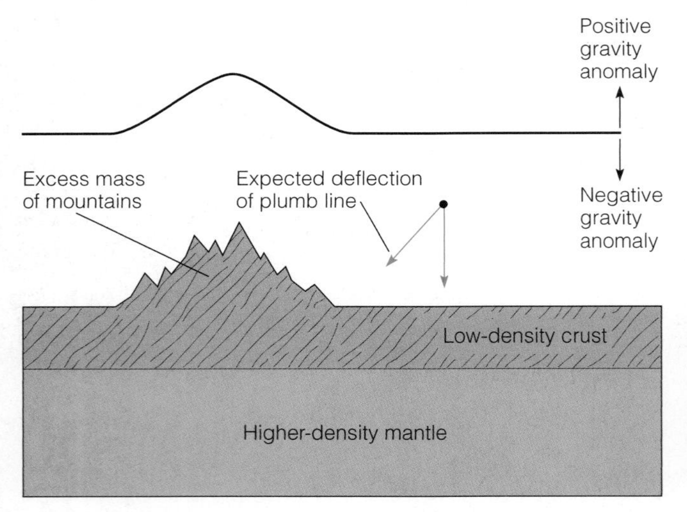

a A plumb line (a cord with a suspended weight) is normally vertical, pointing to Earth's center of gravity. Near a mountain range, the plumb line should be deflected as shown if the mountains are simply thicker, low-density material resting on denser material, and a gravity survey across the mountains would indicate a positive gravity anomaly.

b The actual deflection of the plumb line during the survey in India was less than expected. It was explained by postulating that the Himalayas have a low-density root. A gravity survey, in this case, would show no anomaly because the mass of the mountains above the surface is compensated for at depth by low-density material displacing denser material.

Figure 10.23 Gravity Anomalies

Some of you might realize that crust floating on the mantle raises an apparent contradiction. In Chapter 8, we said that the mantle is a solid because it transmits S-waves, which do not move through a fluid. But according to the principle of isostasy, the mantle behaves as a fluid. When considered in terms of the brief time required for S-waves

Figure 10.24 The Principle of Isostasy An iceberg sinks to an equilibrium level with 10% of its mass above water level. The larger iceberg sinks farther below and rises higher above the water surface than the smaller one. If some of the ice above water level melts, the icebergs will rise to maintain the same proportions of ice above and below water level.

What Would You Do?

As a member of a planning commission, you are charged with developing zoning regulations and building codes for an area with known active faults, steep hills, and deep soils. A number of contractors, as well as developers and citizens in your community, are demanding action because they want to begin several badly needed housing developments. How might geologic maps and an appreciation of geologic structures influence you in this endeavor? Do you think, considering the probable economic gains for your community, that the regulations you draft should be rather lenient or very strict? If the latter, how would you explain why you favored regulations that would involve additional cost for houses?

to pass through it, the mantle is indeed solid. But when subjected to stress over long periods, it yields by flowage; thus, at this time scale, it can be regarded as a viscous fluid.

Isostatic Rebound

What happens when a ship is loaded with cargo and then later unloaded? Of course, it first sinks lower in the water and then rises, but it always finds its equilibrium position. Earth's crust responds similarly to loading and unloading, but much more slowly. For example, if the crust is loaded, as when widespread glaciers accumulate, the crust sinks farther into the mantle to maintain equilibrium. The crust behaves similarly in areas where huge quantities of sediment accumulate.

If loading by glacial ice or sediment depresses Earth's crust farther into the mantle, it follows that when vast glaciers melt or where deep erosion takes place, the crust should rise back up to its equilibrium level. And in fact it does. This phenomenon, known as **isostatic rebound**, is taking place in Scandinavia, which was covered by a thick ice sheet until about 10,000 years ago; it is now rebounding at about 1 m per century. In fact, coastal cities in Scandinavia have rebounded rapidly enough that docks constructed several centuries ago are now far from shore. Isostatic rebound has also occurred in eastern Canada where the crust has risen as much as 100 m in the last 6000 years.

Figure 10.25 shows the response of Earth's continental crust to loading and unloading as mountains form and evolve. Recall that during an orogeny, emplacement of plutons, metamorphism, and general thickening of the crust accompany deformation. However, as the mountains erode, isostatic rebound takes place and the mountains rise, whereas adjacent areas of sedimentation subside (Figure 10.25). If continued long enough, the mountains will disappear and then can be detected only by the plutons and metamorphic rocks that show their former existence.

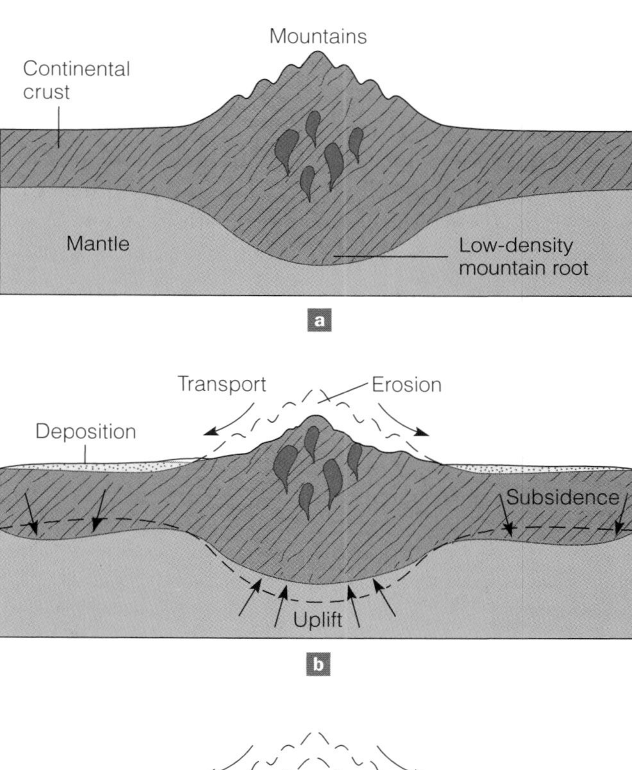

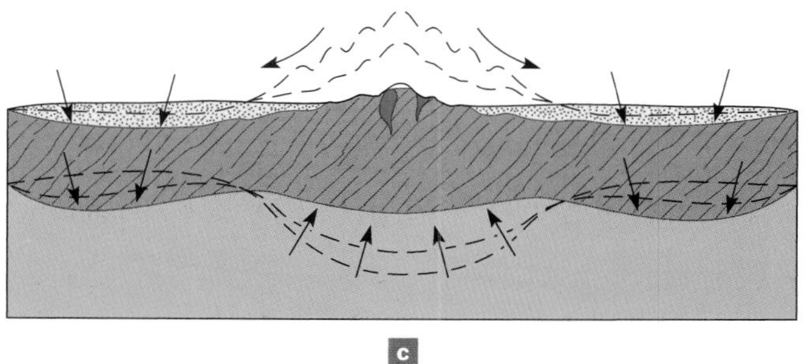

Figure 10.25 Isostatic Rebound A diagrammatic representation showing the isostatic response of the crust to erosion (unloading) and widespread deposition (loading).

Geo-Recap

Chapter Summary

- Folded and fractured rocks have been deformed or strained by applied stresses.
- Stress is compression, tension, or shear. Elastic strain is not permanent, but plastic strain and fracture are, meaning that rocks do not return to their original shape or volume when the deforming forces are removed.
- Strike and dip are used to define the orientation of deformed rock layers. This same concept applies to other planar features such as fault planes.
- Anticlines and synclines are up- and down-arched folds, respectively. They are identified by strike and dip of the folded rocks and the relative ages of rocks in these folds.
- Domes and basins are the circular to oval equivalents of anticlines and synclines, but they are commonly much larger structures.
- The two structures that result from fracture are joints and faults. Joints may open up, but they show no movement parallel with the fracture surface, whereas faults do show movement parallel with the fracture surface.
- Joints are very common and form in response to compression, tension, and shear.
- On dip-slip faults, all movement is up or down the dip of the fault. If the hanging wall moves relatively down it is a normal fault, but if the hanging wall moves up it is a reverse fault. Normal faults result from tension; reverse faults result from compression.
- In strike-slip faults, all movement is along the strike of the fault. These faults are either right-lateral or left-lateral, depending on the apparent direction of offset of one block relative to the other.
- Oblique-slip faults show components of both dip-slip and strike-slip movement.
- A variety of processes account for the origin of mountains. Some involve little or no deformation, but the large mountain systems on the continents resulted from deformation at convergent plate boundaries.
- Subduction of an oceanic plate beneath another oceanic plate or beneath a continental plate causes an orogeny. At an oceanic-oceanic boundary a volcanic island arc intruded by plutons forms, whereas at an oceanic-continental boundary a volcanic arc forms on the continental plate. In both cases deformation and metamorphism occur.
- Some mountain systems are within continents far from a present-day plate boundary. These mountains formed when two continental plates collided and became sutured.
- Geologists now realize that orogenies also involve collisions of terranes with continents.
- Continental crust is characterized as granitic, and it is much thicker and less dense than oceanic crust that is composed of basalt and gabbro.
- According to the principle of isostasy, Earth's crust floats in equilibrium in the denser mantle below. Continental crust stands higher than oceanic crust because it is thicker and less dense.

Important Terms

anticline (p. 249)
basin (p. 251)
compression (p. 247)
continental accretion (p. 264)
deformation (p. 246)
dip (p. 248)
dip-slip fault (p. 255)
dome (p. 251)
elastic strain (p. 247)
fault (p. 253)
fault plane (p. 253)
fold (p. 249)
footwall block (p. 253)
fracture (p. 248)
geologic structure (p. 248)
gravity anomaly (p. 265)
hanging wall block (p. 253)
isostatic rebound (p. 269)
joint (p. 252)
monocline (p. 249)
normal fault (p. 255)
oblique-slip fault (p. 257)
orogeny (p. 258)
plastic strain (p. 248)
principle of isostasy (p. 265)
reverse fault (p. 255)
shear stress (p. 247)
strain (p. 246)
stress (p. 246)
strike (p. 248)
strike-slip fault (p. 256)
syncline (p. 249)
tension (p. 247)
terrane (p. 264)
thrust fault (p. 255)

Review Questions

1. A fault on which the footwall block has moved up relative to the hanging wall block is a ________ fault.
   a. ______ normal;
   b. ______ thrust;
   c. ______ strike-slip;
   d. ______ oblique-slip;
   e. ______ reverse.
2. A time of mountain building during which deformation occurs, accompanied by the emplacement of plutons, metamorphism, and thickening of the crust is a/an
   a. ______ terrane;
   b. ______ monocline;
   c. ______ orogeny;
   d. ______ gravity anomaly;
   e. ______ magnetic reversal.
3. According to the principle of isostasy,
   a. ______ continental crust is denser than oceanic crust;
   b. ______ anticlines and synclines are caused by compression;
   c. ______ the hanging wall block of a fault moves down relative to the footwall block;
   d. ______ the strike of an inclined rock layer is defined by its volume;
   e. ______ the continents are "floating" on the denser mantle.
4. A circular fold with all strata dipping toward the center and the youngest exposed rock layers at the center is a/an
   a. ______ joint;
   b. ______ basin;
   c. ______ horst;
   d. ______ anticline;
   e. ______ monocline.
5. Which one of the following statements is incorrect?
   a. ______ Most orogenies occur at convergent plate boundaries;
   b. ______ Shear stress involves forces acting along the same line, but in opposite directions;
   c. ______ Joint are fractures along which no movement has occurred parallel with the fracture;
   d. ______ The dip of a rock formation is its inclination from horizontal;
   e. ______ Folds and faults are collectively known as geologic structures.
6. A continent–continent collision was responsible for the
   a. ______ Cascade Range;
   b. ______ Basin and Range mountains;
   c. ______ Andes Mountains;
   d. ______ Himalaya Mountains;
   e. ______ Sea of Japan Range.
7. Rocks characterized as ductile
   a. ______ show a great amount of plastic strain;
   b. ______ are the main rocks in terranes;
   c. ______ are found along the crests of anticlines;
   d. ______ fracture easily when in compression;
   e. ______ are found on the hanging wall blocks of faults.
8. The line formed by the intersection of an inclined plane with a horizontal plane is the definition of
   a. ______ stress;
   b. ______ brittle behavior;
   c. ______ strike;
   d. ______ uplift;
   e. ______ jointing.
9. The fault in Figure 10.16e shows both __________ and __________ faulting.
   a. ______ thrust/reverse;
   b. ______ normal/left-lateral strike-slip;
   c. ______ left-lateral strike-slip/monocline;
   d. ______ elastic/dome;
   e. ______ overturned/recumbent.
10. A fold in which one limb has been rotated more than 90° from its original position is said to be
    a. ______ left-lateral;
    b. ______ oblique;
    c. ______ synclinal;
    d. ______ overturned;
    e. ______ inverted.
11. What are the similarities and differences between a basin and a syncline?
12. Describe how time, rock type, pressure, and temperature influence rock deformation.
13. Suppose that rocks were displaced 200 km along a strike-slip fault during a period of 5 million years. What was the average rate of movement per year? Is the average likely to represent the actual rate of displacement on this fault? Explain.
14. Draw a surface view of a plunging syncline and an adjacent plunging anticline. Show strike and dip symbols and indicate on your diagram which rock layer is oldest and which is youngest.
15. Is there any connection between the principle of isostasy and mountain building? If so, what?
16. How would you explain stress and strain to someone unfamiliar with the concepts?
17. What is meant by the elastic limit of rocks, and what happens when rocks are strained beyond their elastic limit?
18. How does a dip-slip fault differ from a strike-slip fault?
19. What is the concept of isostatic rebound? Give an example of how it may take place.
20. What kinds of evidence would indicate that mountain building once took place in an area where mountains are no longer present?

CHAPTER 11

Mass Wasting

OUTLINE

OBJECTIVES

At the end of this chapter, you will have learned that

- It is important to understand the different types of mass wasting because mass wasting affects us all and causes significant destruction.
- Factors such as slope angle, weathering and climate, water content, vegetation, and overloading are interrelated, and all contribute to mass wasting.
- Mass movements can be triggered by such factors as overloading, soil saturation, and ground shaking.
- Mass wasting is categorized as either rapid mass movements or slow mass movements.
- The different types of rapid mass movements are rockfalls, slumps, rock slides, mudflows, debris flows, and quick clays; each type has recognizable characteristics.
- The different types of slow mass movements are earthflows, solifluction, and creep; each type has recognizable characteristics.
- People can minimize the effects of mass wasting by conducting geologic investigations of an area and stabilizing slopes to prevent and ameliorate movement.

Residents of Caracas, Venezuela, clean up the debris from massive flooding and mudslides that devastated large areas of the country during December, 1999.

INTRODUCTION

Triggered by relentless torrential rains that began in December 1999, the floods and mudslides that devastated Venezuela were some of the worst ever to strike that country. Although an accurate death toll is impossible to determine, it is estimated that at least 19,000 people were killed, as many as 150,000 were left homeless, 35,000 to 40,000 homes were destroyed or buried by mudslides, and between $10 and $20 billion in damage was done before the rains and slides abated. It is easy to cite the numbers of dead and homeless, but the human side of the disaster was most vividly brought home by a mother who described standing helplessly by and watching her four small children buried alive in the family car as a raging mudslide carried it away.

Mudslides engulfed and buried not only homes, buildings, and roads, but also entire communities. Some areas were covered with as much as 7 m of mud. In addition, flooding and the accompanying mudslides swept away large parts of many of Venezuela's northern coastal communities, leaving huge areas uninhabitable.

This terrible tragedy illustrates how geology affects all of our lives. The underlying causes of the mudslides in Venezuela can be found anywhere in the world. In fact, *landslides* (a general term for mass movements of earth) cause, on average, between 25 and 50 deaths and more than $2 billion in damage annually in the United States. By being able to recognize and understand how landslides occur and what the results may be, we can find ways to reduce hazards and minimize damage in terms of both human life and property damage.

Mass wasting (also called *mass movement*) is defined as the downslope movement of material under the direct influence of gravity. Most types of mass wasting are aided by weathering and usually involve surficial material. The material moves at rates ranging from almost imperceptible, as in the case of creep, to extremely fast, as in a rockfall or slide. Although water can play an important role, the relentless pull of gravity is the major force behind mass wasting.

Mass wasting is an important geologic process that can occur at any time and almost any place. It is thus important to study this phenomenon because it affects all of us, no matter where we live (Table 11.1). Although all major landslides have natural causes, many smaller ones are the result of human activity and could have been prevented or their damage minimized.

FACTORS THAT INFLUENCE MASS WASTING

When the gravitational force acting on a slope exceeds its resisting force, slope failure (mass wasting) occurs. The resisting forces that help maintain slope stability include the slope material's strength and cohesion, the amount of internal friction between grains, and any external support of the slope (◗ Figure 11.1). These factors collectively define a slope's **shear strength.**

Opposing a slope's shear strength is the force of gravity. Gravity operates vertically, but has a component acting parallel to the slope, thereby causing instability (Figure 11.1).

TABLE 11.1 Selected Landslides, Their Cause, and the Number of People Killed

Date	Location	Type	Deaths
218 B.C.	Alps (European)	Avalanche—destroyed Hannibal's army	18,000
1556	China (Hsian)	Landslides—earthquake triggered	1,000,000
1806	Switzerland (Goldau)	Rock slide	457
1903	Canada (Frank, Alberta)	Rock slide	70
1920	China (Kansu)	Landslides—earthquake triggered	~200,000
1941	Peru (Huaraz)	Avalanche and mudflow	7,000
1962	Peru (Mt. Hauscarán)	Ice avalanche and mudflow	~4,000
1963	Italy (Vaiont Dam)	Landslide—subsequent flood	~2,000
1966	United Kingdom (Aberfan, South Wales)	Debris flow—collapse of mining-waste tip	144
1970	Peru (Mt. Hauscarán)	Rockfall and debris avalanche— earthquake triggered	25,000
1981	Indonesia (West Irian)	Landslide—earthquake triggered	261
1987	El Salvador (San Salvador)	Landslide	1,000
1989	Tadzhikistan	Mudflow—earthquake triggered	274
1994	Colombia (Paez River Valley)	Avalanche—earthquake triggered	>300
1999	Venezuela	Mudflow	>19,000
2005	U.S.A. (Southern California)	Mudflow	10

Source: Data from J. Whittow, *Disasters: The Anatomy of Environmental Hazards* (Athens: University of Georgia Press, 1979); *Geotimes*; *Earth*; and U.S.G.S.

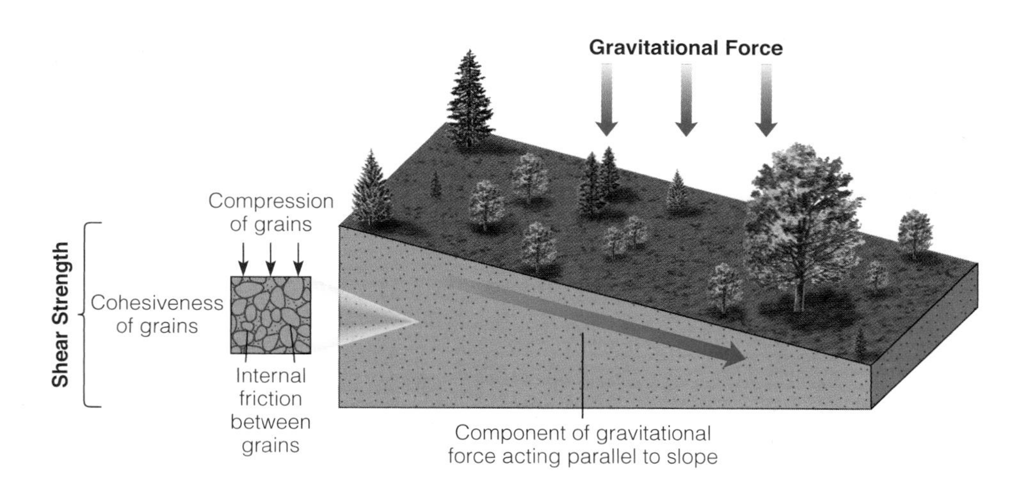

Figure 11.1 Slope Shear Strength A slope's shear strength depends on the slope material's strength and cohesion, the amount of internal friction between grains, and any external support of the slope. These factors promote slope stability. The force of gravity operates vertically, but has a component acting parallel to the slope. When this force, which promotes instability, exceeds a slope's shear strength, slope failure occurs.

The steeper a slope's angle, the greater the component of force acting parallel to the slope, and the greater the chance for mass wasting. The steepest angle that a slope can maintain without collapsing is its *angle of repose.* At this angle, the shear strength of the slope's material exactly counterbalances the force of gravity. For unconsolidated material, the angle of repose normally ranges from 25 to 40 degrees. Slopes steeper than 40 degrees usually consist of unweathered solid rock.

All slopes are in a state of *dynamic equilibrium*, which means that they are constantly adjusting to new conditions. Although we tend to view mass wasting as a disruptive and usually destructive event, it is one of the ways that a slope adjusts to new conditions. Whenever a building or road is constructed on a hillside, the equilibrium of that slope is affected. The slope must then adjust, perhaps by mass wasting, to this new set of conditions.

Many factors can cause mass wasting: a change in slope angle, weakening of material by weathering, increased water content, changes in the vegetation cover, and overloading. Although most of these are interrelated, we will examine them separately for ease of discussion, but we will also show how they individually and collectively affect a slope's equilibrium.

Slope Angle

Slope angle is probably the major cause of mass wasting. Generally speaking, the steeper the slope, the less stable it is. Therefore, steep slopes are more likely to experience mass wasting than gentle ones.

A number of processes can oversteepen a slope. One of the most common is undercutting by stream or wave action (Figure 11.2). This process removes the slope's base, increases the slope angle, and thereby increases the gravitational force acting parallel to the slope. Wave action, especially during storms, often results in mass movements along the shores of oceans or large lakes (Figure 11.3).

Excavations for road cuts and hillside building sites are another major cause of slope failure (Figure 11.4). Grading the slope too steeply or cutting into its side increases the stress in the rock or soil until it is no longer strong enough to remain at the steeper angle, and mass movement ensues.

Such action is analogous to undercutting by streams or waves and has the same result, thus explaining why so many mountain roads are plagued by frequent mass movements.

Weathering and Climate

Mass wasting is more likely to occur in loose or poorly consolidated slope material than in bedrock. As soon as rock is exposed at Earth's surface, weathering begins to disintegrate and decompose it, reducing its shear strength and increasing its susceptibility to mass wasting. The deeper the weathering zone extends, the greater the likelihood of some type of mass movement.

Recall that some rocks are more susceptible to weathering than others and that climate plays an important role in the rate and type of weathering. In the tropics, where temperatures are high and considerable rain falls, the effects of weathering extend to depths of several tens of meters, and mass movements most commonly occur in the deep weathering zone. In arid and semiarid regions, the weathering zone is usually considerably shallower. Nevertheless, intense, localized cloudbursts can drop large quantities of water on an area in a short time. With little vegetation to absorb this water, runoff is rapid and frequently results in mudflows.

Water Content

The amount of water in rock or soil influences slope stability. Large quantities of water from melting snow or heavy rainfall greatly increase the likelihood of slope failure. The additional weight that water adds to a slope can be enough to cause mass movement. Furthermore, water percolating through a slope's material helps to decrease friction between grains, contributing to a loss of cohesion. For example, slopes composed of dry clay are usually quite stable, but

Figure 11.2 Undercutting a Slope's Base by Stream Erosion

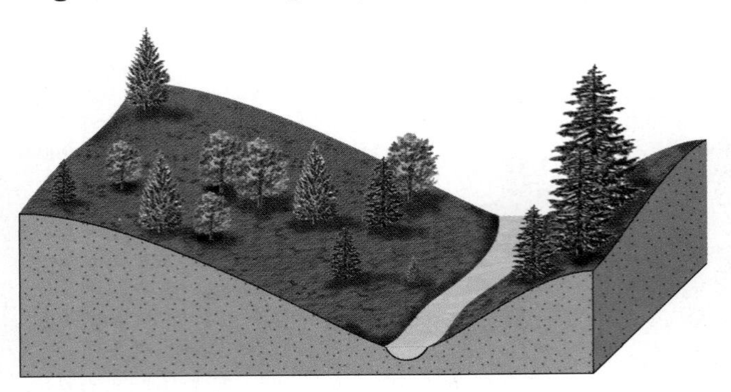

a Undercutting by stream erosion removes a slope's base,

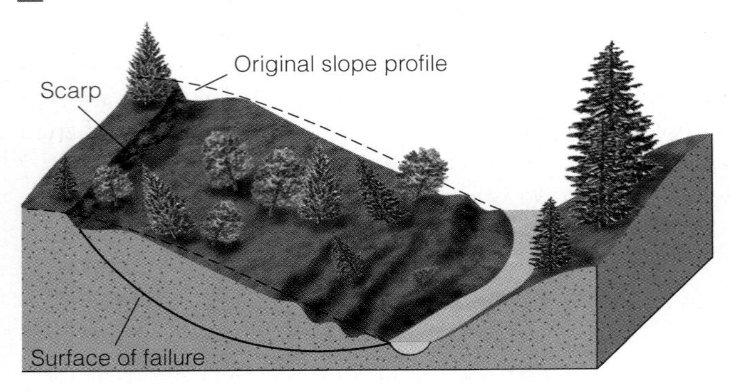

b which increases the slope angle, and can lead to slope failure.

c Undercutting by stream erosion caused slumping along this stream near Weidman, Michigan. Notice the scarp, which is the exposed surface of the underlying material following slumping.

when wetted, they quickly lose cohesiveness and internal friction and become an unstable slurry. This occurs because clay, which can hold large quantities of water, consists of platy particles that easily slide over each other when wet. For this reason, clay beds are frequently the slippery layer along which overlying rock units slide downslope.

Vegetation

Vegetation affects slope stability in several ways. By absorbing the water from a rainstorm, vegetation decreases water saturation of a slope's material that would otherwise lead to a loss of shear strength. Vegetation's root system also helps stabilize a slope by binding soil particles together and holding the soil to bedrock.

The removal of vegetation by either natural or human activity is a major cause of many mass movements. Summer brush and forest fires in southern California frequently leave the hillsides bare of vegetation. Fall rainstorms saturate the ground, causing mudslides that do tremendous damage and cost millions of dollars to clean up. The soils of many hillsides in New Zealand are sliding because deep-rooted native bushes have been replaced by shallow-rooted grasses used for sheep grazing. When heavy rains saturate the soil, the shallow-rooted grasses cannot hold the slope in place, and parts of it slide downhill.

Overloading

Overloading is almost always the result of human activity and typically results from the dumping, filling, or piling up of material. Under natural conditions, a material's load is carried by its grain-to-grain contacts, with the friction between the grains maintaining a slope. The additional weight created by overloading increases the water pressure within the material, which in turn decreases its shear strength, thereby weakening the slope material. If enough material is added, the slope will eventually fail, sometimes with tragic consequences.

Geology and Slope Stability

The relationship between the topography and the geology of an area is important in determining slope stability (Figure 11.5). If the rocks underlying a slope dip in the same direction as the slope, mass wasting is more likely to occur than if the rocks are horizontal or dip in the opposite direction. When the rocks dip in the same direction as the slope, water can percolate along the various bedding planes and decrease the cohesiveness and friction between adjacent rock units. This is particularly true when clay layers are present because clay becomes slippery when wet.

Even if the rocks are horizontal or dip in a direction opposite to that of the slope, joints may dip in the same direction as the slope. Water migrating through them weathers the rock and expands these openings until the weight of the overlying rock causes it to fall.

Figure 11.3 Undercutting a Slope's Base by Wave Action This sea cliff north of Bodega Bay, California, was undercut by waves during the winter of 1997–1998. As a result, part of the land slid into the ocean, damaging several houses.

Parvinder Sethi

Figure 11.4 Highway Excavation

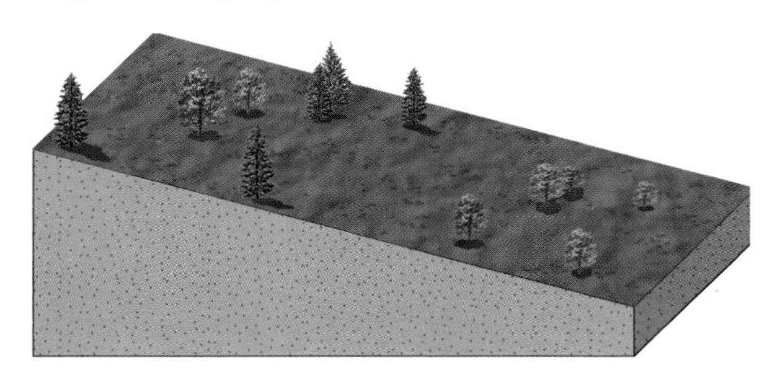

a Highway excavations disturb the equilibrium of a slope by

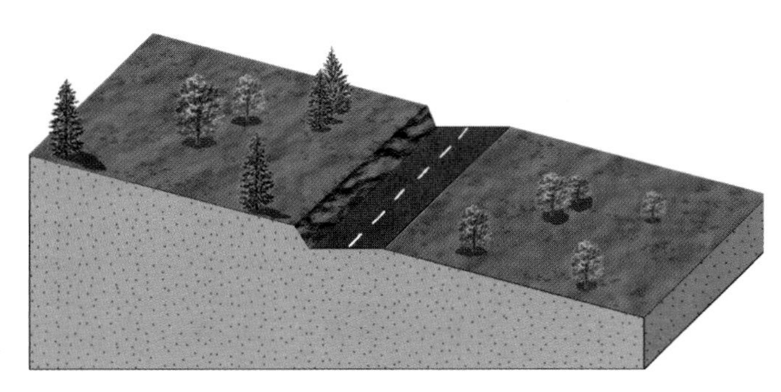

b removing a portion of its support, as well as oversteepening it at the point of excavation, which can result in

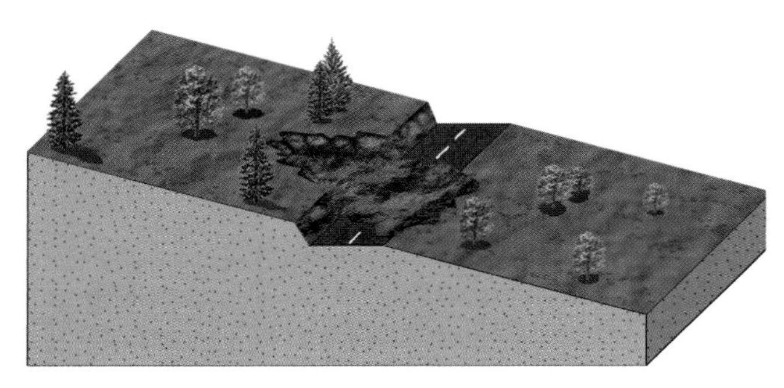

c landslides along the highway.

Courtesy of R. V. Dietrich

d Cutting into the hillside to construct this portion of the Pan-American Highway in Mexico resulted in a rockfall that completely blocked the road.

Figure 11.5 Geology, Slope Stability, and Mass Wasting Rocks dipping in the same direction as a hill's slope are particularly susceptible to mass wasting.

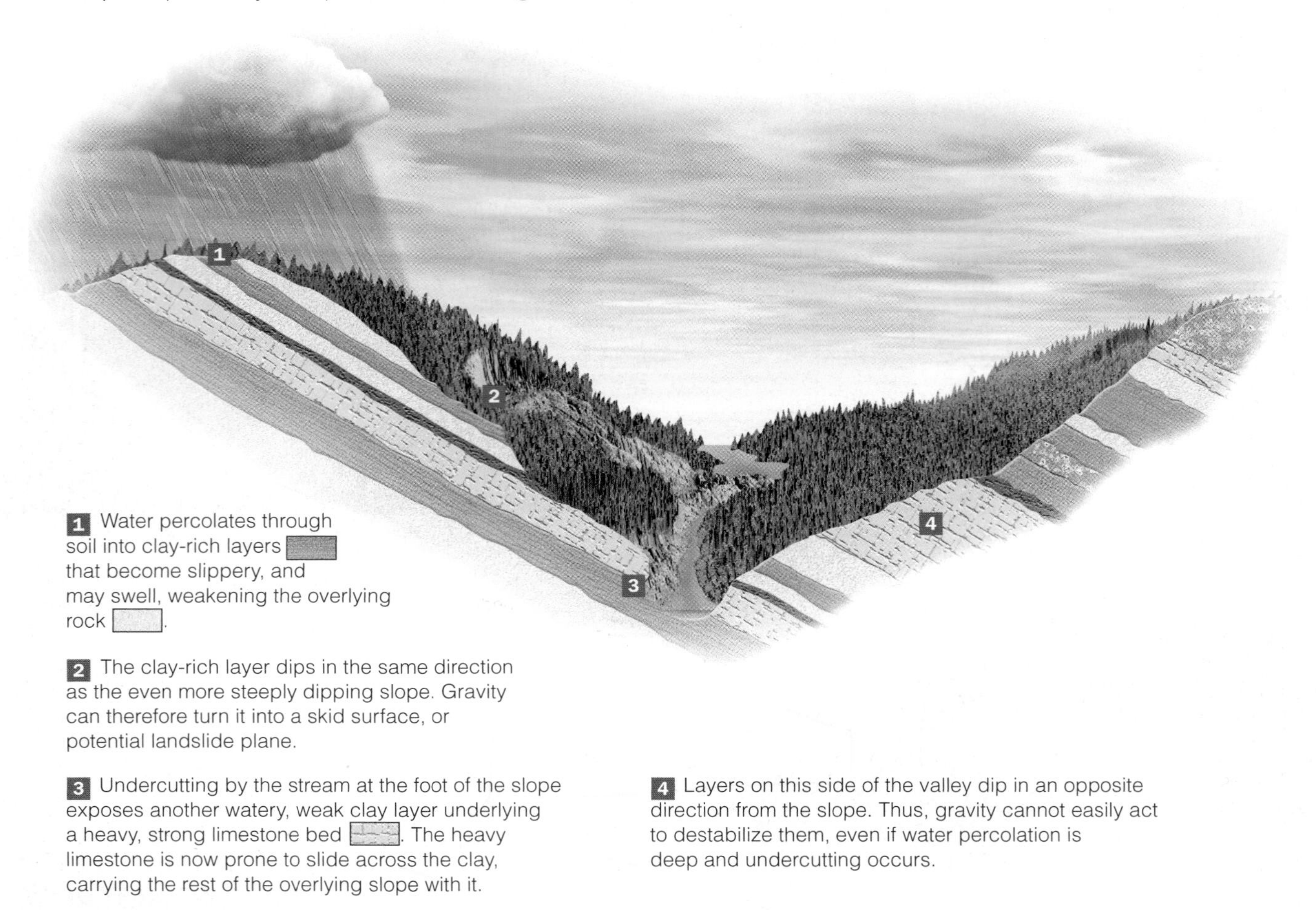

Triggering Mechanisms

The factors discussed thus far all contribute to slope instability. Most, though not all, rapid mass movements are triggered by a force that temporarily disturbs slope equilibrium. The most common triggering mechanisms are strong vibrations from earthquakes and excessive amounts of water from a winter snow melt or a heavy rainstorm (Figure 11.6).

Volcanic eruptions, explosions, and even loud claps of thunder may be enough to trigger a landslide if the slope is sufficiently unstable. Many *avalanches*, which are rapid movements of snow and ice down steep mountain slopes, are triggered by a loud gunshot or, in rare cases, even a person's shout.

TYPES OF MASS WASTING

Mass movements are generally classified on the basis of three major criteria (Table 11.2): (1) rate of movement (rapid or slow); (2) type of movement (primarily falling, sliding, or flowing); and (3) type of material involved (rock, soil, or debris). Even though many slope failures are combinations of different materials and movements, the resulting mass movements are typically classified according to their dominant behavior.

Rapid mass movements involve a visible movement of material. Such movements usually occur quite suddenly, and the material moves quickly downslope. Rapid mass movements are potentially dangerous and frequently result in loss of life and property damage. Most rapid mass movements occur on relatively steep slopes and can involve rock, soil, or debris.

Slow mass movements advance at an imperceptible rate and are usually detectable only by the effects of their movement, such as tilted trees and power poles or cracked foundations. Although rapid mass movements are more dramatic, slow mass movements are responsible for the downslope transport of a much greater volume of weathered material.

Falls

Rockfalls are a common type of extremely rapid mass movement in which rocks of any size fall through the air (Figure 11.7). Rockfalls occur along steep canyons, cliffs, and road cuts and build up accumulations of loose rocks and rock fragments at their base called *talus* (see Figure 6.3b).

Figure 11.6 Landslide Triggered by Heavy Rains, La Conchita, California Heavy winter rains caused this 200,000 m^3 landslide in March, 1995 at La Conchita, California, 120 km northeast of Los Angeles. Although no casualties occurred, nine homes were destroyed or badly damaged. Another landslide 10 years later occurred in the same area and under similar conditions, resulting in 10 deaths.

Rockfalls result from failure along joints or bedding planes in the bedrock and are commonly triggered by natural or human undercutting of slopes, or by earthquakes. Many rockfalls in cold climates are the result of frost wedging. Chemical weathering caused by water percolating through the fissures in carbonate rocks (limestone, dolostone, and marble) is also responsible for many rockfalls.

Rockfalls range in size from small rocks falling from a cliff to massive falls involving millions of cubic meters of debris that destroy buildings, bury towns, and block highways (Figure 11.7b). Rockfalls are a particularly common hazard in mountainous areas where roads have been built by blasting and grading through steep hillsides of bedrock. Anyone who has ever driven through the Appalachians, the Rocky Mountains, or the Sierra Nevada is familiar with the "Watch for Falling Rocks" signs posted to warn drivers of the danger. Slopes that are particularly susceptible to rockfalls are sometimes covered with wire mesh in an effort to prevent dislodged rocks from falling to the road below (Figure 11.8a). Another tactic is to put up wire mesh fences along the base of the slope to catch or slow down bouncing or rolling rocks (Figure 11.8b).

Figure 11.7 Rockfalls

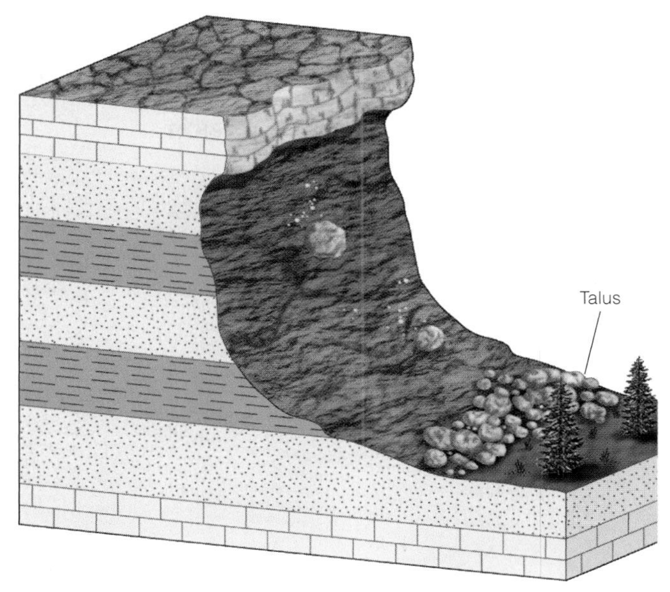

a Rockfalls result from failure along cracks, fractures, or bedding planes in the bedrock and are common features in areas of steep cliffs.

b A huge rockfall closed both lanes of traffic on Highway 70 near Rogers Flat, California, on July 25, 2003. Rocks the size of large dump trucks had to be blasted into smaller pieces to clear the highway. Despite the pavement cracking caused by the falling boulders, geologists determined that the roadbase was undamaged and the road would be safe following cleanup operations.

Slides

A **slide** involves movement of material along one or more surfaces of failure. The type of material may be soil, rock, or a combination of the two, and it may break apart during movement or remain intact. A slide's rate of movement can vary from extremely slow to very rapid (Table 11.2).

Figure 11.8 Minimizing Damage from Rock Falls

Parvinder Sethi

a Wire mesh has been used to cover this steep slope near Narvik in northern Norway. This is a common practice in mountainous areas to prevent rocks from falling on the road.

James S. Monroe

b A wire mesh fence along the base of this hillside of Highway 44 in California has caught many boulders and prevented them from rolling onto the highway.

Two types of slides are generally recognized: (1) slumps or rotational slides, in which movement occurs along a curved surface; and (2) rock or block slides, which move along a more or less planar surface.

A **slump** involves the downward movement of material along a curved surface of rupture and is characterized by the backward rotation of the slump block (Figure 11.9). Slumps usually occur in unconsolidated or weakly consolidated material and range in size from small individual sets, such as occur along stream banks, to massive, multiple sets that affect large areas and cause considerable damage.

Slumps can be caused by a variety of factors, but the most common is erosion along the base of a slope, which removes support for the overlying material. This local steepening may be caused naturally by stream erosion along its banks (Figure 11.2c) or by wave action at the base of a coastal cliff (Figure 11.10).

TABLE 11.2 Classification of Mass Movements and Their Characteristics

Type of Movement	Subdivision	Characteristics	Rate of Movement
Falls	Rockfall	Rocks of any size fall through the air from steep cliffs, canyons, and road cuts	Extremely rapid
Slides	Slump	Movement occurs along a curved surface of rupture; most commonly involves unconsolidated or weakly consolidated material	Extremely slow to moderate
Flows	Rock slide	Movement occurs along a generally planar surface	Rapid to very rapid
	Mudflow	Consists of at least 50% silt- and clay-sized particles and up to 30% water	Very rapid
	Debris flow	Contains larger-sized particles and less water than mudflows	Rapid to very rapid
	Earthflow	Thick, viscous, tongue-shaped mass of wet regolith	Slow to moderate
	Quick clays	Composed of fine silt and clay particles saturated with water; when disturbed by a sudden shock, lose their cohesiveness and flow like a liquid	Rapid to very rapid
	Solifluction	Water-saturated surface sediment	Slow
	Creep	Downslope movement of soil and rock	Extremely slow
Complex movements		Combination of different movement types	Slow to extremely rapid

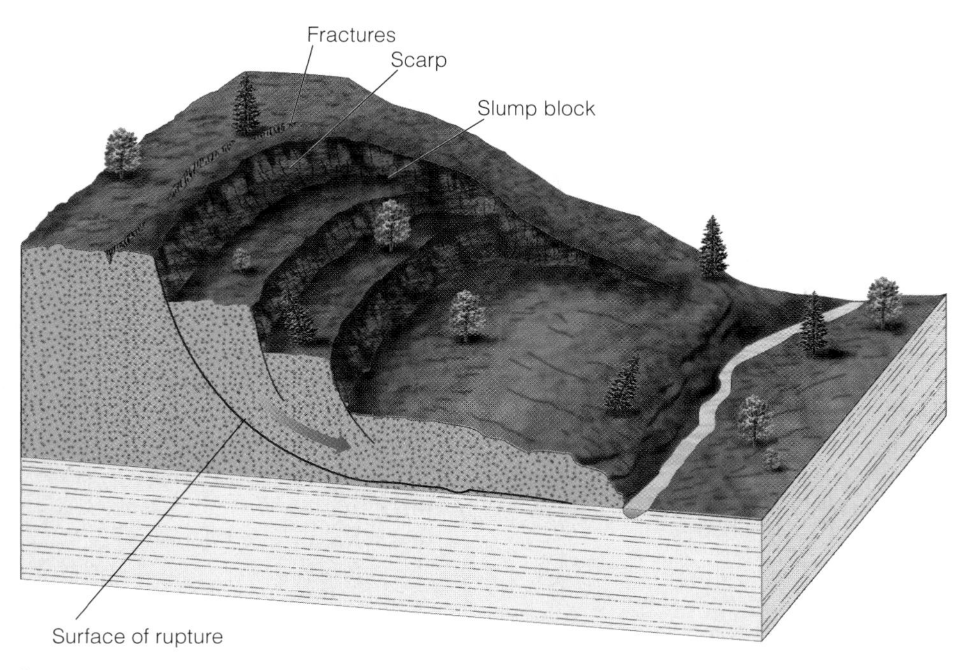

▶ **Figure 11.9 Slumping** In a slump, material moves downward along the curved surface of a rupture, causing the slump block to rotate backward. Most slumps involve unconsolidated or weakly consolidated material and are typically caused by erosion along the slope's base.

Slope oversteepening can also be caused by human activity, such as the construction of highways and housing developments. Slumps are particularly prevalent along highway cuts, where they are generally the most frequent type of slope failure observed.

Although many slumps are merely a nuisance, large-scale slumps in populated areas and along highways can cause extensive damage. Such is the case in coastal southern California where slumping and sliding have been a constant problem, resulting in the destruction of many homes and the closing and relocation of numerous roads and highways (see Geo-Focus on pages 282 and 283).

A **rock** or *block* **slide** occurs when rocks move downslope along a more or less planar surface. Most

▶ **Figure 11.10 Slumping in the Pacific Palisades, Southern California** Undercutting of steep sea cliffs by wave action resulted in massive slumping in the Pacific Palisades area of southern California on March 31 and April 3, 1958. Highway 1 was completely blocked. Note the heavy earth-moving equipment for scale.

Geo-Focus Southern California Landslides

Southern California is no stranger to landslides. La Conchita, Point Fermin, Pacific Palisades, and Laguna Beach are all locations in southern California that have suffered damaging mass movements during the past 50 years. Two regions in particular, La Conchita and Laguna Beach, have been in the news because of the landslides that destroyed numerous homes.

La Conchita is a small community along the coast at the base of a 100-m-high terrace, 120 km northwest of Los Angeles, California. On March 4, 1995, following a period of heavy rains, some residents of this beach community noticed that the steep slope above their homes was slowly moving and that cracks were appearing in the walls of their houses, indicating that the homes were also moving. Shortly thereafter, a 200,000 m^3 slide destroyed or damaged nine homes in its path (Figure 1).

Almost 10 years later, following a week of heavy rainfall in southern California, in which the hillside and previous landslide deposits were saturated with water, another landslide occurred in the same area. This time 10 people were killed and 15 homes were buried under 10 m and 400,000 tons of mud (Figure 2).

What went wrong and why was this situation repeated? The rocks that make up the steep sloped

USGS

Figure 1 La Conchita, California, is located at the base of a steep sloped terrace. Heavy rains and irrigation of an avocado orchard (visible at the top of the terrace) contributed to the landslide that destroyed nine homes in 1995.

Kevork Djansezian/AP/Wide World Photo

Figure 2 La Conchita, California, 10 years later (2005). Here the view is from the east. Similar factors caused another massive landslide in the same area. The landslide can clearly be seen in the center of this photograph, and the scarp and the remains of the 1995 landslide are still visible on the right side, as is the avocado orchard in the foreground.

rock slides take place because the local slopes and rock layers dip in the same direction (Figure 11.11), although they can also occur along fractures parallel to a slope. Rock slides are also common occurrences along the southern California coast. At Point Fermin, seaward-dipping rocks with interbedded slippery clay layers are undercut by waves, causing numerous slides (see Geo-inSight on pages 284 and 285).

terrace behind La Conchita consist of soft, weak, and porous sediments that are not well lithified, and thus are easily weathered and susceptible to mass wasting. In addition, an irrigated avocado orchard sits on top of the hill, contributing to the water percolating through the porous sediments and rocks and contributing to the instability of the hillside. Add in heavy rainfall over an extended period, and you have all of the ingredients for a landslide in the making. An ancient landslide area to begin with, a steep slope that has been undercut at its base by a road, well-saturated sediments decreasing the cohesion of the sediments that hold the hillside together, and continuing rains all contribute to the making of a landslide. And the potential is still there for another landslide, with no guarantee it won't happen again in the next 10 years!

Nick Ut/AP/Wide World Photo

Figure 3 A rock slide on June 1, 2005, destroyed 18 expensive homes and damaged at least 20 others in Laguna Beach, California. Heavy rains combined with unstable underlying geology contributed to this most recent landslide in this area.

Farther south in Laguna Beach, another landslide, in this case a rock slide, destroyed 18 expensive hillside homes and severely damaged approximately 20 others on June 1, 2005 (Figure 3). Just as happened in 1978, the main triggering mechanism was probably unusually heavy winter rains, in this case the second-rainiest season on record. In this area of southern California, the rocks dip in the same direction as the slope and contain numerous clay beds interbedded with porous sandstones. Such conditions, when combined with heavy rainfall, are ideal for rock slides. It should come as no surprise that the area where both the 1978 and 2005 rock slides occurred is also part of an ancient slide complex.

Can anything be done to prevent future landslides? The short answer is probably no. Decreasing the slope, benching the hillside, and making sure that there is sufficient drainage and a good cover of vegetation are all steps that can minimize future mass wasting. But the sad fact is that the geologic conditions are such that future landslides are inevitable as the landscape seeks equilibrium conditions by adjusting its slope. Add in the fact that the coastal terraces of Laguna Beach offer some of the most breathtaking views of the Pacific, and people are willing to pay a premium to live here, and you have the formula for future landslides, loss of life, and property damage.

Farther south in the town of Laguna Beach, residents have been hit by rock slides and mudslides in 1978, 1998, and as recently as 2005, in which numerous homes have been destroyed or damaged and two people killed (Figure 11.12). Just as at Point Fermin, the rocks at Laguna Beach dip about 25 degrees in the same direction as the slope of the canyon walls and contain clay beds that "lubricate" the overlying rock layers, causing the rocks and the houses built on them

Geo-inSight Point Fermin—Slip Sliding Away

Dubbed the "sunken city" by residents of the area, Point Fermin in Southern California is famous for its numerous examples of mass wasting. The area is underlain by fine-grained sedimentary rocks interbedded with diatomite layers and volcanic ash. When these layers get wet, they become slippery and tend to slide easily. The rocks also dip slightly toward the ocean and form steep coastal bluffs that are being undercut by constant wave action at their base. This wave action results in oversteepening of the cliffs, which causes slumping.

Mass wasting began in 1929 with minor slumping in the area. In the early 1940s, water mains in the region were broken and several individual blocks began slumping. Movement largely ceased following this main phase of slumping, but it has continued intermittently until the present, and residents have paid the price for living in an unstable coastal area.

▲ **1.** A map of Southern California, showing the location of Point Fermin and an aerial view at low tide of the sliding that has taken place. Note the numerous slump blocks and oversteepened cliffs. The continuous pounding of waves and surf along the base of the cliffs further erodes and undercuts them, leading to even more slumping and sliding.

▼ **2.** A view of one portion of the Point Fermin slide area, showing the fine-grained sedimentary rocks dipping slightly toward the ocean and the oversteepened cliffs resulting from slumping and sliding in the foreground.

◀ **3.** A view of one slump block shows remnants of a former road and a palm tree still growing as if nothing has happened.

Reed Wicander

▶ **4.** Homes dangerously close to an oversteepened, eroding coastal bluff. Two white drainage pipes that helped drain water out of the ground are dangling from the top of the cliff. It is just a matter of time before these homes are destroyed by the effects of erosion or slumping of the unstable bluff.

Parvinder Sethi

▼ **5.** Creep and minor slumping are evident in this photo. Note the two small slump scarps. The smaller one in the background is mostly grass-covered, whereas the one in the foreground has bare spots and is apparently moving at a slightly faster rate. Notice the effect of creep on the right-hand wall of the house. The bottom part of the wall is moving toward the right of the photo as a result of creep, producing a bend in the wall that can be seen clearly near its base.

Reed Wicander

▲ **6.** Slumping along the oversteepened cliffs at Point Fermin. The abandoned house in Figure 4 is just to the right of this view at the top of the cliff.

Reed Wicander

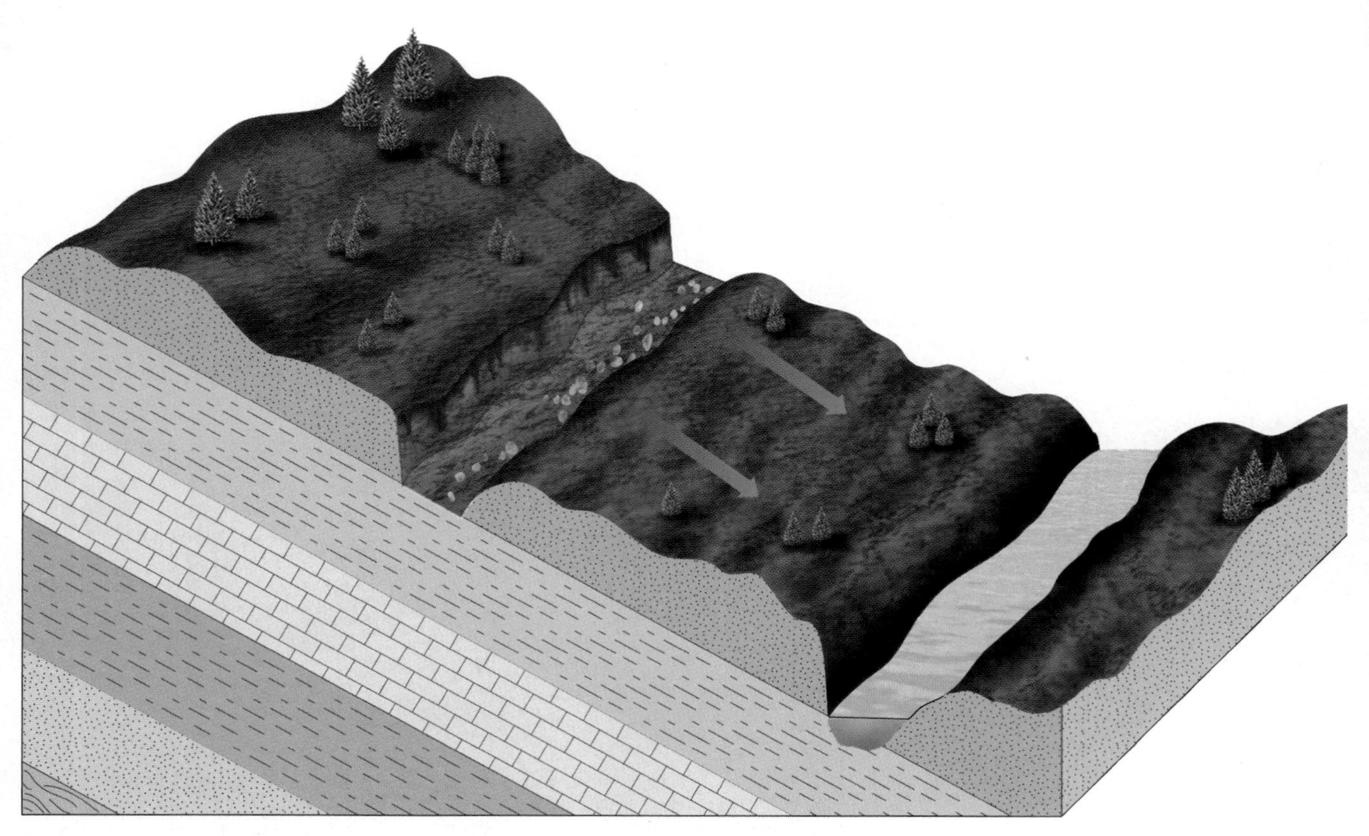

Figure 11.11 Rock Slides Rock slides occur when material moves downslope along a generally planar surface. Most rock slides result when the underlying rocks dip in the same general angle as the slope of the land. Undercutting along the base of the slope and clay layers beneath porous rock or soil layers increase the chance of rock slides.

to slide. Percolating water from heavy rains wets subsurface clayey siltstone, thus reducing its shear strength and helping to activate the slide. In addition, these slides are part of a larger ancient slide complex.

Not all rock slides are the result of rocks dipping in the same direction as a hill's slope. The rock slide at Frank, Alberta, Canada, on April 29, 1903, illustrates how nature and human activity can combine to create a situation with tragic results (Figure 11.13).

It would appear at first glance that the coal-mining town of Frank, lying at the base of Turtle Mountain, was in no danger from a landslide (Figure 11.13). After all, many of the rocks dipped away from the mining valley, unlike the situations at Point Fermin and Laguna Beach. The joints in the massive limestone composing Turtle Mountain, however, dip steeply toward the valley and are essentially parallel with the slope of the mountain itself. Furthermore, Turtle Mountain is supported by weak limestones, shales, and coal layers that underwent slow plastic deformation from the weight of the overlying massive limestone. Coal mining along the base of the valley also contributed to the stress on the rocks by removing some of the underlying support. All of these factors, as well as the frost action and chemical weathering that widened the joints, finally resulted in a massive rock slide. Approximately 40 million m^3 of rock slid down Turtle Mountain along joint planes, killing 70 people and partially burying the town of Frank.

Flows

Mass movements in which material flows as a viscous fluid or displays plastic movement are termed *flows.* Their rate of movement ranges from extremely slow to extremely rapid (Table 11.2). In many cases, mass movements begin as falls, slumps, or slides and change into flows farther downslope.

Of the major mass movement types, **mudflows** are the most fluid and move most rapidly (at speeds up to 80 km/hr). They consist of at least 50% silt- and clay-sized material combined with a significant amount of water (up to 30%). Mudflows are common in arid and semiarid environments where they are triggered by heavy rainstorms that quickly saturate the regolith, turning it into a raging flow of mud that engulfs everything in its path. Mudflows can also occur in mountain regions (Figure 11.14) and in areas covered by volcanic ash where they can be particularly destructive (see Chapter 5). Because mudflows are so fluid, they generally follow preexisting channels until the slope decreases or the channel widens, at which point they fan out.

As urban areas in arid and semiarid climates continue to expand, mudflows and the damage that they create are becoming problems. Mudflows are common, for example, in

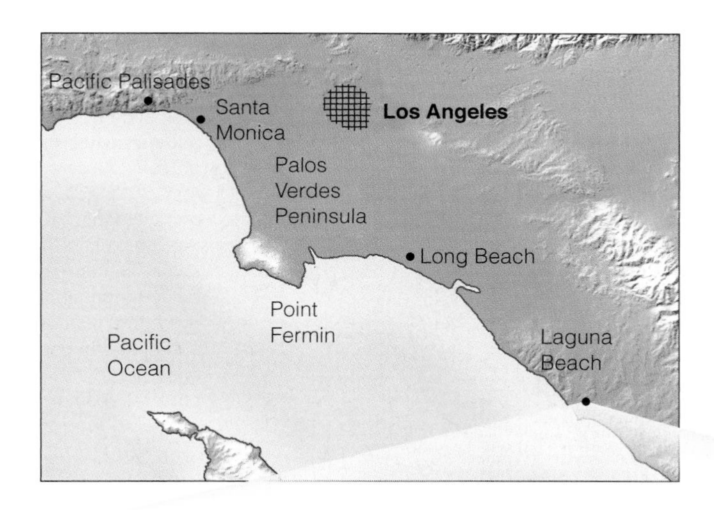

Figure 11.12 Rock Slide, Laguna Beach, California A combination of interbedded clay layers that become slippery when wet, rocks dipping in the same direction as the slope of the sea cliffs, and undercutting of the sea cliffs by wave action activated a rock slide at Laguna Beach, California, that destroyed numerous homes and cars on October 2, 1978. This same area was hit by another rock slide in 2005.

the steep hillsides around Los Angeles where they have damaged or destroyed many homes.

Debris flows are composed of larger particles than mudflows and do not contain as much water. Consequently, they are usually more viscous than mudflows, typically do not move as rapidly, and rarely are confined to preexisting channels. Debris flows can be just as damaging, though, because they can transport large objects (Figure 11.15).

Earthflows move more slowly than either mudflows or debris flows. An earthflow slumps from the upper part of a hillside, leaving a scarp, and flows slowly downslope as a thick, viscous, tongue-shaped mass of wet regolith (Figure 11.16). Like mudflows and debris flows, earthflows can be of any size and are frequently destructive. They occur most commonly in humid climates on grassy, soil-covered slopes following heavy rains.

Some clays spontaneously liquefy and flow like water when they are disturbed. Such **quick clays** have caused serious damage and loss of lives in Sweden, Norway, eastern Canada (Figure 11.17), and Alaska. Quick clays are composed of fine silt and clay particles made by the grinding action of glaciers. Geologists think that these fine sediments were originally deposited in a marine environment where their pore space was filled with saltwater. The ions in saltwater helped establish strong bonds between the clay particles, thus stabilizing and strengthening the clay. When the clays were subsequently uplifted above sea level, the saltwater was flushed out by fresh groundwater, reducing the effectiveness of the ionic bonds between the clay particles and thereby reducing the overall strength and cohesiveness of the clay. Consequently, when the clay is disturbed by a sudden shock or shaking, it essentially turns to a liquid and flows.

Figure 11.13 Rock Slide, Turtle Mountain, Canada

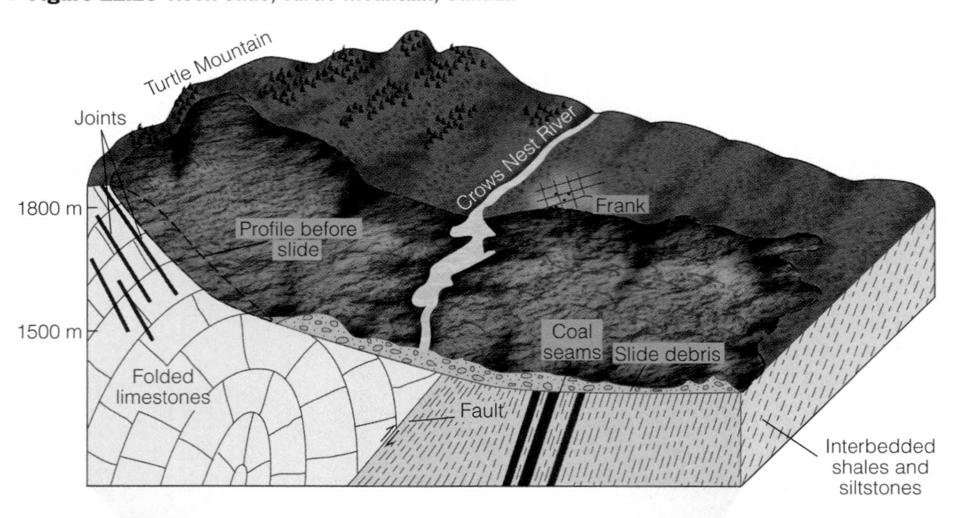

a The tragic Turtle Mountain rock slide that killed 70 people and partially buried the town of Frank, Alberta, Canada, on April 29, 1903, was caused by a combination of factors. These included joints that dipped in the same direction as the slope of Turtle Mountain, a fault partway down the mountain, weak shale and siltstone beds underlying the base of the mountain, and mined-out coal seams.

B. Bradley and the University of Colorado's Geology Department, National Geophysical Data Center

b Results of the 1903 rock slide at Frank.

An excellent example of the damage that can be done by quick clays occurred in the Turnagain Heights area of Anchorage, Alaska, in 1964 (Figure 11.18). Underlying most of the Anchorage area is the Bootlegger Cove Clay, a massive clay unit of poor permeability. Because the Bootlegger Cove Clay forms a barrier that prevents groundwater from flowing through the adjacent glacial deposits to the sea, considerable hydraulic pressure builds up behind the clay. Some of this water has flushed out the saltwater in the clay and has saturated the lenses of sand and silt associated with the clay beds. When the magnitude-8.6 Good Friday earthquake struck on March 27, 1964, the shaking turned parts of the Bootlegger Cove Clay into a quick clay and precipitated a series of massive slides in the coastal bluffs that destroyed

Parvinder Sethi

Figure 11.14 Mudflow, Estes Park, Colorado Mudflows move swiftly downslope, engulfing everything in their path. Note how this mudflow in Rocky Mountain National Park has fanned out at the base of the hill. Also note the small lake adjacent to the mudflow that was formed after this mudflow created a dam across the stream.

most of the homes in the Turnagain Heights subdivision (Figure 11.18b).

Solifluction is the slow downslope movement of water-saturated surface sediment. Solifluction can occur in any climate where the ground becomes saturated with water, but is most common in areas of permafrost.

Permafrost, ground that remains permanently frozen, covers nearly 20% of the world's land surface (Figure 11.19a). During the warmer season when the upper portion of the permafrost thaws, water and surface sediment form a soggy mass that flows by solifluction and produces a characteristic lobate topography (Figure 11.19b).

B. Pipkin, University of Southern California

Figure 11.15 Debris Flow, Ophir Creek, Nevada A debris flow and damaged house in lower Ophir Creek, western Nevada. Note the many large boulders that are part of the debris flow. Debris flows do not contain as much water as mudflows and typically are composed of larger particles.

As might be expected, many problems are associated with construction in a permafrost environment. A good example is what happens when an uninsulated building is constructed directly on permafrost. Heat escapes through the floor, thaws the ground below, and turns it into a soggy, unstable mush. Because the ground is no longer solid, the building settles unevenly into the ground and numerous structural problems result (Figure 11.20).

Creep, the slowest type of flow, is the most widespread and significant mass wasting process in terms of the total amount of material moved downslope and the monetary damage it does annually. Creep involves extremely slow downhill movement of soil or rock. Although it can occur anywhere and in any climate, it is most effective and significant as a geologic agent in humid regions. In fact, it is the most common form of mass wasting in the southeastern United States and the southern Appalachian Mountains.

Because the rate of movement is essentially imperceptible, we are frequently unaware of creep's existence until we notice its effects: tilted trees and power poles, broken streets and sidewalks, or cracked retaining walls or foundations (Figure 11.21). Creep usually involves the whole hillside and probably occurs, to some extent, on any weathered or soil-covered, sloping surface.

Creep is not only difficult to recognize, but also to control. Although engineers can sometimes slow or stabilize creep, many times

Figure 11.16 Earthflow

a Earthflows form tongue-shaped masses of wet regolith that move slowly downslope. They occur most commonly in humid climates on grassy, soil-covered slopes.

b An earthflow near Baraga, Michigan.

Figure 11.17 Quick-Clay Slide, Nicolet, Quebec, Canada The house on the slide (to the right of the bridge and circled in red) traveled several hundred meters with relatively little damage.

the only course of action is to simply avoid the area if at all possible or, if the zone of creep is relatively thin, design structures that can be anchored into the bedrock.

Complex Movements

Recall that many mass movements are combinations of different movement types. When one type is dominant, the movement can be classified as one of those described thus far. If several types are more or less equally involved, however, it is called a **complex movement.**

The most common type of complex movement is the slide-flow, in which there is sliding at the head and then some type of flowage farther along its course. Most slide-flow landslides involve well-defined slumping at the head, followed by a debris flow or earthflow (Figure 11.22). Any combination of different mass movement types is a complex movement.

Figure 11.18 Quick-Clay Slide, Anchorage, Alaska

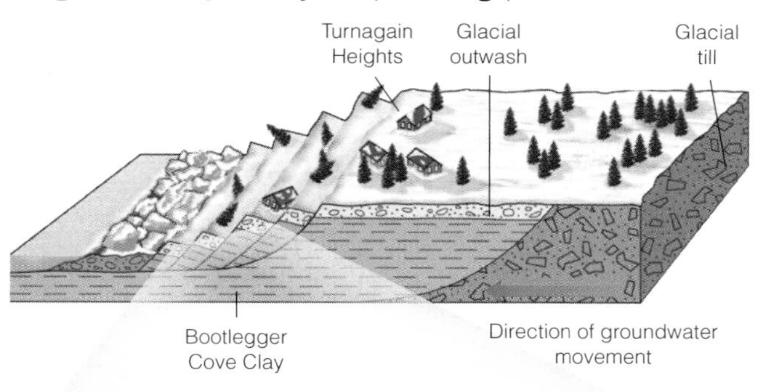

a Ground shaking by the 1964 Alaska earthquake turned parts of the Bootlegger Cove Clay into a quick clay, causing numerous slides.

Alaska Earthquake Collection/USGS

b Low-altitude photograph of the Turnagain Heights subdivision of Anchorage shows some of the numerous landslide fissures that developed, as well as the extensive damage to buildings in the area. The remains of the Four Seasons apartment building can be seen in the background.

A *debris avalanche* is a complex movement that often occurs in very steep mountain ranges. Debris avalanches typically start out as rockfalls when large quantities of rock, ice, and snow are dislodged from a mountainside, frequently as a result of an earthquake. The material then slides or flows down the mountainside, picking up additional surface material and increasing in speed. The 1970 Peru earthquake (Table 11.1) set in motion the debris avalanche that destroyed the towns of Yungay and Ranrahirca, Peru, and killed more than 25,000 people (Figure 11.23).

RECOGNIZING AND MINIMIZING THE EFFECTS OF MASS WASTING

The most important factor in eliminating or minimizing the damaging effects of mass wasting is a thorough geologic investigation of the region in question. In this way, former landslides and areas susceptible to mass movements can be identified and perhaps avoided. By assessing the risks of possible mass wasting before construction begins, engineers can take steps to eliminate or minimize the effects of such events.

Identifying areas with a high potential for slope failure is important in any hazard assessment study; these studies include identifying former landslides, as well as sites of potential mass movement. Scarps, open fissures, displaced or tilted objects, a hummocky surface, and sudden changes in vegetation are some of the features that indicate former landslides or an area susceptible to slope failure. The effects of weathering, erosion, and vegetation may, however, obscure the evidence of previous mass wasting.

Soil and bedrock samples are also studied, in both the field and laboratory, to assess such characteristics as composition, susceptibility to weathering, cohesiveness, and ability to transmit fluids. These studies help geologists and engineers predict slope stability under a variety of conditions.

The information derived from a hazard assessment study can be used to produce *slope-stability maps* of the area (Figure 11.24). These maps allow planners and developers to make decisions about where to site roads, utility lines, and housing or industrial developments based on the relative stability or instability of a particular location. In addition, the maps indicate the extent of an area's landslide problem and the type of mass movement that may occur. This information is important for grading slopes or building structures to prevent or minimize slope-failure damage.

Finally, building codes, which spell out what types of site investigations need to be made and the manner in which structures must be built also help determine how land will be developed and utilized. Of particular interest to us is the section of the *Uniform Building Code* (UBC, Chapter 70) that deals with slopes and the alteration of the landscape. This chapter deals with such geologic situations as the slope angle and direction of the slope in relation to bedding planes, foliation, faults, and joints, for example. In addition, it also specifies surface drainage requirements and compaction of materials in the area under investigation. This section of the *Uniform Building Code* has been widely adopted by many counties and

What Would You Do?

You've found your dream parcel of land in the hills of northern Baja California, where you plan to retire someday. Because you want to make sure that the area is safe to build a house, you decide to do your own geologic investigation of the area to make sure that there aren't any obvious geologic hazards. What specific things would you look for that might indicate mass wasting in the past? Even if there is no obvious evidence of rapid mass wasting, what features would you look for that might indicate a problem with slow types of mass wasting such as creep?

What Would You Do ?

You are a member of a planning board for your seaside community. A developer wants to rezone some coastal property to build 20 condominiums. This would be a boon to the local economy because it would provide jobs and increase the tax base. However, because the area is somewhat hilly and fronts the ocean, you are concerned about how safe the buildings would be. What types of studies would need to be done before any rezoning could take place? Is it possible to build safe structures along a hilly coastline? What specifically would you ask the environmental consulting firm that the planning board has hired to look for in terms of actual or potential geologic hazards if the condominiums are built?

cities in the United States, and has led to stronger local building codes, better land use, and decreased losses to structures built on hillsides and other possibly unstable areas.

Although most large mass movements usually cannot be prevented, geologists and engineers can use various methods to minimize the danger and damage resulting from them. Because water plays such an important role in many landslides, one of the most effective and inexpensive ways to reduce the potential for slope failure or to increase existing slope stability is surface and subsurface drainage of a hillside. Drainage serves two purposes. It reduces the weight of the material likely to slide, and increases the shear strength of the slope material by lowering pore pressure.

Surface waters can be drained and diverted by ditches, gutters, or culverts designed to direct water away from slopes. Drainpipes perforated along one surface and driven into a hillside can help remove subsurface water (Figure 11.25). Finally, planting vegetation on hillsides helps stabilize slopes by holding the soil together and reducing the amount of water in the soil.

Another way to help stabilize a hillside is to reduce its slope. Recall that overloading and oversteepening by grading are common causes of slope failure. Reducing the angle of a hillside decreases the potential for slope failure. Two methods are usually employed to reduce a slope's angle. In the *cut-and-fill* method, material is removed from the upper part of the slope and used as fill at the base, thus providing a flat surface for construction and reducing the slope (Figure 11.26). The second method, which is called *benching*, involves cutting a series of benches or steps into a hillside (Figure 11.27). This process reduces the overall average

Figure 11.19 Permafrost and Solifluction

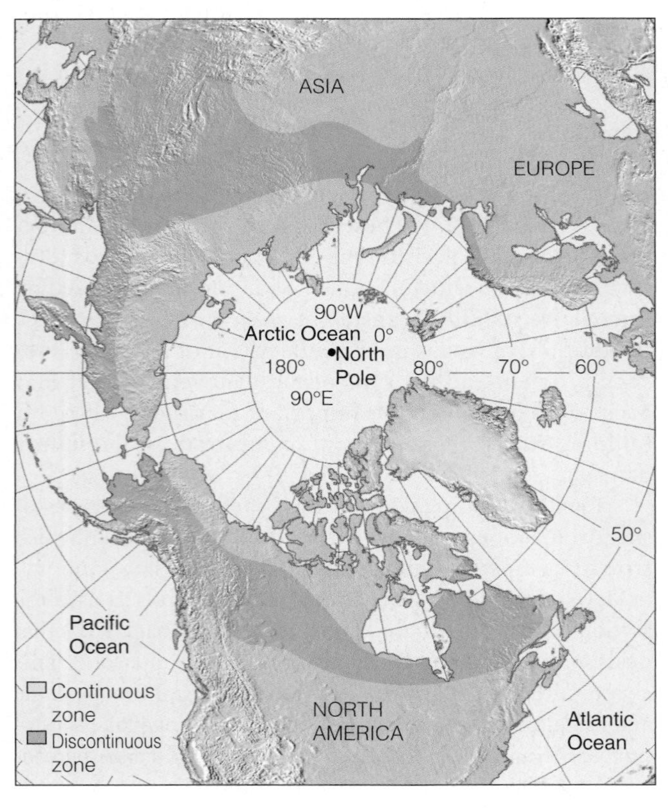

a Distribution of permafrost areas in the Northern Hemisphere.

B. Bradley and the University of Colorado's Geology Department, National Geophysical Data Center

b Solifluction flows near Suslositna Creek, Alaska, show the typical lobate topography that is characteristic of solifluction conditions.

Figure 11.20 Permafrost Damage The house, south of Fairbanks, Alaska, has settled unevenly because the underlying permafrost in fine-grained silts and sands has thawed.

slope, and the benches serve as collecting sites for small landslides or rockfalls that might occur. Benching is most commonly used on steep hillsides in conjunction with a system of surface drains to divert runoff.

In some situations, retaining walls are constructed to provide support for the base of the slope (Figure 11.28). The walls are usually anchored well into bedrock, backfilled with crushed rock, and provided with drain holes to prevent the buildup of water pressure in the hillside.

Rock bolts, similar to those employed in tunneling and mining, are sometimes be used to fasten potentially unstable rock masses into the underlying stable bedrock (Figure 11.29). This technique has been used successfully on the hillsides of Rio de Janeiro, Brazil, and to

Figure 11.21 Creep

a Some evidence of creep: (A) curved tree trunks; (B) displaced monuments; (C) tilted power poles; (D) displaced and tilted fences; (E) roadways moved out of alignment; (F) hummocky surface.

b Trees, bent by creep, Wyoming.

c Creep has bent these sandstone and shale beds of the Haymond Formation near Marathon, Texas.

d Stone wall tilted due to creep in Champion, Michigan.

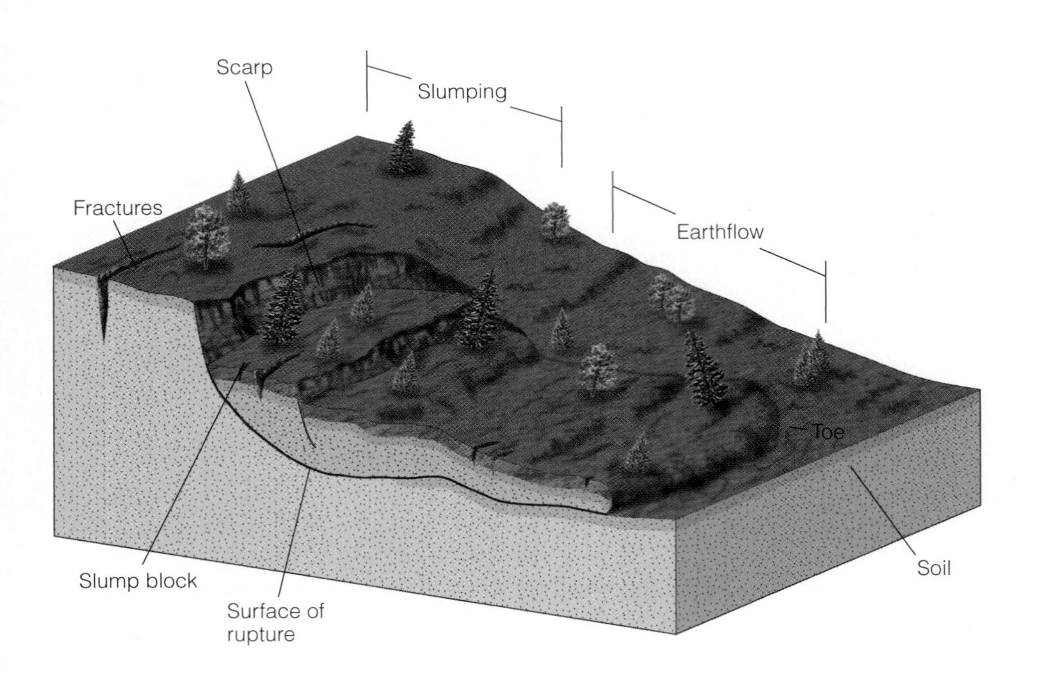

Figure 11.22 Complex Movement A complex movement is one in which several types of mass wasting are involved. In this example, slumpling occurs at the head, followed by an earthflow.

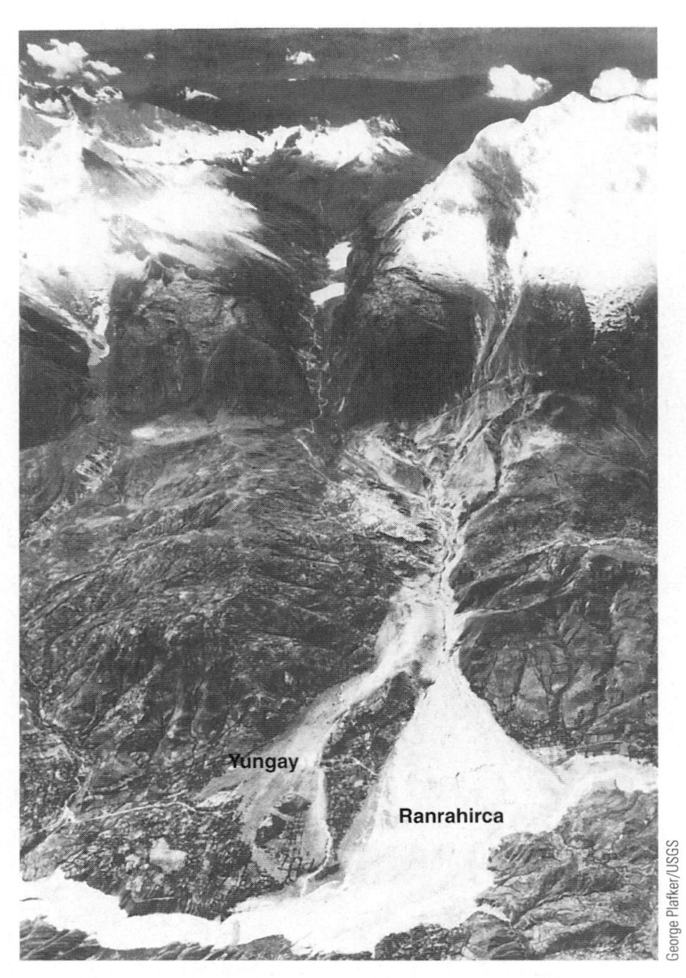

Figure 11.23 Debris Avalanche An earthquake 65 km away triggered this debris avalanche on Nevado Huascarán, Peru, that destroyed the towns of Yungay and Ranrahirca and killed more than 25,000 people.

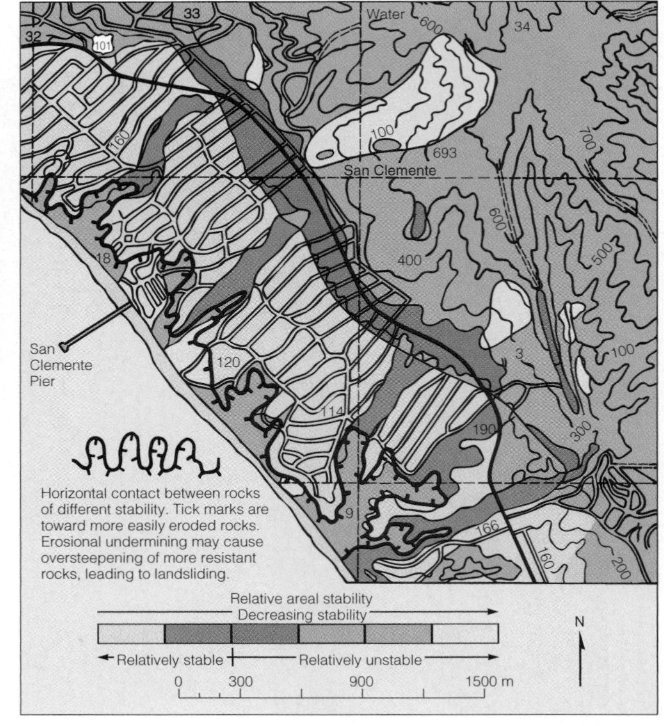

Figure 11.24 Slope-Stability Map This slope-stability map of part of San Clemente, California, shows areas delineated according to relative stability. Such maps help planners and developers make decisions about where to site roads, utility lines, buildings, and other structures.

Figure 11.25 Using Drainpipes to Remove Subsurface Water

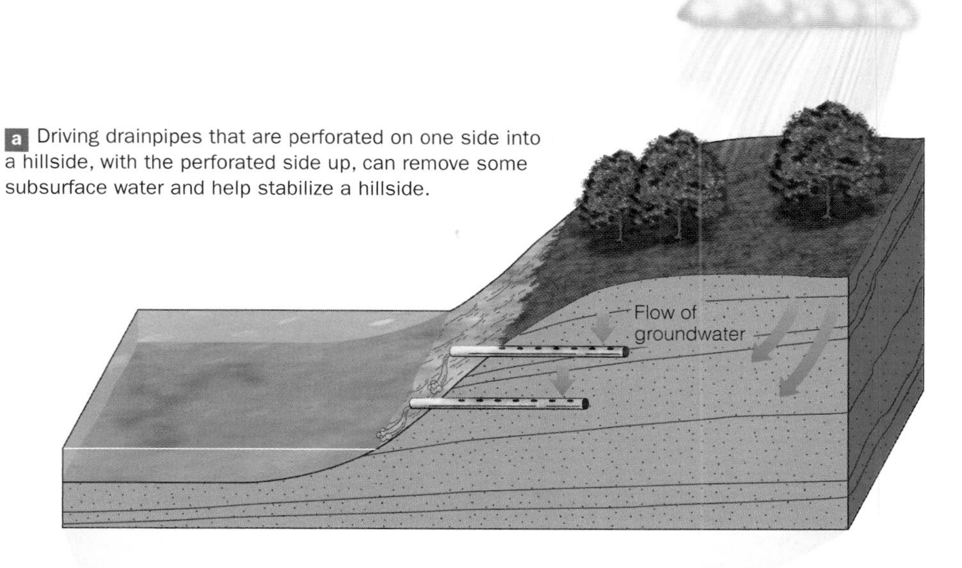

a Driving drainpipes that are perforated on one side into a hillside, with the perforated side up, can remove some subsurface water and help stabilize a hillside.

b A drainpipe driven into the hillside at Point Fermin, California, helps to reduce the amount of subsurface water in these porous beds.

Figure 11.26 Stabilizing a Hillside by the Cut-and-Fill Method One common method used to help stabilize a hillside and reduce its slope is the cut-and-fill method. Material from the steeper upper part of the hillside is removed, thereby decreasing the slope angle, and is used to fill in the base. This provides some additional support at the base of the slope.

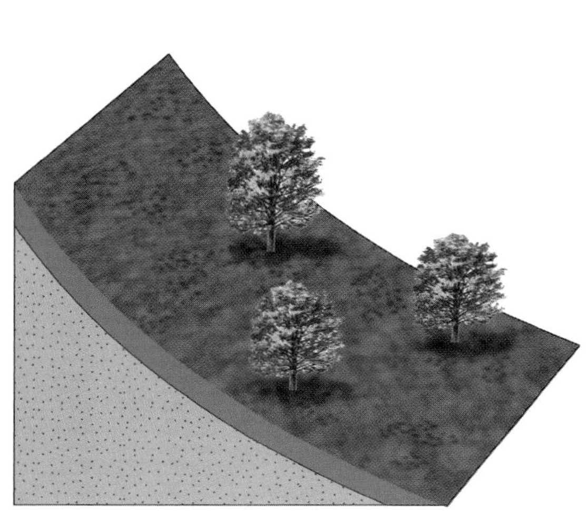

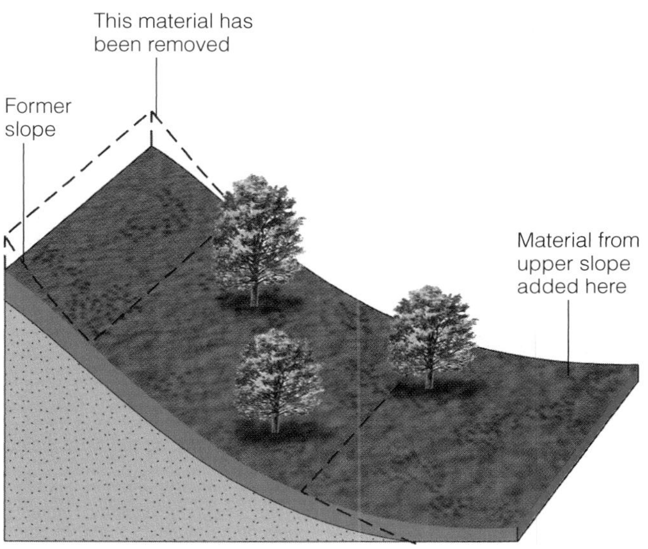

Figure 11.27 Stabilizing a Hillside by Benching

a Another common method used to stabilize a hillside and reduce its slope is benching. This process involves making several cuts along a hillside to reduce the overall slope. Furthermore, individual slope failures are now limited in size, and the material collects on the benches.

b Benching is used in nearly all road cuts and can be clearly seen in this photograph.

Figure 11.28 Retaining Walls Help Reduce Landslides

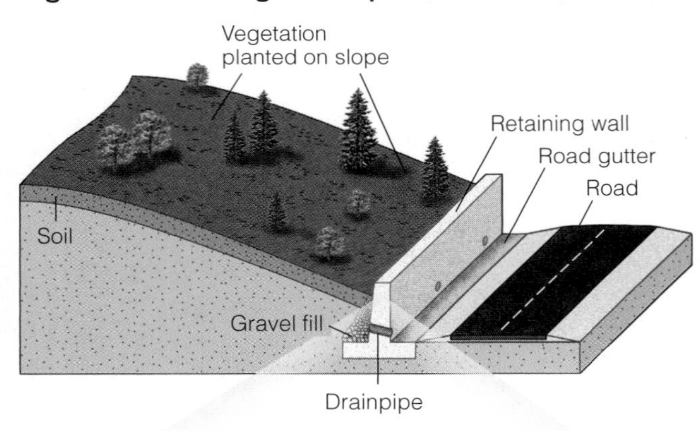

a Retaining walls anchored into bedrock, backfilled with gravel, and provided with drainpipes can support a slope's base and reduce landslides.

b A steel retaining wall built to stabilize the slope and keep falling and sliding rocks off the highway.

help secure the slopes at the Glen Canyon Dam on the Colorado River.

Recognition, prevention, and control of landslide-prone areas are expensive, but not nearly as expensive as the damage can be when such warning signs are ignored or not recognized. Unfortunately, there are numerous examples of landfill and dam collapses that serve as tragic reminders of the price paid in loss of lives and property damage when the warning signs of impending disaster are ignored.

Figure 11.29 Rock Bolts and Wire Mesh Help Reduce Landslides

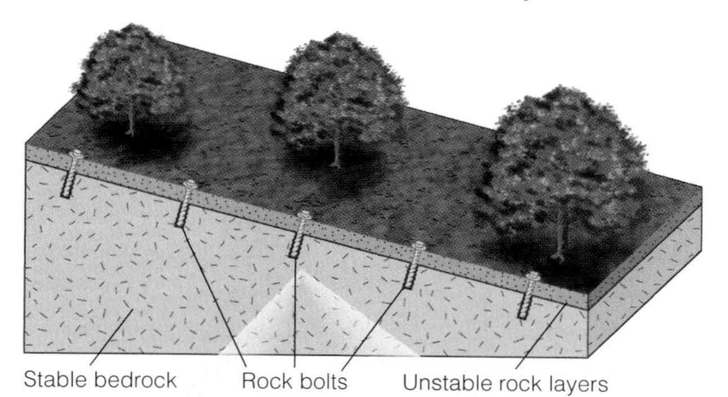

a Rock bolts secured in bedrock can help stabilize a slope and reduce landslides.

b Rock bolts and wire mesh are used to secure rock on a steep hillside in Brisbane, Australia.

Geo-Recap

Chapter Summary

- Mass wasting is the downslope movement of material under the direct influence of gravity. It frequently results in loss of life, as well as millions of dollars in damage annually.
- Mass wasting occurs when the gravitational force acting on a slope exceeds the slope's shear strength (the resisting forces that help maintain slope stability).
- The major factors causing mass wasting include slope angle, weathering and climate, water content, overloading, and removal of vegetation. It is usually several of these factors in combination that result in slope failure.
- Mass movements are generally classified on the basis of their rate of movement (rapid versus slow), type of movement (falling, sliding, or flowing), and type of material (rock, soil, or debris).
- Rockfalls are a common mass movement in which rocks free-fall. They are common along steep canyons, cliffs, and road cuts.
- Two types of slides are generally recognized. Slumps are rotational slides that involve movement along a curved surface and are most common in poorly consolidated or unconsolidated material. Rock slides occur when movement takes place along a more or less planar surface, and they usually involve solid pieces of rock.
- Several types of flows are recognized on the basis of their rate of movement (rapid versus slow), type of material (rock, sediment, or soil), and amount of water.

- Mudflows consist of mostly clay- and silt-sized particles and contain up to 30% water. They are most common in semiarid and arid environments, and generally follow preexisting channels.
- Debris flows are composed of larger particles and contain less water than mudflows.
- Earthflows move more slowly than either debris flows or mudflows, and move downslope as thick, viscous, tongue-shaped masses of wet regolith.
- Quick clays are clays that spontaneously liquefy and flow like water when they are disturbed.
- Solifluction is the slow downslope movement of water-saturated surface material and is most common in areas of permafrost.
- Creep, the slowest type of flow, is the imperceptible downslope movement of soil or rock. It is the most wide-spread of all types of mass wasting.
- Complex movements are combinations of different types of mass movements in which no single type is dominant. Most complex movements involve sliding and flowing.
- The most important factor in reducing or eliminating the damaging effects of mass wasting is a thorough geologic investigation to outline areas susceptible to mass movements.
- Although mass movement cannot be eliminated, its effects can be minimized by building retaining walls, draining excess water, regrading slopes, and planting vegetation.

Important Terms

complex movement (p. 290)
creep (p. 289)
debris flow (p. 287)
earthflow (p. 287)
mass wasting (p. 274)
mudflow (p. 286)
permafrost (p. 289)
quick clay (p. 287)
rapid mass movement (p. 278)
rock slide (p. 281)
rockfall (p. 278)
shear strength (p. 274)
slide (p. 279)
slow mass movement (p. 278)
slump (p. 280)
solifluction (p. 289)

Review Questions

1. Shear strength includes
   a. _____ the strength and cohesion of material;
   b. _____ the amount of internal friction between grains;
   c. _____ gravity;
   d. _____ all of these;
   e. _____ answers a and b.
2. Downslope movement of material along a more or less planar surface is a(n)
   a. _____ slump;
   b. _____ rockfall;
   c. _____ earthflow;
   d. _____ debris flow;
   e. _____ rock slide.
3. The most widespread and costly of all mass wasting processes is
   a. _____ slumps;
   b. _____ creep;
   c. _____ mudflows;
   d. _____ rockfalls;
   e. _____ quick clays.
4. Which of the following factors can actually enhance slope stability?
   a. _____ increasing the slope angle;
   b. _____ vegetation;
   c. _____ overloading;
   d. _____ rocks dipping in the same direction as the slope;
   e. _____ none of these.
5. Which of the following are the most fluid of mass movements?
   a. _____ earthflows;
   b. _____ mudflows;
   c. _____ debris flows;
   d. _____ slumps;
   e. _____ solifluction.
6. Mass wasting can occur
   a. _____ on gentle slopes;
   b. _____ on steep slopes;
   c. _____ in flat-lying areas;
   d. _____ all of these;
   e. _____ none of these.

7. Which of the following is a factor influencing mass wasting?
   a. ______ slope angle;
   b. ______ vegetation;
   c. ______ weathering;
   d. ______ water content;
   e. ______ all of these.
8. Which of the following is a type of mass wasting common in mountainous regions in which talus accumulates?
   a. ______ creep;
   b. ______ solifluction;
   c. ______ rockfalls;
   d. ______ slides;
   e. ______ mudflows.
9. Which of the following helps reduce the slope angle or provides support at the base of a hillside?
   a. ______ cut and fill;
   b. ______ retaining walls;
   c. ______ benching;
   d. ______ all of these;
   e. ______ none of these.
10. Former landslides and areas currently susceptible to slope failure can be identified by which of the following features?
    a. ______ tilted objects;
    b. ______ open fissures;
    c. ______ scarps;
    d. ______ hummocky surfaces;
    e. ______ all of these.
11. Discuss some of the ways slope stability can be maintained in order to reduce the likelihood of mass movements.
12. What potential value would a slope stability map be to a person seeking to purchase new home property? Using the slope stability map in Figure 11.24, locate the line that shows horizontal contact between rocks of different stability. What is the potential for mass wasting along this line, and why?
13. Discuss how topography and the underlying geology contribute to slope failure.
14. Why are slumps such a problem along highways and railroad tracks in areas with relief?
15. How would removing preexisting vegetation tend to affect most slopes in humid regions with respect to mass wasting?
16. Discuss how the different factors that influence mass wasting are interconnected.
17. How could mass wasting be recognized on other planets or moons? What would that tell us about the geology, and perhaps the atmosphere, of the planet or moon on which it occurred?
18. If an area has a documented history of mass wasting that has endangered or taken human life, how should people and governments prevent such events from happening again? Are most large mass wasting events preventable or predictable?
19. Why is it important to know about the different types of mass wasting?
20. Why is creep so prevalent? Why does it do so much damage? What are some of the ways that creep might be controlled?

CHAPTER 12

Running Water

Parvinder Sethi

OUTLINE

OBJECTIVES

At the end of this chapter, you will have learned that

- Running water, one part of the hydrologic cycle, does considerable geologic work.
- Water is continuously cycled from the oceans to land and back to the oceans.
- Running water transports large quantities of sediment and deposits sediment in or adjacent to braided and meandering rivers and streams.
- Alluvial fans (on land) and deltas (in a standing body of water) are deposited when a stream's capacity to transport sediment decreases.
- Flooding is a natural part of stream activity that takes place when a channel receives more water than it can handle.
- The several types of structures to control floods are only partly effective.
- Rivers and streams continuously adjust to changes.
- The concept of a graded stream is an ideal, although many rivers and streams approach the graded condition.
- Most valleys form and change in response to erosion by running water coupled with other geologic processes such as mass wasting.

The Sol Duc Falls in Olympic National Park, Washington. The milky-white color of this water is due to gacial silt.

INTRODUCTION

Mercury, Venus, Earth, and Mars share a similar early history of accretion, differentiation, and volcanism, but Earth is the only terrestrial planet with abundant surface water. The small size and high temperature of Mercury and the runaway greenhouse effect on Venus preclude the possibility of liquid water on these planets. Mars is too small and too cold for liquid water, although it does have some frozen water and trace amounts of water vapor in its atmosphere. However, satellite imagery reveals winding valleys and canyons that were probably eroded by running water during the planet's early history. In contrast, oceans and seas cover approximately 71% of Earth's surface.

The *hydrosphere* consists mostly of water in the oceans (Table 12.1), but it also includes water vapor in the atmosphere, groundwater (see Chapter 13), water frozen in glaciers (see Chapter 14), and water on land in lakes, swamps, bogs, streams, and rivers (see the chapter opening photograph). Our main concern in this chapter is with the small amount of running water confined to channels. In any consideration of the interactions among Earth's systems, it is important to note that running water has a tremendous impact on most of the land surface.

You have probably experienced the power of running water if you have ever swam or canoed in a rapidly flowing stream or river, but to truly appreciate the energy of moving water, you need only read the vivid accounts of floods. For example, at 4:07 p.m. on May 31, 1889, residents of Johnstown, Pennsylvania, heard "a roar like thunder" and within 10 minutes the town was destroyed when an 18-m-high wall of water tore through the town at more than 60 km/hr, sweeping up houses, debris, and entire families (Figure 12.1). According to one account, "Thousands of people desperately tried to escape the wave. Those caught by the wave found themselves swept up in a torrent of oily, muddy water, surrounded by tons of grinding debris. . . . Many became hopelessly entangled in miles of barbed wire from the destroyed wire works."*

When the flood was over, at least 2200 people were dead, some of them victims of a fire that broke out on floating debris on which they escaped the flood. The Johnstown flood, the most deadly river flood in U.S. history, resulted from heavy rainfall and the failure of a dam upstream from the town. This was not the first time that a dam had failed in this country nor would it be the last, but it was certainly the most tragic.

National Park Service

Figure 12.1 Aftermath of the Johnstown, Pennsylvania Flood On May 13, 1889, an 18-m-high wall of water destroyed Johnstown and killed at least 2200 people.

TABLE 12.1 Water on Earth

Location	Volume (km³)	Percent of Total
Oceans	1,327,500,000	97.20
Ice caps and glaciers	29,315,000	2.15
Groundwater	8,442,580	0.625
Freshwater and saline lakes and inland seas	230,325	0.017
Atmosphere at sea level	12,982	0.001
Stream channels	1,255	0.0001

*National Park Service—U.S. Department of Interior, Shiretown Information Service Online.

Every year, floods cause extensive property damage and fatalities, and yet we derive many benefits from running water, even from some floods. In Egypt, before completion of the Aswan High Dam in 1970, farmers depended on annual flooding of the Nile River to replenish their croplands (see Geo-Focus on page 307). Furthermore, running water—that is, water confined to channels—is one source of freshwater for agriculture, industry, domestic use, and recreation, and about 8% of all electricity used in North America is generated by falling water at hydroelectric generating plants. Large waterways throughout the world are avenues of commerce (Figure 12.2), and when Europeans explored the interior of North America, they followed the St. Lawrence, Mississippi, Missouri, and Ohio rivers.

WATER ON EARTH

Most of Earth's 1.33 billion km^3 of water (97.2%) is in the oceans, and nearly all of the rest is frozen in glaciers on land (2.15%) (Table 12.1). That leaves only 0.65% in the atmosphere, groundwater, lakes, swamps, bogs, and a tiny important amount in stream and river channels. Nevertheless, the water in channels is, with few exceptions, the most important geologic agent in modifying the land surface. Only in areas covered by vast glaciers, such as Greenland, and the driest deserts are other agents of erosion, transport, and deposition more important than running water. Even in deserts, however, evidence of running water is common, although channels are dry most of the time.

Much of our discussion of running water is descriptive, but always be aware that streams and rivers are dynamic systems that must continuously respond to change. For example,

Sue Monroe

Figure 12.2 The Danube River at Novi Sad, Serbia The Danube is one of Europe's important waterways used for commerce and recreation.

paving in urban areas increases surface runoff to waterways, and other human activities such as building dams and impounding reservoirs also alter the dynamics of stream and river systems. Natural changes, too, affect the complex interacting parts of stream and river systems.

The Hydrologic Cycle

The connection between precipitation and clouds is obvious, but where does the moisture for rain and snow come from in the first place? In the previous section, we noted that 97.2% of all water on Earth is in the oceans, so you might immediately suspect that the oceans are the ultimate source of precipitation. In fact, water is continuously recycled from the oceans, through the atmosphere, to the continents, and back to the oceans. This **hydrologic cycle,** as it is called (Figure 12.3), is powered by solar radiation and is possible because water changes easily from liquid to gas (water vapor) under surface conditions. About 85% of all water entering the atmosphere comes from a layer about 1 m thick that evaporates from the oceans each year. The remaining 15% comes from water on land, but this water originally came from the oceans as well.

Regardless of its source, water vapor rises into the atmosphere where the complex processes of cloud formation and condensation take place. About 80% of the world's precipitation falls directly back into the oceans, in which case the hydrologic cycle is a three-step process of evaporation, condensation, and precipitation. For the 20% of all precipitation that falls on land, the hydrologic cycle is more complex, involving evaporation, condensation, movement of water vapor from the oceans to land, precipitation, and runoff. Some precipitation evaporates as it falls and reenters the cycle, but about 36,000 km^3 of the precipitation that falls on land returns to the oceans by **runoff,** the surface flow in streams and rivers.

Not all precipitation returns directly to the oceans. Some is temporarily stored in lakes and swamps, snowfields and glaciers, or seeps below the surface where it enters the groundwater system (see Chapter 13). Water might remain in some of these reservoirs for thousands of years, but eventually glaciers melt, lakes and groundwater feed streams and rivers, and this water returns to the oceans. Even the water used by plants evaporates, a process known as *transpiration*, and returns to the atmosphere. In short, all water derived from the oceans eventually makes it back to the oceans and can thus begin the hydrologic cycle again.

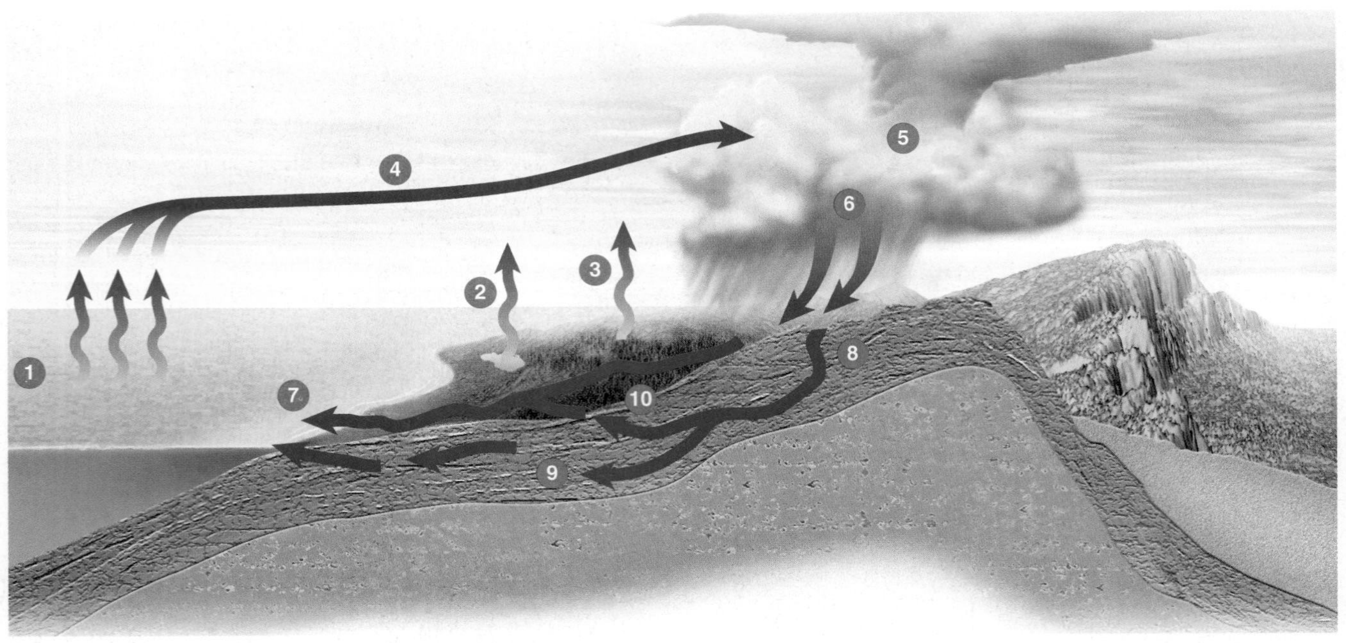

Figure 12.3 Stages of the Hydrologic Cycle Water is recycled from the oceans to land and back to the oceans.

Fluid Flow

Solids are rigid substances that retain their shapes unless deformed by a force, but fluids—that is, liquids and gases—have no strength, so they flow in response to any force, no matter how slight. Liquid water flows downslope in response to gravity, but its flow may be *laminar* or *turbulent.* In laminar flow, lines of flow called streamlines parallel one another with little or no mixing between adjacent layers (Figure 12.4a). All flow is in one direction only, and it remains unchanged through time. In turbulent flow, streamlines intertwine, causing complex mixing within the moving fluid (Figure 12.4b). If we could trace a single water molecule in turbulent flow, it may move in any direction at a particular time although its overall movement is in the direction of flow.

Runoff during a rainstorm depends on the **infiltration capacity,** the maximum rate at which surface materials absorb water. Several factors control the infiltration capacity, including intensity and duration of rainfall. If rain is absorbed as fast as it falls, no surface runoff takes place. Loosely packed dry soil absorbs water faster than tightly packed wet soil, and thus more rain must fall on loose dry soil before runoff begins. Regardless of the initial condition of surface materials, once they are saturated, excess water collects on the surface and, if on a slope, it moves downhill.

RUNNING WATER

The term *running water* applies to any surface water that moves from higher to lower areas in response to gravity. We have already noted that running water is very effective in modifying Earth's land surface by erosion, and that it is the primary geologic process responsible for sediment transport and deposition in many areas. Indeed, it is responsible for the tiniest rills in farmer's fields to scenic wonders such as the Grand Canyon in Arizona, as well as vast deposits such as the Misssissippi River delta.

Sheet Flow and Channel Flow

Even on steep slopes, flow is initially slow and hence causes little or no erosion. As water moves downslope, though, it accelerates and may move by *sheet flow,* a more or less continuous film of water flowing over the surface. Sheet flow is

Figure 12.4 Laminar and Turbulent Flows

a In laminar flow, streamlines are parallel and little mixing takes place in the fluid.

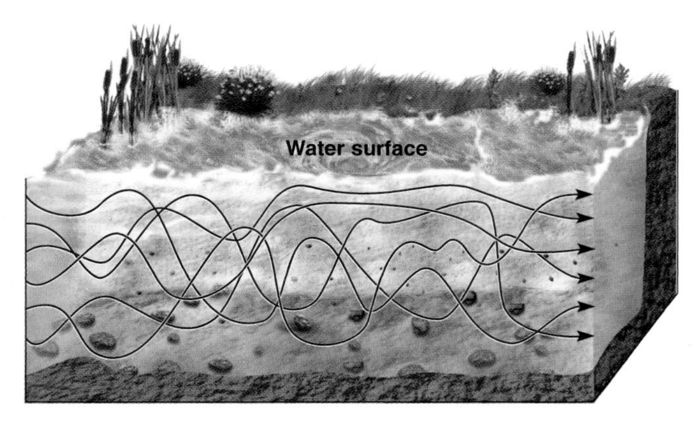

b In turbulent flow, streamlines intertwine, indicating mixing in the fluid.

not confined to depressions, and it accounts for *sheet erosion*, a particular problem on some agricultural lands (see Chapter 6).

In *channel flow*, surface runoff is confined to troughlike depressions that vary in size from tiny rills with a trickling stream of water to the Amazon River in South America, which is 6450 km long and at one place 2.4 km wide and 90 m deep. We describe flow in channels with terms such as *rill*, *brook*, *creek*, *stream*, and *river*, most of which are distinguished by size and volume. Here we use the terms *stream* and *river* more or less interchangeably, although the latter usually refers to a larger body of running water.

Streams and rivers receive water from several sources, including sheet flow and rain that falls directly into their channels. Far more important, though, is the water supplied by soil moisture and groundwater, both of which flow downslope and discharge into waterways. In areas where groundwater is plentiful, streams and rivers maintain a fairly stable flow year-round because their water supply is continuous. In contrast, the amount of water in streams and rivers of arid and semiarid regions fluctuates widely because they depend more on infrequent rainstorms and surface runoff for their water.

Gradient, Velocity, and Discharge

Water in any channel flows downhill over a slope known as its **gradient.** Suppose a river has its headwaters (source) 1000 m above sea level and it flows 500 km to the sea, so it drops vertically 1000 m over a horizontal distance of 500 km. Its gradient is found by dividing the vertical drop by the horizontal distance, which in this example is 1000 m/500 km = 2 m/km (Figure 12.5a). On average, this river drops vertically 2 m for every kilometer along its course.

In this example, we calculated the average gradient for a hypothetical river, but gradients vary not only among channels but even along the course of a single channel. Rivers and streams are steeper in their upper reaches (near their headwaters) where they may have gradients of several tens of meters per kilometer, but they have gradients of only a few centimeters per kilometer where they discharge into the sea.

The **velocity** of running water is a measure of the downstream distance water travels in a given time. It is usually expressed in meters per second (m/sec) or feet per second (ft/sec), and it varies across a channel's width as well as along its length. Water moves more slowly and with greater turbulence near a channel's bed and banks because friction is greater there than it is some distance from these boundaries (Figure 12.5b). Channel shape and roughness also influence flow velocity. Broad, shallow channels and narrow, deep channels have proportionately more water in contact with their perimeters than channels with semicircular cross sections (Figure 12.5c). So, if other variables are the same, water flows faster in a semicircular channel because there is less frictional resistance. As one would expect, rough channels, such as those strewn with boulders, offer more frictional resistance to flow than do channels with a bed and banks composed of sand or mud.

Intuitively, you might think that the gradient is the most important control on velocity—the steeper the gradient, the greater the velocity. In fact, a channel's average velocity actually increases downstream even though its gradient decreases! Keep in mind that we are talking about average velocity for a long segment of a channel, not velocity at a single point. Three factors account for this downstream increase in velocity. First, velocity increases even with decreasing gradient in response to the acceleration of gravity. Second, the upstream reaches of channels tend to be boulder-strewn, broad, and shallow, so frictional resistance to flow is high, whereas downstream segments of the same channels are more semicircular and have banks composed of finer materials. And finally, the number of smaller tributaries joining a larger channel increases downstream. Thus, the total volume of water (discharge) increases, and increasing discharge results in greater velocity.

Figure 12.5 Gradient and Flow Velocity

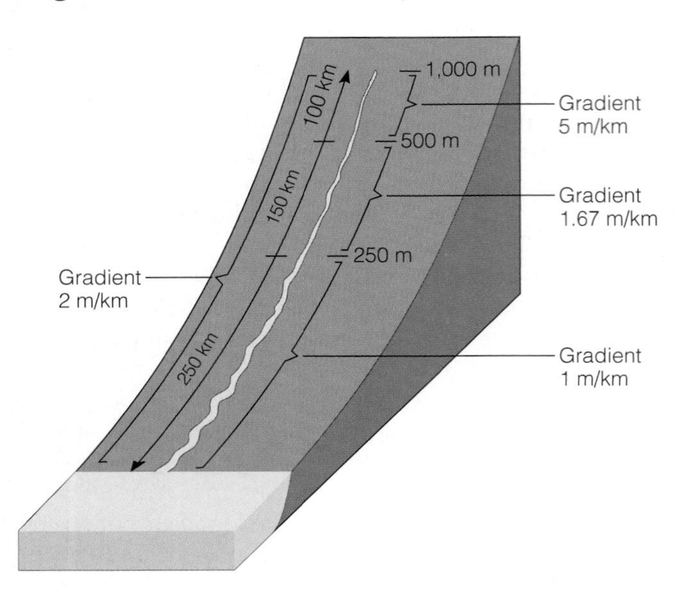

a The average gradient of this stream is 2 m/km; however, gradient can be calculated for any segment of a stream, as shown in this example. Notice that the gradient is steepest in the headwaters area and decreases in a downstream direction.

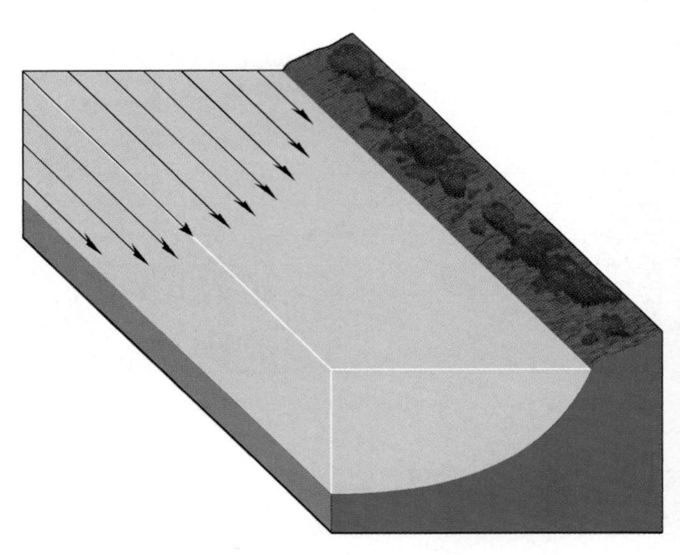

b The maximum flow velocity is near the center and top of a straight channel where the least friction takes place. The arrows are proportional to velocity.

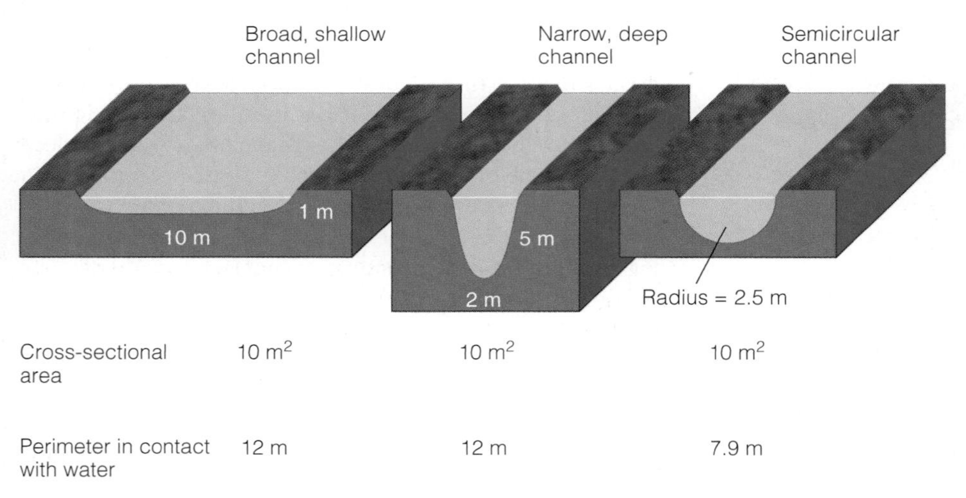

	Broad, shallow channel	Narrow, deep channel	Semicircular channel
Cross-sectional area	10 m^2	10 m^2	10 m^2
Perimeter in contact with water	12 m	12 m	7.9 m

c These three differently shaped channels have the same cross-sectional area; however, the semicircular one has less water in contact with its perimeter and thus less frictional resistance to flow.

Specifically, **discharge** is the volume of water that passes a particular point in a given period of time. Discharge is found from the dimensions of a water-filled channel—that is, its cross-sectional area (A) and flow velocity (V). Discharge (Q) is then calculated with the formula $Q = VA$ and is expressed in cubic meters per second (m^3/sec) or cubic feet per second (ft^3/sec). The Mississippi River has an average discharge of 18,000 m^3/sec, and the average discharge for the Amazon River in South America is 200,000 m^3/sec.

In most rivers and streams, discharge increases downstream as more and more water enters a channel. However, there are a few exceptions. Because of high evaporation rates and infiltration, the flow in some desert waterways actually decreases downstream until the water disappears. And even in perennial rivers and streams, discharge is obviously highest during times of heavy rainfall and at a minimum during the dry season.

HOW DOES RUNNING WATER ERODE AND TRANSPORT SEDIMENT?

Streams and rivers possess two kinds of energy to accomplish their tasks of erosion and transport: potential and kinetic. *Potential energy* is the energy of position, such as the energy of water at high elevation. During stream flow, potential is

Geo-Focus The River Nile and the History of Egypt

Egypt is well known for its remarkable antiquities, such as the pyramids and sphinx at Giza, the tombs in the Valley of the Kings, and numerous temples and monuments. However, Egypt also has some fascinating geologic features, particularly the Nile River. In fact, in no other country has its development and prosperity been so closely tied to a single geologic feature (Figure 1). Although the Nile, at 6825 km long, is the world's longest river, its discharge of 1584 m^3/sec makes it a modest-sized river when compared with the great rivers of the world. Nevertheless, it is the lifeblood of an entire country.

The Nile River rises as the Blue Nile in Ethiopia and the While Nile in Sudan, which join and flow north into Egypt and eventually into the Mediterranean Sea. If it were not for the abundant precipitation in the Ethiopian highlands, Egypt, which is nearly all desert, would have no river at all. Indeed, the Atbara River in Sudan is its only other tributary. Once the Nile enters Egypt it has no tributaries, although a number of gullies called *wadis* do carry water to the river following infrequent rainstorms. The Nile is flanked on the west and east by desert, and the only agricultural land is on the Nile's floodplain (Figure 2) and delta, at the Fayum Depression, and at a few scattered oases.

Although people have occupied what is now Egypt for hundreds of thousands of years, Egyptian civilization took root there about 5000 years ago and persisted for more than 3000 years. Even when ancient Greece flourished, Egyptian civilization was already ancient. But it was a civilization totally dependent on the annual flooding of the Nile, during which fertile deposits accumulated on the floodplain. Indeed, in ancient Egypt, taxes were even determined by the extent of the annual flooding. And, of course, the Nile was also the main avenue of commerce throughout the region.

The Nile is unique in that it has so few tributaries and flows such a great distance through a desert, but in most other respects, it is like any other river. It responds to short- and long-term climatic changes, interruptions in the system where dams are built, and changes in sea level. During the Late Miocene Epoch, about 6 or 7 million years ago, base level for the Nile was much lower than it is now because the Mediterranean Sea was closed off from the Atlantic at the Strait of Gibraltar. As a result, the sea dried up, leaving a vast plain more than 3000 m below sea level. The Nile and other rivers around the Mediterranean basin responded by cutting deep canyons as they adjusted to this new lower base level. By the Pliocene Epoch (about 5 million years ago), however, the barrier at the Strait of Gibraltar was breached and the Mediterranean began to refill, thus raising base level.*

During the time of deep erosion, the Nile cut a canyon about 10–20 km wide and 2500 m deep near Cairo. Even 465 km upstream at Aswan, the canyon was cut 200 m below present-day sea level. As sea level rose, the ancient river canyon became an arm of the sea, but by about 3.3 million years ago, the river began depositing gravel, sand, and mud in its canyon and it took on an appearance much like it has today.

Since the completion of the Aswan High Dam in 1970, flooding on the Nile has been drastically reduced, so now it does not deposit mud on its floodplain each year. As a result, Egyptian farmers have had to apply chemical fertilizers to maintain agricultural productivity.

*Actually, the isolation of the Mediterranean from the Atlantic and its refilling probably occurred several times.

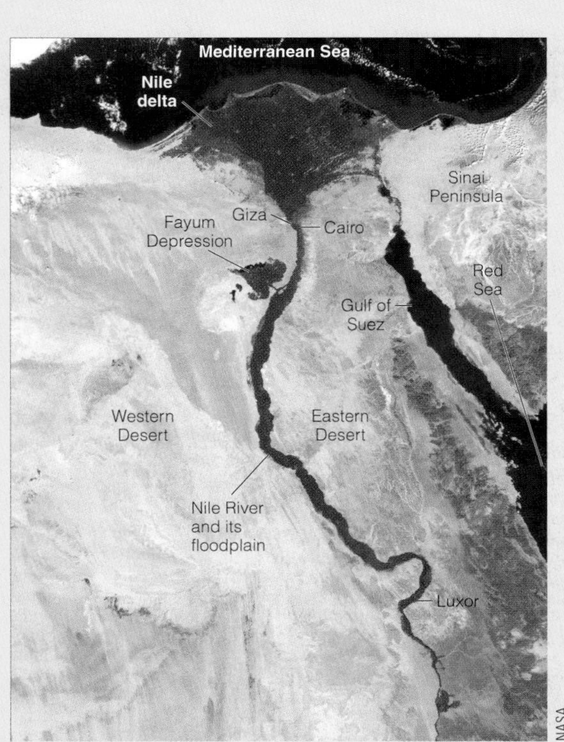

Figure 1 Satellite image of the Nile River as it flows north through Egypt. The only arable land is on the river's floodplain and its delta and at a few scattered oases. The floodplain measures about 20 km across at its widest point.

Figure 2 At this location, between Luxor and Aswan, the floodplain of the Nile River is only a few hundred meters wide. The view is toward the Eastern Desert, which begins a short distance from the river.

converted to *kinetic energy*, which is the energy of motion. Much of this kinetic energy is used up in fluid turbulence, but some is available for erosion and transport. You already know that erosion involves the removal of weathered materials from their source area, so the materials transported by a stream include a load of solid particles (mud, sand, and gravel) and a dissolved load.

Because the **dissolved load** of a stream is invisible, it is commonly overlooked, but it is an important part of the total sediment load. Some of it is acquired from a stream's bed and banks where soluble rocks such as limestone are present, but much of it is carried into waterways by sheet flow and by groundwater. A stream's solid load is made up of particles ranging from clay sized (>1/256 mm) to huge boulders, much of it supplied by mass wasting, but some is eroded directly from a stream's bed and banks (◗ Figure 12.6). The direct impact of running water, **hydraulic action,** is sufficient to set particles in motion, just as the stream from a garden hose gouges out a hole in soil.

Running water carrying sand and gravel erodes by **abrasion,** as exposed rock is worn and scraped by the impact of these particles (◗ Figure 12.7a). Circular to oval depressions called *potholes* in streambeds are one manifestation of abrasion (Figure 12.7b, c). These depressions form where swirling currents with sand and gravel eroded the rock.

Once materials are eroded, they are transported for some distance from their source and eventually deposited. The dissolved load is transported in the water itself, but the load of solid particles moves as *suspended load* or *bed load.* The **suspended load** consists of the smallest particles of silt and clay, which are kept suspended above the channel's bed by fluid turbulence (◗ Figure 12.8). It is the suspended load of streams and rivers that gives their water its murky appearance.

The **bed load** of larger particles, mostly sand and gravel, cannot be kept suspended by fluid turbulence so that it is transported along the bed. However, some of the sand may be temporarily suspended by currents that swirl the streambed and lift grains into the water. The grains move forward with the water, but also settle and finally come to rest and then again move by the same process of intermittent bouncing and skipping, a phenomenon known as *saltation* (Figure 12.8). Particles too large to be even temporarily suspended are transported by traction; that is, they simply roll or slide along a channel's bed.

◗ Figure 12.6 Hydraulic Action and Erosion

a This stream acquires some of its sediment by eroding its banks.

b Some of the sediment in the Snake River of Idaho comes from these talus cones that accumulate as a result of mass wasting.

Figure 12.7 Abrasion by Running Water Carrying Sand and Gravel

a The rocks just above water level have been smoothed and polished by abrasion.

b These potholes in the bed of the Chippewa River in Ontario, Canada, are about 1 m across. Two potholes have merged to form a larger composite pothole.

c These stones from a pothole measure 7–9 cm across. They are spherical and smooth because of abrasion.

DEPOSITION BY RUNNING WATER

Some of the sediment now being deposited in the Gulf of Mexico by the Mississippi River came from sources as distant as Pennsylvania, Minnesota, and Alberta, Canada. Transport might be lengthy, but deposition eventually takes place. Some deposits accumulate along the way in channels, on adjacent floodplains, or where rivers and streams discharge from mountains onto nearby lowlands or where they flow into lakes or seas.

Rivers and streams constantly erode, transport, and deposit sediment, but they do most of their geologic work when they flood. Consequently, their deposits, collectively called **alluvium,** do not represent the day-to-day activities of running water, but rather the periodic sedimentation that takes place during floods. Recall from Chapter 6 that sediments accumulate in *depositional environments*

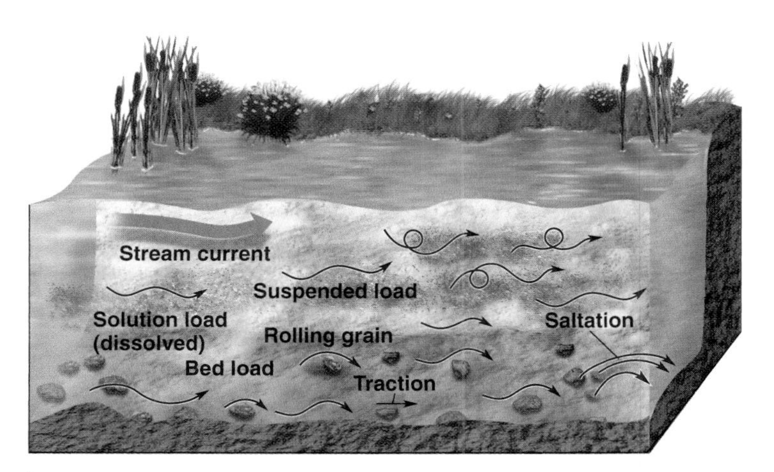

Figure 12.8 Sediment Transport Sediment transport as bed load, suspended load, and dissolved load. Flow velocity is highest near the surface, but gravel- and sand-sized particles are too large to be lifted far from the streambed so they make up the bed load, whereas silt and clay are in the suspended load.

characterized as continental, transitional, and marine (see Figure 6.17). Deposits of rivers and streams are found mostly in the first two of these settings; however, much of the detrital sediment found on continental margins is derived from the land and transported to the oceans by running water.

The Deposits of Braided and Meandering Channels

A **braided stream** has an intricate network of dividing and rejoining channels separated from one another by sand and gravel bars (Figure 12.9). Seen from above, the channels resemble the complex strands of a braid. Braided channels develop when the sediment supply exceeds the transport capacity of running water, resulting in the deposition of sand and gravel bars. During high-water stages, the bars are submerged, but when the water is low, they are exposed and divide a single channel into multiple channels. Braided streams have broad, shallow channels and are characterized as bed-load transport streams because they transport and deposit mostly sand and gravel.

Braided channels are common in arid and semiarid regions with sparse vegetation and surface materials that are easily eroded. So much sediment is released from melting glaciers that rivers and streams discharging from them are also commonly braided (see Chapter 14).

Meandering streams have a single sinuous channel with broadly looping curves known as *meanders* (Figure 12.10). Channels of meandering streams are semicircular in cross section along straight reaches, but markedly asymmetric at meanders, where they vary from shallow to deep across the meander. The deeper side of the channel is known as the *cut bank* because greater velocity and fluid turbulence erode it. In contrast, flow velocity is at a minimum on the opposite bank, which slopes gently into the channel. As a result of this unequal distribution of flow velocity across meanders, the cut bank erodes and a **point bar** is deposited on the gently sloping inner bank. (Figure 12.11).

Meanders commonly become so sinuous that the thin neck of land between adjacent ones is cut off during a flood. Many of the floors of valleys with meandering channels are marked by crescent-shaped **oxbow lakes,** which are simply cutoff meanders (Figures 12.10 and 12.12). Oxbow lakes may persist for a long time, but they eventually fill with organic matter and fine-grained sediments carried by floods.

Figure 12.9 Braided Streams and Their Deposits

James S. Monroe

a A braided stream in Denali National Park, Alaska.

James S. Monroe

b This small braided stream is near Grindelwald, Switzerland. The deposits of both streams are mostly sand and gravel.

Floodplain Deposits

Channels periodically receive more water than they can accommodate, so they overflow their banks and spread across adjacent low-lying, relatively flat **floodplains** (Figure 12.10). Floodplain sediments might be sand and gravel that accumulated when meandering streams deposited a succession of point bars as they migrated laterally (Figure 12.13a). More commonly, however, fine-grained sediments, mostly mud, are dominant on floodplains. During a flood, a stream overtops its banks and water pours onto the floodplain, but as it does so, its velocity and depth rapidly decrease. As a result, ridges of sandy alluvium known as **natural levees** are deposited along the channel margins, and mud is carried beyond the natural levees into the floodplain where it settles from suspension (Figure 12.13b, c).

Deltas

Where a river or stream flows into a standing body of water, such as a lake or the ocean, its flow velocity rapidly diminishes and any sediment in transport is deposited. Under some circumstances, this deposition creates a **delta,** an alluvial deposit that causes the shoreline to build outward into the lake or sea, a process called *progradation.* The simplest prograding deltas have a characteristic vertical sequence of *bottomset beds* overlain successively by *foreset beds* and *topset beds* (Figure 12.14). This vertical sequence develops when a river or stream enters another body of water where the finest sediment (silt and clay) is carried some distance out into the lake or sea; there it settles to form bottomset beds. Nearer shore, foreset beds are deposited as gently inclined layers, and topset beds, consisting of the coarsest sediments, are deposited in a network of *distributary channels* traversing the top of the delta (Figure 12.14).

Small deltas in lakes may have the three-part sequence described above, but deltas deposited along seacoasts are

What Would You Do?

Given what is known about the dynamics of running water in channels, it is remarkable that houses are still built on the cut banks of meandering rivers. No doubt, property owners think these locations provide good views because they sit high above the adjacent channel. Explain why you would or would not build a house in such a location. What recommendations would you make to a planning commission on land use for areas as described here? Are there any specific zoning regulations or building codes you might favor?

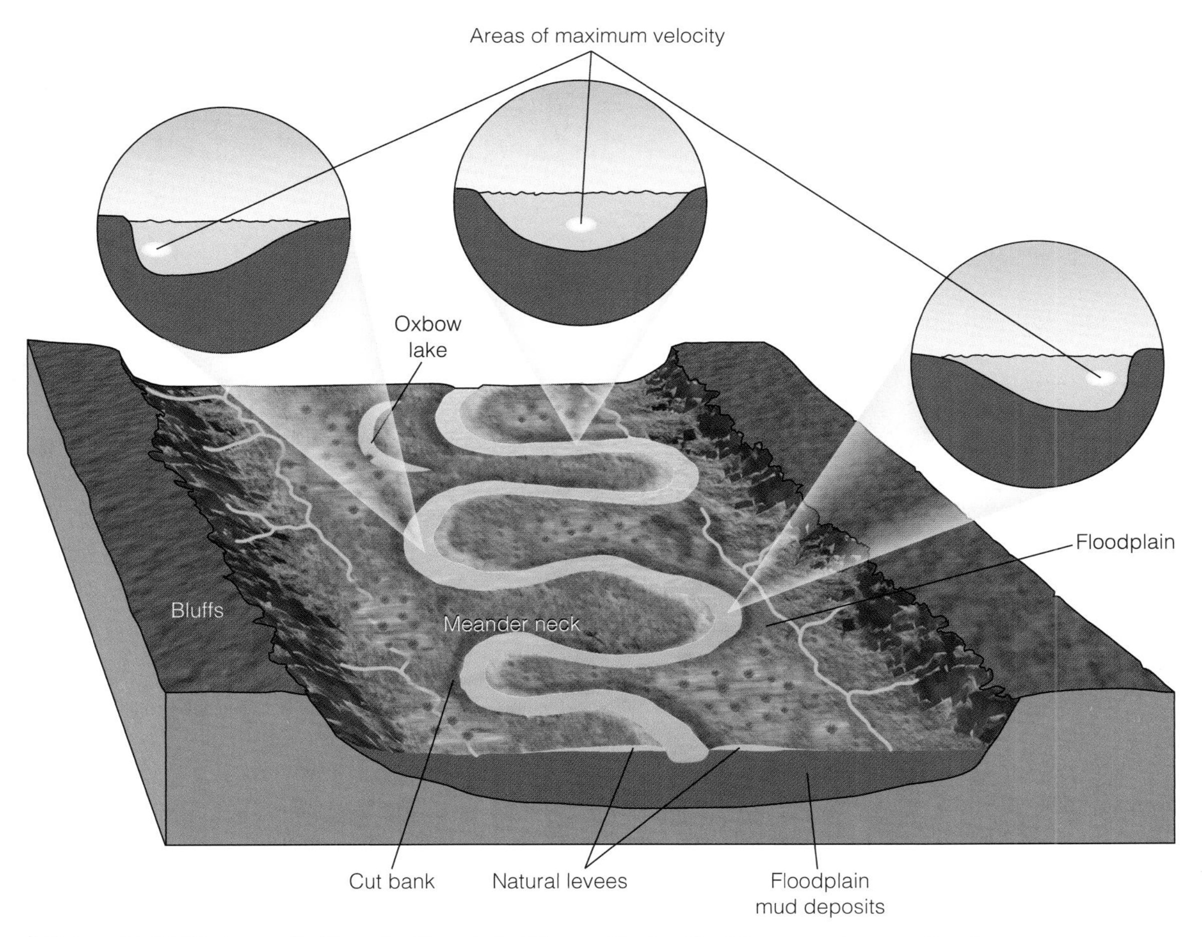

Figure 12.10 Diagrammatic View of a Meandering Stream The length of the arrows is proportional to flow velocity.

a Two point bars of sand in Otter Creek in Yellowstone National Park in Wyoming. Note that the point bars are inclined into the deeper part of the channel.

b Point bar of gravel in the Middle Fork of the San Joaquin River at Devil's Postpile National Monument in California.

Figure 12.11 Point Bars Point bars form on the gently sloping side of meanders where flow velocity is lowest.

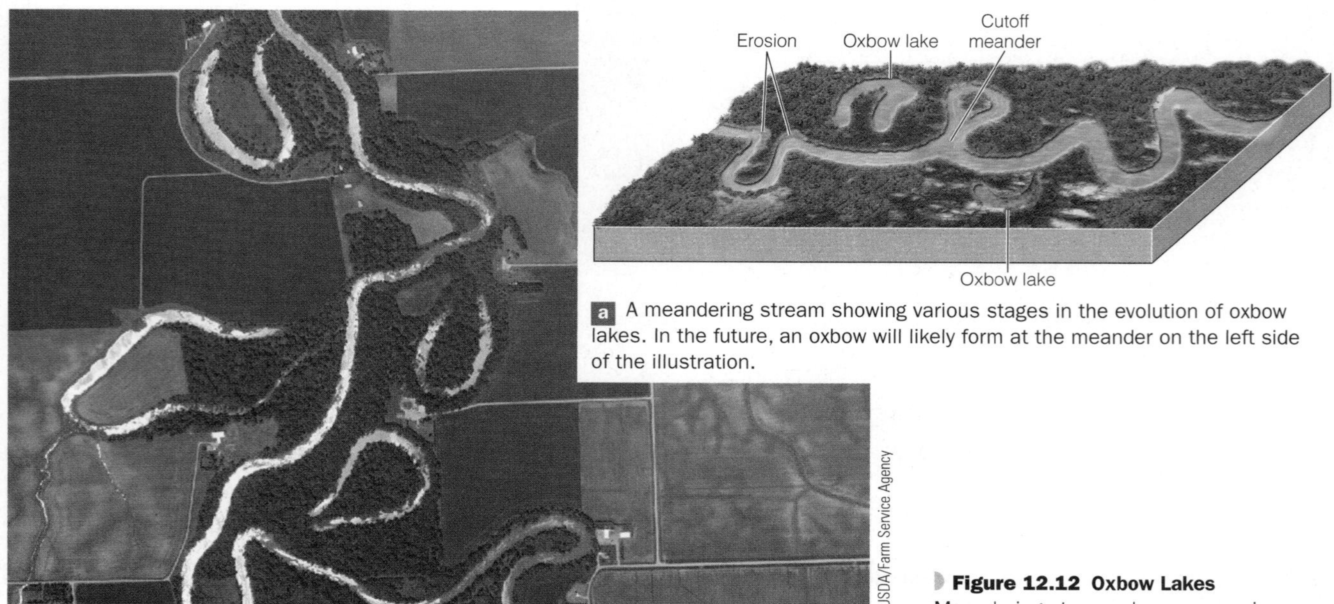

a A meandering stream showing various stages in the evolution of oxbow lakes. In the future, an oxbow will likely form at the meander on the left side of the illustration.

b Oxbow lakes along the Red River in Minnesota.

Figure 12.12 Oxbow Lakes Meandering streams become so sinuous that meanders get cut off.

Figure 12.13 Floodplain Deposits

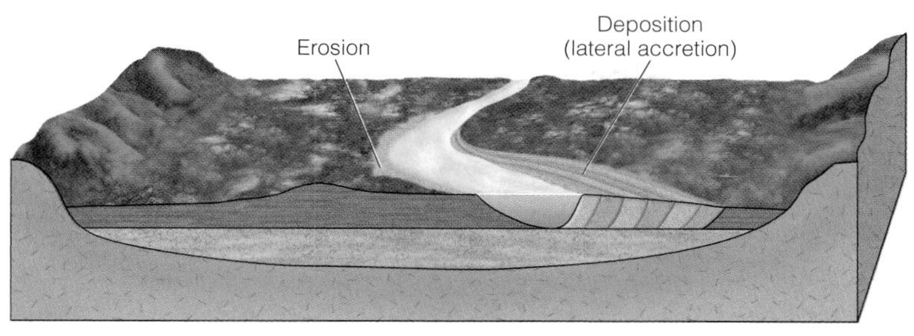

a Floodplain deposits formed by lateral accretion of point bars are made up mostly of sand.

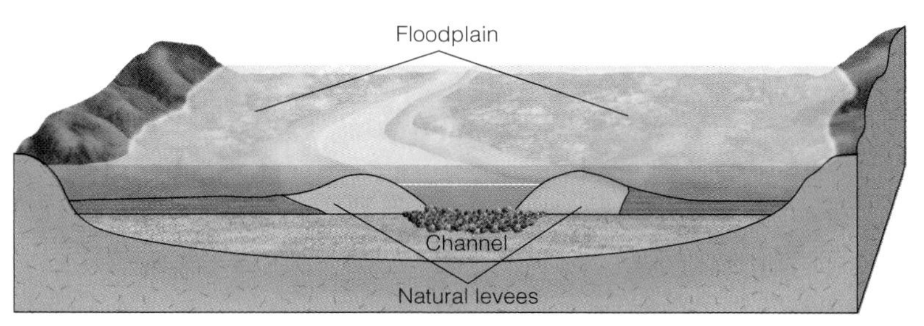

b The origin of vertical accretion deposits. During floods, streams deposit natural levees, and silt and mud settle from suspension on the floodplain.

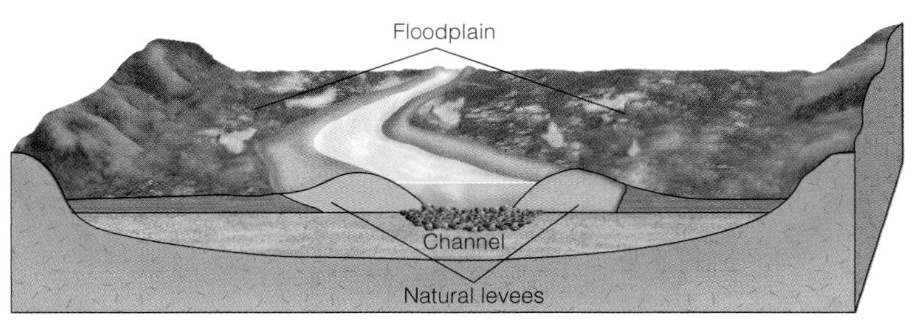

c After flooding.

much larger, far more complex, and considerably more important as potential areas of natural resources. In fact, depending on the relative importance of stream (or river), wave, and tide processes, geologists identify three main types of marine deltas (Figure 12.15). *Stream-dominated deltas* have long fingerlike sand bodies, each deposited in a distributary channel that progrades far seaward. The Mississippi delta is a good example. In contrast, the Nile delta in Egypt is *wave dominated.* It also has distributary channels, but the seaward margin of the delta consists of islands reworked by waves, and the entire margin of the delta progrades. *Tide-dominated deltas* are continuously modified into tidal sand bodies that parallel the direction of tidal flow.

Alluvial Fans

Fan-shaped deposits of alluvium on land known as **alluvial fans** form best on lowlands with adjacent highlands in arid and semiarid regions where little vegetation exists to stabilize surface materials (Figure 12.16). During periodic rainstorms, surface materials are quickly saturated and surface runoff is funneled into a mountain canyon leading to adjacent lowlands. In the mountain canyon, the runoff is confined so that it cannot spread laterally, but when it discharges onto the lowlands, it quickly spreads out, its velocity diminishes, and deposition ensues. Repeated episodes of sedimentation result in the accumulation of a fan-shaped body of alluvium.

Deposition by running water in the manner just described is responsible for many alluvial fans. In this case, they are composed mostly of sand and gravel, both of which contain a variety of sedimentary structures. In some cases, though, the water flowing through a canyon picks up so much sediment that it becomes a viscous debris flow. Consequently, some alluvial fans consist mostly of debris-flow

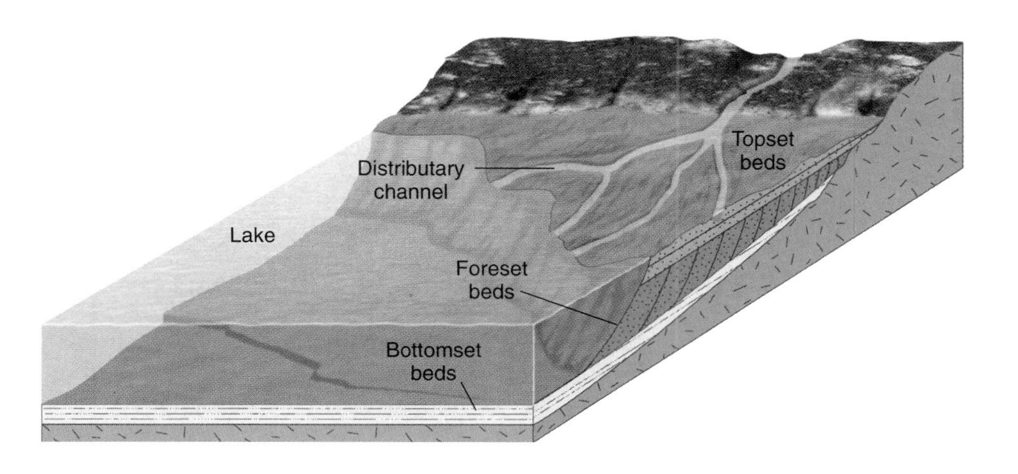

Figure 12.14 Prograding Delta Internal structure of the simplest type of prograding delta. Small deltas in lakes may have this structure, but marine deltas are much more complex.

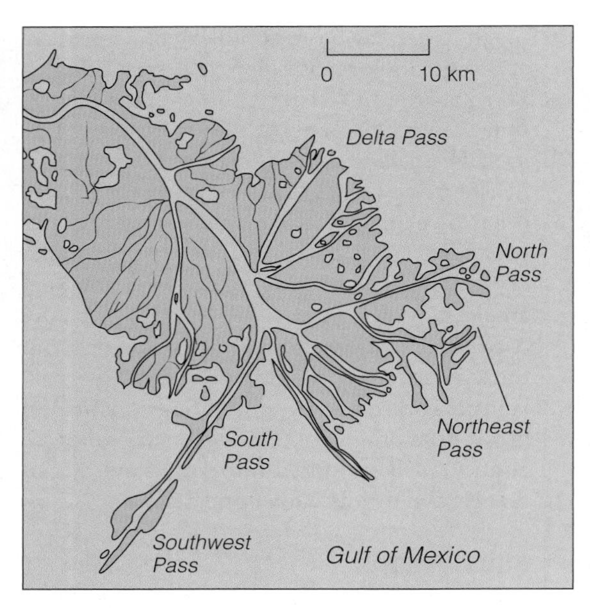

a The Mississippi River delta on the U.S. Gulf Coast is stream dominated.

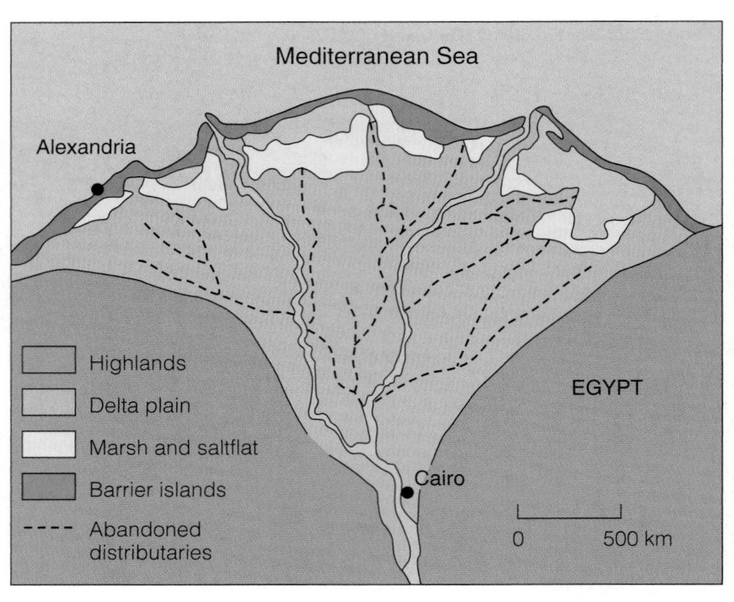

b The Nile delta of Egypt is wave dominated.

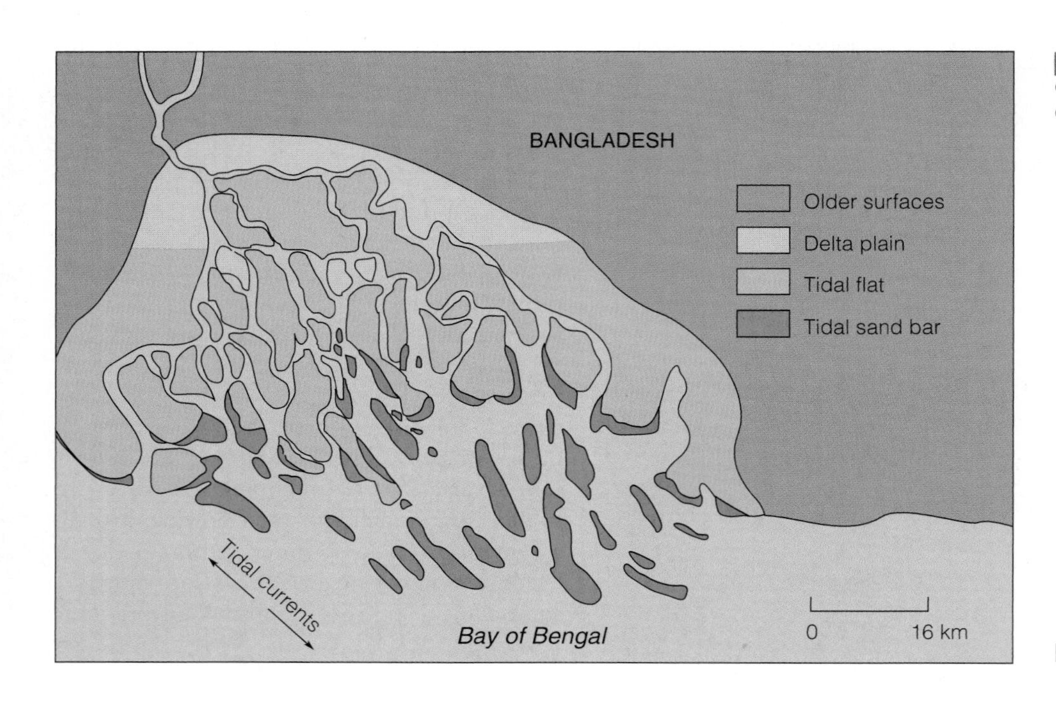

c The Ganges–Brahmaputra delta of Bangledesh is tide dominated.

Figure 12.15 Marine Deltas

deposits that show little or no layering. Of course, the dominant type of deposition can change through time, so a particular fan might have both types of deposits.

CAN FLOODS BE PREDICTED AND CONTROLLED?

When a river or stream receives more water than its channel can handle, it floods, occupying part or all of its floodplain. Indeed, floods are so common that unless they cause considerable property damage or fatalities, they rarely rate more than a passing notice in the news. Widespread catastrophic flooding in the United States took place in 1993 (see Geo-inSight on pages 316 and 317), but since then, several other areas in North America and elsewhere have experienced serious flooding. As this was being written in August 2007, another tragic flood was taking place in Southeast Asia, particularly in India and Bangladesh. More than 2100 people have perished and millions have been displaced by swollen rivers, overflowing from monsoon rains. As displaced people flee to high ground, so does the wildlife, including poisonous snakes that have claimed dozens of lives.

People have tried to control floods for thousands of years. Common practices are to construct dams that impound reservoirs and to build levees along stream banks (Figure 12.17a, b). Levees raise the banks of a stream, thereby restricting flow during floods. Unfortunately, deposition within the channel raises the streambed, making the levees useless unless they, too, are raised. Levees along the banks of the Huang He in China caused the streambed to rise more than 20 m above its surrounding floodplain in 4000 years. When the Huang He breached its levees in 1887, more than 1 million people were killed. Sacramento, California, lying at the junction of two rivers, is among the most flood-prone cities in the United States. Some of the levees that protect the city are 150 years old and in poor condition; the cost of repairing them has risen to as much as $250,000 per 100 m.

Dams and levees alone are insufficient to control large floods, so floodways are also used in many areas. A floodway is a channel constructed to divert part of the excess water in a stream around populated areas or areas of economic importance (Figure 12.17c). Some communities build *floodwalls* to protect them from floods. Floodwalls have gates that permit access to the waterway, but can be closed when the water rises (Figure 12.17d). Reforestation of cleared land also reduces the potential for flooding because vegetated soil helps prevent runoff by absorbing more water.

When flood-control projects are well planned and constructed, they are functional. What many people fail to realize is that these projects are designed to contain floods of a given size; should larger floods occur, rivers spill onto floodplains anyway. Furthermore, dams occasionally collapse, and reservoirs eventually fill with sediment unless dredged. In short, flood-control projects not only are initially expensive, but also require constant, costly maintenance.

What Would You Do?

The largest part of many states' mineral revenues comes from sand and gravel, most of which is used in construction. You become aware of a sand and gravel deposit that you can acquire for a small investment. How would the proximity of the deposit to potential markets influence your decision? Assuming the deposit was stream deposited, would it be important to know whether deposition took place in braided or meandering channels? How could you tell one from the other?

As for predicting floods, the best that can be done is to monitor streams, evaluate their past behavior, and anticipate floods of a given size in a specified period. Most people have heard of 10-year floods, 20-years floods, and so on, but how are such determinations made? The U.S. Geologic Survey, as well as state agencies, record and analyze stream behavior through time and anticipate floods of a specified size. So a 20-year flood, for example, is the period during which a flood of a given magnitude can be expected. It does not mean that the river in question will have a flood of that size every 20 years, only that over a long period of time, it will average 20 years. Or we can say that the chances of a 10-year flood taking place in any one year are 1 in 10 (1/10). In fact, it is possible that two 10-year floods could take place in successive years, but then not occur again for several decades.

DRAINAGE SYSTEMS

Thousands of waterways, which are parts of larger drainage systems, flow directly or indirectly into the oceans. The only exceptions are some rivers and streams that flow into desert basins surrounded by higher areas. But even these are parts of larger systems consisting of a main channel with all its

Figure 12.16 Alluvial Fans and Their Deposits

Parvinder Sethi

a These alluvial fans at the base of the Panamint Range in Death Valley, California, were deposited where streams discharged from mountain canyons onto adjacent lowlands.

James S. Monroe

b Ancient alluvial fan deposits in Montana are made up of sand and gravel.

Geo-inSight The Flood of '93

Although several floods take place each year that cause damage, injuries, and fatalities, the last truly vast river flooding in North America occurred during June and July of 1993. Now called the Flood of '93, it was responsible for 50 deaths and 70,000 were left homeless. Extensive property damage occured in several states, but particularly hard hit were Missouri and Iowa (see chart). Unusual behavior of the jet stream and the convergence of air masses over the Midwest were responsible for numerous thunderstorms that caused the flooding.

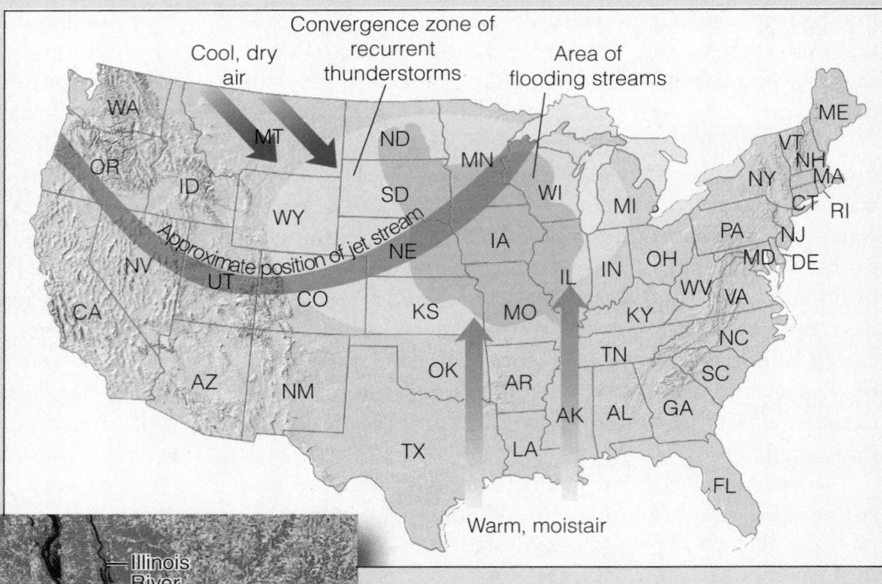

▶ **1.** The dominant weather pattern for June and July 1993. The jet stream remained over the Midwest during the summer rather than shifting north over Canada as it usually does. Thunderstorms developed in the convergence zone where warm, moist air, and cool, dry air met.

Illinois River
Mississippi River
x
Missouri River

◀ **2.** Satellite images of the Mississippi, Missouri, and Illinois rivers near the juncture of three rivers during the drought of 1988 (left), and during the flood of 1993 (below). The "x" marks the site of Portage des Sioux, Missouri (see facing page).

Landsat Imagery courtesy of Earth Observation Satellite Co.

▲ **3.** Portage des Sioux, St. Charles County, Missouri, on July 16, 1993. The channel of the Mississippi River is on the far right.

▲ **4.** Flood waters in Portage des Sioux covered the 5.5-m-high pedestal of this statue on the bank of the Mississippi River.

◀ **5.** Breached levee on the Mississippi River near Davenport, Iowa. This is one of 800 levees that failed or was overtopped during the flood.

▶ **6.** Damage caused by the Flood of '93. Compiled by the U.S. Army Corps of Engineers, figures are rounded to the nearest $1000.

State	Residential	Agricultural	Other	Total
Illinois	$ 176,833,000	$ 166,502,000	$ 409,020,000	$ 752,355,000
Iowa	57,827,000	1,030,030,000	334,835,000	1,422,692,000
Kansas	35,829,000	855,849,000	176,162,000	1,067,840,000
Minnesota	16,940,000	694,041,000	286,540,000	997,521,000
Missouri	405,175,000	540,666,000	1,255,191,000	2,201,032,000
Nebraska	18,584,000	120,521,000	67,899,000	207,004,000
North Dakota	10,138,000	35,039,000	25,806,000	70,983,000
South Dakota	21,919,000	276,218,000	195,408,000	493,545,000
Wisconsin	17,747,000	133,835,000	135,917,000	287,499,000
Total—All States	760,992,000	3,852,701,000	2,886,778,000	7,500,471,000

Figure 12.17 Flood Control Dams and reservoirs, levees, floodways, and floodwalls are some of the structures used to control floods.

James S. Monroe

a Oroville Dam in California, at 235 m high, is the highest dam in the United States. It helps control floods, provides water for irrigation, and produces electricity at its power plant.

James S. Monroe

b This levee, an artificial embankment along a waterway, helps protect nearby areas from floods. A university campus lies out of view just to the right of the levee.

Sue Monroe

c This floodway carries excess water from a river (not visible) around a small community.

James S. Monroe

d This floodwall on the bank of the Danube River at Mohács, Hungary, helps protect the city from floods.

tributaries—that is, streams that contribute water to another stream. The Mississippi River and its tributaries such as the Ohio, Missouri, Arkansas, and Red rivers and thousands of smaller ones, or any other drainage system for that matter, carry runoff from an area known as a **drainage basin**. A topographically high area called a **divide** separates a drainage basin from adjoining ones (Figure 12.18). The continental divide along the crest of the Rocky Mountains in North America, for instance, separates drainage in opposite directions; drainage to the west goes to the Pacific, whereas drainage to the east eventually reaches the Gulf of Mexico.

The arrangements of channels within an area are classified as types of **drainage patterns.** The most common is *dendritic drainage*, which consists of a network of channels resembling tree branching (Figure 12.19a). It develops on gently sloping surfaces composed of materials that respond more or less homogeneously to erosion, such as areas underlain by nearly horizontal sedimentary rocks.

In dendritic drainage, tributaries join larger channels at various angles, but *rectangular drainage* is characterized by right-angle bends and tributaries joining larger channels at right angles (Figure 12.19b). Such regularity in channels is strongly controlled by geologic structures, particularly regional joint systems that intersect at right angles.

Trellis drainage, consisting of a network of nearly parallel main streams with tributaries joining them at right angles, is common in some parts of the eastern United States. In Virginia and Pennsylvania, erosion of folded sedimentary rocks developed a landscape of alternating ridges on resistant rocks and valleys underlain by easily eroded rocks. Main waterways follow the valleys, and short tributaries flowing from the nearby ridges join the main channels at nearly right angles (Figure 12.19c).

In *radial drainage*, streams flow outward in all directions from a central high point, such as a large volcano (Figure 12.19d). Many of the volcanoes in the Cascade

Range of western North America have radial drainage patterns.

In all of the types of drainage mentioned so far, some kind of pattern is easily recognized. *Deranged drainage*, in contrast, is characterized by irregularity, with streams flowing into and out of swamps and lakes, streams with only a few short tributaries, and vast swampy areas between channels (Figure 12.19e). This kind of drainage developed recently and has not yet formed a fully organized drainage system. In parts of Minnesota, Wisconsin, and Michigan, where glaciers obliterated the previous drainage, only 10,000 years have elapsed since the glaciers melted. As a result, drainage systems have not fully developed and large areas remain undrained.

THE SIGNIFICANCE OF BASE LEVEL

Base level is the lowest limit to which a stream or river can erode. With the exception of streams that flow into closed depressions in deserts, all others are restricted ultimately to sea level. That is, they can erode no lower than sea level because they must have some gradient to maintain flow. So *ultimate base level* is sea level, which is simply the lowest level of erosion for any waterway that flows into the sea (Figure 12.20). Ultimate base level applies to an entire stream or river system, but channels may also have a *local* or *temporary base level*. For example, a local base level may be a lake or another stream, or where a stream or river flows across particularly resistant rocks and a waterfall develops: (Figure 12.20).

Ultimate base level is sea level, but suppose that sea level dropped or rose with respect to the land, or suppose that the land rose or subsided? In these cases, base level would change and bring about changes in stream and river systems. During the Pleistocene Epoch (Ice Age), sea level was about 130 m lower than it is now, and streams adjusted by eroding deeper valleys and extending well out onto the continental shelves. Rising sea level at the end of the Ice Age accounted for a rising base level, decreased stream gradients, and deposition within channels.

Natural changes, such as fluctuations in sea level during the Pleistocene, alter the dynamics of rivers and streams, but

Figure 12.18 Drainage Basins

a Small drainage basins are separated by divides (dashed lines), which are along the crests of ridges between channels (solid lines).

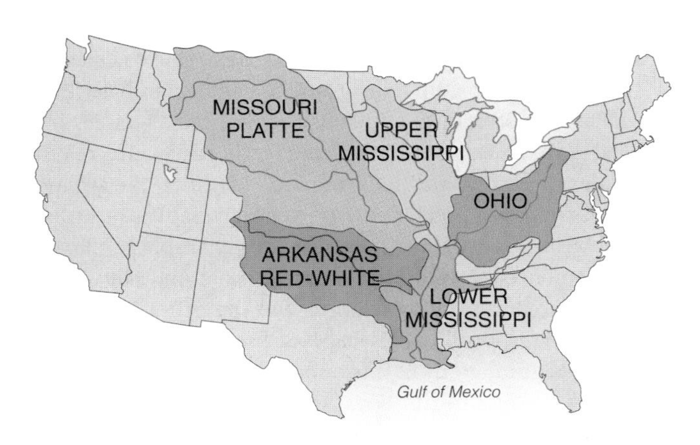

b The drainage basin of the Mississippi River and its main tributaries.

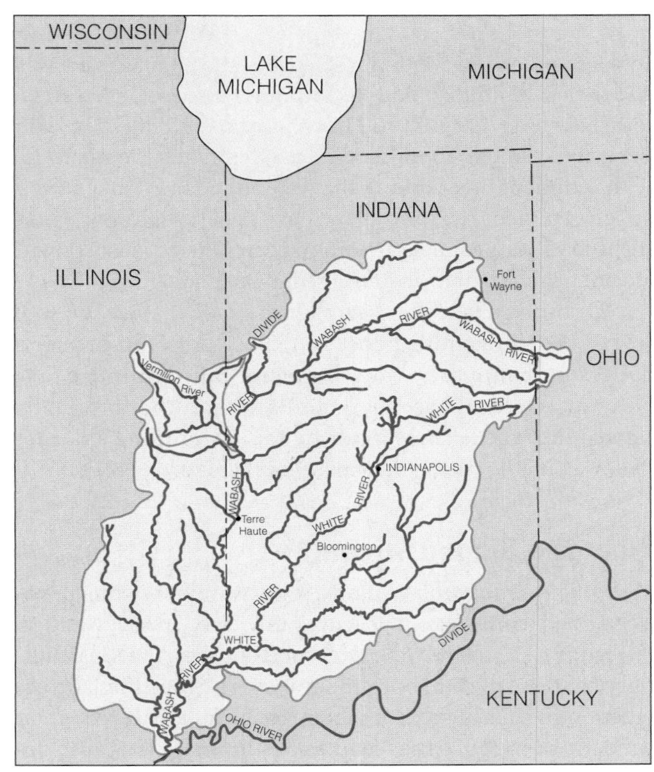

c A detailed view of the Wabash River's drainage basin, a tributary of the Ohio River. All tributary streams within the drainage basin, such as the Vermilion River, have their own smaller drainage basins. Divides are shown by red lines.

Figure 12.19 Drainage Patterns

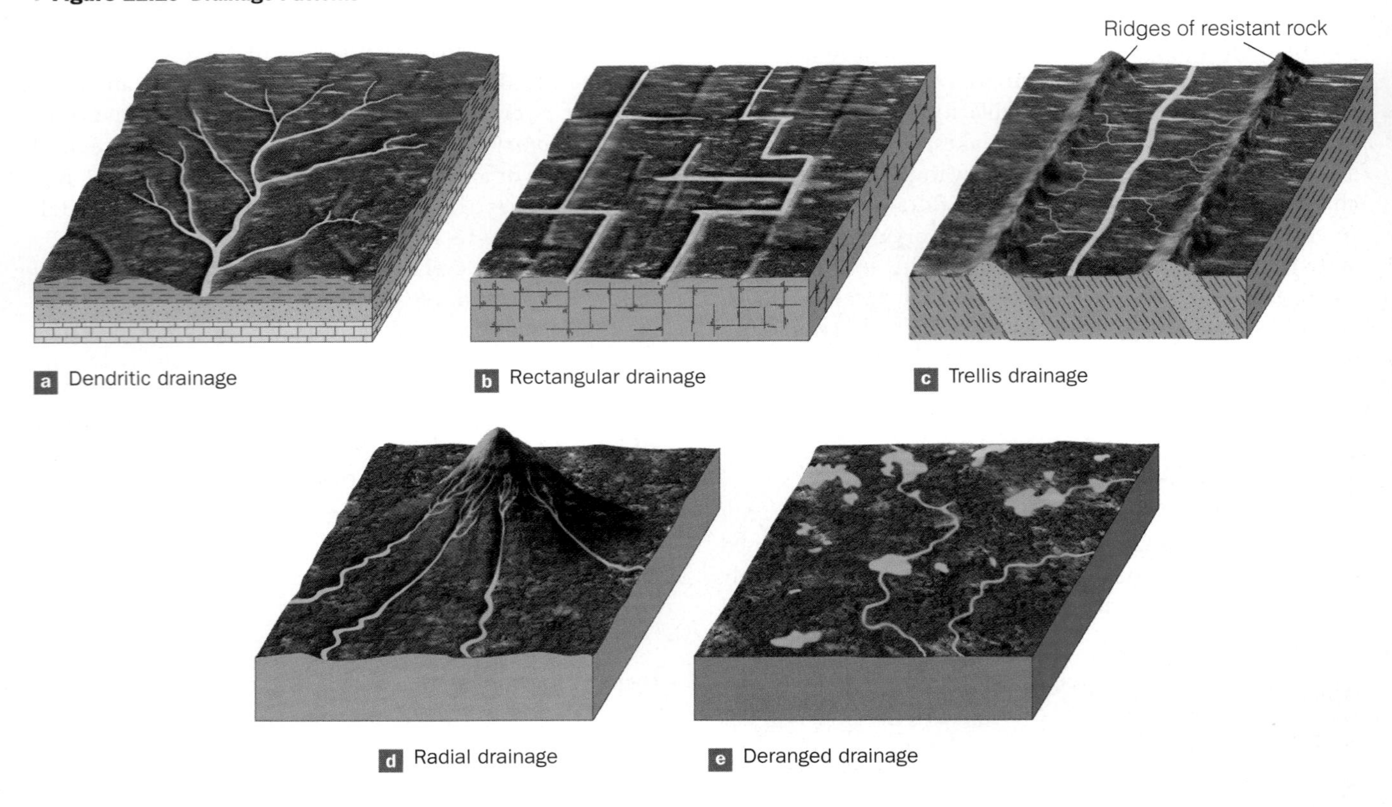

so does human intervention. Geologists and engineers are well aware that building a dam to impound a reservoir creates a local base level (Figure 12.21a). A stream entering a reservoir deposits sediment, so unless dredged, reservoirs eventually fill with sediment. In addition, the water discharged at a dam is largely sediment free, but still possesses energy to carry a sediment load. As a result, a stream may erode vigorously downstream from a dam to acquire a sediment load.

Draining a lake may seem like a small change and well worth the time and expense to expose dry land for agriculture or commercial development. But draining a lake eliminates a local base level, and a stream that originally flowed into the lake responds by rapidly eroding a deeper valley as it adjusts to a new base level (Figure 12.21b).

What Is a Graded Stream?

The *longitudinal profile* of any waterway shows the elevations of a channel along its length as viewed in cross section (Figure 12.22). For some rivers and streams, the longitudinal profile is smooth, but others show irregularities such as lakes and waterfalls, all of which are local base levels. Over time, these irregularities tend to be eliminated because deposition takes place where the gradient is insufficient to maintain sediment transport, and erosion decreases the gradient where it is steep. So, given enough time, rivers and streams develop a smooth, concave longitudinal profile of equilibrium, meaning that all parts of the system dynamically adjust to one another.

A **graded stream** is one with an equilibrium profile in which a delicate balance exists among gradient, discharge, flow velocity, channel shape, and sediment load so that neither significant erosion nor deposition takes place within its channel (Figure 12.22b). Such a delicate balance is rarely attained, so the concept of a graded stream is an ideal. Nevertheless, the graded condition is closely approached in many streams, although only temporarily and not necessarily along their entire lengths.

Even though the concept of a graded stream is an ideal, we can anticipate the response of a graded stream to changes that alter its equilibrium. For instance, a change in base level would cause a stream to adjust as previously discussed. Increased rainfall in a stream's drainage basin would result in greater discharge and flow velocity. In short, the stream would now possess greater energy—energy that must be dissipated within the stream system by, for example, a change from a semicircular to a broad, shallow channel that would dissipate more energy by friction. On the other hand, the stream may respond by eroding a deeper valley, effectively reducing its gradient until it is once again graded.

Vegetation inhibits erosion by stabilizing soil and other loose surface materials. So a decrease in vegetation in a drainage basin might lead to higher erosion rates, causing more sediment to be washed into a stream than it can

Figure 12.20 Base Level

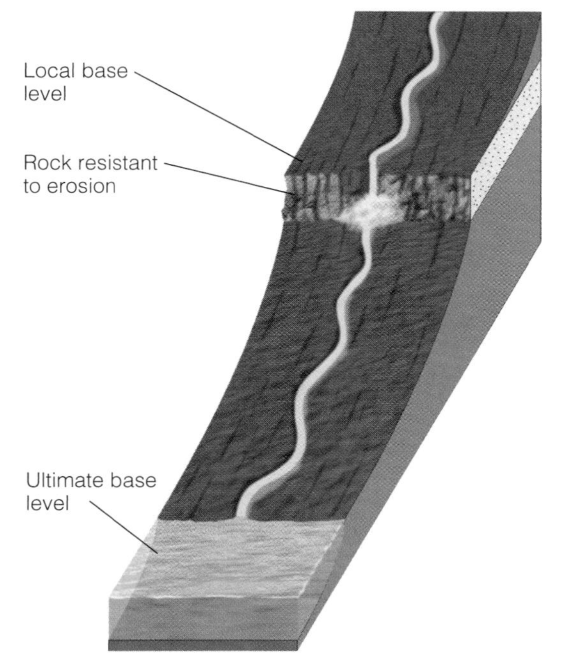

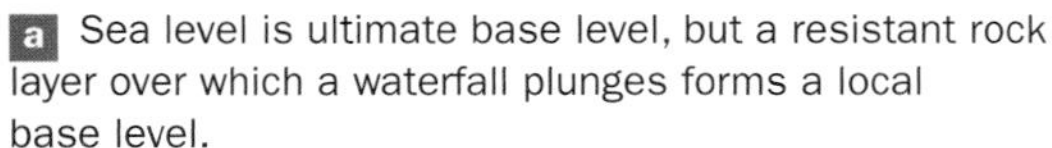

a Sea level is ultimate base level, but a resistant rock layer over which a waterfall plunges forms a local base level.

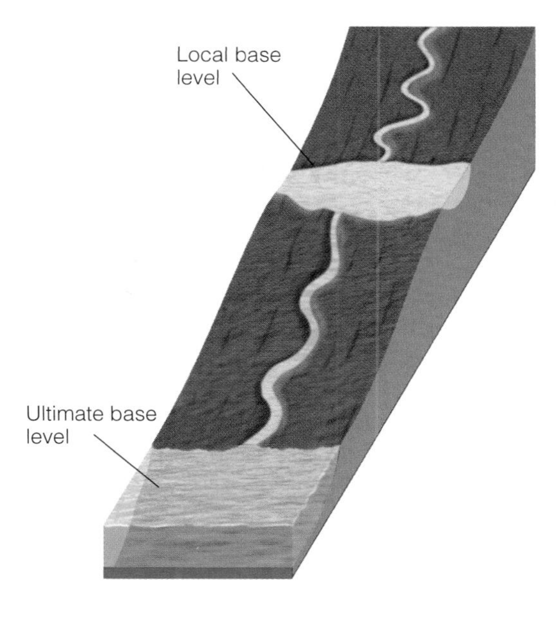

b Local base level where a stream flows into a lake.

Figure 12.21 Local Base Levels

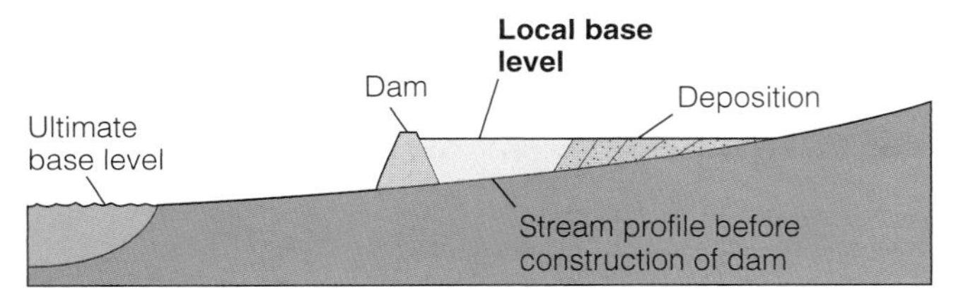

a A stream deposits much of its sediment load where it flows into a reservoir.

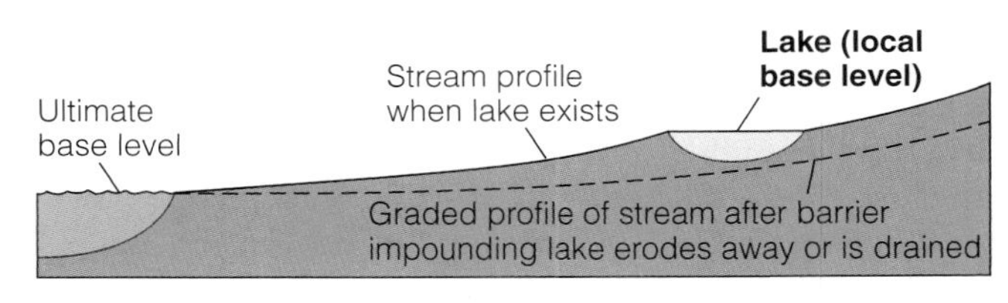

b Local base levels disappear as streams erode their beds into graded profiles matching the ultimate base level.

Figure 12.22 Longitudinal Profiles of Streams

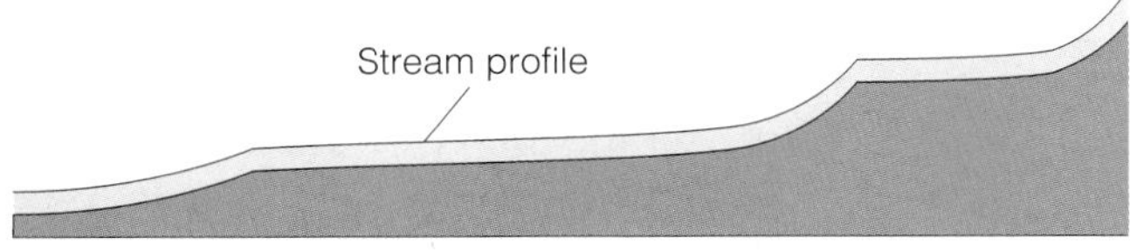

a An ungraded stream has irregularities in its longitudinal profile.

Erosion
Erosion
Erosion
Deposition
Deposition

b Erosion and deposition along the course of a stream eliminate irregularities and cause it to develop the smooth, concave profile typical of a graded stream.

effectively carry. Accordingly, the stream may respond by deposition within its channel, which increases its gradient until it is sufficiently steep to transport the greater sediment load.

THE EVOLUTION OF VALLEYS

Low areas on land known as **valleys** are bounded by higher land, and most of them have a river or stream running their length, with tributaries draining the nearby high areas. Valleys are common landforms, and with few exceptions, they form and evolve in response to erosion by running water, although other processes, especially mass wasting, contribute. The shapes and sizes of valleys vary from small, steep-sided *gullies* to those that are broad with gently sloping valley walls (Figure 12.23). Steep-walled, deep valleys of vast size are *canyons*, and particularly narrow and deep ones are *gorges*.

A valley might start to erode where runoff has sufficient energy to dislodge surface materials and excavate a small rill. Once formed, a rill collects more runoff and becomes deeper and wider and continues to do so until a full-fledged valley develops. Processes that contribute to valley formation include downcutting, lateral erosion, headward erosion, and sheetwash. Mass wasting processes are also important.

Downcutting takes place when a river or stream has more energy than it needs to transport sediment, so some of its excess energy is used to deepen its valley. If downcutting were the only process operating, valleys would be narrow and steep sided. In most cases, though, the valley walls are undercut, a process called *lateral erosion*, creating unstable slopes that may fail by *mass wasting*. Furthermore, erosion by sheetwash and erosion by tributary streams carry materials from the valley walls into the main stream in the valley.

Valleys not only become deeper and wider, but also become longer by *headward erosion*, a phenomenon involving erosion by entering runoff at the upstream end of a valley (Figure 12.24a). Continued headward erosion may result in *stream piracy*, the breaching of a drainage divide and diversion of part of the drainage of another stream (Figure 12.24b). Once stream piracy takes place, both drainage systems must adjust to these new conditions; one system now has greater discharge and the potential to do more erosion and sediment transport, whereas the other is diminished in its ability to accomplish these tasks.

According to one concept, stream erosion of an area uplifted above sea level yields a distinctive series of landscapes. When erosion begins, streams erode downward; their valleys are deep, narrow, and V-shaped, and their profiles have a number of irregularities (Figure 12.25a). As streams cease eroding downward, they start eroding laterally, thereby establishing a meandering pattern and a broad floodplain (Figure 12.25b). Finally, with continued erosion, a vast, rather featureless plain develops (Figure 12.25c).

Many streams do indeed show the features typical of these stages. For instance, the Colorado River flows through the Grand Canyon and closely matches the features in the initial stage shown in Figure 12.25a. Streams in many areas approximate the second stage of development, and certainly the lower Mississippi closely resembles the last stage. Nevertheless, the idea of the sequential development of stream-eroded landscapes has been largely abandoned because there is no reason to think that streams necessarily follow this idealized progression. Indeed, a stream on a gently sloping surface near sea level could develop features of the last stage very early in its history. In addition, as long as the rate of uplift exceeds the rate of downcutting, a stream will continue to erode downward and be confined to a narrow canyon.

Figure 12.23 Gullies and Valleys

a Gullies are small valleys, but they are narrow and deep. This gully measures about 15 m across.

b This valley has steep walls that descend to a narrow valley bottom.

Stream Terraces

Adjacent to many channels are erosional remnants of floodplains that formed when the streams were flowing at a higher level. These **stream terraces** consist of a fairly flat upper surface and a steep slope descending to the level of the lower, present-day floodplain (Figure 12.26). Some streams have several steplike surfaces above their present-day floodplains, indicating that terraces formed several times.

Figure 12.24 Two Stages in the Evolution of a Valley

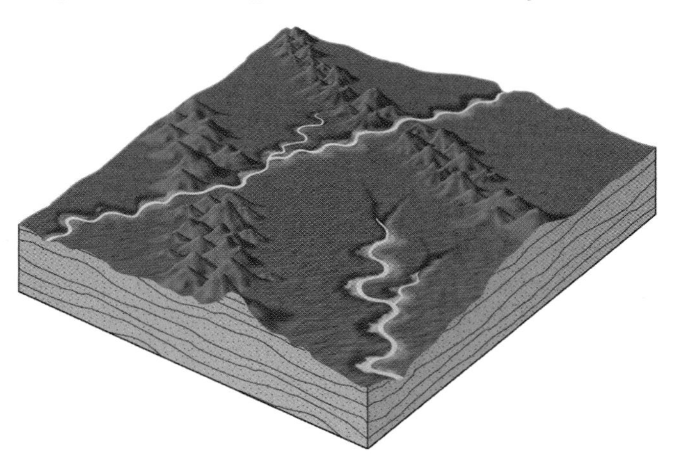

a The stream widens its valley by lateral erosion and mass wasting while simultaneously extending its valley by headward erosion.

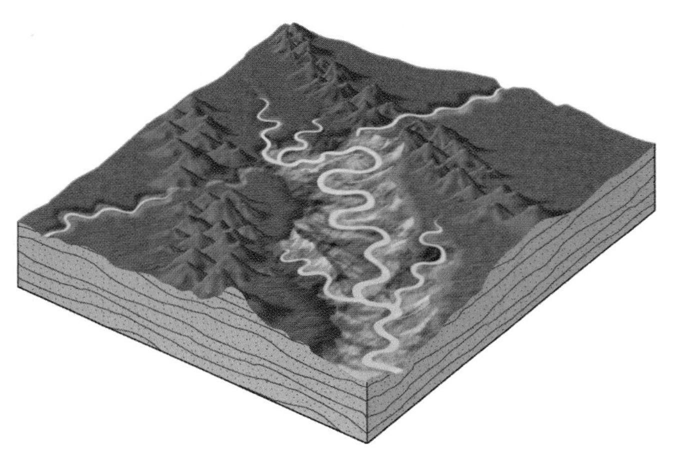

b As the larger stream continues to erode headward, stream piracy takes place when it captures some of the drainage of the smaller stream. Notice also that the valley is wider in (b) than it was in (a).

Figure 12.25 Idealized Stages in the Development of a Stream and Its Associated Landform

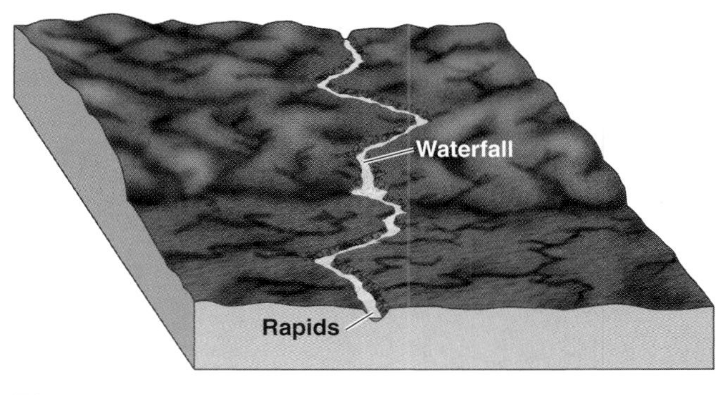

a Initial stage

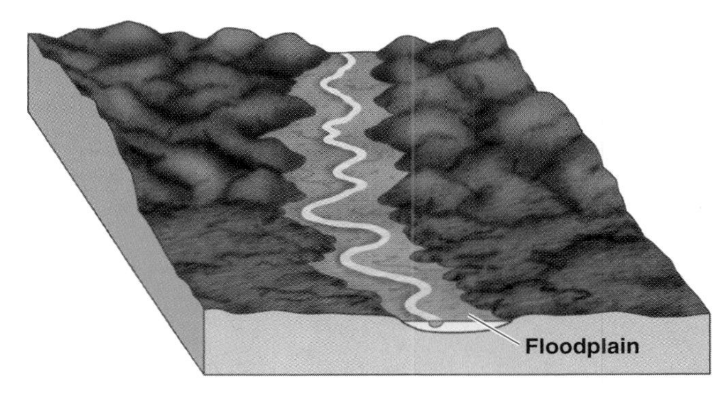

b Intermediate stage

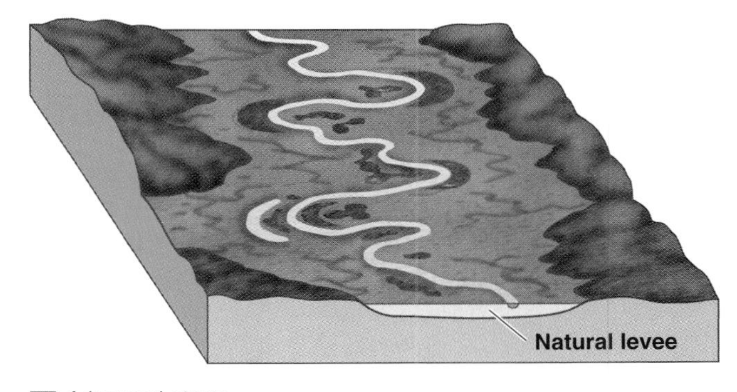

c Advanced stage

Although all stream terraces result from erosion, they are preceded by an episode of floodplain formation and sediment deposition. Subsequent erosion causes the stream to cut downward until it is once again graded (Figure 12.26). Then it begins to erode laterally and establishes a new floodplain at a lower level. Several such episodes account for the multiple terrace levels adjacent to some channels.

Renewed erosion and the formation of stream terraces are usually attributed to a change in base level. Either uplift of the land that a stream flows over or lowering of sea level yields a steeper gradient and increased flow velocity, thus initiating an episode of downcutting. When the stream reaches a level at which it is once again graded, downcutting ceases. Although changes in base level no doubt account for many stream terraces, greater runoff in a stream's drainage basin can also result in the formation of terraces.

Incised Meanders

Some streams are restricted to deep, meandering canyons cut into bedrock, where they form features called **incised meanders.** For example, the Colorado River in Utah occupies a meandering canyon more than 600 m deep (Figure 12.27). Streams restricted by rock walls usually cannot erode laterally; thus, they lack a floodplain and occupy the entire width of the canyon floor.

It is not difficult to understand how a stream can cut downward into rock, but how a stream forms a meandering pattern in bedrock is another matter. Because lateral erosion is inhibited once downcutting begins, one must infer that the meandering course was established when the stream flowed across an area covered by alluvium. For example, suppose that a stream near base level has established a meandering pattern. If the land that the stream flows over is uplifted, then erosion begins and the meanders become incised into the underlying bedrock.

Figure 12.26 Origin of Stream Terraces

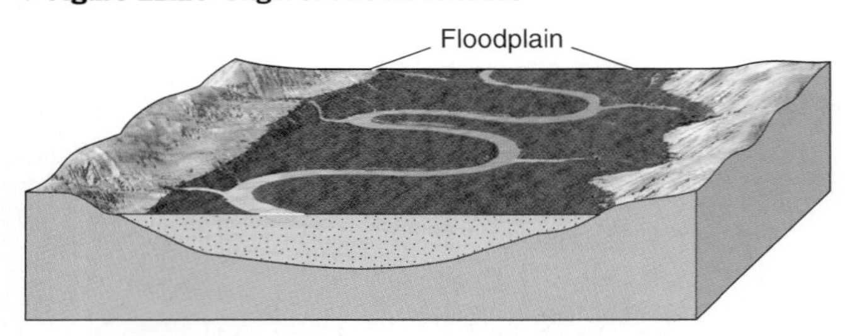

a A stream has a broad floodplain.

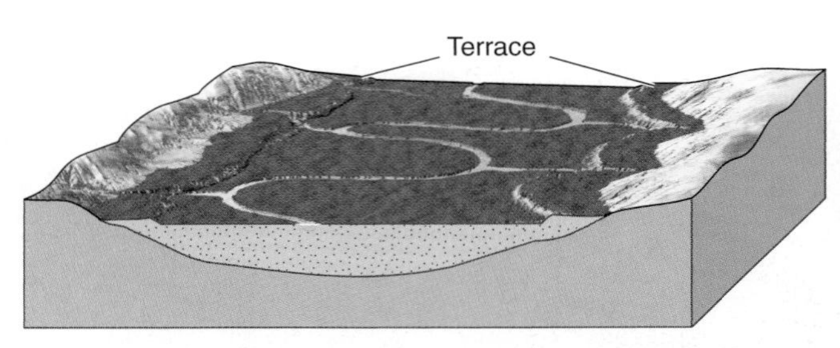

b The stream erodes downward and establishes a new floodplain at a lower level. Remnants of its old, higher floodplain are stream terraces.

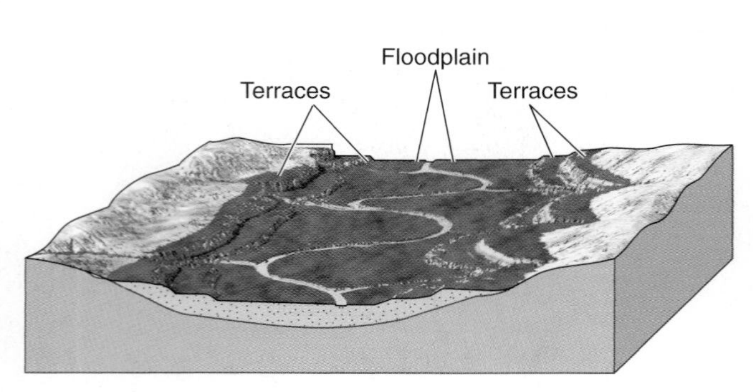

c Another level of stream terraces forms as the stream erodes downward again.

d Stream terraces along the Madison River in Montana.

Superposed Streams

Water flows downhill in response to gravity, so the direction of flow in streams and rivers is determined by topography. Yet a number of waterways seem, at first glance, to have defied this fundamental control. For instance, the Delaware, Potomac, and Susquehanna rivers in the eastern United States flow in valleys that cut directly through ridges that lie in their paths. These are examples of **superposed streams,** all of which once flowed on a surface at a higher level, but as they eroded downward, they eroded into resistant rocks and cut narrow canyons or what geologists call *water gaps* (Figure 12.28). The Jefferson River in Montana was superposed on a ridge that lay in its path (Figure 12.28c).

A water gap has a stream flowing through it, but if the stream is diverted elsewhere, perhaps by stream piracy, the abandoned gap is then called a *wind gap*. The Cumberland Gap in Kentucky is a good example; it was the avenue through which settlers migrated from Virginia to Kentucky from 1790 until well into the 1800s. Furthermore, several water gaps and wind gaps played important strategic roles during the Civil War (1861–1865).

Figure 12.27 Incised Meanders The Colorado River at Dead Horse State Park in Utah is incised to a depth of 600 m.

Figure 12.28 Origin of a Superposed Stream

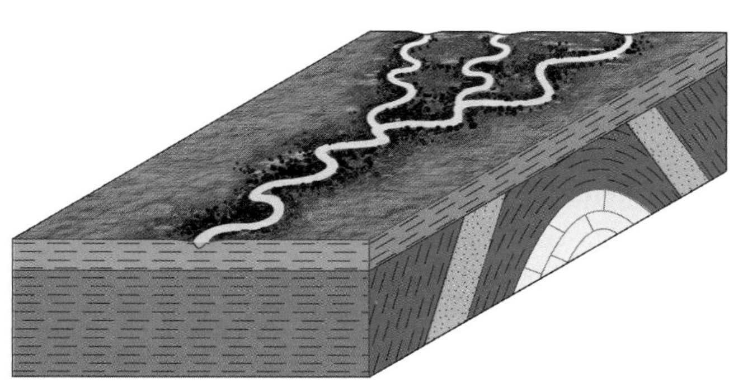

a As a stream erodes down and removes the surface layers of rock, it is lowered onto ridges that form when resistant rocks in the underlying structure are exposed.

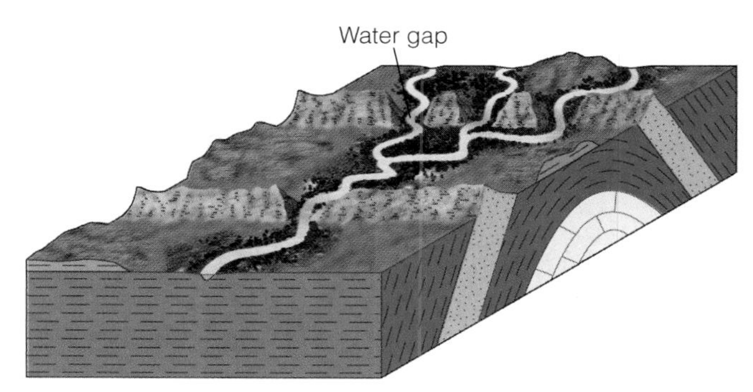

b The narrow valleys through the ridges are water gaps.

c View of a water gap cut by the Jefferson River in Montana.

Geo-Recap

Chapter Summary

- Water continuously evaporates from the oceans, rises as water vapor, condenses, and falls as precipitation. About 20% of this precipitation falls on land and eventually returns to the oceans, mostly by surface runoff.
- Running water moves by laminar flow, in which streamlines parallel one another, and by turbulent flow, in which streamlines complexly intertwine. Almost all flow in channels is turbulent.
- Runoff takes place by sheet flow, a thin, more or less continuous sheet of water, and by channel flow, confined to long, troughlike stream and river channels.
- The vertical drop in a given distance, or the gradient, for a channel varies from steep in its upper reaches to more gentle in its lower reaches.
- Flow velocity and discharge are related, so that if either changes, the other changes as well.
- Erosion by running water takes place by hydraulic action, abrasion, and solution.
- The bed load in channels is made up of sand and gravel, whereas suspended load consists of silt- and clay-sized particles. Running water also transports a dissolved load.
- Braided waterways have a complex of dividing and rejoining channels, and their deposits are mostly sheets of sand and gravel.
- A single sinuous channel is typical of meandering streams that deposit mostly mud, with subordinate point-bar deposits of sand or, more rarely, gravel.

- Broad, flat floodplains adjacent to channels are the sites of oxbow lakes, which are simply abandoned meanders.
- An alluvial deposit at a river's mouth is a delta. Some deltas conform to the three-part division of bottomset, foreset, and topset beds, but large marine deltas are much more complex and are characterized as stream-, wave-, or tide-dominated.
- Alluvial fans are fan-shaped deposits of sand and gravel on land that form best in semiarid regions. They form mostly by deposition from running water, but debris flows are also important.
- Rivers and streams carry runoff from their drainage basins, which are separated from one another by divides.
- Sea level is ultimate base level, the lowest level to which streams or rivers can erode. Local base levels may be lakes or where streams or rivers flow across resistant rocks.
- Graded streams tend to eliminate irregularities in their channels, so they develop a smooth, concave profile of equilibrium.
- A combination of processes, including downcutting, lateral erosion, sheetwash, mass wasting, and headward erosion, are responsible for the origin and evolution of valleys.
- Stream terraces and incised meanders usually form when a stream or river that was formerly in equilibrium begins a new episode of downcutting.

Important Terms

abrasion (p. 308)
alluvial fan (p. 313)
alluvium (p. 309)
base level (p. 319)
bed load (p. 308)
braided stream (p. 310)
delta (p. 311)
discharge (p. 306)
dissolved load (p. 308)
divide (p. 318)
drainage basin (p. 318)
drainage pattern (p. 318)
floodplain (p. 310)
graded stream (p. 320)
gradient (p. 305)
hydraulic action (p. 308)
hydrologic cycle (p. 303)
incised meanders (p. 323)
infiltration capacity (p. 304)
meandering stream (p. 310)
natural levee (p. 311)
oxbow lake (p. 310)
point bar (p. 310)
runoff (p. 303)
stream terrace (p. 322)
superposed stream (p. 324)
suspended load (p. 308)
valley (p. 321)
velocity (p. 305)

Review Questions

1. Base level is
   a. _______ the lowest level to which a stream can erode;
   b. _______ the distance from a river's headwaters to where it discharges into the sea;
   c. _______ a stream profile with numerous irregularities such as lakes and waterfalls;
   d. _______ a type of river with a single, sinuous channel;
   e. _______ the amount of water moving past a given point in a specified amount of time.
2. A board, flat area adjacent to a river is its
   a. _______ oxbow lake;
   b. _______ floodplain;
   c. _______ natural levee;
   d. _______ delta;
   e. _______ point bar.
3. Running water carrying sand and gravel effectively erodes by
   a. _______ solution;
   b. _______ sheetwash;
   c. _______ deposition;
   d. _______ piracy;
   e. _______ abrasion.
4. The type of drainage that is likely to develop on a composite volcano is
   a. _______ incised;
   b. _______ meandering;
   c. _______ radial;
   d. _______ alluvial;
   e. _______ solutional.
5. A ridge of alluvium along the banks of a river that forms during floods is a/an
   a. _______ tide-dominated delta;
   b. _______ natural levee;
   c. _______ alluvial fan;
   d. _______ incised meander;
   e. _______ drainage basin.
6. Drainage basins are separated from one another by a/an
   a. _______ bed load;
   b. _______ infiltration capacity;
   c. _______ base level;
   d. _______ divide;
   e. _______ natural levee.

7. The recycling of water from the oceans to the land and back to the oceans is known as the
   a. _______ superposed drainage;
   b. _______ erosion capacity;
   c. _______ hydrologic cycle;
   d. _______ infiltration rate;
   e. _______ lateral accretion depth.
8. A deposit usually composed of sand that is deposited on the gently sloping side of a river meander is called
   a. _______ an oxbow lake;
   b. _______ the dissolved load;
   c. _______ a floodplain;
   d. _______ a delta;
   e. _______ a point bar.
9. The suspended load of a stream is made up of
   a. _______ silt and clay;
   b. _______ dissolved material;
   c. _______ sand and gravel;
   d. _______ organic matter;
   e. _______ runoff.
10. Dividing the vertical drop of a stream by the horizontal distance it flows gives the
   a. _______ discharge;
   b. _______ alluvial capacity;
   c. _______ gradient;
   d. _______ profile of equilibrium;
   e. _______ velocity.
11. Calculate the daily discharge of a river 148 m wide and 2.6 m deep, with a flow velocity of 0.3 m/sec.
12. Diagram and describe how an ideal small delta forms.
13. Explain how the hydrologic cycle works.
14. Why is Earth the only terrestrial planet with abundant surface water?
15. Describe the processes whereby running water erodes.
16. The discharge of most rivers and streams increases downstream, but in a few cases, it actually decreases and they eventually disappear. Why? Give an example.
17. Explain how stream terraces form and what they indicate about the dynamics of a stream or river system.
18. About 10.75 km^3 of sediment erodes from the continents each year, and the volume of the continents above sea level is 93,000,000 km^3. Thus the continents should erode to sea level in just over 8,600,000 years. The figures given are reasonably accurate, but something is seriously wrong with the conclusion that the continents will be leveled in so short a time. Explain.
19. A river with headwaters 2000 m above sea level flows 1500 km to the sea. What is its gradient? Do you think that your calculated gradient is valid for all segments of this river? Explain
20. Explain how a superposed stream develops. Also describe the landforms that result from superposition.

CHAPTER 13

Groundwater

Parvinder Sethi

OUTLINE

Introduction

Groundwater and the Hydrologic Cycle

Porosity and Permeability

The Water Table

Groundwater Movement

Springs, Water Wells, and Artesian Systems

Groundwater Erosion and Deposition

GEO-INSIGHT: The Burren Area of Ireland

Modifications of the Groundwater System and Its Effects

GEO-FOCUS: Arsenic and Old Lace

Hydrothermal Activity

Geo-Recap

OBJECTIVES

At the end of this chapter, you will have learned that

- Groundwater is one reservoir of the hydrologic cycle and accounts for approximately 22% of the world's supply of freshwater.
- Porosity and permeability are largely responsible for the amount, availability, and movement of groundwater.
- The water table separates the zone of aeration from the underlying zone of saturation and is a subdued replica of the overlying land surface.
- Groundwater moves downward because of the force of gravity.
- In an artesian system, groundwater is confined and builds up high hydrostatic pressure.
- Groundwater is an important agent of both erosion and deposition and is responsible for karst topography and a variety of cave features.
- Modifications of the groundwater system may result in lowering of the water table, saltwater incursion, subsidence, and contamination.

A variety of cave deposits such as stalactites, stalagmites, and curtains can be seen in caves such as this one in the Shenandoah Caverns, in Virginia.

- Hot springs and geysers result when groundwater is heated, typically in regions of recent volcanic activity.
- Geothermal energy is a desirable and relatively nonpolluting alternative form of energy.

INTRODUCTION

Within the limestone region of western Kentucky lies the largest cave system in the world. In 1941, approximately 51,000 acres were set aside and designated as Mammoth Cave National Park. In 1981, it became a World Heritage Site.

From ground level, the topography of the area is unimposing, with gently rolling hills. Beneath the surface, however, are more than 540 km of interconnected passageways whose spectacular geologic features have been enjoyed by millions of cave explorers and tourists.

During the War of 1812, approximately 180 metric tons of saltpeter, used in the manufacture of gunpowder, was mined from Mammoth Cave. At the end of the war, the saltpeter market collapsed, and Mammoth Cave was developed as a tourist attraction, easily overshadowing the other caves in the area. During the next 150 years, the discovery of new passageways and links to other caverns helped establish Mammoth Cave as the world's premier cave and the standard against which all others were measured.

The formation of the caves themselves began approximately 3 million years ago when groundwater began dissolving the region's underlying St. Genevieve Limestone to produce a complex network of openings, passageways, and huge chambers that constitute present-day Mammoth Cave. Flowing through the various caverns is the Echo River, a system of streams that eventually joins the Green River at the surface.

The colorful cave deposits are the primary reason that millions of tourists have visited Mammoth Cave over the years. Hanging down from the ceiling and growing up from the floor are spectacular icicle-like structures, as well as columns and curtains in a variety of colors. Moreover, intricate passageways connect rooms of various sizes. The cave is also home to more than 200 species of insects and other animals, including about 45 blind species.

In addition to the beautiful caves, caverns, and cave deposits produced by groundwater movement, groundwater is also an important natural resource. Although groundwater constitutes only 0.6% of the world's water, it is, nonetheless, a significant source of freshwater for agriculture, industry, and domestic users. More than 65% of the groundwater used in the United States each year goes for irrigation, with industrial use second, followed by domestic needs. Furthermore, groundwater provides 80% of the water used for rural livestock and domestic use, as well as providing 40% of public water supplies. In fact, of the largest 100 cities in the United States, 34 depend solely on local groundwater supplies. These demands have severely depleted the groundwater supply in many areas and have led to such problems as ground subsidence and saltwater contamination. In other areas, pollution from landfills, toxic waste, and agriculture has rendered the groundwater supply unsafe.

As the world's population and industrial development expand, the demand for water, particularly groundwater, will increase. Not only must new groundwater sources be located, but once found, these sources must be protected from pollution and managed properly to ensure that users do not withdraw more water than can be replenished. It is therefore important that people become aware of what a valuable resource groundwater is, so that they can ensure that future generations have a clean and adequate supply of this water source.

GROUNDWATER AND THE HYDROLOGIC CYCLE

Groundwater, water that fills open spaces in rocks, sediment, and soil beneath Earth's surface, is one reservoir in the hydrologic cycle, accounting for approximately 22% (8.4 million km^3) of the world's supply of freshwater (see Table 12.1). Like all other water in the hydrologic cycle, the ultimate source of groundwater is the oceans; however, its more immediate source is the precipitation that infiltrates the ground and seeps down through the voids in soil, sediment, and rocks. Groundwater may also come from water infiltrating from streams, lakes, swamps, artificial recharge ponds, and water-treatment systems.

Regardless of its source, groundwater moving through the tiny openings between soil and sediment particles and the spaces in rocks filters out many impurities such as disease-causing microorganisms and many pollutants. However, not all soils and rocks are good filters, and sometimes so much undesirable material may be present that it contaminates the groundwater. Groundwater movement and its recovery at wells depend on two critical aspects of the materials that it moves through: *porosity and permeability*.

TABLE 13.1 Porosity Values for Different Materials

Material	Percentage Porosity
Unconsolidated sediment	
Soil	55
Gravel	20–40
Sand	25–50
Silt	35–50
Clay	50–70
Rocks	
Sandstone	5–30
Shale	0–10
Solution activity in limestone, dolostone	10–30
Fractured basalt	5–40
Fractured granite	10

Source: U.S. Geological Survey, Water Supply Paper 2220 (1983) and others.

POROSITY AND PERMEABILITY

Porosity and permeability are important physical properties of Earth materials and are largely responsible for the amount, availability, and movement of groundwater. Water soaks into the ground because soil, sediment, and rock have open spaces or pores. **Porosity** is the percentage of a material's total volume that is pore space. Porosity most often consists of the spaces between particles in soil, sediment, and sedimentary rocks, but other types of porosity include cracks, fractures, faults, and vesicles in volcanic rocks (◗ Figure 13.1).

Porosity varies among different rock types and is dependent on the size, shape, and arrangement of the material composing the rock (Table 13.1). Most igneous and metamorphic rocks, as well as many limestones and dolostones, have very low porosity because they consist of tightly interlocking crystals. Their porosity can be increased, however, if they have been fractured or weathered by groundwater. This is particularly true for massive limestone and dolostone, whose fractures can be enlarged by acidic groundwater.

By contrast, detrital sedimentary rocks composed of well-sorted and well-rounded grains can have high porosity because any two grains touch at only a single point, leaving relatively large open spaces between the grains (Figure 13.1a). Poorly sorted sedimentary rocks, on the other hand, typically have low porosity because smaller grains fill in the spaces between the larger grains, further reducing porosity (Figure 13.1b). In addition, the amount of cement between grains can decrease porosity.

Porosity determines the amount of groundwater that Earth materials can hold, but it does not guarantee that the water can be easily extracted. So, in addition to being porous, Earth materials must have the capacity to transmit fluids, a property known as **permeability**. Thus, both porosity and permeability play important roles in groundwater movement and recovery.

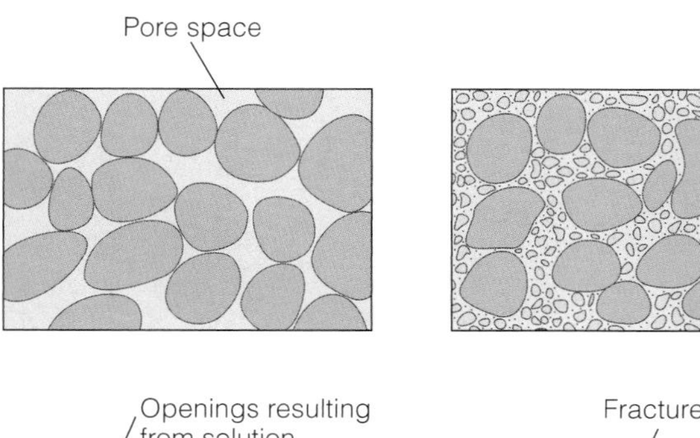

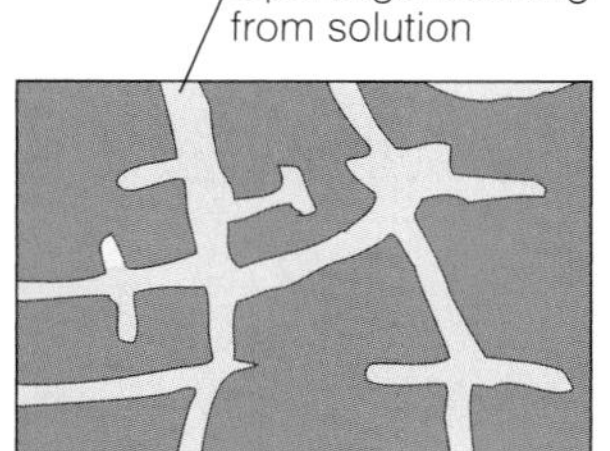

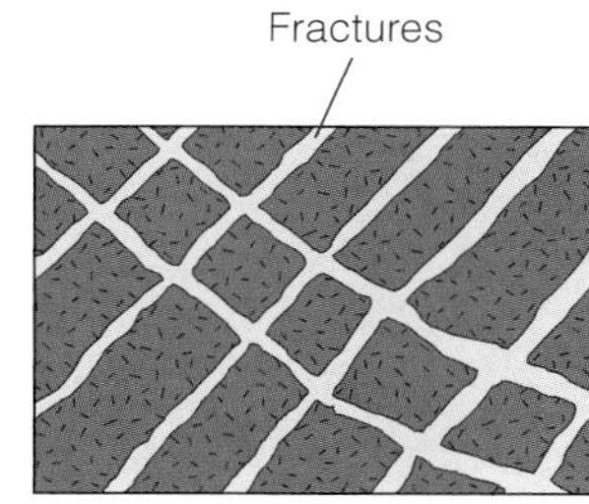

◗ **Figure 13.1 Porosity** A rock's porosity depends on the size, shape, and arrangement of the material composing the rock.

Permeability is dependent not only on porosity, but also on the size of the pores or fractures and their interconnections. For example, deposits of silt or clay are typically more porous than sand or gravel, but they have low permeability because the pores between the particles are very small, and molecular attraction between the particles and water is great, thereby preventing movement of the water. In contrast, the pore spaces between grains in sandstone and conglomerate are much larger, and molecular attraction on the water is therefore low. Chemical and biochemical sedimentary rocks, such as limestone and dolostone, and many igneous and metamorphic rocks that are highly fractured can also be very permeable provided that the fractures are interconnected.

The contrasting porosity and permeability of familiar substances are well demonstrated by sand versus clay. Pour some water on sand and it rapidly sinks in, whereas water poured on clay simply remains on the surface. Furthermore, wet sand dries quickly, but once clay absorbs water, it may take days to dry out because of its low permeability. Neither sand nor clay makes a good substance in which to grow crops or gardens, but a mixture of the two, plus some organic matter in the form of humus, makes an excellent soil for farming and gardening (see Chapter 6).

A permeable layer transporting groundwater is an *aquifer*, from the Latin *aqua*, "water." The most effective aquifers are deposits of well-sorted and well-rounded sand and gravel. Limestones in which fractures and bedding planes have been enlarged by solution are also good aquifers. Shales and many igneous and metamorphic rocks make poor aquifers because they are typically impermeable, unless fractured. Rocks such as these and any other materials that prevent the movement of groundwater are *aquicludes*.

THE WATER TABLE

Some of the precipitation on land evaporates, and some enters streams and returns to the oceans by surface runoff; the remainder seeps into the ground. As this water moves down from the surface, a small amount adheres to the material it moves through and halts its downward progress. With the exception of this *suspended water*, however, the rest seeps further downward and collects until it fills all of the available pore spaces. Thus, two zones are defined by whether their pore spaces contain mostly air, the **zone of aeration**, or mostly water, the underlying **zone of saturation**. The surface that separates these two zones is the **water table** (◗ Figure 13.2).

The base of the zone of saturation varies from place to place, but usually extends to a depth where an impermeable layer is encountered or to a depth where confining pressure closes all open space. Extending irregularly upward a few centimeters to several meters from the zone of saturation is the *capillary fringe*. Water moves upward in this region because of surface tension, much as water moves upward through a paper towel.

In general, the configuration of the water table is a subdued replica of the overlying land surface; that is, it rises beneath hills

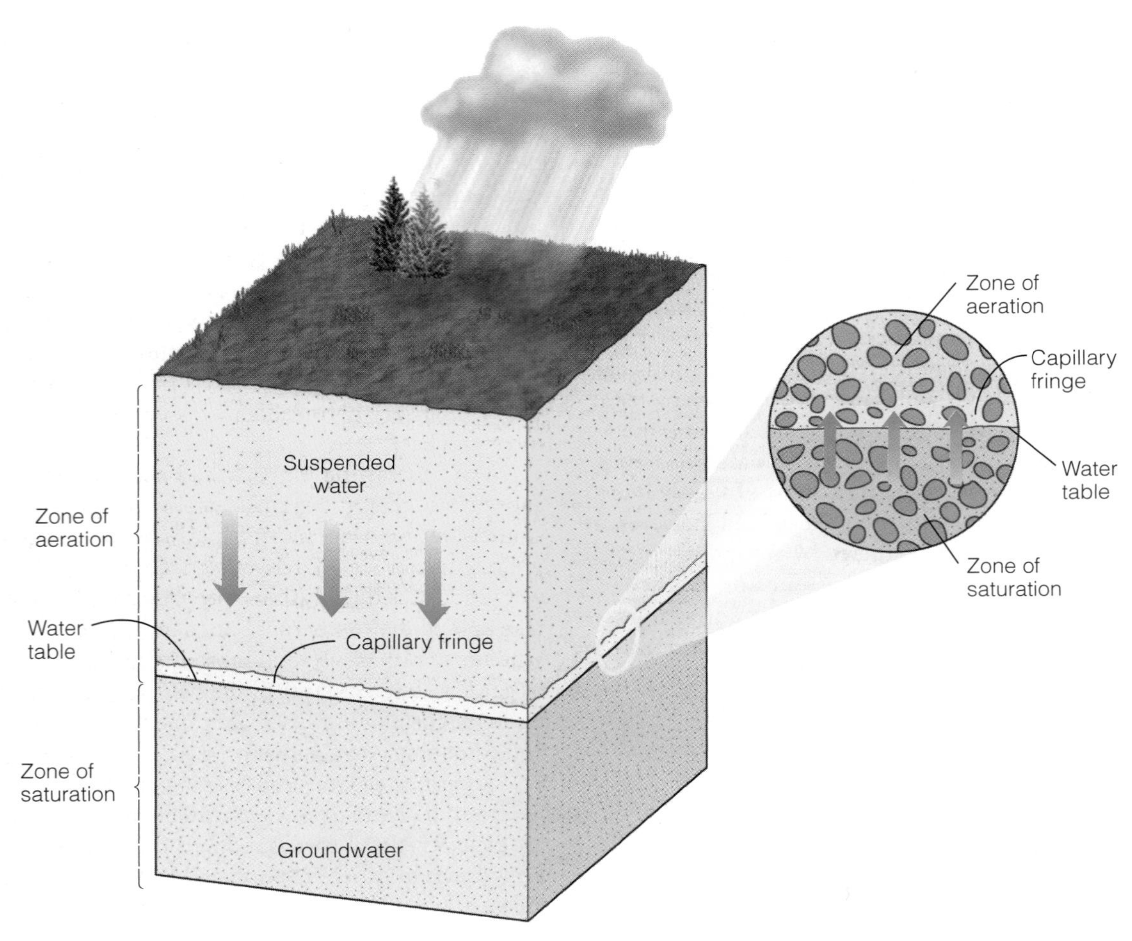

Figure 13.2 Water Table The zone of aeration contains both air and water within its pore spaces, whereas all pore spaces in the zone of saturation are filled with groundwater. The water table is the surface separating the zones of aeration and saturation. Within the capillary fringe, water rises by surface tension from the zone of saturation into the zone of aeration.

and has its lowest elevations beneath valleys. Several factors contribute to the surface configuration of a region's water table, including regional differences in amount of rainfall, permeability, and rate of groundwater movement. During periods of high rainfall, groundwater tends to rise beneath hills because it cannot flow fast enough into adjacent valleys to maintain a level surface. During droughts, the water table falls and tends to flatten out because it is not being replenished. In arid and semiarid regions, the water table is usually quite flat regardless of the overlying land surface.

GROUNDWATER MOVEMENT

Gravity provides the energy for the downward movement of groundwater. Water entering the ground moves through the zone of aeration to the zone of saturation (Figure 13.3). When water reaches the water table, it continues to move through the zone of saturation

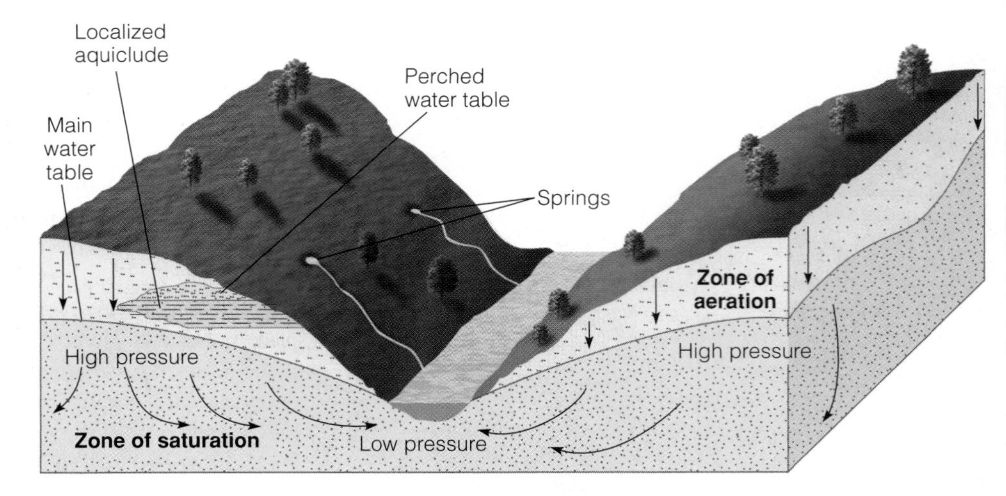

Figure 13.3 Groundwater Movement Groundwater moves down through the zone of aeration to the zone of saturation. Then some of it moves along the slope of the water table, and the rest moves through the zone of saturation from areas of high pressure toward areas of low pressure. Some water might collect over a local aquiclude, such as a shale layer, thus forming a perched water table.

from areas where the water table is high toward areas where it is lower, such as streams, lakes, or swamps. Only some of the water follows the direct route along the slope of the water table. Most of it takes longer curving paths down and then enters a stream, lake, or swamp from below, because it moves from areas of high pressure toward areas of lower pressure within the saturated zone.

Groundwater velocity varies greatly and depends on many factors. Velocities range from 250 m per day in some extremely permeable material to less than a few centimeters per year in nearly impermeable material. In most ordinary aquifers, the average velocity of groundwater is a few centimeters per day.

SPRINGS, WATER WELLS, AND ARTESIAN SYSTEMS

You can think of the water in the zone of saturation much like a reservoir whose surface rises or falls depending on additions as opposed to natural and artificial withdrawals. *Recharge*—that is, additions to the zone of saturation—comes from rainfall or melting snow, or water might be added artificially at wastewater-treatment plants or recharge ponds constructed for just this purpose. But if groundwater is discharged naturally or withdrawn at wells without sufficient recharge, the water table drops just as a savings account diminishes if withdrawals exceed deposits. Withdrawals from the groundwater system take place where groundwater flows laterally into streams, lakes, or swamps, where it discharges at the surface as *springs*, and where it is withdrawn from the system at water wells.

Springs

Places where groundwater flows or seeps out of the ground as **springs** have always fascinated people. The water flows out of the ground for no apparent reason and from no readily identifiable source. So it is not surprising that springs have long been regarded with superstition and revered for their supposed medicinal value and healing powers. Nevertheless, there is nothing mystical or mysterious about springs.

Although springs can occur under a wide variety of geologic conditions, they all form in basically the same way (Figure 13.4a). When percolating water reaches the water table or an impermeable layer, it flows laterally, and if this flow intersects the surface, the water discharges as a spring (Figure 13.4b). The Mammoth Cave area in Kentucky is underlain by fractured limestones whose fractures have been enlarged into caves by solution activity. In this geologic environment, springs occur where the fractures and caves intersect the ground surface, allowing groundwater to exit onto the surface. Most springs are along valley walls where streams have cut valleys below the regional water table.

Springs can also develop wherever a perched water table intersects the surface (Figure 13.3). A *perched water table* may occur wherever a local aquiclude is present within a larger aquifer, such as a lens of shale within sandstone. As water migrates through the zone of aeration, it is stopped by the local aquiclude, and a localized zone of saturation "perched"

Figure 13.4 Springs Springs form wherever laterally moving groundwater intersects Earth's surface.

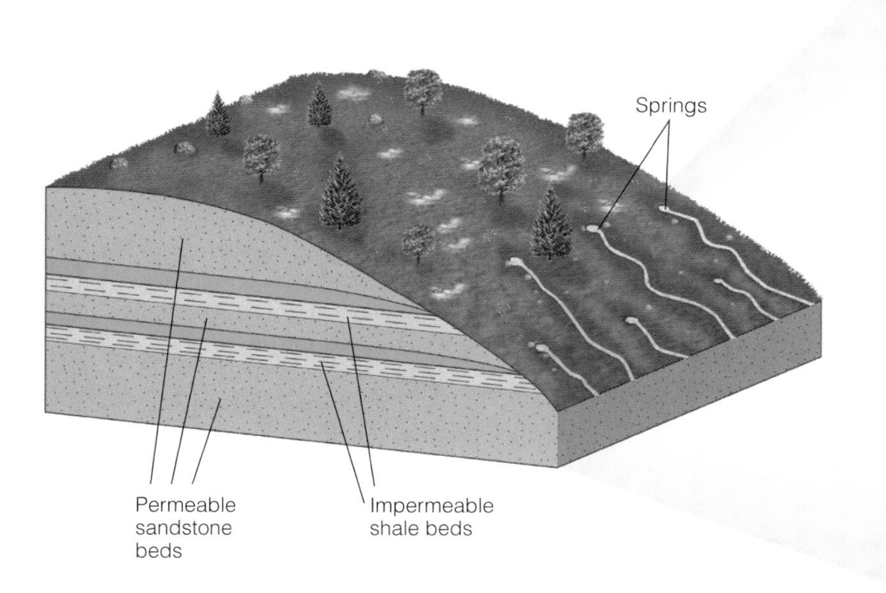

a Most commonly, springs form when percolating water reaches an impermeable layer and migrates laterally until it seeps out at the surface.

b Thunder River Spring in the Grand Canyon, Arizona, issues from rocks along a wall of the Grand Canyon. Water percolating downward through permeable rocks is forced to move laterally when it encounters an impermeable zone, and thus gushes out along this cliff. Notice the vegetation parallel to and below the springs, indicating enough water flows from springs along the cliff wall to support the vegetation.

above the main water table forms. Water moving laterally along the perched water table may intersect the surface to produce a spring.

Water Wells

Water wells are openings made by digging or drilling down into the zone of saturation. Once the zone of saturation has been penetrated, water percolates into the well, filling it to the level of the water table. A few wells are free flowing (see the next section); however, for most, the water must be brought to the surface by pumping.

In some parts of the world, water is raised to the surface with nothing more than a bucket on a rope or a hand-operated pump. In many parts of the United States and Canada, one can see windmills from times past that used wind power to pump water. Most of these are no longer in use, having been replaced by more efficient electric pumps.

When groundwater is pumped from a well, the water table in the area around the well is lowered, forming a **cone of depression** (Figure 13.5). A cone of depression forms because the rate of water withdrawal from the well exceeds the rate of water inflow to the well, thus lowering the water table around the well. A cone of depression's gradient, that is, whether it is steep or gentle, depends to a great extent on the permeability of the aquifer being pumped. A highly permeable aquifer produces a gentle gradient in the cone of depression, whereas a low-permeability aquifer results in a steep cone of depression because water cannot easily flow to the well to replace the water being withdrawn.

The formation of a cone of depression does not normally pose a problem for the average domestic well, provided that the well is drilled deep enough into the zone of saturation. However, the tremendous amounts of water used by industry and for irrigation may create a large cone of depression that lowers the water table sufficiently to cause shallow wells in the immediate area to go dry (Figure 13.5). This situation is not uncommon and frequently results in lawsuits by the owners of the shallow dry wells.

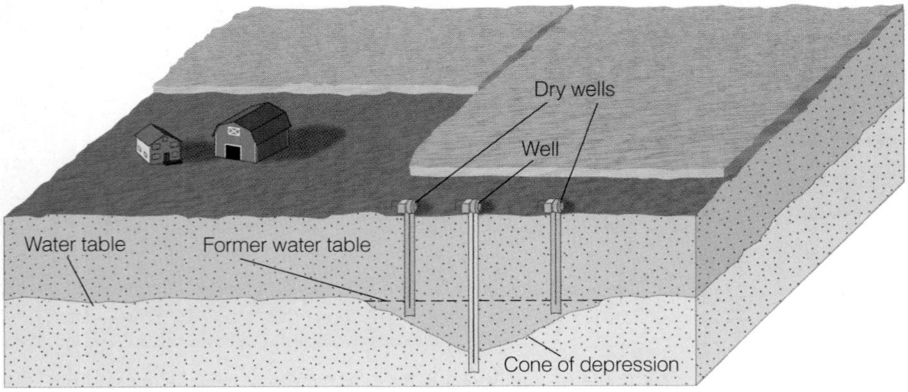

Lowering of the regional water table, because more groundwater is being withdrawn than is being replenished, is becoming a serious problem in many areas, particularly in the southwestern United States where rapid growth has placed tremendous demands on the groundwater system. As mentioned earlier, some of the largest cities in the United States depend entirely on groundwater for their municipal needs. Furthermore, some of the largest agricultural states are withdrawing groundwater from regional aquifers that are not being sufficiently replenished. Unrestricted withdrawal of groundwater cannot continue indefinitely, and the rising costs and decreasing supply of groundwater should soon limit the growth of some regions in the United States, such as the Southwest.

Artesian Systems

The word *artesian* comes from the French town and province of Artois (called Artesium during Roman times) near Calais, where the first European artesian well was drilled in 1126 and is still flowing today. The term **artesian system** can be applied to any system in which groundwater is confined and builds up high hydrostatic (fluid) pressure (Figure 13.6). Water in such a system is able to rise above the level of the aquifer if a well is drilled through the confining layer, thereby reducing the pressure and forcing the water upward. An artesian system can develop when an aquifer is confined above and below by aquicludes, the rock sequence is usually tilted to build up hydrostatic pressure, and the aquifer is exposed at the surface, thus enabling it to be recharged.

The elevation of the water table in the recharge area and the distance of the well from the recharge area determine the height to which artesian water rises in a well. The surface defined by the water table in the recharge area, called the *artesian-pressure surface,* is indicated by the sloping dashed line in Figure 13.6. If there were no friction in the aquifer, well water from an artesian aquifer would rise exactly to the elevation of the artesian-pressure surface. Friction, however, slightly reduces the pressure of the aquifer water and consequently the level to which artesian water rises, which is why the pressure surface slopes.

An artesian well will flow freely at the ground surface only if the wellhead is at an elevation below the artesian-pressure surface. In this situation, the water flows out of the well because it rises toward the artesian-pressure surface, which is at a higher elevation than the wellhead. In a nonflowing artesian well, the wellhead is above the artesian-pressure surface, and the water will rise in the well only as high as the artesian-pressure surface.

One of the best-known artesian systems in the United States underlies South Dakota and extends southward to central Texas. The majority of the artesian water from this

Figure 13.5 Cone of Depression A cone of depression forms whenever water is withdrawn from a well. If water is withdrawn faster than it can be replenished, the cone of depression will grow in depth and circumference, lowering the water table in the area and causing nearby shallow wells to go dry.

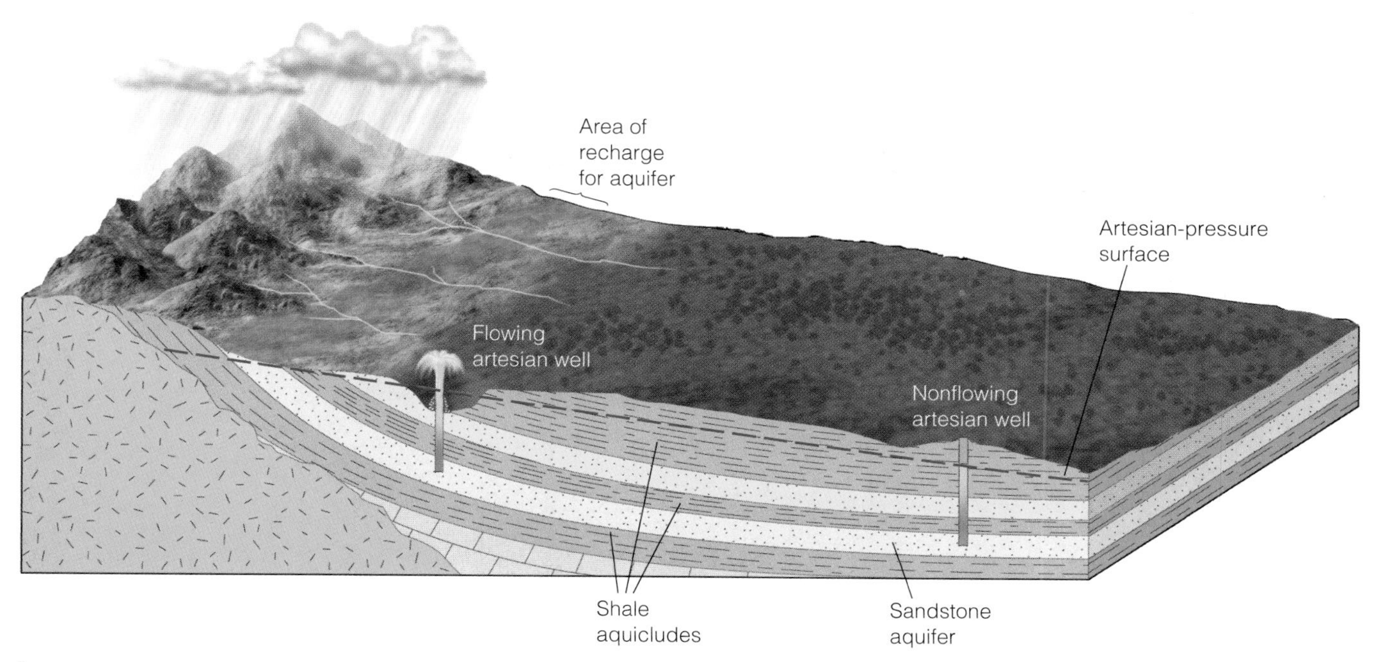

Figure 13.6 Artesian System An artesian system must have an aquifer confined above and below by aquicludes, the aquifer must be exposed at the surface, and the rock units are typically tilted as to build up hydrostatic pressure within the aquifer. The elevation of the water table in the recharge area, which is indicated by a sloping dashed line (the artesian-pressure surface), defines the highest level to which well water can rise. A wellhead below the elevation of the artesian-pressure surface will be free-flowing because the water will rise toward the artesian-pressure surface, which is at a higher elevation than the wellhead. Conversely, a well will be nonflowing if the elevation of the wellhead is at or above that of the artesian-pressure surface.

system is used for irrigation. The aquifer of this artesian system, the Dakota Sandstone, is recharged where it is exposed along the margins of the Black Hills of South Dakota. The hydrostatic pressure in this system was originally great enough to produce free-flowing wells and to operate waterwheels. However, because of the extensive use of this groundwater for irrigation over the years, the hydrostatic pressure in many of the wells is so low that they are no longer free-flowing and the water must be pumped.

As a final comment on artesian systems, we should mention that it is not unusual for advertisers to tout the quality of artesian water as somehow being superior to other groundwater. Some artesian water might in fact be of excellent quality, but its quality is not dependent on the fact that water rises above the surface of an aquifer. Rather, its quality is a function of dissolved minerals and any introduced substances, so artesian water really is no different from any other groundwater. The myth of its superiority probably arises from the fact that people have always been fascinated by water that flows freely from the ground.

GROUNDWATER EROSION AND DEPOSITION

When rainwater begins to seep into the ground, it immediately starts to react with the minerals it contacts, weathering them chemically. In an area underlain by soluble rock, groundwater is the principal agent of erosion and is responsible for the formation of many major features of the landscape.

Limestone, a common sedimentary rock composed primarily of the mineral calcite ($CaCO_3$), underlies large areas of Earth's surface (Figure 13.7). Although limestone is practically insoluble in pure water, it readily dissolves if a small amount of acid is present. Carbonic acid (H_2CO_3) is a weak acid that forms when carbon dioxide combines with water ($H_2O + CO_2 \rightarrow H_2CO_3$). Because the atmosphere contains a small amount of carbon dioxide (0.03%) and carbon dioxide is also produced in soil by the decay of organic matter, most groundwater is slightly acidic. When groundwater percolates through the various openings in limestone, the slightly acidic water readily reacts with the calcite to dissolve the rock by forming soluble calcium bicarbonate, which is carried away in solution (see Chapter 6).

Sinkholes and Karst Topography

In regions underlain by soluble rock, the ground surface may be pitted with numerous depressions that vary in size and shape. These depressions, called **sinkholes** or merely *sinks*, mark areas with underlying soluble rock (Figure 13.8). Most sinkholes form in one of two ways. The first is when soluble rock below the soil is dissolved by seeping water, and openings in the rock are enlarged and filled in by the overlying soil. As the groundwater continues to dissolve the rock, the soil is eventually removed, leaving shallow depressions with gently sloping sides. When adjacent sinkholes merge, they form a network of larger, irregular, closed depressions called *solution valleys*.

Geo-inSight The Burren Area of Ireland

The Burren region in northwest County Clare, Ireland, covers more than 260 square kilometers and is one of the finest examples of karst topography in Europe. Although the Burren landscape is frequently referred to as lunar-like because it is seemingly barren and lifeless, it is actually teeming with life and is world-famous for its variety of vegetation. Because of the heat-retention capacity of the massive limestones and soil trapped in the vertical joints of bare limestone pavement, and extremely diverse community of plants abounds.

▼ **2.** The present landcape is best described as glaciated karst. Like most of Ireland, the Burren was covered by a warm, shallow sea some 340 million years ago. As much as 780 m of interbedded marine limestones and shales were deposited at this time. These rocks were then covered by nearly 330 m of sandstones, slitstones, and shales. During the Pleistocene Epoch, glaciers stripped off most of the detrital rocks, thus exposing the underlying limestones to weathering and the humic acids produced by localized vegetation. Together, they have been the driving force producing the distinctive karst topography we find today.

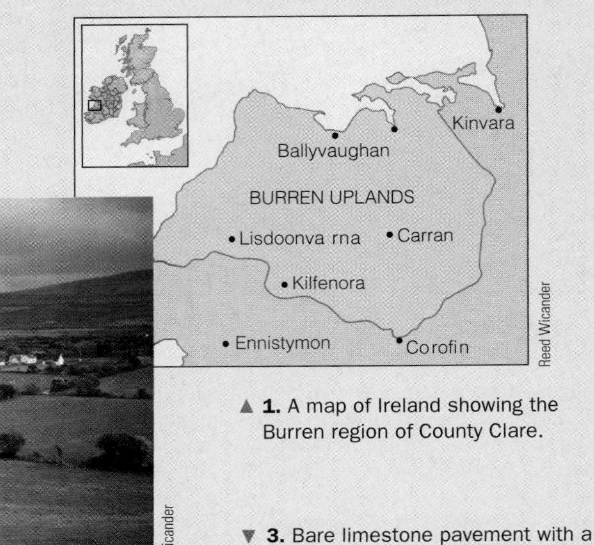

▲ **1.** A map of Ireland showing the Burren region of County Clare.

▼ **3.** Bare limestone pavement with a small wedge tomb from the Neolithic period—approximately 6000 years ago—in the background.

▲ **4.** Bare limestone pavements typically display a blocky appearance. The network of vertical cracks is the result of weathering of the joint pattern produced in the limestone during uplift of the region.

◀ **5.** A close-up shows the characteristic karren weathering pattern produced by solution of the limestone. Karren is used to describe the various microsolutional features of limestone pavement.

Reed Wicander

Reed Wicander

◄ **6.** Despite a lack of significant soil cover, a profusion of plants live in the soil trapped in the cracks, joints, and solution cavities of the limestone beds in the Burren.

▼ **7.** Perhaps the most famous Irish Neolithic dolmen is the Poulnabrone portal tomb which dates from about 5800 years ago. The name Poulnabrone literally means “the hole of the sorrows.” The capstone of the Poulnabrone dolmen rests on two 1.8 m-high portal stones to create a chamber in a low circular cairn, 9 m in diameter.

Reed Wicander

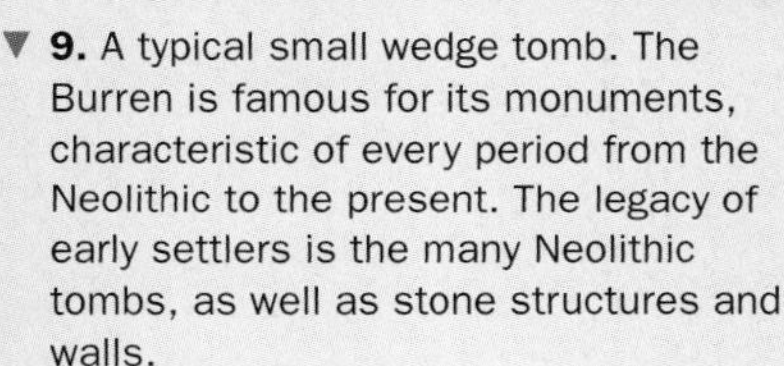

▼ **9.** A typical small wedge tomb. The Burren is famous for its monuments, characteristic of every period from the Neolithic to the present. The legacy of early settlers is the many Neolithic tombs, as well as stone structures and walls.

▲ **8.** Excavation of the Poulnabrone dolmen revealed that the tomb contained the remains of up to 22 people, buried over a period of six centuries, as well as a polished stone axe, two stone disc beads, flint and chert arrowheads and scrapers, and more than 60 shards of pottery.

Reed Wicander

Reed Wicander

► **10.** A Burren limestone wall. Burren limestone has been used for centuries in the construction of tombs, forts, houses, and walls. Today, it is in demand for limestone-finished houses, walls, and garden features.

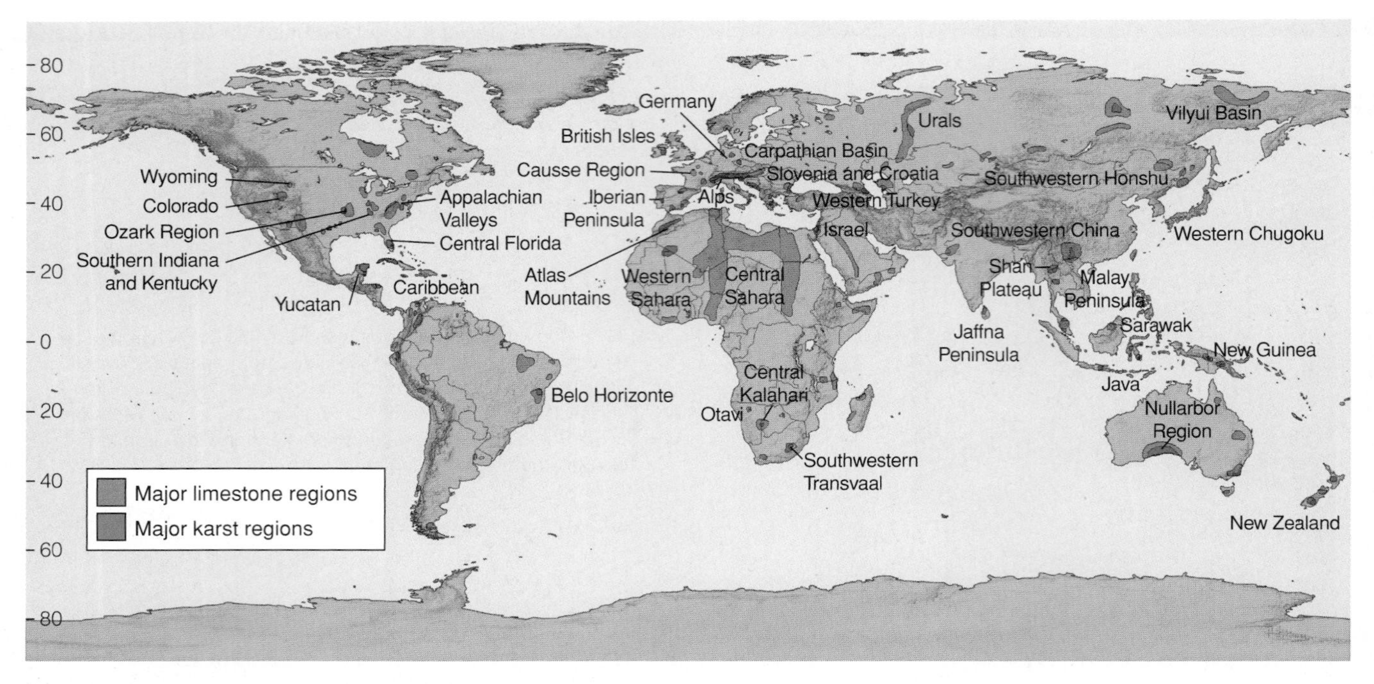

Figure 13.7 Major Limestone and Karst Areas of the World Distribution of the major limestone and karst areas of the world. Karst topography develops largely by groundwater erosion in areas underlain by soluble rocks.

Figure 13.8 Sinkholes

a This sinkhole formed on May 8 and 9, 1981, in Winter Park, Florida. It formed in previously dissolved limestone following a drop in the water table. The 100-m-wide, 35-m-deep sinkhole destroyed a house, numerous cars, and a municipal swimming pool.

b A small sinkhole in Montana, now occupied by a lake. The water enters the lake from a hot spring so it remains warm year-round. In fact, tropical fish have been introduced into the lake and thus live in an otherwise inhospitable climate.

Sinkholes also form when a cave's roof collapses, usually producing a steep-sided crater. Sinkholes formed in this way are a serious hazard, particularly in populated areas. In regions prone to sinkhole formation, extensive geologic and hydrogeologic investigation must be performed to determine the depth and extent of underlying cave systems prior to any site development. This is necessary to ensure that the underlying rocks are thick enough to support planned structures.

Karst topography, or simply *karst*, develops largely by groundwater erosion in many areas underlain by soluble rocks (Figure 13.9). The name *karst* is derived from the plateau region of the border area of Slovenia, Croatia, and northeastern Italy where this type of topography is well developed. In the United States, regions of karst topography include large areas of southwestern Illinois, southern Indiana, Kentucky, Tennessee, northern Missouri, Alabama, and central and northern Florida (Figure 13.7).

Karst topography is characterized by numerous caves, springs, sinkholes, solution valleys, and disappearing streams (Figure 13.9). *Disappearing streams* are so named because they typically flow only a short distance at the surface and then disappear into a sinkhole. The water continues flowing underground through fractures or caves until it surfaces again at a spring or other stream.

Karst topography varies from the spectacular high-relief landscapes of China to the subdued and pockmarked landforms of Kentucky (Figure 13.10). Common to all karst topography, though, is the presence of thick-bedded, readily soluble rock at the surface or just below the soil, and enough water for solution activity to occur (see Geo-inSight on pages 336 and 337). Karst topography is therefore typically restricted to humid and temperate climates.

Caves and Cave Deposits

Caves are perhaps the most spectacular examples of the combined effects of weathering and erosion by groundwater. As groundwater percolates through carbonate rocks, it dissolves and enlarges fractures and openings to form a complex interconnecting system of crevices, caves, caverns, and underground streams. A **cave** is usually defined as a naturally formed subsurface opening that is generally connected to the surface and is large enough for a person to enter. A *cavern* is a very large cave or a system of interconnected caves.

More than 17,000 caves are known in the United States. Most of them are small, but some are large and spectacular. Some of the more famous ones are Mammoth Cave, Kentucky, Carlsbad Caverns, New Mexico, Lewis and Clark Caverns, Montana, Lehman Cave, Nevada, and Meramec Caverns, Missouri, which Jesse James and his outlaw band often used as a hideout. The United States has many famous caves, but so has Canada, including 536-m-deep Arctomys Cave in Mount Robson Provincial Park, British Columbia, the deepest known cave in North America.

Caves and caverns form as a result of the dissolution of carbonate rocks by weakly acidic groundwater (Figure 13.11). Groundwater percolating through the zone of aeration slowly dissolves the carbonate rock and enlarges its fractures and bedding planes. On reaching the water table, the groundwater migrates toward the region's surface streams. As the groundwater moves through the zone of saturation, it continues to dissolve the rock and gradually forms a system of horizontal passageways through which the dissolved rock is carried to the streams. As the surface streams erode deeper valleys, the water table drops in response to the lower elevation of the streams. The water that flowed through the system of horizontal passageways now percolates to the lower water table where a new system of passageways begins to form. The abandoned channelways form an interconnecting system of caves and caverns. Caves eventually become unstable and collapse, littering the floor with fallen debris.

When most people think of caves, they think of the seemingly endless variety of colorful and

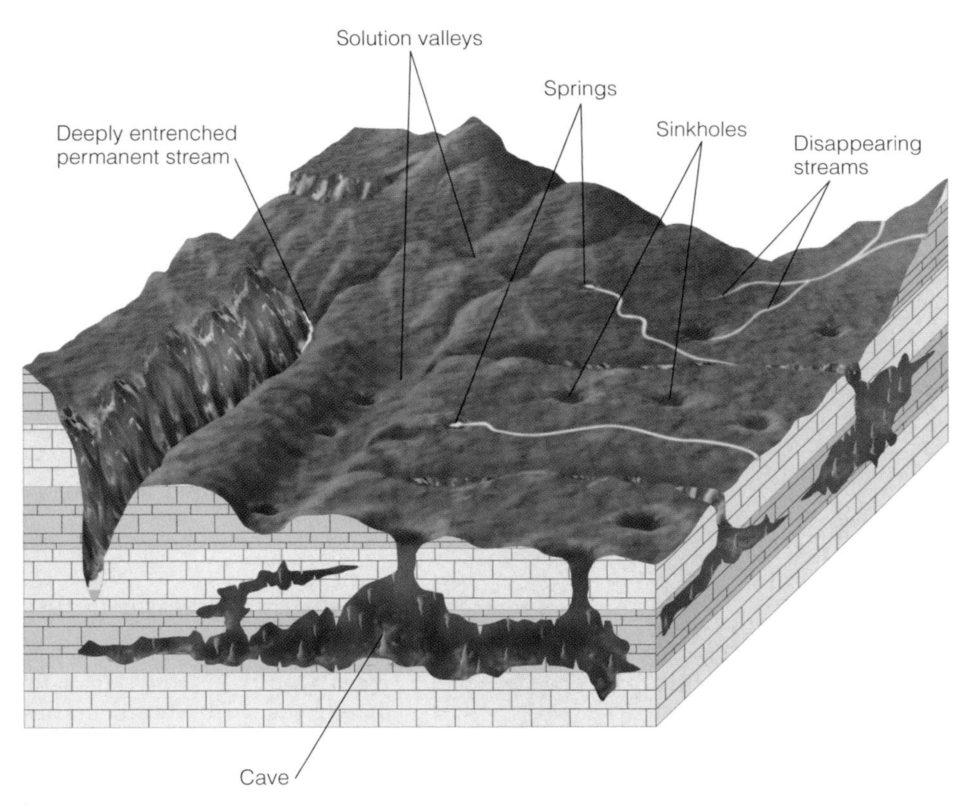

Figure 13.9 Features of Karst Topography Erosion of soluble rock by groundwater produces karst topography. Features commonly found include solution valleys, springs, sinkholes, and disappearing streams.

What Would You Do?

Your well periodically goes dry and never produces the amount of water you would like or need. Is your well too shallow, is it drilled into Earth materials with low permeability, or are nearby irrigation pumps causing your problems? What can you do to resolve these problems?

Figure 13.10 Karst Landscape in Kunming, China, and Bowling Green, Kentucky

Reed Wicander

a The Stone Forest, 125 km southeast of Kunming, China, is a high-relief karst landscape formed by the dissolution of carbonate rocks.

John S. Shelton

b Solution valleys, sinkholes, and sinkhole lakes dominate the subdued karst topography east of Bowling Green, Kentucky.

bizarre-shaped deposits found in them. Although a great many different types of cave deposits exist, most form in essentially the same manner and are collectively known as *dripstone*. As water seeps into a cave, some of the dissolved carbon dioxide in the water escapes, and a small amount of calcite is precipitated. In this manner, the various dripstone deposits are formed (Figure 13.11c).

Stalactites are icicle-shaped structures hanging from cave ceilings that form as a result of precipitation from dripping water (Figure 13.12). With each drop of water, a thin layer of calcite is deposited over the previous layer, forming a cone-shaped projection that grows down from the ceiling. The water that drips from a cave's ceiling also precipitates a small amount of calcite when it hits the floor. As additional calcite is deposited, an upward-growing projection called a *stalagmite* forms (Figure 13.12). If a stalactite and stalagmite meet, they form a *column*. Groundwater seeping from a crack in a cave's ceiling may form a vertical sheet of rock called a *drip curtain*, and water flowing across a cave's floor may produce *travertine terraces* (Figure 13.11c).

MODIFICATIONS OF THE GROUNDWATER SYSTEM AND ITS EFFECTS

Groundwater is a valuable natural resource that is rapidly being exploited with seemingly little regard to the effects of overuse and misuse. Currently, approximately 20% of all water used in the United States is groundwater. This percentage is rapidly increasing, and unless this resource is used more wisely, sufficient amounts of clean ground water will not be available in the future. Modifications of the groundwater system may have many consequences, including (1) lowering of the water table, causing wells to dry up; (2) saltwater incursion; (3) subsidence; and (4) contamination.

Lowering the Water Table

Withdrawing groundwater at a significantly greater rate than it is replaced by either natural or artificial recharge can have serious effects. For example, the High Plains aquifer is one of the most important aquifers in the United States. It

Figure 13.11 Cave Formation

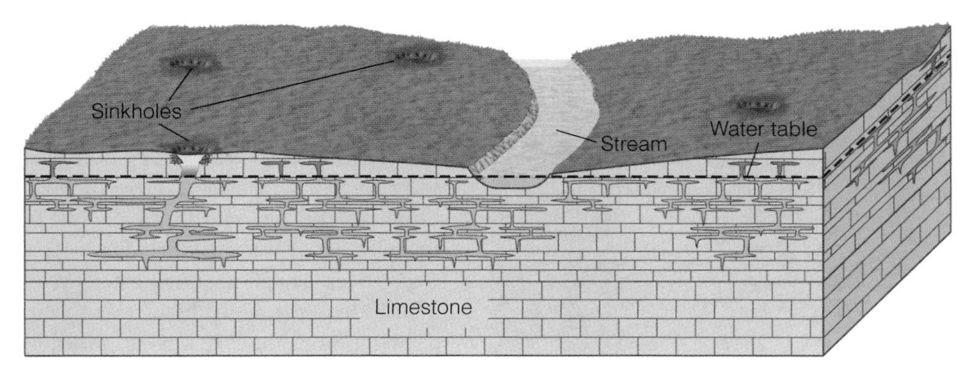

a As groundwater percolates through the zone of aeration and flows through the zone of saturation, it dissolves the carbonate rocks and gradually forms a system of passageways.

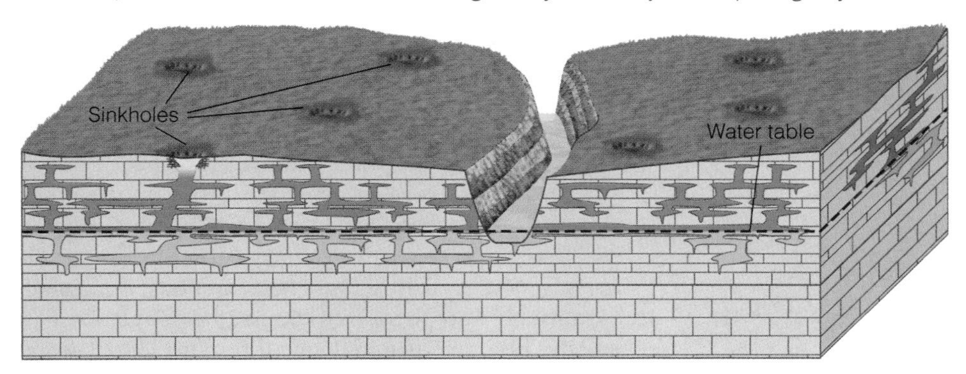

b Groundwater moves along the surface of the water table, forming a system of horizontal passageways through which dissolved rock is carried to the surface streams, thus enlarging the passageways.

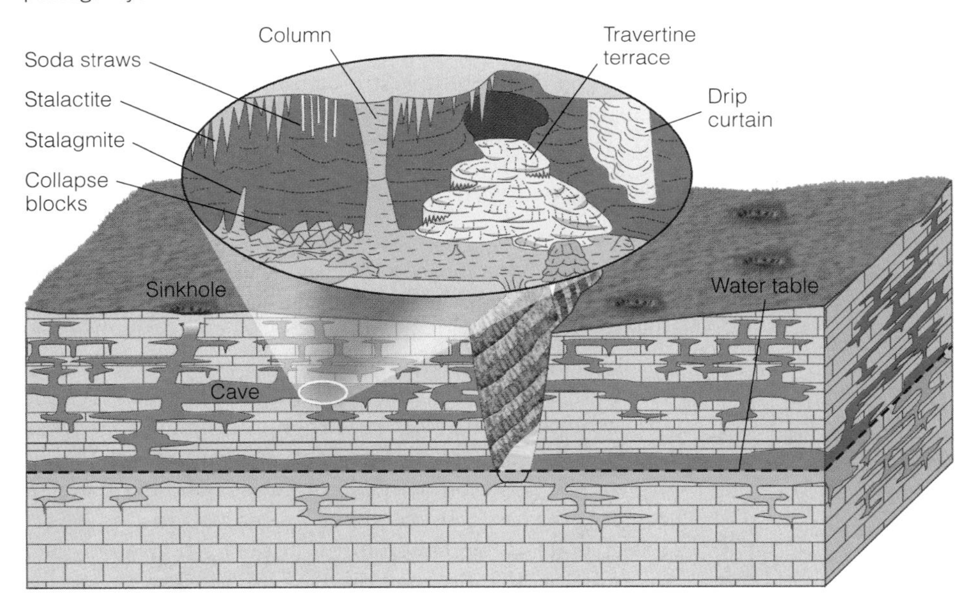

c As the surface streams erode deeper valleys, the water table drops and the abandoned channelways form an interconnecting system of caves and caverns.

underlies more than 450,000 km^2, including most of Nebraska, large parts of Colorado and Kansas, portions of South Dakota, Wyoming, and New Mexico, as well as the panhandle regions of Oklahoma and Texas, and accounts for approximately 30% of the groundwater used for irrigation in the United States (Figure 13.13).

Significant withdrawal of groundwater from the High Plains aquifer for irrigation began in the 1950s, and by 1980, the water table had dropped an average of 3 meters. Irrigation from the High Plains aquifer is largely responsible for the region's agricultural productivity, including a significant percentage of the nation's corn, cotton, and wheat, and half of U.S. beef cattle.

Although the High Plains aquifer has contributed to the high productivity of the region, it cannot continue to provide the quantities of water that it has in the past. Based on the present rate of groundwater withdrawal, a 1982 Department of Commerce study estimated that by 2020, about a fourth of the High Plains aquifer water will have been withdrawn. In some parts of the High Plains, from 2 to 100 times more water is being pumped annually than is being recharged, causing a substantial drop in the water table in many areas (Figure 13.13). It must be noted that much of the aquifer's water infiltrated during wetter glacial climates more than 10,000 years ago. Consequently, most of the water being pumped is fossil water that is not being replenished at anywhere near the same rate when it formed during the Pleistocene Epoch.

What will happen to this region's economy if long-term withdrawal of water from the High Plains aquifer continues to greatly exceed its recharge rate, and the aquifer can no longer supply the quantities of water necessary for irrigation? Solutions range from going back to farming without irrigation to diverting water from other regions such as the Great Lakes. Most users of the aquifer realize that they cannot continue to withdraw the quantities of groundwater that they have in the past, and thus are turning to greater conservation, monitoring of the aquifer, and using

Figure 13.12 Cave Deposits Stalactites are the icicle-shaped structures hanging from a cave's ceiling, whereas the upward-pointing structures on the floor are stalagmites. Columns result when stalactites and stalagmites meet. All three structures are visible in Shenandoah Caverns, Virginia, USA.

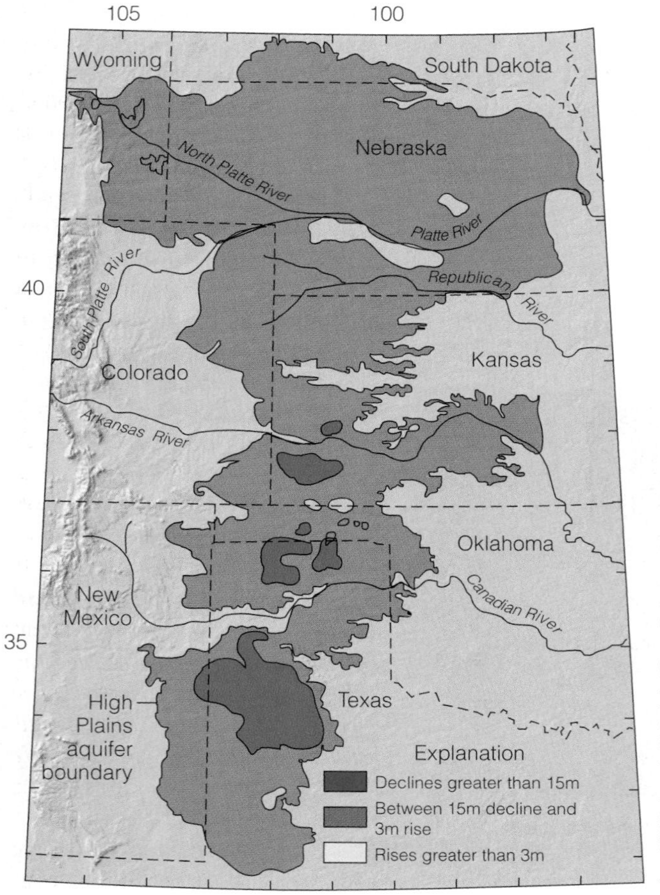

Figure 13.13 High Plains Aquifer The geographic extent of the High Plains aquifer and changes in water level from predevelopment through 1993. Irrigation from the High Plains aquifer is largely responsible for the region's agricultural productivity.

new technologies to try to better balance withdrawal with recharge rates.

Another excellent example of what we might call deficit spending with regard to groundwater took place in California during the drought of 1987–1992. During that time, the state's aquifers were overdrawn at a rate of 10 million acre-feet per year (an acre-foot is the amount of water that covers 1 acre, 1 foot deep). In short, during each year of the drought, California was withdrawing more than 12 km^3 of groundwater more than was being replaced. Unfortunately, excessive depletion of the groundwater reservoir has other consequences, such as subsidence involving sinking or settling of the ground surface (discussed in a later section).

Water supply problems certainly exist in many areas, but on the positive side, water use in the United States actually declined during the five years following 1980 and has remained nearly constant since then, even though the population has increased. This downturn in demand resulted largely from improved techniques in irrigation, more efficient industrial water use, and a general public awareness of water problems coupled with conservation practices. Nevertheless, the rates of withdrawal of groundwater from some aquifers still exceeds their rates of recharge, and population growth in the arid to semiarid Southwest is continuing to put significant demands on an already limited water supply.

Saltwater Incursion

The excessive pumping of groundwater in coastal areas has resulted in *saltwater incursion* such as occurred on Long Island, New York, during the 1960s. Along coastlines where permeable rocks or sediments are in contact with the ocean, the fresh groundwater, being less dense than seawater, forms a lens-shaped body above the underlying saltwater (Figure 13.14a). The weight of the freshwater exerts pressure on the underlying saltwater. As long as rates of recharge equal rates of withdrawal, the contact between the fresh groundwater and the seawater remains the same. If excessive pumping occurs, however, a deep cone of depression forms in the fresh groundwater (Figure 13.14b). Because some of the pressure from the overlying freshwater has been removed, saltwater forms a *cone of ascension* as it rises to fill the pore space that formerly contained freshwater. When this occurs, wells become contaminated with saltwater and remain contaminated until recharge by freshwater restores the former level of the fresh-groundwater water table.

Saltwater incursion is a major problem in many rapidly growing coastal communities. As the population in these

Figure 13.14 Saltwater Incursion

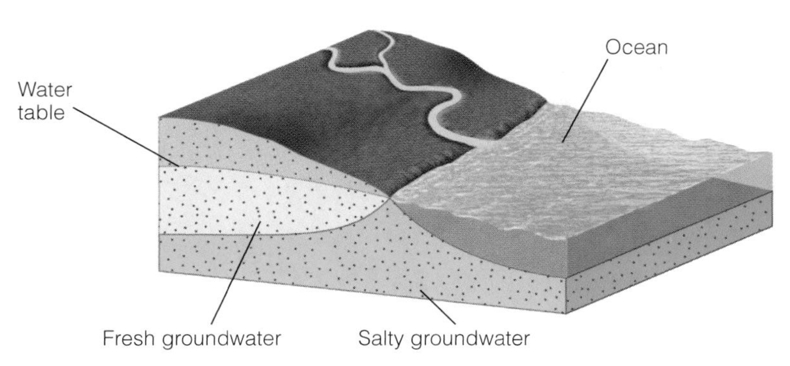

a Because freshwater is not as dense as saltwater, it forms a lens-shaped body above the underlying saltwater.

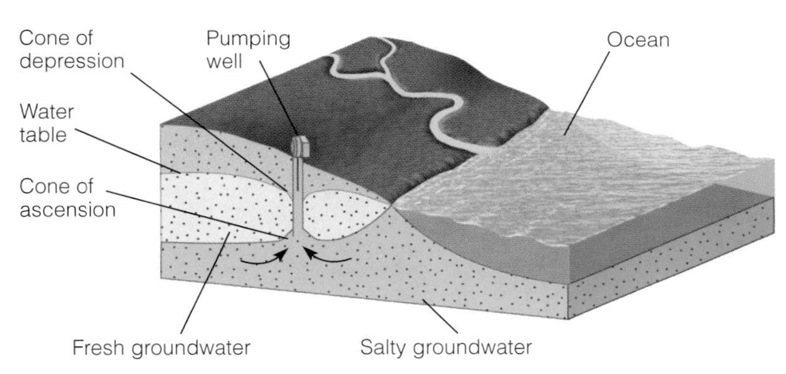

b If excessive pumping occurs, a cone of depression develops in the fresh groundwater, and a cone of ascension forms in the underlying salty groundwater, which may result in saltwater contamination of the well.

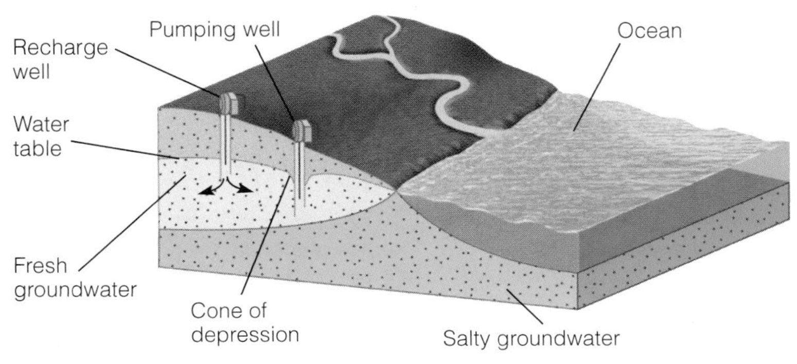

c Pumping water back into the groundwater system through recharge wells can help lower the interface between the fresh groundwater and the salty groundwater and reduce saltwater incursion.

areas grows, greater demand for groundwater creates an even greater imbalance between withdrawal and recharge.

To counteract the effects of saltwater incursion, recharge wells are often drilled to pump water back into the groundwater system (Figure 13.14c). Recharge ponds that allow large quantities of fresh surface water to infiltrate the groundwater supply may also be constructed. Both of these methods are used successfully on Long Island to mitigate the saltwater incursion problem that has persisted for several decades.

Subsidence

As excessive amounts of groundwater are withdrawn from poorly consolidated sediments and sedimentary rocks, the water pressure between grains is reduced, and the weight of the overlying materials causes the grains to pack more closely together, resulting in *subsidence* of the ground. As more and more groundwater is pumped to meet the increasing needs of agriculture, industry, and population growth, subsidence is becoming more prevalent.

The San Joaquin Valley of California is a major agricultural region that relies largely on groundwater for irrigation. Between 1925 and 1977, groundwater withdrawals in parts of the valley caused subsidence of nearly 9 m (Figure 13.15). Other areas in the United States that have experienced subsidence due to groundwater withdrawal are New Orleans, Louisiana, and Houston, Texas, both of which have subsided more than 2 m, and Las Vegas, Nevada, where 8.5 m of subsidence has taken place (Table 13.2).

Looking elsewhere in the world, the tilt of the Leaning Tower of Pisa in Italy is partly due to groundwater withdrawal (Figure 13.16). The tower started tilting soon after construction began in 1173 because of differential compaction of the foundation. During the 1960s, the city of Pisa withdrew ever-larger amounts of groundwater, causing the ground to subside further; as a result, the tilt of the tower increased until it was in danger of falling over. Strict control of groundwater withdrawal, stabilization of the foundation, and recent renovations have reduced the amount of tilting to about 1 mm per year, thus ensuring that the tower should stand for several more centuries.

A spectacular example of continuing subsidence is taking place in Mexico City, which is built on a former lake bed. As groundwater is removed for the increasing needs of the city's 17.8 million people, the water table has been lowered up to 10 m. As a result, the fine-grained lake deposits are compacting, and Mexico City is slowly and unevenly subsiding. Its opera house has settled more than 3 m, and half of the first floor is now below ground level. Other parts of the city have subsided as much as 7.5 m, creating similar problems for other structures. The fact that 72% of the city's water comes from the aquifer beneath the metropolitan area ensures that problems of subsidence will continue.

The extraction of oil can also cause subsidence. Long Beach, California, has subsided 9 m as a result of many decades of oil production. More than $100 million in damages was done to the pumping, transportation, and harbor facilities in this area because of subsidence and encroachment of the sea (Figure 13.17). Once water was pumped back into the oil reservoir, thus stabilizing it, subsidence virtually stopped.

USGS

Figure 13.15 Subsidence in the San Joaquin Valley, California The dates on this power pole dramatically illustrate the amount of subsidence in the San Joaquin Valley, California. Because of groundwater withdrawals and subsequent sediment compaction, the ground subsided nearly 9 m between 1925 and 1977. For a time, surface water use reduced subsidence, but during the drought of 1987 to 1992, it started again as more groundwater was withdrawn.

Sarah Stone/Getty Images

Figure 13.16 The Leaning Tower of Pisa, Italy The tilting is partly the result of subsidence due to the removal of groundwater. Strict control of groundwater withdrawal, recent stabilization of the foundation, and renovation of the structure itself have ensured that the Leaning Tower will continue leaning for many more centuries.

TABLE 13.2 Subsidence of Cities and Regions Due Primarily to Groundwater Removal

Location	Maximum Subsidence (m)	Area Affected (km^2)
Mexico City, Mexico	8.0	25
Long Beach and Los Angeles, California	9.0	50
Taipei Basin, Taiwan	1.0	100
Shanghai, China	2.6	121
Venice, Italy	0.2	150
New Orleans, Louisiana	2.0	175
London, England	0.3	295
Las Vegas, Nevada	8.5	500
Santa Clara Valley, California	4.0	600
Bangkok, Thailand	1.0	800
Osaka and Tokyo, Japan	4.0	3,000
San Joaquin Valley, California	9.0	9,000
Houston, Texas	2.7	12,100

Source: Data from R. Dolan and H. G. Goodell, "Sinking Cities," American Scientist 74 (1986): 38–47; and J. Whittow, *Disasters: The Anatomy of Environmental Hazards* (Athens: University of Georgia Press, 1979).

Long Beach Department of Oil Properties

Figure 13.17 Oil Field Subsidence, Long Beach, California The withdrawal of petroleum from the Long Beach, California, oil field resulted in up to 9 m of ground subsidence in some areas because of sediment compaction. In this photograph, note that the ground has settled around the well stems (the white "posts"), leaving the wellheads up above the ground. The levee on the left edge of the photograph was built to keep seawater in the adjacent marina from flooding the oil field. It was not until water was pumped back into the reservoir to replace the extracted petroleum that ground subsidence finally ceased.

Groundwater Contamination

A major problem facing our society is the safe disposal of the numerous pollutant byproducts of an industrialized economy. We are becoming increasingly aware that streams, lakes, and oceans are not unlimited reservoirs for waste and that we must find new safe ways to dispose of pollutants.

The most common sources of groundwater contamination are sewage, landfills, toxic waste disposal sites, and agriculture. Once pollutants get into the groundwater system, they spread wherever groundwater travels, which can make their containment difficult (see Geo-Focus on pages 346 and 347). Furthermore, because groundwater moves so slowly, it takes a long time to cleanse a groundwater reservoir once it has become contaminated.

In many areas, septic tanks are the most common way of disposing of sewage. A septic tank slowly releases sewage into the ground, where it is decomposed by oxidation and microorganisms and filtered by the sediment as it percolates through the zone of aeration. In most situations, by the time the water from the sewage reaches the zone of saturation, it has been cleansed of any impurities and is safe to use (Figure 13.18a). If the water table is close to the surface or if the rocks are very permeable, however, water entering the zone of saturation may still be contaminated and unfit to use.

Landfills are also potential sources of groundwater contamination (Figure 13.18b). Not only does liquid waste seep into the ground, but rainwater also carries dissolved chemicals and other pollutants down into the groundwater reservoir. Unless the landfill is carefully designed and lined with an impermeable layer such as clay, many toxic compounds such as paints, solvents, cleansers, pesticides, and battery acid will find their way into the groundwater system.

Toxic waste sites where dangerous chemicals are either buried or pumped underground are an increasing source of groundwater contamination. The United States alone must dispose of several thousand metric tons of hazardous chemical waste per year. Unfortunately, much of this waste has been, and still is, being improperly dumped and is contaminating the surface water, soil, and groundwater.

Examples of indiscriminate dumping of dangerous and toxic chemicals can be found in every state. Perhaps the most famous is Love Canal, near Niagara Falls, New York. During the 1940s, the Hooker Chemical Company dumped approximately 19,000 tons of chemical waste into Love Canal. In 1953, Hooker covered one of the dump sites with dirt and sold it for one dollar to the Niagara Falls Board of Education, which built an elementary school and playground on the site. Heavy rains and snow during the winter of 1976–1977 raised the water table and turned the area into a muddy swamp in the spring of 1977. Mixed with the mud were thousands of toxic, noxious chemicals that formed puddles in the playground, oozed into people's basements, and covered gardens and lawns. Trees, lawns, and gardens began to die, and many of the residents of the area suffered from serious illnesses. The cost of cleaning up the Love Canal site and relocating its residents exceeded $100 million, and the site and neighborhood are now vacant.

Groundwater Quality

Finding groundwater is rather easy because it is present beneath the land surface nearly everywhere, although the depth to the water table varies considerably. But just finding water is not enough. Sufficient amounts in porous and permeable materials must be located if groundwater is to be withdrawn for agricultural, industrial, or domestic use. The availability of groundwater was important in the westward expansion in both Canada and the United States, and now more than one-third of all water for irrigation comes from the groundwater system. As we mentioned in the Introduction, groundwater provides 80% of the water used for rural livestock and domestic use in rural America and it is the primary source of water for a number of large cities. Furthermore, as one would expect, quality is more important here than it is for most other purposes.

If we discount contamination by humans from landfills, septic systems, toxic waste sites, and industrial effluents, groundwater quality is mostly a function of (1) the kinds of materials that make up an aquifer, (2) the residence time of

What Would You Do?

Americans generate tremendous amounts of waste. Some of this waste, such as battery acid, paint, cleaning agents, insecticides, and pesticides, can easily contaminate the groundwater system. Your community is planning to construct a city dump to contain waste products, but it simply wants to dig a hole, dump waste in, and then bury it. What do you think of this plan? Are you skeptical of this plan's merits, and if so, what would you suggest to remedy any potential problems?

Geo-Focus Arsenic and Old Lace

Many people probably learned that arsenic is a poison from either reading or seeing the play *Arsenic and Old Lace*, written by Joseph Kesselring. In the play, the elderly Brewster sisters poison lonely old men by adding a small amount of arsenic to their homemade elderberry wine.

Arsenic is a naturally occurring toxic element found in the environment, and several types of cancer have been linked to arsenic in water. Arsenic also harms the central and peripheral nervous systems and may cause birth defects and reproductive problems. In fact, because of arsenic's prevalence in the environment and its adverse health effects, Congress included it in the amendments to the Safe Drinking Water Act in 1996. Arsenic gets into the groundwater system mainly as arsenic-bearing minerals dissolve in the natural weathering process of rocks and soils.

A map published in 2001 by the U.S. Geological Survey (USGS) shows the extent and concentration of arsenic in the nation's groundwater supply (Figure 1). The highest concentrations of groundwater arsenic were found throughout the West and in parts of the Midwest and Northeast. Although the map is not intended to provide specific information for individual wells or even a locality within a

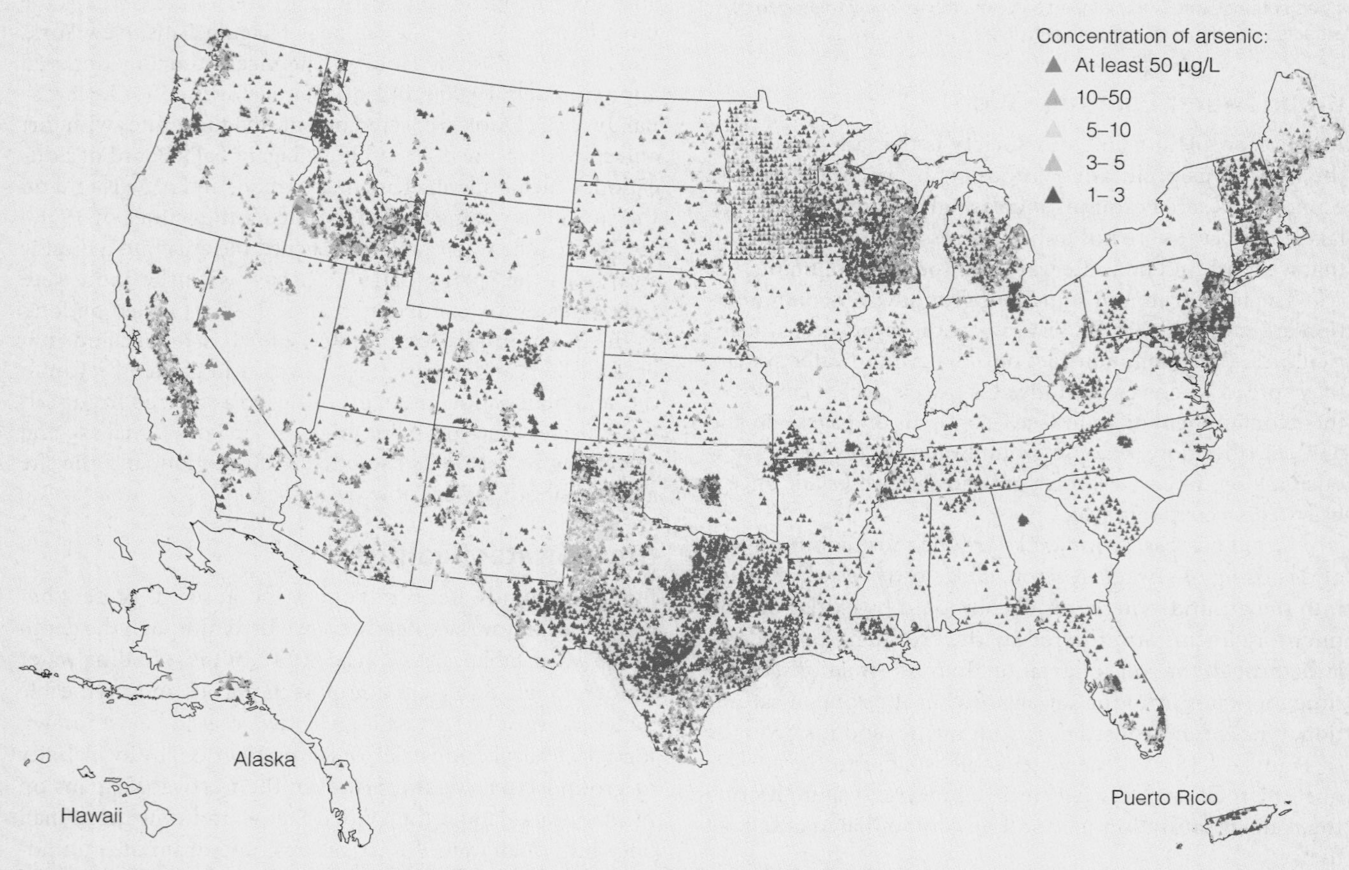

Figure 1 Arsenic concentrations for 31,350 groundwater samples collected from 1973 to 2001.

water in an aquifer, and (3) the solubility of rocks and minerals. These factors account for the amount of dissolved materials in groundwater, such as calcium, iron, fluoride, and several others. Most pose no health problems, but some have undesirable effects such as deposition of minerals in water pipes and water heaters, an offensive taste or smell, or they may stain clothing and fixtures or inhibit the effectiveness of detergents.

A good example that everyone is familiar with is *hard water*, which is a problem in many areas, especially those underlain by limestone and dolostone. Hard water is caused by dissolved county, it helps researchers and policymakers identify areas of high concentration so that they can make informed decisions about water use. We should point out, however, that a high degree of local variability in the amount of arsenic in the groundwater can be caused by local geology, type of aquifer, depth of well, and other factors. The only way to learn the arsenic concentration in any well is to have it tested.

What is considered a safe level of arsenic in drinking water? In 2001, the U.S. Environmental Protection Agency (U.S. EPA, or EPA) lowered the maximum level of arsenic permitted in drinking water from 50 μg (50 micrograms) arsenic per liter to 10 μg of arsenic per liter. This is the same figure used by the World Health Organization.

From the data in Figure 1, additional maps were created. Figure 2 shows arsenic concentrations found in at least 25% of groundwater samples per county. Based on these data, approximately 10% of the samples in the USGS study exceed the new standard of 10 μg of arsenic per liter of drinking water.

Public water supply systems that exceed the existing EPA arsenic standard are required to either treat the water to remove the arsenic or find an alternative supply. Although reducing the acceptable level of arsenic in drinking water will surely increase the cost of water to consumers, it will also decrease their exposure to arsenic and the possible adverse health effects associated with this toxic element.

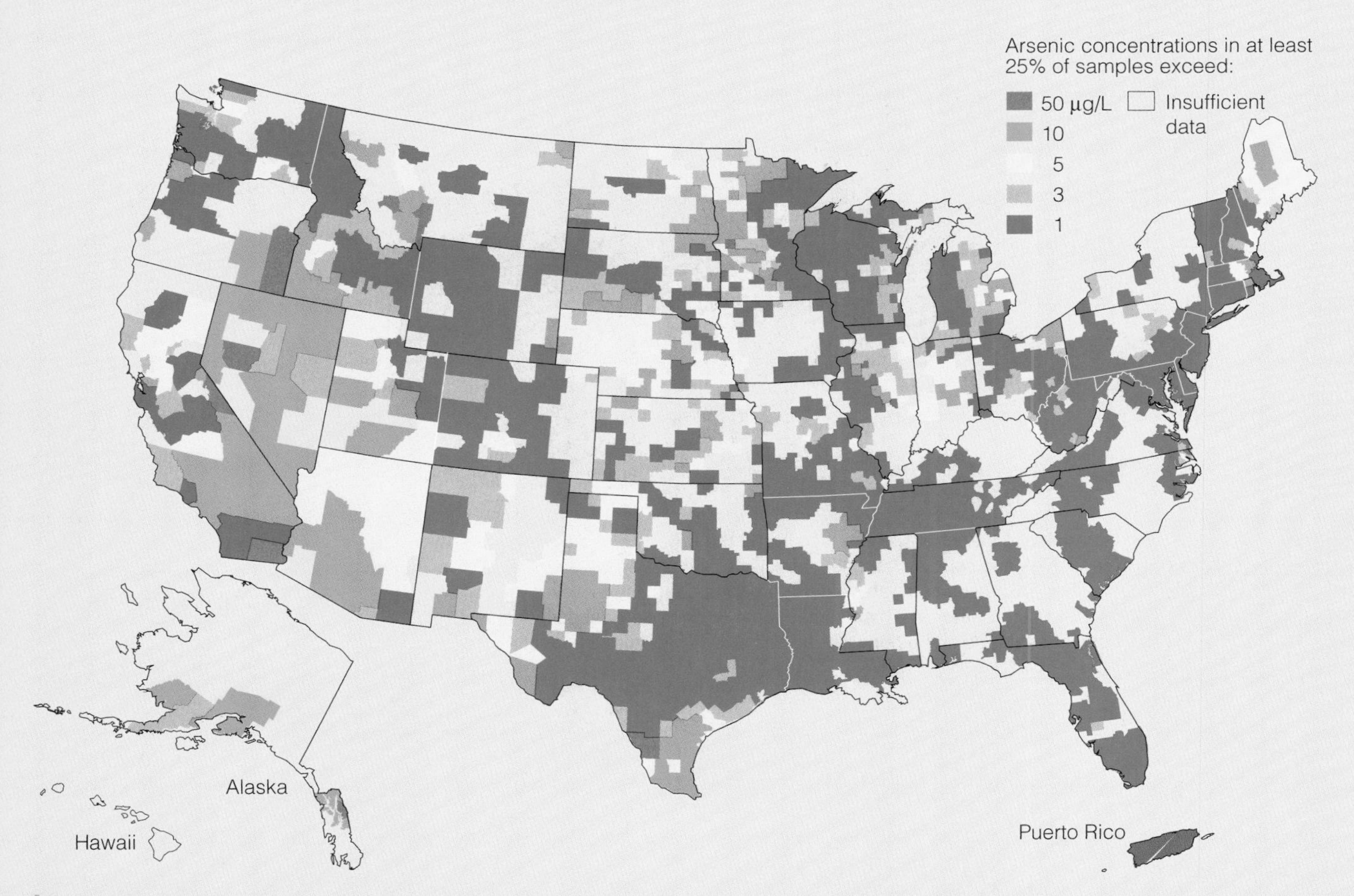

Figure 2 Arsenic concentrations found in at least 25% of groundwater samples per county. The map is based on 31,350 groundwater samples collected between 1973 and 2001.

calcium (Ca^{+2}) and magnesium (Mg^{+2}) ions. Water containing less than 60 milligrams of Ca^{+2} and Mg^{+2} per liter (mg/L) is considered soft, whereas 61–120 mg/L indicates moderately hard water, values from 121 to 180 mg/L characterize hard water, and any water with more than 180 mg/L is very hard.

One of the negative aspects of hard water is the precipitation of scale (Ca and Mg salts) in water pipes, water heaters, dishwashers, and even on glasses and dinnerware. To remedy this problem, many households have a water softener, whereby calcium and magnesium ions in the water are

Figure 13.18 Groundwater Contamination

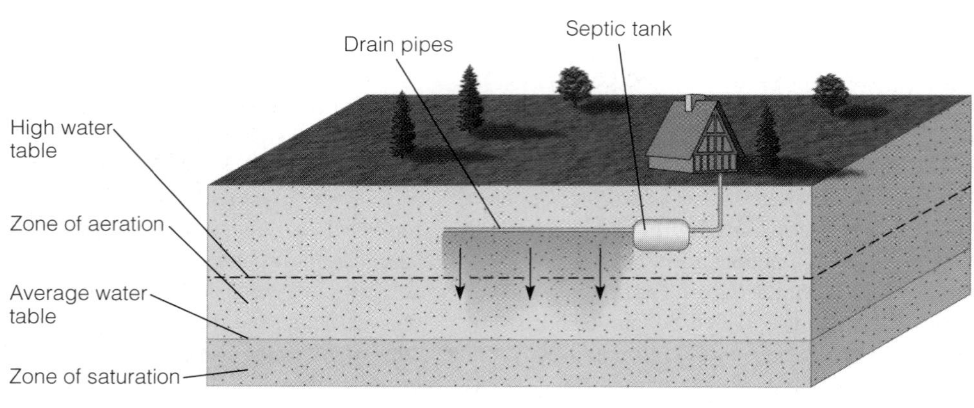

a A septic system slowly releases sewage into the zone of aeration. Oxidation, bacterial degradation, and filtering usually remove impurities before they reach the water table. However, if the rocks are very permeable or the water table is too close to the septic system, contamination of the groundwater can result.

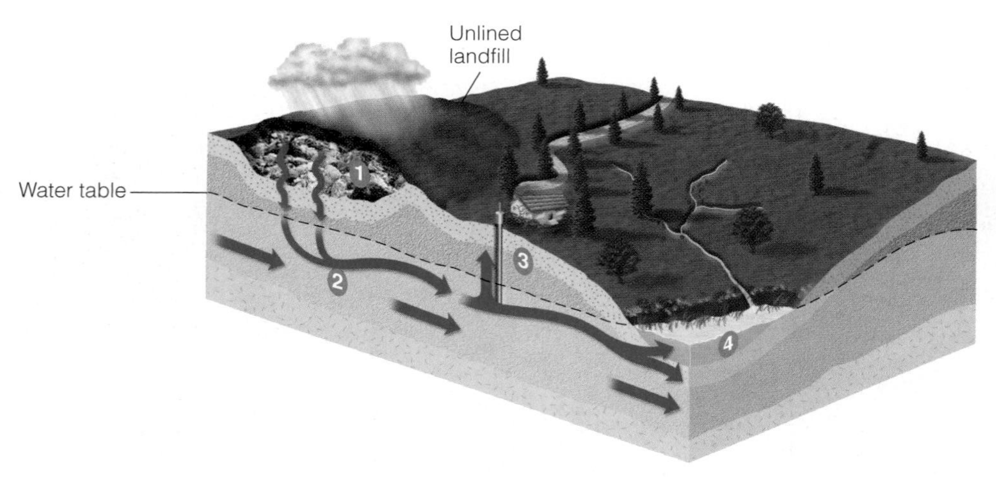

b Unless there is an impermeable barrier between a landfill and the water table, pollutants can be carried into the zone of saturation and contaminate the groundwater supply: (1) Infiltrating water leaches contaminates from the landfill; (2) the polluted water enters the water table and moves away from the landfill; (3) wells may tap the polluted water and thus contaminate drinking water supplies; and (4) the polluted water may emerge into streams and other water bodies downslope from the landfill.

replaced by sodium (Na^+) ions through the use of an ion exchanger or a mineral sieve. Thus, the amount of calcium and magnesium is reduced and the water is more desirable for most domestic purposes. However, people on low-sodium diets, such as those with hypertension (high blood pressure), are cautioned not to drink softened water because it contains more sodium.

Not all dissolved materials in groundwater are undesirable, at least in small quantities. Fluoride (F^-), for instance, if present in amounts of 1.0 to 1.5 parts per million (ppm), combines with the calcium phosphate in teeth and makes them more resistant to decay. However, too much fluoride—more than 4.0 ppm—gives children's teeth a dark, blotchy appearance.

Fluoride in natural waters is rare, so few communities benefit from its presence. However, many cities and towns add fluorine to their drinking water, and many dentists routinely use fluoride treatments, such that the population as a whole has seen a reduction in cavities and tooth decay. Despite the beneficial effect of fluoridation in reducing tooth decay, there has been an increase in opposition to fluoridating municipal water supplies because of possible health risks.

HYDROTHERMAL ACTIVITY

Hydrothermal is a term referring to hot water. Some geologists restrict the meaning to include only water heated by magma, but here we use it to refer to any hot subsurface water and the surface activity that results from its discharge. One manifestation of hydrothermal activity in areas of active or recently active volcanism is the discharge of gases, such as steam, at vents known as *fumeroles* (see Figure 5.2). Of more immediate concern here, however, is the groundwater that rises to the surface as *hot springs* or *geysers*. It may be heated by its proximity to magma or by Earth's geothermal gradient because it circulates deeply.

Hot Springs

A **hot spring** (also called a *thermal spring* or *warm spring*) is any spring in which the water temperature is higher than 37°C, the temperature of the human body (Figure 13.19a). Some hot springs are much hotter, with temperatures up to the boiling point in many instances (Figure 13.19b). Another type of hot spring, called a *mud pot*, results when chemically altered rocks yield clays that bubble as hot water and steam rise through them (Figure 13.19c). Of the approximately 1100 known hot springs in the United States, more than 1000 are in the far West, with the others in the Black Hills of South Dakota, Georgia, the Ouachita region of Arkansas, and the Appalachian region.

Hot springs are also common in other parts of the world. One of the most famous is in Bath, England, where shortly after the Roman conquest of Britain in A.D. 43, numerous bathhouses and a temple were built around the hot springs (Figure 13.20).

Figure 13.19 Hot Springs

Parvinder Sethi

a One of the more colorful hot-springs in Yellowstone National Park, the Morning Glory hot-spring is fringed with multicolored mats of heat-loving, cyanobacteria and algal mats. Each color represents a certain temperature range that allows for specific bacterial species to thrive in this extreme environment.

James S. Monroe

b The water in this hot spring at Bumpass Hell in Lassen Volcanic National Park, California, is boiling.

James S. Monroe

c Mud pot at the Sulfur Works, also in Lassen Volcanic National Park.

James S. Monroe

d The U.S. Park Service warns of the dangers in hydrothermal areas, but some people ignore the warnings and are injured or killed.

The heat for most hot springs comes from magma or cooling igneous rocks. The geologically recent igneous activity in the western United States accounts for the large number of hot springs in that region. The water in some hot springs, however, circulates deep into Earth, where it is warmed by the normal increase in temperature, the geothermal gradient. For example, the spring water of Warm Springs, Georgia, is heated in this manner. This hot spring was a health and bathing resort long before the Civil War (1861–1865); later, with the establishment of the Georgia Warm Springs Foundation, it was used to help treat polio victims.

Figure 13.20 Bath, England One of the many bathhouses in Bath, England, that were built around hot springs shortly after the Roman conquest in A.D. 43.

Figure 13.21 Old Faithful Geyser Old Faithful Geyser in Yellowstone National Park, Wyoming, is one of the world's most famous geysers, erupting faithfully every 30 to 90 minutes and spewing water 32 to 56 m high.

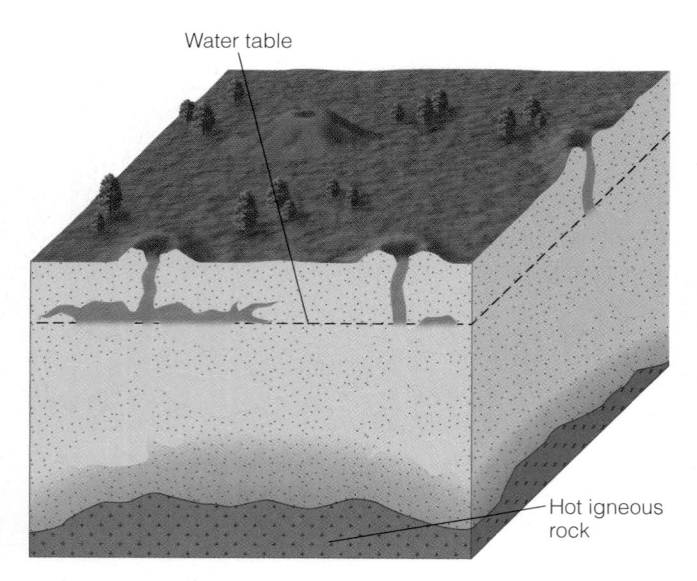

a The eruption of a geyser starts when groundwater percolates down into a network of interconnected openings and is heated by the hot igneous rocks. The water near the bottom of the fracture system is under higher pressure than the water near the top and consequently must be heated to a higher temperature before it will boil.

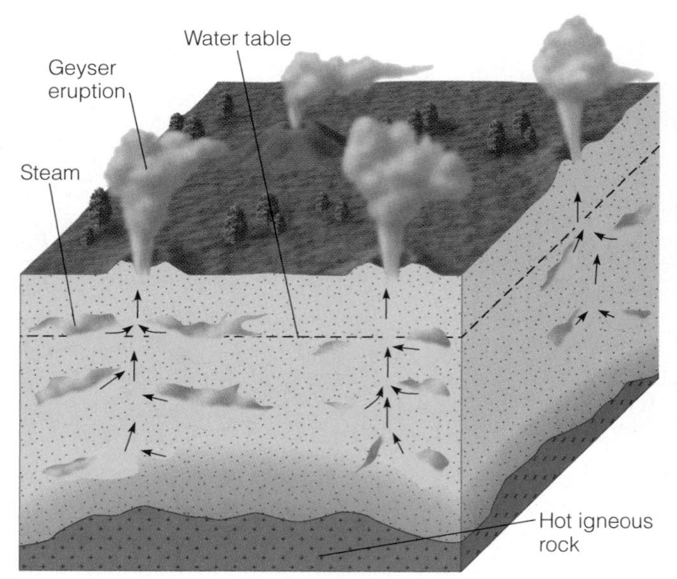

b Any rise in the temperature of the water above its boiling point or a drop in pressure will cause the water to change to steam, which quickly pushes the water above it up and out of the ground, producing a geyser eruption.

Figure 13.22 Anatomy of a Geyser

Geysers

Hot springs that intermittently eject hot water and steam with tremendous force are known as **geysers**. The word comes from the Icelandic *geysir*, "to gush" or "to rush forth." One of the most famous geysers in the world is Old Faithful in Yellowstone National Park in Wyoming (Figure 13.21). With a thunderous roar, it erupts a column of hot water and steam every 30 to 90 minutes. Other well-known geyser areas are found in Iceland and New Zealand.

Geysers are the surface expression of an extensive underground system of interconnected fractures within hot igneous rocks (Figure 13.22). Groundwater percolating down into the network of fractures is heated as it comes into

Figure 13.23 Hot-Spring Deposits in Yellowstone National Park, Wyoming

Parvinder Sethi

a Minerva Terrace, formed when calcium-carbonate-rich hot-spring water cooled, precipitating travertine.

Reed Wicander

b Liberty Cap is a geyserite mound formed by numerous geyser eruptions of silicon-dioxide-rich hot-spring water.

contact with the hot rocks. Because the water near the bottom of the fracture system is under higher pressure than the water near the top, it must be heated to a higher temperature before it will boil. Thus, when the deeper water is heated to near the boiling point, a slight rise in temperature or a drop in pressure, such as from escaping gas, will instantly change it to steam. The expanding steam quickly pushes the water above it out of the ground and into the air, producing a geyser eruption. After the eruption, relatively cool groundwater starts to seep back into the fracture system where it heats to near its boiling temperature and the eruption cycle begins again. Such a process explains how geysers can erupt with some regularity.

Hot spring and geyser water typically contains large quantities of dissolved minerals because most minerals dissolve more rapidly in warm water than in cold water. Because of this high mineral content, some believe that the waters of many hot springs have medicinal properties. Numerous spas and bathhouses have been built at hot springs throughout the world to take advantage of these supposed healing properties.

When the highly mineralized water of hot springs or geysers cools at the surface, some of the material in solution is precipitated, forming various types of deposits. The amount and type of precipitated minerals depend on the solubility and composition of the material that the groundwater flows through. If the groundwater contains dissolved calcium carbonate ($CaCO_3$), then *travertine* or *calcareous tufa* (both of which are varieties of limestone) are precipitated. Spectacular examples of hot spring travertine deposits are found at Pamukhale in Turkey and at Mammoth Hot Springs in Yellowstone National Park (Figure 13.23a). Groundwater containing dissolved silica will, upon reaching the surface, precipitate a soft, white, hydrated mineral called *siliceous sinter* or *geyserite*, which can accumulate around a geyser's opening (Figure 13.23b).

Figure 13.24 The Geysers, Sonoma County, California Steam rising from one of the geothermal power plants at The Geysers in Sonoma County, California. Steam from wells drilled into this geothermal region, about 120 km north of San Francisco, is piped directly to electricity-generating turbines to produce electricity that is distributed throughout the area.

Geothermal Energy

Geothermal energy is any energy produced from Earth's internal heat. In fact, the term *geothermal* comes from *geo*, "Earth," and *thermal*, "heat." Several forms of internal heat are known, such as hot dry rocks and magma, but so far, only hot water and steam are used.

Approximately 1–2% of the world's current energy needs could be met by geothermal energy. In those areas where it is plentiful, geothermal energy can supply most, if not all, of the energy needs, sometimes at a fraction of the cost of other types of energy. Some of the countries currently using geothermal energy in one form or another are Iceland, the United States, Mexico, Italy, New Zealand, Japan, the Philippines, and Indonesia.

In the United States, the first commercial geothermal electricity-generating plant was built in 1960 at The Geysers, about 120 km north of San Francisco, California. Here, wells were drilled into the numerous near-vertical fractures underlying the region. As pressure on the rising groundwater decreases, the water changes to steam, which is piped directly to electricity-generating turbines and generators (Figure 13.24).

As oil reserves decline, geothermal energy is becoming an attractive alternative, particularly in parts of the western United States, such as the Salton Sea area of southern California, where geothermal exploration and development have begun.

Geo-Recap

Chapter Summary

- Groundwater is part of the hydrologic cycle and an important natural resource. It consists of all subsurface water trapped in the pores and other open spaces in rocks, sediment, and soil.
- Porosity is the percentage of a material's total volume that is pore space. Permeability is the capacity to transmit fluids. Permeability is dependent on porosity, but also on the size of the pores or fractures and their interconnections.
- The water table is the surface separating the zone of saturation (in which pores are filled with water) from the overlying zone of aeration (in which pores are filled with air and water). The water table is a subdued replica of the overlying land surface in most places.
- Groundwater moves slowly downward under the influence of gravity through the pore spaces in the zone of aeration to the zone saturation. Some of it moves along the surface of the water table, and the rest moves from areas of high pressure to areas of low pressure.
- Springs are found wherever the water table intersects the surface. Some springs are the result of a perched water table—that is, a localized aquiclude within an aquifer and above the regional water table.
- Water wells are openings made by digging or drilling down into the zone of saturation. When water is pumped from a well, the water table in the area around the well is lowered, forming a cone of depression.
- In an artesian system, confined groundwater builds up high hydrostatic pressure. For an artesian system to develop, an aquifer must be confined above and below by aquicludes with the rock units usually tilted so as to build up hydrostatic pressure, and the aquifer must be exposed at the surface so it can be recharged.
- Karst topography largely develops by groundwater erosion in many areas underlain by soluble rocks, and is characterized by sinkholes, caves, solution valleys, and disappearing streams.
- Caves form when groundwater in the zone of saturation weathers and erodes soluble rock such as limestone. Cave deposits, called dripstone, result from the precipitation of calcite.
- Modifications of the groundwater system can cause serious problems such as lowering of the water table, saltwater incursion, subsidence, and contamination.
- Groundwater contamination by humans is becoming a serious problem and can result from landfills, septic systems, toxic waste sites, and industrial effluents, all of which affect the quality of the groundwater.
- Hydrothermal refers to hot water, typically heated by magma, but also resulting from Earth's geothermal gradient as it circulates deeply beneath the surface. Manifestations of hydrothermal activity include fumaroles, hot springs, and geysers.
- Geothermal energy is energy produced from Earth's internal heat and comes from the steam and hot water trapped within Earth's crust. It is a relatively nonpolluting from of energy that is used as a source of heat and to generate electricity.

Important Terms

artesian system (p. 334)
cave (p. 339)
cone of depression (p. 334)
geothermal energy (p. 352)
geyser (p. 351)
groundwater (p. 330)
hot spring (p. 348)
hydrothermal (p. 348)
karst topography (p. 339)
permeability (p. 331)
porosity (p. 331)
sinkhole (p. 335)
spring (p. 333)
water table (p. 331)
water well (p. 334)
zone of aeration (p. 331)
zone of saturation (p. 331)

Review Questions

1. Two features typical of areas of karst topography are
   a. ______ geysers and hot springs;
   b. ______ hydrothermal activity and springs;
   c. ______ saltwater incursion and pollution;
   d. ______ sinkholes and disappearing streams;
   e. ______ dripstone and a cone of depression.
2. The water table is a surface separating the
   a. ______ zone of porosity from the underlying zone of permeability;
   b. ______ capillary fringe from the underlying zone of aeration;
   c. ______ capillary fringe from the underlying zone of saturation;
   d. ______ zone of aeration from the underlying zone of saturation;
   e. ______ zone of saturation from the underlying zone of aeration.
3. The term hydrothermal refers to
   a. ______ calcareous tufa;
   b. ______ groundwater contamination;
   c. ______ hot water;
   d. ______ sinkhole formation;
   e. ______ artesian wells.
4. A cone of depression forms when
   a. ______ a stream flows into a sinkhole;
   b. ______ water in the zone of aeration is replaced by water from the zone of saturation;
   c. ______ a spring forms where a perched water table intersects the surface;
   d. ______ water is withdrawn faster than it can be replaced;
   e. ______ the ceiling of a cave collapses, forming a steep-sided crater.
5. The water in hot springs and geysers
   a. ______ is thought by some people to have curative properties;
   b. ______ is noncorrosive;
   c. ______ contain large quantities of dissolved minerals;
   d. ______ answers a and b;
   e. ______ answers a and c.
6. Which of the following is not an example of groundwater erosion?
   a. ______ karst topography;
   b. ______ stalactites;
   c. ______ sinkholes;
   d. ______ caves;
   e. ______ caverns.
7. When water is pumped from wells in some coastal areas, a problem arises that is known as
   a. ______ artesian recharge;
   b. ______ dripstone deposition;
   c. ______ geothermal depression;
   d. ______ saltwater incursion;
   e. ______ permeability decrease.
8. The capacity of a material to transmit fluids is
   a. ______ porosity;
   b. ______ permeability;
   c. ______ solubility;
   d. ______ aeration quotient;
   e. ______ saturation.
9. A perched water table
   a. ______ occurs wherever there is a localized aquiclude;
   b. ______ is frequently the site of springs;
   c. ______ lacks a zone of aeration;
   d. ______ answers a and b;
   e. ______ answers b and c.
10. What is the correct order, from highest to lowest usage of groundwater in the United States?
    a. ______ agricultural, industrial, domestic;
    b. ______ industrial, domestic, agricultural;
    c. ______ domestic, agricultural, industrial;
    d. ______ agricultural, domestic, industrial;
    e. ______ industrial, agricultural, domestic.
11. What is different between an artesian system and a water well? Is there any difference between the water obtained from an artesian system and the water from a water well?
12. Describe the configuration of the water table beneath a humid area and beneath an arid region. Why are the configurations different?
13. Explain how some Earth materials can be porous yet not permeable. Give an example.
14. Explain how groundwater weathers and erodes Earth materials.
15. Why does groundwater move so much slower than surface water?
16. Explain how saltwater incursion takes place and why it is a problem in coastal areas.

17. Discuss the role of groundwater in the hydrologic cycle.
18. Why should we be concerned about how fast the groundwater supply is being depleted in some areas?
19. Describe some ways to quantitatively measure the rate of groundwater movement.
20. One concern geologists have about burying nuclear waste in present-day arid regions such as Nevada is that the climate may change during the next several thousand years and become more humid, thus allowing more water to percolate through the zone of aeration. Why is this a concern? What would the average rate of groundwater movement have to be during the next 5000 years to reach buried canisters containing radioactive waste buried at a depth of 400 m?

CHAPTER 14

Glaciers and Glaciation

Peter Essick/Aurora/Getty Images

OUTLINE

OBJECTIVES

At the end of this chapter, you will have learned that

- Moving bodies of ice on land known as glaciers cover about 10% of Earth's land surface.
- During the Pleistocene Epoch (Ice Age), glaciers were much more widespread than they are now.
- Water frozen in glaciers constitutes one reservoir in the hydrologic cycle.
- In any area with a yearly net accumulation of snow, the snow is first converted to granular ice known as firn and eventually into glacial ice.
- The concept of the glacial budget is important to understanding the dynamics of any glacier.
- Glaciers move by a combination of basal slip and plastic flow, but several factors determine their rates of movement, and under some conditions they may move rapidly.
- Glaciers effectively erode, transport, and deposit sediment, thus accounting for the origin of several distinctive landforms.

This image shows the Unteraar Glacier in Switzerland, which is a shrunken remnant of the much larger glacier that occupied this valley during the Little Ice Age. The Little Ice Age lasted from about 1500 into the 1800s, during which temperatures were cooler, and glaciers moved much farther down their valleys than they do now.

- As a result of glaciation during the Ice Age, Earth's crust was depressed into the mantle and has since risen, sea level fluctuated widely, and lakes were present in areas that are now quite arid.
- A current widely and accepted theory explaining the onset of ice ages points to irregularities in Earth's rotation and orbit.

INTRODUCTION

Scientists know that Earth's surface temperatures have increased during the last few decades, and even though the evidence for a connection between greenhouse gases, especially carbon dioxide (CO_2), and global warming is increasing, the human contribution to the problem has not been fully resolved. We know from the geologic record that an Ice Age took place between 1.8 million and 10,000 years ago, and since that time, Earth has experienced several entirely natural climatic changes. During the Holocene Maximum about 6000 years ago, the average temperatures were slightly higher than they are now, and some of today's arid regions were much more humid. The Sahara Desert of North Africa had sufficient precipitation to support lush vegetation, swamps, and lakes. Indeed, Egypt's only arable land today is along the Nile River (see the Geo-Focus in Chapter 12), but until only a few thousands of years ago, much of North Africa was covered by grasslands.

After the Holocene Maximum was a time of cooler temperatures, but from about A.D. 1000 to 1300, Europe experienced the Medieval Warm Period during which wine grapes grew 480 km farther north than they do now. Then a cooling trend beginning in about A.D. 1300 led to the **Little Ice Age**, which lasted from 1500 to the middle or late 1800s. The Little Ice Age was a time of expansion of glaciers (Figure 14.1); cooler, wetter summers; colder winters; and the persistence of sea ice for long periods in Greenland, Iceland, and the Canadian Arctic islands. And because of the cooler, wetter summers, growing seasons were shorter, which accounts for several widespread famines.

During the coldest part of the Little Ice Age (1680–1730), the growing season in England was five weeks shorter than during the 1900s, and in 1695 Iceland was surrounded by sea ice for most of the year. Occasionally, Eskimos following the southern edge of the sea ice paddled their kayaks as far south as Scotland, and the canals in Holland froze over in some winters. In 1607, the first Frost Fair was held in London, England, on the Thames River, which began to freeze over nearly every winter. In the late 1700s, New York Harbor froze over, and 1816 is known as the "year without a summer" when unusually cold temperatures persisted into June and July in New England and northern Europe. (The eruption of Tambora in 1815 contributed to the cold spring and summer of 1816.)

Many of you have probably heard of the Ice Age and have some idea of what a glacier is, but it is doubtful that you know much about the dynamics of glaciers, how they form, and what may cause ices ages. In any case, *glaciers* are moving bodies of ice on land that are particularly effective at erosion, sediment transport, and deposition. They deeply scour the

Figure 14.1 Glacier Expansion During the Little Ice Age

Reprinted with permission from T. H. Van Andel. *New Views on an Old Planet*, p. 175 (Table 11.1) © 1989 Cambridge University Press.

a Samuel Birmann (1793–1847) painted this view in Switzerland, titled *Unterer Grindelwald,* in 1826 when the glacier extended onto the valley floor in the foreground.

Sue Monroe

b The same glacier today. Its terminus is hidden behind a rock projection at the lower end of its valley.

surfaces they move over, producing many easily recognizable landforms, and they deposit huge amounts of sediment, much of it important sources of sand and gravel. Glaciers today cover about 10% of Earth's land surface, but during the Ice Age they were much more widespread.

Studying glaciers and their possible causes may clarify some aspects of long-term climatic change and the debate on global warming (see Geo-Focus on pages 364 and 365). Glaciers are very sensitive to even short-term changes in climate, so scientists closely monitor them to see whether they advance, remain stationary, or retreat.

THE KINDS OF GLACIERS

Geologists define a **glacier** as a moving body of ice on land that flows downslope or outward from an area of accumulation. We will discuss how glaciers flow in a later section. Our definition of a glacier excludes frozen seawater as in the North Polar region, and sea ice that forms yearly adjacent to Greenland and Iceland. Drifting icebergs are not glaciers either, although they may have come from glaciers that flowed into lakes or the sea. The critical points in the definition are *moving* and *on land*. Accordingly, permanent snowfields in high mountains, though on land, are not glaciers because they do not move. All glaciers share several characteristics, but they differ enough in size and location for geologists to define two specific types, valley glaciers and continental glaciers, and several subvarieties.

Valley Glaciers

Valley glaciers are confined to mountain valleys where they flow from higher to lower elevations (Figure 14.2), whereas continental glaciers cover vast areas, they are not confined by the underlying topography, and they flow outward in all directions from areas of snow and ice accumulation. We use the term **valley glacier,** but some geologists prefer the synonyms *alpine glacier* and *mountain glacier*. Valley glaciers commonly have tributaries, just as streams do, thereby forming a network of glaciers in an interconnected system of mountain valleys.

Valley glaciers are common in the mountains of western North America, especially Alaska and Canada, as well as the Andes in South America, the Alps in Europe, the Alps of New Zealand, and the Himalayas in Asia. Even a few of the loftier mountains in Africa, though near the equator, are high enough to support small glaciers. In fact, Australia is the only continent that has no glaciers.

A valley glacier's shape is obviously controlled by the shape of the valley it occupies, so it tends to be a long, narrow tongue of moving ice. Valley glaciers that flow into the ocean are called *tidewater glaciers*; they differ from other valley glaciers only in that their terminus is in the sea rather than on land (Figure 14.2b).

Valley glaciers are small compared with the much more extensive continental glaciers, but even so, they may be as large as several kilometers across, 200 km long, and hundreds of meters thick. Some glaciers in Alaska are up to 1500 m thick. Erosion and deposition by valley glaciers were responsible for much of the spectacular scenery in several U.S. and Canadian national parks, notably Yosemite, Glacier, and Banff-Jasper.

Figure 14.2 Valley Glaciers

Journal of Engineering Mechanics, Virginia Polytechnic Institute and State University

a A valley glacier in Alaska. Notice the tributaries that unite to form a larger glacier.

Sue Monroe

b Wellesley Glacier in Alaska flows into the sea, so it is a tidewater glacier.

Continental Glaciers

Continental glaciers, also known as *ice sheets*, are vast, covering at least 50,000 km^2, and they are unconfined by topography; that is, their shape and movement are not controlled by the underlying landscape. Valley glaciers are long, narrow tongues of ice that conform to the shape of the valley they occupy, and the existing slope determines their direction of flow. In contrast, continental glaciers flow outward in all directions from a central area or areas of accumulation in response to variations in ice thickness.

In Earth's two areas of continental glaciation, Greenland and Antarctica, the ice is more than 3000 m thick and covers all but the highest mountains (Figure 14.3a, b). The continental glacier in Greenland covers about 1,800,000 km^2, and in Antarctica the East and West Antarctic Glaciers merge to form a continuous ice sheet covering more than 12,650,000 km^2. The glaciers in Antarctica flow into the sea, where the buoyant effect of water causes the ice to float in vast *ice shelves*; the Ross Ice Shelf alone covers more than 547,000 km^2 (Figure 14.3a).

Although valley glaciers and continental glaciers are easily differentiated by size and location, geologists also recognize an intermediate variety called an **ice cap**. Ice caps are similar to, but smaller than, continental glaciers, covering less than 50,000 km^2. The 6000-km^2 Penny Ice Cap on Baffin Island, Canada (Figure 14.3c), and the Juneau Icefield in Alaska and Canada at about 3900 km^2 are good examples. Some ice caps form when valley glaciers grow and overtop the divides and passes between adjacent valleys and coalesce to form a continuous ice cover. They also form on fairly flat terrain in Iceland and some of the islands in the Canadian Arctic.

GLACIERS—MOVING BODIES OF ICE ON LAND

We use the term **glaciation** to indicate all glacial activity, including the origin, expansion, and retreat of glaciers, as well as their impact on Earth's surface. Presently, glaciers

Figure 14.3 Continental Glaciers and Ice Caps

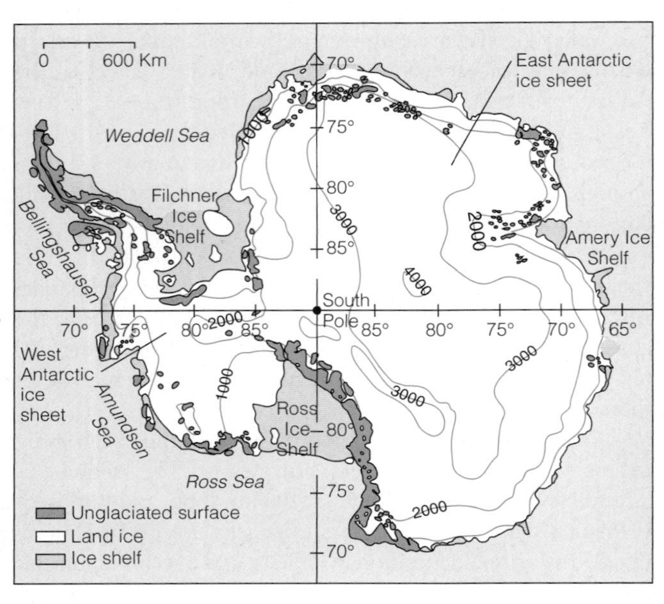

a The West and East Antarctic ice sheets merge to form a nearly continuous ice cover that averages 2160 m thick. The blue lines are lines of equal thickness.

Slide 397, Slide Set B, *Physical Geology*, 1992.

b View of part of the Antarctic ice sheet. Notice the *nunatak*, which is a peak extending above the glacial ice.

James S. Monroe

c View of the Penny Ice Cap on Baffin Island, Canada. It covers about 6000 km^2.

cover nearly 15 million km^2, or about 10% of Earth's land surface. As a matter of fact, if all glacial ice were in the United States and Canada, it would form a continuous ice cover about 1.5 km thick!

At first glance, glaciers appear static. Even briefly visiting a glacier may not dispel this impression because, although glaciers move, they usually do so slowly. Nevertheless, they do move and just like other geologic agents such as running water, glaciers are dynamic systems that continuously adjust to changes. For example, a glacier may flow slower or more rapidly depending on decreased or increased amounts of snow or the absence or presence of water at its base. And glaciers may expand or contract depending on climatic changes.

Glaciers—Part of the Hydrologic Cycle

In Chapter 1, you learned that one of Earth's systems, the hydrosphere, consists of all surface water, including water frozen in glaciers. So glaciers make up one reservoir in the hydrologic cycle where it is stored for long periods, but even this water eventually returns to its original source, the oceans (see Figure 12.3). Glaciers at high latitudes, as in Alaska, northern Canada, and Scandinavia, flow directly into the oceans where they melt, or icebergs break off (a process known as calving) and drift out to sea where they eventually melt. At low latitudes or areas remote from the oceans, glaciers flow to lower elevations where they melt and the liquid water enters the groundwater system (another reservoir in the hydrologic cycle) or it returns to the seas by surface runoff.

In addition to melting, glaciers lose water by *sublimation*, when ice changes to water vapor without an intermediate liquid phase. Sublimation is not an exotic process; it occurs in the freezer compartment of a refrigerator. Because of sublimation, the older ice cubes at the bottom of the container are much smaller than the more recently formed ones. In any case, the water vapor so derived enters the atmosphere where it may condense and fall as rain or snow, but in the long run, this water also returns to the oceans.

Figure 14.4 Glacial Ice

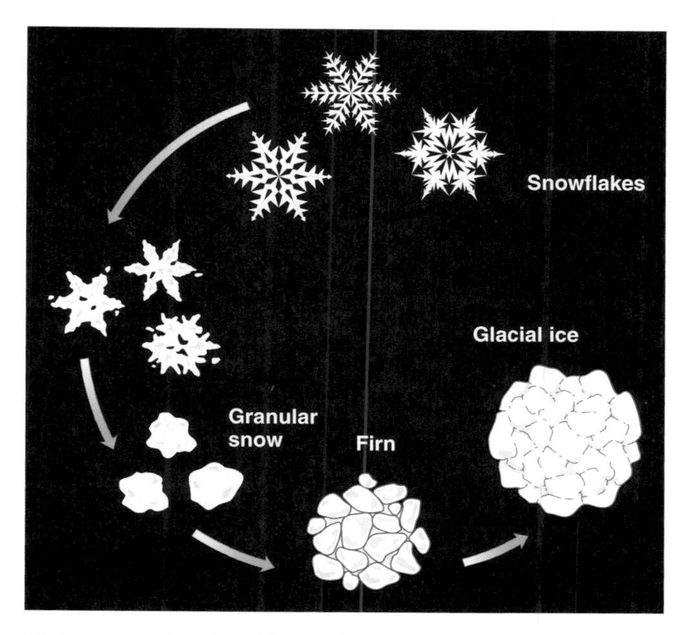

a The conversion of freshly fallen snow to firn and then to glacial ice.

b This iceberg in Portage Lake in Alaska shows the blue color of glacial ice. The longer wavelengths of white light are absorbed by the ice, but blue (short wavelength) is transmitted into the ice and scattered, accounting for the blue color.

How Do Glaciers Originate and Move?

Ice is a crystalline solid with characteristic physical properties and a specific chemical composition, and thus is a mineral. Accordingly, glacial ice is a type of metamorphic rock, but one that is easily deformed. Glaciers form in any area where more snow falls than melts during the warmer seasons and a net accumulation takes place. Freshly fallen snow has about 80% air-filled pore space and 20% solids, but it compacts as it accumulates, partially thaws, and refreezes, converting to a granular type of snow known as **firn**. As more snow accumulates, the firn is buried and further compacted and recrystallized until it is transformed into **glacial ice**, consisting of about 90% solids and 10% air (Figure 14.4).

Now you know how glacial ice forms, but we still have not addressed how glaciers move. At this time, it is useful to recall some terms from Chapter 10. Remember that *stress* is force per unit area and *strain* is a change in the shape or volume of solids. When accumulating snow and ice reach a critical thickness of about 40 m, the stress on the ice at depth is great enough to induce **plastic flow**, a type of permanent deformation involving no fracturing. Glaciers move mostly by plastic flow, but they may also slide over their underlying surface by **basal slip** (Figure 14.5). Liquid water facilitates basal slip because it reduces friction between a glacier and the surface over which it moves.

The total movement of a glacier in a given time is a consequence of plastic flow and basal slip, although the former occurs continuously, whereas the latter varies depending on the season, latitude, and elevation. Indeed, if a glacier is solidly frozen to the surface below, as in the case of many polar environments, it moves only by plastic flow. Furthermore,

What Would You Do?

Suppose you are a high school earth science teacher trying to explain to students that ice is a mineral and rock, and how a solid like ice can flow like a fluid. Furthermore, you explain that the upper 40 m or so of a glacier is brittle and fractures, whereas ice below that depth simply flows when subjected to stress. Now that your students are thoroughly confused, how will you explain and demonstrate that ice can behave like a solid yet show properties of fluid flow?

basal slip is far more important in valley glaciers as they flow from higher to lower elevations, whereas continental glaciers need no slope for flow. Although glaciers move by plastic flow, the upper 40 m or so of ice behaves like a brittle solid and fractures if subjected to stress. Large crevasses commonly develop in glaciers where they flow over an increase in slope of the underlying surface or where they flow around a corner (Figure 14.6). In either case, the ice is stretched (subjected to tension) and crevasses open, which extend down to the zone of plastic flow. In some cases, a glacier descends over such a steep precipice that crevasses break up the ice into a jumble of blocks and spires, and an icefall develops.

Distribution of Glaciers

As you might suspect, the amount of snowfall and temperature are important factors in determining where glaciers form. Parts of northern Canada are cold enough to support glaciers but receive too little snowfall, whereas some mountain areas in California receive huge amounts of snow but are too warm for glaciers. Of course, temperature varies with elevation and latitude, so we would expect to find glaciers in high mountains and at high latitudes, if these areas receive enough snow.

Many small glaciers are present in the Sierra Nevada of California, but only at elevations exceeding 3900 m. In fact, the high mountains in California, Oregon, and Washington all have glaciers because they receive so much snow. Mount Baker in Washington had almost 29 m of snow during the winter of 1998–1999, and average accumulations of 10 m or more are common in many parts of these mountains.

Glaciers are also found in the mountains along the Pacific Coast of Canada, which also receive considerable snowfall, and of course, they are farther north. Some of the higher peaks in the Rocky Mountains in both the United States and Canada also support glaciers. At even higher latitudes, as in Alaska, northern Canada, and Scandinavia, glaciers exist at sea level.

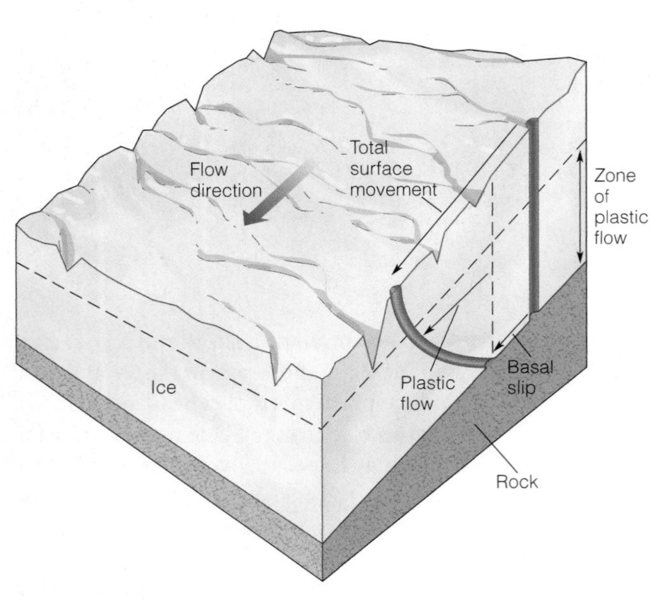

Figure 14.5 Part of a Glacier Showing Movement by a Combination of Plastic Flow and Basal Slip Plastic flow involves internal deformation within the ice, whereas basal slip is sliding over the underlying surface. If a glacier is solidly frozen to its bed, it moves only by plastic flow. Notice that the top of the glacier moves farther in a given time than the bottom does.

THE GLACIAL BUDGET—ACCUMULATION AND WASTAGE

Just as a savings account grows and shrinks as funds are deposited and withdrawn, a glacier expands and contracts in response to accumulation and wastage. We describe a glacier's behavior in terms of a **glacial budget**, which is essentially a balance sheet of accumulation and wastage. For instance, the upper part of a valley glacier is a **zone of accumulation**, where additions exceed losses and the surface is perennially snow covered. In contrast, the lower part of the same glacier is a **zone of wastage**, where losses from melting, sublimation, and calving of icebergs exceed the rate of accumulation (Figure 14.7).

At the end of winter, a glacier's surface is completely covered with the accumulated seasonal snowfall. During the spring and summer, the snow begins to melt, first at lower elevations and then progressively higher up the glacier. The elevation to which snow recedes during a wastage season is the *firn limit* (Figure 14.7). You can easily identify the zones of accumulation and wastage by noting the location of the firn limit.

The firn limit on a glacier may change yearly, but if it does not change or shows only minor fluctuations, the glacier has a balanced budget. That is, additions in the zone of accumulation are exactly balanced by losses in the zone of wastage, and the distal end, or terminus, of the glacier remains stationary (Figure 14.7a). If the firn limit moves up the glacier, indicating a negative budget, the glacier's terminus retreats (Figure 14.7b). If the firn limit moves down the glacier, however, the glacier has a positive budget, additions exceed losses, and its terminus advances (Figure 14.7c).

Even though a glacier may have a negative budget and a retreating terminus, the glacial ice continues to move toward the terminus by plastic flow and basal slip. If a negative budget persists long enough, though, the glacier continues to recede and it thins until it is no longer thick enough to maintain flow. It then ceases moving and becomes a *stagnant glacier*; if wastage continues, the glacier eventually disappears.

Figure 14.6 Crevasses Crevasses are common in the upper parts of glaciers when the ice is subjected to tension.

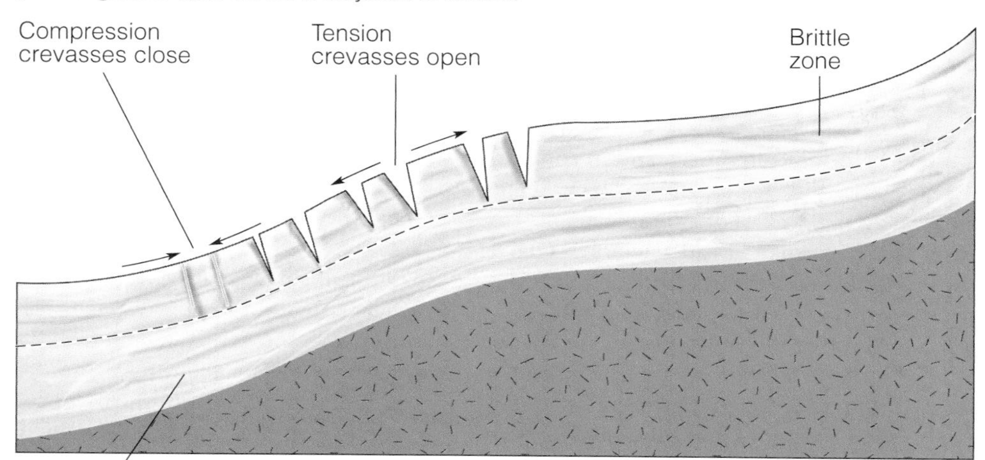

a Crevasses open where the brittle part of a glacier is stretched as it moves over a steeper slope in its valley.

b These crevasses are on the Byron Glacier in Kenai Fjords National Park and Preserve, near Seward, Alaska. Note the two hikers for scale.

Also, notice in Figure 14.7a that a valley glacier with a balanced budget transports and deposits sediment at its terminus as a *terminal moraine*. At a later time, the same glacier may have a negative budget, in which case its terminus retreats and perhaps becomes stabilized again if the budget is once more balanced. It then deposits another moraine; however, since the glacier's terminus has retreated, this one is known as a *recessional moraine* (Figure 14.7b). We will discuss moraines and other glacial landforms later in this chapter.

We used a valley glacier as an example, but the same budget considerations control the flow of ice caps and continental glaciers as well. The entire Antarctic ice sheet is in the zone of accumulation, but it flows into the ocean where wastage occurs.

How Fast Do Glaciers Move?

In general, valley glaciers move more rapidly than continental glaciers, but the rates for both vary from centimeters to tens of meters per day. Valley glaciers moving down steep slopes flow more rapidly than glaciers of comparable size on gentle slopes, assuming that all other variables are the same. The main glacier in a valley glacier system contains a greater volume of ice and thus has a greater discharge and flow velocity than its tributaries (Figure 14.2a). Temperature exerts a seasonal control on valley glaciers because, although plastic flow remains rather constant year-round, basal slip is more important during warmer months when meltwater is abundant.

Flow rates also vary within the ice itself. For example, flow velocity increases downslope in the zone of accumulation until the firn limit is reached; from that point, the velocity becomes progressively lower toward the glacier's terminus. Valley glaciers are similar to streams, in that the valley walls and floor cause frictional resistance to flow, so the ice in contact with the walls and floor moves more slowly than the ice some distance away (Figure 14.8).

Notice in Figure 14.8 that flow velocity in the interior of a glacier increases upward until the top few tens of meters of ice are reached, but little or no additional increase occurs after that point. This upper ice layer constitutes the rigid part of the glacier that is moving as a result of basal slip and plastic flow below.

Geo-Focus Glaciers and Global Warming

As you know, global warming is a phenomenon involving warming of Earth's atmosphere during the last 100 years or so. Many scientists think that the concentration of greenhouse gases (carbon dioxide, methane, and nitrous oxide) as a result of human activities, especially the combustion of fossil fuels, is the cause of global warming. There are, however, dissenters who acknowledge that Earth's surface temperatures have increased but attribute the increase to normal climatic variations. Needless to say, the issue has not been resolved.

Whatever the cause of climate change, no one doubts that glaciers are good indicators of short-term variations in climate. According to one estimate, there are about 160,000 glaciers outside Antarctica and Greenland, with Alaska alone having several tens of thousands. What do valley glaciers tell us about climate? Remember that the behavior of valley glaciers depends on their glacial budget, which is in turn controlled by temperature and precipitation.

It is true that not very many of the 160,000 or so glaciers on Earth have been studied, but those that have reveal an alarming trend: Many are retreating, and in some cases they have nearly or completely disappeared. For example, of the approximately 150 glaciers in Glacier National Park in Montana in 1850, only a few very small ones remain. Many of the valley glaciers in Alaska at lower elevations are much smaller than they were a few decades ago, and so it goes just about everywhere glaciers are studied, including those in the Cascade Range of the Pacific Northwest.

Recall from Chapter 5 that the Cascade Range is made up of several large composite volcanoes and hundreds of smaller volcanic cones and vents (see Chapter 5). All of the higher peaks in the range have glaciers, although some are very small and most are retreating. When Mount St. Helens erupted in May 1980, all of its 12 glaciers were destroyed or at least considerably diminished. But by 1982, the lava dome in the crater had cooled sufficiently for snow to accumulate yearly and it is now as much as 190 m thick. In any case, it is thick enough so that pressure on the snow at depth converts it to glacial ice, and "giant cracks in the ice, called crevasses, and other flow features, indicate that the ice body is transforming into a glacier"* (Figure 1).

Three factors account for the birth of this new glacier. First, the Cascade Range receives huge amounts of snowfall. Second, the crater provides protection for the accumulating snow. And third, rockfalls from the crater walls help insulate the forming glacier. This new glacier formed in little more than 20 years, and it is the only one in the continental United States that is advancing.

The story for the other Cascade Range glaciers is not so comforting. Glacier Peak in Washington, which last erupted in 1880, has more than a dozen glaciers, all of which are retreating, and Whitechuck Glacier will soon be inactive (Figure 2). During the Little Ice Age, Whitechuck Glacier had northern and southern branches, each with a separate accumulation zone, that merged to form a single glacier. It covered about

*U.S. Forest Service, *Volcano Review*, Summer 2002, contribution by Charlie Anderson, Director of the International Glaciospeleological Survey (http://glaciercaves.com/html/birtho_1.HTM).

Continental glaciers ordinarily flow at a rate of centimeters to meters per day. One reason continental glaciers move comparatively slowly is that they exist at higher latitudes and are frozen to the underlying surface most of the time, which limits the amount of basal slip. But some basal slip does occur even beneath the Antarctic ice sheet, although most of its movement is by plastic flow. Nevertheless, some parts of continental glaciers manage to achieve extremely high flow rates. Near the margins of the Greenland ice sheet, the ice is forced between mountains in what are called *outlet glaciers*. In some of these outlets, flow velocities exceed 100 m per day.

In parts of the continental glacier covering West Antarctica, scientists have identified ice streams in which flow rates are considerably higher than in adjacent glacial ice. Drilling has revealed a 5-m-thick layer of water-saturated sediment beneath these ice streams, which acts to facilitate movement of the ice above. Some geologists think that geothermal heat from subglacial volcanism melts the underside of the ice, thus accounting for the layer of water-saturated sediment.

Glacial Surges

A **glacial surge** is a short-lived episode of accelerated flow in a glacier during which its surface breaks into a maze of crevasses and its terminus advances noticeably. Glacial surges are best documented in valley glaciers, although they also take place in ice caps and perhaps in continental glaciers. In 1995, for instance, a huge ice shelf at the

4.8 km^2, but a rapid episode of retreat began in 1930 and now it covers only about 0.9 km^2 and has thinned to about 35 m thick. In fact, by 2002, the glacier's northern branch was gone, and what had been an ice-filled valley is now filled with rubble. The south branch may persist for a few more decades, but it, too, is retreating and thinning.

Matt Logan/USGS

Figure 1 View of a lava dome and the newly formed glacier in the crater of Mount St. Helens on April 19, 2005. Notice the ash on the surface of the glacier, which also shows large crevasses.

Corbis

Figure 2 Whitechuck Glacier on Glacier Peak in Washington State. The south branch of the glacier on the right has a small accumulation area, but the north branch has none.

northern end of the Antarctic Peninsula broke apart and several ice streams from the Antarctic ice sheet surged toward the ocean.

During a surge, a glacier may advance several tens of meters per day for weeks or months and then return to its normal flow rate. Surging glaciers constitute only a tiny proportion of all glaciers, and none of these are in the United States outside of Alaska. Even in Canada, they are found only in the Yukon Territory and the Queen Elizabeth Islands.

The fastest glacial surge ever recorded was in 1953 in the Kutiah Glacier in Pakistan; the glacier advanced 12 km in three months, for an average daily rate of about 130 m. In 1986 the terminus of the Hubbard Glacier in Alaska began advancing at about 10 m per day, and in 1993 Alaska's Bering Glacier advanced more than 1.5 km in just three weeks.

The onset of a glacial surge is heralded by a thickened bulge in the upper part of a glacier that begins to move toward the terminus at several times the glacier's normal velocity. When the bulge reaches the terminus, it causes rapid movement and displacement of the terminus by as much as 20 km. Surges are also probably related to accelerated rates of basal slip rather than more rapid plastic flow. One theory holds that thickening in the zone of accumulation with concurrent thinning in the zone of wastage increases the glacier's slope and accounts for accelerated flow. But another theory holds that pressure on

Figure 14.7 Response of a Hypothetical Glacier to Changes in Its Budget

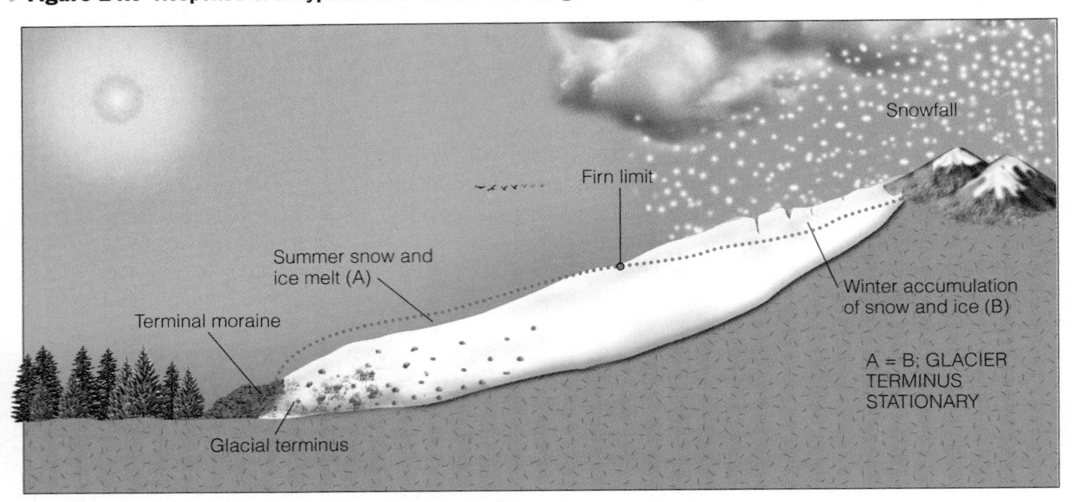

a Winter accumulation (B) and summer snow and ice melt (A) are equal. That is, additions and losses are equal so the glacier's terminus remains stationary. The terminal moraine is deposited at the terminus of a glacier.

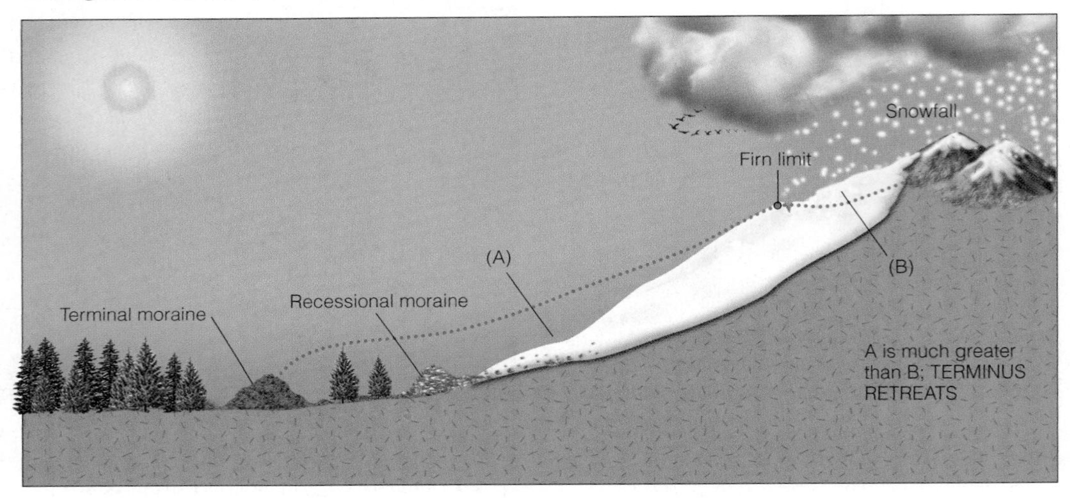

b Summer snow and ice melt (A) are much greater than winter accumulation (B), and the glacier's terminus retreats, although the glacier continues to move by plastic flow and basal slip. The recessional moraine is deposited at the glacier's new terminus.

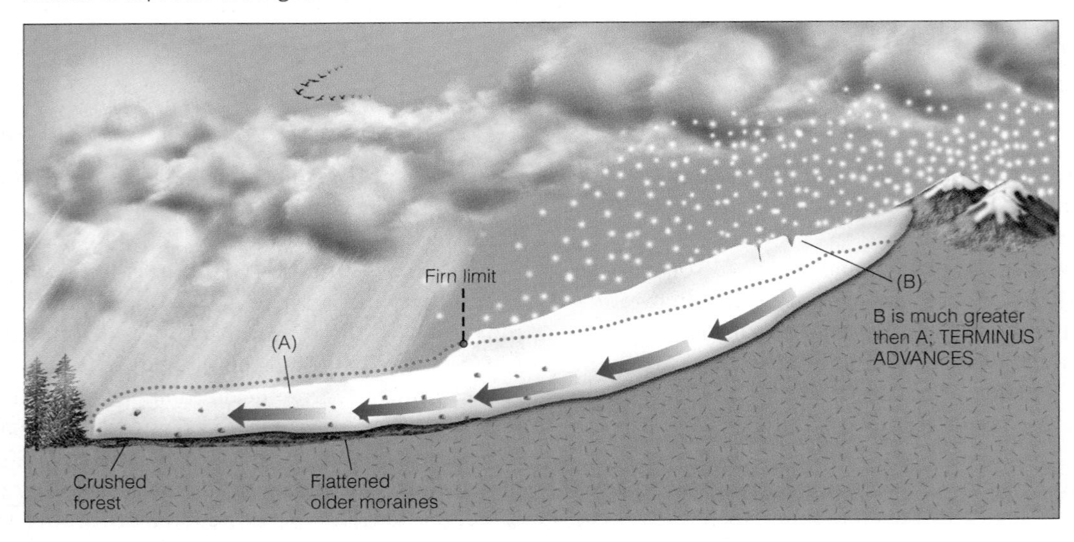

c Winter accumulation (B) is much greater than summer snow and ice melt (A), so the glacier's terminus advances. As it does so, it overrides and modifies its previously deposited moraines.

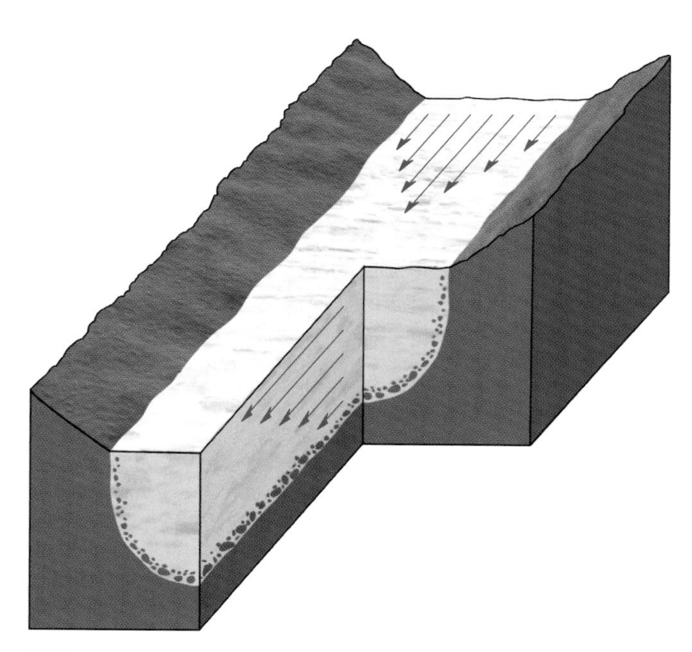

Figure 14.8 **Flow Velocity in a Valley Glacier** Flow velocity in a valley glacier varies both horizontally and vertically. Velocity is greatest at the top center of the glacier because friction with the walls and floor of the trough slows the flow adjacent to these boundaries. The lengths of the arrows in the figure are proportional to velocity.

soft sediment beneath a glacier squeezes fluids through the sediment, thereby allowing the overlying glacier to slide more effectively.

EROSION AND TRANSPORT BY GLACIERS

As moving solids, glaciers erode, transport, and eventually deposit huge quantities of sediment and soil. Indeed, they have the capacity to transport any size of sediment, including boulders the size of houses, as well as clay-sized particles. Important processes of erosion include bulldozing, plucking, and abrasion.

Although *bulldozing* is not a formal geologic term, it is fairly self-explanatory; glaciers shove or push unconsolidated materials in their paths. This effective process was aptly described in 1744 during the Little Ice Age by an observer in Norway:

> When at times [the glacier] pushes forward a great sound is heard, like that of an organ and it pushes in front of it unmeasurable masses of soil, grit and rocks bigger than any house could be, which it then crushes small like sand.*

Plucking, also called *quarrying*, results when glacial ice freezes in the cracks and crevices of a bedrock projection and eventually pulls it loose. One manifestation of plucking is a landform called a *roche moutonnée*, a French term for "rock sheep." As shown in Figure 14.9, a glacier smooths the "upstream" side of an obstacle, such as a small hill, and plucks pieces of rock from the "downstream" side by repeatedly freezing to and pulling away from the obstacle.

Bedrock over which sediment-laden glacial ice moves is effectively eroded by **abrasion** and develops a **glacial polish,** a smooth surface that glistens in reflected light (Figure 14.10a). Abrasion also yields **glacial striations,** consisting of rather straight scratches rarely more than a few millimeters deep on rock surfaces. Abrasion thoroughly pulverizes rocks, yielding an aggregate of clay- and silt-sized particles that have the consistency of flour—hence, the name *rock flour*. Rock flour is so common in streams discharging from glaciers that the water has a milky appearance (Figure 14.10b).

Continental glaciers derive sediment from mountains projecting through them, and wind-blown dust settles on their surfaces, but most of their sediment comes from the surface they move over. As a result, most sediment is transported in the lower part of the ice sheet. In contrast, valley glaciers carry sediment in all parts of the ice, but it is concentrated at the base and along the margins (Figure 14.11). Some of the marginal sediment is derived by abrasion and plucking, but much of it is supplied by mass wasting processes, as when soil, sediment, or rock falls or slides onto the glacier's surface.

Erosion by Valley Glaciers

When mountain ranges are eroded by valley glaciers, they take on a unique appearance of angular ridges and peaks in the midst of broad, smooth valleys with near-vertical walls. The erosional landforms produced by valley glaciers are easily recognized and enable us to appreciate the tremendous erosive power of moving ice. See Geo-inSight on pages 370 and 371 which features U-shaped glacial troughs, fjords, hanging valleys, cirques, arêtes, and horns.

U-Shaped Glacial Troughs A **U-shaped glacial trough** is one of the most distinctive features of valley glaciation. Mountain valleys eroded by running water are typically V-shaped in cross section; that is, they have valley walls that descend to a narrow valley bottom (Figure 14.12a). In contrast, valleys scoured by glaciers are deepened, widened, and straightened so that they have very steep or vertical walls, but

What Would You Do?

After carefully observing the same valley glacier for several years, you conclude (1) that its terminus has retreated at least 1 km and (2) that debris on its surface have obviously moved several hundreds of meters toward the glacier's terminus. Can you think of an explanation for these observations? Also, why do you think studying glaciers might have some implications for the debate on global warming?

*Quoted in C. Officer and J. Page, *Tales of the Earth* (New York: Oxford University Press, 1993), p. 99.

Figure 14.9 Roche Moutonnée A French term meaning "rock sheep," a roche moutonnée is a bedrock projection that was shaped by glacial abrasion and plucking.

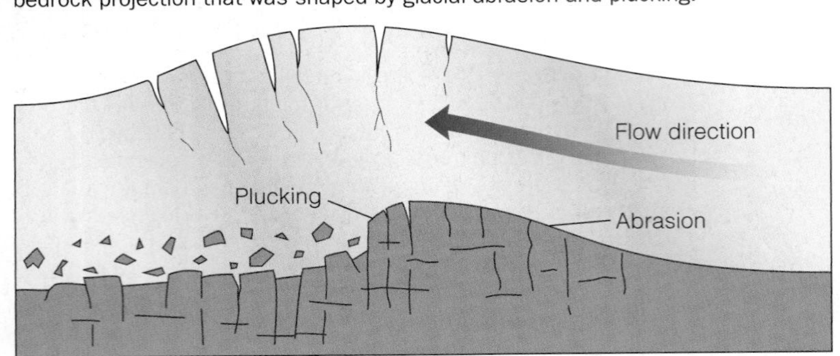

a Origin of a roche moutonnée. A glacier abrades and polishes the "upstream" side of a bedrock projection and shapes its "downstream" side by plucking.

b A roche moutonnée in Montana. The glacier that formed this feature moved from right to left.

James S. Monroe

Figure 14.10 Glacial Striations, Polish, and Rock Flour

James S. Monroe

a Glacial polish and striations, the straight scratches, on basalt at Devils Postpile National Monument in California.

James S. Monroe

b The water in this stream in Switzerland is discolored by rock flour, small particles produced by glacial abrasion.

broad, rather flat valley floors; thus they exhibit a U-shaped profile (Figure 14.12b).

Many glacial troughs contain triangular-shaped *truncated spurs*, which are cutoff or truncated ridges that extend into the preglacial valley (Figure 14.12c). Another common feature is a series of steps or rock basins in the valley floor where the glacier eroded rocks of varying resistance; many of the basins now contain small lakes.

Figure 14.11 Sediment Transport by Valley Glaciers Debris on the surface of the Mendenhall Glacier in Alaska. The largest boulder is about 2 m across. Notice the icefall in the background. The person left of center provides scale.

During the Pleistocene, when glaciers were more extensive, sea level was as much as 130 m lower than at present, so glaciers flowing into the sea eroded their valleys below present sea level. When the glaciers melted at the end of the Pleistocene, sea level rose and the ocean filled the lower ends of the glacial troughs so that now they are long, steep-walled embayments called **fiords**.

Fiords are restricted to high latitudes where glaciers exist at low elevations, such as Alaska, western Canada, Scandinavia, Greenland, southern New Zealand, and southern Chile. Lower sea level during the Pleistocene was not entirely responsible for the formation of all fiords. Unlike running water, glaciers can erode a considerable distance below sea level. In fact, a glacier 500 m thick can stay in contact with the seafloor and effectively erode it to a depth of about 450 m before the buoyant effects of water cause the glacial ice to float! The depth of some fiords is impressive; some in Norway and southern Chile are about 1300 m deep.

Hanging Valleys Waterfalls form in several ways, but some of the world's highest and most spectacular are found in recently glaciated areas. Bridalveil Falls in Yosemite National Park, California, plunge from a **hanging valley**, which is a tributary valley whose floor is at a higher level than that of the main valley. Where the two valleys meet, the mouth

Figure 14.12 Erosional Landforms Produced by Valley Glaciers

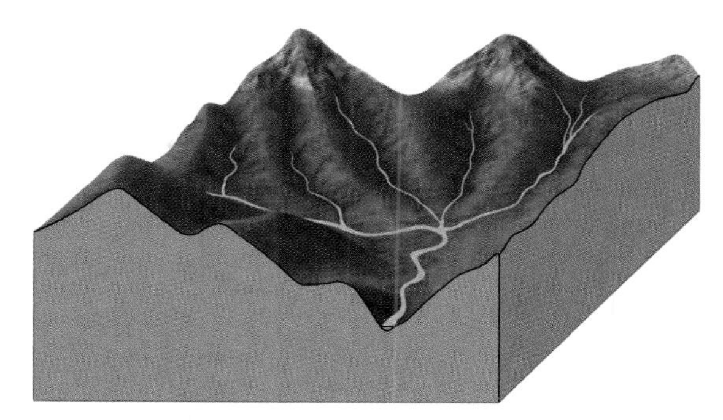

a A mountain area before glaciation.

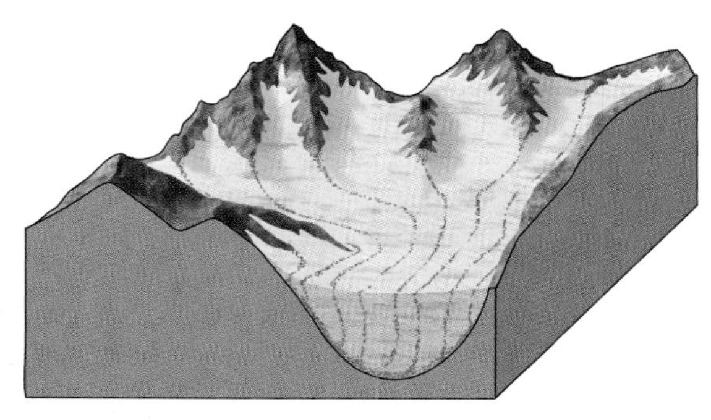

b The same area during the maximum extent of valley glaciers.

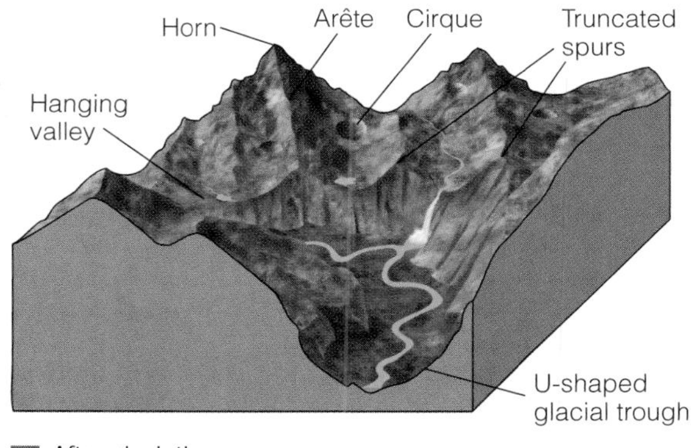

c After glaciation.

of the hanging valley is perched far above the main valley's floor (Figure 14.12c). Accordingly, streams flowing through hanging valleys plunge over vertical or steep precipices.

Although not all hanging valleys form by glacial erosion, many do. As Figure 14.12 shows, the large glacier in the main valley vigorously erodes, whereas the smaller glaciers in tributary valleys are less capable of erosion. When the glaciers disappear, the smaller tributary valleys remain as hanging valleys.

Cirques, Arêtes, and Horns Perhaps the most spectacular erosional landforms in areas of valley glaciation are at the upper ends of glacial troughs and along the divides that separate adjacent glacial troughs (see Geo-inSight on pages 370 and 371. Valley glaciers form and move out from steep-walled, bowl-shaped depressions called **cirques** at the upper end of their troughs

Geo-inSight Valley Glaciers and Erosion

Valley glaciers effectively erode and produce several easily recognized landforms. Where glaciers move through mountain valleys, the valleys are deepened and widened, giving them a distinctive U-shaped profile. The peaks and ridges rising above valley glaciers are also eroded, and they become jagged and angular. Much of the spectacular scenery in Grand Teton National Park, Wyoming, Yosemite National Park, California, and Glacier National Park, Montana, resulted from erosion by valley glaciers. In fact, valley glaciers remain active in some of the mountains of western North America, especially in Alaska and Canada.

James S. Monroe

James S. Monroe

▶ **1.** ▲ **2.** Part of southwestern Greenland (right) where valley glaciers merged to form an ice cap. Should these glaciers melt, several landforms like those in Figure 14.12c would be present. The Teton Range in Wyoming (above) acquired its angular peaks and ridges and broadly rounded valleys as a result of erosion by valley glaciers.

James S. Monroe

James S. Monroe

James S. Monroe

▲ **3.** ▲ **4.** ▶ **5.** U-shaped glacial troughs. The glacial trough above is in northern Montana, whereas the one above right is in southern Germany. The lake is impounded behind a glacial deposit known as an end moraine. The steep-walled glacial trough in Norway (right) extends below sea level, so it is a fiord.

James S. Monroe

◀ **6.** The bowl-shaped depression on Mount Wheeler in Great Basin National Park, Nevada, is a cirque. It has steep walls on three sides and opens out into a glacial trough.

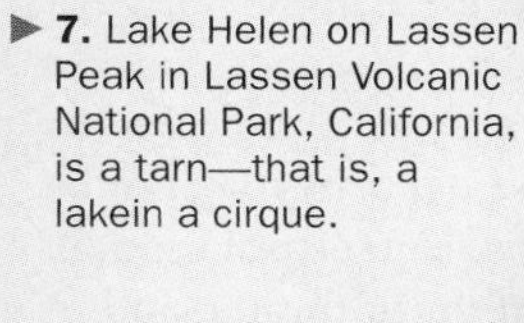

▶ **7.** Lake Helen on Lassen Peak in Lassen Volcanic National Park, California, is a tarn—that is, a lakein a cirque.

James S. Monroe

Sue Monroe

▲ **8.** Nevada Falls in Yosemite National Park, California, plunges 181 m from a hanging valley. The valley in the foreground is a hugeU-shaped glacial trough.

James S. Monroe

Swiss National Tourist Office

▲ **9.** ◀ **10.** This view on the Jungfrau (left) in Switzerland shows two small glaciers, a cirque headwall, and an arête.The Matterhorn (above) in Switzerland is a well-known horn.

14.12c). Cirques are typically steep-walled on three sides, but one side is open and leads into the glacial trough. Some cirques slope continuously into the glacial trough, but many have a lip or threshold at their lower end.

The details of cirque origin are not fully understood, but they probably form by erosion of a preexisting depression on a mountain side. As snow and ice accumulate in the depression, frost wedging and plucking, combined with glacial erosion, enlarge and transform the head of a steep mountain valley into a typical amphitheater-shaped cirque. Tension in the upper part of the glacier may reduce the erosive power of the ice on the immediate, downslope side of a cirque, leaving a lip or threshold in the valley floor after the ice melts away. Small lakes of meltwater, called *tarns*, often form on the floors of cirques behind such thresholds.

Cirques become wider and are cut deeper into mountainsides by headward erosion as a result of abrasion, plucking, and several mass wasting processes. For example, part of a steep cirque headwall may collapse while frost wedging continues to pry loose rocks that tumble downslope, so a combination of processes erode a small mountainside depression into a large cirque.

Arêtes—narrow, serrated ridges—form in two ways. In many cases, cirques form on opposite sides of a ridge, and headward erosion reduces the ridge until only a thin partition of rock remains (Figure 14.12c). The same effect occurs when erosion in two parallel glacial troughs reduces the intervening ridge to a thin spine of rock.

The most majestic of all mountain peaks are **horns**, steep-walled, pyramidal peaks formed by headward erosion of cirques. For a horn to form, a mountain peak must have at least three cirques on its flanks, all of which erode headward (Figure 14.12c). Excellent examples of horns are Mount Assiniboine in the Canadian Rockies, the Grand Teton in Wyoming, and the most famous of all, the Matterhorn in Switzerland.

Continental Glaciers and Erosional Landforms

Areas eroded by continental glaciers tend to be smooth and rounded because these glaciers bevel and abrade high areas that project into the ice. Rather than yielding the sharp, angular landforms typical of valley glaciation, they produce a landscape of subdued topography interrupted by rounded hills because they bury landscapes entirely during their development.

In a large part of Canada, particularly the vast Canadian Shield region, continental glaciers have stripped off the soil and unconsolidated surface sediment, revealing extensive exposures of striated and polished bedrock. These areas have deranged drainage (see Figure 12.19e), numerous lakes and swamps, low relief, extensive bedrock exposures, and little or no soil. They are referred to as *ice-scoured plains* (Figure 14.13). Similar though smaller bedrock exposures are also widespread in the northern United States from Maine through Minnesota.

Allan Kellelheim/Mary Pat Ziter, JLM Visuals

Figure 14.13 An Ice-Scoured Plain in Northwest Territories of Canada This low-relief surface is an ice-scoured plain in the Northwest Territories of Canada. Numerous lakes, little or no soil, and extensive bedrock exposures are typical of these areas eroded by continental glaciers.

DEPOSITS OF GLACIERS

Glacial Drift

What kinds of glacial deposits do geologists recognize, and are they as distinctive as the erosional features? Both valley and continental glaciers deposit sediment as **glacial drift**, a general term for all deposits resulting from glacial activity. A vast sheet of Pleistocene glacial drift is present in the northern tier of the United States and adjacent parts of Canada. Smaller but similar deposits are found where valley glaciers existed or remain active. The appearance of these deposits may not be as inspiring as some landforms resulting from glacial erosion, but they are important as reservoirs of groundwater, and in many areas, they are exploited for their sand and gravel.

One conspicuous aspect of glacial drift is rock fragments of various sizes that were obviously not derived from the underlying bedrock. These **glacial erratics**, as they are called, were derived from some distant source and transported to their present location (Figure 14.14). Some erratics are gigantic. For instance, the Madison Boulder in New Hampshire and Daggett Rock in Maine weigh about 4550 and 7270 metric tons, respectively. The glacial erratic shown in Figure 14.14b is not the world's largest, but it is one of many in a narrow belt of erratics stretching more than 640 km from their source.

As noted, *glacial drift* is a general term, and geologists define two types of drift: till and stratified drift. **Till** consists of sediments deposited directly by glacial ice. They are not

Figure 14.14 Glacial Erratics

a A glacial erratic in the making. This boulder on the surface of the Mendenhall Glacier in Alaska will eventually be deposited far from its source.

James S. Monroe

Royal Alberta Museum

b This is the Airdrie erratic near Calgary, Alberta, Canada. It is about 14 m long and 7.6 m high. The world's largest erratic is about one-third again as large as this one.

sorted by particle size or density, and they show no stratification. The till of both valley and continental glaciers is similar, but that of continental glaciers is much more extensive and usually has been transported much farther.

As opposed to till, **stratified drift** is layered—that is, stratified—and it invariably exhibits some degree of sorting by particle size. As a matter of fact, most stratified drift is actually layers of sand and gravel or mixtures thereof that accumulated in braided stream channels. In Chapter 12, we mentioned that streams issuing from melting glaciers are commonly braided because they receive more sediment than they can effectively transport.

Landforms Composed of Till

Landforms composed of till include several types of *moraines* and elongated hills know as *drumlins*.

End Moraines The terminus of any glacier may become stabilized in one position for some period of time, perhaps a few years or even decades. Stabilization of the ice front does not mean that the glacier has ceased to flow, only that it has a balanced budget (Figure 14.7). When an ice front is stationary, flow within the glacier continues, and any sediment transported within or upon the ice is dumped as a pile of rubble at the glacier's terminus (Figure 14.7 and Figure 14.15). These deposits are **end moraines,** which continue to grow as long as the ice front remains stationary. End moraines of valley glaciers are crescent-shaped ridges of till spanning the valley occupied by the glacier. Those of continental glaciers similarly parallel the ice front but are much more extensive.

Following a period of stabilization, a glacier may advance or retreat, depending on changes in its budget. If it advances, the ice front overrides and modifies its former moraine. If it has a negative budget, though, the ice front retreats toward the zone of accumulation. As the ice front recedes, till is deposited as it is liberated from the melting ice and forms a layer of **ground moraine**. Ground moraine has an irregular, rolling topography, whereas end moraine consists of long ridgelike accumulations of sediment.

After a glacier has retreated for some time, its terminus may once again stabilize, and it deposits another end moraine. Because the ice front has receded, such moraines are called **recessional moraines** (Figure 14.7b). During the Pleistocene

Figure 14.15 End Moraine

End moraine

James S. Monroe

a An end moraine deposited by a valley glacier. This particular end moraine is also a terminal moraine because it is the one most distant from the glacier's source.

James S. Monroe

b Closeup of an end moraine. Notice that the deposit is not sorted by particle size, and it shows no layering or stratification.

Epoch, continental glaciers in the mid-continent region extended as far south as southern Ohio, Indiana, and Illinois. Their outermost end moraines, marking the greatest extent of the glaciers, go by the special name **terminal moraine** (valley glaciers also deposit terminal moraines). As the glaciers retreated from the positions where their terminal moraines were deposited, they temporarily stopped retreating numerous times and deposited dozens of recessional moraines.

Lateral and Medial Moraines Valley glaciers transport considerable sediment along their margins, much of it abraded and plucked from the valley walls, but a significant amount falls or slides onto the glacier's surface by mass wasting processes. In any case, this sediment is transported and deposited as long ridges of till called **lateral moraines** along the margin of the glacier (Figure 14.16).

Where two lateral moraines merge, as when a tributary glacier flows into a larger glacier, a **medial moraine** forms (Figure 14.16b). A large glacier will often have several dark stripes of sediment on its surface, each of which is a medial moraine. One can determine how many tributaries a valley glacier has by the number of its medial moraines.

Drumlins In many areas where continental glaciers deposited till, the till has been reshaped into elongated hills known as **drumlins** (Figure 14.17). Some drumlins are as large as 50 m high and 1 km long, but most are much smaller. From the side, a drumlin looks like an inverted spoon, with the steep end on the side from which the glacial ice advanced and the gently sloping end pointing in the direction of ice movement. Drumlins are rarely found as single, isolated hills; instead, they occur in *drumlin fields* that contain hundreds or thousands of drumlins. Drumlin fields are found in several states and Ontario, Canada, but perhaps the finest example is near Palmyra, New York.

According to one hypothesis, drumlins form when till beneath a glacier is reshaped into streamlined hills as the ice moves over it by plastic flow. Another hypothesis holds that huge floods of glacial meltwater modify till into drumlins.

Landforms Composed of Stratified Drift

Stratified drift is a type of glacial deposit that exhibits sorting and layering, both indications that it was deposited by running water. Stratified drift is deposited by streams discharging from both valley and continental glaciers, but as you would expect, it is more extensive in areas of continental glaciation.

Outwash Plains and Valley Trains Glaciers discharge meltwater laden with sediment most of the time, except perhaps

Figure 14.16 Types of Moraines A moraine is a mound or ridge of unstratified till. The types shown here—lateral, medial, and end moraines—are defined by their position.

a Lateral moraine and end moraine deposited by a glacier on Baffin Island in Canada.

Cristopher J. Morris/Corbis

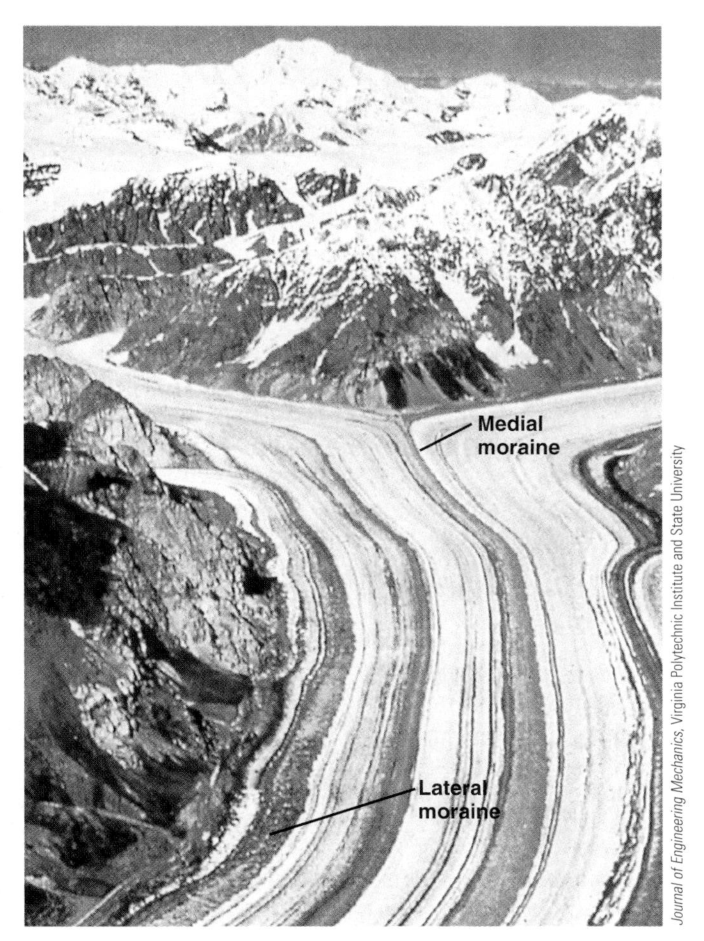

b Lateral and medial moraines on the Bernard Glacier in the St. Elias Mountains in Alaska.

Journal of Engineering Mechanics, Virginia Polytechnic Institute and State University

during the coldest months. This meltwater forms a series of braided streams that radiate out from the front of continental glaciers over a wide region. So much sediment is supplied to these streams that much of it is deposited within their channels as sand and gravel bars. The vast blanket of sediment so formed is an **outwash plain**.

Valley glaciers also discharge large amounts of meltwater and, like continental glaciers, have braided streams extending from them. However, these streams are confined to the lower parts of glacial troughs, and their long, narrow deposits of stratified drift are known as **valley trains** (Figure 14.18a).

Outwash plains, valley trains, and some moraines commonly contain numerous circular to oval depressions, many of which contain small lakes. These depressions are *kettles* that form when a retreating glacier leaves a block of ice that is subsequently partly or wholly buried (Figures 14.17 and 14.18b). When the ice block eventually melts, it leaves a depression; if the depression extends below the water table, it becomes the site of a small lake. Some outwash plains have so many kettles that they are called *pitted outwash plains.*

Kames and Eskers **Kames** are conical hills of stratified drift up to 50 m high (Figure 14.17 and Figure 14.19a). Many form when a stream deposits sediment in a depression on a glacier's surface; as the ice melts, the deposit is lowered to the land surface. Kames also form in cavities within or beneath stagnant ice.

Long sinuous ridges of stratified drift, many of which meander and have tributaries, are **eskers** (Figures 14.17 and 14.19b). Most eskers have sharp crests and sides that slope at about 30 degrees. Some are as high as 100 m and can be traced for more than 500 km. The sorting and stratification of the sediments in eskers clearly indicate deposition by running water. The features of ancient eskers and observations of present-day glaciers show that they form in tunnels beneath stagnant ice. Excellent examples of eskers can be seen at Kettle Moraine State Park in Wisconsin and in several other states, but the most extensive eskers in the world are in northern Canada.

Deposits in Glacial Lakes

Some lakes in areas of glaciation formed as a result of glaciers scouring out depressions; others occur where a stream's drainage was blocked; and others are the result of water accumulating behind moraines or in kettles. Regardless of how they formed, glacial lakes, like all lakes, are areas of deposition. Sediment may be carried into them and deposited as small deltas, but of special interest are the fine-grained deposits. Mud deposits in glacial lakes are commonly finely laminated (having layers less than 1 cm thick) and consist of alternating light and dark layers known as *varves* (Figure 14.20a), which represents an annual episode of deposition. The light layer formed during the spring and summer and consists of silt and clay; the dark layer formed during the winter when the smallest particles of clay and organic matter settled from suspension as the lake froze over. The number of varves indicates how many years a glacial lake has existed.

Another distinctive feature of glacial lakes with varves is *dropstones* (Figure 14.20b). These are pieces of gravel, some of boulder size, in otherwise very fine-grained deposits. The presence of varves indicates that currents and

Figure 14.17 Development of Features Associated with Past Continental Glaciation

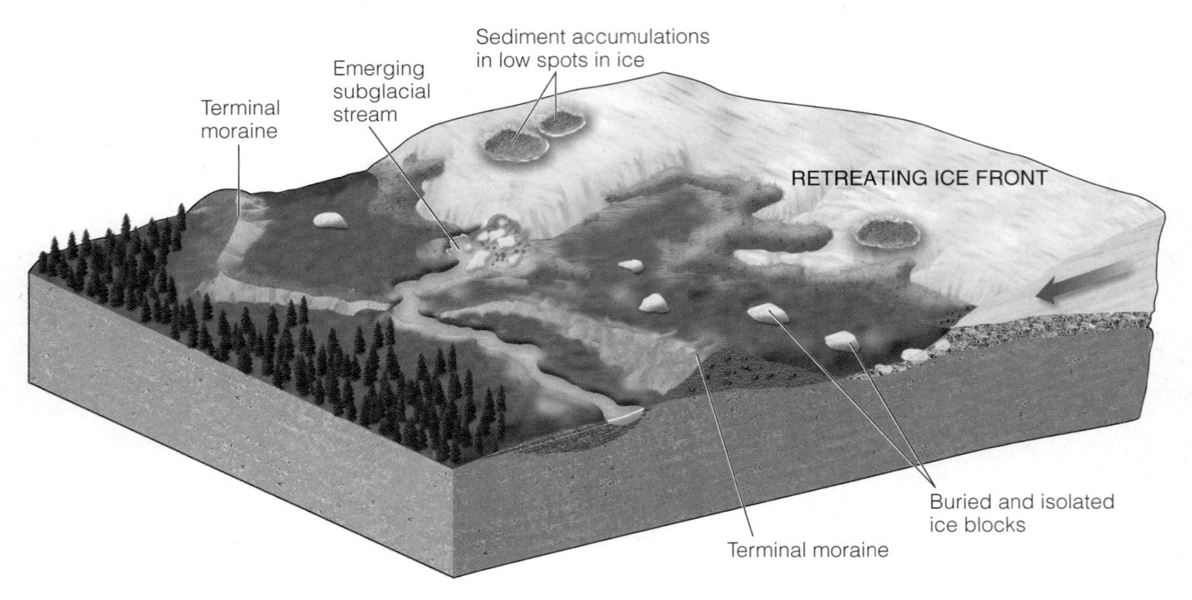

a This retreating continental glacier once covered a larger area as indicated by the terminal moraines.

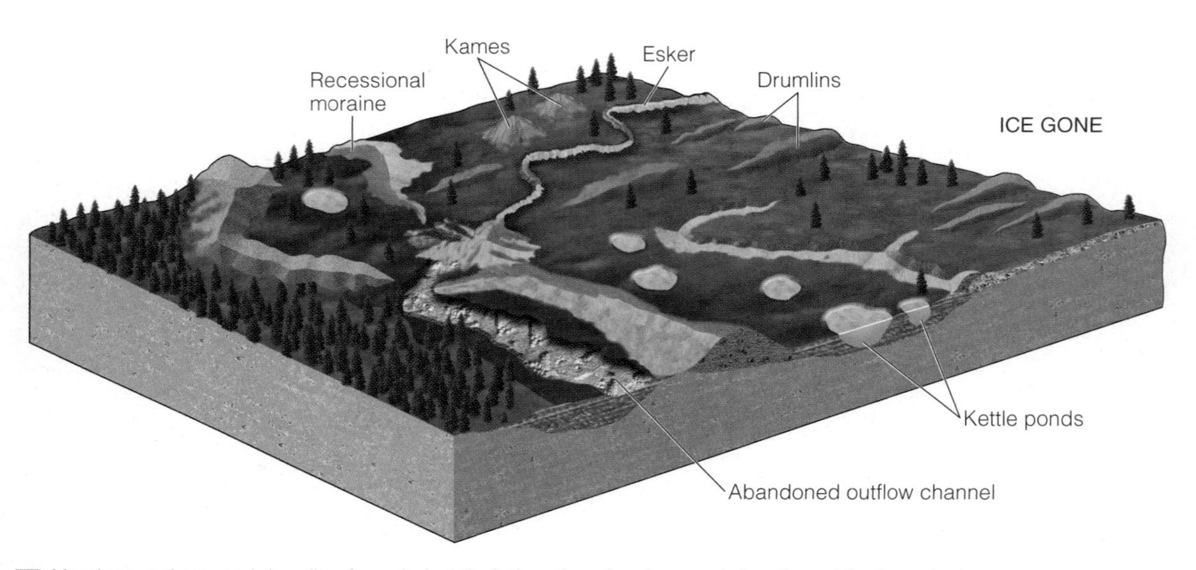

b Moraines, eskers, and drumlins form during glaciation, though eskers and drumlins originate under the ice cover. Kames and kettles develop at the end of glaciation.

turbulence in these lakes were minimal; otherwise, clay and organic matter would not have settled from suspension. How then can we account for dropstones in a low-energy environment? Most of them were probably carried into the lakes by icebergs that eventually melted and released sediment contained in the ice.

WHAT CAUSES ICE AGES?

We discussed the conditions necessary for a glacier to form earlier in this chapter: More snow falls than melts during the warm season, thus accounting for a net accumulation of snow and ice over the years. But this really does not address the broader question of what causes ice ages—that is, times of much more extensive glaciation. Actually, we need to address not only what causes ice ages, but also why there have been so few episodes of widespread glaciation in all of Earth history. Only during the Late Proterozoic Eon, the Pennsylvanian and Permian periods, and the Pleistocene Epoch has Earth had glaciers on a grand scale. Additionally, widespread glaciation occurred at least four times during the Pleistocene, with each glacial episode separated by a long *interglacial stage* during which glaciers were restricted in their distribution (see Chapter 23).

Figure 14.18 Valley Train and a Kettle

James S. Monroe

a A valley train in Alaska made up of stratified drift.

James S. Monroe

b A kettle in a moraine in Alaska.

Figure 14.19 Kames and Eskers

Courtesy of B.M.C. Pape

a This small hill in Wisconsin is a kame.

Tom Bean/Corbis

b An esker near Dahlen, North Dakota.

For more than a century, scientists have attempted to develop a comprehensive theory explaining all aspects of ice ages, but they have not yet been completely successful. One reason for their lack of success is that the climatic changes responsible for glaciation, the cyclic occurrence of glacial–interglacial episodes, and short-term events such as the Little Ice Age operate on vastly different time scales.

Only a few periods of glaciation are recognized in the geologic record, each separated from the others by long intervals of mild climate. Such long-term climatic changes probably result from slow geographic changes related to plate tectonic activity. Moving plates carry continents to high latitudes where glaciers exist, provided they receive enough precipitation as snow. Plate collisions, the subsequent uplift of vast areas far above sea level, and the changing atmospheric and oceanic circulation patterns caused by the changing shapes and positions of plates also contribute to long-term climate change.

The Milankovitch Theory

Changes in Earth's orbit as a cause of intermediate-term climatic events were first proposed during the mid-1800s, but the idea was made popular during the 1920s by the Serbian astronomer Milutin Milankovitch. He proposed that minor irregularities in Earth's rotation and orbit are sufficient to alter the amount of solar radiation received at any given latitude and hence bring about climate changes. Now called the **Milankovitch theory**, it was initially ignored but has received renewed interest since the 1970s and is now widely accepted.

Figure 14.20 Varves and a Dropstone in Glacial Deposits

a Each of these pairs of dark and light layers make up a varve, an annual deposit in a glacial lake.

b These varves have a dropstone that was probably liberated from floating ice.

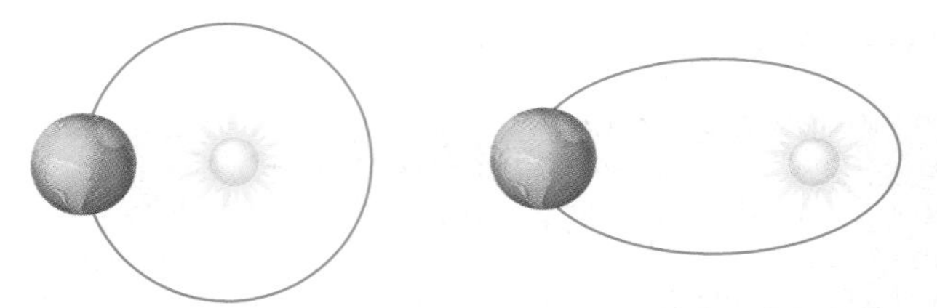

a Earth's orbit varies from nearly a circle (left) to an ellipse (right) and back again in about 100,000 years.

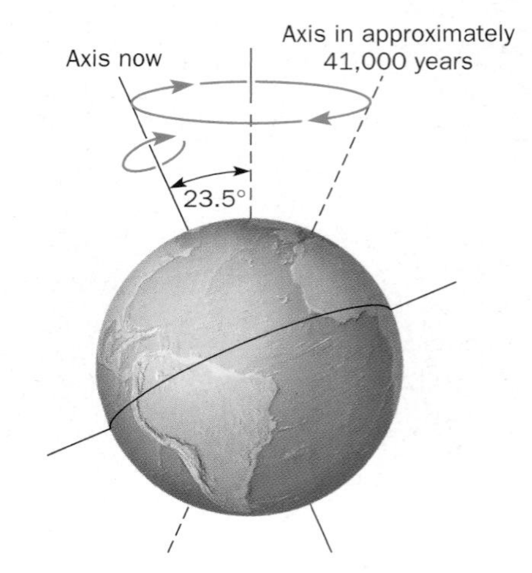

b Earth moves around its orbit while rotating on its axis, which is tilted to the plane of its orbit around the Sun at 23.5 degrees and points to the North Star. Earth's axis of rotation slowly moves and traces out a cone in space.

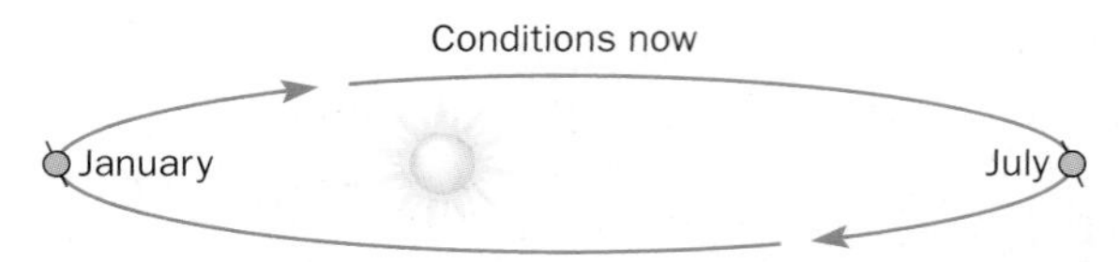

c At present, Earth is closest to the Sun in January (top), when Northern Hemisphere experiences winter.

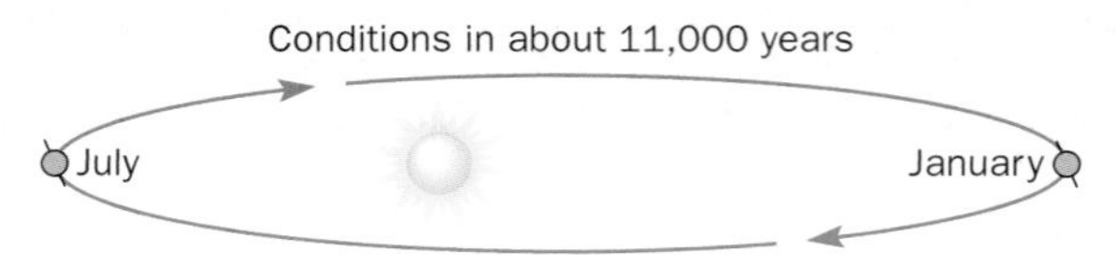

d In about 11,000 years, however, as a result of precession Earth will be closer to the Sun in July (bottom), when summer occurs in the Northern Hemisphere.

Figure 14.21 According to the Milankovitch Theory, Minor Irregularities in Earth's Rotation and Orbit May Affect Climatic Changes

Milankovitch attributed the onset of the Pleistocene Ice Age to variations in three aspects of Earth's orbit. The first is *orbital eccentricity*, which is the degree to which Earth's orbit around the sun changes over time (Figure 14.21a). When the orbit is nearly circular, both the Northern and Southern Hemispheres have similar contrasts between the seasons. However, if the orbit is more elliptic, hot summers and cold winters will occur in one hemisphere, while warm summers and cool winters will take place in the other hemisphere. Calculations indicate a roughly 100,000-year cycle between times of maximum eccentricity, which corresponds closely to the 20 warm–cold climatic cycles that took place during the Pleistocene.

Milankovitch also pointed out that the angle between Earth's axis and a line perpendicular to the plane of Earth's orbit shifts about 1.5 degrees from its current value of 23.5 degrees during a 41,000-year cycle (Figure 14.21b).

Although changes in *axial tilt* have little effect on equatorial latitudes, they strongly affect the amount of solar radiation received at high latitudes and the duration of the dark period at and near Earth's poles. Coupled with the third aspect of Earth's orbit, precession of the equinoxes, high latitudes might receive as much as 15% less solar radiation, certainly enough to affect glacial growth and melting.

Precession of the equinoxes, the last aspect of Earth's orbit that Milankovitch cited, refers to a change in the time of the equinoxes. At present, the equinoxes take place on about March 21 and September 21 when the Sun is directly over the equator. But as Earth rotates on its axis, it also wobbles as its axial tilt varies 1.5 degrees from its current value, thus changing the time of the equinoxes. Taken alone, the time of the equinoxes has little climatic effect, but changes in Earth's axial tilt also change the times of *aphelion* and *perihelion*, which are, respectively, when Earth is farthest from and closest to the Sun during its orbit (Figure 14.21c). Earth is now at perihelion, closest to the Sun, during Northern Hemisphere winters, but in about 11,000 years perihelion will be in July. Accordingly, Earth will be at aphelion, farthest from the Sun, in January and have colder winters.

Continuous variations in Earth's orbit and axial tilt cause the amount of solar heat received at any latitude to vary slightly through time. The total heat received by the planet changes little, but according to Milankovitch, and now many scientists agree, these changes cause complex climatic variations and provided the triggering mechanism for the glacial–interglacial episodes of the Pleistocene.

Short-Term Climatic Events

Climatic events with durations of several centuries, such as the Little Ice Age, are too short to be accounted for by plate tectonics or Milankovitch cycles. Several hypotheses have been proposed, including variations in solar energy and volcanism.

Variations in solar energy could result from changes within the Sun itself or from anything that would reduce the amount of energy Earth receives from the Sun. The latter could result from the solar system passing through clouds of interstellar dust and gas or from substances in the atmosphere reflecting solar radiation back into space. Records kept over the past 90 years indicate that during this time the amount of solar radiation has varied only slightly. Although variations in solar energy may influence short-term climatic events, such a correlation has not been demonstrated.

During large volcanic eruptions, tremendous amounts of ash and gases are spewed into the atmosphere, where they reflect incoming solar radiation and thus reduce atmospheric temperatures. Small droplets of sulfur gases remain in the atmosphere for years and can have a significant effect on climate. Several large-scale volcanic events have occurred, such as the 1815 eruption of Tambora, and are known to have had climatic effects. However, no relationship between periods of volcanic activity and periods of glaciation has yet been established.

Geo-Recap

Chapter Summary

- Glaciers currently cover about 10% of the land surface and contain about 2.15% of all water on Earth.
- A glacier forms when winter snowfall exceeds summer melt and therefore accumulates year after year. Snow is compacted and converted to glacial ice, and when the ice is about 40 m thick, pressure causes it to flow.
- Glaciers move by plastic flow and basal slip.
- Valley glaciers are confined to mountain valleys and flow from higher to lower elevations, whereas continental glaciers cover vast areas and flow outward in all directions from a zone of accumulation.
- The behavior of a glacier depends on its budget, which is the relationship between accumulation and wastage. If a glacier has a balanced budget, its terminus remains stationary; a positive or negative budget results in the advance or retreat of the terminus, respectively.
- Glaciers move at varying rates depending on slope, discharge, and season. Valley glaciers tend to flow more rapidly than continental glaciers.
- Glaciers effectively erode and transport because they are solids in motion. They are particularly effective at eroding soil and unconsolidated sediment, and they can transport any size sediment supplied to them.
- Continental glaciers transport most of their sediment in the lower part of the ice, whereas valley glaciers may carry sediment in all parts of the ice.
- Erosion of mountains by valley glaciers yields several sharp, angular landforms including cirques, arêtes, and horns. U-shaped glacial troughs, fiords, and hanging valleys are also products of valley glaciation.
- Continental glaciers abrade and bevel high areas, producing a smooth, rounded landscape known as an ice-scoured plain.
- Depositional landforms include moraines, which are ridgelike accumulations of till. The several types of moraines are terminal, recessional, lateral, and medial.
- Drumlins are composed of till that was apparently reshaped into streamlined hills by continental glaciers or floods.

- Stratified drift in outwash plains and valley trains consists of sand and gravel deposited by meltwater streams issuing from glaciers. Ridges known as eskers, and conical hills called kames are also composed of stratified drift.
- Major glacial intervals separated by tens or hundreds of millions of years probably occur as a result of the changing positions of tectonic plates, which in turn cause changes in oceanic and atmospheric circulation patterns.
- Currently, the Milankovitch theory is widely accepted as the explanation for glacial–interglacial intervals.
- The reasons for short-term climatic changes, such as the Little Ice Age, are not understood. Two proposed causes are changes in the amount of solar energy received by Earth and volcanism.

Important Terms

abrasion (p. 367)
arête (p. 372)
basal slip (p. 361)
cirque (p. 369)
continental glacier (p. 360)
drumlin (p. 374)
end moraine (p. 373)
esker (p. 375)
fiord (p. 369)
firn (p. 361)
glacial budget (p. 362)
glacial drift (p. 372)
glacial erratic (p.372)
glacial ice (p. 361)
glacial polish (p. 367)
glacial striations (p. 367)
glacial surge (p. 364)
glaciation (p. 360)
glacier (p. 359)
ground moraine (p. 373)
hanging valley (p. 369)
horn (p. 372)
ice cap (p. 360)
kame (p. 375)
lateral moraine (p. 374)
Little Ice Age (p. 358)
medial moraine (p. 374)
Milankovitch theory (p. 377)
outwash plain (p. 375)
plastic flow (p. 361)
recessional moraine (p. 373)
stratified drift (p. 373)
terminal moraine (p. 374)
till (p. 372)
U-shaped glacial trough (p. 367)
valley glacier (p. 359)
valley train (p. 375)
zone of accumulation (p. 362)
zone of wastage (p. 362)

Review Questions

1. The zone of accumulation on a glacier is
   a. ______ the area where firn is converted to glacial ice;
   b. ______ where additions exceed losses;
   c. ______ at a depth of about 40 meters;
   d. ______ where most basal slip takes place;
   e. ______ an erosional landform.
2. All sediment deposited directly by glacial ice is called
   a. ______ varve;
   b. ______ esker;
   c. ______ outwash;
   d. ______ till;
   e. ______ kame.
3. Which one of the following is a feature caused by valley glacier erosion?
   a. ______ valley train;
   b. ______ dropstone;
   c. ______ terminal moraine;
   d. ______ glacial recession;
   e. ______ hanging valley.
4. Glaciers move mostly by
   a. ______ plastic flow;
   b. ______ surging;
   c. ______ lateral compression;
   d. ______ abrasion;
   e. ______ glacial polish.
5. When freshly fallen snow compacts and partly melts and refreezes, it forms granular ice known as
   a. ______ drift;
   b. ______ firn;
   c. ______ cirque;
   d. ______ drumlin;
   e. ______ till.
6. If a glacier has a negative budget
   a. ______ its terminus retreats toward the zone of accumulation;
   b. ______ its firn limit moves toward the glacier's terminus;
   c. ______ it flows much more rapidly for a short time;
   d. ______ its zone of wastage disappears;
   e. ______ it advances and modifies previously deposited moraines.
7. The only two areas where continental glaciers are present today are
   a. ______ Canada and Alaska;
   b. ______ Montana and Wyoming;
   c. ______ Greenland and Antarctica;
   d. ______ Scandinavia and Siberia;
   e. ______ Iceland and Baffin Island.

8. A bowl-shaped depression on a mountainside at the upper end of a valley glacier is a/an
   a. ______ horn;
   b. ______ cirque;
   c. ______ valley train;
   d. ______ arête;
   e. ______ end moraine.
9. An ice-scoured plain is a/an
   a. ______ subdued landscape resulting from erosion by a continental glacier;
   b. ______ vast area covered by outwash deposits;
   c. ______ deposit made up of alternating light and dark layers of clay;
   d. ______ region with numerous kames, drumlins, and eskers;
   e. ______ mountain range that has been deeply eroded by valley glaciers.
10. Boulders deposited by glaciers far from their source are known as
   a. ______ firns;
   b. ______ horns;
   c. ______ erratics;
   d. ______ varves;
   e. ______ arêtes.
11. How is it possible for glacial ice, which is solid, to flow?
12. Draw side views of a drumlin and a roche moutonnée, indicate the direction of ice flow, and explain how each forms.
13. A valley glacier has a cross sectional area of 400,000 m^2 and a flow velocity of 2 m per day. How long will it take for 1 km^3 of ice to move past a given point?
14. Explain how a glacier can erode below sea level but streams and rivers cannot.
15. How does the Milankovitch theory explain the onset of Pleistocene episodes of glaciation?
16. What kinds of evidence would indicate that an ice-free area was once covered by glacial ice?
17. What are glacial surges and what might cause one to occur?
18. Explain in terms of a glacial budget how a once-active glacier becomes stagnant.
19. How do valley glaciers, continental glaciers, and ice caps differ?
20. Identify the glacial features indicated in the image below and explain how they form.

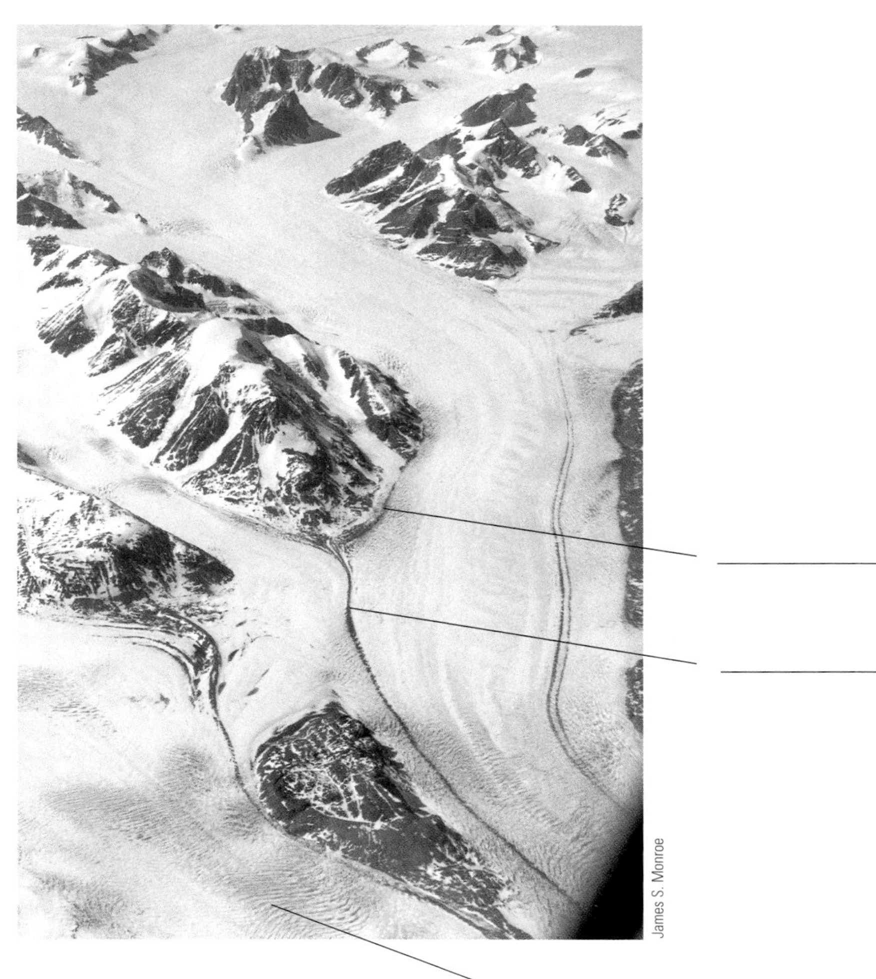

CHAPTER 15

The Work of Wind and Deserts

George Gerster/The National Audubon Society/Photo Researchers, Inc.

OUTLINE

OBJECTIVES

At the end of this chapter, you will have learned that

- Wind transports sediment and modifies the landscape through the processes of abrasion and deflation.
- Dunes and loess are the result of wind depositing material.
- Dunes form when wind flows over and around obstructions.
- The four major dune types are barchan, longitudinal, transverse, and parabolic.
- Loess is formed from wind-blown silt and clay, and is derived from three main sources: deserts, Pleistocene glacial outwash deposits, and river floodplains in semiarid regions.
- The global pattern of air-pressure belts and winds is responsible for Earth's atmospheric circulation patterns.
- Deserts are dry and receive less than 25 cm of rain per year, have high evaporation rates, typically have poorly developed soils, and are mostly or completely devoid of vegetation.
- The majority of deserts are found in the dry climates of the low and middle latitudes.
- Deserts have many distinctive landforms produced by both wind and running water.

The Saharan community of El Gedida in western Egypt is slowly being overwhelmed by advancing sand.

INTRODUCTION

During the past several decades, deserts have been advancing across millions of acres of productive land, destroying rangeland, croplands, and even villages. Such expansion, estimated at 70,000 km^2 per year, has exacted a terrible toll in human suffering. Because of the relentless advance of deserts, hundreds of thousands of people have died of starvation or been forced to migrate as "environmental refugees" from their homelands to camps, where the majority are severely malnourished. This expansion of deserts into formerly productive lands is called **desertification** and is a major problem in many countries.

Most regions undergoing desertification lie along the margins of existing deserts where a delicately balanced ecosystem serves as a buffer between the desert on one side and a more humid environment on the other. These regions have limited potential to adjust to increasing environmental pressures from natural causes as well as human activity. Ordinarily, desert regions expand and contract gradually in response to natural processes such as climatic change, but much recent desertification has been greatly accelerated by human activities.

In many areas, the natural vegetation has been cleared as crop cultivation has expanded into increasingly drier fringes to support growing populations. Because grasses are the dominant natural vegetation in most fringe areas, raising livestock is a common economic activity. However, increasing numbers of livestock in many areas have greatly exceeded the land's ability to support them. Consequently, the vegetation cover that protects the soil has diminished, causing the soil to crumble and be stripped away by wind and water, which results in increased desertification.

One particularly hard-hit area of desertification is the Sahel of Africa (a belt 300–1100 km wide, lying south of the Sahara). Because drought is common in the Sahel, the region can support only a limited population of livestock and humans. Unfortunately, expanding humans (30 million people in 1980, increasing to an estimated 50 million in 2000) and animal populations and more intensive agriculture have increased the demands on the lands. Plagued with periodic droughts, this region has suffered tremendously as crops have failed and livestock has overgrazed the natural vegetation, resulting in thousands of deaths, displaced people, and the encroachment of the Sahara.

The tragedy of the Sahel and prolonged droughts in other desert fringe areas remind us of the delicate equilibrium of ecosystems in such regions. Once the fragile soil cover has been removed by erosion, it takes centuries for new soil to form (see Chapter 6).

There are many important reasons to study deserts and the processes that are responsible for their formation. First, deserts cover large regions of Earth's surface. More than 40% of Australia is desert, and the Sahara occupies a vast part of northern Africa. Although deserts are generally sparsely populated, some desert regions are experiencing an influx of people, such as Las Vegas, Nevada, the high desert area of southern California, and various locations in Arizona. Many of these places already have problems with population growth and the strains it places on the environment, particularly the need for greater amounts of groundwater (see Chapter 13).

Furthermore, with the current debate about global warming, it is important to understand how desert processes operate and how global climate changes affect the various Earth systems and subsystems. By understanding how desertification operates, people can take steps to eliminate or reduce the destruction done, particularly in terms of human suffering.

Learning about the underlying causes of climate change by examining ancient desert regions may provide insight into the possible duration and severity of future climatic changes. This can have important ramifications in decisions about whether burying nuclear waste in a desert, such as Yucca Mountain, Nevada, is as safe as some claim and is in our best interests as a society.

As an example, more than 6000 years ago, the Sahara was a fertile savannah supporting a diverse fauna and flora, including humans. Then the climate changed, and the area became a desert. How did this happen? Will this region change back again in the future? These are some of the questions geoscientists hope to answer by studying deserts.

And last, many agents and processes that have shaped deserts do not appear to be limited to our planet. Features found on Mars, especially as seen in images transmitted by the *Spirit* and *Opportunity* rovers, are apparently the result of the same wind-driven processes that operate on Earth.

SEDIMENT TRANSPORT BY WIND

Wind is a turbulent fluid and therefore transports sediment in much the same way as running water. Although wind typically flows at a greater velocity than water, it has a lower density and thus can carry only clay- and silt-size particles as *suspended load.* Sand and larger particles are moved along the ground as *bed load.*

Bed Load

Sediments too large or heavy to be carried in suspension by water or wind are moved as bed load either by *saltation* or by rolling and sliding. As we discussed in Chapter 12, saltation is the process by which a portion of the bed load moves by intermittent bouncing along a streambed. Saltation also occurs on land. Wind starts sand grains rolling and lifts and carries some grains short distances before they fall back to the surface. As the descending sand grains hit the surface, they strike other grains, causing them to bounce along by saltation (Figure 15.1). Wind-tunnel experiments show that once sand grains begin moving, they continue to move, even if the wind drops below the speed necessary to start them moving! This happens because once saltation begins, it sets off a chain reaction of collisions between sand grains that keeps the grains in constant motion.

Saltating sand usually moves near the surface, and even when winds are strong, grains are rarely lifted higher than about a meter. If the winds are very strong, these wind-whipped grains can cause extensive abrasion. A car's paint can

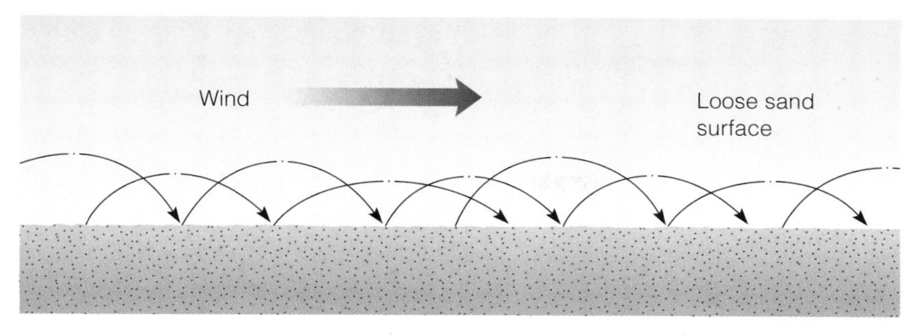

Figure 15.1 Saltation Most sand is moved near the ground surface by saltation. Sand grains are picked up by the wind and carried a short distance before falling back to the ground, where they usually hit other grains, causing them to bounce and move in the direction of the wind.

be removed by sandblasting in a short time, and its windshield will become completely frosted and translucent from pitting.

Suspended Load

Silt- and clay-sized particles constitute most of a wind's suspended load. Even though these particles are much smaller and lighter than sand-sized particles, wind usually starts the latter moving first. The reason for this phenomenon is that a very thin layer of motionless air lies next to the ground where the small silt and clay particles remain undisturbed. The larger sand grains, however, stick up into the turbulent air zone where they can be moved. Unless the stationary air layer is disrupted, the silt and clay particles remain on the ground, providing a smooth surface.

This phenomenon can be observed on a dirt road on a windy day. Unless a vehicle travels over the road, little dust is raised even though it is windy. When a vehicle moves over the road, it breaks the calm boundary layer of air and disturbs the smooth layer of dust, which is picked up by the wind and forms a dust cloud in the vehicle's wake.

In a similar manner, when a sediment layer is disturbed, silt- and clay-sized particles are easily picked up and carried in suspension by the wind, creating clouds of dust or even dust storms. Once these fine particles are lifted into the atmosphere, they may be carried thousands of kilometers from their source. For example, large quantities of fine dust from the southwestern United States were blown eastward and fell on New England during the dust storms of the 1930s (see Chapter 6 Geo-Focus on pages 148 and 149).

WIND EROSION

Although wind action produces many distinctive erosional features and is an extremely efficient sorting agent, running water is responsible for most erosional landforms in arid regions, even though stream channels are typically dry. Wind erodes material in two ways: abrasion and deflation.

Abrasion

Abrasion involves the impact of saltating sand grains on an object and is analogous to sandblasting. The effects of abrasion are usually minor because sand, the most common agent of abrasion, is rarely carried more than a meter above the surface. Rather than creating major erosional features, wind abrasion typically modifies existing features by etching, pitting, smoothing, or polishing. Nonetheless, wind abrasion can produce many strange-looking and bizarre-shaped features (Figure 15.2).

Ventifacts are a common product of wind abrasion; these are stones whose surfaces have been polished, pitted, grooved, or faceted by the wind (Figure 15.3). If the wind blows from different directions, or if the stone is moved, the ventifact will have multiple facets. Ventifacts are most common in deserts, yet they can form wherever stones are exposed to saltating sand grains, as on beaches in humid regions and some outwash plains in New England.

Yardangs are larger features than ventifacts and also result from wind erosion (Figure 15.4). They are elongated, streamlined ridges that look like an overturned ship's hull. Yardangs are typically found in clusters aligned parallel to the prevailing winds. They probably form by differential erosion in which depressions, parallel to the direction of wind, are carved out of a rock body, leaving sharp, elongated ridges.

O. Alamany & W. Vicens/Corbis

Figure 15.2 Wind Abrasion Wind abrasion has formed these bizarre-shaped structures by eroding the lower part of the exposed limestone in Desierto Libico, Egypt.

Figure 15.3 Ventifacts

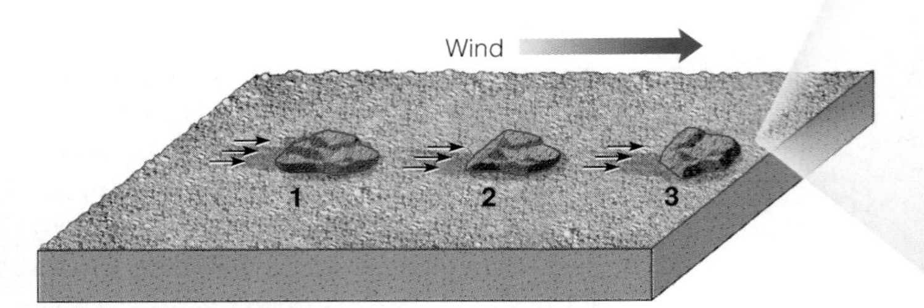

a A ventifact forms when wind-borne particles **(1)** abrade the surface of a rock, **(2)** forming a flat surface. If the rock is moved, **(3)** additional flat surfaces are formed.

Marli Bryant Miller

b Large ventifacts lying on desert pavement in Death Valley National Monument, California.

Marion A. Whitney

Figure 15.4 Yardang A profile view of a streamlined yardang in the Roman playa deposits of the Kharga Depression, Egypt. Yardangs form by wind erosion.

These ridges may then be further modified by wind abrasion into their characteristic shape. Although yardangs are fairly common desert features, interest in them was renewed when images radioed back from Mars showed that they are also widespread features on the Martian surface.

Deflation

Another important mechanism of wind erosion is **deflation,** which is the removal of loose surface sediment by the wind. Among the characteristic features of deflation in many arid and semiarid regions are *deflation hollows* or *blowouts* (Figure 15.5). These shallow depressions of variable dimensions result from differential erosion of surface materials. Ranging in size from several kilometers in diameter and tens of meters deep to small depressions only a few meters wide and less than a meter deep, deflation hollows are common in the southern Great Plains region of the United States.

In many dry regions, the removal of sand-sized and smaller particles by wind leaves a surface of pebbles, cobbles, and boulders. As the wind removes the fine-grained material from the surface, the effects of gravity and occasional heavy rain, and even the swelling of clay grains, rearrange the remaining coarse particles into a mosaic of close-fitting rocks called **desert pavement** (Figure 15.3b and Figure 15.6). Once desert pavement forms, it protects the underlying material from further deflation.

WIND DEPOSITS

Although wind is of minor importance as an erosional agent, it is responsible for impressive deposits, which are primarily of two types. The first, *dunes,* occur in several distinctive types, all of which consist of sand-sized particles that are usually deposited near their source. The second is *loess,* which consists of layers of wind-blown silt and clay deposited over large areas downwind and commonly far from their source.

Figure 15.5 **Deflation Hollow** A deflation hollow, the low area, between two sand dunes in Death Valley, California. Deflation hollows result when loose surface sediment is differentially removed by wind.

and around an obstruction, resulting in the deposition of sand grains, which accumulate and build up a deposit of sand. As they grow, these sand deposits become self-generating in that they form ever-larger wind barriers that further reduce the wind's velocity, resulting in more sand deposition and growth of the dune.

Most dunes have an asymmetrical profile, with a gentle windward slope and a steeper downwind or leeward slope that is inclined in the direction of the prevailing wind (Figure 15.8a). Sand grains move up the gentle windward slope by saltation and accumulate on the leeward side, forming an angle of 30 to 34 degrees from the horizontal, which is the angle of repose of dry sand. When this angle is exceeded by accumulating sand, the slope collapses, and the sand slides down the leeward slope, coming to rest at its base. As sand moves from a dune's windward side and periodically slides down its

The Formation and Migration of Dunes

The most characteristic features in sand-covered regions are **dunes,** which are mounds or ridges of wind-deposited sand (Figure 15.7). Dunes form when wind flows over

Figure 15.6 Desert Pavement

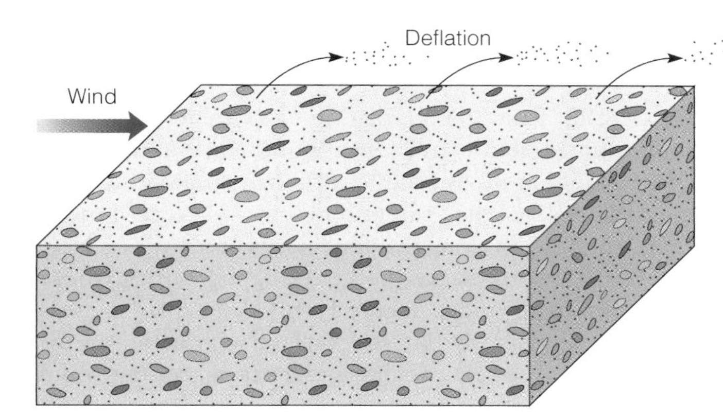

a Fine-grained material is removed by wind,

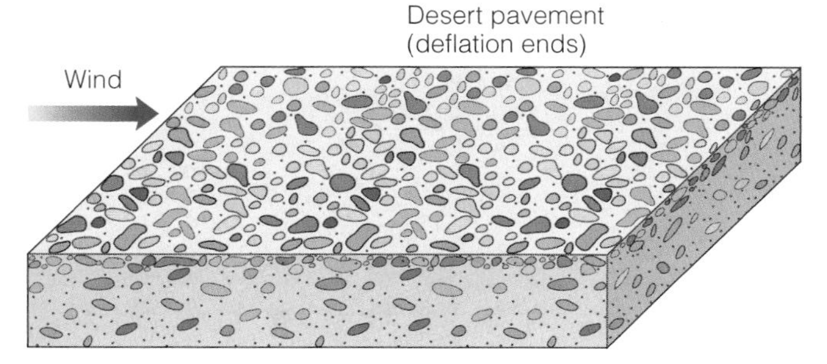

b leaving a concentration of larger particles that form desert pavement.

c Desert pavement in the Mojave Desert, California. Numerous ventifacts are visible in this photo. Desert pavement prevents further erosion and transport of a desert's surface materials by forming a protective layer of close-fitting, larger rocks.

Figure 15.7 **Sand Dunes** Large sand dunes in Death Valley, California. The prevailing wind direction is from left to right, as indicated by the sand dunes in which the gentle windward side is on the left and the steeper leeward slope is on the right.

leeward slope, the dune slowly migrates in the direction of the prevailing wind (Figure 15.8b). When preserved in the geologic record, dunes help geologists determine the prevailing direction of ancient winds (Figure 15.9).

Dune Types

Geologists recognize four major dune types (barchan, longitudinal, transverse, and parabolic), although intermediate forms also exist. The size, shape, and arrangement of dunes result from the interaction of such factors as sand supply, the direction and velocity of the prevailing wind, and the amount of vegetation. Although dunes are usually found in deserts, they can also develop wherever sand is abundant, such as along the upper parts of many beaches.

Barchan dunes are crescent-shaped dunes whose tips point downwind (Figure 15.10). They form in areas that have a generally flat, dry surface with little vegetation, a limited supply of sand, and a nearly constant wind

Figure 15.8 Dune Migration

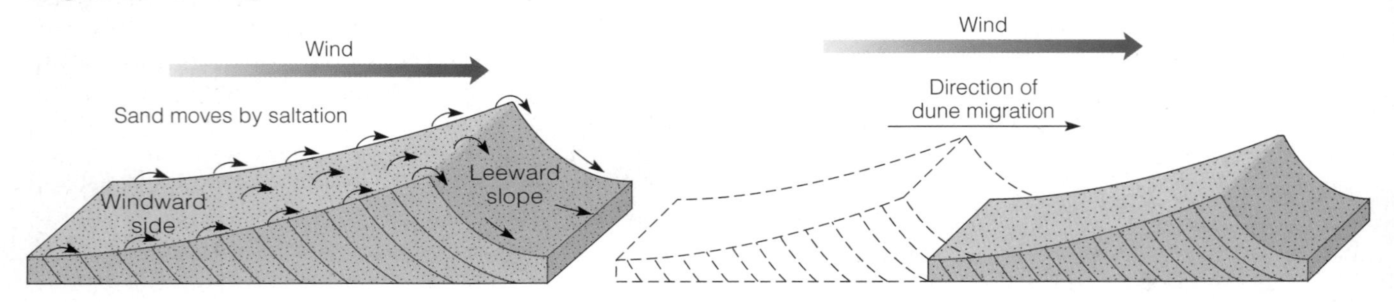

a Profile of a sand dune.

b Dunes migrate when sand moves up the windward side and slides down the leeward slope. Such movement of the sand grains produces a series of cross-beds that slope in the direction of wind movement.

Figure 15.9 **Cross-Bedding** Ancient cross-bedding in sandstone beds in Zion National Park, Utah, helps geologists determine the prevailing direction of the wind that formed these ancient sand dunes.

direction. Most barchans are small, with the largest reaching about 30 m high. Barchans are the most mobile of the major dune types, moving at rates that can exceed 10 m per year.

Longitudinal dunes (also called *seif dunes*) are long, parallel ridges of sand aligned generally parallel to the direction of the prevailing winds; they form where the sand supply is somewhat limited (Figure 15.11). Longitudinal dunes result when winds converge from slightly different directions to produce the prevailing wind. They range in height from about 3 m to more than 100 m, and some stretch for more than 100 km. Longitudinal dunes are especially well developed in central Australia, where they cover nearly one-fourth of the continent. They also cover extensive areas in Saudi Arabia, Egypt, and Iran.

Transverse dunes form long ridges perpendicular to the prevailing wind direction in areas that have abundant sand and little or no vegetation (Figure 15.12). When viewed from the air, transverse dunes have a wavelike appearance and are therefore sometimes called *sand seas.* The crests of transverse dunes can be as high as 200 m, and the dunes may be as wide as 3 km. Some transverse dunes develop a clearly distinguishable barchan form and may separate into individual barchan dunes along the edges of the dune field where there is less sand. Such intermediate-form dunes are known as *barchanoid dunes.*

Parabolic dunes are most common in coastal areas with abundant sand, strong onshore winds, and a partial cover of vegetation (Figure 15.13). Although parabolic dunes have a crescent shape like barchan dunes, their tips point upwind. Parabolic dunes form when the vegetation cover is broken and deflation produces a deflation hollow or blowout. As the wind transports the sand out of the depression, it builds up on the convex downwind dune crest. The central part of the dune is excavated by the wind, while vegetation holds the ends and sides fairly well in place.

Another type of dune commonly found in the deserts of North Africa and Saudi Arabia is the *star dune,* so named because of its resemblance to a multipointed star (Figure 15.14). Star dunes are among the tallest in the world, rising, in some cases, more than 100 m above the surrounding desert plain. They consist of pyramidal hills of sand, from which radiate several ridges of sand, and they develop where the wind direction is variable. Star dunes can remain stationary for centuries at a time and have served as desert landmarks for nomadic peoples.

Loess

Wind-blown silt and clay deposits composed of angular quartz grains, feldspar, micas, and calcite are known as **loess.** The distribution of loess shows that it is derived from three main sources: deserts, Pleistocene glacial outwash deposits, and the floodplains of rivers in semiarid regions. Loess must be stabilized by moisture and vegetation in order to accumulate. Consequently, loess is not found in deserts, even though deserts provide much of its material. Because of its unconsolidated nature, loess is easily eroded, and as a result, eroded loess areas are characterized by steep cliffs and rapid lateral and headward stream erosion (Figure 15.15).

At present, loess deposits cover approximately 10% of Earth's land surface and 30% of the United States. The most extensive and thickest loess deposits are found in northeast China, where accumulations greater than 30 m thick are common. The extensive deserts in central Asia are the source for this loess. Other important loess deposits are on the North European Plain from Belgium eastward to Ukraine, in Central Asia, and the Pampas of Argentina. In the United States, loess deposits are found in the Great Plains, the Midwest, the Mississippi River Valley, and eastern Washington state.

Loess-derived soils are some of the world's most fertile (Figure 15.15). It is therefore not surprising that the world's major grain-producing regions correspond to large loess deposits, such as the North European Plain, Ukraine, and the Great Plains of North America.

Figure 15.10 Barchan Dunes

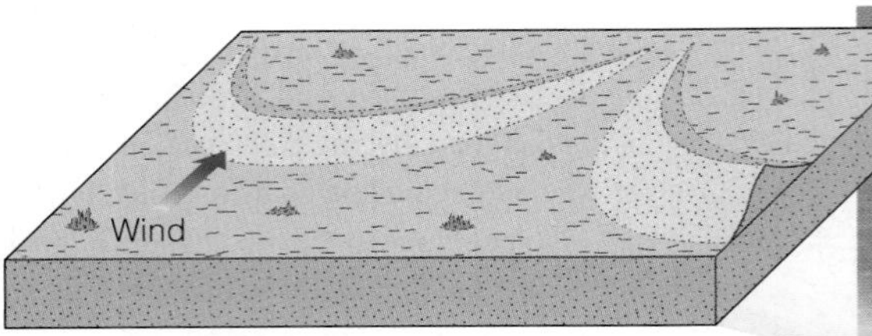

a Barchan dunes form in areas that have a limited amount of sand, a nearly constant wind direction, and a generally flat, dry surface with little vegetation. The tips of barchan dunes point downward.

b An aerial view of several barchan dunes west of the Salton Sea, California. The prevailing wind direction is from left to right, as indicated by the barchan dune's tips pointing toward the right.

c A ground-level view of several barchan dunes.

Figure 15.11 Longitudinal Dunes

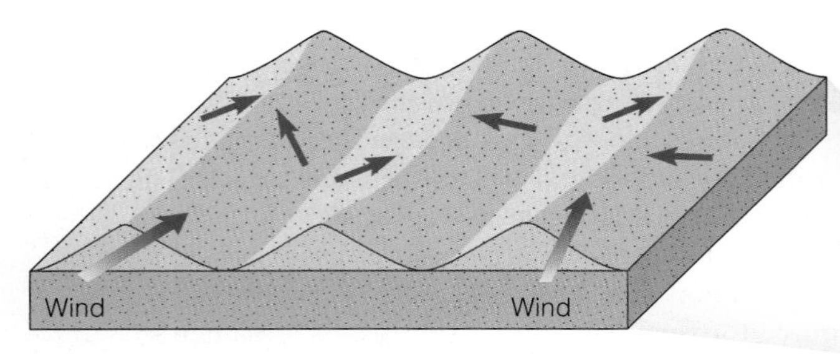

a Longitudinal dunes form long, parallel ridges of sand aligned roughly parallel to the prevailing wind direction. They typically form where sand supplies are limited.

b Longitudinal dunes, 15 m high, in the Gibson Desert, west central Australia. The bright blue areas between the dunes are shallow pools of rainwater, and the darkest patches are areas where the Aborigines have set fires to encourage the growth of spring grasses.

AIR-PRESSURE BELTS AND GLOBAL WIND PATTERNS

To understand the work of wind and the distribution of deserts, we need to consider the global pattern of air-pressure belts and winds, which are responsible for Earth's atmospheric circulation patterns. Air pressure is the density of air exerted on its surroundings (that is, its weight). When air is heated, it expands and rises, reducing its mass for a given volume and causing a decrease in air pressure. Conversely, when air is cooled, it contracts and air pressure increases. Therefore, those areas of Earth's surface that

Figure 15.12 Transverse Dunes

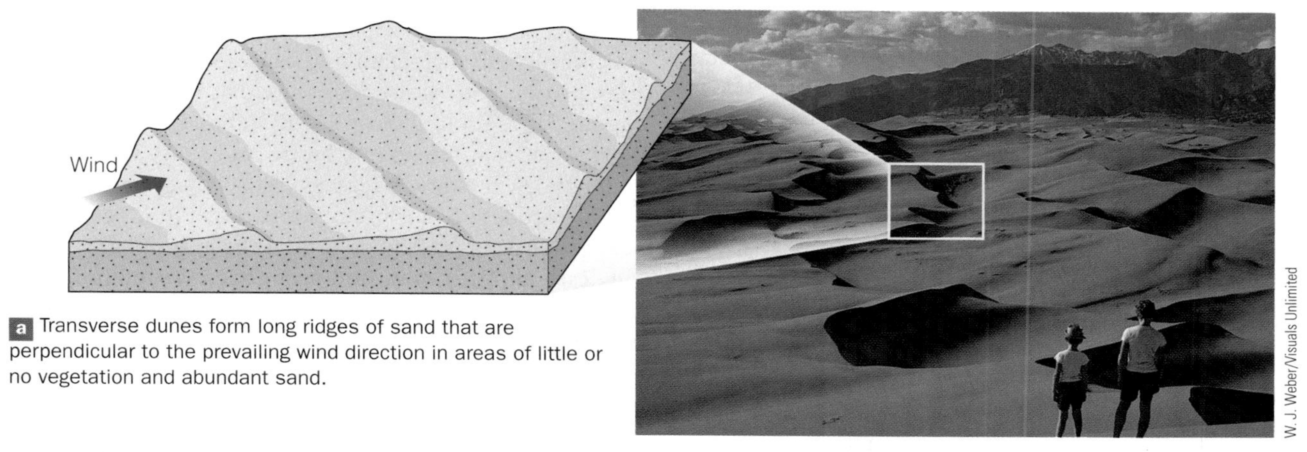

a Transverse dunes form long ridges of sand that are perpendicular to the prevailing wind direction in areas of little or no vegetation and abundant sand.

b Transverse dunes, Great Sand Dunes National Monument, Colorado. The prevailing wind direction is from lower left to upper right.

receive the most solar radiation, such as the equatorial regions, have low air pressure, whereas the colder areas, such as the polar regions, have high air pressure.

Air flows from high-pressure zones to low-pressure zones. If Earth did not rotate, winds would move in a straight line from one zone to another. Because Earth rotates, however, winds are deflected to the right of their direction of motion (clockwise) in the Northern Hemisphere and to the left of their direction of motion (counterclockwise) in the Southern Hemisphere. This deflection of air between latitudinal zones resulting from Earth's rotation is known as the **Coriolis effect.** The combination of latitudinal pressure differences and the

Figure 15.13 Parabolic Dunes

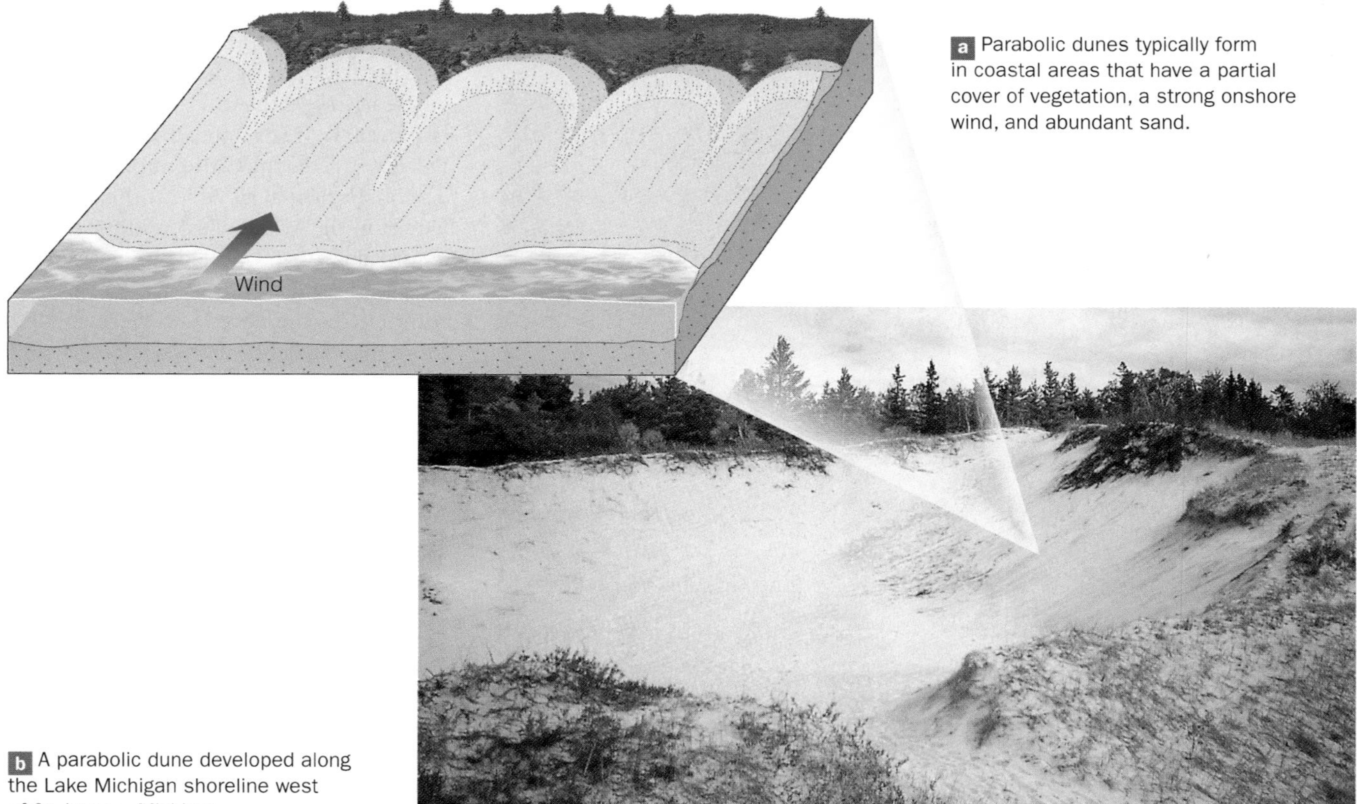

a Parabolic dunes typically form in coastal areas that have a partial cover of vegetation, a strong onshore wind, and abundant sand.

b A parabolic dune developed along the Lake Michigan shoreline west of St. Ignace, Michigan.

Figure 15.14 Star Dunes

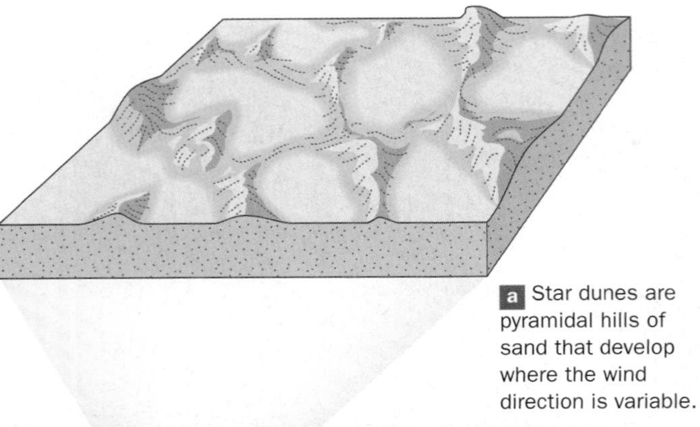

a Star dunes are pyramidal hills of sand that develop where the wind direction is variable.

Nigel J. Dennis/Photo Researchers Inc.

b A ground-level view of star dunes in Namib-Naukluft Park, Namibia.

Coriolis effect produces a worldwide pattern of east–west–oriented wind belts (Figure 15.16).

Earth's equatorial zone receives the most solar energy, which heats the surface air and causes it to rise. As the air rises, it cools and releases moisture that falls as rain in the equatorial region (Figure 15.16). The rising air is now much drier as it moves northward and southward toward each pole. By the time it reaches 20 to 30 degrees north and south latitudes, the air has become cooler and denser and begins to descend. Compression of the atmosphere warms the descending air mass and produces a warm, dry, high-pressure area, the perfect conditions for the formation of the low-latitude deserts of the Northern and Southern Hemispheres (Figure 15.17).

THE DISTRIBUTION OF DESERTS

Dry climates occur in the low and middle latitudes where the potential loss of water by evaporation may exceed the yearly precipitation (Figure 15.17). Dry climates cover 30% of Earth's land surface and are subdivided into semiarid and arid regions. *Semiarid regions* receive more precipitation than arid regions, yet are moderately dry. Their soils are usually well developed and fertile, and support a natural grass cover. *Arid regions,* generally described as **deserts,** are dry; they receive less than 25 cm of rain per year, have high evaporation rates, typically have poorly developed soils, and are mostly or completely devoid of vegetation.

The majority of the world's deserts are in the dry climates of the low and middle latitudes (Figure 15.17). In North America, most of the southwestern United States and northern Mexico are characterized by this hot, dry climate, whereas in South America, this climate is primarily restricted to the Atacama Desert of coastal Chile and Peru. The Sahara in northern Africa, the Arabian Desert in the Middle East, and the majority of Pakistan and western India form the largest essentially unbroken desert environment in the Northern Hemisphere. More than 40% of Australia is desert, and most of the rest of it is semiarid.

The remaining dry climates of the world are found in the middle and high latitudes, mostly within continental interiors in the Northern Hemisphere (Figure 15.17). Many of these areas are dry because of their remoteness from moist maritime air and the presence

Lowell Georgia/Corbis

Figure 15.15 Terraced Wheat Fields in the Loess Soil at Tangwa Village, China Because of the unconsolidated nature of loess, many farmers live in hillside caves that they carved from the loess.

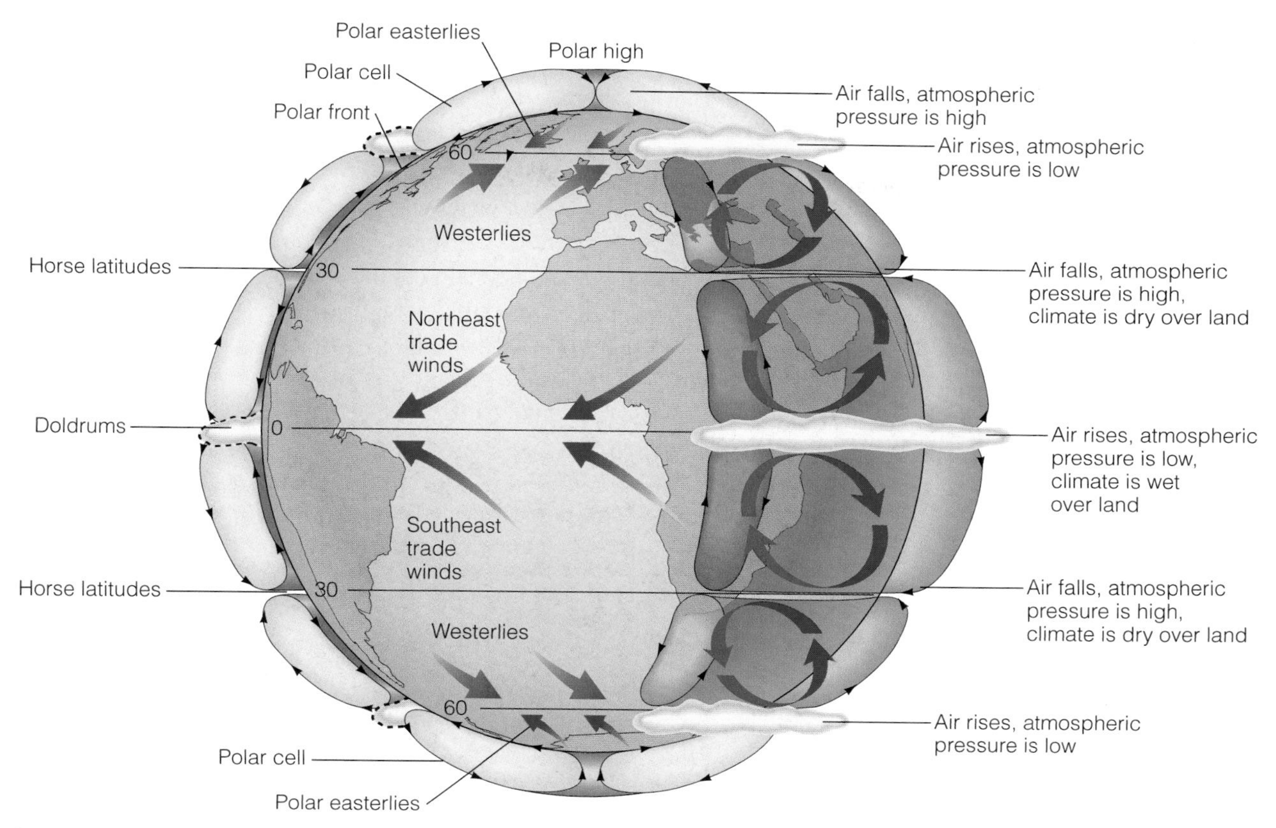

Figure 15.16 The General Circulation Pattern of Earth's Atmosphere Air flows from high-pressure zones to low-pressure zones, and the resulting winds are deflected to the right of their direction of movement (clockwise) in the Northern Hemisphere and to the left of their direction of movement (counterclockwise) in the Southern Hemisphere. This deflection of air between latitudinal zones resulting from Earth's rotation is known as the Coriolis effect.

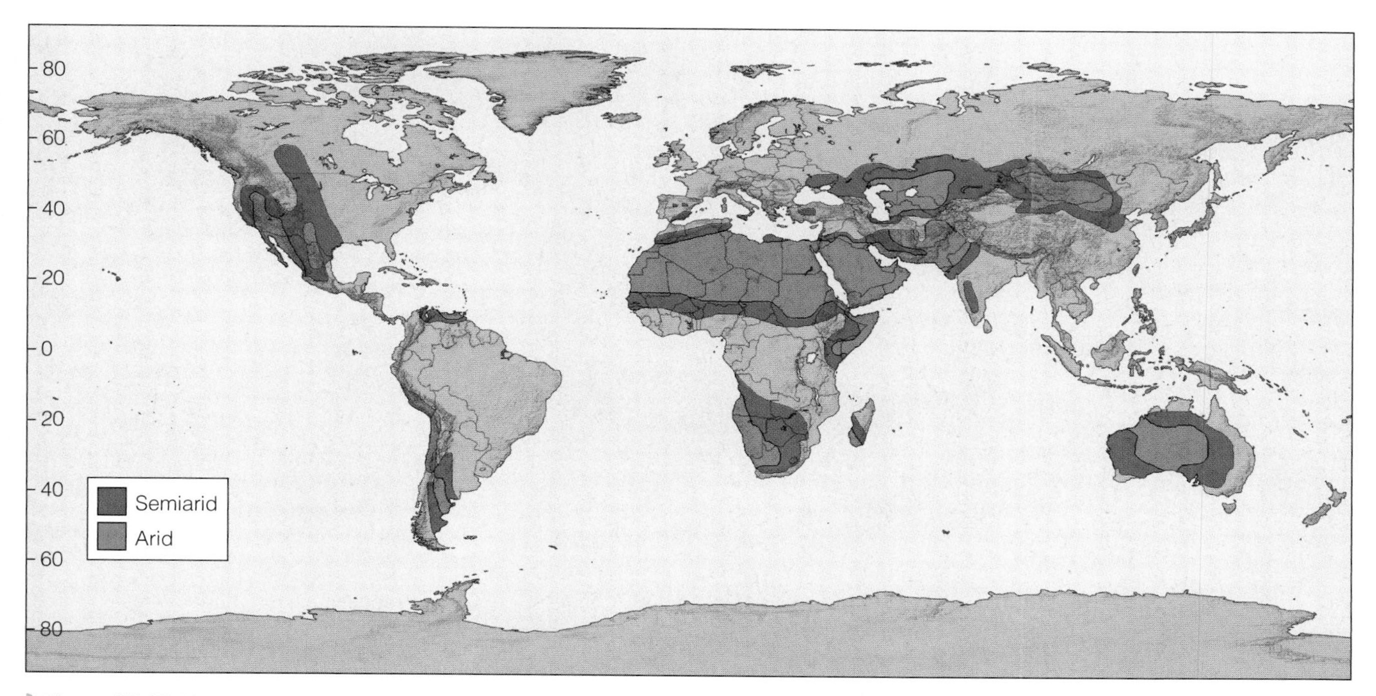

Figure 15.17 The Distribution of Earth's Arid and Semiarid Regions Semiarid regions receive more precipitation than arid regions, yet they are still moderately dry. Arid regions, generally described as deserts, are dry and receive less than 25 cm of rain per year. The majority of the world's deserts are located in the dry climates of the low and middle latitudes.

Geo-inSight Rock Art for the Ages

Rock art includes rock paintings (where paints made from natural pigments are applied to a rock surface) and petroglyphs (from the Greek petro, meaning "rock," and glyph, meaning "carving or engraving"), which are the abraded, pecked, incised, or scratched marks made by humans on boulders, cliffs, and cave walls.

Rock art has been found on every continent except Antarctica and is a valuable archaeological resource that provides graphic evidence of the cultural, social, and religious relationships and practices of ancient peoples. The oldest known rock art was made by hunters in western Europe and dates back to the Pleistocene Epoch. Africa has more rock art sites than any other continent. The oldest known African rock art, found in the southern part of the continent, is estimated to be 27,000 years old.

Petroglyphs are a fragile and nonrenewable cultural resource that cannot be replaced if they are damaged or destroyed. A commitment to their preservation is essential so that future generations can study them, as well as enjoy their beauty and mystery.

In the arid Southwest and Great Basin of North America where rock art is plentiful, rock paintings and petroglyphs extend back to about 2000 B.C. Here, rock art can be divided into two categories. *Representational art* deals with life-forms such as humans, birds, snakes, and human-like supernatural beings. Rarely exact replicas, they are more or less stylized versions of the beings depicted. *Abstract art,* in contrast, bears no resemblance to any real-life images.

▲ **1.** Various petroglyphs exposed at an outcrop along Cub Creek Road in Dinosaur National Monument, Utah.

Reed Wicander

◀ **2.** A human-like petroglyph, an example of representational art, exposed at an outcrop along Cub Creek Road in Dinosaur National Monument, Utah. Note the contrast between the fresh exposure of the rock where the upper part of the petroglyph's head has been removed, the weathered brown surface of the rest of the petroglyph, and the black rock varnish coating the rock surface.

Petroglyphs are the most common form of rock art in North America and are made by pecking, incising, or scratching the rock surface with a tool harder than the rock itself. In arid regions, many rock surfaces display a patination, or thin brown or black coating, known as rock varnish (Figure 15.20). When this coating is broken by pecking, incising, or scratching, the underlying lighter clolored natural rock surface provides an excellent contrast for the petroglyphs.

Petroglyphs are expecially abundant in the Southwest and Great Basin area, where they occur by the thosands, having been made by Native Americans from many cultures during the past several thousand years. Petroglyhs can be seen in many of the U.S. national parks and monuments, such as Petrified Forest National Park, Arizona, Dinosaur National Monument, Utah, Canyonlands National Park, Utah, and Petroglyph National Monument. New Mexico—to name a few.

Reed Wicander

▲ **3.** Examples of both representational and abstract art are displayed in these petroglyphs from Arizona.

Parvinder Sethi

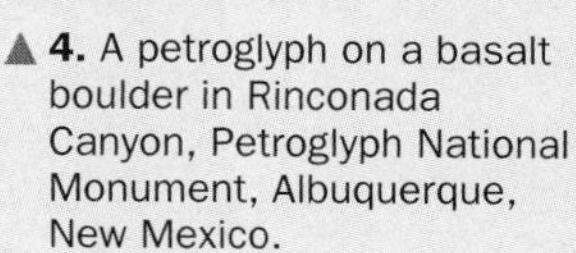

▲ **4.** A petroglyph on a basalt boulder in Rinconada Canyon, Petroglyph National Monument, Albuquerque, New Mexico.

▼► **5.** Examples of rock art from Tassili n'Aijer, Algeria.

Thomas Abertocrombine/National Geographic/Getty images

Thomas Abertocrombine/National Geographic/Getty images

▼ **6.** Rock art found in a rock shelter open to visitors in Uluru, Australia. These rock paintings are the work of Anagu artists, local Aboriginal people, and are created for religious and ceremonial expression, as well as for teaching and storytelling. Each abstract symbol can have many levels of meaning and can help illustrate a story.

Reed Wicander

◄ **7.** The paints used in this rock art are made from natural mineral substances mixed with water and sometimes with animal fat. Red, yellow, and orange pigments come from iron-stained clays, whereas the white pigments come from either ash or the mineral calcite. The black colors come from charcoal.

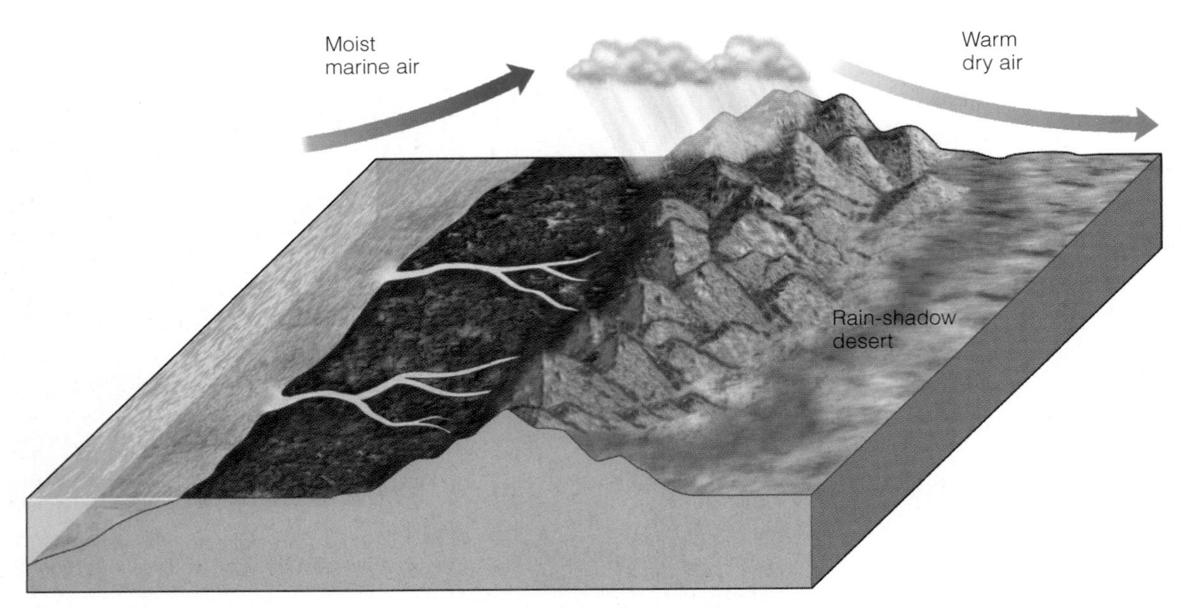

Figure 15.18 Rain-Shadow Deserts Many deserts in the middle and high latitudes are rain-shadow deserts, so named because they form on the leeward side of mountain ranges. When moist marine air moving inland meets a mountain range, it is forced upward, where it cools and forms clouds that produce rain. This rain falls on the windward side of the mountains. The air descending on the leeward side is much warmer and drier, producing a rain-shadow desert.

of mountain ranges that produce a **rain-shadow desert** (Figure 15.18). When moist marine air moves inland and meets a mountain range, it is forced upward. As it rises, it cools, forming clouds and producing precipitation that falls on the windward side of the mountains. The air that descends on the leeward side of the mountain range is much warmer and drier, producing a rain-shadow desert.

Three widely separated areas are included within the mid-latitude dry-climate zone (Figure 15.17). The largest is the central part of Eurasia, extending from just north of the Black Sea eastward to north-central China. The Gobi Desert in China is the largest desert in this region. The Great Basin area of North America is the second largest mid-latitude dry-climate zone and results from the rain shadow produced by the Sierra Nevada. This region adjoins the southwestern deserts of the United States that formed as a result of the low-latitude subtropical high-pressure zone. The smallest of the mid-latitude dry-climate areas is the Patagonian region of southern and western Argentina. Its dryness results from the rain-shadow effect of the Andes. The remainder of the world's deserts are found in the cold but dry high latitudes, such as Antarctica.

CHARACTERISTICS OF DESERTS

To people who live in humid regions, deserts may seem stark and inhospitable. Instead of a landscape of rolling hills and gentle slopes with an almost continuous cover of vegetation, deserts are dry, have little vegetation, and consist of nearly continuous rock exposures, desert pavement, or sand dunes. Yet despite the great contrast between deserts and more humid areas, the same geologic processes are at work, only operating under different climatic conditions.

Temperature, Precipitation, and Vegetation

The heat and dryness of deserts are well known. Many of the deserts of the low latitudes have average summer temperatures that range between 32° and 38°C. It is not uncommon for some low-elevation inland deserts to record daytime highs of 46° to 50°C for weeks at a time. The highest temperature ever recorded was 58°C in El Azizia, Libya, on September 13, 1922.

During the winter months when the Sun's angle is lower and there are fewer daylight hours, daytime temperatures average between 10° and 18°C. Winter nighttime lows can be cold, with frost and freezing temperatures common in the more poleward deserts. Winter daily temperature fluctuations in low-latitude deserts are among the greatest in the world, ranging between 18° and 35°C. Temperatures have been known to fluctuate from below 0°C to higher than 38°C in a single day!

The dryness of the low-latitude deserts results primarily from the year-round dominance of the subtropical high-pressure belt, whereas the dryness of the mid-latitude deserts is due to their isolation from moist marine winds and the rain-shadow effect created by mountain ranges. The dryness of both is accentuated by their high temperatures.

Although deserts are defined as regions that receive, on average, less than 25 cm of rain per year, the amount of rain

Charlie Ott, The National Audubon Society Collection/Photo Researchers, Inc.

Figure 15.19 Desert Vegetation Desert vegetation is typically sparse, widely spaced, and characterized by slow growth rates. The vegetation shown here in Organ Pipe National Monument, Arizona, includes saguaro and cholla cacti, paloverde trees, and jojoba bushes and is characteristic of the vegetation found in the Sonoran Desert of North America.

that falls each year is unpredictable and unreliable. It is not uncommon for an area to receive more than an entire year's average rainfall in one cloudburst and then to receive little rain for several years. Thus, yearly rainfall averages can be misleading.

Deserts display a wide variety of vegetation (Figure 15.19). Although the driest deserts, or those with large areas of shifting sand, are almost devoid of vegetation, most deserts support at least sparse plant cover. Compared to the vegetation in humid areas, desert vegetation may appear monotonous. A closer examination, however, reveals an amazing diversity of plants that have evolved the ability to live in the near absence of water.

Desert plants are widely spaced, typically small, and grow slowly. Their stems and leaves are usually hard and waxy to minimize water loss by evaporation and to protect the plant from sand erosion. Most plants have a widespread shallow root system to absorb the dew that forms each morning in all but the driest deserts and to help anchor the plant in what little soil there may be. In extreme cases, many plants lie dormant during particularly dry years and spring to life after the first rain shower with a beautiful profusion of flowers.

Weathering and Soils

Mechanical weathering is dominant in desert regions. Daily temperature fluctuations and frost wedging are the primary forms of mechanical weathering (see Chapter 6). The breakdown of rocks by roots and from salt crystal growth is of minor importance. Some chemical weathering does occur, but its rate is greatly reduced by aridity and the scarcity of organic acids produced by the sparse vegetation. Most chemical weathering takes place during the winter months when there is more precipitation, particularly in the midlatitude deserts.

An interesting feature seen in many deserts is a thin red, brown, or black shiny coating on the surface of many rocks (see Geo-inSight on pages 394 and 395). This coating, called *rock varnish,* is composed of iron and manganese oxides (Figure 15.20). Because many of the varnished rocks contain little or no iron and manganese oxides, the varnish is thought to result either from wind-blown iron and manganese dust that settles on the ground or from the precipitated waste of microorganisms.

Desert soils, if developed, are usually thin and patchy because the limited rainfall and the resultant scarcity of vegetation reduce the efficiency of chemical weathering and hence soil formation. Furthermore, the sparseness of the vegetative cover enhances wind and water erosion of what little soil actually forms.

Mass Wasting, Streams, and Groundwater

When traveling through a desert, most people are impressed by such wind-formed features as moving sand, sand dunes, and sand and dust storms. They may also notice the dry washes and dry streambeds. Because of the lack of running water, most people would conclude that wind is the most important erosional agent in deserts. They would be wrong! Running water, even though it occurs infrequently,

Reed Wicander

Figure 15.20 Rock Varnish The shiny black coating on this rock, exposed at Castle Valley, Utah, is rock varnish. It is composed of iron and manganese oxides.

Geo-Focus Windmills and Wind Power

Whoosh, whoosh, whoosh. Ah, the gentle sound of a windmill's blades turning in the wind. The image most people associate with windmills is one of a pastoral landscape in Holland dominated by a classic Dutch windmill crafted of wood (Figure 1), or perhaps Don Quixote tilting at windmills in the famous novel *Don Quixote de la Mancha* by Miguel Cervantes. Today, instead of a whoosh, whoosh, whoosh, the sound of modern electricity-generating windmills in a wind farm is more like a woomph, woomph, woomph (Figure 2).

As early as 5000 B.C., people began to harness the power of wind to propel boats along the Nile River. The Chinese used windmills to pump water for irrigating crops as long ago as 200 B.C. Wind power was used in the Middle Ages in Europe, particularly in Holland, where windmills have played an important role in society. Windmills were first used to grind corn, which is where the term "windmill" originally came from. Later windmills were used to drain lakes and marshes from the low-lying districts, and to saw timber. Settlers in the United States in the late 19th and early 20th centuries used this technology to pump water and generate electricity in the Great Plains.

With the application of stream power and industrialization in Europe and later the United States, the use of windmills rapidly declined. However, industrialization led to the development of

Larry Lee/Corbis

Figure 1 Five traditional Dutch windmills lined up along a canal at Kinderdiik, the Netherlands.

Parvinder Sethi

Figure 2 A windmill farm in Alameda County, California. California has more windmills than any other state in the United States.

causes most of the erosion in deserts. The dry conditions and sparse vegetation characteristic of deserts enhance water erosion. If you look closely, you can see the evidence of erosion and transportation by running water nearly everywhere except in areas covered by sand dunes.

Most of a desert's average annual rainfall of 25 cm or less comes in brief, heavy, localized cloudbursts. During these times, considerable erosion takes place because the ground cannot absorb all of the rainwater. With so little vegetation to hinder the flow of water, runoff is rapid, especially on moderately to steeply sloping surfaces, resulting in flash floods and sheet flows. Dry stream channels quickly fill with raging torrents of muddy water and mudflows, which carve out steep-sided gullies and overflow their banks. During these times, a tremendous amount of sediment is rapidly transported and deposited far downstream.

Although water is the major erosive agent in deserts today, it was even more important during the Pleistocene Epoch when these regions were more humid (see Chapter 23). During that time, many of the major topographic features of deserts were forming. Today that topography is modified by wind and infrequently flowing streams.

Most desert streams are poorly integrated and flow only intermittently. Many of them never reach the sea because the larger and more efficient windmills exclusively designed to generate electricity. Denmark began using such windmills as early as 1890, and other countries soon followed suit. Interest in electricity-generated windmills has always mirrored the price of fossil fuels. When the price of petroleum and coal is low, it is cheaper to use these fuels to generate electricity. When the price of fossil fuels goes up, interest in wind power also increases. Today, wind power is an important source of electricity, both in the United States and elsewhere around the world, particularly in Europe. As wind turbine technology has increased the efficiency of wind-generated electricity, the price has decreased significantly, so that wind farms can now compete in many areas with traditional fossil fuel-burning power plants.

How do windmills produce electricity? Simply stated, wind turbines (the term commonly used to describe electricity-producing windmills) convert the kinetic energy (the energy of an object due to motion) of the wind into mechanical power, in this case, the generation of electricity. This electricity is sent to the local power grid where it is distributed throughout the area.

In order to be effective, numerous wind turbines are clustered together in wind farms that are located in areas with relatively strong, steady winds. The number of turbines on wind farms can range from several to thousands, as in California (Figure 2). In addition to wind farms on land, wind farms offshore are becoming more popular because they are out of sight and people cannot hear the blades turning. In fact, wind farms in the North Sea already provide approximately 20% of Denmark's energy needs. In the United States, wind power production currently accounts for about 9% of the total power produced by all means.

What are the advantages and disadvantages of wind power? First of all, the wind is a free, renewable energy source, so it cannot be used up. It is also a clean source of energy that doesn't pollute the water or atmosphere, or contribute to greenhouse gases. Thus, it reduces the consumption of fossil fuels. The land on which windmills are sited can still be used for farming and ranching, thereby increasing the productivity of the land and providing an additional source of income to the landowner who leases the land to utilities. In addition, wind farms can benefit the local economy of rural and remote areas by supplying wind energy for local consumption.

There are some disadvantages to wind-generated electricity. The major disadvantage is that wind doesn't always blow with sufficient strength to be totally reliable, thereby necessitating backup generation. Furthermore, good wind sites are frequently located in remote areas, far from the areas where large quantities of electricity are needed, or in coastal areas, where land is expensive and local residents do not want large wind turbines as neighbors. The initial start-up cost of a wind farm is usually higher than the cost of a conventional power plant. However, as the cost of wind power has decreased because of better technology, wind-generated electricity can compete favorably with traditional power plants in many areas. The "not-in-my-backyard" opposition to wind frames can make siting a wind farm difficult. The major objection to wind farms is the noise generated by the turbines, although as the windmills are built taller and the turbines are more efficient at noise reduction, that is not the major concern it once was. Of course, aesthetics still plays a role in many people's objection to wind farms, and the fact that some birds are killed by the rotating blades is another issue. Studies have indicated, however, that the impact of wind turbines on bird mortality and injury is less than that of many other structures, such as buildings, power lines, and communication towers.

As the price of fossil fuel continues to rise, and our dependency on foreign sources of oil increases, the use of a centuries-old staple, the windmill, albeit modernized, will continue to gain in popularity and use.

water table is usually far deeper than the channels of most streams, so they cannot draw upon groundwater to replace water lost to evaporation and absorption into the ground. This type of drainage in which a stream's load is deposited within the desert is called *internal drainage* and is common in most arid regions.

Although most deserts have internal drainage, some deserts have permanent through-flowing streams, such as the Nile (see Chapter 12 Geo-Focus on page 307) and Niger rivers in Africa, the Rio Grande and Colorado River in the southwestern United States, and the Indus River in Asia. These streams can flow through desert regions because their headwaters are well outside the desert and water is plentiful enough to offset losses resulting from evaporation and infiltration. Demands for greater amounts of water from the Colorado River for agriculture and domestic use, however, are leading to increased salt concentrations in its lower reaches and causing political problems between the United States and Mexico.

Wind

Although running water does most of the erosional work in deserts, wind can also be an effective geologic agent capable

What Would You Do?

As an expert in desert processes, you have been assigned the job of teaching the first astronaut crew that will explore Mars all about deserts and their landforms. The reason is that many Martian features display evidence of having formed as a result of wind processes, and many Martian landforms are the same as those found in deserts on Earth. Describe how you would teach the astronauts to recognize wind-formed features and where on Earth you would take the astronauts to show them the types of landforms they may find on Mars.

of producing a variety of distinctive erosional (Figures 15.2 and 15.4) and depositional features (Figures 15.10 through 15.14). Wind is effective in transporting and depositing unconsolidated sand-,silt-, and dust-sized particles. Contrary to popular belief, most deserts are not sand-covered wastelands, but rather vast areas of rock exposures and desert pavement. Sand-covered regions, or sandy deserts, constitute less than 25% of the world's deserts. The sand in these areas has accumulated primarily by the action of wind.

Wind is not only an effective erosional and depositional agent in deserts, it is also becoming an important resource in generating electricity in many parts of the world (see GeoFocus on pages 398 and 399). The same wind that erodes, transports, and deposits materials is increasingly being harnessed to produce electricity for an energy-hungry world.

DESERT LANDFORMS

Because of differences in temperature, precipitation, and wind, as well as the underlying rocks and recent tectonic events, landforms in arid regions vary considerably. Running water, although infrequent in deserts, is responsible for producing and modifying many distinctive landforms found there.

After an infrequent and particularly intense rainstorm, excess water not absorbed by the ground may accumulate in low areas and form *playa lakes* (Figure 15.21a). These lakes are temporary, lasting from a few hours to several months. Most of them are shallow and have rapidly shifting boundaries as water flows in or leaves by evaporation and seepage into the ground. The water is often very saline.

When a playa lake evaporates, the dry lake bed is called a **playa** or *salt pan* and is characterized by mud cracks and precipitated salt crystals (Figure 15.21b, c). Salts in some playas are thick enough to be mined commercially. For example, borates have been mined in Death Valley, California, for more than 100 years.

Other common features of deserts, particularly in the Basin and Range Province of the United States, are alluvial fans and bajadas. **Alluvial fans** form when sediment-laden streams flowing out from the generally straight, steep mountain fronts deposit their load on the relatively flat desert floor. Once beyond the mountain front where no valley

Figure 15.21 Playas and Playa Lakes

John S. Shelton

a A playa lake formed after a rainstorm filled Croneis Dry Lake, Mojave Desert, California. Playa lakes are ephemeral features, lasting from a few hours to several months.

John S. Shelton

b A playa is the dry lake bed that remains after the water in a playa lake evaporates. Racetrack Playa, Death Valley, California. The Inyo Mountains can be seen in the background.

John Karachewski

c Salt deposits and salt ridges cover the floor of this playa in the Mojave Desert, California. Salt crystals and mud cracks are characteristic features of playas.

walls confine streams, the sediment spreads out laterally, forming a gently sloping and poorly sorted fan-shaped sedimentary deposit (Figure 15.22). Alluvial fans are similar in origin and shape to deltas (see Chapter 12), but are formed entirely on land. Alluvial fans may coalesce to form a *bajada,* a broad alluvial apron that typically has an undulating surface resulting from the overlap of adjacent fans (Figure 15.23).

Large alluvial fans and bajadas are frequently important sources of groundwater for domestic and agricultural use. Their outer portions are typically composed of fine-grained sediments suitable for cultivation, and their gentle slopes allow good drainage of water. Many alluvial fans and bajadas are also the sites of large towns and cities, such as San Bernardino, California; Salt Lake City, Utah; and Teheran, Iran.

What Would You Do?

You have been asked to testify before a congressional committee charged with determining whether the National Science Foundation should continue to fund research devoted to the study of climate changes during the Cenozoic Era. Your specialty is desert landforms and the formation of deserts. What arguments would you make to convince the committee to continue funding research on ancient climates?

Parvinder Sethi

Figure 15.22 Alluvial Fan A ground view of an alluvial fan, Death Valley, California. Alluvial fans form when sediment-laden streams flowing out from a mountain deposit their load on the desert floor, forming a gently sloping, fan-shaped, sedimentary deposit.

Most mountains in desert regions, including those of the Basin and Range Province, rise abruptly from gently sloping surfaces called pediments. **Pediments** are erosional bedrock surfaces of low relief that slope gently away from mountain bases (Figure 15.24). Most pediments are covered by a thin layer of debris, alluvial fans, or bajadas.

The origin of pediments has been the subject of much controversy. Most geologists agree that they are erosional features developed on bedrock in association with the erosion and retreat of a mountain front (Figure 15.24a). The

Marli Bryant Miller

Figure 15.23 Bajada Coalescing alluvial fans forming a bajada at the base of these mountains in Death Valley, California.

Figure 15.24 Pediment

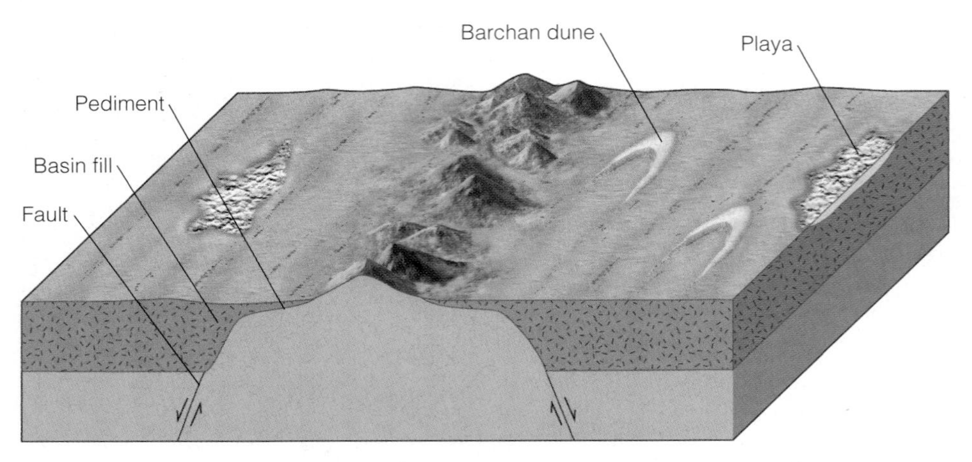

a Pediments are erosional bedrock surfaces formed by erosion along a mountain front.

b A pediment north of Mesquite, Nevada.

disagreement concerns how the erosion has occurred. Although not all geologists would agree, it appears that pediments are produced by the combined activities of lateral erosion by streams, sheet flooding, and various weathering processes along the retreating mountain front. Thus, pediments grow at the expense of the mountains, and they will continue to expand as the mountains are eroded away or partially buried.

Rising conspicuously above the flat plains of many deserts are isolated steep-sided erosional remnants called *inselbergs*, a German word meaning "island mountain." Inselbergs have survived for a longer period of time than other mountains because of their greater resistance to weathering. Uluṟu (formerly known as Ayers Rock) is an excellent example of an inselberg (see Chapter 17 Geo-inSight on pages 456 and 457).

Other easily recognized erosional remnants common to arid and semiarid regions are mesas and buttes (Figure 15.25). A **mesa** is a broad, flat-topped erosional remnant bounded on all sides by steep slopes. Continued weathering and stream erosion form isolated pillar-like structures known as **buttes.** Buttes and mesas consist of relatively easily weathered sedimentary rocks capped by nearly horizontal, resistant rocks such as sandstone, limestone, or basalt. They form when the resistant rock layer is breached, which allows rapid erosion of the less resistant underlying sediment. One of the best-known areas of mesas and buttes in the United States is Monument Valley on the Arizona–Utah border (Figure 15.25).

Figure 15.25 Buttes and Mesas

Royalty-Free/Corbis

a Several mesas and buttes can be seen in this aerial view of Monument Valley Navajo Tribal Park, Navajo Indian Reservation, Arizona. The mesas are the broad, flat-topped structures, one of which is prominent in the foreground, whereas the buttes are more pillar-like structures.

Parvinder Sethi

b Left Mitten Butte and Right Mitten Butte in Monument Valley on the border of Arizona and Utah.

Geo-Recap

Chapter Summary

- Desertification is the expansion of deserts into formerly productive lands. It destroys croplands and rangelands, causing massive starvation and forcing hundreds of thousands of people from their homelands.
- Wind transports sediment in suspension or as bed load. Suspended load is the material that is carried in suspension by water or wind. Silt- and clay-sized particles constitute most of a wind's suspended load. Bed load is the material that is too large or heavy to be carried in suspension by water or wind. Particles are moved along the surface by saltation, rolling, or sliding.
- Wind erodes material by either abrasion or deflation. Abrasion is the impact of saltating sand grains on an object. Ventifacts are common products of wind abrasion.
- Deflation is the removal of loose surface material by wind. Deflation hollows resulting from differential erosion of surface material are common features of many deserts, as is desert pavement, which effectively protects the underlying surface from additional deflation.
- The two major wind deposits are dunes and loess. Dunes are mounds or ridges of wind-deposited sand that form when wind flows over and around an obstruction, resulting in the deposition of sand grains, which accumulate and build up a deposit of sand. Loess is wind-blown silt and clay deposits composed of angular quartz grains, feldspar, micas, and calcite.
- The four major dune types are barchan, longitudinal, transverse, and parabolic. The amount of sand available, prevailing wind direction, wind velocity, and amount of vegetation determine which type of dune will form.
- Loess is derived from deserts, Pleistocene glacial outwash deposits, and river floodplains in semiarid regions. Loess covers approximately 10% of Earth's land surface and weathers to a rich, productive soil.
- The winds of the major air-pressure belts, oriented east-west, result from the rising and cooling of air. The winds are deflected clockwise in the Northern Hemisphere and counterclockwise in the Southern Hemisphere by the Coriolis effect to produce Earth's global wind patterns.
- Dry climates, located in the low and middle latitudes where the potential loss of water by evaporation exceeds the yearly precipitation, cover 30% of Earth's land surface and are subdivided into semiarid and arid regions. Semiarid regions receive more precipitation than arid regions, yet are moderately dry. Arid regions, generally described as deserts, are dry and receive less than 25 cm of rain per year.
- The majority of the world's deserts are in the low-latitude dry-climate zone between 20 and 30 degrees north and south latitudes. Their dry climate results from a high-pressure belt of descending dry air. The remaining deserts are in the middle latitudes, where their distribution is related to the rain-shadow effect, and in the dry polar regions.
- Deserts are characterized by high temperatures, little precipitation, and sparse plant cover. Rainfall is unpredictable and, when it does occur, tends to be intense and of short duration.
- Mechanical weathering is the dominant form of weathering in deserts, and coupled with slow rates of chemical weathering, results in poorly developed soils.
- Running water is the major agent of erosion in deserts and was even more important during the Pleistocene, when wetter climates resulted in humid conditions.
- Wind is an erosional agent in deserts and is very effective in transporting and depositing unconsolidated fine-grained sediments.
- Important desert landforms include playas, which are dry lakebeds; when temporarily filled with water, they form playa lakes. Alluvial fans are fan-shaped sedimentary deposits that may coalesce to form bajadas.
- Pediments are erosional bedrock surfaces of low relief that slope gently away from mountain bases.
- Inselbergs are isolated steep-sided erosional remnants that rise above the surrounding desert plains. Buttes and mesas are, respectively, pinnacle-like and flat-topped erosional remnants with steep sides.

Important Terms

abrasion (p. 385)
alluvial fan (p. 400)
barchan dune (p. 388)
butte (p. 402)
Coriolis effect (p. 391)
deflation (p. 386)
desert (p. 392)
desert pavement (p. 386)
desertification (p. 384)
dune (p. 387)
loess (p. 389)
longitudinal dune (p. 389)
mesa (p. 402)
parabolic dune (p. 389)
pediment (p. 401)
playa (p. 400)
rain-shadow desert (p. 396)
transverse dune (p. 389)
ventifact (p. 385)

Review Questions

1. The primary process by which bed load is transported is
   a. ______ suspension;
   b. ______ abrasion;
   c. ______ saltation;
   d. ______ precipitation;
   e. ______ answers a and c.
2. The Coriolis effect causes wind to be deflected
   a. ______ to the right in the Northern Hemisphere and to the left in the Southern Hemisphere;
   b. ______ to the left in the Northern Hemisphere and to the right in the Southern Hemisphere;
   c. ______ only to the left in both hemispheres;
   d. ______ only to the right in both hemispheres;
   e. ______ not at all.
3. The dominant form of weathering in deserts is ______, and the soils are ______.
   a. ______ mechanical, thick;
   b. ______ mechanical, thin;
   c. ______ mechanical, limited;
   d. ______ chemical, thick;
   e. ______ chemical, thin.
4. The major agent of erosion in deserts today is______, while during the Pleistocene Epoch, it was______.
   a. ______ wind, running water;
   b. ______ wind, wind;
   c. ______ running water, wind;
   d. ______ running water, running water;
   e. ______ wind, glaciers.
5. Which of the following is a crescent-shaped dune whose tips point downwind?
   a. ______ barchan;
   b. ______ longitudinal;
   c. ______ parabolic;
   d. ______ transverse;
   e. ______ barchanoid.
6. An important source of groundwater for domestic and agricultural use in arid regions is
   a. ______ alluvial fans;
   b. ______ playa lakes;
   c. ______ bajadas;
   d. ______ answers a and b;
   e. ______ answers a and c.
7. Where are the thickest and most extensive loess deposits in the world?
   a. ______ United States;
   b. ______ Pampas of Argentina;
   c. ______ Belgium;
   d. ______ Ukraine;
   e. ______ northeast China.
8. Coalescing alluvial fans form
   a. ______ buttes;
   b. ______ mesas;
   c. ______ bajadas;
   d. ______ inselbergs;
   e. ______ playas.
9. Which of the following is a feature produced by wind erosion?
   a. ______ playa;
   b. ______ loess;
   c. ______ dune;
   d. ______ yardang;
   e. ______ none of these.
10. The dry lakebeds found in many deserts are
    a. ______ playas;
    b. ______ bajadas;
    c. ______ inselbergs;
    d. ______ pediments;
    e. ______ mesas.
11. Is it possible for Mars to have the same type of sand dunes as Earth? What does that tell us about the climate and geology of Mars?
12. Using what you now know about deserts, their location, how they form, and the various landforms found in them, how can you determine where deserts may have existed in the past?
13. As noted in the text, some large cities are built on alluvial fans. What are the advantages and disadvantages of such an arrangement?
14. Why are so many desert rock formations red?
15. Why are low-latitude deserts so common?
16. If deserts are dry regions in which mechanical weathering predominates, why are so many of their distinctive landforms the result of running water and not of wind?
17. Why is desert pavement important in a desert environment?
18. How do dunes form and migrate? Why is dune migration a problem in some areas?
19. Much of the recent desertification has been greatly accelerated by human activity. Can you think of anything that can be done to slow the process?
20. What is loess, and why is it important?

CHAPTER 16

Shorelines and Shoreline Processes

OUTLINE

OBJECTIVES

At the end of this chapter, you will have learned that

- Wind-generated waves and their associated nearshore currents effectively modify shorelines by erosion and deposition.
- The gravitational attraction by the Moon and Sun and Earth's rotation are responsible for the rhythmic daily rise and fall of sea level known as tides.
- Seacoasts and lakeshores are both modified by waves and nearshore currents, but seacoasts also experience tides, which are insignificant in even the largest lakes.
- Distinctive erosional and depositional landforms such as wave-cut platforms, spits, and barrier islands are found along shorelines.
- The concept of a nearshore sediment budget considers equilibrium, losses, and gains in the amount of sediment in a coastal area.
- Several types of coasts are recognized based on criteria such as deposition and erosion and fluctuations in sea level.

This part of the Pacific Coast in California is rocky and rugged.

- Shoreline management is complicated by rising sea level, and structures originally far from the shoreline are now threatened or have already been destroyed.

INTRODUCTION

No doubt you know that a **shoreline** is the area of land in contact with the sea or a lake; however, we can expand this definition by noting that ocean shorelines include the land between low tide and the highest level on land affected by storm waves. How does a *shoreline* differ from a *coast*? Actually, the terms are commonly used interchangeably, but coast is a more inclusive term that includes the shoreline as well as an area of indefinite width both seaward and landward of the shoreline. For instance, in addition to the shoreline area, a coast includes nearshore sandbars and islands, and areas on land such as wind-blown sand dunes, marshes, and cliffs (Figure 16.1a).

Our main concern in this chapter is ocean shorelines, or seashores, but waves and nearshore currents are also effective in large lakes (Figure 16.1b). Waves and nearshore currents are certainly more vigorous along seashores, and even the largest lakes have insignificant tides. Lake Superior has a tidal rise and fall of only about 2.5 cm, whereas tidal fluctuations on seashores may be several meters.

You already know that the hydrosphere consists of all water on Earth, most of which is in the oceans. In this enormous body of water, wave energy is transferred through the water to shorelines where it has a tremendous impact. Accordingly, understanding the geologic processes operating on shorelines is important to many people. Indeed, many centers of commerce and much of Earth's population are concentrated in a narrow band at or near shorelines. In addition, coastal communities such as Myrtle Beach, South Carolina, Fort Lauderdale, Florida, and Padre Island, Texas, depend heavily on tourists visiting their beaches.

Geologists, oceanographers, coastal engineers, and marine biologists, among others, are interested in the dynamics of shorelines. So are elected officials and city planners of coastal communities because they must be familiar with shoreline processes in order to develop policies and zoning regulations that serve the public while protecting the fragile shoreline environment. Understanding shorelines is especially

Figure 16.1 Seashores and Lakeshores

James S. Monroe

a The shoreline of this part of the U.S. Pacific Coast consists of the area from about where the waves start to break to the base of the sea cliffs. The coast, however, extends farther seaward and also includes the sea cliffs and an area some distance inland.

James S. Monroe

b This feature called Miner's Castle in Lake Superior in Michigan formed by erosion just as similar features do along seashores. The turret of Miner's Castle on the right collapsed on April 13, 2006.

important now because in many areas sea level is rising, so buildings that were far inland are now in peril or have already been destroyed. Furthermore, hurricanes expend much of their energy on shorelines, resulting in extensive coastal flooding, numerous fatalities, and widespread property damage.

The study of shorelines provides another excellent example of systems interactions—in this case, between part of the hydrosphere and the solid Earth. The atmosphere is also involved in transferring energy from wind to water, thereby causing waves, which in turn generate nearshore currents. And, of course, the gravitational attraction of the Moon and Sun on ocean waters is responsible for the rhythmic rise and fall of tides. As dynamic systems, shorelines continuously adjust to any change that takes place, such as increased wave energy or an increase or decrease in sediment supply.

TIDES, WAVES, AND NEARSHORE CURRENTS

In contrast to other geologic agents such as running water, wind, and glaciers, which operate over vast areas, shoreline processes are restricted to a narrow zone at any particular time. However, shorelines might migrate landward or seaward depending on changing sea level or uplift or subsidence of coastal regions. During a rise in sea level, for instance, the shoreline migrates landward, and then wave, tide, and nearshore-current activity shifts landward as well; during times when sea level falls, just the opposite takes place. Recall from Chapter 6 that during marine transgressions and regressions, beach and nearshore sediments are deposited over vast regions (see Figure 6.22).

In the marine realm, several biological, chemical, and physical processes are operating continuously. Organisms change the local chemistry of seawater and contribute their skeletons to nearshore sediments, and temperature and salinity changes and internal waves occur in the oceans. However, the processes most important for modifying shorelines are purely physical ones, especially waves, tides, and nearshore currents. We cannot totally discount some of these other processes, though; offshore reefs composed of the skeletons of organisms, for instance, may protect a shoreline from most of the energy of waves.

Tides

The surface of the oceans rises and falls twice daily in response to the gravitational attraction of the Moon and Sun. These regular fluctuations in the ocean's surface, or **tides**, result in most seashores having two daily high tides and two low tides as sea level rises and falls anywhere from a few centimeters to more than 15 m (Figure 16.2). A complete tidal cycle includes a *flood tide* that progressively covers more and more of a nearshore area until high tide is reached, followed by *ebb tide*, during which the nearshore area is once again exposed (Figure 16.2). These regular fluctuations in sea level constitute one largely untapped source of energy as do waves, ocean currents, and temperature differences in seawater (see Geo-Focus on pages 416 and 417).

Both the Moon and the Sun have sufficient gravitational attraction to exert tide-generating forces strong enough to deform the solid body of Earth, but they have a much greater influence on the oceans. The Sun is 27 million times more massive than the Moon, but it is 390 times as far from Earth, and its tide-generating force is only 46% as strong as that of the Moon. Accordingly, the tides are dominated by the Moon, but the Sun plays an important role as well.

If we consider only the Moon acting on a spherical, water-covered Earth, its tide-generating forces produce two bulges on the ocean surface (Figure 16.3a). One bulge points toward the Moon because it is on the side of Earth where the Moon's gravitational attraction is greatest. The other bulge is on the opposite side of Earth; it points away from the Moon because of centrifugal force due to Earth's rotation, and the Moon's gravitational attraction is less. These two bulges always point toward and away from the Moon (Figure 16.3a), so as Earth rotates and the Moon's position changes, an observer at a particular shoreline location experiences the rhythmic rise and fall of tides twice daily, but the heights of two successive high tides may vary depending on the Moon's inclination with respect to the equator.

The Moon revolves around Earth every 28 days, so its position with respect to any latitude changes slightly each day. That is, as the Moon moves in its orbit and Earth rotates on its axis, it takes the Moon 50 minutes longer each day to return to the same position it was in the previous day. Thus, an observer would experience a high tide at 1:00 P.M. on one day, for example, and at 1:50 P.M. on the following day.

Tides are also complicated by the combined effects of the Moon and the Sun. Even though the Sun's tide-generating force is weaker than the Moon's, when the two are aligned every two weeks, their forces added together generate *spring tides* about 20% higher than average tides (Figure 16.3b). When the Moon and Sun are at right angles to each another, also at two-week intervals, the Sun's tide-generating force cancels some of the Moon's, and *neap tides* about 20% lower than average occur (Figure 16.3c).

Tidal ranges are also affected by shoreline configuration. Broad, gently sloping continental shelves as in the Gulf of Mexico have low tidal ranges, whereas steep, irregular shorelines experience much greater rise and fall of tides. Tidal ranges are greatest in some narrow, funnel-shaped bays and inlets. The Bay of Fundy in Nova Scotia has a tidal range of 16.5 m, and ranges greater than 10 m occur in several other areas.

Tides have an important impact on shorelines because the area of wave attack constantly shifts onshore and offshore as the tides rise and fall. Tidal currents themselves, however, have little modifying effect on shorelines, except in narrow passages where tidal current velocity is great enough to erode and transport sediment. Indeed, if it were not for strong tidal currents, some passageways would be blocked by sediments deposited by nearshore currents.

Waves

You can see **waves**, or oscillations of a water surface, on all bodies of water, but they are best developed in the oceans

a Low tide.

b High tide.

Figure 16.2 Low and High Tides Low tide a and high tide b in Turnagain Arm, part of Cook Inlet in Alaska. The tidal range here is about 10 m. Turnagain Arm is a huge fiord now being filled with sediment carried in by rivers. Notice the mudflats in a.

where they have their greatest impact on seashores. In fact, waves are directly or indirectly responsible for most erosion, sediment transport, and deposition in coastal areas. Wave terminology is illustrated with a typical series of waves in Figure 16.4a. A **crest**, as you would expect, is the highest part of a wave, whereas the low area between crests is a **trough**. The distance from crest to crest (or trough to trough) is the *wavelength,* and the vertical distance from trough to crest is *wave height.* You can calculate the speed at which a wave advances, called celerity (C), by the formula

$$C = L/T$$

where L is wavelength and T is wave period—that is, the time it takes for two successive wave crests, or troughs, to pass a given point.

The speed of wave advance (C) is actually a measure of the velocity of the wave form rather than the speed of the molecules of water in a wave. In fact, water waves are somewhat similar to waves moving across a grass-covered field; the grass moves forward and back as the wave passes, but the individual blades of grass remain in their original position. When waves move across a water surface, the water moves in circular orbits but shows little or no net forward movement (Figure 16.4a). Only the wave form moves forward, and as it does it transfers energy in the direction of wave movement.

The diameters of the orbits that water follows in waves diminish rapidly with depth, and at a depth of about one-half wavelength ($L/2$), called **wave base**, they are essentially zero. Thus, at a depth exceeding wave base, the water and seafloor or lake floor are unaffected by surface waves (Figure 16.4a). Wave base is an important consideration in some aspects of shoreline modification; we will explore it more fully in later sections of this chapter.

Wave Generation Several processes generate waves. Landslides into the oceans and lakes displace water and generate waves, and so do faulting and volcanic eruptions. Some waves so formed are huge and might be devastating to coastal areas, but most geologic work on shorelines is accomplished by wind-generated waves, especially storm waves. When wind blows over water—that is, one fluid (air) moves over another fluid (water)—friction between the two transfers energy to the water, causing the water surface to oscillate.

In areas where waves are generated, such as beneath a storm center at sea, sharp-crested, irregular waves called *seas* develop. Seas have various heights and lengths, and one wave cannot be easily distinguished from another. But as seas move out from their area of generation, they are sorted into broad *swells* with rounded, long crests and all are about the same size.

The harder and longer the wind blows, the larger are the waves. Wind velocity and duration, however, are not the only

a Tidal bulges if only the Moon caused them.

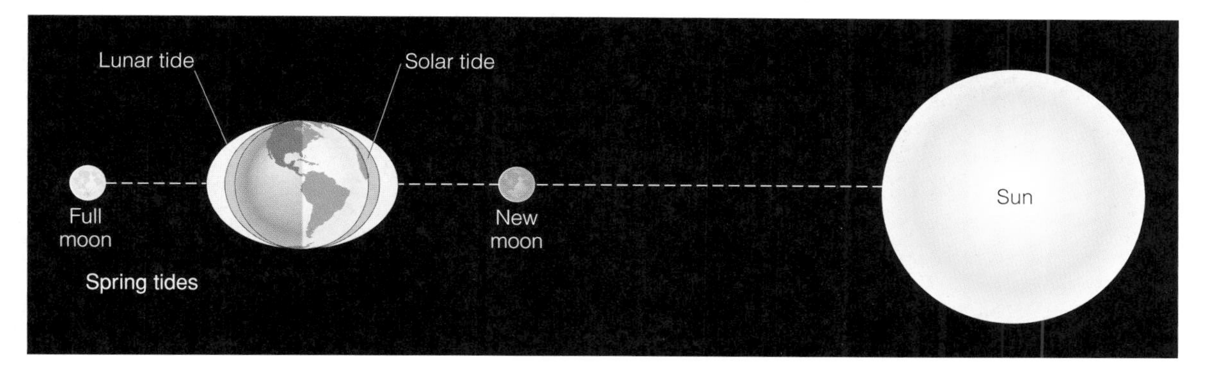

b When the Moon is new or full, the solar and lunar tides reinforce one another, causing spring tides, the highest high tides and lowest low tides.

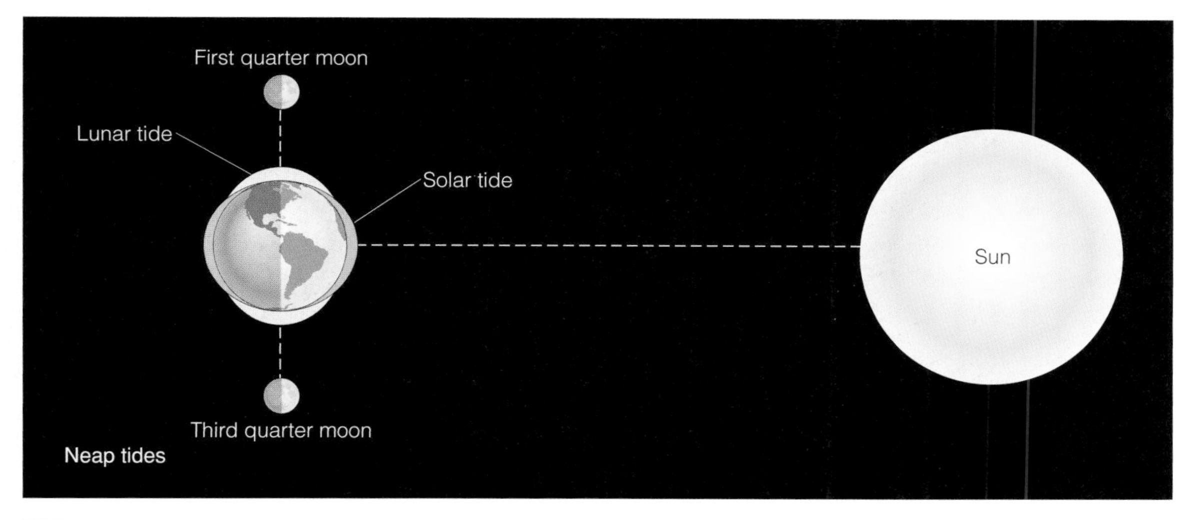

c During the Moon's first and third quarters, the Moon, Sun, and Earth form right angles, causing neap tides, the lowest high tides and highest low tides.

Figure 16.3 Tidal Bulges The gravitational attraction of the Moon and Sun cause tides. The sizes of the tidal bulges are greatly exaggerated.

factors that control wave size. High-velocity wind blowing over a small pond will never generate large waves regardless of how long it blows. In fact, waves on ponds and most lakes appear only while the wind is blowing; once the wind stops, the water quickly smooths out. In contrast, the surface of the ocean is always in motion, and waves with heights of 34 m have been recorded during storms in the open sea.

The reason for the disparity between wave sizes on ponds and lakes and on the oceans is the **fetch**, which is the distance the wind blows over a continuous water surface. Fetch is limited by the available water surface, so on ponds and lakes it corresponds to their length or width, depending on wind direction. To produce waves of greater length and height, more energy must be transferred from wind to water; hence large waves form beneath large storms at sea.

Waves with different lengths, heights, and period may merge, making them smaller or larger. Under some circumstances, two wave crests merge to form *rogue waves* that are three or four times higher than the average. These waves can rise unexpectedly out of an otherwise comparatively calm sea and threaten even the largest ships. During the last two decades, more than 200 supertankers and container ships have been lost at sea, many of them apparently hit by these huge waves. As recently as April 16, 2005, the Norwegian cruise

Figure 16.4 Waves and Wave Terminology

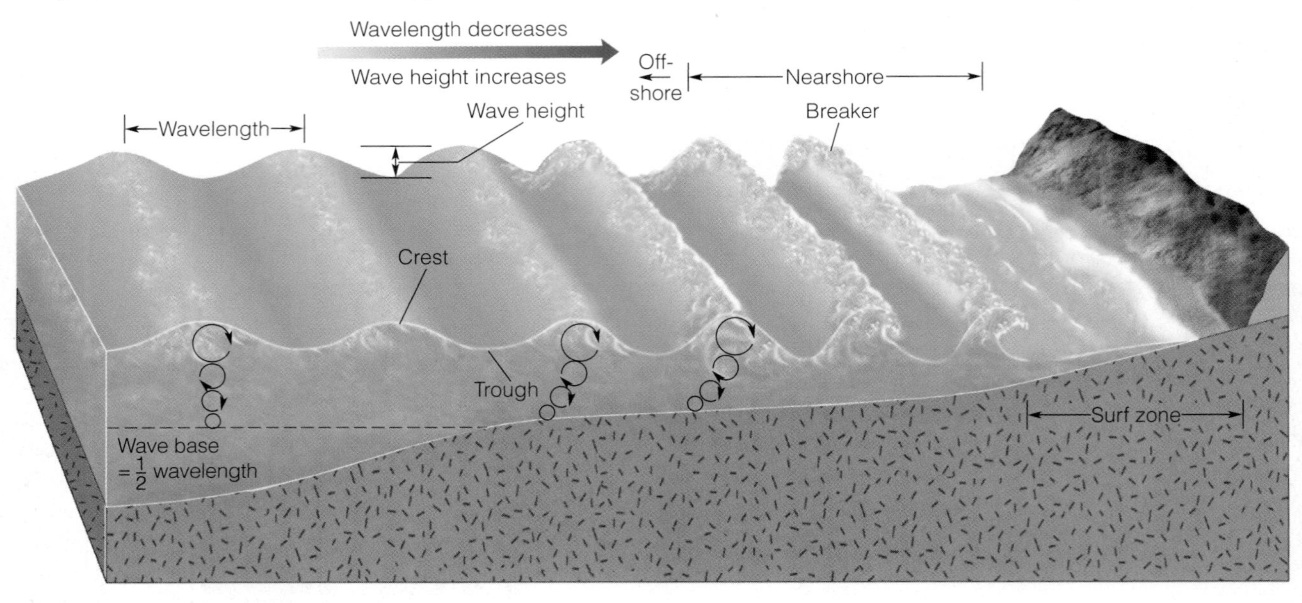

a Waves and the terminology applied to them.

Tom Garrison

b When swells in the deep water move toward the shore, the orbital motion of water within them is disrupted as they enter water shallower than wave base. Wavelength decreases while wave height increases, causing the waves to oversteepen and plunge forward as breakers.

Sue Monroe

c For scale, notice the surfers waiting for this wave that is beginning to break.

ship *Dawn* was damaged by a 21-m-high rogue wave that flooded more than 60 cabins and injured four passengers.

Shallow-Water Waves and Breakers Swells moving out from an area of wave generation lose little energy as they travel long distances across the ocean. In these deepwater swells, the water surface oscillates and water moves in circular orbits, but little net displacement of water takes place in the direction of wave travel (Figure 16.4a). Of course, wind blows some water from wave crests, thus forming whitecaps with foamy white crests, and surface currents transport water great distances; but deepwater waves themselves accomplish little actual water movement. When these waves enter progressively shallower water, however, the wave shape changes and water is displaced in the direction of wave advance.

Broad, undulating deepwater waves are transformed into sharp-crested waves as they enter shallow water. This transformation begins at a water depth corresponding to wave base—that is, one-half wavelength (Figure 16.5a). At this point, the waves "feel" the seafloor, and the orbital motion of water within the waves is disrupted (Figure 16.4a). As waves continue moving shoreward, the speed of wave advance and wavelength decrease, but wave height increases. Thus, as they enter shallow water, waves become oversteepened as the wave crest advances faster than the wave form, and eventually the crest plunges forward as a **breaker** (Figure 16.4c). Breaking waves might be several times higher than their deepwater counterparts, and when they break, they expend their kinetic energy on the shoreline.

The waves described above are the classic *plunging breakers* that crash onto shorelines with steep offshore slopes, such as those on the north shore of Oahu in the Hawaiian Islands (Figure 16.5b). In contrast, shorelines where the offshore slope is more gentle usually have *spilling breakers,* where the waves build up slowly and the wave's crest spills down the wave front (Figure 16.5c). Whether the breakers are plunging or spilling, the water rushes onto the shore and then returns seaward to become part of another breaking wave.

Nearshore Currents

The area extending seaward from the upper limit of the shoreline to just beyond the area of breaking waves is conveniently designated as the *nearshore zone.* Within the nearshore zone are the breaker zone and a surf zone, where water from breaking waves rushes forward and then flows seaward as backwash. The nearshore zone's width varies depending on the length of approaching waves because long waves break at a greater depth, and thus farther offshore, than do short waves. Incoming waves are responsible for two types of currents in the nearshore zone: *longshore currents* and *rip currents.*

What Would You Do?

While swimming at your favorite beach, you suddenly notice that you are far down the beach from the point where you entered the water. Furthermore, the size of the waves you were swimming in has diminished considerably. You decide to swim back to shore and then walk back to your original staring place, but no matter how hard you swim, you are carried farther and farther away from the shore. Assuming that you survive this incident, explain what happened and what you did to remedy the situation.

Wave Refraction and Longshore Currents Deepwater waves have long, continuous crests, but rarely are their crests parallel with the shoreline (Figure 16.6). In other words, they seldom approach a shoreline head-on, but rather at some angle. Thus, one part of a wave enters shallow water where it encounters wave base and begins breaking before other parts of the same wave. As a wave begins to break, its velocity diminishes, but the part of the wave still in deep

Figure 16.5 Wave Base and Breakers

James S. Monroe

a The waves in this lake have a wavelength of about 2 m, so we can infer that wave base is 1 m deep. Where the waves encounter wave base, they stir up the bottom sediment, thus accounting for the nearshore turbid water.

Tom Servais/Surfer Magazine

b A plunging breaker on the north shore of Oahu, Hawaii.

Tom Servais/Surfer Magazine

c A spilling breaker.

water races ahead until it too encounters wave base. The net effect of this oblique approach is that waves bend so that they more nearly parallel the shoreline, a phenomenon known as **wave refraction** (Figure 16.6).

Even though waves are refracted, they still usually strike the shoreline at some angle, causing the water between the breaker zone and the beach to flow parallel to the shoreline. These **longshore currents**, as they are called, are long and narrow and flow in the same general direction as the approaching waves. These currents are particularly important in transporting and depositing sediment in the nearshore zone.

Rip Currents Waves carry water into the nearshore zone, so there must be a mechanism for mass transfer of water back out to sea. One way in which water moves seaward from the nearshore zone is in **rip currents**, narrow surface currents that flow out to sea through the breaker zone (Figure 16.7). Surfers commonly take advantage of rip currents for an easy ride out beyond the breaker zone, but these currents pose a danger to inexperienced swimmers. Some rip currents flow at several kilometers per hour, so if a swimmer is caught in one, it is useless to try to swim directly back to shore. Instead, because rip currents are narrow and usually nearly perpendicular to the shore, one can swim parallel to the shoreline for a short distance and then turn shoreward with little difficulty.

Rip currents are circulating cells fed by longshore currents that increase in velocity from midway between each rip current (Figure 16.7a). When waves approach a shoreline, the amount of water builds up until the excess moves out to sea through the breaker zone.

Relief on the seafloor plays an important role in determining the location of rip currents. They commonly develop where wave heights are lower than in adjacent areas, and differences in wave height are controlled by variations in water depth. For instance, if waves move over a depression, the height of the waves over the depression tends to be less than in adjacent areas, forming the ideal environment for rip currents.

SHORELINE EROSION

Beaches, the most familiar coastal landforms, are absent, poorly developed, or restricted to protected areas on seacoasts where erosion rather than deposition predominates. Erosion creates steep or vertical slopes known as *sea cliffs.* During storms, these cliffs are pounded by waves (hydraulic action), worn by the impact of sand and gravel (abrasion) (Figure 16.8), and more or less continuously eroded by dissolution involving the chemical breakdown of rocks by the solvent action of seawater. Tremendous energy from waves is concentrated on the bases of sea cliffs and is most effective on those composed of sediments or highly fractured rocks. In any case, the net effect of these processes is erosion of the sea cliff and retreat of the cliff face landward.

Wave-Cut Platforms

Wave intensity and the resistance of shoreline materials to erosion determine the rate at which a sea cliff retreats landward. A sea cliff of glacial drift on Cape Cod, Massachusetts, erodes as much as 30 m per century, and some parts of the White Cliffs of Dover in England retreat landward at more than 100 m per century. By comparison, sea cliffs of dense igneous or metamorphic rocks erode and retreat much more slowly.

Sea cliffs erode mostly as a result of hydraulic action and abrasion at their bases. As a sea cliff is undercut by erosion, the upper part is left unsupported and susceptible to mass wasting processes. Thus, sea cliffs retreat little by little, and as they do, they leave a beveled surface called a **wave-cut platform** that slopes gently seaward (Figure 16.9). Broad wave-cut platforms exist in many areas, but invariably the water over them is shallow because

Figure 16.6 Wave Refraction Wave refraction (wave crests are indicated by dashed lines). These waves are refracted as they enter shallow water and more nearly parallel the shoreline. The waves generate a longshore current that flows in the direction of wave approach, from upper left to lower right (arrow) in this example.

Figure 16.7 **Rip Currents**

a Rip currents are fed on each side by currents moving parallel to the shoreline.

b Suspended sediment, indicated by discolored water, is being carried seaward in these rip currents.

c Sign along the California coast warning swimmers of dangerous rip currents, higher than normal (sleeper) waves, and backwash.

the abrasive planing action of waves is effective to a depth of only about 10 m. The sediment eroded from sea cliffs is transported seaward until it reaches deeper water at the edge of the wave-cut platform. There it is deposited and forms a *wave-built platform*, which is a seaward extension of the wave-cut platform (Figure 16.9). Wave-cut platforms now above sea level are known as **marine terraces** (Figure 16.9b).

Sea Caves, Arches, and Stacks

Sea cliffs do not retreat uniformly because some of the materials of which they are composed are more resistant to erosion than others. **Headlands** are seaward-projecting parts of the shoreline that are eroded on both sides by wave refraction (Figure 16.10). *Sea caves* form on opposite sides of a headland, and if these caves join, they form a *sea arch* (Figure 16.10a, b). Continued erosion causes the span of an arch to collapse, creating isolated *sea stacks* on wave-cut platforms (Figure 16.10a, c).

In the long run, shoreline processes tend to straighten an initially irregular shoreline. Wave refraction causes more wave energy to be expended on headlands and less on embayments. Thus, headlands erode, and some of the sediment yielded by erosion is deposited in the embayments.

DEPOSITION ALONG SHORELINES

In a previous section, we mentioned that longshore currents are effective at transporting sediment, and indeed they are. In fact, we can think of the area from the breaker zone to the upper limit of wave swash as a "river" that flows along the shoreline. Unlike rivers on land, though, its direction of flow changes if waves approach from a different direction. Nevertheless, the analogy is apt, and just like rivers on land, a longshore current's capacity for transport varies with flow velocity and water depth.

Wave refraction and the resulting longshore currents are the primary agents of sediment transport and deposition on shorelines, but tides also play a role because, as they rise and

Geo-Focus: Energy from the Oceans

If we could harness the energy of waves, ocean currents, temperature differences in oceanic waters, and tides, an almost limitless, largely nonpolluting energy supply would be ensured. Unfortunately, ocean energy is diffuse, meaning that the energy for a given volume of water is small and thus difficult to concentrate and use. Of the several sources of ocean energy, only tides show much promise for the near future.

Ocean thermal energy conversion (OTEC) exploits the temperature difference between surface waters and those at depth to run turbines and generate electricity. The amount of energy from this source is enormous, but a number of practical problems must be solved. For one thing, large quantities of water must be circulated through a power plant, which requires large surface areas devoted to this purpose. Despite several decades of research, only a few have been tested in Hawaii and Japan.

Ocean currents also possess energy that might be tapped to generate electricity. Unfortunately, these currents flow at only a few kilometers per hour at most, whereas hydroelectric power plants on land rely on water moving rapidly from higher to lower elevations. Furthermore, ocean currents cannot be dammed, their energy is diffuse, and any power plant would have to contend with unpredictable changes in flow direction.

Harnessing wave energy to generate electricity is not a new idea, and, in fact, it is used on a limited scale. Any facility using this form of energy would obviously have to be designed to withstand the effects of storms and saltwater corrosion. No large-scale wave-energy plants are operating, but the Japanese have developed devices to power lighthouses and buoys, and a facility with a capacity to provide power to approximately 300 homes began operating in Scotland in September 2000.

Tidal power has been used for centuries in some coastal areas to run mills, but its use at present for electrial generation is limited. One limitation is that the tidal range must be at last 5 m, and there must also be a coastal region where water can be stored following high tide. Suitable sites for tidal power plants are limited not only by tidal range but also by location.

Many areas along the U.S. Gulf Coast would certainly benefit from a tidal power plant, but the tidal range of generally less than 1 m precludes the possibility of development. However, an area with an appropriate tidal range in some remote area such as southern Chile or the Arctic islands of Canada offers little potential because of the great distance from population centers. Accordingly, in North America, only a few areas show much potential for developing tidal energy.

The idea behind tidal power is simple, although putting it into practice is not easy. First, a dam with sluice gates to regulate water flow must be built across the entrance to a bay or estuary. When the water level has risen sufficiently high during flood tide, the sluice gates are closed. Water held on the landward side of the dam is then released and electricity is generated

fall, the position of wave attack shifts onshore and offshore. Rip currents play no role in shoreline deposition, but they do transport fine-grained sediment (silt and clay) offshore through the breaker zone.

Beaches

Beaches are the most familiar coastal landforms, attracting millions of visitors each year and providing the economic base for many communities. Depending on shoreline materials and wave intensity, beaches may be discontinuous, existing as only *pocket beaches* in protected areas such as embayments, or they may be continuous for long distances.

By definition, a **beach** is a deposit of unconsolidated sediment extending landward from low tide to a change in topography, such as a line of sand dunes, a sea cliff, or the point where permanent vegetation begins. Typically, a beach has several component parts (see Geo-inSight on pages 424–425), including a **backshore** that is usually dry, being covered by water only during storms or exceptionally high tides. The backshore consists of one or more **berms**, platforms composed of sediment deposited by waves; the berms are nearly horizontal or slope gently landward. The sloping area below a berm exposed to wave swash is the **beach face**. The beach face is part of the **foreshore**, an area covered by water during high tide but exposed during low tide.

Some of the sediment on beaches is derived from weathering and wave erosion of the shoreline, but most of it is transported to the coast by streams and redistributed along the shoreline by longshore currents. As we noted, waves usually strike beaches at some angle, causing the sand grains to move up the beach face at a similar angle; as the sand grains are carried seaward in the backwash, however, they move perpendicular to the long axis of the beach. Thus, individual sand grains move in a zigzag pattern in the direction of longshore currents. This movement is not restricted to the beach; it extends seaward to the outer edge of the breaker zone.

In an attempt to widen a beach or prevent erosion, shoreline residents often build *groins*, structures that project seaward at right angles from the shoreline. Groins interrupt the flow of longshore currents, causing sand deposition on their upcurrent sides and widening of the beach at that just as it is at a hydroelectric dam. Actually, a tidal power plant can operate during both flood and ebb tides (Figure 1).

The first tidal power-generating facility was constructed in 1966 at the La Rance River estuary in France. In North America, a much smaller tidal power plant has been operating in the Bay of Fundy, Nova Scotia, where the tidal range, the greatest in the world, exceeds 16 m.

Although tidal power shows some promise, it will not solve our energy needs even if developed to its fullest potential. Most analysts think that only 100 to 150 sites worldwide have sufficiently high tidal ranges and the appropriate coastal configuration to exploit this energy resource. This, coupled with the facts that construction costs are high and tidal energy systems can have disastrous effects on the ecology (biosphere) of estuaries, makes it unlikely that tidal energy will ever contribute more than a small percentage of all energy production.

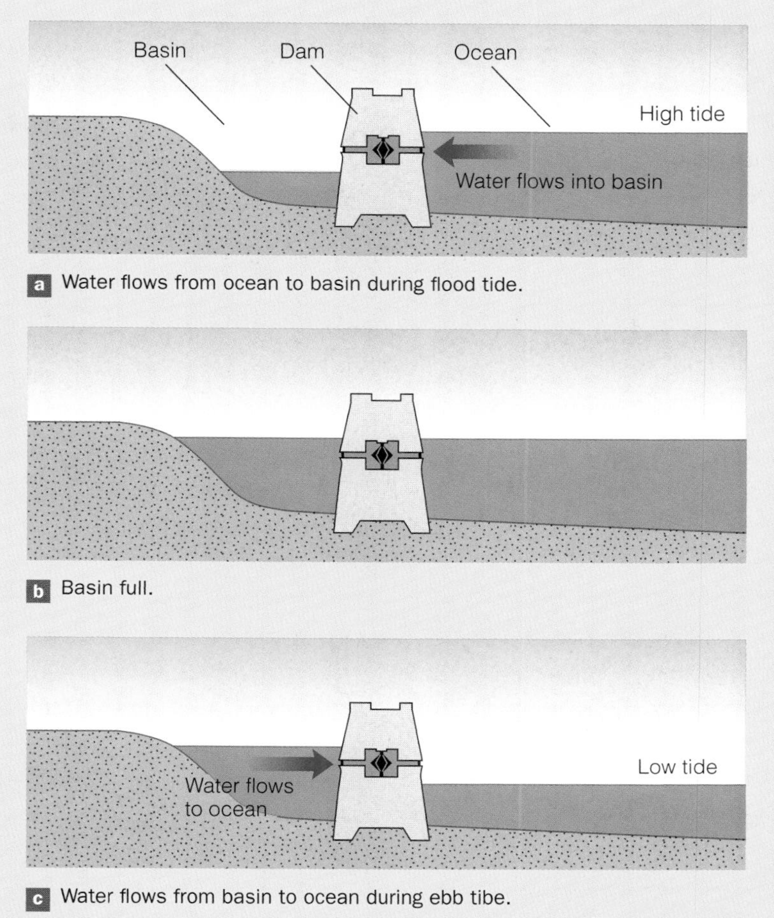

Figure 1 Rising and falling tides produce electricity by spinning turbines connected to generators, just as at hydroelectric plants. This view in the foreground is a cross section showing how water flows into and out of the basin, but the basin would actually be closed off here by land.

location. However, erosion inevitably occurs on the downcurrent side of a groin.

Quartz is the most common mineral in most beach sands, but there are some notable exceptions. For example, the black sand beaches of Hawaii are composed of sand- and gravel-sized basalt rock fragments or small grains of volcanic glass, and some Florida beaches are composed of the fragmented calcium carbonate shells of marine organisms. In short, beaches are composed of whatever material is available; quartz is most abundant simply because it is available in most areas and is the most durable and stable of the common rock-forming minerals.

Seasonal Changes in Beaches

The loose grains on beaches are constantly moved by waves, but the overall configuration of a beach remains unchanged as long as equilibrium conditions persist. We can think of the beach profile consisting of a berm or berms and a beach face, shown on pages 424 and 425, as a profile of equilibrium; that is, all parts of the beach are adjusted to the prevailing conditions of wave intensity, nearshore currents, and materials composing the beach.

Tides and longshore currents affect the configuration of beaches to some degree, but storm waves are by far the most important agent modifying their equilibrium profile. In many areas, beach profiles change with the seasons; so we recognize *summer beaches* and *winter beaches*, each of which is adjusted to the conditions prevailing at those times. Summer beaches are sand covered and have a wide berm, a gently sloping beach face, and a smooth offshore profile. Winter beaches, in contrast, tend to be coarser grained and steeper; they have a small berm or none at all, and their offshore profiles reveal sandbars paralleling the shoreline (Figure 16.11a).

Seasonal changes in beach profiles are related to changing wave intensity. During the winter, energetic storm waves erode the sand from beaches and transport it offshore where it is stored in sandbars (Figure 16.11b, c). The same sand that was eroded from a beach during the winter returns the next summer when it is driven onshore by more gentle swells. The

Figure 16.8 Wave Erosion by Abrasion and Hydraulic Action

a The rocks in the lower part of this image on a small island in the Irish Sea have been smoothed by abrasion, but the rocks higher up are out of the reach of waves.

b These waves pounding the California coast carry sand and gravel and effectively erode by abrasion, that is, the impact of solid particles against shoreline materials. In many areas sea cliffs undercut by abrasion fail by mass wasting processes (see Figure 11.3).

volume of sand in the system remains more or less constant; it simply moves farther offshore or onshore depending on wave energy.

The terms *winter beach* and *summer beach*, though widely used, are somewhat misleading. A winter beach profile can develop at any time if there is a large storm, and likewise a summer beach profile can develop during a prolonged winter calm period.

Spits, Baymouth Bars, and Tombolos

Beaches are the most familiar depositional features of coasts, but spits, baymouth bars, and tombolos are common, too. In fact, these features are simply continuations of a beach. A **spit,** for instance, is a fingerlike projection of a beach into a body of water such as a bay, and a **baymouth bar** is a spit that has grown until it completely closes off a bay from the open sea (Figure 16.12). Both are composed of sand, more rarely gravel, that was transported and deposited by longshore currents where they weakened as they entered the deeper water of a bay's opening. Some spits are modified by waves so that their free ends are curved; they go by the name *hook* or *recurved spit* (Figure 16.12a)

A rarer type of spit is a **tombolo** that extends out from the shoreline to an island (Figure 16.13). A tombolo forms on the shoreward side of an island as wave refraction around the island creates converging currents that turn seaward and deposit a sandbar. So, in contrast to spits and baymouth bars, which are usually nearly parallel with the shoreline, the long axes of tombolos are nearly at right angles to the shoreline.

Spits, baymouth bars, and tombolos are most common along irregular seashores, but they can also be found in large lakes. Regardless of their setting, spits and baymouth bars, especially, constitute a continuing problem where bays must be kept open for pleasure boating, commercial shipping, or both. Obviously, a bay closed off by a sandbar is of little use for either endeavor, so a bay must be regularly dredged or protected from deposition by longshore

Figure 16.9 Origin of a Wave-Cut Platform

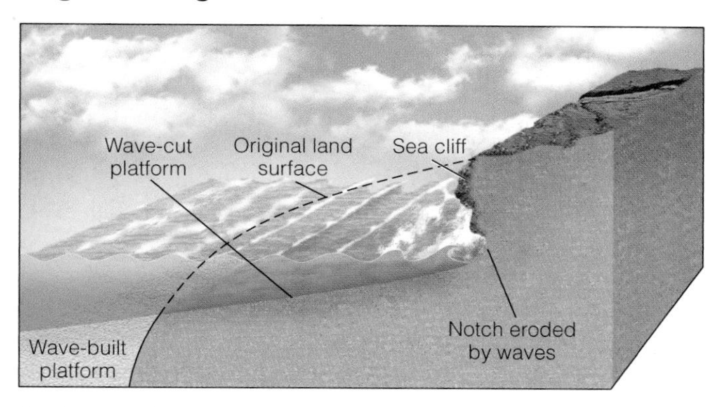

a Wave erosion causes a sea cliff to migrate landward, leaving a gently sloping surface, called a wave-cut platform. A wave-built platform originates by deposition at the seaward margin of the wave-cut platform.

b This gently sloping surface along the coast of California is a marine terrace. Notice the sea stacks rising above the terrace.

Figure 16.10 Erosion of a Headland

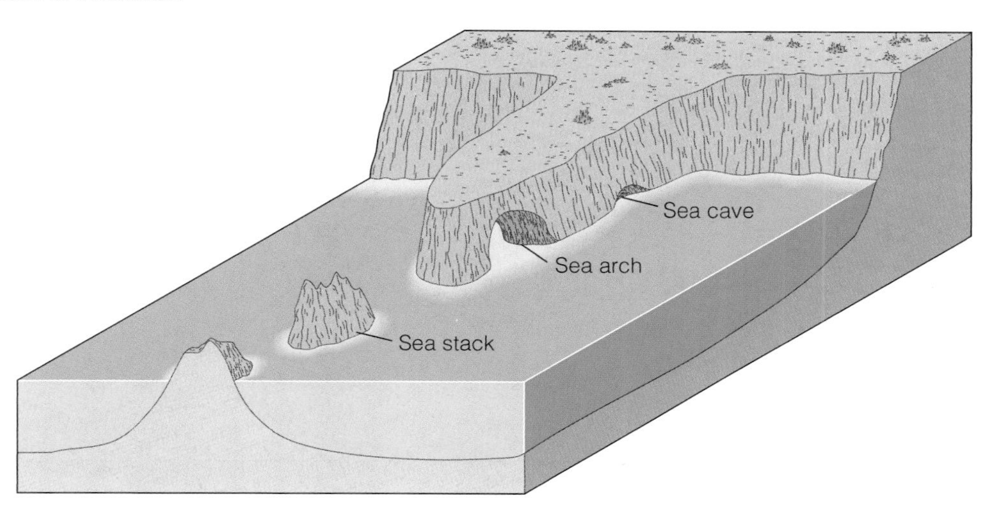

a Erosion of a headland and the origin of a sea cave, a sea arch, and sea stacks.

b This sea stack in Australia has an arch that developed in it.

c Sea stacks at Shell Beach along the California coast.

What Would You Do?

While visiting a barrier island, you notice some fine-grained, organic-rich deposits along the beach. In fact, there is so much organic matter that the sediments are black and have a foul odor. Their presence on the beach does not seem to make sense because you know that sediments like these were almost certainly deposited in a marsh on the landward (opposite) side of the island. How can you explain your observations?

currents. In some areas, *jetties* are constructed that extend seaward (or lakeward) to interrupt the flow of longshore currents and thus protect the opening to a bay.

Barrier Islands

Long, narrow islands of sand lying a short distance offshore from the mainland are **barrier islands** (Figure 16.14). On their seaward sides, they are smoothed by waves, but their landward margins are irregular because storm waves carry sediment over the island and deposit it in a lagoon where it is little modified by further wave activity. The component parts of a barrier island include a beach, wind-blown sand dunes, and a marshy area on their landward sides.

Everyone agrees that barrier islands form on gently sloping continental shelves where abundant sand is available and where both wave energy and the tidal range are low. In fact, these are the reasons that so many are along the United States' Atlantic and Gulf Coasts. But even though it is well known where barrier islands form, the details of their origin are still unresolved. According to one model, they formed as spits that became detached from land, whereas another model holds that they formed as beach ridges that subsequently subsided.

Most barrier islands are migrating landward as a result of erosion on their seaward sides and deposition on their landward sides. This is a natural part of barrier island

Figure 16.11 Seasonal Changes in Beaches

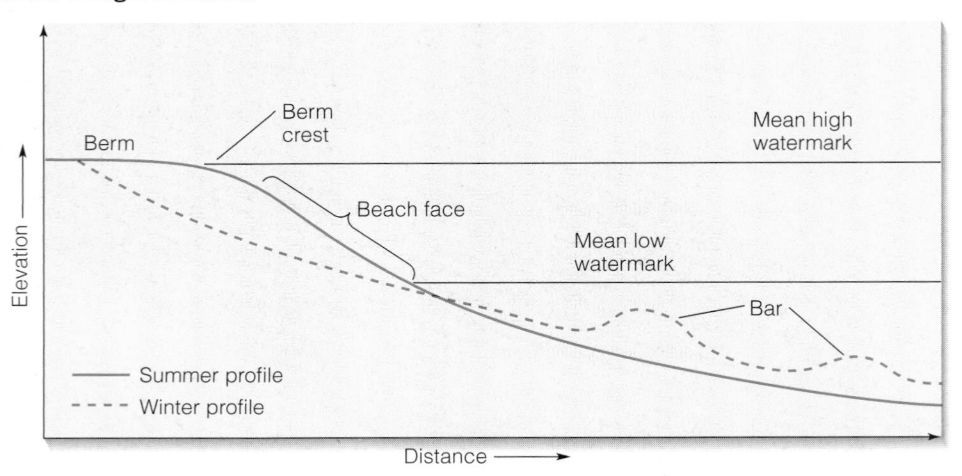

a Seasonal changes in beach profiles.

b **c**

San Gregorio State Beach, California. These photographs were taken from nearly the same place, but **c** was taken two years after **b**. Much of the change from **b** to **c** can be accounted for by beach erosion during 1997–1998 winter storms.

Figure 16.12 Spits and Baymouth Bars

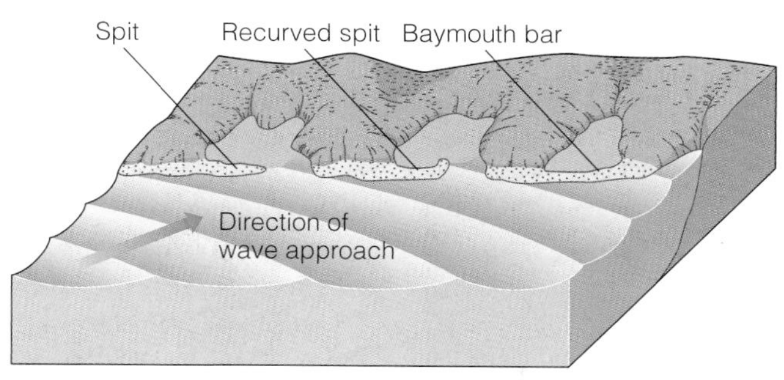

a Spits form where longshore currents deposit sand in deeper water, as at the entrance to a bay. A baymouth bar is simply a spit that has grown until it extends across the mouth of a bay.

b A spit at the mouth of the Russian River near Jenner, California.

c Rodeo Beach north of San Francisco, California, is a baymouth bar.

Figure 16.13 Tombolos and Breakwaters

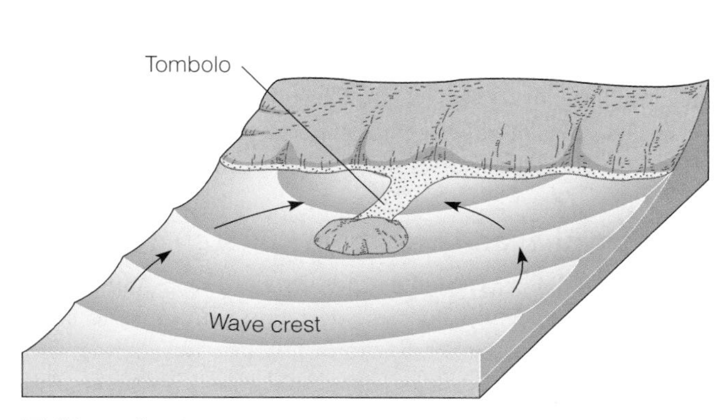

a Wave refraction around an island and the origin of a tombolo.

b A small tombolo along the Pacific coast of the United States.

Figure 16.14 Barrier Islands

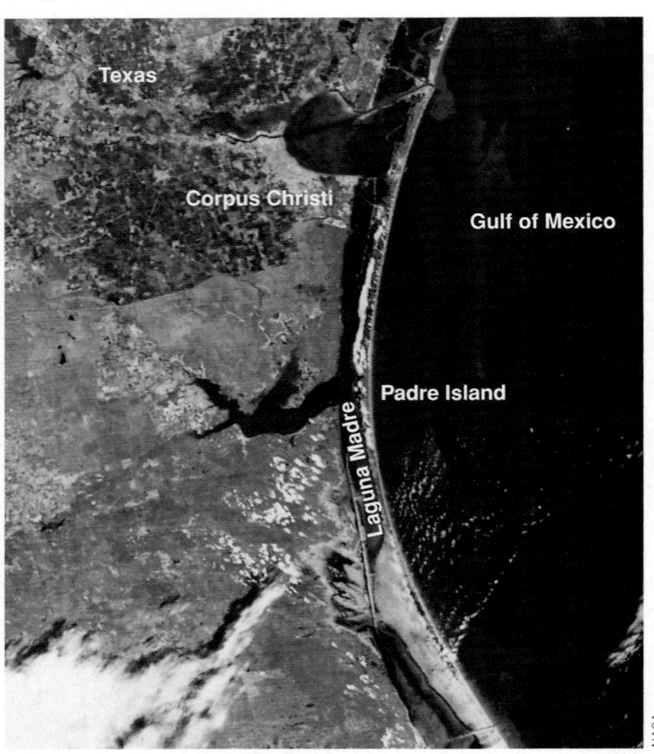

a View from space of the barrier islands along the Gulf Coast of Texas. Notice that a lagoon up to 20 km wide separates the long, narrow barrier islands from the mainland.

b Aerial view of Padre Island on the Texas Gulf Coast. Laguna Madre is on the left, and the Gulf of Mexico is on the right.

evolution, and it takes place rather slowly. However, it takes place fast enough to cause many problems for island residents and communities.

THE NEARSHORE SEDIMENT BUDGET

We can think of the gains and losses of sediment in the nearshore zone in terms of a **nearshore sediment budget** (Figure 16.15). If a nearshore system has a balanced budget, sediment is supplied to it as fast as it is removed, and the volume of sediment remains more or less constant, although sand may shift offshore and onshore with the changing seasons (Figure 16.11). A positive budget means that gains exceed losses, whereas a negative budget means that losses exceed gains. If a negative budget prevails long enough, a nearshore system is depleted and beaches may disappear.

Erosion of sea cliffs provides some sediment to beaches, but in most areas probably no more than 5–10% of the total sediment comes from this source. There are exceptions, though; almost all of the sediment on the beaches of Maine is derived from the erosion of shoreline rocks. Most sediment on typical beaches is transported to the shoreline by streams and then redistributed along the shoreline by longshore currents. Thus, longshore currents also play a role in the nearshore sediment budget because they continuously move sediment into and away from beach systems.

The primary ways that a nearshore system loses sediment are offshore transport, wind, and deposition in submarine canyons. Offshore transport mostly involves fine-grained sediment carried seaward where it eventually settles in deeper water. Wind is an important process because it removes sand from beaches and blows it inland where it piles up as sand dunes.

If the heads of submarine canyons are nearshore, huge quantities of sand are funneled into them and deposited in deeper water. La Jolla and Scripps submarine canyons off the coast of southern California funnel off an estimated 2 million m^3 of sand each year. In most areas, however, submarine canyons are too far offshore to interrupt the flow of sand in the nearshore zone.

It should be apparent from this discussion that if a nearshore system is in equilibrium, its incoming supply of sediment exactly offsets its losses. Such a delicate balance tends to continue unless the system is somehow disrupted. One change that affects this balance is the construction of dams across the streams that supply sand. Once dams have been built, all sediment from the upper reaches of the drainage systems is trapped in reservoirs and thus cannot reach the shoreline.

TYPES OF COASTS

Coasts are difficult to classify because of variations in the factors that control their development and variations in their composition and configuration. Rather than attempt to categorize all coasts, we shall simply note that two types of coasts have already been discussed: those dominated by deposition and those dominated by erosion.

Depositional and Erosional Coasts

Depositional coasts, such as the U.S. Gulf Coast, are characterized by an abundance of detrital sediment and such depositional landforms as wide sandy beaches, deltas, and barrier islands. In contrast, erosional coasts are steep and

Figure 16.15 The Nearshore Sediment Budget The long-term sediment budget can be assessed by considering inputs versus outputs. If inputs and outputs are equal, the system is in a steady state, or equilibrium. If outputs exceed inputs, however, the beach has a negative budget and erosion occurs. Accretion takes place when the beach has a positive budget with inputs exceeding outputs.

irregular and typically lack well-developed beaches, except in protected areas (see Geo-inSight on pages 424 and 425). They are further characterized by sea cliffs, wave-cut platforms, and sea stacks. Many of the coasts along the West Coast of North America fall into this category.

The following section will examine coasts in terms of their changing relationship to sea level. But note that although some coasts, such as those in southern California, are described as emergent (uplifted), these same coasts may be erosional as well. In other words, coasts commonly possess features that allow them to be classified in more than one way.

Submergent and Emergent Coasts

If sea level rises with respect to the land or the land subsides, coastal regions are flooded and said to be **submergent** or *drowned* **coasts** (Figure 16.16). Much of the East Coast of North America from Maine southward through South Carolina was flooded during the rise in sea level following the Pleistocene Epoch, so it is extremely irregular. Recall that during the expansion of glaciers during the Pleistocene, sea level was as much as 130 m lower than at present, and that streams eroded their valleys more deeply and extended across continental shelves. When sea level rose, the lower ends of these valleys were drowned, forming *estuaries* such as Delaware and Chesapeake bays (Figure 16.16). Estuaries are simply the seaward ends of river valleys where seawater and freshwater mix. The divides between adjacent drainage systems on submergent coasts project seaward as broad headlands or a line of islands.

Submerged coasts are also present at higher latitudes where Pleistocene glaciers flowed into the sea. When sea level rose, the lower ends of the glacial troughs were drowned, forming fiords (see Chapter 14 Geo-inSight on pages 370 and 371).

Geo-inSight Shoreline Processes and Beaches

Beaches are the most familiar depositional landforms along shorelines. They are found in a variety of sizes and shapes, with long sandy beaches typical of the East and Gulf Coasts, and smaller, mostly protected beaches along the West Coast. The sand on most beaches is primarily quartz, but there are some notable exceptions, shell sand beaches in Florida and black sand beaches in Hawaii. Beaches are dynamic systems where waves, tides, and marine currents constantly bring about change.

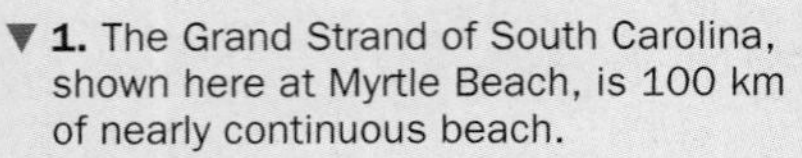

▼ **1.** The Grand Strand of South Carolina, shown here at Myrtle Beach, is 100 km of nearly continuous beach.

Micheal Slear

Parvinder Sethi

▲ **2.** Small pocket beach at Big Sur, California.

▼ **3.** Diagram of a beach showing its component parts.

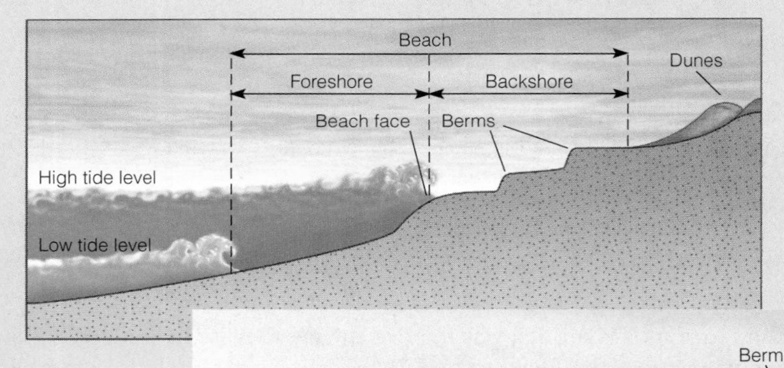

James S. Monroe

► **4.** The backshore area of a pocket beach along the Pacific coast. Note that the berm ends at the rocks on the right and the beach face slopes steeply seaward.

James S. Monroe

▲ **5.** The origin of beach cusps like these at Mykanos, Greece, is poorly known, but they are very common.

▼ **6,** ▶ **7.** Longshore currents transport sediment along the shoreline, a phenomenon called longshore drift, between the breaker zone and the upper limit of wave action. These groins (right) at Cape May, New Jersey, interrupt the flow of longshore currents. Sand is trapped on their upcurrent sides, whereas erosion takes place on their downcurrent sides. The longshore currents moved toward the top of the image.

Jhon S. Shelton

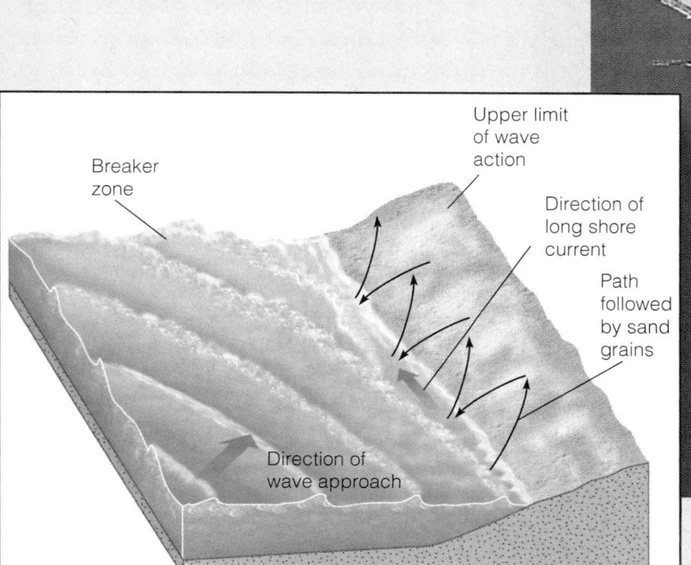

◀ **8.** Although most beaches are made up predominantly of quartz sand, there are exceptions. This black sand beach on Maui in Hawaii is made up of small basalt rock fragments and particles of obsidian that formed when lava flowed into the sea.

Sue Monroe

Sue Monroe

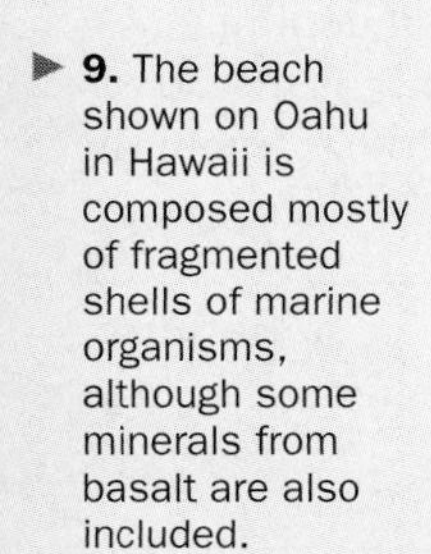

▶ **9.** The beach shown on Oahu in Hawaii is composed mostly of fragmented shells of marine organisms, although some minerals from basalt are also included.

James S. Monroe

▲ **10.** This California beach is made up of quartz and sand- and gravel-sized rock fragments.

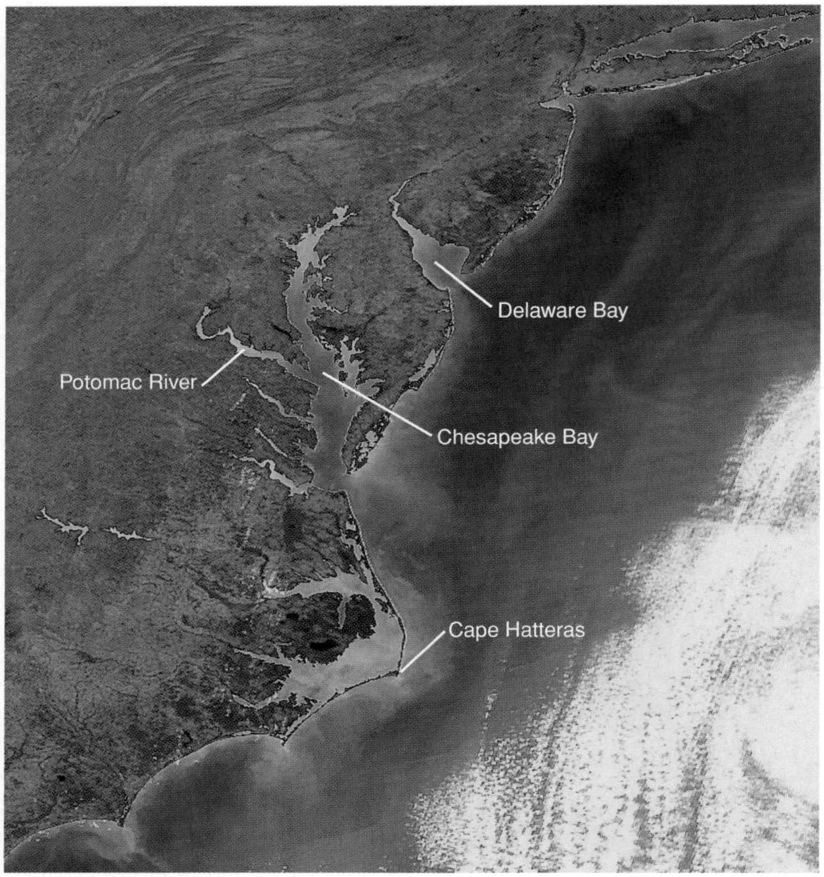

Figure 16.16 Submergent Coasts Submergent coasts tend to be extremely irregular, with estuaries such as Chesapeake and Delaware bays. They formed when the East Coast of the United States was flooded as sea level rose following the Pleistocene Epoch.

Emergent coasts are found where the land has risen with respect to sea level (Figure 16.17). Emergence takes place when water is withdrawn from the oceans, as occurred during the Pleistocene expansion of glaciers. Presently, coasts are emerging as a result of isostasy or tectonism. In northeastern Canada and the Scandinavian countries, for instance, the coasts are irregular because isostatic rebound is elevating formerly glaciated terrain from beneath the sea.

Coasts that form in response to tectonism, on the other hand, tend to be straighter because the seafloor topography being exposed by uplift is smooth. The west coasts of North and South America are rising as a consequence of plate tectonics. Distinctive features of these coasts are marine terraces (Figures 16.9b and 16.17), which are wave-cut platforms now elevated above sea level. Uplift in these areas appears to be episodic rather than continuous, as indicated by the multiple levels of terraces in some places. In southern California, several terrace levels are present, each of which probably represents a period of tectonic stability followed by uplift. The highest terrace is now about 425 m above sea level.

STORM WAVES AND COASTAL FLOODING

What causes most fatalities during hurricanes? Many people think it is strong wind, but actually, coastal flooding caused by storm waves and heavy rainfall are more dangerous. Coastal flooding during hurricanes results when large waves are driven onshore, and by as much as 60 cm of rain in as little as 24 hours. In addition, as a hurricane moves over the ocean, low atmospheric pressure causes the ocean surface to bulge upward as much as 0.5 m. When the eye of the storm reaches shoreline, the bulge, coupled with wind-driven waves, piles up in a **storm surge** that may rise several meters above normal high tide and inundate areas far inland.

In 1900, hurricane-driven waves surged inland, eventually covering the entire island that Galveston, Texas, was built on. Buildings and other structures near the seashore were battered to pieces and "great beams and railway ties were lifted by the [waves] and driven like battering rams into dwellings and business houses."* Between 6000 and 8000 people died. In an effort to protect the city from future storms, a huge seawall was constructed, and the entire city was elevated to the level of the top of the seawall (Figure 16.18).

Although Galveston, Texas, has been largely protected from more recent storm surges, the same is not true for several other areas. In 1989, Charleston, South Carolina, and nearby areas were flooded by a storm surge generated by Hurricane Hugo, which caused 21 deaths and more than $7 billion in property damages. Bangladesh is even more susceptible to storm surges; in 1970, 300,000 people drowned, and in 1991, another 130,000 were lost. Large-scale coastal flooding took place when Hurricane Isabel hit the Outer Banks of North Carolina in 2003. And in 2004, four hurricanes hit Florida and parts of the Gulf Coast, causing widespread wind damage and coastal flooding.

The examples just cited were certainly disasters of great magnitude, but the U.S. Gulf Coast was hardest hit in 2005, first in August by Hurricane Katrina and then again in September by Hurricane Rita. When Hurricane Katrina roared ashore on August 29, high winds, a huge storm surge, and coastal flooding destroyed nearly everything in an area

*L. W. Bates, Jr., "Galveston—A City Built upon Sand," *Scientific American*, 95 (1906), p. 64.

Sue Monroe

Figure 16.17 An Emergent Coast in California Emergent coasts tend to be steep and straighter than submergent coasts. Notice the several sea stacks and the sea arch. Also, a marine terrace is visible in the distance.

of 230,000 km^2. Gulfport and Biloxi, Mississippi were mostly leveled, but most of the public's attention has focused on the damage in New Orleans, Louisiana (Figure 16.19).

Even when New Orleans was founded in 1718, engineers warned that building a community on a swampy, subsiding parcel of land between Lake Pontchartrain to the north and the Mississippi River to the south was risky; now most of the city lies below sea level. In any case, New Orleans was nearly surrounded by levees (earthen embankments) and floodwalls (structures of concrete and steel) to protect it from the lake and river.

When Hurricane Katrina came ashore, the levees initially held, but on the next day some of the floodwalls were breached and about 80% of the city was flooded. And because New Orleans is mostly below sea level, the floodwaters could not drain out naturally. In fact, the city has 22 pumping stations to remove water from normal rainstorms, but as the city flooded, the pumps were overwhelmed, and when the electricity failed, the pumps were useless. All in all, Hurricane Katrina was the most expensive natural disaster in U.S. history; the property damages exceeded $100 billion and more than 1800 people died. Then, in September, Hurricane Rita hit the Gulf Coast, mostly in Texas, and caused about 120 more fatalities and nearly $12 billion in damages.

Scientists, engineers, and some politicians had warned of a Katrina-type disaster for New Orleans for many years. They were aware that during a large hurricane the levees or floodwalls would likely fail. In fact, the political leaders in Louisiana had pleaded for years for funds to strengthen the levees, but for one reason or another, the funds were never allocated. "What do we do now?" is a question that will be debated for many years. Another factor to consider is that coastal flooding is exacerbated by rising sea level that results from global warming. Most of you reading this book will not become geologists, but perhaps you will be engineers, city planners, members of planning commissions, politicians, or simply concerned citizens of coastal communities and will have to deal with these or similar problems.

Rosenberg Library, Galveston, Texas

Figure 16.18 Seawall Construction Construction on this seawall to protect Galveston, Texas, from storm waves began in 1902. Notice that the wall is curved to deflect waves upward.

HOW ARE COASTAL AREAS MANAGED AS SEA LEVEL RISES?

During the last century, sea level rose about 12 cm worldwide, and all indications are that it will continue to rise. The absolute rate of sea-level rise in a shoreline region depends on two factors. The first is the volume of water in the ocean basins, which is increasing as a result of glacial ice melting and the

Figure 16.19 Hurricane Katrina, 2005

NOAA

a On August 29, 2005, the Hurricane Katrina made landfall along the Gulf Coast. In this image, the eye of the storm is passing just to the east of New Orleans, Louisiana.

Denis Paquin/AP/World Wide Photos

c Destruction caused by the storm surge from Hurricane Katrina. This image shows a man in Biloxi, Mississippi, trying to find his house.

Manahem Kahana/AFP/Getty Images

b Although many areas were hit hard by Katrina, New Orleans was extensively flooded when the floodwalls failed that were built to protect the city from Lake Pontchartrain and the Mississippi River.

Sue Monroe

d This is a repaired section of floodwall in New Orleans. This barrier is along one of the canals leading into the city from Lake Pontchartrain. Failure of the floodwalls during and following Hurricane Katrina caused flooding of about 80% of New Orleans.

thermal expansion of near-surface seawater. Many scientists think that sea level will continue to rise because of global warming caused by increasing concentrations of greenhouse gases in the atmosphere.

The second factor that controls sea level is the rate of uplift or subsidence of a coastal area. In some areas, uplift is occurring fast enough that sea level is actually falling with respect to the land. In other areas, sea level is rising while the coastal region is simultaneously subsiding, resulting in a net change in sea level of as much as 30 cm per century. Perhaps such a "slow" rate of sea level change seems insignificant; after all, it amounts to only a few millimeters per year. But in gently sloping coastal areas, as in the eastern United States from New Jersey southward, even a slight rise in sea level will eventually have widespread effects.

Many of the nearly 300 barrier islands along the East and Gulf Coasts of the United States are migrating landward as sea level rises (Figure 16.20). Landward migration of barrier islands would pose few problems if it were not for the numerous communities, resorts, and vacation homes located on them. Moreover, barrier islands are not the only threatened areas. For example, Louisiana's coastal wetlands, an important wildlife habitat and seafood-producing area, are currently

Figure 16.20 Barrier Island Migration

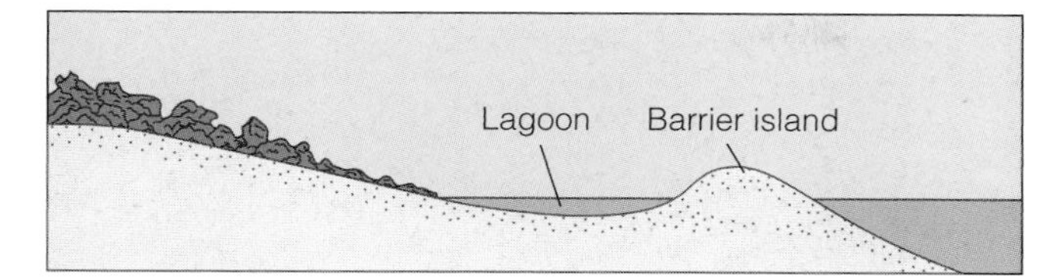

a A barrier island.

b A barrier island migrates landward as sea level rises and storm waves carry sand from its seaward side into its lagoon.

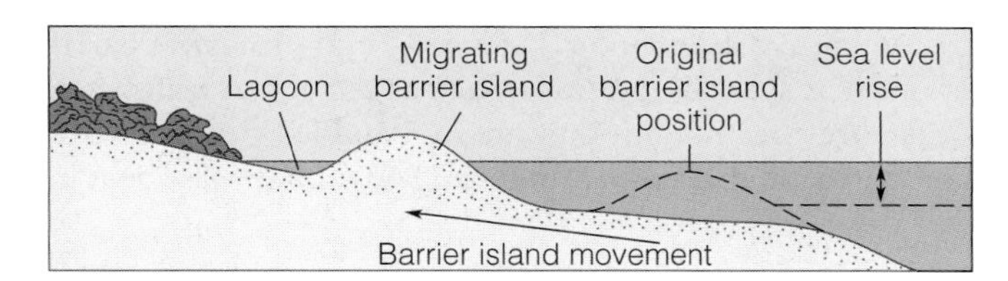

c Over time, the entire island shifts toward the land.

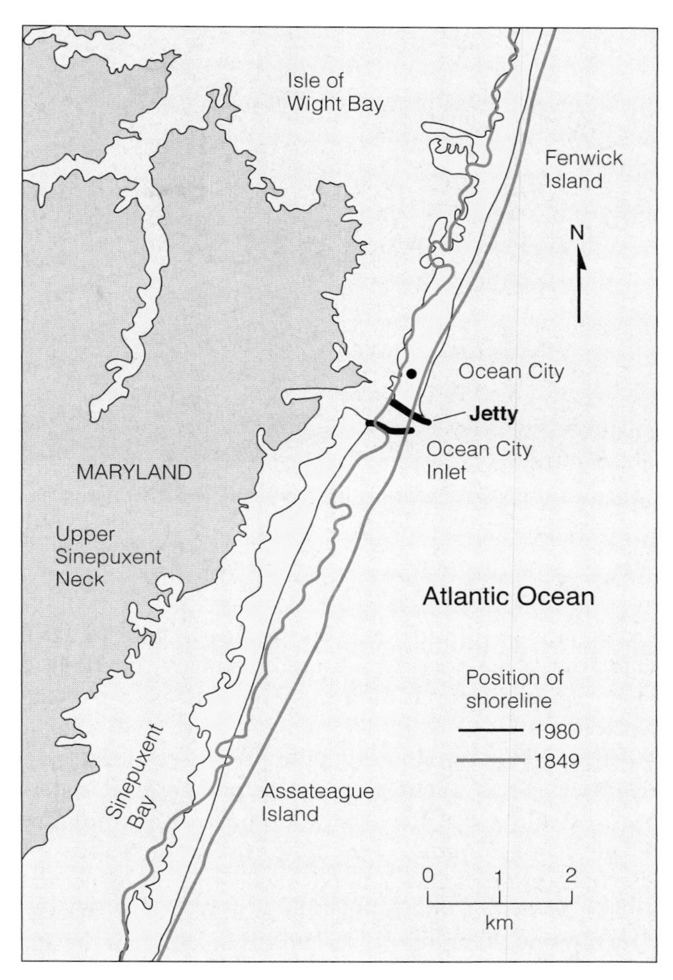

d Jetties were constructed during the 1930s to protect the inlet at Ocean City, Maryland, but they disrupted the net southerly longshore drift and Assateague Island, starved of sediment, has migrated 500 m landward. Beginning in the fall of 2002, beach sand was artificially replenished in an effort to stabilize the island.

being lost at a rate of about 90 km^2 per year. Much of this loss results from sediment compaction, but rising sea level exacerbates the problem.

Another consequence of Hurricane Katrina was its effect on the Chandeleur Islands, off of Louisiana's southeast coast (Figure 16.21). These are barrier islands that absorb some of the impact of approaching storms, especially storm surges. When Hurricane Katrina swept across this area, the islands were battered and reduced to small shoals (Figure 16.21).

Rising sea level also directly threatens many beaches upon which communities depend for revenue. The beach at Miami Beach, Florida, for instance, was disappearing at an alarming rate until the Army Corps of Engineers began replacing the eroded beach sand. The problem is even more serious in other countries. A rise in sea level of only 2 m would inundate large areas of the U.S. East and Gulf Coasts, but would cover 20% of the entire country of Bangladesh. Other problems associated with rising sea level include increased coastal flooding during storms and saltwater incursions that may threaten groundwater supplies (see Chapter 13).

Armoring shorelines with *seawalls* (embankments of reinforced concrete or stone) (Figure 16.18) and using *riprap* (piles of stones, Figure 16.22a) protect beachfront structures, but both are initially expensive and during large storms are commonly damaged or destroyed. Seawalls do afford some protection and are seen in many coastal areas along the oceans and large lakes, but some states, including North and South Carolina, Rhode Island, Oregon, and Maine, no longer allow their construction. The futility of artificially maintaining beaches is aptly shown by efforts to protect homes on a South Carolina barrier island. After each spring tide, heavy equipment builds a sand berm to

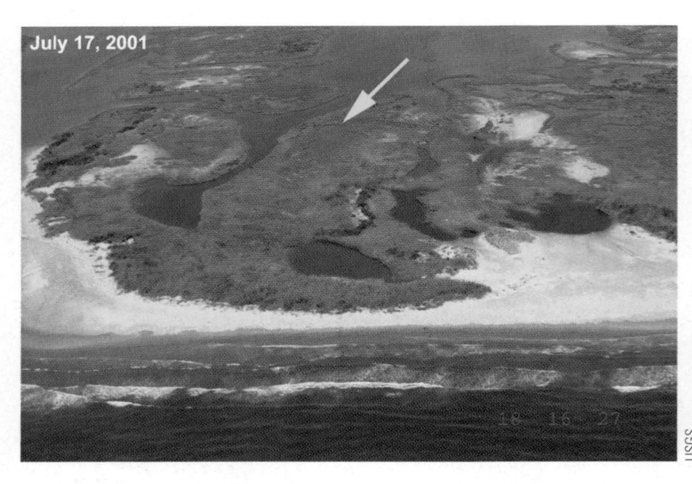

USGS

USGS

Figure 16.21 The Chandeleur Islands Before and After Hurricane Katrina

protect homes from the next spring tide (Figure 16.22b), only to see the berm disappear and then have to rebuild it in a never-ending cycle of erosion and expensive artificial replacement.

Because we can do nothing to prevent sea level from rising, engineers, scientists, planners, and political leaders must examine what they can do to prevent or minimize the effects of shoreline erosion. At present, only a few viable options exist. One is to put strict controls on coastal development. North Carolina, for example, permits large structures no closer to the shoreline than 60 times the annual erosion rate. Although a growing awareness of shoreline processes has resulted in similar legislation elsewhere, some states have virtually no restrictions on coastal development.

Regulating coastal development is commendable, but it has no impact on existing structures and coastal communities. A general retreat from the shoreline may be possible, but expensive, for individual dwellings and small communities, but it is impractical for large population centers. Such communities as Atlantic City, New Jersey, Miami Beach, Florida, and Galveston, Texas, have adopted one of two strategies to combat coastal erosion. One is to build protective barriers such as seawalls. Seawalls, such as the one at Galveston, Texas, are effective, but they are tremendously expensive to construct and maintain. More than $50 million was spent in

Figure 16.22 Riprap and Sand Berm

Sue Monroe

a Riprap made up of large pieces of basalt was piled on this beach to protect a luxury hotel just to the left of this image.

Courtesy of Dr. Stanley Riggs

b Heavy equipment builds a berm, an embankment of sand, on the seaward side of beach homes on the Isle of Palms, South Carolina, to protect them from waves. The berm must be rebuilt every two weeks after each spring tide.

just five years to replenish the beach and build a seawall at Ocean City, Maryland. Furthermore, barriers retard erosion only in the area directly behind them; Galveston Island west of its seawall has been eroded back about 45 m.

Another option, adopted by both Atlantic City, New Jersey, and Miami Beach, Florida, is to pump sand onto the beaches to replace that lost to erosion. This, too, is expensive as the sand must be replenished periodically because erosion is a continuing process. In many areas, groins are constructed to preserve beaches, but unless additional sand is artificially supplied to the beaches, longshore currents invariably erode sand from the downcurrent sides of the groins.

Geo-Recap

Chapter Summary

- Shorelines are continuously modified by the energy of waves and longshore currents and, to a limited degree, by tidal currents.
- The gravitational attraction of the Moon and Sun causes the ocean surface to rise and fall as tides twice daily in most shoreline areas.
- Waves are oscillations on water surfaces that transmit energy in the direction of wave movement. Surface waves affect the water and seafloor only to wave base, which is equal to half the wavelength.
- Little or no net forward motion of water occurs in waves in the open sea. When waves enter shallow water, they are transformed into waves in which water moves in the direction of wave advance.
- Wind-generated waves, especially storm waves, are responsible for most geologic work on shorelines, but waves can also be generated by faulting, volcanic explosions, and rockfalls.
- Breakers form where waves enter shallow water and disrupt the orbital motion of water particles. The waves become oversteepened and plunge forward or spill onto the shoreline, thus expending their kinetic energy.
- Waves approaching a shoreline at an angle generate a longshore current. These currents are capable of considerable erosion, transport, and deposition.
- Narrow surface currents called rip currents carry water from the nearshore zone seaward through the breaker zone.
- Many shorelines are characterized by erosion. Such shorelines have sea cliffs and wave-cut platforms. Other features commonly present include sea caves, sea arches, and sea stacks. Depositional coasts, on the other hand, are characterized by long sandy beaches, deltas, and barrier islands.
- Beaches, the most common shoreline depositional features, are continuously modified by nearshore processes, and their profiles generally exhibit seasonal changes.
- Spits, baymouth bars, and tombolos all form and grow as a result of longshore current transport and deposition.
- Barrier islands are nearshore sediment deposits of uncertain origin. They parallel the mainland but are separated from it by a lagoon.
- The volume of sediment, or nearshore sediment budget, in a nearshore system remains rather constant unless the system is somehow disrupted, as when dams are built across streams that supply sand to the system.
- Submergent and emergent coasts are defined on the basis of their relationship to changes in sea level.
- Coastal flooding during storms by waves and storm surge is an ongoing problem in many areas.

Important Terms

backshore (p. 416)
barrier island (p. 420)
baymouth bar (p. 418)
beach (p. 416)
beach face (p. 416)
berm (p. 416)
breaker (p. 412)
crest (wave) (p. 410)
emergent coast (p. 426)
fetch (p. 411)
foreshore (p. 416)
headland (p. 415)
longshore current (p. 414)
marine terrace (p. 415)

nearshore sediment budget (p. 422)
rip current (p. 414)
shoreline (p. 408)
spit (p. 418)
storm surge (p. 426)
submergent coast (p. 423)
tide (p. 409)
tombolo (p. 418)
trough (wave) (p. 410)
wave (p. 409)
wave base (p. 410)
wave-cut platform (p. 414)
wave refraction (p. 414)

Review Questions

1. A type of spit that connects a beach with an offshore island is a
   a. ______ headland;
   b. ______ tombolo;
   c. ______ sea stack;
   d. ______ baymouth bar;
   e. ______ breaker.
2. The distance wind blows over a continuous water surface is the
   a. ______ fetch;
   b. ______ wavelength;
   c. ______ trough;
   d. ______ rip current;
   e. ______ berm.
3. A marine terrace is
   a. ______ a type of submergent coast;
   b. ______ made up of the beach face and a berm;
   c. ______ the highest level reached by storm waves;
   d. ______ a wave-cut platform above sea level;
   e. ______ the area exposed during neap tides.
4. Although there are some exceptions, most beaches receive most of their sediment from
   a. ______ erosion of reefs;
   b. ______ shoreline rocks;
   c. ______ coastal emergence;
   d. ______ the deep seafloor;
   e. ______ streams and rivers.
5. A long narrow sand deposit separated from the mainland by a lagoon is a
   a. ______ tidal delta;
   b. ______ beach cusp;
   c. ______ barrier island;
   d. ______ wave-built platform;
   e. ______ marine terrace.
6. The most common depositional landforms along seashores are
   a. ______ beaches;
   b. ______ sea arches;
   c. ______ headlands;
   d. ______ wave-dominated deltas;
   e. ______ drift.
7. Erosion caused by the wearing action of water carrying sand and gravel is known as
   a. ______ solution;
   b. ______ cavitation;
   c. ______ abrasion;
   d. ______ impacting;
   e. ______ hydrolysis.
8. Longshore currents are generated by
   a. ______ water flowing offshore through the breaker zone;
   b. ______ waves striking the shore at an angle;
   c. ______ the distance wind blows over water;
   d. ______ submarine currents in the deep sea;
   e. ______ erosion by abrasion and hydraulic action.
9. The time it takes for two successive wave crests (or troughs) to pass a given point is known as the
   a. ______ wave depth;
   b. ______ wave celerity;
   c. ______ wave period;
   d. ______ wave fetch;
   e. ______ wave height.
10. Water is driven onshore by ___________ during hurricanes.
    a. ______ rogue waves;
    b. ______ storm surges;
    c. ______ low tides;
    d. ______ hydraulic action;
    e. ______ wave base.
11. Explain what wave base is and how it affects waves as they enter shallow water.
12. How are waves responsible for transporting sediment long distances along shorelines?
13. Why does an observer at a shoreline location experience two high and two low tides daily?
14. How do rising sea level and coastal development, including developments on barrier islands, complicate efforts to control shoreline erosion?
15. While driving along North America's West Coast, you see a surface sloping gently toward the sea with several masses of rock rising above it. How would you explain the origin of this feature to your children?

16. How does a spit and a baymouth bar form?
17. Explain why waves are so small in ponds and lakes but much larger ones develop in the oceans.
18. As a member of a planning commission for a coastal region where shoreline erosion is a continuing problem, what recommendations would you make to remedy or at least mitigate the problem?
19. What are rip currents? If caught in one what would you do?
20. A hypothetical shoreline has a balanced budget, but a dam is built on the river supplying sand to the system and a wall is built to protect sea cliffs from erosion. Explain what is likely to happen to the beaches in this area.

CHAPTER 17

Geologic Time: Concepts and Principles

OUTLINE

OBJECTIVES

At the end of this chapter, you will have learned that

- The concept of geologic time and its measurements have changed throughout human history.
- The principle of uniformitarianism is fundamental to geology.
- Relative dating—placing geologic events in a sequential order—provides a means to interpret geologic history.
- The three types of unconformities—disconformities, angular unconformities, and nonconformities—are erosional surfaces separating younger from older rocks and represent significant intervals of geologic time for which we have no record at a particular location.
- Time equivalency of rock units can be demonstrated by various correlation techniques.
- Absolute dating methods are used to date geologic events in terms of years before present.

The Grand Canyon, Arizona. Major John Wesley Powell led two expeditions down the Colorado River and through the canyon in 1869 and 1871. He was struck by the seemingly limitless time represented by the rocks exposed in the canyon walls and by the recognition that these rock layers, like the pages in a book, contain the geologic history of this region.

- The most accurate radiometric dates are obtained from igneous rocks.
- The geologic time scale evolved primarily during the 19th century through the efforts of many people.
- Stratigraphic terminology includes units based on content and units related to geologic time.
- Geologic time is an important element in the study of climate change.

INTRODUCTION

In 1869, Major John Wesley Powell, a Civil War veteran who lost his right arm in the battle of Shiloh, led a group of hardy explorers down the uncharted Colorado River through the Grand Canyon. With no maps or other information, Powell and his group ran the many rapids of the Colorado River in fragile wooden boats, hastily recording what they saw. Powell wrote in his diary that "all about me are interesting geologic records. The book is open and I read as I run."

From this initial reconnaissance, Powell led a second expedition down the Colorado River in 1871. This second trip included a photographer, a surveyor, and three topographers. Members of the expedition made detailed topographic and geologic maps of the Grand Canyon area, as well as the first photographic record of the region.

Probably no one has contributed as much to the understanding of the Grand Canyon as Major Powell. In recognition of his contributions, the Powell Memorial was erected on the South Rim of the Grand Canyon in 1969 to commemorate the 100th anniversary of this history-making first expedition.

Most tourists today, like Powell and his fellow explorers in 1869, are astonished by the seemingly limitless time represented by the rocks exposed in the walls of the Grand Canyon. For most visitors, viewing a 1.5-km-deep cut into Earth's crust is the only encounter they'll ever have with the magnitude of geologic time. When standing on the rim and looking down into the Grand Canyon, we are really looking far back in time, all the way back to the early history of our planet. In fact, more than 1 billion years of history are preserved in the rocks of the Grand Canyon.

Vast periods of time set geology apart from most of the other sciences, and an appreciation of the immensity of geologic time is fundamental to understanding the physical and biological history of our planet. In fact, understanding and accepting the magnitude of geologic time are major contributions that geology has made to the sciences.

Besides providing an appreciation for the immensity of geologic time, why is the study of geologic time important? One of the most valuable lessons you will learn in this chapter is how to reason and apply the fundamental geologic principles in solving geologic problems. The logic used in applying these principles to interpret the geologic history of an area involves basic reasoning skills that can be transferred to and used in almost any profession or discipline.

HOW IS GEOLOGIC TIME MEASURED?

In some respects, time is defined by the methods used to measure it. Geologists use two different frames of reference when discussing geologic time. **Relative dating** is placing geologic events in a sequential order as determined from their position in the geologic record. Relative dating will not tell us how long ago a particular event took place, only that one event preceded another. A useful analogy for relative dating is a television guide that does not list the times that programs are shown. You cannot tell what time a particular program will be shown, but by watching a few shows and checking the guide, you can determine whether you have missed the show or how many shows are scheduled before the one you want to see.

The various principles used to determine relative dating were discovered hundreds of years ago, and since then, they have been used to construct the *relative geologic time scale* (◗ Figure 17.1). Furthermore, these principles are still widely used by geologists today.

Absolute dating provides specific dates for rock units or events expressed in years before the present. In our analogy of the television guide, the time when the programs are actually shown would be the absolute dates. In this way, you not only can determine whether you have missed a show (relative dating), but also know how long it will be until a show you want to see will be shown (absolute dating).

Radiometric dating is the most common method of obtaining absolute ages. Dates are calculated from the natural decay rates of various radioactive elements present in trace amounts in some rocks. It was not until the discovery of radioactivity near the end of the 19th century that absolute ages could be accurately applied to the relative geologic time scale. Today, the geologic time scale is really a dual scale: a relative scale based on rock sequences with radiometric dates expressed as years before the present (Figure 17.1).

EARLY CONCEPTS OF GEOLOGIC TIME AND THE AGE OF EARTH

The concept of geologic time and its measurement have changed throughout human history. Many early Christian scholars and clerics tried to establish the date of creation by analyzing historical records and the genealogies found in Scripture. One of the most influential scholars was James Ussher (1581–1656), Archbishop of Armagh, Ireland, who, based upon Old Testament genealogy, asserted that God created Earth on Sunday, October 23, 4004 B.C. In 1701, an authorized version of the Bible made this date accepted Church doctrine. For nearly a century thereafter, it was considered heresy to assume that Earth and all of its features were more than about 6000 years old. Thus, the idea of a very young Earth provided the basis for most Western chronologies of Earth history prior to the 18th century.

During the 18th and 19th centuries, several attempts were made to determine Earth's age on the basis of scientific evidence rather than revelation. The French zoologist Georges Louis de Buffon (1707–1788) assumed that Earth gradually cooled to its present condition from a molten beginning.

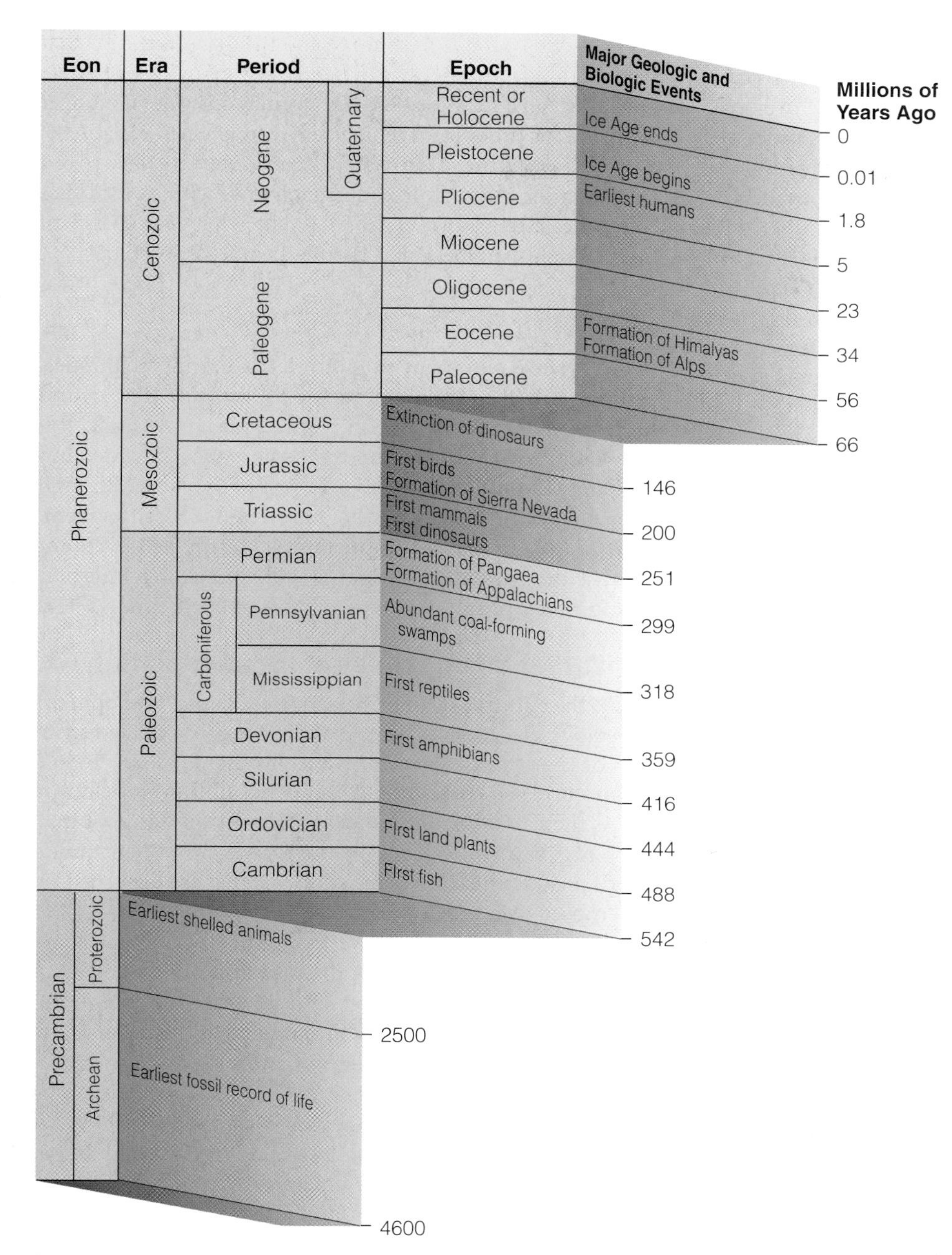

Figure 17.1 The Geologic Time Scale Some of the major geologic and biologic events are indicated for the various eras, periods, and epochs. Dates are from Gradstein, F., Ogg, J., and Smith, A., *A Geologic Timescale 2004* (Cambridge, UK: Cambridge University Press, 2005), Figure 1.2.

To simulate this history, he melted iron balls of various diameters and allowed them to cool to the surrounding temperature. By extrapolating their cooling rate to a ball the size of Earth, he determined that Earth was at least 75,000 years old. Although this age was much older than that derived from Scripture, it was vastly younger than we now know our planet to be.

Other scholars were equally ingenious in attempting to calculate Earth's age. For example, if deposition rates could be determined for various sediments, geologists reasoned that they could calculate how long it would take to deposit any rock layer. They could then extrapolate how old Earth was from the total thickness of sedimentary rock in its crust. Rates of deposition vary, however, even for the same type of rock. Furthermore, it is impossible to estimate how much of a rock has been removed by erosion, or how much a rock sequence has been reduced by compaction. As a result of these variables, estimates of Earth's age ranged from younger than 1 million years to older than 2 billion years.

Another attempt to determine Earth's age involved ocean salinity. Scholars assumed that Earth's ocean waters were originally fresh and that their present salinity was the result of dissolved salt being carried into the ocean basins by streams. Knowing the volume of ocean water and its salinity, John Joly, a 19th-century Irish geologist, measured the amount of salt currently in the world's streams. He then calculated that it would have taken at least 90 million years for the oceans to reach their present salinity level. This was still much younger than the now-accepted age of 4.6 billion years for Earth, mainly because Joly had no way to calculate either how much salt had been recycled or the amount of salt stored in continental salt deposits and seafloor clay deposits.

Besides trying to determine Earth's age, the naturalists of the 18th and 19th centuries were formulating some of the fundamental geologic principles that are used in deciphering Earth history. From the evidence preserved in the geologic record, it was clear to them that Earth is very old and that geologic processes have operated over long periods of time. A good example of geologic processes operating over long periods of time to produce a spectacular landscape is the evolution of Uluru and Kata Tjuta (see Geo-inSight on pages 456 and 457).

JAMES HUTTON AND THE RECOGNITION OF GEOLOGIC TIME

Many consider the Scottish geologist James Hutton (1726–1797) to be the father of modern geology. His detailed studies and observations of rock exposures and present-day geologic processes were instrumental in establishing the *principle of uniformitarianism* (see Chapter 1), the concept that the same processes seen today have operated over vast amounts of time. Because Hutton relied on known processes to account for Earth history, he concluded that Earth must be very old and wrote that "we find no vestige of a beginning, and no prospect of an end."

Unfortunately, Hutton was not a particularly good writer, so his ideas were not widely disseminated or accepted. In 1830, Charles Lyell published a landmark book, *Principles of Geology*, in which he championed Hutton's concept of uniformitarianism. Instead of relying on catastrophic events to explain various Earth features, Lyell recognized that imperceptible changes brought about by present-day processes could, over long periods of time, have tremendous cumulative effects. Through his writings, Lyell firmly established uniformitarianism as the guiding principle of geology. Furthermore, the recognition of virtually limitless amounts of time was also necessary for, and instrumental in, the acceptance of Darwin's 1859 theory of evolution (see Chapter 18).

After establishing that present-day processes have operated over vast periods of time, geologists were nevertheless nearly forced to accept a very young age for Earth when a highly respected English physicist, Lord Kelvin (1824–1907), claimed, in a paper written in 1866, to have destroyed the uniformitarian foundation of geology. Starting with the generally accepted belief that Earth was originally molten, Kelvin assumed that it has gradually been losing heat and that, by measuring this heat loss, he could determine its age.

Kelvin knew from deep mines in Europe that Earth's temperature increases with depth, and he reasoned that Earth is losing heat from its interior. By knowing the size of Earth, the melting temperatures of rocks, and the rate of heat loss, Kelvin calculated the age at which Earth was entirely molten. From these calculations, he concluded that Earth could not be older than 400 million years nor younger than 20 million years. This wide discrepancy in age reflected uncertainties in average temperature increases with depth and the various melting points of Earth's constituent materials.

After establishing that Earth was very old and that present-day processes operating over long periods of time account for geologic features, geologists were in a quandary. If they accepted Kelvin's dates, they would have to abandon the concept of seemingly limitless time that was the underpinning of uniformitarian geology and one of the foundations of Darwinian evolution and squeeze events into a shorter time frame.

Kelvin's reasoning and calculations were sound, but his basic premises were false, thereby invalidating his conclusions. Kelvin was unaware that Earth has an internal heat source, radioactivity, that has allowed it to maintain a fairly constant temperature through time.* His 40-year campaign for a young Earth ended with the discovery of radioactivity near the end of the 19th century, and the insight in 1905 that natural radioactive decay can be used in many cases to date how long ago a rock formed. His calculations were no longer valid and his proof for a geologically young Earth collapsed.

Although the discovery of radioactivity destroyed Kelvin's arguments, it provided geologists with a clock that could measure Earth's age and validate what geologists had long thought—namely, that Earth was indeed very old!

*Actually, Earth's temperature has decreased through time because the original amount of radioactive materials has been decreasing and thus is not supplying as much heat. However, the temperature is decreasing at a rate considerably slower than would be required to lend any credence to Kelvin's calculations.

RELATIVE DATING METHODS

Before the development of radiometric dating techniques, geologists had no reliable means of absolute dating and therefore depended solely on relative dating methods. Recall that relative dating places events in sequential order, but does not tell us how long ago an event took place. Although the principles of relative dating may now seem self-evident, their discovery was an important scientific achievement because they provided geologists with a means to interpret geologic history and develop a relative geologic time scale.

Fundamental Principles of Relative Dating

The 17th century was an important time in the development of geology as a science because of the widely circulated writings of the Danish anatomist Nicolas Steno (1638–1686). Steno observed that when streams flood, they spread out across their floodplains and deposit layers of sediment that bury organisms dwelling on the floodplain. Subsequent floods produce new layers of sediments that are deposited or superposed over previous deposits.

When lithified, these layers of sediment become sedimentary rock. Thus, in an undisturbed succession of sedimentary rock layers, the oldest layer is at the bottom and the youngest layer is at the top. This **principle of superposition** is the basis for relative-age determinations of strata and their contained fossils (Figure 17.2a).

Steno also observed that, because sedimentary particles settle from water under the influence of gravity, sediment is deposited in essentially horizontal layers, thus illustrating the **principle of original horizontality** (Figure 17.2a). Therefore, a sequence of sedimentary rock layers that is steeply inclined from the horizontal must have been tilted after deposition and lithification (Figure 17.2b).

Steno's third principle, the **principle of lateral continuity**, states that sediment extends laterally in all directions until it thins and pinches out or terminates against the edge of the depositional basin (Figure 17.2a).

James Hutton is credited with discovering the **principle of cross-cutting relationships.** Based on his detailed studies and observations of rock exposures in Scotland, Hutton recognized that an igneous intrusion or fault must be younger than the rocks it intrudes or displaces (Figure 17.3).

Although this principle illustrates that an intrusive igneous structure is younger than the rocks it intrudes, the association of sedimentary and igneous rocks may cause problems in relative dating. Buried lava flows and sills look very similar in a sequence of strata (Figure 17.4). A buried lava flow, however, is older than the rocks above it (principle of superposition),

Figure 17.2 The Principles of Original Horizontality, Superposition, and Lateral Continuity

a The sedimentary rocks of Bryce Canyon National Park, Utah, were originally deposited horizontally in a variety of continental environments (principle of original horizontality). The oldest rocks are at the bottom of this highly dissected landscape, and the youngest rocks are at the top, forming the rims (principle of superposition). The exposed rock layers extend laterally in all directions for some distance (principle of lateral continuity).

b These shales and limestones of the Postolonnec Formation, at Postolonnec Beach, Crozon Peninsula, France, were originally deposited horizontally, but have been significantly tilted since their formation.

Figure 17.3 The Principle of Cross-Cutting Relationships

a A dark gabbro dike cuts across granite in Acadia National Park, Maine. The dike is younger than the granite that it intrudes.

b A small fault (arrows show direction of movement) cuts across, and thus displaces, tilted sedimentary beds along the Templin Highway in Castaic, California. The fault is therefore younger than the youngest beds that are displaced.

Figure 17.4 Differentiating Between a Buried Lava Flow and a Sill In frames a and b below, the ages of sedimentary strata are shown by numbering from 1–6, with (1) being the oldest sedimentary rock layer in each frame, and (6) the youngest. Remember that the lava flow took place *during* deposition of an ordinary sedimentary sequence, whereas the sill intruded the sedimentary strata *after* they had accumulated. The notes in both frames highlight the physical features that one would look for to distinguish between a lava flow and a sill, which can look quite similar in an outcrop.

a A buried lava flow has baked underlying bed 2 when it flowed over it. Clasts of the lava were deposited along with other sediments during deposition of bed 3. The lava flow is younger than bed 2 and older than beds 3, 4, and 5.

Baking zones on both sides of sill
Sill
Inclusions of rock from both layers above (3) and below (2) may exist in the sill
6
5
4
3
2
1

b The rock units above and below the sill have been baked, indicating that the sill is younger than beds 2 and 3, but its age relative to beds 4–6 cannot be determined.

c This buried lava flow in Yellowstone National Park, Wyoming, displays columnar joining. A baked zone is present below, and not above, the igneous structure, indicating that it is a buried lava flow and not a sill.

whereas a sill, resulting from later igneous intrusion, is younger than all of the beds below it and younger than the immediately overlying bed as well.

To resolve such relative-age problems as these, geologists look to see whether the sedimentary rocks in contact with the igneous rocks show signs of baking or alteration by heat (see the section on contact metamorphism in Chapter 7). A sedimentary rock that shows such effects must be older than the igneous rock with which it is in contact. In Figure 17.4, for example, a sill produces a zone of baking immediately above and below it because it intruded into previously existing sedimentary rocks. A lava flow, in contrast, bakes only those rocks below it.

Another way to determine relative ages is by using the **principle of inclusions**. This principle holds that inclusions, or fragments of one rock contained within a layer of another, are older than the rock layer itself. The batholith shown in Figure 17.5a contains sandstone inclusions, and the sandstone unit shows the effects of baking. Accordingly, we conclude that the sandstone is older than the batholith. In Figure 17.5b, however, the sandstone contains granite rock pieces, indicating that the batholith was the source rock for the inclusions and is therefore older than the sandstone.

Fossils have been known for centuries (see Chapter 18), yet their utility in relative dating and geologic mapping was not fully appreciated until the early 19th century. William Smith (1769–1839), an English civil engineer involved in surveying and building canals in southern England, independently recognized the principle of superposition by reasoning that the fossils at the bottom of a sequence of strata are older than those at the top of the sequence. This recognition served as the basis for the **principle of fossil succession** or the *principle of faunal and floral succession*, as it is sometimes called (Figure 17.6).

According to this principle, fossil assemblages succeed one another through time in a regular and predictable order. The validity and successful use of this principle depend on three points: (1) Life has varied through time, (2) fossil assemblages are recognizably different from one another, and (3) the relative ages of the fossil assemblages can be determined. Observations of fossils in older versus younger strata clearly demonstrate that life-forms have changed. Because this is true, fossil assemblages (point 2) are recognizably different. Furthermore, superposition can be used to demonstrate the relative ages of the fossil assemblages.

Unconformities

Our discussion so far has been concerned with vertical relationships among conformable strata—that is, sequences of rock in which deposition was more or less continuous. A bedding plane between strata may represent a depositional break of anywhere from minutes to tens of years, but it is inconsequential in the context of geologic time. However, in some sequences of strata, surfaces known as **unconformities** may be present, representing times of nondeposition, erosion, or both. Unconformities encompass long periods of geologic time, perhaps millions or tens of millions of years.

What Would You Do?

You have been chosen to be part of the first astronaut crew to land on Mars. You were selected because you are a geologist, and therefore your primary responsibility is to map the geology of the landing site area. An important goal of the mission is to work out the geologic history of the area. How will you go about this? Will you be able to use the principles of relative dating? How will you correlate the various rock units? Will you be able to determine absolute ages? How will you do this?

Accordingly, the geologic record is incomplete wherever an unconformity is present, just as a book with missing pages is incomplete, and the interval of geologic time not represented by strata is called a *hiatus* (Figure 17.7).

The general term *unconformity* encompasses three specific types of surfaces. First, a **disconformity** is a surface of erosion or nondeposition separating younger from older rocks, both of which are parallel with one another (Figure 17.8). Unless the erosional surface separating the older from the younger parallel beds is well defined or distinct, the disconformity frequently resembles an ordinary bedding plane. Hence, many disconformities are difficult to recognize and must be identified on the basis of fossil assemblages.

Second, an **angular unconformity** is an erosional surface on tilted or folded strata over which younger strata were deposited (Figure 17.9). The strata below the unconformable surface generally dip more steeply than those above, producing an angular relationship.

The angular unconformity illustrated in Figure 17.9b is probably the most famous in the world. It was here at Siccar Point, Scotland, that James Hutton realized that severe upheavals had tilted the lower rocks and formed mountains that were then worn away and covered by younger, flat-lying rocks. The erosional surface between the older tilted rocks and the younger flat-lying strata meant that a significant gap existed in the geologic record. Although Hutton did not use the term *unconformity*, he was the first to understand and explain the significance of such discontinuities in the geologic record.

A **nonconformity** is the third type of unconformity. Here, an erosional surface cut into metamorphic or igneous rocks is covered by sedimentary rocks (Figure 17.10). This type of unconformity closely resembles an intrusive igneous contact with sedimentary rocks. The principle of inclusions (Figure 17.5) is helpful in determining whether the relationship between the underlying igneous rocks and the overlying sedimentary rocks is the result of an intrusion or erosion. A nonconformity is also marked in many places by an ancient zone of weathering, or even a reddened, brick-like soil horizon, or paleosol. In the case of an intrusion, the igneous rocks are younger, whereas in the case of erosion, the sedimentary rocks are younger. Being able to distinguish between a nonconformity and an intrusive contact is important because they represent different sequences of events.

Figure 17.5 The Principle of Inclusions

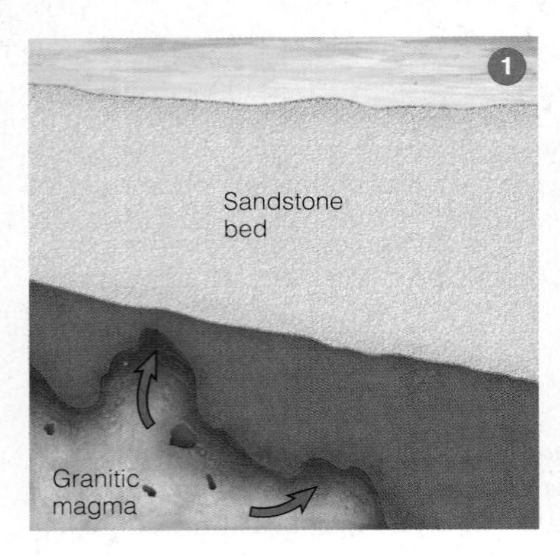

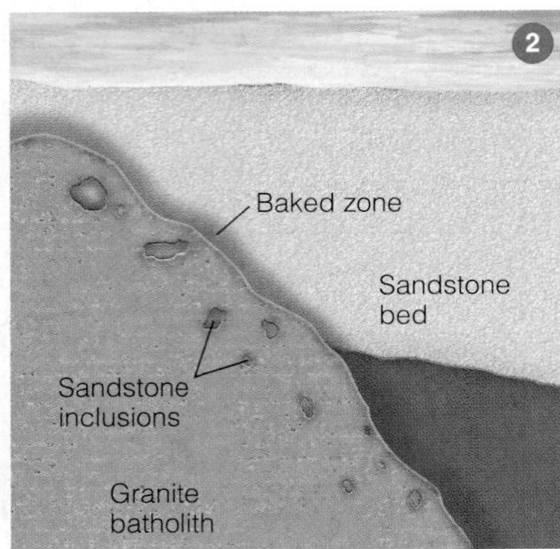

a The sandstone is older than the granite batholith because there are inclusions of sandstone inside the granite. The sandstone also shows evidence of having been baked along its contact with the granite batholith when the granitic magma intruded the overlying sedimentary beds.

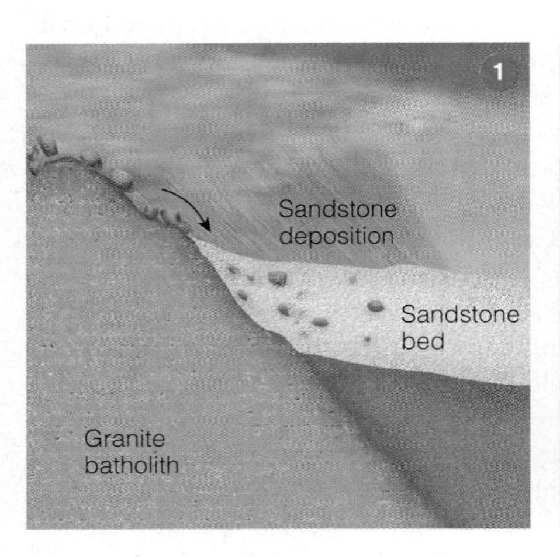

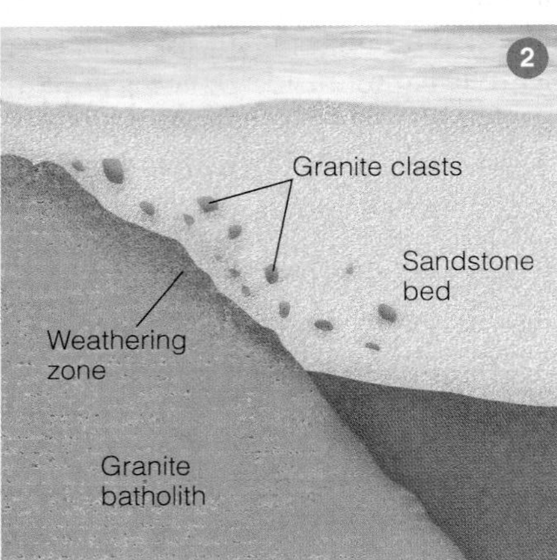

b The sandstone is younger than the granite batholith because it contains pieces (clasts) of granite. The granite is also weathered along the contact with the sandstone, indicating that it was the source of the granite clasts, and must therefore be older than the sandstone.

c Outcrop in northern Wisconsin showing basalt inclusions (dark gray) in granite (white). Accordingly, the basalt inclusions are older than the granite.

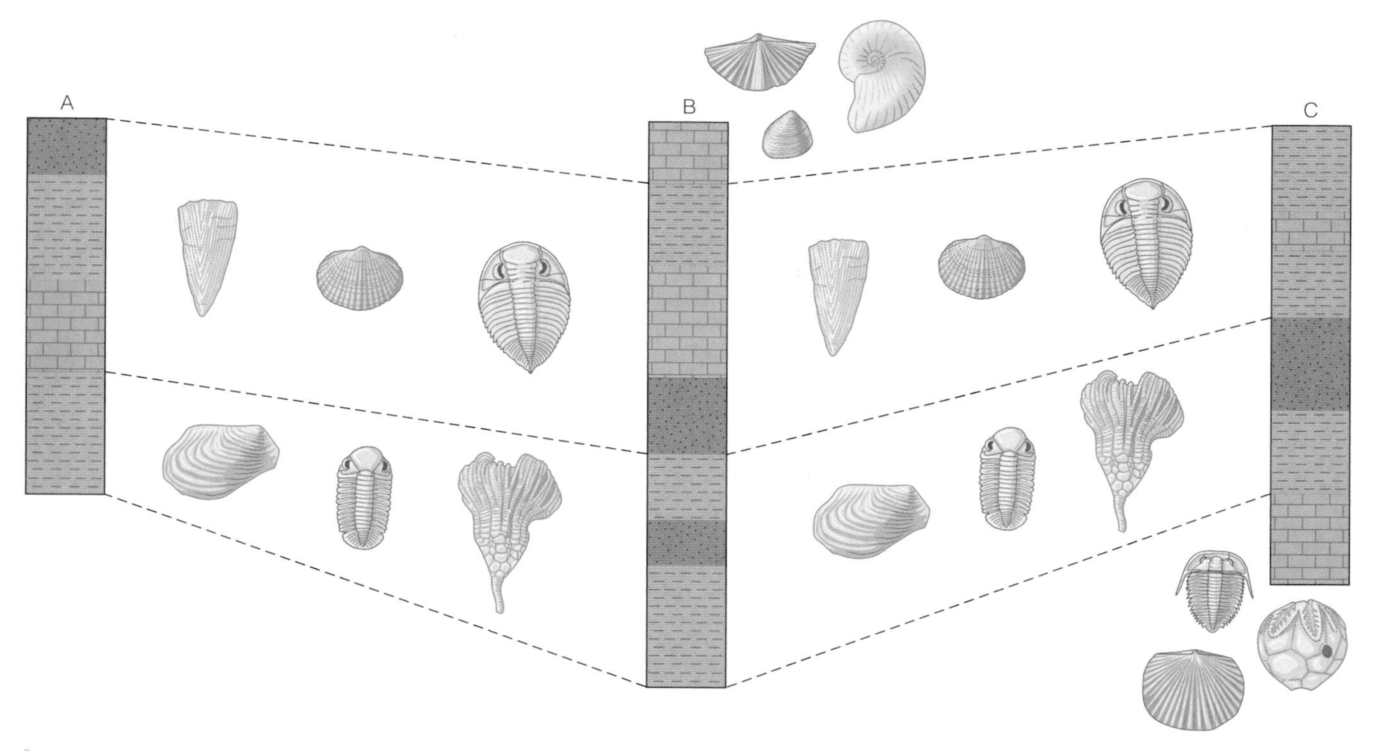

Figure 17.6 **The Principle of Fossil Succession** This generalized diagram shows how geologists use the principle of fossil succession to identify strata of the same age in different areas. The rocks in the three sections encompassed by the dashed lines contain similar fossils and are therefore the same age. Note that the youngest rocks in this region are in section B, whereas the oldest rocks are in section C.

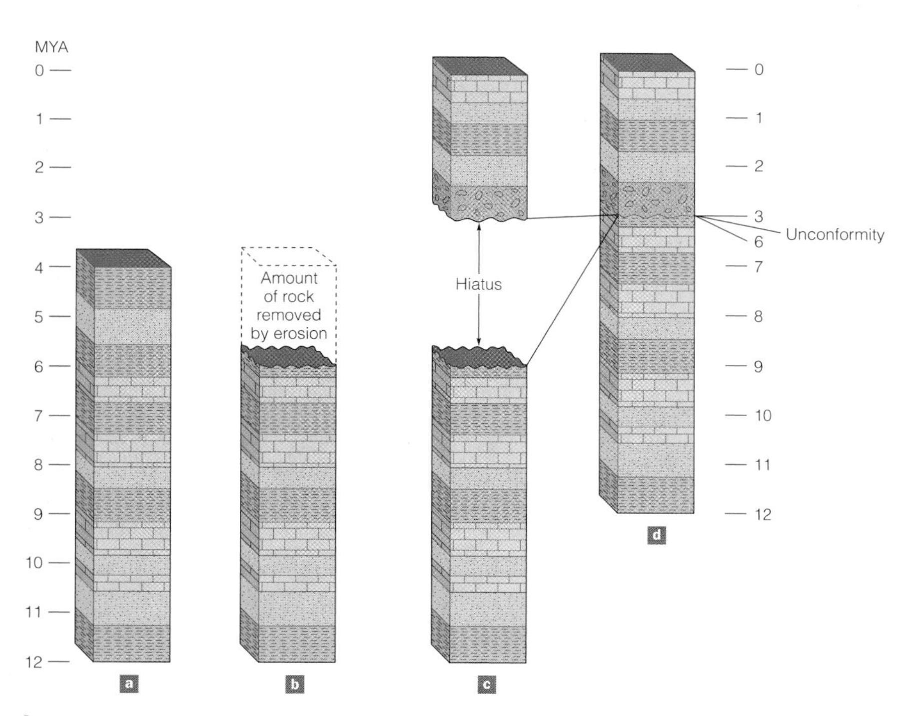

Figure 17.7 The Development of a Hiatus and an Unconformity a Deposition began 12 million years ago (mya) and continued more or less uninterrupted until 4 mya. b Between 3 and 4 mya, an episode of erosion occurred. During that time, some of the strata deposited earlier was eroded. c A hiatus of 3 million years thus exists between the older strata and the strata that formed during a renewed episode of deposition that began 3 mya. d The actual stratigraphic record as seen in an outcrop today. The unconformity is the surface separating the strata and represents a major break in our record of geologic time.

Applying the Principles of Relative Dating

We can decipher the geologic history of the area represented by the block diagram in Figure 17.11 by applying the various relative-dating principles just discussed. The methods and logic used in this example are the same as those applied by 19th-century geologists in constructing the geologic time scale.

According to the principles of superposition and original horizontality, beds A–G were deposited horizontally; then either they were tilted, faulted (H), and eroded, or after deposition, they were faulted (H), tilted, and then eroded (Figure 17.12a–c). Because the fault cuts beds A–G, it must be younger than the beds according to the principle of cross-cutting relationships.

Beds J–L were then deposited horizontally over this erosional surface, producing an angular unconformity (I) (Figure 17.12d). Following deposition of these three beds, the entire sequence was intruded by a dike (M), which, according to the principle of

Figure 17.8 Formation of a Disconformity

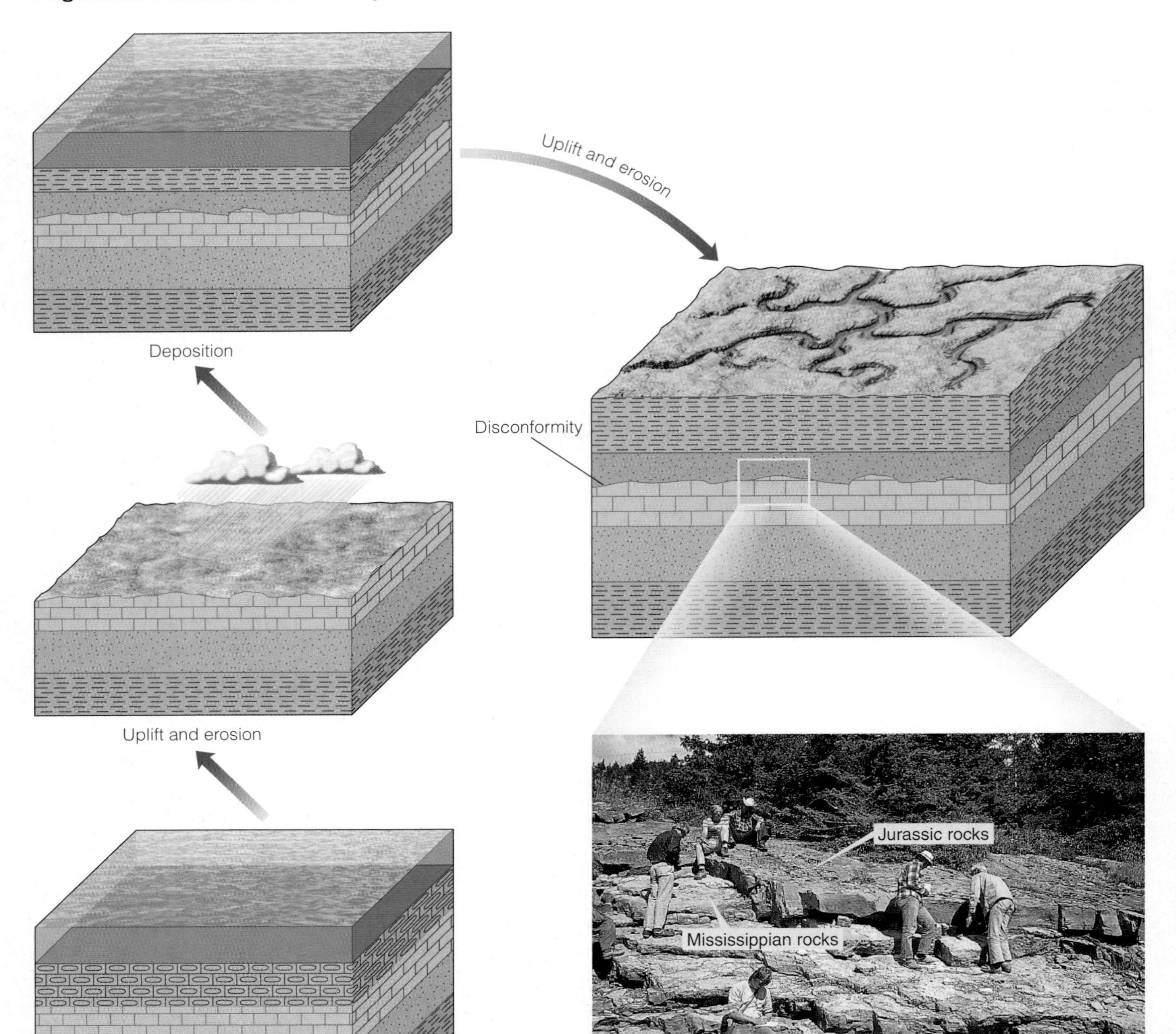

a Formation of a disconformity.

b Disconformity between Mississippian and Jurassic strata in Montana. The geologist at the upper left is sitting on Jurassic strata, and his right foot is resting on Mississippian rocks. This disconformity represent approximately 165 million years.

cross-cutting relationships, must be younger than all of the rocks that it intrudes (Figure 17.12e).

The entire area was then uplifted and eroded; next, beds P and Q were deposited, producing a disconformity (N) between beds L and P and a nonconformity (O) between the igneous intrusion M and the sedimentary bed P (Figure 17.12f, g). We know that the relationship between igneous intrusion M and the overlying sedimentary bed P is a nonconformity because of the inclusions of M in P (principle of inclusions).

At this point, there are several possibilities for reconstructing the geologic history of this area. According to the principle of cross-cutting relationships, dike R must be younger than bed Q because it intrudes into it. It could have intruded anytime *after* bed Q was deposited; however, we cannot determine whether R was formed right after Q, right after S, or after T was formed. For purposes of this

Figure 17.9 Formation of an Angular Unconformity

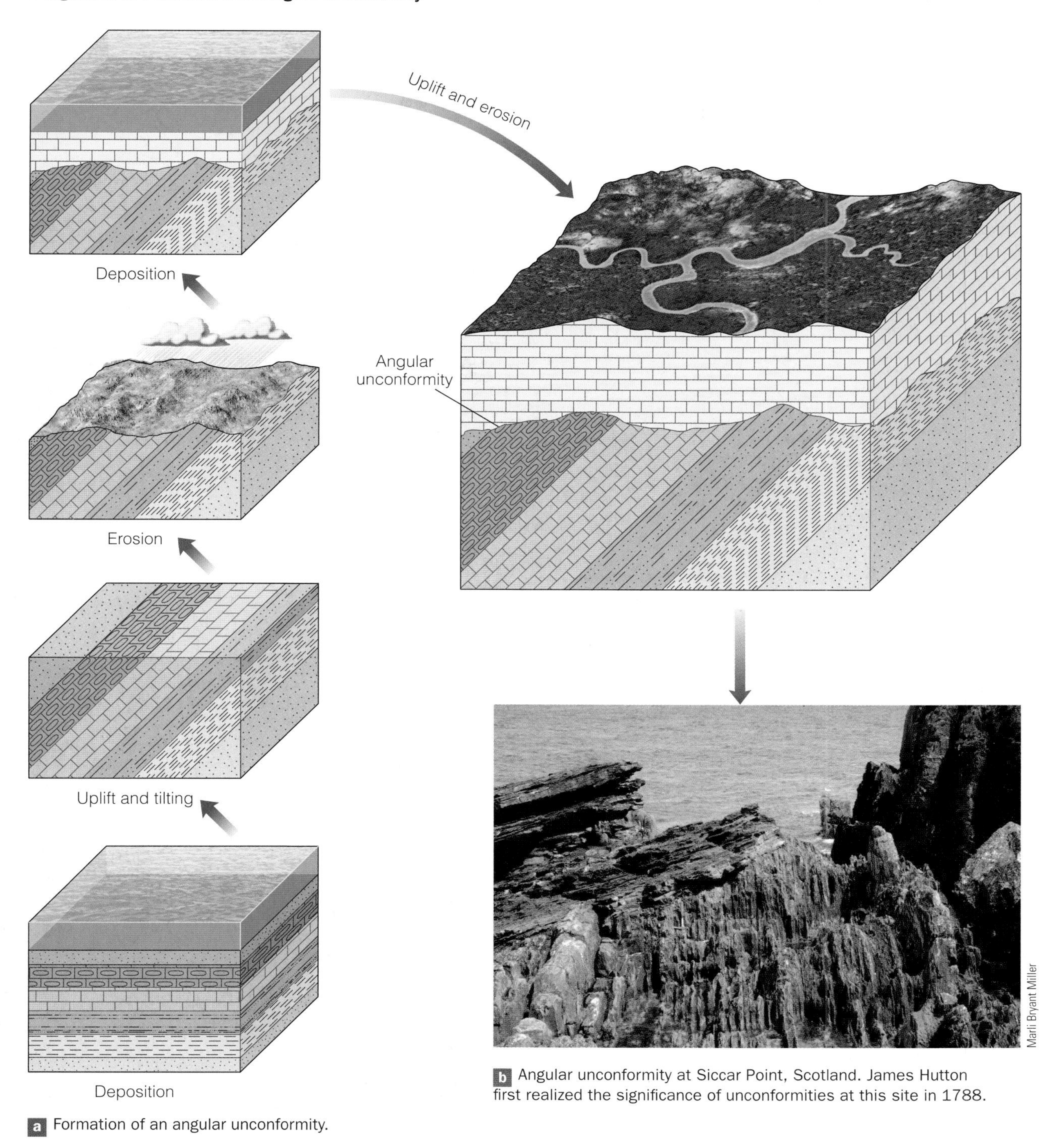

a Formation of an angular unconformity.

b Angular unconformity at Siccar Point, Scotland. James Hutton first realized the significance of unconformities at this site in 1788.

history, we will say that it intruded after the deposition of bed Q (Figure 17.12g, h).

Following the intrusion of dike R, lava S flowed over bed Q, followed by the deposition of bed T (Figure 17.12i, j). Although the lava flow (S) is not a sedimentary unit, the principle of superposition still applies because it flowed onto the surface, just as sediments are deposited on Earth's surface.

We have established a relative chronology for the rocks and events of this area by using the principles of relative dating. Remember, however, that we have no way of knowing how many years ago these events occurred unless we can obtain radiometric dates for the igneous rocks. With these dates, we can establish the range of absolute ages between which the different sedimentary units were deposited and also determine how much time is represented by the unconformities.

Figure 17.10 Formation of a Nonconformity

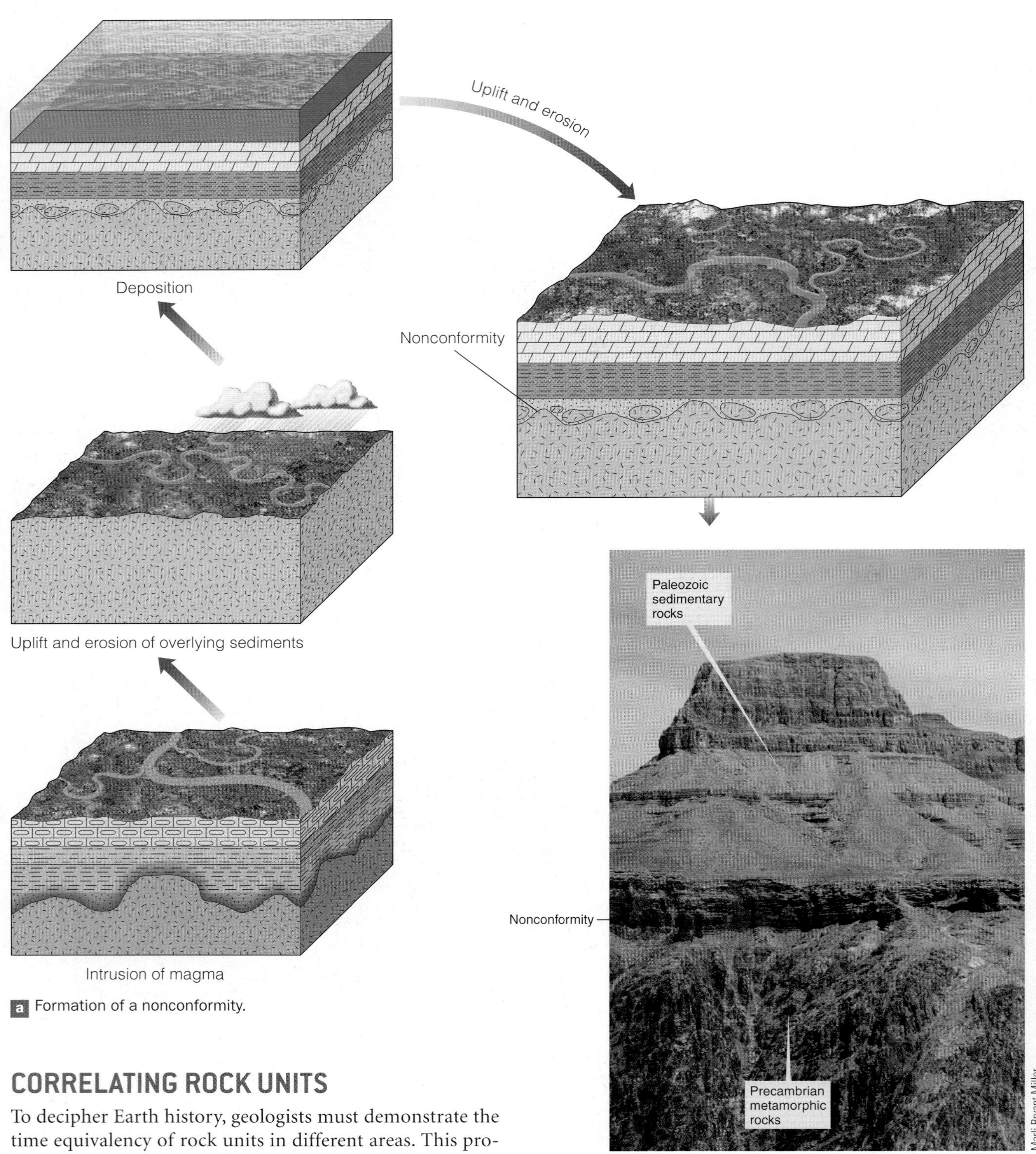

a Formation of a nonconformity.

b Nonconformity between Precambrian metamorphic rocks and overlying Paleozoic sedimentary rocks in the Grand Canyon, Arizona.

CORRELATING ROCK UNITS

To decipher Earth history, geologists must demonstrate the time equivalency of rock units in different areas. This process is known as **correlation**.

If surface exposures are adequate, units may simply be traced laterally (principle of lateral continuity), even if occasional gaps exist (Figure 17.13). Other criteria used to correlate units are similarity of rock type, position in a sequence, and key beds. *Key beds* are units, such as coal beds or volcanic ash layers, that are sufficiently distinctive to allow identification of the same unit in different areas (Figure 17.13).

Generally, no single location in a region has a geologic record of all events that occurred during its history; therefore, geologists must correlate from one area to another to determine the complete geologic history of the region.

An excellent example is the history of the Colorado Plateau (Figure 17.14). This region provides a record of events occurring over approximately 2 billion years. Because of the forces of erosion, the entire record is not preserved at any single location. Within the walls of the Grand Canyon are rocks of the Precambrian and Paleozoic eras, whereas Paleozoic and Mesozoic Era rocks are found in Zion National Park, and Mesozoic and Cenozoic Era rocks are exposed in Bryce Canyon (Figure 17.14). By correlating the uppermost rocks at one location with the lowermost equivalent rocks of another area, geologists can decipher the history of the entire region.

Although geologists match up rocks on the basis of similar rock type and superposition, correlation of this type can be done only in a limited area where beds can be traced from one site to another. To correlate rock units over a large area or to correlate age-equivalent units of different composition, fossils and the principle of fossil succession must be used.

Fossils are useful as relative time indicators because they are the remains of organisms that lived for a certain length of time during the geologic past. Fossils that are easily identified, are geographically widespread, and existed for a rather short interval of geologic time are particularly useful. Such fossils are **guide fossils** or *index fossils* (Figure 17.15). The trilobite *Paradoxides* and the brachiopod *Atrypa* meet these criteria and are therefore good guide fossils. In contrast, the brachiopod *Lingula* is easily identified and widespread, but its long geologic range of Ordovician to Recent makes it of little use in correlation.

What Would You Do?

You are a member of a regional planning commission that is considering a plan for constructing what is said to be a much-needed river dam that will create a recreational lake. Opponents of the dam project have come to you with a geologic report and map showing that a fault underlies the area of the proposed dam, and the fault trace can be clearly seen at the surface. The opponents say that the fault may be active, and thus someday it will move suddenly, bursting the dam and sending a wall of water downstream. You seek the advice of a local geologist who has worked in the dam area. She tells you that she found a lava flow covering the fault, less than a kilometer from the proposed dam project site. Can you use this information, along with a radiometric date from the lava flow, to help convince the opponents that the fault has not moved in any direction (vertically or laterally) in the recent past? How would you do this, and what type of reasoning would you use?

Because most fossils have fairly long geologic ranges, geologists construct *concurrent range zones* to determine the age of the sedimentary rocks containing the fossils. Concurrent range zones are established by plotting the overlapping geologic ranges of two or more fossils that have different geologic ranges (Figure 17.16). The first and last occurrences of fossils are used to determine zone boundaries. Correlating concurrent range zones is probably the most accurate method of determining time equivalence.

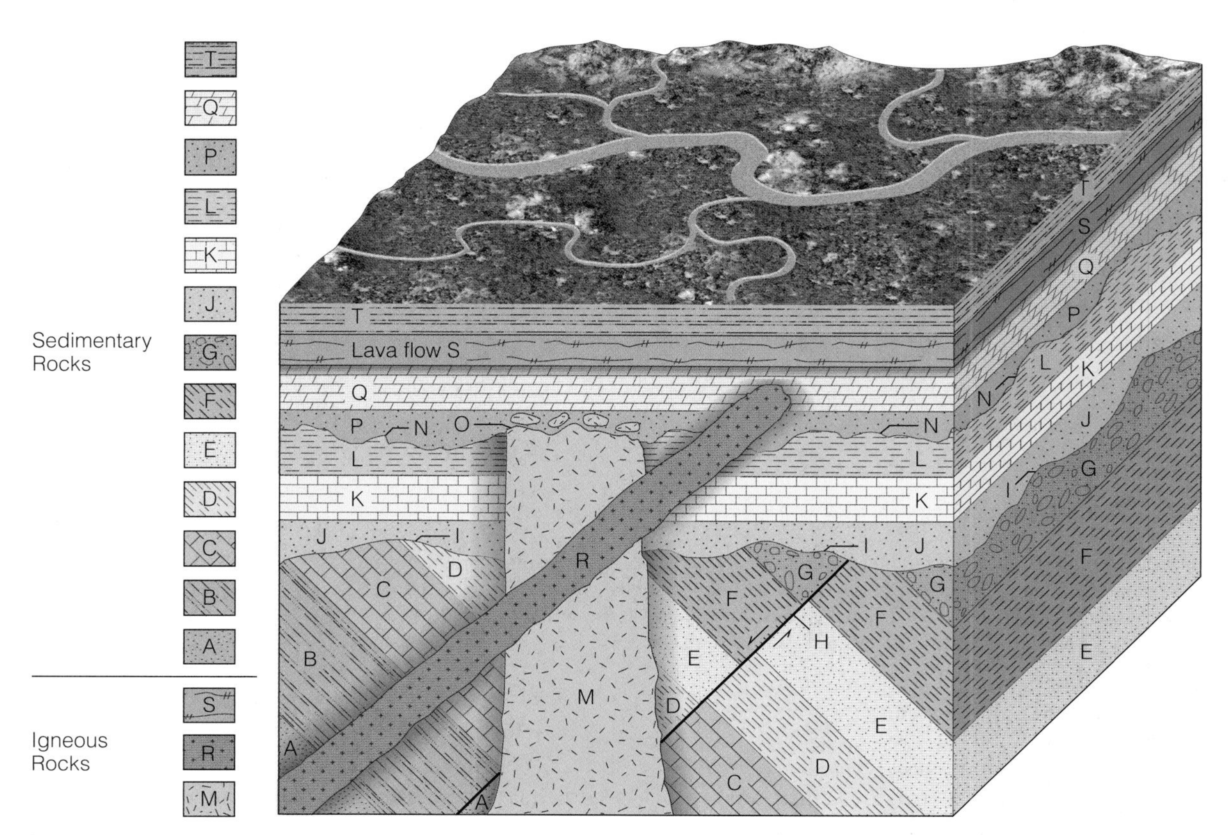

Figure 17.11 Block Diagram of a Hypothetical Area A block diagram of a hypothetical area in which the various relative dating principles can be applied to determine its geologic history. See Figure 17.12 to see how the geologic history was determined using relative dating principles.

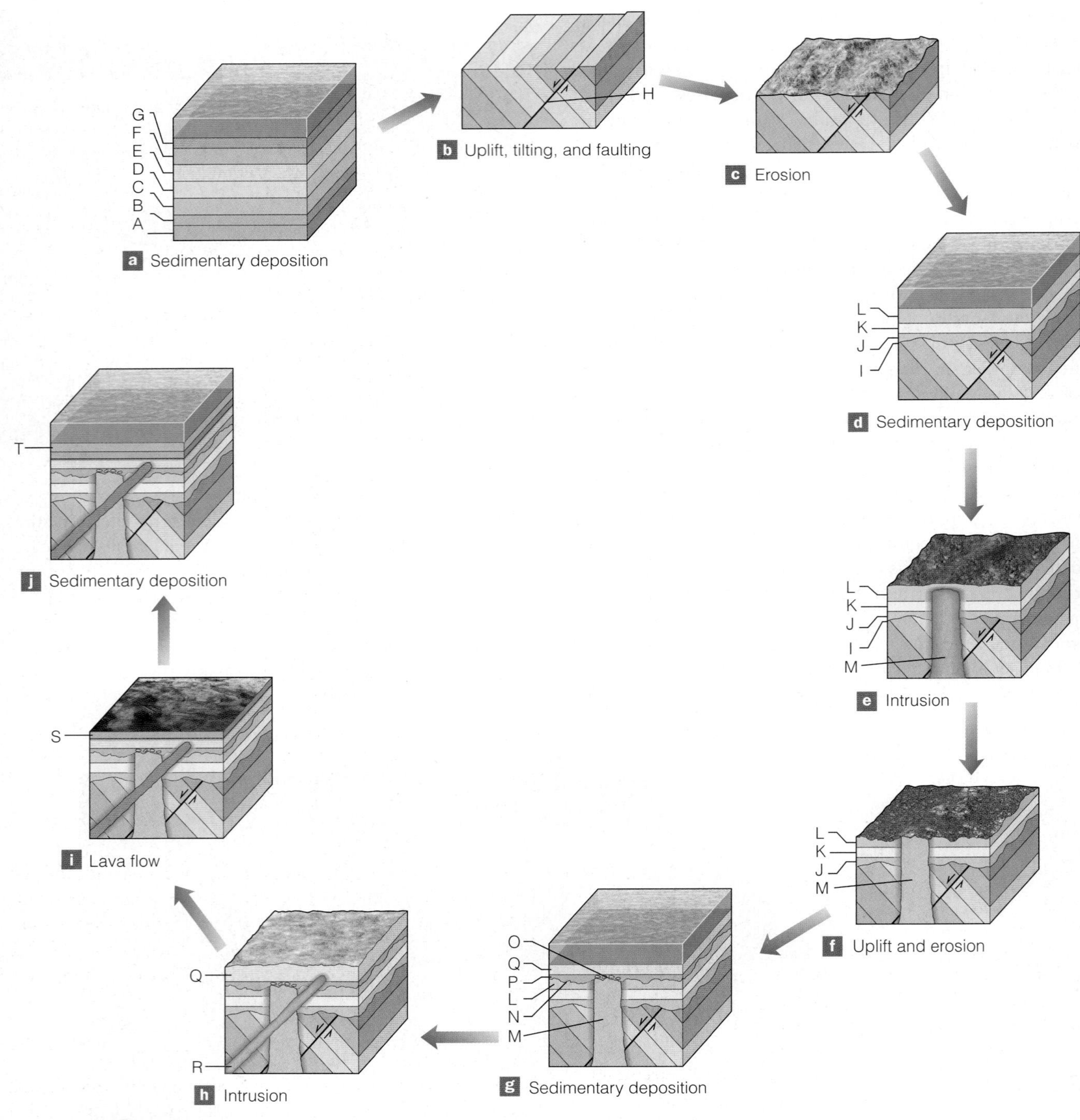

Figure 17.12 Using Relative Dating Principles to Interpret the Geologic History of a Hypothetical Area **a** Beds A–G are deposited, **b** The preceding beds are titled and faulted. **c** Erosion. **d** Beds J–L are deposited, producing an angular unconformity (I). **e** The entire sequence is intruded by a dike. **f** The entire sequence is uplifted and eroded. **g** Beds P and Q are deposited, producing a disconformity (N) and a nonconformity (O). **h** Dike R intrudes. **i** Lava S flows over bed Q, baking it. **j** Bed T is deposited.

Subsurface Correlation

In addition to surface geology, geologists are interested in subsurface geology because it provides additional information about geologic features beneath Earth's surface. A variety of techniques and methods are used to acquire and interpret data about the subsurface geology of an area.

When drilling is done for oil or natural gas, cores or rock chips called *well cuttings* are commonly recovered from the drill hole. These samples are studied under the microscope and reveal such important information as rock type, porosity (the amount of pore space), permeability (the ability to transmit fluids), and the presence of oil stains. In addition, the samples

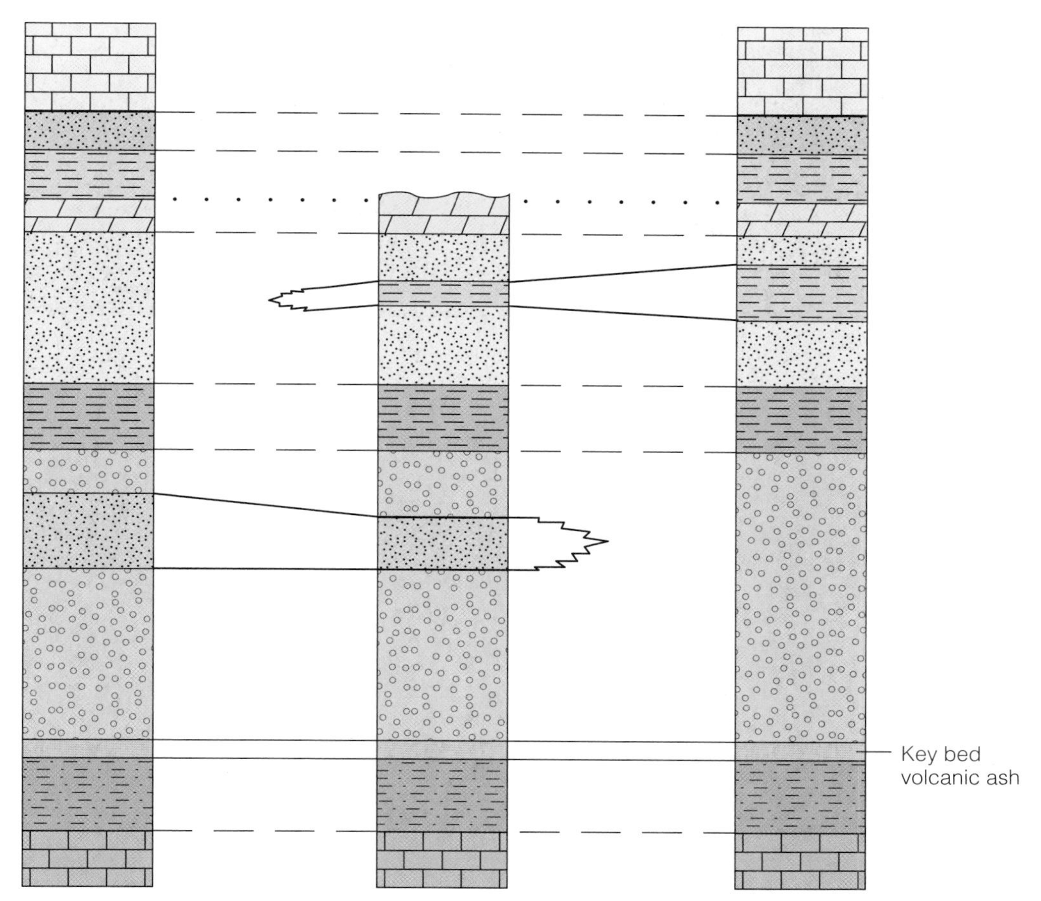

Figure 17.13 Correlating Rock Units In areas of adequate exposure, rock units can be traced laterally, even if occasional gaps exist, and correlated on the basis of similarity in rock type and position in a sequence. Rocks can also be correlated by a key bed—in this case, volcanic ash.

can be processed for a variety of microfossils that aid in determining the geologic age of the rock and the environment of deposition.

Geophysical instruments may be lowered down the drill hole to record such rock properties as electrical resistivity and radioactivity, thus providing a record or *well log* of the rocks penetrated. Cores, well cuttings, and well logs are all extremely useful in making subsurface correlations (Figure 17.17).

Subsurface rock units may also be detected and traced by the study of seismic profiles. Energy pulses, such as those from explosions, travel through rocks at a velocity determined by rock density, and some of this energy is reflected from various horizons (contacts between contrasting layers) back to the surface, where it is recorded (see Figure 9.3). Seismic stratigraphy is particularly useful in tracing units in areas such as the continental shelves, where it is very expensive to drill holes and other techniques have limited use.

ABSOLUTE DATING METHODS

Although most of the isotopes of the 92 naturally occurring elements are stable, some are radioactive and spontaneously decay to other more stable isotopes of elements, releasing energy in the process. The discovery in 1903 by Pierre and Marie Curie that radioactive decay produces heat meant that geologists finally had a mechanism for explaining Earth's internal heat that did not rely on residual cooling from a molten origin. Furthermore, geologists now had a powerful tool to date geologic events accurately and to verify the long time periods postulated by Hutton and Lyell.

Atoms, Elements, and Isotopes

As discussed in Chapter 3, all matter is made up of chemical elements, each composed of extremely small particles called *atoms*. The *nucleus* of an atom is composed of *protons* (positively charged particles) and *neutrons* (neutral particles), with *electrons* (negatively charged particles) encircling it (see Figure 3.2). The number of protons defines an element's *atomic number* and helps determine its properties and characteristics.

The combined number of protons and neutrons in an atom is its *atomic mass number*. However, not all atoms of the same element have the same number of neutrons in their nuclei. These variable forms of the same element are called *isotopes* (see Figure 3.4). Most isotopes are stable, but some are unstable and spontaneously decay to a more stable form. It is the decay rate of unstable isotopes that geologists measure to determine the absolute ages of rocks.

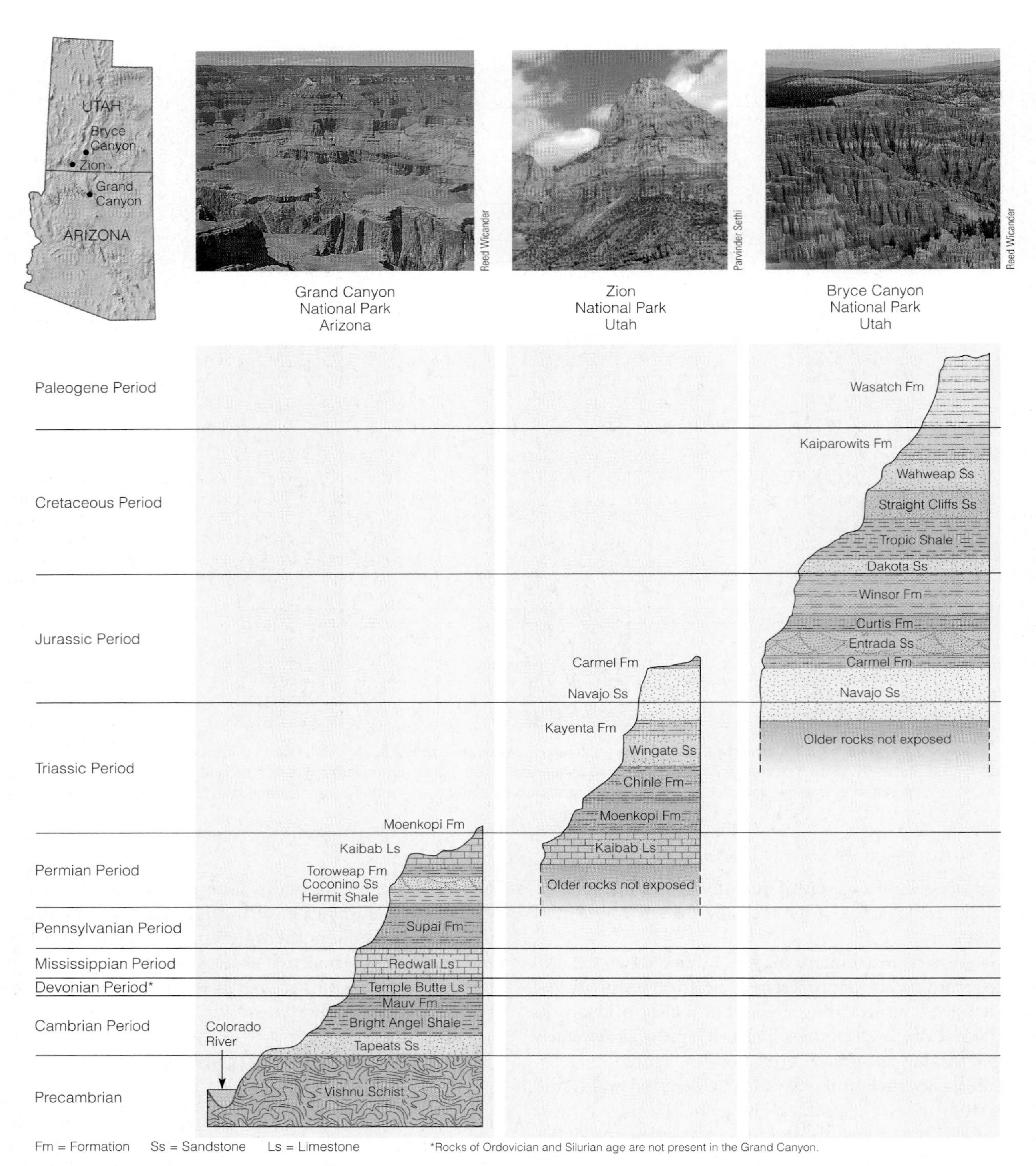

Figure 17.14 Correlation of Rock Units within the Colorado Plateau At each location, only a portion of the geologic record of the Colorado Plateau is exposed. By correlating the youngest rocks at one exposure with the oldest rocks at another exposure, geologists can determine the entire history of the region. For example, the rocks forming the rim of the Grand Canyon, Arizona, are the Kaibab Limestone and Moenkopi Formation—the youngest rocks exposed in the Grand Canyon. The Kaibab Limestone and Moenkopi Formation are the oldest rocks exposed in Zion National Park, Utah, and the youngest rocks are the Navajo Sandstone and Carmel Formation. The Navajo Sandstone and Carmel Formation are the oldest rocks exposed in Bryce Canyon National Park, Utah. By correlating the Kaibab Limestone and Moenkopi Formation between the Grand Canyon and Zion National Park, geologists have extended the geologic history from the Precambrian to the Jurassic. And by correlating the Navajo Sandstone and Carmel Formation between Zion and Bryce Canyon National Parks, geologists can extend the geologic history through the Paleogene Period. Thus, by correlating the rock exposures between these areas and applying the principle of superposition, geologists can reconstruct the geologic history of the region.

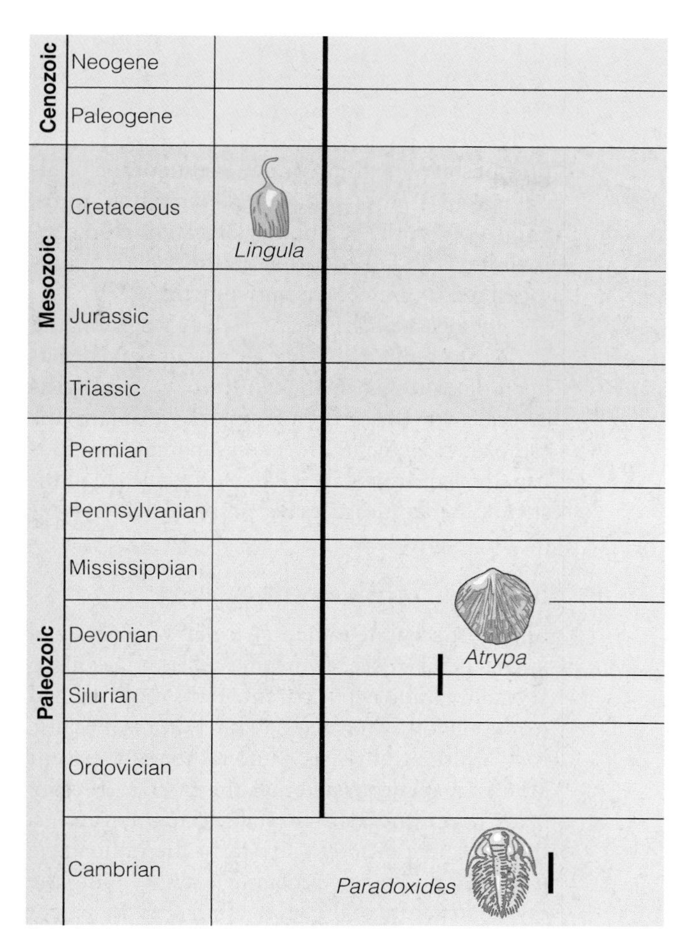

Figure 17.15 Guide Fossils Comparison of the geologic ranges (heavy vertical lines) of three marine invertebrate animals. *Lingula* is of little use in correlation because it has such a long range. But *Atrypa* and *Paradoxides* are good guide fossils because both are widespread, easily identified, and have short geologic ranges. Thus, both can be used to correlate rock units that are widely separated and to establish the relative age of a rock that contains them.

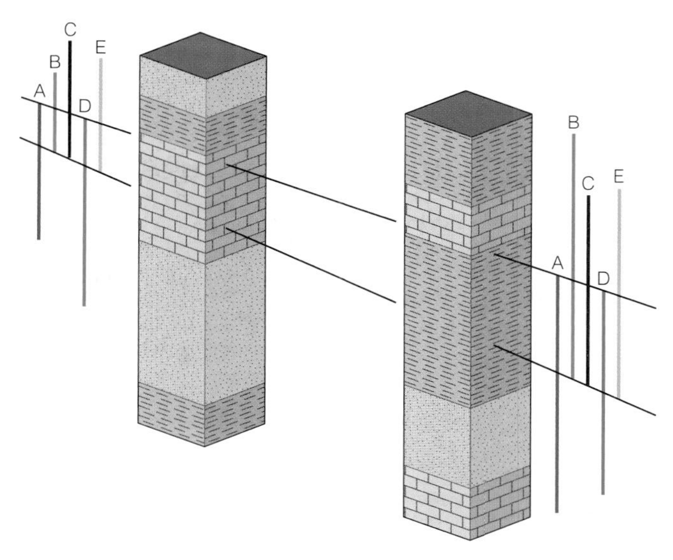

Figure 17.16 Correlation of Two Sections Using Concurrent Range Zones This concurrent range zone was established by the overlapping geologic ranges of fossils symbolized here by the letters A through E. The concurrent range zone is of shorter duration than any of the individual fossil geologic ranges. Correlating by concurrent range zones is probably the most accurate method of determining time equivalence.

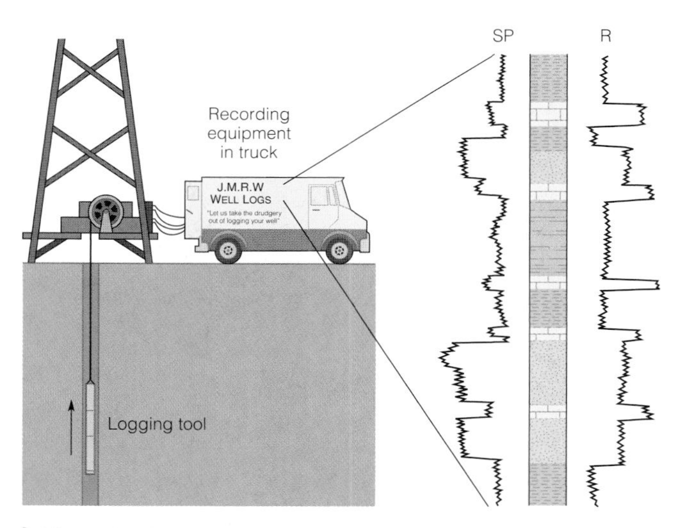

Figure 17.17 Well Logs A schematic diagram showing how well logs are made. As the logging tool is withdrawn from the drill hole, data are transmitted to the surface, where they are recorded and printed as a well log. The curve labeled SP in this diagrammatic electric log is a plot of self-potential (electrical potential caused by different conductors in a solution that conducts electricity) with depth. The curve labeled R is a plot of electrical resistivity with depth. Electric logs yield information about the rock type and fluid content of subsurface formations. Electric logs are also used to correlate from well to well.

Radioactive Decay and Half-Lives

Radioactive decay is the process whereby an unstable atomic nucleus is spontaneously transformed into an atomic nucleus of a different element. Scientists recognize three types of radioactive decay, all of which result in a change of atomic structure (Figure 17.18). In *alpha decay*, 2 protons and 2 neutrons are emitted from the nucleus, resulting in the loss of 2 atomic numbers and 4 atomic mass numbers. In *beta decay*, a fast-moving electron is emitted from a neutron in the nucleus, changing that neutron to a proton and consequently increasing the atomic number by 1, with no resultant atomic mass number change. *Electron capture* is when a proton captures an electron from an electron shell and thereby converts to a neutron, resulting in the loss of 1 atomic number, but not changing the atomic mass number.

Some elements undergo only one decay step in the conversion from an unstable form to a stable form. For example, rubidium 87 decays to strontium 87 by a single beta emission, and potassium 40 decays to argon 40 by a single electron capture. Other radioactive elements undergo several decay steps. Uranium 235 decays to lead 207 by 7 alpha and 6 beta steps, whereas uranium 238 decays to lead 206 by 8 alpha and 6 beta steps (Figure 17.19).

Figure 17.18 Three Types of Radioactive Decay

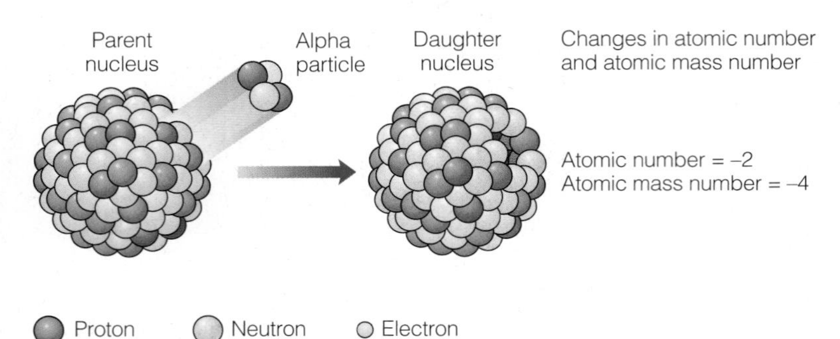

a Alpha decay, in which an unstable parent nucleus emits 2 protons and 2 neutrons.

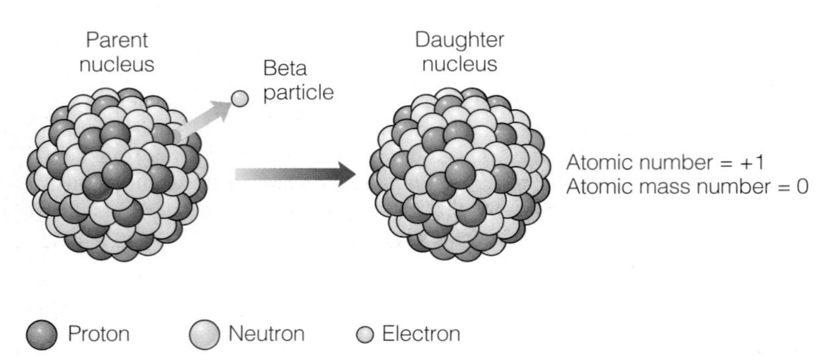

b Beta decay, in which an electron is emitted from the nucleus.

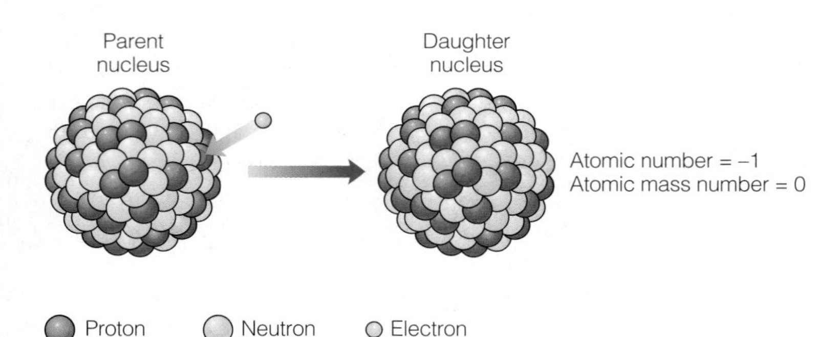

c Electron capture, in which a proton captures an electron and is thereby converted to a neutron.

When we discuss decay rates, it is convenient to refer to them in terms of half-lives. The **half-life** of a radioactive element is the time it takes for half of the atoms of the original unstable *parent element* to decay to atoms of a new, more stable *daughter element.* The half-life of a given radioactive element is constant and can be precisely measured. Half-lives of various radioactive elements range from less than a billionth of a second to 49 billion years.

Radioactive decay occurs at a geometric rate rather than a linear rate. Therefore, a graph of the decay rate produces a curve rather than a straight line (Figure 17.20). For example, an element with *1,000,000* parent atoms will have *500,000* parent atoms and *500,000* daughter atoms after one half-life. After two half-lives, it will have *250,000* parent atoms (one-half of the previous parent atoms, which is equivalent to one-fourth of the original parent atoms) and *750,000* daughter atoms. After three half-lives, it will have *125,000* parent atoms (one-half of the previous parent atoms, or one-eighth of the original parent atoms) and *875,000* daughter atoms, and so on, until the number of parent atoms remaining is so few that they cannot be accurately measured by present-day instruments.

By measuring the parent–daughter ratio and knowing the half-life of the parent (which has been determined in the laboratory), geologists can calculate the age of a sample that contains the radioactive element. The parent–daughter ratio is usually determined by a *mass spectrometer*, an instrument that measures the proportions of atoms of different masses.

Sources of Uncertainty

The most accurate radiometric dates are obtained from igneous rocks. As magma cools and begins to crystallize, radioactive parent atoms are separated from previously formed daughter atoms. Because they are the right size, some radioactive parent atoms are incorporated into the crystal structure of certain minerals. The stable daughter atoms, though, are a different size from the radioactive parent atoms and consequently cannot fit into the crystal structure of the same mineral as the parent atoms. Therefore, a mineral crystallizing in cooling magma will contain radioactive parent atoms but no stable daughter atoms (Figure 17.21). Thus, the time that is being measured is the time of crystallization of the mineral that contains the radioactive atoms and not the time of formation of the radioactive atoms.

Except in unusual circumstances, sedimentary rocks cannot be radiometrically dated because one would be measuring the age of a particular mineral rather than the time that it was deposited as a sedimentary particle. One of the few instances in which radiometric dates can be obtained on sedimentary rocks is when the mineral glauconite is present. Glauconite is a greenish mineral containing potassium 40, which decays to argon 40 (Table 17.1). It forms in certain marine environments as a result of chemical reactions with clay minerals during the conversion of sediments to sedimentary rock. Thus, glauconite forms when the sedimentary rock forms, and a radiometric date indicates the time of the sedimentary rock's origin. Being a gas, however, the daughter product argon can easily escape from a mineral. Therefore, any date obtained from glauconite, or any other mineral containing the potassium 40 and argon 40 pair, must be considered a minimum age.

To obtain accurate radiometric dates, geologists must be sure that they are dealing with a *closed system*, meaning that neither parent nor daughter atoms have been added or

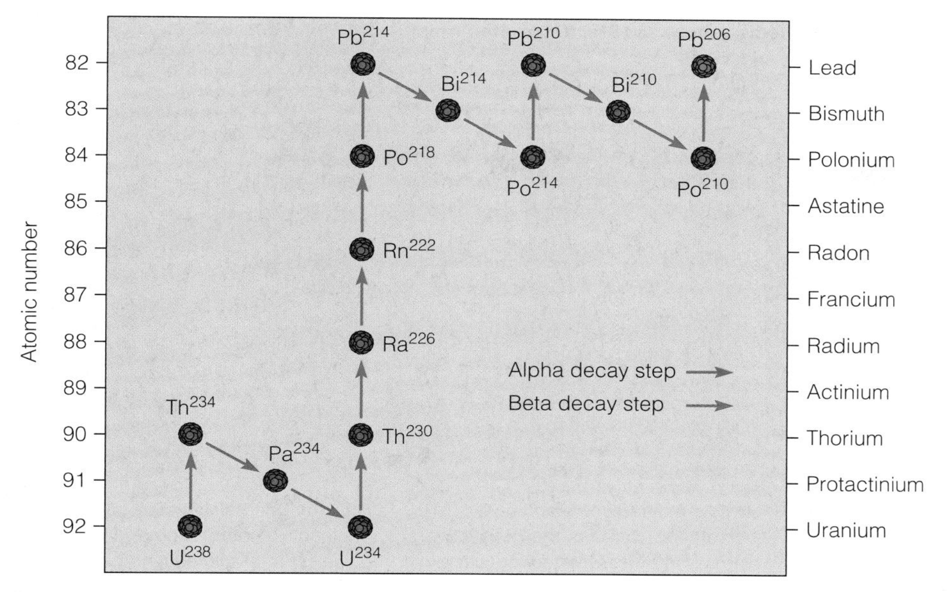

Figure 17.19 Radioactive Decay Series for Uranium 238 to Lead 206 Radioactive uranium 238 decays to its stable daughter product, lead 206, by 8 alpha and 6 beta decay steps. A number of different isotopes are produced as intermediate steps in the decay series.

removed from the system since crystallization, and that the ratio between them results only from radioactive decay. Otherwise, an inaccurate date will result. If daughter atoms have leaked out of the mineral being analyzed, the calculated age will be too young; if parent atoms have been removed, the calculated age will be too old.

Leakage may take place if the rock is heated or subjected to intense pressure as can sometimes occur during metamorphism. If this happens, some of the parent or daughter atoms may be driven from the mineral being analyzed, resulting in an inaccurate age determination. If the daughter product was completely removed, then one would be measuring the time since metamorphism (a useful measurement itself) and not the time since crystallization of the mineral (Figure 17.22).

Because heat and pressure affect the parent–daughter ratio, metamorphic rocks are difficult to date accurately. Remember that although the resulting parent–daughter ratio of the sample being analyzed may have been affected by heat, the decay rate of the parent element remains constant, regardless of any physical or chemical changes.

To obtain an accurate radiometric date, geologists must make sure that the sample is fresh and unweathered, and that it has not been subjected to high temperature or intense pressures after crystallization. Furthermore, it is sometimes possible to cross-check the radiometric date obtained by measuring the parent–daughter ratio of two different radioactive elements in the same mineral.

For example, naturally occurring uranium consists of both uranium 235 and uranium 238 isotopes. Through various decay steps, uranium 235 decays to lead 207, whereas uranium 238 decays to lead 206 (Figure 17.19). If the minerals that contain both uranium isotopes have remained closed systems, the ages obtained from each parent–daughter ratio should agree closely. If they do, they are said to be *concordant*, thus reflecting the time of crystallization of the

Figure 17.20 Uniform, Linear Change Compared to Geometric Radioactive Decay

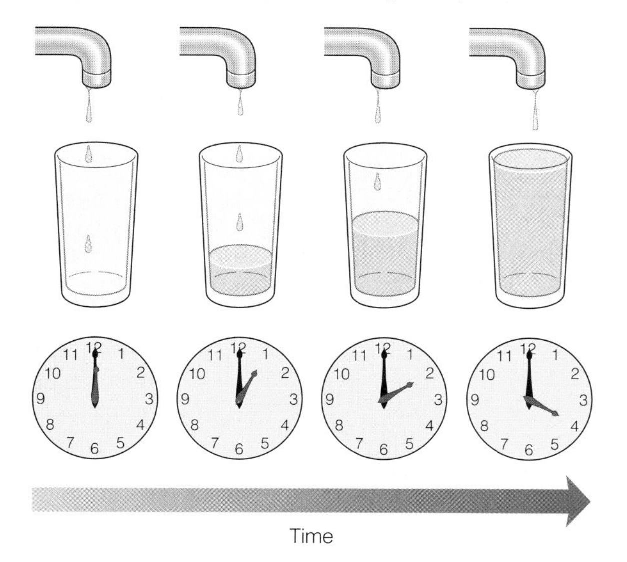

a Uniform, linear change is characteristic of many familiar processes. In this example, water is being added to a glass at a constant rate.

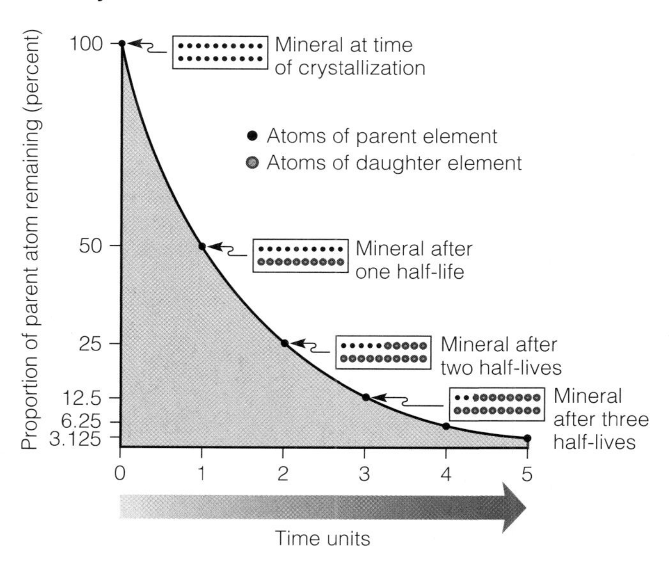

b A geometric radioactive decay curve, in which each time unit represents one half-life, and each half-life is the time it takes for half of the parent element to decay to the daughter element.

Figure 17.21 Crystallization of Magma Containing Radioactive Parent and Stable Daughter Atoms

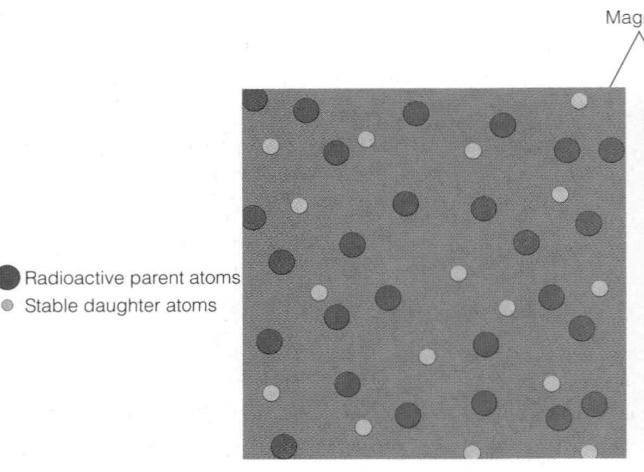

a Magma contains both radioactive parent atoms and stable daughter atoms. The radioactive parent atoms are larger than the stable daughter atoms.

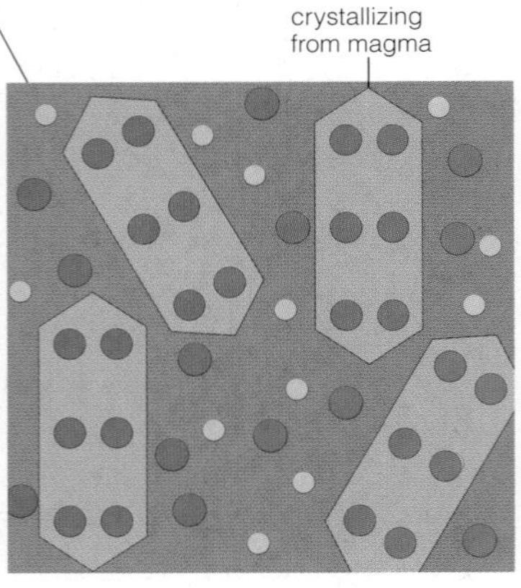

b As magma cools and begins to crystallize, some of the radioactive atoms are incorporated into certain minerals because they are the right size and can fit into the crystal structure. In this example, only the larger radioactive parent atoms fit into the crystal structure. Therefore, at the time of crystallization, minerals in which the radioactive parent atoms can fit into the crystal structure will contain 100% radioactive parent atoms and 0% stable daughter atoms.

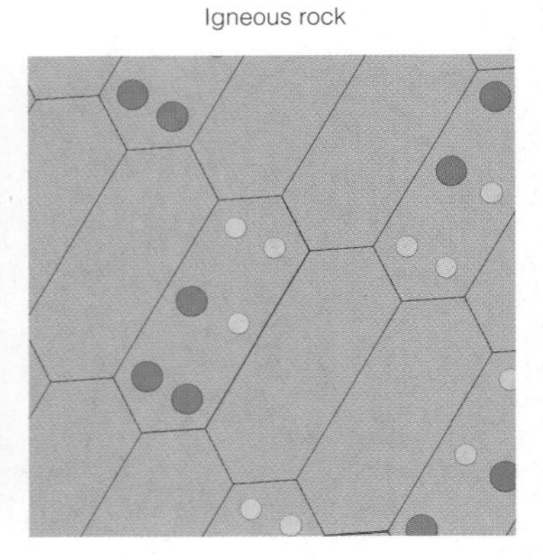

c After one half-life, 50% of the radioactive parent atoms will have decayed to stable daughter atoms, such that those minerals that had radioactive parent atoms in their crystal structure will now have 50% radioactive parent atoms and 50% stable daughter atoms.

magma. If the ages do not closely agree, then they are said to be *disconcordant*, and other samples must be taken and ratios measured to see which, if either, date is correct.

Recent advances and the development of new techniques and instruments for measuring various isotope ratios have enabled geologists to analyze not only increasingly smaller samples, but with a greater precision than ever before. Presently, the measurement error for many radiometric dates is typically less than 0.5% of the age, and in some cases is even better than 0.1%. Thus, for a rock 540 million years old (near the beginning of the Cambrian Period), the possible error could range from nearly 2.7 million years to less than 540,000 years.

Long-Lived Radioactive Isotope Pairs

Table 17.1 shows the five common, long-lived parent–daughter isotope pairs used in radiometric dating. Long-lived pairs have half-lives of millions or billions of years. All of these were present when Earth formed and are still present in measurable quantities. Other shorter-lived radioactive isotope pairs have decayed to the point that only small quantities near the limit of detection remain.

The most commonly used isotope pairs are the uranium–lead and thorium–lead series, which are used principally to date ancient igneous intrusives, lunar samples, and some

TABLE 17.1 Five of the Principal Long-Lived Radioactive Isotope Pairs Used in Radiometric Dating

ISOTOPES Parent	Daughter	Half-Life of Parent (years)	Effective Dating Range (years)	Minerals and Rocks That Can Be Dated
Uranium 238	Lead 206	4.5 billion	10 million to 4.6 billion	Zircon Uraninite
Uranium 235	Lead 207	704 million		
Thorium 232	Lead 208	14 billion		
Rubidium 87	Strontium 87	48.8 billion	10 million to 4.6 billion	Muscovite Biotite Potassium feldspar Whole metamorphic or igneous rock
Potassium 40	Argon 40	1.3 billion	100,000 to 4.6 billion	Glauconite Hornblende Muscovite Whole volcanic rock Biotite

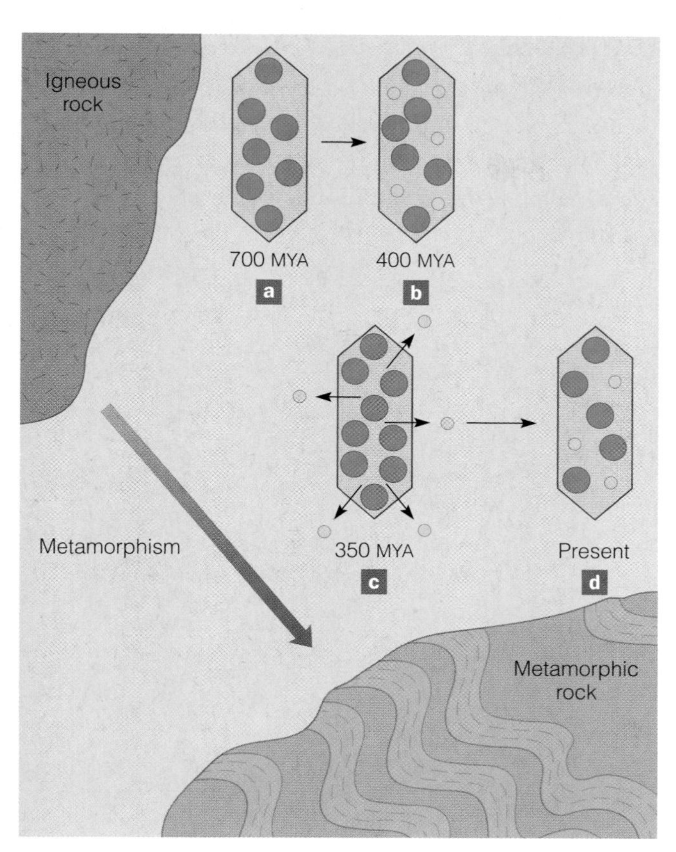

Figure 17.22 Effects of Metamorphism on Radiometric Dating The effect of metamorphism in driving out daughter atoms from a mineral that crystallized 700 million years ago (mya). The mineral is shown immediately after crystallization a, then at 400 million years b, when some of the parent atoms had decayed to daughter atoms. Metamorphism at 350 mya c, drives the daughter atoms out of the mineral into the surrounding rock. If the rock has remained a closed chemical system throughout its history, dating the mineral today d, yields the time of metamorphism, whereas dating the whole rock provides the time of its crystallization, 700 mya.

meteorites. The rubidium–strontium pair is also used for very old samples and has been effective in dating the oldest rocks on Earth, as well as meteorites.

The potassium–argon method is typically used for dating fine-grained volcanic rocks from which individual crystals cannot be separated; hence, the whole rock is analyzed. Because argon is a gas, great care must be taken to ensure that the sample has not been subjected to heat, which would allow argon to escape; such a sample would yield an age that is too young. Other long-lived radioactive isotope pairs exist, but they are rather rare and are used only in special situations.

Fission Track Dating

The emission of atomic particles that result from the spontaneous decay of uranium within a mineral damages its crystal structure. The damage appears as microscopic linear tracks that are visible only after the mineral has been etched with hydrofluoric acid, an acid so powerful that its vapors can destroy one's sense of smell without careful handling. The age of the sample is determined from the number of fission tracks present and the amount of uranium that the sample contains: the older the sample, the greater the number of tracks (Figure 17.23).

Fission track dating is of particular interest to archaeologists and geologists because the technique can be used to date samples ranging from only a few hundred to hundreds of millions of years old. It is most useful for dating samples between approximately 40,000 and 1.5 million years ago, a period for which other dating techniques are not always particularly suitable. One of the problems in fission track dating occurs when the rocks have later been subjected to high temperatures. If this happens, the damaged crystal structures are repaired by annealing, and consequently the tracks disappear. In such instances, the calculated age will be younger than the actual age.

Radiocarbon and Tree-Ring Dating Methods

Carbon is an important element in nature and is one of the basic elements found in all forms of life. It has three isotopes; two of these, carbon 12 and 13, are stable, whereas carbon 14 is radioactive (see Figure 3.4). Carbon 14 has a half-life of 5730 years plus or minus 30 years. The **carbon-14 dating technique** is based on the ratio of carbon 14 to carbon 12 and is generally used to date formerly living material.

The short half-life of carbon 14 makes this dating technique practical only for specimens younger than about 70,000 years. Consequently, the carbon-14 dating method is especially useful in archaeology and has greatly helped unravel the events of the latter portion of the Pleistocene Epoch. For example, carbon-14 dates of maize from the Tehuacan Valley of Mexico have forced archaeologists to rethink their ideas of where the first center for maize domestication in Mesoamerica arose. Carbon-14 dating is also helping to answer the question of when humans began populating North America.

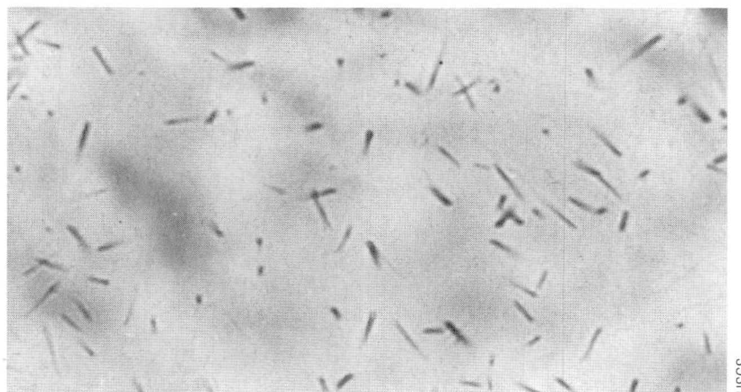

Figure 17.23 Fission Track Dating Each fission track (about 16 microns [= 16/1000 mm] long) in this apatite crystal is the result of the radioactive decay of a uranium atom. The apatite crystal, which has been etched with hydrofluoric acid to make the fission tracks visible, comes from one of the dikes at Shiprock, New Mexico, and has a calculated age of 27 million years.

Geo-inSight Uluru and Kata Tjuta

Rising majestically above the surrounding flat desert of central Australia are Uluru and Kata Tjuta. Uluru and Kata Tjuta are the aboriginal names for what most people know as Ayers Rock and The Olgas. The history of Uluru and Kata Tjuta began approximately 550 million years ago when a huge mountain range formed in what is now central Australia. It subsequently eroded, and vast quantities of gravel were transported by streams and deposited along its base to form large alluvial fans. Marine sediments then covered the alluvial fans and the entire region was uplifted by tectonic forces between 400 and 300 million years ago and then subjected to weathering.

The spectacular and varied rock shapes of Uluru and Kata Tjuta are the result of millions of years of weathering and erosion by water and, to a lesser extent, wind acting on the fractures formed during uplift. Differences in the composition and texture of the rocks also played a role in sculpting these colorful and magnificent structures.

▶ **1.** Location map of Uluru and Kata Tjuta, Australia.

◀ **2.** Aerial view of Uluru with Kata Tjuta in the background. Contrary to popular belief, Uluru is not a giant boulder. Rather, it is the exposed portion of the nearly vertically tilted Uluru Arkose. The caves, cavems, and depressions visible on the northeastern side are the result of weathering.

Reed Wicander

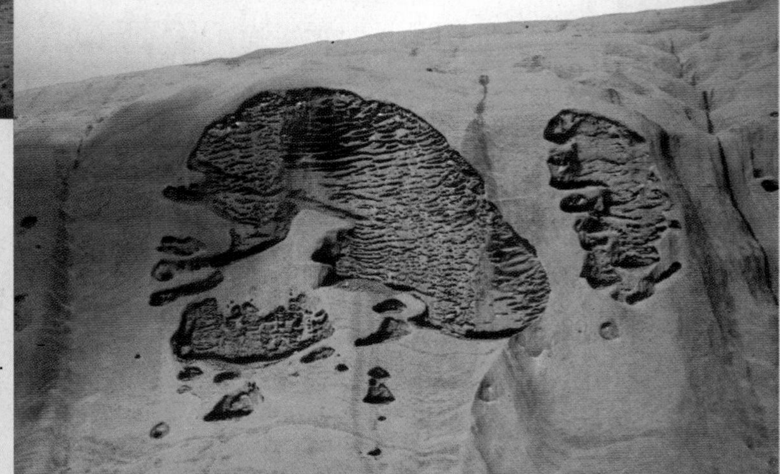

Reed Wicander

▶ **3.** A close-up view of the brain-and honeycomb-like small caves seen on the northeastern side of Uluru.

Reed Wicander

◀ **4.** Uluru at sunset. The near-vertical tilting of the sedimentary beds of the Uluru Arkose that make up Uluru can be seen clearly. Differential weathering of the sedimentary layers has produced the distinct parallel ridges and other features characteristic of Uluru.

▶ **5.** Aerial view of Kata Tjuta with Uluru in the background. Kata Tjuta is composed of the Mount Currie Conglomerate, a coarse-grained and poorly sorted conglomerate. The sediments that were lithified into the Mount Currie Conglomerate were deposited, like the Uluru Arkose, as an alluvial fan beginning approximately 550 million years ago.

Reed Wicander

Reed Wicander

◀ **6.** A view of the rounded domes of Kata Tjuta and typical vegetation as seen from within one of its canyons. The red color of the rocks is the result of oxidation of iron in the sediments.

▶ **7.** An aerial close-up view of Kata Tjuta. The distinctive dome shape of the rocks is the result of weathering and erosion of the Mount Currie Conglomerate. In addition to weathering, the release of pressure on the buried rocks when they were exposed at the surface by tectonic forces contributed to the characteristic rounded shapes of Kata Tjuta.

Carbon 14 is constantly formed in the upper atmosphere when cosmic rays, which are high-energy particles (mostly protons), strike the atoms of upper-atmospheric gases, splitting their nuclei into protons and neutrons. When a neutron strikes the nucleus of a nitrogen atom (atomic number 7, atomic mass number 14), it may be absorbed into the nucleus and a proton emitted. Thus, the atomic number of the atom decreases by 1, whereas the atomic mass number stays the same. Because the atomic number has changed, a new element, carbon 14 (atomic number 6, atomic mass number 14), is formed. The newly formed carbon 14 is rapidly assimilated into the carbon cycle and, along with carbon 12 and 13, is absorbed in a nearly constant ratio by all living organisms (Figure 17.24). When an organism dies, however, carbon 14 is not replenished, and the ratio of carbon 14 to carbon 12 decreases as carbon 14 decays back to nitrogen by a single beta decay step (Figure 17.24).

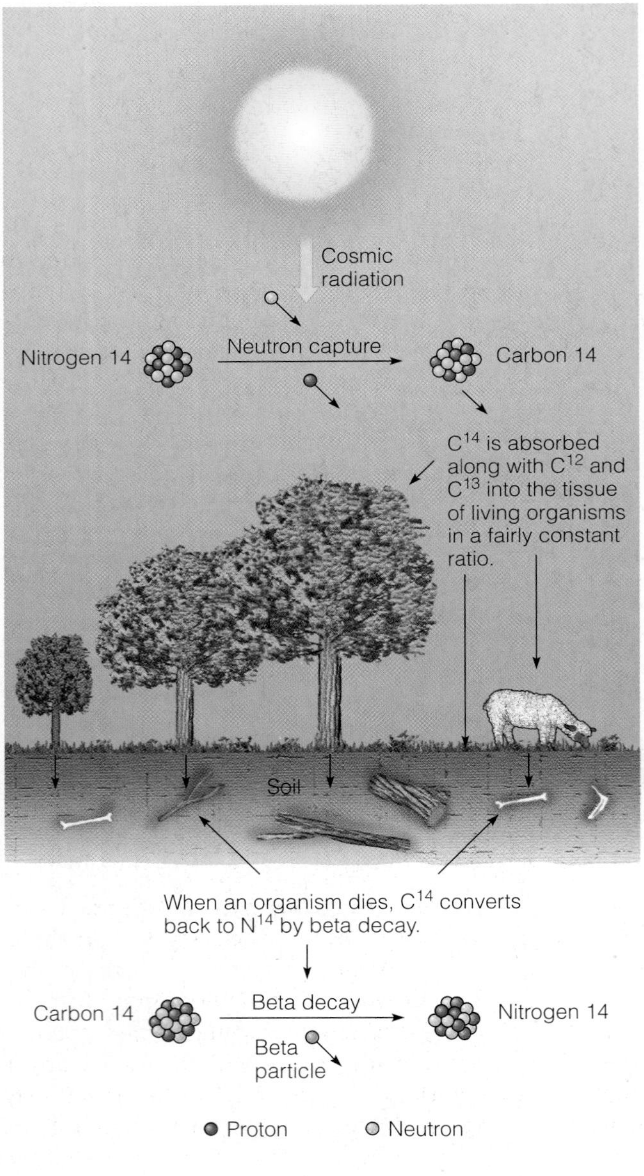

Figure 17.24 Carbon-14 Dating Method The carbon cycle showing the formation of carbon 14 in the upper atmosphere, its dispersal and incorporation into the tissues of all living organisms, and its decay back to nitrogen 14 by beta decay.

Currently, the ratio of carbon 14 to carbon 12 is remarkably constant both in the atmosphere and in living organisms. There is good evidence, however, that the production of carbon 14, and thus the ratio of carbon 14 to carbon 12, has varied somewhat during the past several thousand years. This was determined by comparing ages established by carbon-14 dating of wood samples with ages established by counting annual tree rings in the same samples. As a result, carbon-14 ages have been corrected to reflect such variations in the past.

Tree-ring dating is another useful method for dating geologically recent events. The age of a tree can be determined by counting the growth rings in the lower part of the trunk. Each ring represents one year's growth, and the pattern of wide and narrow rings can be compared among trees to establish the exact year in which the rings were formed. The procedure of matching ring patterns from numerous trees and wood fragments in a given area is called *cross-dating*.

By correlating distinctive tree-ring sequences from living and nearby dead trees, scientists can construct a time scale that extends back approximately 14,000 years (Figure 17.25). By matching ring patterns to the composite ring scale, wood samples whose ages are not known can be accurately dated.

The applicability of tree-ring dating is somewhat limited because it can be used only where continuous tree records are found. It is therefore most useful in arid regions, particularly the southwestern United States, where trees live a very long time.

DEVELOPMENT OF THE GEOLOGIC TIME SCALE

The geologic time scale is a hierarchical scale in which the 4.6-billion-year history of Earth is divided into time units of varying duration (Figure 17.1). It did not result from the work of any one individual, but rather evolved, primarily during the 19th century, through the efforts of many people.

By applying relative-dating methods to rock outcrops, geologists in England and western Europe defined the major geologic time units without the benefit of radiometric dating techniques. Using the principles of superposition and fossil succession, they correlated various rock exposures and pieced together a composite geologic section. This composite section is, in effect, a relative time scale because the rocks are arranged in their correct sequential order.

By the beginning of the 20th century, geologists had developed a relative geologic time scale, but did not yet have any absolute dates for the various time–unit boundaries. Following the discovery of radioactivity near the end of the 19th century, radiometric dates were added to the relative geologic time scale (Figure 17.1).

Because sedimentary rocks, with rare exceptions, cannot be radiometrically dated, geologists have had to

Figure 17.25 Tree-Ring Dating Method In the cross-dating method, tree-ring patterns from different woods are compared to establish a ring-width chronology backward in time.

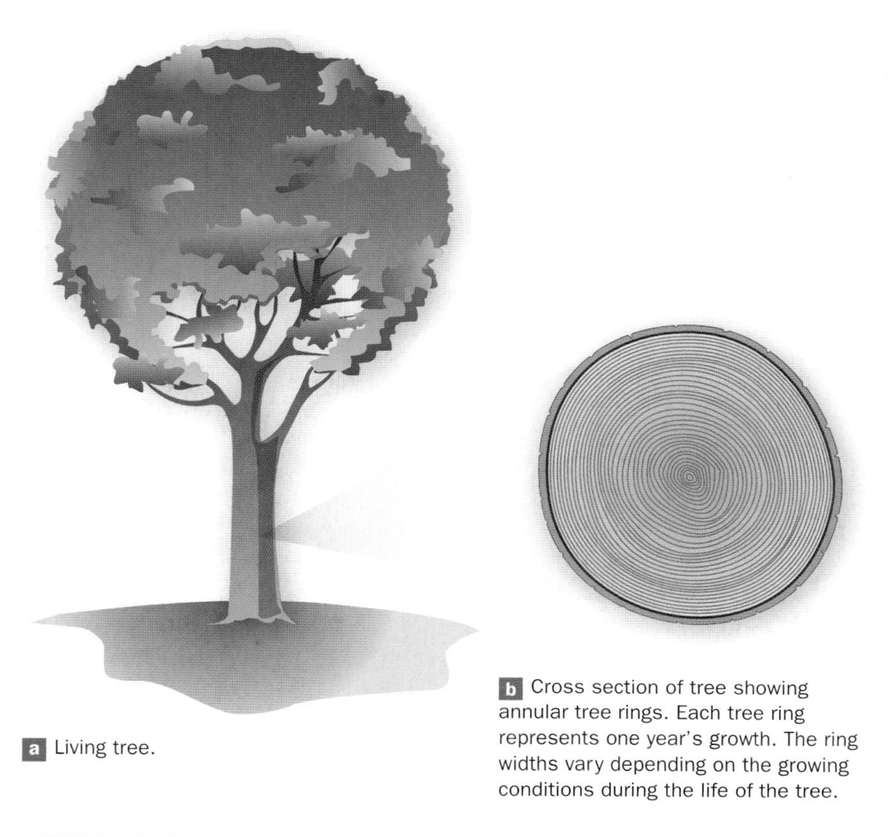

a Living tree.

b Cross section of tree showing annular tree rings. Each tree ring represents one year's growth. The ring widths vary depending on the growing conditions during the life of the tree.

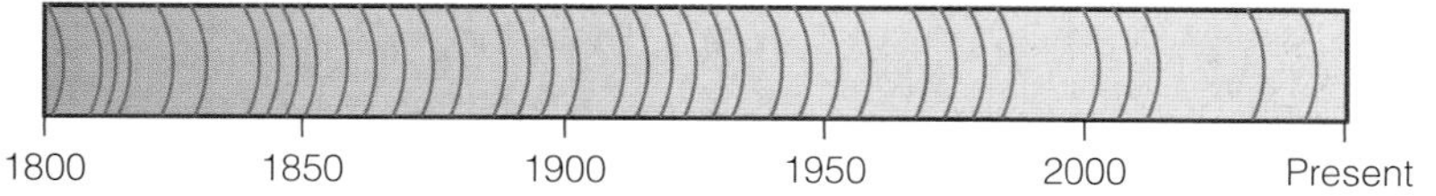

c By matching up tree ring patterns from numerous trees in a region, a master tree ring chronology can be constructed. Note that these tree rings are not to scale, but are diagrammatic to show how a master tree ring chronology would work.

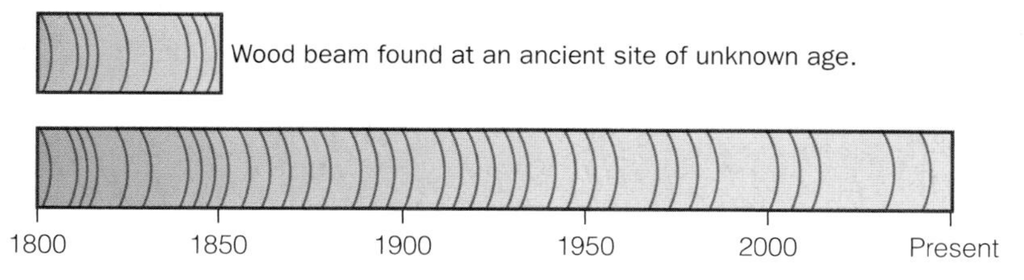

d Matching the tree ring pattern from a wood beam found at an ancient site to the same pattern of the master tree ring chronology for that region, the age of the site can be determined, at least to the time the tree died or was cut down.

rely on interbedded volcanic rocks and igneous intrusions to apply absolute dates to the boundaries of the various subdivisions of the geologic time scale (Figure 17.26). An ashfall or lava flow provides an excellent marker bed that is a time-equivalent surface, supplying a minimum age for the sedimentary rocks below and a maximum age for the rocks above. Ashfalls are particularly useful because they may fall over both marine and nonmarine sedimentary environments and can provide a connection between these different environments.

Thousands of absolute ages are now known for sedimentary rocks of known relative ages, and these absolute dates have been added to the relative time scale. In this way, geologists have been able to determine both the absolute ages of the various geologic periods and their durations (Figure 17.1). In fact, the dates for the era, period, and epoch boundaries of the geologic time scale are still being refined as more accurate dating methods are developed and new exposures dated. The ages shown in Figures 1.17 and 17.1 are the most recently published ages as of 2004.

STRATIGRAPHY AND STRATIGRAPHIC TERMINOLOGY

The recognition of a relative geologic time scale brought some order to *stratigraphy* (the study of the composition, origin, areal distribution, and age relationships of layered rocks); however, problems still remained because many sedimentary rock units are time transgressive. This means that they were deposited during one geologic period in a particular area, and during another period elsewhere (see Figure 6.22). Therefore, modern stratigraphic terminology includes two fundamentally different kinds of units to deal with both rocks and time: those defined by their content and those related to geologic time (Table 17.2).

Units defined by their content include **lithostratigraphic** and **biostratigraphic units**. Lithostratigraphic (*lith-* and *litho-* are prefixes meaning "stone" or "stonelike") units are defined by the physical attributes of the rocks, such as rock type (for example, sandstone or limestone), with no consideration of time of origin. The basic lithostratigraphic unit is the *formation*, which is a mappable body of rock with distinctive upper and lower boundaries (Figure 17.27). A formation may consist of a single rock type (e.g., the Redwall Limestone, Figure 17.14), or a variety of related rock types (e.g., the Morrison Formation, see Figure 22.17). Formations are commonly subdivided into smaller units known as *members* and *beds*, and they may be parts of larger units known as *groups* and *supergroups* (Table 17.2).

Geo-Focus Denver's Weather—280 Million Years Ago!

With all of the concern about global climate change, it might be worthwhile to step back a bit and look at climate change from a geologic perspective. We're all aware that some years are hotter than others and some years we have more rain, but generally things tend to average out over time. We know that it will be hot in the summer in Arizona, and it will be very cold in Minnesota in the winter. We also know that scientists, politicians, and concerned people everywhere are debating whether humans are partly responsible for the global warming that Earth seems to be experiencing.

What about long-term climate change? We know that Earth has experienced periods of glaciation in the past—for instance, during the Precambrian, the end of the Ordovician Period, and most recently during the Pleistocene Epoch. Earth has also undergone large-scale periods of aridity, such as during the end of the Permian and beginning of the Triassic periods. Such long-term climatic changes are probably the result of slow geographic changes related to plate tectonic activity. Not only are continents carried into higher and lower latitudes, but their movement affects ocean circulation and atmospheric circulation patterns, which in turn affect climate, and result in climate changes.

Even though we can't physically travel back in time, geologists can reconstruct what the climate was like in the past. The distribution of plants and animals is controlled, in part, by climate. Plants are particularly sensitive to climate change and many can only live in particular environments. The fossils of plants and animals can tell us something about the environment and climate at the time that these organisms were living. Furthermore, climate-sensitive sedimentary rocks can be used to interpret past climatic conditions. Desert dunes are typically well sorted and exhibit large-scale cross-bedding. Coals form in freshwater swamps where climatic conditions promote abundant plant growth. Evaporites such as rock salt result when evaporation exceeds precipitation, such as in desert regions or along hot, dry shorelines. Tillites (glacial sediments) result from glacial activity and indicate cold, wet environments. So by combining all relevant geologic and paleontologic information, geologists can reconstruct what the climate was like in the past and how it has changed over time at a given locality.

In a recently published book titled *Ancient Denvers: Scenes from the Past 300 Million Years of the Colorado Front Range* by Kirk R. Johnson depicts what Denver, Colorado, looked like at 13 different time periods in the past. The time slices begin during the Pennsylvanian Period, 300 million years ago, and end with a view of the Front Range amid a spreading wave of houses on the southern edge of metropolitan Denver.

The information for piecing together Denver's geologic past was derived mainly from a 688-m-deep well drilled by the Denver Museum of Nature and Science beneath Kiowa, Colorado, in 1999. Using the information gleaned from the rocks recovered from the well, plus additional geologic evidence from other parts of the area, museum scientists and artists were able to reconstruct Denver's geologic past.

Beginning 300 million years ago (Pennsylvanian Period), the Denver area had coastlines on its eastern and western borders and a mountain range (not the Rocky Mountains of today). The climate was mostly temperate with lots of seedless vascular plants, such as ferns, as well as very tall scale trees related to the modern horsetail rush. Huge insects such as millipedes, cockroaches, and dragonflies shared this region with relatively small fin-backed reptiles and a variety of amphibians.

By 280 million years ago, the area was covered by huge sand seas, much like the Sahara is today (Figure 1). This change in climate and landscape was the result of the formation of Pangaea. As the continents collided, arid and semiarid conditions prevailed over much of the supercontinent, and the Denver area was no exception.

During the late Jurassic (150 million years ago), herds of plant-eating dinosaurs such as *Apatosaurus* roamed throughout the Denver area, feasting on the succulent and abundant vegetation.

Figure 1 Denver as it appeared 280 million years ago. As a result of the collision of continents and the formation of Pangaea, the world's climate was generally arid, and Denver was no exception. Denver was probably covered by great seas of sand, much as the Sahara is today.

Grasses and flowering plants had not yet evolved, so the dinosaurs ate the ferns and gymnosperms that were abundant at this time.

As a result of rising sea level, Denver was covered by a warm, shallow sea 70 million years ago (late Cretaceous). Marine reptiles such as plesiosaurs and mosasaurs ruled these seas, while overhead, pterosaurs soared through the skies, looking for food (Figure 2).

Beginning approximately 66 million years ago, the Rocky Mountains began to form as tectonic forces started a mountain-building episode known as the *Laramide orogeny* that resulted in the present-day Rocky Mountains. Dinosaurs still roamed the land around Denver, and flowering plants began their evolutionary history.

By 55 million years ago (Eocene), the world was in the grip of an intense phase of global warming. A subtropical rainforest with many trees that would be recognizable today filled the landscape. Primitive mammals were becoming more abundant, and many warm-climate-loving animals could be found living north of the Arctic Circle.

Although ice caps still covered portions of North America, mammoths and other mammals wandered among the plains of Denver 16,000 years ago (Figure 3). Mastodons, horses, bison, lions, and giant ground sloths, to name a few, all lived in this region, and their fossils can be found in the sedimentary rocks from this area.

What was once a rainforest, desert, warm, shallow sea, and mountainous region is now home to thousands of people. What the Denver region will be like in the next several million years is anyone's guess. Whereas humans can affect change, what change we will cause is open to debate. Certainly, the same forces that have shaped the Denver area in the past will continue to determine its future. With the rise of humans and technology, we, as a species, will also influence what future Denvers will be like. Let us hope that the choices we make are good ones.

Figure 2 Pterosaurs (flying reptiles) soar over Denver 70 million years ago. At this time, Denver was below a warm, shallow sea that covered much of western North America. Marine reptiles such as plesiosaurs and mosasaurs swam in these seas in search of schools of fish.

Figure 3 Mammoths and camels wander on the prairie during a summer day 16,000 years ago. Whereas much of northern North America was covered by an ice sheet, Denver sported pine trees and prairie grass. This area was home to a large variety of mammals, including mammoths, camels, horses, bison, and giant ground sloths. Humans settled in this area approximately 11,000 years ago, hunting the plentiful game that was available.

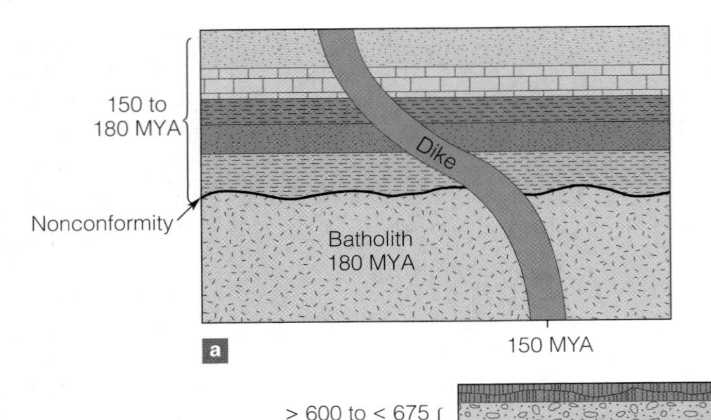

Figure 17.26 Determining Absolute Dates for Sedimentary Rocks The absolute ages of sedimentary rocks can be determined by dating associated igneous rocks. In a and b, sedimentary rocks are bracketed by rock bodies for which absolute ages have been determined.

Figure 17.27 Lithostratigraphic Units The Madison Group in Montana consists of two formations, the Lodgepole Formation and the overlying Mission Canyon Formation. The Mission Canyon Formation is the rock unit exposed on the skyline. The underlying Lodgepole Formation are the rocks that are mostly covered by vegetation on the slopes below.

A body of strata recognized only on the basis of its fossil content is a biostratigraphic unit, the boundaries of which do not necessarily correspond to those of lithostratigraphic boundaries (Figure 17.28). The fundamental biostratigraphic unit is the *biozone*. Several types of biozones are recognized, one of which, the *concurrent range zone*, was discussed in the section on correlation (Figure 17.16).

The category of units expressing or related to geologic time includes **time-stratigraphic units** (also known as chronostratigraphic units) and **time units** (Table 17.2). Time-stratigraphic units consist of rocks deposited during a particular interval of geologic time. The *system*, the basic time-stratigraphic unit, is based on a stratotype consisting of rocks in an area where the system was first described. Systems are recognized beyond their stratotype area by their fossil content.

Time units are simply designations for certain parts of geologic time. The basic time unit is the *period*; however, two or more periods may be designated as an *era*, and two or more eras constitute an *eon*. Periods also consist of shorter designations such as *epoch* and *age*. The time units known as period, epoch, and age correspond to the time-stratigraphic units known as system, series, and stage, respectively (Table 17.2).

TABLE 17.2 Classification of Stratigraphic Units

UNITS DEFINED BY CONTENT		UNITS EXPRESSING OR RELATED TO GEOLOGIC TIME	
Lithostratigraphic Units	**Biostratigraphic Units**	**Time-Stratigraphic Units**	**Time Units**
Supergroup	Biozones	Eonothem	Eon
Group		Erathem	Era
Formation		System	Period
Member		Series	Epoch
Bed		Stage	Age

For example, the Cambrian Period is defined as the time during which strata of the Cambrian System were deposited.

GEOLOGIC TIME AND CLIMATE CHANGE

Given the debate concerning global warming and its possible implications, it is extremely important to be able to reconstruct past climatic regimes as accurately as possible (see Geo-Focus on pages 460 and 461). To model how Earth's climate system has responded to changes in the past and to use that information for simulations of future climate scenarios, geologists must have a geologic calendar that is as precise and accurate as possible.

New dating techniques with greater precision are providing geologists with more accurate dates for when and how long past climate changes occurred. The ability to accurately determine when past climate changes occurred helps geologists correlate these changes with regional and global geologic events to see if there are any possible connections.

One interesting method that is becoming more common in reconstructing past climates is to analyze stalagmites from caves. Recall that stalagmites are icicle-shaped structures rising from a cave floor and formed of calcium carbonate precipitated from evaporating water. A stalagmite therefore records a layered history because each newly precipitated layer of calcium carbonate is younger than the previously precipitated layer (Figure 17.29). Thus, a stalagmite's layers are oldest in the center at its base and progressively younger as they move outward (principle of superposition). Using techniques based on ratios of uranium 234 to thorium 230, geologists can achieve very precise radiometric dates on individual layers of a stalagmite. This technique enables geologists to determine the age of materials much older than they can date by the carbon-14 method, and it is reliable back to approximately 500,000 years.

A study of stalagmites from Crevice Cave in Missouri revealed a history of climatic and vegetation change in the midcontinent region of the United States during the interval between 75,000 and 25,000 years ago. Dates obtained from the Crevice Cave stalagmites were correlated with major changes in vegetation and average temperature fluctuations obtained from carbon-13 and oxygen-18 isotope profiles, to reconstruct a detailed picture of climate changes during this time period.

Thus, precise dating techniques in stalagmite studies provide an accurate chronology that allows geologists to model climate systems of the past and perhaps to determine what causes global climate changes and their duration. Without these sophisticated dating techniques and others like them, geologists would not be able to make precise correlations and accurately reconstruct past environments and climates. By analyzing past environmental and climate changes and their duration, geologists hope they can use these data, sometime in the near future, to predict and possibly modify regional climate changes.

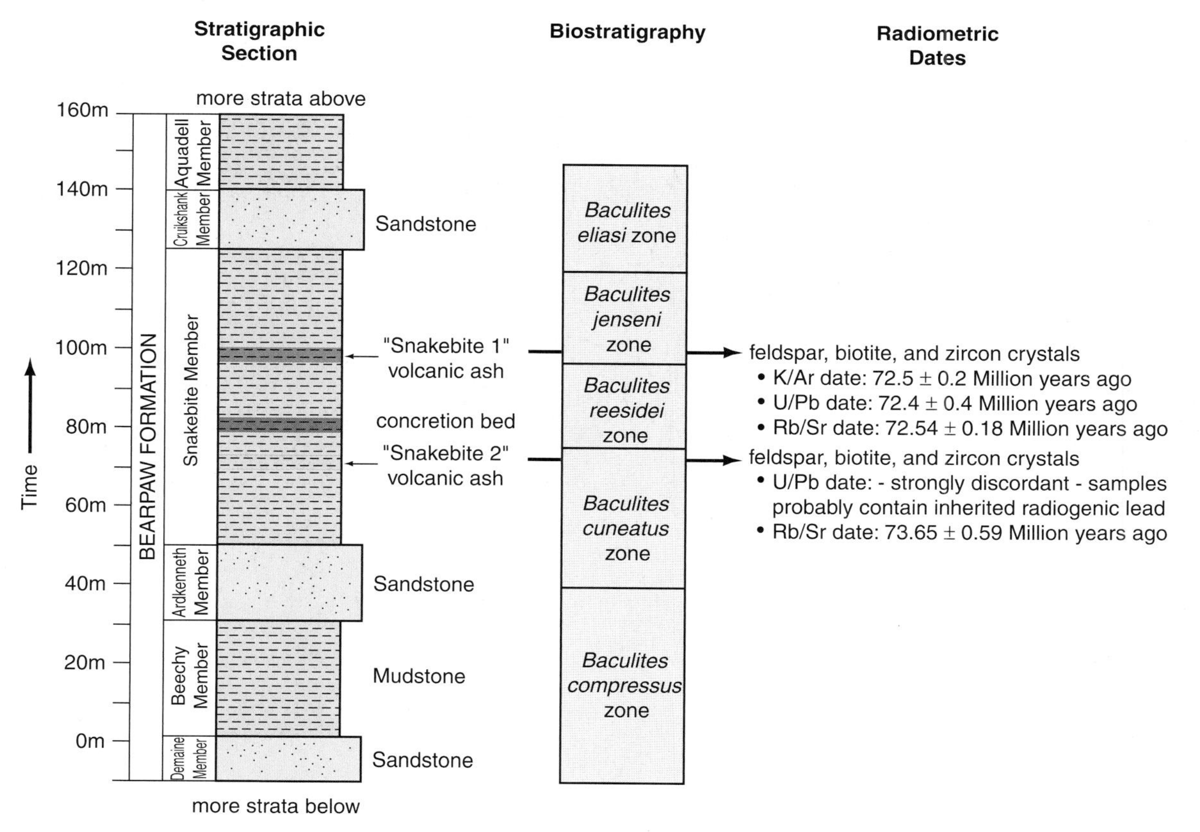

Figure 17.28 Rocks and Fossils of the Bearpaw Formation in Saskatchewan, Canada The column on the left shows formation and members that are lithostratigraphic units. Notice that the biozone boundaries do not correspond with lithostratigraphic boundaries. The absolute ages for the two volcanic ash layers indicate that the *Baculites reesidei* zone is approximately 72 to 73 million years old.

Figure 17.29 Stalagmites and Climate Change

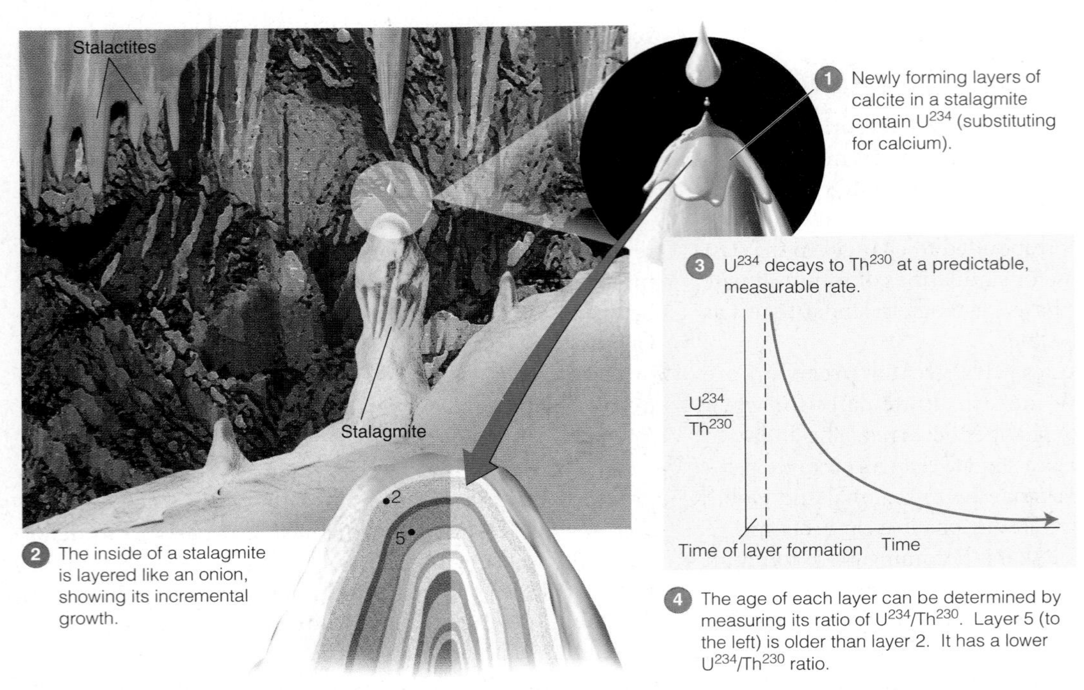

a Stalagmites are icicle-shaped structures rising from the floor of a cave and are formed by the precipitation of calcium carbonate from evaporating water. A stalagmite is thus layered, with the oldest layer in the center and the youngest layers on the outside. Uranium 234 frequently substitutes for the calcium ion in the calcium carbonate of the stalagmite. Uranium 234 decays to thorium 230 at a predictable and measurable rate. Therefore, the age of each layer of the stalagmite can be dated by measuring its ratio of uranium 234 to thorium 230.

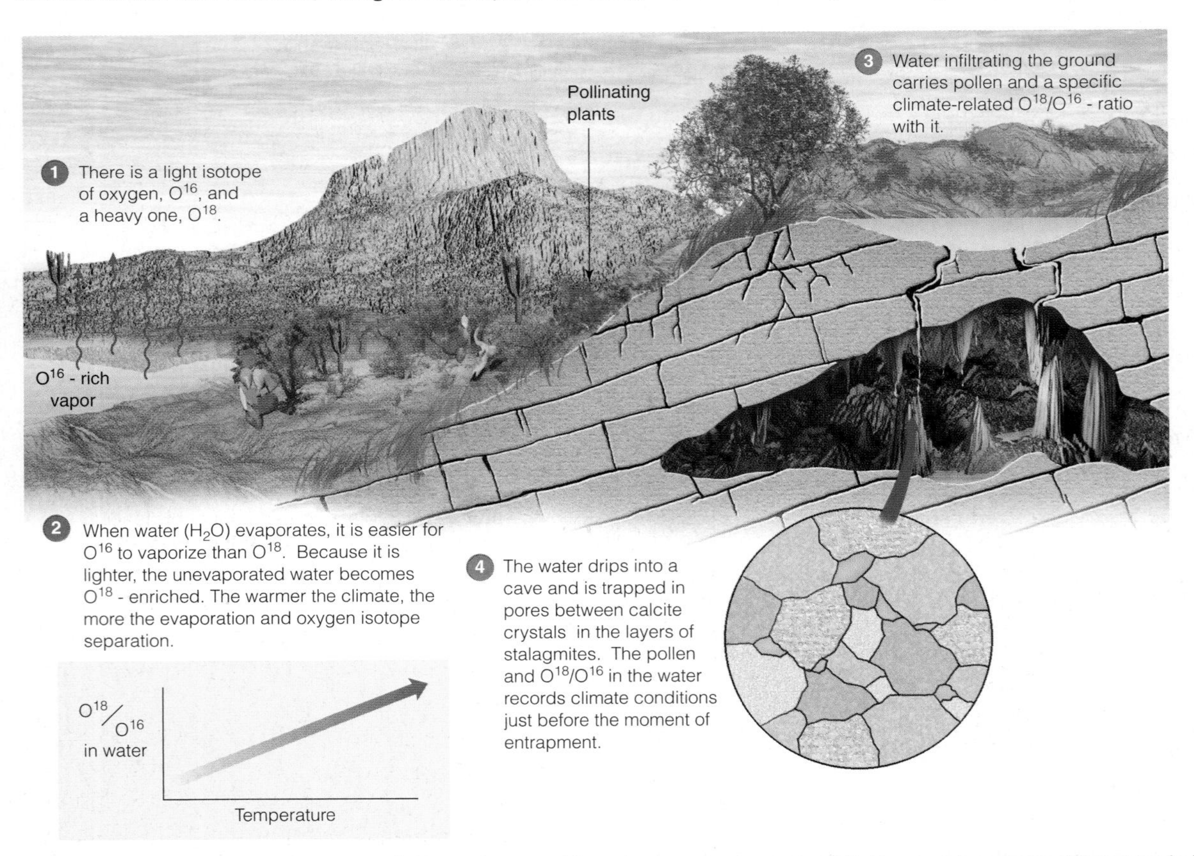

b There are two isotopes of oxygen, a light one, oxygen 16, and a heavy one, oxygen 18. Because the oxygen 16 isotope is lighter than the oxygen 18 isotope, it vaporizes more readily than the oxygen 18 isotope when water evaporates. Therefore, as the climate becomes warmer, evaporation increases, and the O^{18}/O^{16} ratio becomes higher in the remaining water. Water in the form of rain or snow percolates into the ground and becomes trapped in the pores between the calcite forming the stalagmites.

Figure 17.29 Stalagmites and Climate Change (*continued*)

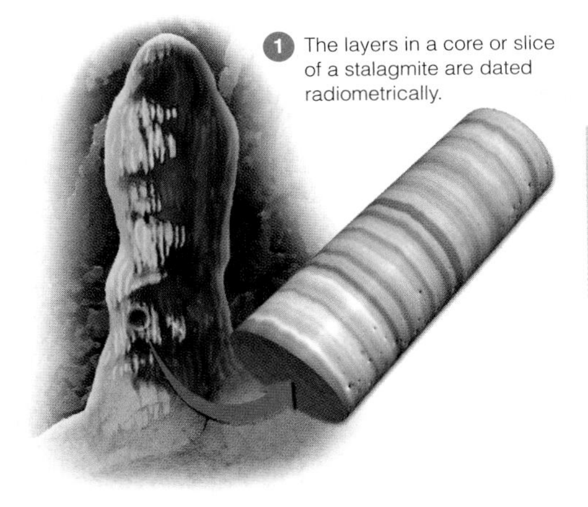

1 The layers in a core or slice of a stalagmite are dated radiometrically.

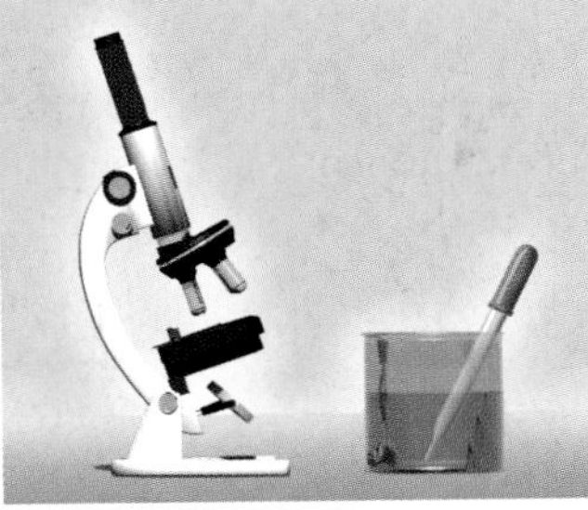

2 The pore water is analyzed in each layer for O^{18}/O^{16} and species of plants (from the pollen).

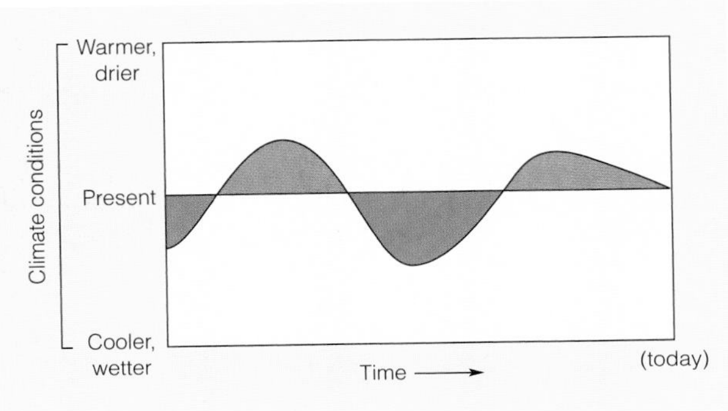

3 A record of climatic change is put together for the area of the caves.

c The layers of a stalagmite can be dated by measuring the U^{234}/Th^{230} ratio, and the O^{18}/O^{16} ratio determined for the pore water trapped in each layer. Thus, a detailed record of climatic change for the area can be determined by correlating the climate of the area as determined by the O^{18}/O^{16} ratio to the time period determined by the U^{234}/Th^{230} ratio.

Geo-Recap

Chapter Summary

- Time is defined by the methods used to measure it. Relative dating places geologic events in sequential order as determined from their position in the geologic record. Absolute dating provides specific dates for geologic rock units or events, expressed in years before the present.
- During the 18th and 19th centuries, attempts were made to determine Earth's age based on scientific evidence rather than revelation. Although some attempts were ingenious, they yielded a variety of ages that now are known to be much too young.
- James Hutton, considered by many to be the father of modern geology, thought that present-day processes operating over long periods of time could explain all of the geologic features of Earth. His observations were instrumental in establishing the principle of uniformitarianism and the fact that Earth was much older than earlier scientists thought.
- Uniformitarianism, as articulated by Charles Lyell, soon became the guiding principle of geology. It holds that the laws of nature have been constant through time and that the same processes operating today have operated in the past, although not necessarily at the same rates.
- Besides uniformitarianism, the principles of superposition, original horizontally, lateral continuity, cross-cutting relationships, inclusions, and fossil succession are basic for determining relative geologic ages and for interpreting Earth history.
- An unconformity is a surface of erosion, nondeposition, or both separating younger strata from older strata.

These surfaces encompass long periods of geologic time for which there is no geologic record at that location.

- Three types of unconformities are recognized. A disconformity separates younger from older sedimentary strata that are parallel to each other. An angular unconformity is an erosional surface on tilted or folded rocks, over which younger sedimentary rocks were deposited. A nonconformity is an erosional surface cut into igneous or metamorphic rocks and overlain by younger sedimentary rocks.
- Correlation is the demonstration of time equivalency of rock units in different areas. Similarity of rock type, position within a rock sequence, key beds, and fossil assemblages can all be used to correlate rock units.
- Radioactivity was discovered during the late 19th century, and soon afterward, radiometric dating techniques enabled geologists to determine absolute ages for rock units and geologic events.
- Absolute dates for rocks are usually obtained by determining how many half-lives of a radioactive parent element have elapsed since the sample originally crystallized. A half-life is the time it takes for one-half of the original unstable radioactive parent element to decay to a new, more stable daughter element.
- The most accurate radiometric dates are obtained from long-lived radioactive isotope pairs in igneous rocks. The most reliable dates are those obtained by using at least two different radioactive decay series in the same rock.
- Carbon-14 dating can be used only for organic matter such as wood, bones, and shells and is effective back to approximately 70,000 years ago. Unlike the long-lived isotopic pairs, the carbon-14 dating technique determines age by the ratio of radioactive carbon-14 to stable carbon-12.
- The geologic time scale was developed primarily during the 19th century through the efforts of many people. It was originally a relative geologic time scale, but with the discovery of radioactivity and the development of radiometric dating methods, absolute age dates were added at the beginning of the 20th century. Since then, refinement of the time-unit boundaries has continued.
- Stratigraphic terminology includes two fundamentally different kinds of units: those based on content and those related to geologic time.
- To reconstruct past climate changes and link them to possible causes, geologists must have a geologic calendar that is as precise and accurate as possible. Thus, they must be able to date geologic events and the onset and duration of climate changes as precisely as possible.

Important Terms

absolute dating (p. 436)
angular unconformity (p. 441)
biostratigraphic unit (p. 459)
carbon-14 dating technique (p. 455)
correlation (p. 446)
disconformity (p. 441)
fission track dating (p. 455)
guide fossil (p. 447)
half-life (p. 452)
lithostratigraphic unit (p. 459)
nonconformity (p. 441)
principle of cross-cutting relationships (p. 438)
principle of fossil succession (p. 441)
principle of inclusions (p. 441)
principle of lateral continuity (p. 438)
principle of original horizontality (p. 438)
principle of superposition (p. 438)
radioactive decay (p. 451)
relative dating (p. 436)
time-stratigraphic unit (p. 462)
time unit (p. 462)
tree-ring dating (p. 458)
unconformity (p. 441)

Review Questions

1. Because of the heat and pressure exerted during metamorphism, daughter atoms were driven out of a mineral being analyzed for a radiometric date. The date obtained from this mineral will therefore be _______ its actual age of formation.
   a. _____ younger than;
   b. _____ older than;
   c. _____ the same as;
   d. _____ it can't be determined;
   e. _____ none of these.
2. Placing geologic events in sequential or chronological order as determined by their position in the geologic record is
   a. _____ absolute dating;
   b. _____ correlation;
   c. _____ historical dating;
   d. _____ relative dating;
   e. _____ uniformitarianism.
3. If a radioactive element has a half-life of 16 million years, what fraction of the original amount of parent material will remain after 96 million years?
   a. _____ $^1/_2$;
   b. _____ $^1/_{16}$;
   c. _____ $^1/_{32}$;
   d. _____ $^1/_4$;
   e. _____ $^1/_{64}$.

4. If a flake of biotite within a sedimentary rock (such as a sandstone) is radiometrically dated, the date obtained indicates when
   a. ______ the biotite crystal formed;
   b. ______ the sedimentary rock formed;
   c. ______ the parent radioactive isotope formed;
   d. ______ the daughter radioactive isotope(s) formed;
   e. ______ none of these.
5. In which type of radioactive decay are two protons and two neutrons emitted from the nucleus?
   a. ______ alpha decay;
   b. ______ beta decay;
   c. ______ electron capture;
   d. ______ fission track;
   e. ______ none of these.
6. What is being measured in radiometric dating is
   a. ______ the time when a radioactive isotope formed;
   b. ______ the time of crystallization of a mineral containing an isotope;
   c. ______ the amount of the parent isotope only;
   d. ______ when the dated mineral became part of a sedimentary rock;
   e. ______ when the stable daughter isotope was formed.
7. How many half-lives are required to yield a mineral with 625,000,000 atoms of thorium 232 and 19,375,000,000 atoms of lead 235?
   a. ______ 1;
   b. ______ 2;
   c. ______ 3;
   d. ______ 4;
   e. ______ 5.
8. As carbon 14 decays back to nitrogen in radiocarbon dating, what isotopic ratio decreases?
   a. ______ nitrogen 14 to carbon 12;
   b. ______ carbon 14 to carbon 12;
   c. ______ carbon 13 to carbon 12;
   d. ______ nitrogen 14 to carbon 14;
   e. ______ nitrogen 14 to carbon 13.
9. Considering that the half-life of uranium 235 is 704 million years, what fraction of the original uranium 235 will remain after 2,816,000,000 years?
   a. ______ $^1/_2$;
   b. ______ $^1/_4$;
   c. ______ $^1/_8$;
   d. ______ $^1/_{16}$;
   e. ______ $^1/_{32}$.
10. A lithostratigraphic unit made up of two or more formations is a
    a. ______ group;
    b. ______ facies;
    c. ______ range zone;
    d. ______ member;
    e. ______ disconformity.
11. What is the difference between relative dating and absolute dating?
12. In some places where disconformities are particularly difficult to discern from a physical point of view, how could you use the principle of fossil succession to recognize a disconformity? How could the principle of inclusions be used to recognize a nonconformity?
13. An igneous rock was radiometrically dated using the uranium 235 to lead 207 and potassium 40 to argon 40 isotope pairs. The isotope pairs yielded distinctly different ages. What possible explanation could be offered for this result? What would you do to rectify the discrepancy in ages?
14. Can the various principles of relative dating be used to reconstruct the geologic history of Mars? Which principles might not apply to interpreting the geologic history of another planet?
15. Why were Lord Kelvin's arguments and calculations so compelling, and what was the basic flaw in his assumption? What do you think the course of geology would have been if radioactivity had not been discovered?
16. When geologists reconstruct the geologic history of an area, why is it important for them to differentiate between a sill and a lava flow? How could you tell the difference between a sill and a lava flow at an outcrop if both structures consisted of basalt? What features would you look for in an outcrop to positively identify the structure as either a sill or a lava flow?
17. If you wanted to calculate the absolute age of an intrusive body, what information would you need?
18. Why do igneous rocks yield the most accurate radiometric dates? Why can't sedimentary rocks be dated radiometrically? What problems are encountered in dating metamorphic rocks?
19. What is the major difference between the carbon-14 dating technique and the techniques used for the five common, long-lived radioactive isotope pairs?
20. Given the current debate over global warming and the many short-term consequences for humans, can you visualize how the world might look in 100,000 years or even 10 million years? Use what you've learned about plate tectonics and the direction and rate of movement of plates, as well as how plate movement and global warming will affect ocean currents, weather patterns, weathering rates, and other factors, to make your prediction. Do you think such short-term changes can be extrapolated to long-term trends in trying to predict what Earth will be like using a geologic time perspective?

APPENDIX A

ENGLISH-METRIC CONVERSION CHART

	English Unit	Conversion Factor	Metric Unit	Conversion Factor	English Unit
Length	Inches (in.)	2.54	Centimeters (cm)	0.39	Inches (in.)
	Feet (ft)	0.305	Meters (m)	3.28	Feet (ft)
	Miles (mi)	1.61	Kilometers (km)	0.62	Miles (mi)
Area	Square inches (in.^{2})	6.45	Square centimeters (cm^2)	0.16	Square inches (in.^{2})
	Square feet (ft^2)	0.093	Square meters (m^2)	10.8	Square feet (ft^2)
	Square miles (mi^2)	2.59	Square kilometers (km^2)	0.39	Square miles (mi^2)
Volume	Cubic inches (in.^{3})	16.4	Cubic centimeters (cm^3)	0.061	Cubic inches (in.^{3})
	Cubic feet (ft^3)	0.028	Cubic meters (m^3)	35.3	Cubic feet (ft^3)
	Cubic miles (mi^3)	4.17	Cubic kilometers (km^3)	0.24	Cubic miles (mi^3)
Weight	Ounces (oz)	28.3	Grams (g)	0.035	Ounces (oz)
	Pounds (lb)	0.45	Kilograms (kg)	2.20	Pounds (lb)
	Short tons (st)	0.91	Metric tons (t)	1.10	Short tons (st)
Temperature	Degrees Fahrenheit (°F)	−32° × 0.56	Degrees centrigrade (Celsius) (°C)	× 1.80 + 32°	Degrees Fahrenheit (°F)

Examples:

10 inches = 25.4 centimeters; 10 centimeters = 3.9 inches
100 square feet = 9.3 square meters; 100 square meters = 1080 square feet
50°F = 10.1°C; 50°C = 122°F

APPENDIX B

TOPOGRAPHIC MAPS

Nearly everyone has used a map of one kind or another and is probably aware that a map is a scaled-down version of the area depicted. For a map to be of any use, however, one must understand what is shown on a map and how to read it. A particularly useful type of map for geologists, and people in many other professions, is a *topographic map*, which shows the three-dimensional configuration of Earth's surface on a two-dimensional sheet of paper.

Maps showing relief—differences in elevation in adjacent areas—are actually models of Earth's surface. Such maps are available for some areas, but they are expensive, difficult to carry, and impossible to record data on. Thus, paper sheets that show relief by using lines of equal elevation known as *contours* are most commonly used. Topographic maps depict (1) relief, which includes hills, mountains, valleys, canyons, and plains; (2) bodies of water such as rivers, lakes, and swamps; (3) natural features such as forests, grasslands, and glaciers; and (4) various cultural features, including communities, highways, railroads, land boundaries, canals, and power transmission lines.

Topographic maps known as *quadrangles* are published by the U.S. Geological Survey (USGS). The area depicted on a topographic map is identified by referring to the map's name in the upper right and lower right corners, which is usually derived from some prominent geographic feature (Lincoln Creek Quadrangle, Idaho) or community (Mt. Pleasant Quadrangle, Michigan). In addition, most maps have a state outline map along the bottom margin, and shown within the outline is a small black rectangle indicating the part of the state represented by the map.

Contours

Contour lines, or simply contours, are lines of equal elevation used to show topography. Think of contours as the lines formed where imaginary horizontal planes intersect Earth's surface at specific elevations. On maps, contours are brown, and every fifth contour, called an *index contour*, is darker than adjacent ones and labeled with its elevation (Figure B1). Elevations on most USGS topographic maps are in feet, although a few use meters; in either case, the specified elevation is above or below mean sea level. Because contours are defined as lines of equal elevation, they cannot divide or cross one another, although they will converge and appear to join in areas with vertical or overhanging cliffs. Notice in Figure B1 that where contours cross a stream they form a V that points upstream toward higher elevations.

The vertical distance between contours is the *contour interval*. If an area has considerable relief, a large contour interval is used, perhaps 80 or 100 feet, whereas a small interval such as 5, 10, or 20 feet is used in areas with little relief. The values recorded on index contours are always multiples of the map's contour interval, shown at the bottom of the map. For instance, if a map has a contour interval of 10 feet, index contour values such as 3600, 3650, and 3700 feet might be shown (Figure B1). In addition to contours, specific elevations are shown at some places on maps and may be indicated by a small X, next to which is a number. A specific elevation might also be shown adjacent to the designation BM (benchmark), a place where the elevation and location are precisely known.

Contour spacing depends on slope, so in areas with steep slopes, contours are closely spaced because there is a considerable increase in elevation in a short distance. In contrast, if slopes are gentle, contours are widely spaced (Figure B1). Furthermore, if contour spacing is uniform, the slope angle remains constant, but if spacing changes, the slope angle changes. However, one must be careful in comparing slopes on maps with different contour intervals or different scales.

Topographic features such as hills, valleys, plains, and so on are easily shown by contours. For instance, a hill is shown by a concentric pattern of contours with the highest elevation in the central part of the pattern. All contours must close on themselves, but they may do so beyond the confines of a particular map. A concentric contour pattern also might show a closed depression, but in this case, special contours with short bars perpendicular to the contour pointing toward the central part of the depression are used (Figure B1).

Map Scales

All maps are scaled-down versions of the areas shown, so to be of any use, they must have a scale. Highway maps, for example, commonly have a scale such as "1 inch equals 10 miles," by which one can readily determine distances. Two types of scales are used on topographic maps. The first and most easily understood is a graphic scale, which is simply a bar subdivided into appropriate units of length (Figure B1). This scale appears at the bottom center of the map and may show miles, feet, kilometers, or meters. Indeed, graphic scales on USGS topographic maps generally show both English and metric distance units.

A ratio or fractional scale, which represents the degree of reduction of the area depicted, appears above the graphic scale. On a map with a ratio scale of 1:24,000, for instance, the area shown is 1/24,000th the size of the actual land area (Figure B1). Another way to express this relationship is to say that any unit of length on the map equals 24,000 of the same units on the ground. Thus, 1 inch on the map equals 24,000 inches on the ground, which is more meaningful if

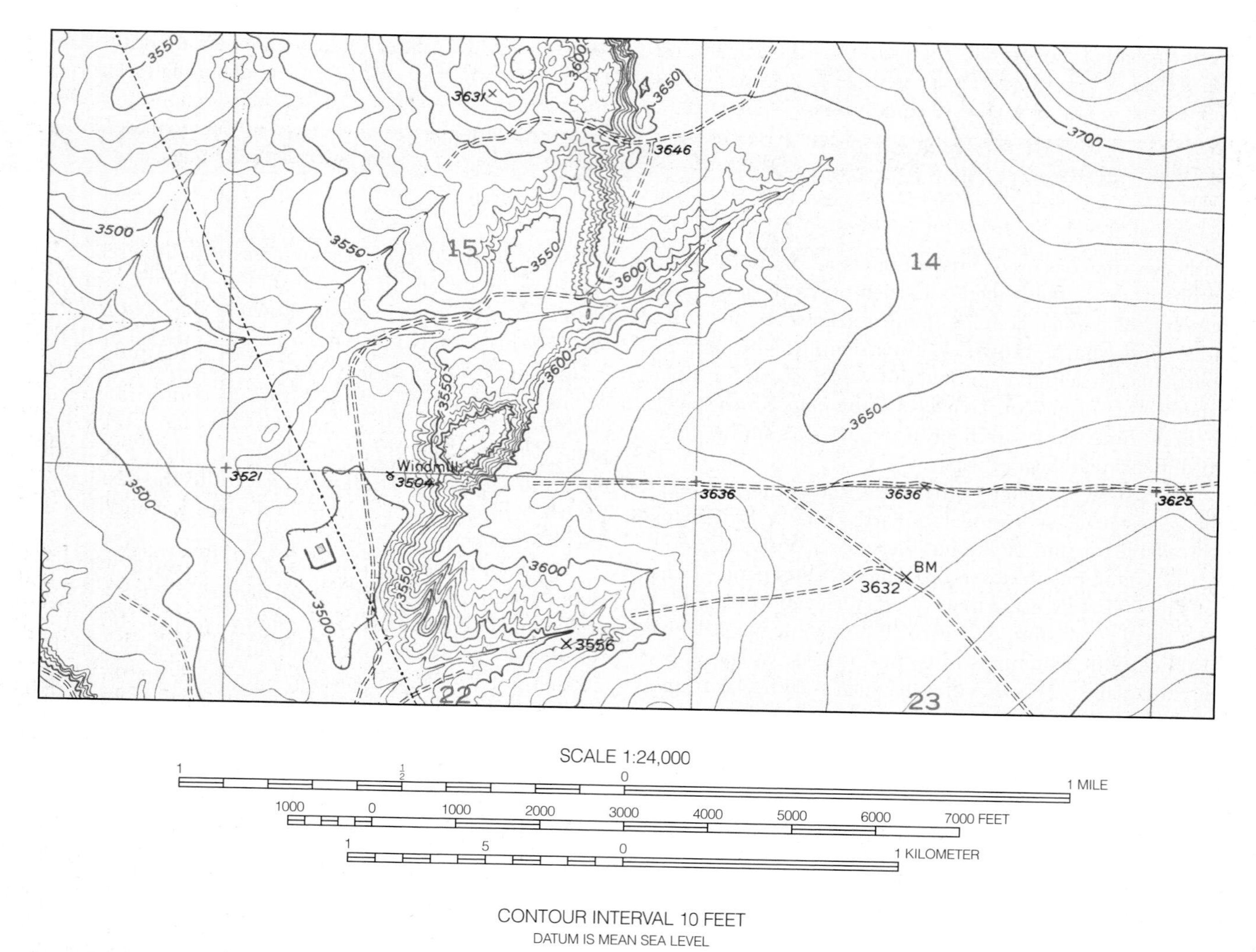

Figure B1 Part of the Bottomless Lakes Quadrangle, New Mexico, which has a contour interval of 10 feet; every fifth contour is darker and labeled with its elevation. Notice that contours are widely spaced where slopes are gentle and more closely spaced where they are steeper, as in the central part of the map. Hills are shown by contours that close on themselves, whereas depressions are indicated by contours with hachure marks pointing toward the center of the depression. The dashed blue lines on the map represent intermittent streams; notice that where contours cross a stream's channel they form a V that points upstream.

one converts inches to feet, making 1 inch equal to 2000 feet. A few maps have scales of 1:63,360, which converts to 1 inch equals 5280 feet, or 1 inch equals 1 mile.

USGS topographic maps are published in a variety of scales such as 1:50,000, 1:62,500, 1:125,000, and 1:250,000. One should also realize that large-scale maps cover less area than small-scale maps, and the former show much more detail than the latter. For example, a large-scale map (1:24,000) shows more surface features in greater detail than does a small-scale map (1:125,000) for the same area.

Map Locations

Location on topographic maps can be determined in two ways. First, the borders of maps correspond to lines of latitude and longitude. Latitude is measured north and south of the equator in degrees, minutes, and seconds, whereas the same units are used to designate longitude east and west of the prime meridian, which passes through Greenwich, England. Maps depicting all areas within the United States are noted in north latitude and west longitude. Latitude and longitude are noted in degrees and minutes at the corners of maps, but usually only minutes and seconds are shown along the margins. Many USGS topographic maps cover 7 1/2 or 15 minutes of latitude and longitude and are thus referred to as 7 ½- and 15-minute quadrangles.

Beginning in 1812, the General Land Office (now known as the Bureau of Land Management) developed a standardized method for accurately defining the location of property in the United States. This method, known as the General Land Office Grid System, has been used for all states except those along the eastern seaboard (except Florida), parts of Ohio, Tennessee, Kentucky, West Virginia, and Texas.

As new land acquired by the United States was surveyed, the surveyors laid out north–south lines they called *principal meridians* and east–west lines known as *base lines.* These intersecting lines form a set of coordinates for locating specific pieces of property. The basic unit in the General Land Office Grid System is the *township*, an area measuring 6 miles on a side and thus covering 36 square miles (◗ Figure B2). Townships are numbered north and south of base lines and are designated as T.1N., T.1S., and so on. Rows of townships known as *ranges* are numbered east and west of principal meridians—R.2W and R.4E, for example. Note in Figure B2 that each township has a unique designation of township and range numbers.

Townships are subdivided into 36 1-square-mile (640-acre) *sections* numbered from 1 to 36. Because of surveying errors and the adjustments necessary to make a grid system conform to Earth's curved surface, not all sections are exactly 1 mile square. Nevertheless, each section can be further subdivided into half sections and quarter sections designated NE 1/4, NW 1/4, SE 1/4, and SW 1/4, and each quarter section can be further divided into quarter-quarter sections. To show the complete designation for an area, the smallest unit is noted first (quarter-quarter section) followed by quarter section, section number, township, and range. For example, the area shown in Figure B2 is the NW 1/4, SW 1/4, Sec. 34, T.2N., R.3W.

Because only a few principal meridians and base lines were established, they do not appear on most topographic maps. Nevertheless, township and range numbers are printed along the margins of 7 1/2- and 15-minute quadrangles, and a grid consisting of red land boundaries depicts sections. In addition, each section number is shown in red within the map. However, small-scale maps show only township and range.

Where to Obtain Topographic Maps

Many people find topographic maps useful. Land use planners, personnel in various local, state, and federal agencies, as well as engineers and real estate developers might use these maps for a variety of reasons. In addition, hikers, backpackers, and others interested in exploring undeveloped areas commonly use topographic maps because trails are shown by black dashed lines. Furthermore, map users can readily determine their location by interpreting the topographic features depicted by contours, and they can anticipate the type of terrain they will encounter during off-road excursions.

Topographic maps for local areas are available at some sporting goods stores, at National Park Visitor Centers, and from some state geologic surveys. Free index maps showing the names and locations of all quadrangles for each state are available from the USGS to anyone uncertain of which specific map is needed. Any published topographic map can be

◗ **Figure B2** The General Land Office Grid System. Each 36-square-mile township is designated by township and range numbers. Townships are subdivided into sections, which can be further subdivided into quarter sections and quarter-quarter sections.

purchased from two main sources. For maps of areas east of the Mississippi River, write to

Branch of Distribution
U.S. Geological Survey
1200 S. Eads Street
Arlington, Virginia 22202

Maps for areas west of the Mississippi River can be obtained from

Branch of Distribution
U.S. Geological Survey
Box 25286 Federal Center
Denver, Colorado 80225

ANSWERS

MULTIPLE-CHOICE QUESTIONS

Chapter 1
1. c; 2. c; 3. b; 4. d; 5. e; 6. e; 7. c; 8. d; 9. c; 10. c.

Chapter 2
1. b; 2. b; 3. b; 4. c; 5. a; 6. e; 7. a; 8. c; 9. a; 10. c.

Chapter 3
1. b; 2. c; 3. d; 4. e; 5. a; 6. c; 7. c; 8. b; 9. a; 10. c.

Chapter 4
1. b; 2. d; 3. b; 4. a; 5. a; 6. e; 7. c; 8. a; 9. c; 10. e.

Chapter 5
1. c; 2. a; 3. b; 4. d; 5. c; 6. b; 7. e; 8. a; 9. d; 10. c.

Chapter 6
1. b; 2. c; 3. a; 4. e; 5. a; 6. c; 7. a; 8. c; 9. b; 10. a.

Chapter 7
1. b; 2. d; 3. e; 4. d; 5. b; 6. c; 7. e; 8. d; 9. c; 10. d.

Chapter 8
1. c; 2. d; 3. d; 4. c; 5. d; 6. d; 7. b; 8. b; 9. c; 10. c.

Chapter 9
1. c; 2. a; 3. b; 4. e; 5. c; 6. d; 7. c; 8. b; 9. c; 10. b.

Chapter 10
1. a; 2. c; 3. e; 4. b; 5. b; 6. d; 7. a; 8. c; 9. b; 10. d.

Chapter 11
1. e; 2. e; 3. b; 4. b; 5. b; 6. d; 7. e; 8. c; 9. d; 10. e.

Chapter 12
1. a; 2. b; 3. e; 4. c; 5. b; 6. d; 7. c; 8. e; 9. a; 10. c.

Chapter 13
1. d; 2. d; 3. c; 4. d; 5. e; 6. b; 7. d; 8. b; 9. d; 10. a.

Chapter 14
1. b; 2. d; 3. e; 4. a; 5. b; 6. a; 7. c; 8. b; 9. a; 10. c.

Chapter 15
1. c; 2. a; 3. b; 4. d; 5. a; 6. e; 7. e; 8. c; 9. d; 10. a.

Chapter 16
1. b; 2. a; 3. d; 4. e; 5. c; 6. a; 7. c; 8. b; 9. c; 10. b.

Chapter 17
1. a; 2. d; 3. e; 4. a; 5. a; 6. b; 7. e; 8. b; 9. d; 10. a.

ANSWERS

SELECTED SHORT-ANSWER QUESTIONS

Chapter 1

15. The Big Bang is a model for the origin and evolution of the universe. According to the Big Bang, the universe originated about 14 billion years ago when all matter and energy were compressed into an infinitely small high-temperature and high-density state in which both time and space were set at zero. This was followed by expansion, cooling, and a less dense state.

    The evidence for the universe beginning about 14 billion years ago is the fact that the universe is expanding and it is permeated by a background radiation of 2.7 Kelvin above absolute zero. This background radiation is thought to be the fading afterglow of the Big Bang.

16. Earth is considered a dynamic planet because it has been continuously changing during its 4.6-billion-year existence. Furthermore, both internal and external forces have been at work during Earth's existence shaping and reshaping both its external and internal features.

    Earth's three concentric layers consist, from the center to the surface, of the *core*, with a calculated density of 10–13 g/cm^3 and a composition of largely iron with a small amount of nickel, and divided into a solid inner region and apparently liquid outer portion; *mantle*, with a density of 3.3–5.7 g/cm^3, and thought to be composed largely of peridotite; and a *crust* which consists of a relatively thick (20–90 km) continental crust with an average density of 2.7 g/cm^3, and a granitic composition, and a thinner (5–10 km) oceanic crust, with a density of 3.0 g/cm^3, and a composition of basalt.

Chapter 2

15. Los Angeles and San Francisco are approximately 550 km apart. 550 km = 55,000,000 cm (100 cm/m × 1000 m/km × 550 km = 55,000,000 cm). Movement along the San Andreas fault averages 5.5 cm/year. The time it will take before Los Angeles is opposite San Francisco is 10,000,000 years (55,000,000 cm/5.5 cm/yr. = 10,000,000 years).
18. The present distribution of plants and animals is controlled largely by climate and geographic barriers. The movement of plates has created and destroyed ocean basins, as well as suturing together continental blocks, breaking apart continents, and moving continents around the globe. This movement has affected weather patterns and oceanic current patterns, and also created barriers for the migration of plants and animals. When continental masses are separated, newly formed ocean basins act as barriers to the migration of land animals. The formation of land bridges, such as the Isthmus of Panama, resulting from a convergent plate boundary, create barriers for ocean species, leading to the evolution of new species.

Chapter 3

11. The atomic number equals the number of protons in the nucleus of an atom of an element, whereas the atomic mass number is found by adding together the number of protons and neutrons in the nucleus.
16. All minerals are crystalline, meaning that their atoms are arranged in a specific three-dimensional framework. A crystal is a geometric shape with planar faces, sharp corners and straight edges; it is the external manifestation of a mineral's internal structure. However, not all minerals display well developed crystals because they may grow in proximity and form an interlocking mass of minerals in which individual crystals are not apparent.

Chapter 4

12. Dikes, sills, and laccoliths are all igneous intrusions, but dikes and sills are tabular with the former being discordant and the latter concordant. A laccolith is similar to a sill in that it is concordant but its central part bulges upward giving it a mushroom-like geometry.
19. The size of minerals in igneous rocks is related to their cooling history. In volcanic (extrusive) rocks the minerals tend to be too small to see without high magnification because lava cools quickly allowing little time for minerals to grow. The minerals in plutonic (intrusive) rocks are generally easily seen without magnification because magma in plutons cools slowly thereby allowing enough time for large minerals to grow. There are, however, exceptions to these generalizations, though. Large minerals might develop in the interiors of thick lava flows, and small minerals may form in very shallow igneous intrusions.

Chapter 5

13. Shield volcanoes have gentle slopes because the lava erupted from them is basaltic which flows easily because of its low viscosity. A cinder cone is made up of pyroclastic materials that resemble cinders which are blasted into the air and settle around the volcano's vent. Because the "cinders" tend to be angular they accumulate in steep-sided piles.

18. Geologists monitor several physical and chemical aspects of volcanoes to better anticipate eruptions. For example, they monitor gas emissions, changes in hot spring activity, temperature, bulging as magma is injected into a volcano, as well as changes in the local magnetic and electrical fields. They are also keenly aware of volcanic tremor and a volcano's eruptive history.

Chapter 6

13. Sedimentary rocks preserve evidence of how they were deposited in the form of ripple marks, mud cracks, cross-bedding, and fossils. The only requirement for geologists to interpret ancient sedimentary rocks is that this evidence has been preserved in the geologic record. If it has, geologists use their knowledge of present-day processes and deposits to make inferences about ancient deposits.
20. The lowermost layer of sedimentary rocks in the diagram originated on land (dinosaur bones) by stream deposition (lenses of sandstone) and perhaps floodplain deposits (mudstone and siltstone). The next three layers upward look exactly like the sequence expected during a marine transgression as shown in Figure 6.22. The uppermost layer was also deposited on land (mammal tracks) and it has the characteristics you would expect in desert dunes.

Chapter 7

11. Contact metamorphism occurs wherever a magma flows over the land surface as a lava flow, or is intruded into crustal rocks such as along mid-ocean ridges, or is an igneous intrusion. Contact metamorphism causes thermal alteration to the affected rock body.
13. The metamorphic facies represented by 200°C and 4 kbar of pressure is the *prehnite-pumpellyite* facies. At 12 kbar of pressure and 200°C, the metamorphic facies is *blueschist*. A change in depth from 15 km to 42 km is needed to effect the pressure change from 4 to 12 kbars.

Chapter 8

13. The difference between an earthquake's focus and epicenter is that the focus is the location within Earth's lithosphere where fracturing begins—that is, the point at which energy is first released, whereas the epicenter is the point on Earth's surface directly above the focus. An earthquake's epicenter is the location usually reported by the news media because it is a geographic location which can be identified by longitude and latitude coordinates.
15. Insurance companies use the Modified Mercalli Intensity Scale because it measures actual damage done to structures and areas. Therefore, insurance companies can determine the type of damage to expect from a given earthquake in a particular area based on the amount and type of damage that resulted from previous earthquakes in that area, and thus adjust their rates accordingly.

Chapter 9

12. To find the average rate of movement in cm/yr first find the number of cm in 1000 km (1 km = 1000 m; 1 m = 100 cm). Thus, $1000 \times 1000 \times 100 = 100{,}000{,}000$. Divide distance by time (100,000,000/30,000,000) and the rate of movement is 3.3 cm/yr.
15. In some mountain ranges geologists identified ophiolites, sequences of rocks made up of what they thought were upper mantle and oceanic crust long before they made direct observations of the same rocks along fractures on the seafloor. Of course they already knew from deep-sea drilling that the upper oceanic crust was made up of pillow lava and sheeted dikes, so they inferred that the gabbro and peridotite parts of these ophiolite sequences represented the lower oceanic crust and upper mantle, respectively.

Chapter 10

16. Stress is the force applied to a given area of rock, usually expressed in kilograms per square centimeter (kg/cm^2). For instance, the stress exerted by a person walking on an ice-covered pond is a function of the person's weight and the area beneath his or her feet. Strain is a measure of the amount of deformation (change in shape or volume or both) as a result of stress. In our ice-covered pond example, the strain is the bending or fracturing of the ice caused by stress.
18. On a dip-slip fault all movement is in the direction of the fault plane's dip; that is, the rocks in the hanging wall block have moved up (reverse fault) or down (normal fault) relative to the footwall block. In contrast, all movement on a strike-slip fault is horizontal, that is, in the direction of the fault plane's strike.

Chapter 11

11. Slope stability can be maintained by a variety of measures, depending on the underlying geology and topography. Such measures as reducing the slope angle of a hillside, surface and subsurface water drainage, construction of retaining walls, and even rock bolts, all can help in maintaining or increasing slope stability.
14. Slumps are usually the result of erosion along the base of a slope, thus increasing the slope angle. The construction of highways and railroad beds in areas of relief, such as hillsides, causes the slope to become oversteepened near its base, resulting in increased pressure along the highway or railroad bed, and increasing the likelihood of slope failure, or slumping.

Chapter 12

15. Running water effectively erodes by hydraulic action (the impact of water especially on soil and sediment), abrasion (the wearing away of rock by the impact of solid parcels such as sand), and by solution. Hydraulic action and abrasion are especially effective, but most

solution takes place in the weathering environment rather than in stream channels.

18. The conclusion that the continents would be leveled in 8,600,000 years is unwarranted because it assumes static continents (no uplift or mountain building) and a constant erosion rate. Erosion rates are complicated, but we can safely say that if the continents were static and progressively lowered by erosion, the erosion rate would necessarily decrease as the continents became lower. As a result the time to level continents would be vastly longer than 8,600,000 years.

Chapter 13

13. Porosity is the percentage of a material's total volume that is pore space. Permeability is the capacity to transmit fluids. Permeability is dependent not only on porosity, but also on the size of the pores or fractures in a rock and their interconnections. Therefore, if a rock is fractured, such as a basalt or quartzite, the total volume of open or pore space may be large, but if the fractures aren't connected, fluid cannot flow through the rock, and permeability will be low or nonexistent.

20. In arid regions, such as Nevada, the water table is very low, in some cases tens and hundreds of meters below the surface. Furthermore, rainfall is scant, and what little does fall and percolates into the ground usually evaporates before it gets very far. Thus, the likelihood of water reaching deeply buried canisters containing radioactive nuclear waste material where it could corrode the containers or become polluted by the radioactive material is virtually nonexistent for the foreseeable future.

    However, if the climate should change and the region become humid, water from rainfall will freely percolate through the zone of aeration into the zone of saturation, and eventually cause the water table to rise, possibly to the level of the buried radioactive nuclear waste material, where the groundwater system could then become polluted by the radioactive material.

    The average rate of groundwater movement during the next 5000 years would have to be 8 cm per year to reach radioactive waste canisters buried at a depth of 400 m. (400 m = 40,000 cm [100 cm/m × 400 m = 40,000 cm]; 40,000 cm/5000 years = 8 cm/year)

Chapter 14

14. The depth to which running water can erode is determined by sea level because the upstream part of a stream cannot be lower than where it discharges into the sea. Glaciers, because of their density, can stay in contact with the seafloor and erode some distance below sea level. For instance, a 500-m-thick glacier could effectively erode about 450 m below sea level before the buoyant effect of seawater would cause it to float.

18. A glacier's budget is essentially a balance sheet of gains and losses. If gains exceed losses, the glacier has a positive budget, it actively flows, and its terminus advances. However, should losses exceed gains, the glacier has a negative budget and its terminus will retreat toward its area of accumulation although it still actively flows. If a negative budget persists long enough, the glacier will recede and thin until it no longer flows and it will eventually disappear.

Chapter 15

14. Many desert rock formations are red in color because of the large amount of iron in the rock formation. Iron exposed to the atmosphere becomes oxidized, which produces a red color.

17. Desert pavement is important in a desert environment because it prevents further erosion and transport of a desert's surface materials by forming a protective layer of close-fitting larger rocks.

Chapter 16

15. The gently sloping surface is a marine terrace with old sea stacks rising above it. It formed initially as a wave-cut platform, an erosional coastal landform with resistant masses of rocks known as sea stacks. Subsequent uplift of the coastal area elevated the platform above sea level.

17. No matter how long or hard the wind blows, large waves will not develop on small ponds and lakes because of the fetch, the distance the wind blows over a continuous water surface. In the oceans, though, fetch is unrestricted and huge wind-generated waves develop if the wind blows hard and for a long time.

Chapter 17

11. The difference between relative and absolute dating is that relative dating is placing geologic events in a sequential order as determined from their position in the geologic record without any regard to how long ago an event took place Absolute dating, on the other hand, provides specific dates for rock units or events expressed in years before the present.

16. In reconstructing the geologic history of an area, it is important to differentiate between a sill and a lava flow because they formed differently in relation to the rocks found in association with them. A sill is a tabular igneous intrusion and thus is younger than the rocks it intrudes. A lava flow moving over the land surface will be younger than rocks that it is flowing over, and will be older than any rocks deposited on it after it cools and solidifies.

    To identify a sill, one should look for baking (contact metamorphism) of the rocks immediately above and below the sill. A lava flow, however, will only show the effects of baking in the rocks below it, because the rocks above it were deposited on the solidified flow.

GLOSSARY

aa Lava flow with a surface of rough, angular blocks and fragments.

abrasion The process whereby rock is worn smooth by the impact of sediment transported by running water, glaciers, waves, or wind.

absolute dating Using various radioactive decay dating techniques to assign ages in years before the present to rocks. (*See also* relative dating)

abyssal plain Vast flat area on the seafloor adjacent to the continental rise of a passive continental margin.

active continental margin A continental margin with volcanism and seismicity at the leading edge of a continental plate where oceanic lithosphere is subducted. (*See also* passive continental margin)

alluvial fan A cone-shaped accumulation of mostly sand and gravel deposited where a stream flows from a mountain valley onto an adjacent lowland.

alluvium A collective term for all detrital sediment transported and deposited by running water.

angular unconformity An unconformity below which older rocks dip at a different angle (usually steeper) than overlying strata. (*See also* disconformity and nonconformity)

anticline A convex upward fold in which the oldest exposed rocks coincide with the fold axis and all strata dip away from the axis.

aphanitic texture A texture in igneous rocks in which individual mineral grains are too small to be seen without magnification; results from rapid cooling of magma and generally indicates an extrusive origin.

arête A narrow, serrated ridge between two glacial valleys or adjacent cirques.

artesian system A confined groundwater system with high hydrostatic pressure that causes water to rise above the level of the aquifer.

aseismic ridge A ridge or broad area rising above the seafloor that lacks seismic activity.

ash Pyroclastic materials that measure less than 2 mm.

assimilation A process whereby magma changes composition as it reacts with country rock.

asthenosphere The part of the mantle that lies below the lithosphere; it behaves plastically and flows slowly.

atom The smallest unit of matter that retains the characteristics of an element.

atomic mass number The number of protons plus neutrons in the nucleus of an atom.

atomic number The number of protons in the nucleus of an atom.

aureole A zone surrounding a pluton in which contact metamorphism took place.

backshore That part of a beach that is usually dry, being water covered only by storm waves or especially high tides.

barchan dune A crescent-shaped sand dune with its tips pointing downwind.

barrier island A long, narrow island of sand parallel to a shoreline but separated from the mainland by a lagoon.

basal slip Movement involving a glacier sliding over its underlying surface.

basalt plateau A plateau built up by horizontal or nearly horizontal overlapping lava flows that erupted from fissures.

base level The level below which a stream or river cannot erode; sea level is ultimate base level.

basin A circular fold in which all strata dip inward toward a central point and the youngest exposed strata are in the center.

batholith An irregularly shaped, discordant pluton with at least 100 km^2 of surface area.

baymouth bar A spit that has grown until it closes off a bay from the open sea or lake.

beach Any deposit of sediment extending landward from low tide to a change in topography or where permanent vegetation begins.

beach face The sloping area of a beach that is exposed to wave swash.

bed An individual layer of rock, especially sediment or sedimentary rock. (*See also* strata)

bed load That part of a stream's sediment load, mostly sand and gravel, transported along its bed.

berm A platform of sediment with a steeply sloping seaward face deposited by waves; some beaches have no berm, whereas others may have several.

Big Bang A model for the evolution of the universe in which a dense, hot state was followed by expansion, cooling, and a less dense state.

biochemical sedimentary rock Any sedimentary rock produced by the chemical activities of organisms. (*See also* chemical sedimentary rock)

biostratigraphic unit An association of sedimentary rocks defined by its fossil content.

black smoker A type of submarine hydrothermal vent that emits a black plume of hot water colored by dissolved minerals.

bonding The process whereby atoms join to other atoms.

Bowen's reaction series A series of minerals that form in a specific sequence in cooling magma or lava; originally proposed to explain the origin of intermediate and felsic magma from mafic magma.

braided stream A stream with multiple dividing and rejoining channels.

breaker A wave that steepens as it enters shallow water until its crest plunges forward.

butte An isolated, steep-sided, pinnacle-like hill formed when resistant cap rock is breached allowing erosion of less resistant underlying rocks.

caldera A large, steep-sided, oval to circular depression usually formed when a volcano's summit collapses into a partially drained underlying magma chamber.

carbon 14 dating technique Absolute dating technique relying on the ratio of C14 to C12 in an organic substance; useful back to about 70,000 years ago.

carbonate mineral A mineral with the carbonate radical $(CO_3)^{-2}$, as in calcite ($CaCO_3$) and dolomite [$CaMg(CO_3)^2$].

carbonate rock Any rock, such as limestone and dolostone, made up mostly of carbonate minerals.

cave A natural subsurface opening generally connected to the surface and large enough for a person to enter.

cementation The process whereby minerals crystallize in the pore spaces of sediment and bind the loose particles together.

chemical sedimentary rock Sedimentary rock made up of minerals that were dissolved during chemical weathering and later precipitated from seawater, more rarely lake water, or extracted from solution by organisms. (*See also* biochemical sedimentary rock)

chemical weathering The decomposition of rocks by chemical alteration of parent material.

cinder cone A small, steep-sided volcano made up of pyroclastic materials resembling cinders that accumulate around a vent.

circum-Pacific belt A zone of seismic and volcanic activity and mountain building that nearly encircles the Pacific Ocean basin.

cirque A steep-walled, bowl-shaped depression on a mountainside at the upper end of a glacial valley.

cleavage Breakage along internal planes of weakness in mineral crystals.

columnar jointing The phenomenon of forming columns bounded by fractures in some igneous rocks as they cooled and contracted.

compaction Reduction in the volume of a sedimentary deposit that results from its own weight and the weight of any additional sediment deposited on top of it.

complex movement A combination of different types of mass movements in which no single type is dominant; usually involves sliding and flowing.

composite volcano (stratovolcano) A volcano composed of lava flows and pyroclastic layers, typically of intermediate composition, and mudflows.

compound Any substance resulting from the bonding of two or more different elements (e.g., water, H_2O, and quartz, SiO_2).

compression Stress resulting when rocks are squeezed by external forces directed toward one another.

concordant pluton Intrusive igneous body whose boundaries parallel the layering in the country rock. (*See also* discordant pluton)

cone of depression A cone-shaped depression around a well where water is pumped from an aquifer faster than it can be replaced.

contact (thermal) metamorphism Metamorphism of country rock adjacent to a pluton.

continental–continental plate boundary A convergent plate boundary along which two continental lithospheric plates collide.

continental drift The theory that the continents were joined into a single landmass that broke apart with the various fragments (continents) moving with respect to one another.

continental glacier A glacier that covers a vast area (at least 50,000 km^2) and is not confined by topography; also called an ice sheet.

continental margin The area separating the part of a continent above sea level from the deep seafloor.

continental rise The gently sloping part of the continental margin between the continental slope and the abyssal plain.

continental shelf The very gently sloping part of the continental margin between the shoreline and the continental slope.

continental slope The relatively steeply inclined part of the continental margin between the continental shelf and the continental rise or between the continental shelf and an oceanic trench.

convergent plate boundary The boundary between two plates that move toward each other.

core The interior part of Earth beginning at a depth of 2900 km that probably consists mostly of iron and nickel.

Coriolis effect The apparent deflection of a moving object from its anticipated course because of Earth's rotation. Winds and oceanic currents are deflected clockwise in the Northern Hemisphere and counterclockwise in the Southern Hemisphere.

correlation Demonstration of the physical continuity of rock units or biostratigraphic units, or demonstration of time equivalence as in time-stratigraphic correlation.

country rock Any preexisting rock that has been intruded by a pluton or altered by metamorphism.

covalent bond A chemical bond formed by the sharing of electrons between atoms.

crater An oval to circular depression at the summit of a volcano resulting from the eruption of lava, pyroclastic materials, and gases.

creep A widespread type of mass wasting in which soil or rock moves slowly downslope.

crest The highest part of a wave.

Cretaceous Interior Seaway A Late Cretaceous arm of the sea that effectively divided North America into two large landmasses.

cross-bedding A type of bedding in which layers are deposited at an angle to the surface on which they accumulate, as in sand dunes.

crust Earth's outermost layer; the upper part of the lithosphere that is separated from the mantle by the Moho; divided into continental and oceanic crust.

crystal A naturally occurring solid of an element or compound with a specific internal structure that is manifested externally by planar faces, sharp corners, and straight edges.

crystal settling The physical separation and concentration of minerals in the lower part of a magma chamber or pluton by crystallization and gravitational settling.

crystalline solid A solid in which the constituent atoms are arranged in a regular, three-dimensional framework.

Curie point The temperature at which iron-bearing minerals in cooling magma or lava attain their magnetism.

debris flow A type of mass wasting that involves a viscous mass of soil, rock fragments, and water that moves downslope; debris flows have larger particles than mudflows and contain less water.

deflation The removal of sediment and soil by wind.

deformation A general term for any change in shape or volume, or both, of rocks in response to stress; involves folding and fracturing.

delta An alluvial deposit formed where a stream or river flows into the sea or a lake.

density The mass of an object per unit volume; usually expressed in grams per cubic centimeter (g/cm^3).

depositional environment Any site such as a floodplain or beach where physical, biologic, and chemical processes yield a distinctive kind of sedimentary deposit.

desert Any area that receives less than 25 cm of rain per year and that has a high evaporation rate.

desert pavement A surface mosaic of close-fitting pebbles, cobbles, and boulders found in many dry regions; results from wind erosion of sand and smaller particles.

desertification The expansion of deserts into formerly productive lands.

detrital sedimentary rock Sedimentary rock made up of the solid particles (detritus) of preexisting rocks.

differential pressure Pressure that is not applied equally to all sides of a rock body.

differential weathering Weathering that occurs at different rates on rocks, thereby yielding an uneven surface.

dike A tabular or sheetlike discordant pluton.

dip A measure of the maximum angular deviation of an inclined plane from horizontal.

dip-slip fault A fault on which all movement is parallel with the dip of the fault plain. (*See also* normal fault and reverse fault)

discharge The volume of water in a stream or river moving past a specific point in a given interval of time; expressed in cubic meters per second (m^3/sec) or cubic feet per second (ft^3/sec).

disconformity An unconformity above and below which the rock layers are parallel. (*See also* angular unconformity and nonconformity)

discontinuity A boundary across which seismic wave velocity or direction of travel changes abruptly, such as the mantle–core boundary.

discordant pluton Pluton with boundaries that cut across the layering in the country rock. (*See also* concordant pluton)

dissolved load The part of a stream's load consisting of ions in solution.

divergent plate boundary The boundary between two plates that are moving apart.

divide A topographically high area that separates adjacent drainage basins.

dome A rather circular geologic structure in which all rock layers dip away from a central point and the oldest exposed rocks are at the dome's center.

drainage basin The surface area drained by a stream or river and its tributaries.

drainage pattern The regional arrangement of channels in a drainage system.

drumlin An elongate hill of till formed by the movement of a continental glacier or by floods.

dune A mound or ridge of wind-deposited sand.

dynamic metamorphism Metamorphism in fault zones where rocks are subjected to high differential pressure.

earthflow A mass wasting process involving the downslope movement of water-saturated soil.

earthquake Vibrations caused by the sudden release of energy, usually as a result of displacement of rocks along faults.

elastic rebound theory An explanation for the sudden release of energy that causes earthquakes when deformed rocks fracture and rebound to their original undeformed condition.

elastic strain A type of deformation in which the material returns to its original shape when stress is relaxed.

electron A negatively charged particle of very little mass that encircles the nucleus of an atom.

electron shell Electrons orbit an atom's nucleus at specific distances in electron shells.

element A substance composed of atoms that all have the same properties; atoms of one element can change to atoms of another element by radioactive decay, but otherwise they cannot be changed by ordinary chemical means.

emergent coast A coast where the land has risen with respect to sea level.

end moraine A pile or ridge of rubble deposited at the terminus of a glacier. (*See also* terminal moraine and recessional moraine)

erosion The removal of weathered materials from their source area by running water, wind, glaciers, and waves.

esker A long, sinuous ridge of stratified drift deposited by running water in a tunnel beneath stagnant ice.

evaporite Any sedimentary rock, such as rock salt, formed by inorganic chemical precipitation of minerals from evaporating water.

Exclusive Economic Zone (EEZ) An area extending 371 km seaward from the coast of the United States and its possessions in which the United States claims rights to all resources.

exfoliation dome A large, rounded dome of rock resulting when concentric layers of rock are stripped from the surface of a rock mass.

fault A fracture along which rocks on opposite sides of the fracture have moved parallel with the fracture surface.

fault plane A fault surface that is more or less planar.

felsic magma Magma with more than 65% silica and considerable sodium, potassium, and aluminum, but little calcium, iron, and magnesium. (*See also* intermediate magma and mafic magma)

ferromagnesian silicate Any silicate mineral that contains iron, magnesium, or both. (*See also* nonferromagnesian silicate)

fetch The distance the wind blows over a continuous water surface.

fiord An arm of the sea extending into a glacial trough eroded below sea level.

firn Granular snow formed by partial melting and refreezing of snow; transitional material between snow and glacial ice.

fission track dating The absolute dating process in which small linear tracks (fission tracks) resulting from alpha decay are counted in mineral crystals.

fissure eruption A volcanic eruption in which lava or pyroclastic materials issue from a long, narrow fissure (crack) or group of fissures.

floodplain A low-lying, flat area adjacent to a channel that is partly or completely water-covered when a stream or river overflows its banks.

fluid activity An agent of metamorphism in which water and carbon dioxide promote metamorphism by increasing the rate of chemical reactions.

focus The site within Earth where an earthquake originates and energy is released.

fold A type of geologic structure in which planar features in rock layers such as bedding and foliation have been bent.

foliated texture A texture in metamorphic rocks in which platy and elongate minerals are aligned in a parallel fashion.

footwall block The block of rock that lies beneath a fault plane. (*See also* hanging wall block)

foreshore That part of a beach covered by water at high tide but exposed during low tide.

fracture A break in rock resulting from intense applied pressure.

frost action The disaggregation of rocks by repeated freezing and thawing of water in cracks and crevasses.

geologic structure Any feature in rocks that results from deformation, such as folds, joints, and faults.

geologic time scale A chart arranged so that the designation for the earliest part of geologic time appears at the bottom followed upward by progressively younger time designations.

geology The science concerned with the study of Earth materials (minerals and rocks), surface and internal processes, and Earth history.

geothermal energy Energy that comes from steam and hot water trapped within Earth's crust.

geothermal gradient Earth's temperature increase with depth; it averages 25°C/km near the surface but varies from area to area.

geyser A hot spring that periodically ejects hot water and steam.

glacial budget The balance between expansion and contraction of a glacier in response to accumulation versus wastage.

glacial drift A collective term for all sediment deposited directly by glacial ice (till) and by meltwater streams (outwash).

glacial erratic A rock fragment carried some distance from its source by a glacier and usually deposited on bedrock of a different composition.

glacial ice Water in the solid state within a glacier; forms as snow partially melts and refreezes and compacts so that it is transformed first to firn and then to glacial ice.

glacial polish A smooth, glistening rock surface formed by the movement of sediment-laden ice over bedrock.

glacial striation A straight scratch rarely more than a few millimeters deep on a rock caused by the movement of sediment-laden glacial ice.

glacial surge A time of greatly accelerated flow in a glacier. Commonly results in displacement of the glacier's terminus by several kilometers.

glaciation Refers to all aspects of glaciers, including their origin, expansion, and retreat, and their impact on Earth's surface.

glacier A mass of ice on land that moves by plastic flow and basal slip.

***Glossopteris* flora** A Late Paleozoic association of plants found only on the Southern Hemisphere continents and India; named for its best-known genus, *Glossopteris*.

graded bedding Sedimentary layer in which a single bed shows a decrease in grain size from bottom to top.

graded stream A stream that has an equilibrium profile in which a delicate balance exists among gradient, discharge, flow velocity, channel characteristics, and sediment load so that neither significant deposition nor erosion takes place within its channel.

gradient The slope over which a stream or river flows expressed in m/km or ft/mi.

gravity anomaly A departure from the expected force of gravity; anomalies may be positive or negative.

ground moraine The layer of sediment released from melting ice as a glacier's terminus retreats.

groundwater Underground water stored in the pore spaces of soil, sediment, and rock.

guide fossil Any easily identified fossil with an extensive geographic distribution and short geologic range useful for determining the relative ages of rocks in different areas.

guyot A flat-topped seamount of volcanic origin rising more than 1 km above the seafloor.

half-life The time necessary for half of the original number of radioactive atoms of an element to decay to a stable daughter product; for example, the half-life for potassium 40 is 1.3 billion years.

hanging valley A tributary glacial valley whose floor is at a higher level than that of the main glacial valley.

hanging wall block The block of rock that overlies a fault plane. (*See also* footwall block)

hardness A term used to express the resistance of a mineral to abrasion.

headland Part of a shoreline commonly bounded by cliffs that extends out into the sea or a lake.

heat An agent of metamorphism.

horn A steep-walled, pyramid-shaped peak formed by the headward erosion of at least three cirques.

hot spot Localized zone of melting below the lithosphere that probably overlies a mantle plume.

hot spring A spring in which the water temperature is warmer than the temperature of the human body (37°C).

hydraulic action The removal of loose particles by the power of moving water.

hydrologic cycle The continuous recycling of water from the oceans, through the atmosphere, to the continents, and back to the oceans, or from the oceans, through the atmosphere, and back to the oceans.

hydrolysis The chemical reaction between hydrogen (H^+) ions and hydroxyl (OH^-) ions of water and a mineral's ions.

hydrothermal A term referring to hot water as in hot springs and geysers.

hypothesis A provisional explanation for observations that is subject to continual testing. If well supported by evidence, a hypothesis may be called a theory.

ice cap A dome-shaped mass of glacial ice that covers less than 50,000 km^2.

igneous rock Any rock formed by cooling and crystallization of magma or lava or the consolidation of pyroclastic materials.

incised meander A deep, meandering canyon cut into bedrock by a stream or river.

index mineral A mineral that forms within specific temperature and pressure ranges during metamorphism.

infiltration capacity The maximum rate at which soil or sediment absorbs water.

intensity The subjective measure of the kind of damage done by an earthquake as well as people's reaction to it.

intermediate magma Magma with a silica content between 53% and 65% and an overall composition intermediate between mafic and felsic magma.

ion An electrically charged atom produced by adding or removing electrons from the outermost electron shell.

ionic bond A chemical bond resulting from the attraction between positively and negatively charged ions.

isostasy See principle of isostasy.

isostatic rebound The phenomenon in which unloading of the crust causes it to rise until it attains equilibrium.

joint A fracture along which no movement has occurred or where movement is perpendicular to the fracture surface.

Jovian planet Any of the four planets (Jupiter, Saturn, Uranus, and Neptune) that resemble Jupiter. All are large and have low mean densities, indicating that they are composed mostly of lightweight gases, such as hydrogen and helium, and frozen compounds, such as ammonia and methane.

kame Conical hill of stratified drift originally deposited in a depression on a glacier's surface.

karst topography Landscape consisting of numerous caves, sinkholes, and solution valleys formed by groundwater solution of rocks such as limestone and dolostone.

laccolith A concordant pluton with a mushroom-like geometry.

lahar A mudflow composed of pyroclastic materials such as ash.

lateral moraine Ridge of sediment deposited along the margin of a valley glacier.

laterite A red soil, rich in iron or aluminum, or both, resulting from intense chemical weathering in the tropics.

lava Magma that reaches Earth's surface.

lava dome A bulbous, steep-sided mountain formed by viscous magma moving upward through a volcanic conduit.

lava flow A stream of magma flowing over Earth's surface.

lava tube A tunnel beneath the solidified surface of a lava flow through which lava moves; also, the hollow space left when the lava within a tube drains away.

lithification The process of converting sediment into sedimentary rock by compaction and cementation.

lithosphere Earth's outer, rigid part, consisting of the upper mantle, oceanic crust, and continental crust.

lithostatic pressure Pressure exerted on rocks by the weight of overlying rocks.

lithostratigraphic unit A body of rock, such as a formation, defi ned solely by its physical attributes.

Little Ice Age An interval from about 1500 to the mid- to late-1800s during which glaciers expanded to their greatest historic extent.

loess Wind-blown deposit of silt and clay.

longitudinal dune A long ridge of sand generally parallel to the direction of the prevailing wind.

longshore current A current resulting from wave refraction found between the breaker zone and a beach that flows parallel to the shoreline.

Love wave (L-wave) A surface wave in which the individual particles of material move only back and forth in a horizontal plane perpendicular to the direction of wave travel.

luster The appearance of a mineral in reflected light. Luster is metallic or nonmetallic, although the latter has several subcategories.

mafic magma Magma with between 45% and 52% silica and proportionately more calcium, iron, and magnesium than intermediate and felsic magma.

magma Molten rock material generated within Earth.

magma chamber A reservoir of magma within Earth's upper mantle or lower crust.

magma mixing The process whereby magmas of different composition mix together to yield a modified version of the parent magmas.

magnetic anomaly Any deviation, such as a change in average strength, in Earth's magnetic field.

magnetic field The area in which magnetic substances are affected by lines of magnetic force emanating from Earth.

magnetic reversal The phenomenon involving the complete reversal of the north and south magnetic poles.

magnetism A physical phenomenon resulting from moving electricity and the spin of electrons in some solids in which magnetic substances are attracted toward one another.

magnitude The total amount of energy released by an earthquake at its source. (*See also* Richter Magnitude Scale)

mantle The thick layer between Earth's crust and core.

mantle plume A cylindrical mass of magma rising from the mantle toward the surface; recognized at the surface by a hot spot, an area such as the Hawaiian Islands where volcanism takes place.

marine regression The withdrawal of the sea from a continent or coastal area, resulting in the emergence of the land as sea level falls or the land rises with respect to sea level.

marine terrace A wave-cut platform now above sea level.

marine transgression The invasion of a coastal area or a continent by the sea, resulting from a rise in sea level or subsidence of the land.

mass wasting The downslope movement of Earth materials under the influence of gravity.

meandering stream A stream that has a single, sinuous channel with broadly looping curves.

mechanical weathering Disaggregation of rocks by physical processes that yields smaller pieces that retain the composition of the parent material.

medial moraine A moraine carried on the central surface of a glacier; formed where two lateral moraines merge.

Mediterranean belt A zone of seismic and volcanic activity extending through the Mediterranean region of southern Europe and eastward to Indonesia.

mesa A broad, flat-topped erosional remnant bounded on all sides by steep slopes.

metamorphic facies A group of metamorphic rocks characterized by particular minerals that formed under the same broad temperature and pressure conditions.

metamorphic grade The degree to which parent rocks have undergone metamorphic change; the higher the grade the greater the change, as in low-, medium-, and high-grade.

metamorphic rock Any rock that has been changed from its original condition by heat, pressure, and the chemical activity of fluids, as in marble and slate.

metamorphic zone The region between lines of equal metamorphic intensity known as isograds.

metamorphism The phenomenon of changing rocks subjected to heat, pressure, and fluids so that they are in equilibrium with a new set of environmental conditions.

Milankovitch theory An explanation for the cyclic variations in climate and the onset of ice ages as a result of irregularities in Earth's rotation and orbit.

mineral A naturally occurring, inorganic, crystalline solid that has characteristic physical properties and a narrowly defined chemical composition.

Modified Mercalli Intensity Scale A scale with values from I to XII used to characterize earthquakes based on damage.

Mohorovičić discontinuity (Moho) The boundary between Earth's crust and mantle.

monocline A bend or flexure in otherwise horizontal or uniformly dipping rock layers.

mud crack A crack in clay-rich sediment that forms in response to drying and shrinkage.

mudflow A flow consisting mostly of clay- and silt-sized particles and up to 30% water that moves downslope under the influence of gravity.

native element A mineral composed of a single element, such as gold.

natural levee A ridge of sandy alluvium deposited along the margins of a channel during floods.

nearshore sediment budget The balance between additions and losses of sediment in the nearshore zone.

neutron An electrically neutral particle found in the nucleus of an atom.

nonconformity An unconformity in which stratified sedimentary rocks overlie an erosion surface cut into igneous or metamorphic rocks. (*See also* angular unconformity and disconformity)

nonferromagnesian silicate A silicate mineral that has no iron or magnesium. (*See also* ferromagnesian silicate)

nonfoliated texture A metamorphic texture in which there is no discernable preferred orientation of minerals.

normal fault A dip-slip fault on which the hanging wall block has moved downward relative to the footwall block. (*See also* reverse fault)

nucleus The central part of an atom consisting of protons and neutrons.

nuée ardente A fast moving, dense cloud of hot pyroclastic materials and gases ejected from a volcano.

oblique-slip fault A fault showing both dip-slip and strike-slip movement.

oceanic–continental plate boundary A convergent plate boundary along which oceanic lithosphere is subducted beneath continental lithosphere.

oceanic–oceanic plate boundary A convergent plate boundary along which two oceanic plates collide and one is subducted beneath the other.

oceanic ridge A mostly submarine mountain system composed of basalt found in all ocean basins.

oceanic trench A long, narrow feature restricted to active continental margins and along which subduction occurs.

ooze Deep-sea sediment composed mostly of shells of marine animals and plants.

ophiolite A sequence of igneous rocks representing a fragment of oceanic lithosphere; composed of peridotite overlain successively by gabbro, sheeted basalt dikes, and pillow lava.

organic evolution See theory of evolution.

orogeny An episode of mountain building involving deformation, usually accompanied by igneous activity, and crustal thickening.

outwash plain The sediment deposited by meltwater discharging from a continental glacier's terminus.

oxbow lake A cutoff meander filled with water.

oxidation The reaction of oxygen with other atoms to form oxides or, if water is present, hydroxides.

pahoehoe A type of lava flow with a smooth ropy surface.

paleomagnetism Residual magnetism in rocks, studied to determine the intensity and direction of Earth's past magnetic field.

Pangaea The name Alfred Wegener proposed for a supercontinent consisting of all Earth's landmasses at the end of the Paleozoic Era.

parabolic dune A crescent-shaped dune with its tips pointing upwind.

parent material The material that is chemically and mechanically weathered to yield sediment and soil.

passive continental margin A continental margin within a tectonic plate as in the East Coast of North America where little seismic activity and no volcanism occur; characterized by a broad continental shelf and a continental slope and rise.

pedalfer Soil formed in humid regions with an organic-rich A horizon and aluminum-rich clays and iron oxides in horizon B.

pediment An erosion surface of low relief gently sloping away from the base of a mountain range.

pedocal Soil characteristic of arid and semiarid regions with a thin A horizon and a calcium carbonate–rich B horizon.

pelagic clay Brown or red deep-sea sediment composed of clay-sized particles.

permafrost Ground that remains permanently frozen.

permeability A material's capacity to transmit fluids.

phaneritic texture Igneous rock texture in which minerals are easily visible without magnification.

pillow lava Bulbous masses of basalt, resembling pillows, formed when lava is rapidly chilled under water.

plastic flow The flow that takes place in response to pressure and causes deformation with no fracturing.

plastic strain Permanent deformation of a solid with no failure by fracturing.

plate An individual segment of the lithosphere that moves over the asthenosphere.

plate tectonic theory The theory holding that large segments of Earth's outer part (lithospheric plates) move relative to one another.

playa A dry lakebed found in deserts.

pluton An intrusive igneous body that forms when magma cools and crystallizes within the crust, such as a batholith or sill.

plutonic (intrusive igneous) rock Igneous rock that formed from magma intruded into or formed in place within the crust.

point bar The sediment body deposited on the gently sloping side of a meander loop.

porosity The percentage of a material's total volume that is pore space.

porphyritic texture An igneous texture with minerals of markedly different sizes.

pressure release A mechanical weathering process in which rocks that formed under pressure expand on being exposed at the surface.

principle of cross-cutting relationships A principle holding that an igneous intrusion or fault must be younger than the rocks it intrudes or cuts across.

principle of fossil succession A principle holding that fossils, and especially groups or assemblages of fossils, succeed one another through time in a regular and predictable order.

principle of inclusions A principle holding that inclusions or fragments in a rock unit are older than the rock unit itself; for example, granite inclusions in sandstone are older than the sandstone.

principle of isostasy The theoretical concept of Earth's crust "floating" on a dense underlying layer. (*See also* isostatic rebound)

principle of lateral continuity A principle holding that rock layers extend outward in all directions until they terminate.

principle of original horizontality According to this principle, sediments are deposited in horizontal or nearly horizontal layers.

principle of superposition A principle holding that in a vertical sequence of undeformed sedimentary rocks, the relative ages of the rocks can be determined by their position in the sequence—oldest at the bottom followed by successively younger layers.

principle of uniformitarianism A principle holding that we can interpret past events by understanding present-day processes, based on the idea that natural processes have always operated in the same way.

proton A positively charged particle found in the nucleus of an atom.

P-wave A compressional, or push-pull, wave; the fastest seismic wave and one that can travel through solids, liquids, and gases; also called a primary wave.

P-wave shadow zone An area between 103 and 143 degrees from an earthquake focus where little P-wave energy is recorded by seismographs.

pyroclastic (fragmental) texture A fragmental texture characteristic of igneous rocks composed of pyroclastic materials.

pyroclastic materials Fragmental substances, such as ash, explosively ejected from a volcano.

pyroclastic sheet deposit Vast, sheetlike deposit of felsic pyroclastic materials erupted from fissures.

quick clay A clay deposit that spontaneously liquefies and flows like water when disturbed.

radioactive decay The spontaneous change of an atom to an atom of a different element by emission of a particle from its nucleus (alpha and beta decay) or by electron capture.

rain-shadow desert A desert found on the lee side of a mountain range because precipitation falls mostly on the windward side of the range.

rapid mass movement Any kind of mass wasting that involves a visible downslope displacement of material.

Rayleigh wave (R-wave) A surface wave in which individual particles of material move in an elliptical path within a vertical plane oriented in the direction of wave movement.

recessional moraine An end moraine that forms when a glacier's terminus retreats, then stabilizes, and a ridge or mound of till is deposited. (*See also* end moraine and terminal moraine)

reef A mound-like, wave-resistant structure composed of the skeletons of organisms.

reflection The return to the surface of some of a seismic wave's energy when it encounters a boundary separating materials of different density or elasticity.

refraction The change in direction and velocity of a seismic wave when it travels from one material into another of different density or elasticity.

regional metamorphism Metamorphism that occurs over a large area, resulting from high temperatures, tremendous pressure, and the chemical activity of fluids within the crust.

regolith The layer of unconsolidated rock and mineral fragments and soil that covers most of the land surface.

relative dating The process of determining the age of an event as compared to other events; involves placing geologic events in their correct chronological order, but does not involve consideration of when the events occurred in number of years ago. (*See also* absolute dating)

reserve The part of the resource base that can be extracted economically.

resource A concentration of naturally occurring solid, liquid, or gaseous material in or on Earth's crust in such form and amount that economic extraction of a commodity from the concentration is currently or potentially feasible.

reverse fault A dip-slip fault on which the hanging wall block has moved upward relative to the footwall block. (*See also* normal fault)

Richter Magnitude Scale An open-ended scale that measures the amount of energy released during an earthquake.

rip current A narrow surface current that flows out to sea through the breaker zone.

ripple mark Wavelike (undulating) structure produced in granular sediment, especially sand, by unidirectional wind and water currents or by oscillating wave currents.

rock A solid aggregate of one or more minerals, as in limestone and granite, or a consolidated aggregate of rock fragments, as in conglomerate, or masses of rocklike materials, such as coal and obsidian.

rock cycle A group of processes through which Earth materials may pass as they are transformed from one major rock type to another.

rockfall A type of extremely fast mass wasting in which rocks fall through the air.

rock-forming mineral Any mineral common in rocks that is important in their identification and classification.

rock slide Rapid mass wasting in which rocks move downslope along a more or less planar surface.

runoff The surface flow in streams and rivers.

salt crystal growth A mechanical weathering process in which salt crystals growing in cracks and pores disaggregate rocks.

scientific method A logical, orderly approach that involves gathering data, formulating and testing hypotheses, and proposing theories.

seafloor spreading The theory that the seafloor moves away from spreading ridges and is eventually consumed at subduction zones.

seamount A submarine volcanic mountain rising at least 1 km above the seafloor.

sediment Loose aggregate of solids derived by weathering from preexisting rocks, or solids precipitated from solution by inorganic chemical processes or extracted from solution by organisms.

sedimentary facies Any aspect of a sedimentary rock unit that makes it recognizably different from adjacent sedimentary rocks of the same or approximately the same age.

sedimentary rock Any rock composed of sediment, such as limestone and sandstone.

sedimentary structure Any feature in sedimentary rock that formed at or shortly after the time of deposition, such as cross-bedding, animal burrows, and mud cracks.

seismic profiling A method in which strong waves generated at an energy source penetrate the layers beneath the seafloor. Some of the energy is reflected back from various layers to the surface, making it possible to determine the nature of the layers.

seismograph An instrument that detects, records, and measures the various waves produced by earthquakes.

seismology The study of earthquakes.

shear strength The resisting forces that help maintain a slope's stability.

shear stress The result of forces acting parallel to one another but in opposite directions; results in deformation by displacement of adjacent layers along closely spaced planes.

shield volcano A dome-shaped volcano with a low, rounded profile built up mostly by overlapping basalt lava flows.

shoreline The area between mean low tide and the highest level on land affected by storm waves.

silica A compound of silicon and oxygen.

silica tetrahedron The basic building block of all silicate minerals; consists of one silicon atom and four oxygen atoms.

silicate A mineral that contains silica, such as quartz (SiO_2).

sill A tabular or sheetlike concordant pluton.

sinkhole A depression in the ground that forms by the solution of the underlying carbonate rocks or by the collapse of a cave roof.

slide Mass wasting involving movement of material along one or more surfaces of failure.

slow mass movement Mass movement that advances at an imperceptible rate and is usually detectable only by the effects of its movement.

slump Mass wasting that takes place along a curved surface of failure and results in the backward rotation of the slump mass.

soil Regolith consisting of weathered materials, water, air, and humus that can support vegetation.

soil degradation Any process leading to a loss of soil productivity; may involve erosion, chemical pollution, or compaction.

soil horizon A distinct soil layer that differs from other soil layers in texture, structure, composition, and color.

solar nebula theory A theory for the evolution of the solar system from a rotating cloud of gas.

solifluction Mass wasting involving the slow downslope movement of water-saturated surface materials; especially the flow at high elevations or high latitudes where the flow is underlain by frozen soil.

solution A reaction in which the ions of a substance become dissociated in a liquid and the solid substance dissolves.

specific gravity The ratio of a substance's weight, especially a mineral, to an equal volume of water at 4°C.

spheroidal weathering A type of chemical weathering in which corners and sharp edges of rocks weather more rapidly than flat surfaces, thus yielding spherical shapes.

spit A fingerlike projection of a beach into a body of water such as a bay.

spring A place where groundwater flows or seeps out of the ground.

stock An irregularly shaped discordant pluton with a surface area smaller than 100 km^2.

stoping A process in which rising magma detaches and engulfs pieces of the country rock.

storm surge The surge of water onto a shoreline as a result of a bulge in the ocean's surface beneath the eye of a hurricane and wind-driven waves.

strain Deformation caused by stress.

strata (singular, stratum) Refers to layering in sedimentary rocks. (*See also* bed)

stratified drift Glacial deposits that show both stratification and sorting.

stream terrace An erosional remnant of a floodplain that formed when a stream was flowing at a higher level.

stress The force per unit area applied to a material such as rock.

strike The direction of a line formed by the intersection of an inclined plane and a horizontal plane.

strike-slip fault A fault involving horizontal movement of blocks of rock on opposite sides of a fault plane.

submarine canyon A steep-walled canyon best developed on the continental slope, but some extend well up onto the continental shelf.

submarine fan A cone-shaped sedimentary deposit that accumulates on the continental slope and rise.

submarine hydrothermal vent A crack or fissure in the seafloor through which superheated water issues. (*See also* black smoker)

submergent coast A coast along which sea level rises with respect to the land or the land subsides.

superposed stream A stream that once flowed on a higher surface and eroded downward into resistant rocks while maintaining its course.

suspended load The smallest particles (silt and clay) carried by running water, which are kept suspended by fluid turbulence.

S-wave A shear wave that moves material perpendicular to the direction of travel, thereby producing shear stresses in the material it moves through; also known as a secondary wave; S-waves travel only through solids.

S-wave shadow zone Those areas more than 103 degrees from an earthquake focus where no S-waves are recorded.

syncline A down-arched fold in which the youngest exposed rocks coincide with the fold axis and all strata dip toward the axis.

system A combination of related parts that interact in an organized fashion; Earth systems include the atmosphere, hydrosphere, biosphere, and solid Earth.

talus Accumulation of coarse, angular rock fragments at the base of a slope.

tension A type of stress in which forces act in opposite directions but along the same line, thus tending to stretch an object.

terminal moraine An end moraine consisting of a ridge or mound of rubble marking the farthest extent of a glacier. (*See also* end moraine *and* recessional moraine)

terrestrial planet Any of the four innermost planets (Mercury, Venus, Earth, and Mars). They are all small and have high mean densities, indicating that they are composed of rock and metallic elements.

theory An explanation for some natural phenomenon that has a large body of supporting evidence. To be scientific, a theory must be testable — for example, plate tectonic theory.

thermal convection cell A type of circulation of material in the asthenosphere during which hot material rises, moves laterally, cools and sinks, and is reheated and continues the cycle.

thermal expansion and contraction A type of mechanical weathering in which the volume of rocks changes in response to heating and cooling.

thrust fault A type of reverse fault in which a fault plane dips less than 45 degrees.

tide The regular fluctuation of the sea's surface in response to the gravitational attraction of the Moon and Sun.

till All sediment deposited directly by glacial ice.

time-stratigraphic unit A body of strata that was deposited during a specific interval of geologic time; for example, the Devonian System (a time-stratigraphic unit) was deposited during the Devonian Period.

time unit Any of the units such as eon, era, period, epoch, and age used to refer to specific intervals of geologic time.

tombolo A type of spit that extends out from the shoreline and connects the mainland with an island.

transform fault A fault along which one type of motion is transformed into another; commonly displaces oceanic ridges; on land, recognized as a strike-slip fault, such as the San Andreas fault.

transform plate boundary Plate boundary along which plates slide past one another and crust is neither produced nor destroyed.

transverse dune A ridge of sand with its long axis perpendicular to the wind direction.

tree-ring dating The process of determining the age of a tree or wood in a structure by counting the number of annual growth rings.

trough The lowest point between wave crests.

tsunami A large sea wave that is usually produced by an earthquake, but can also result from submarine landslides and volcanic eruptions.

turbidity current A sediment-water mixture, denser than normal seawater, that flows downslope to the deep seafloor.

unconformity A break in the geologic record represented by an erosional surface separating younger strata from older rocks. (*See also* nonconformity, disconformity, and angular unconformity)

U-shaped glacial trough A valley with steep or vertical walls and a broad, rather flat floor formed by the movement of a glacier through a stream valley.

valley A linear depression bounded by higher areas such as ridges or mountains.

valley glacier A glacier confined to a mountain valley or an interconnected system of mountain valleys.

valley train A long, narrow deposit of stratified drift confined within a glacial valley.

velocity A measure of distance traveled per unit of time, as in the flow velocity in a stream or river.

ventifact A stone with a surface polished, pitted, grooved, or faceted by wind abrasion.

vesicle A small hole or cavity formed by gas trapped in cooling lava.

viscosity A fluid's resistance to flow.

volcanic ash Pyroclastic materials that measure less than 2 mm.

volcanic explosivity index (VEI) A semi-quantative scale for the size of a volcanic eruption based on evaluation of criteria such as volume of material explosively erupted and height of eruption cloud.

volcanic (extrusive igneous) rock An igneous rock formed when magma is extruded onto Earth's surface where it cools and crystallizes, or when pyroclastic materials become consolidated.

volcanic neck An erosional remnant of the material that solidified in a volcanic pipe.

volcanic pipe The conduit connecting the crater of a volcano with an underlying magma chamber.

volcanic tremor Ground motion lasting from minutes to hours, resulting from magma moving beneath the surface, as opposed to the sudden jolts produced by most earthquakes.

volcanism The processes whereby magma and its associated gases rise through the crust and are extruded onto the surface or into the atmosphere.

volcano A hill or mountain formed around a vent as a result of the eruption of lava and pyroclastic materials.

water table The surface that separates the zone of aeration from the underlying zone of saturation.

water well A well made by digging or drilling into the zone of saturation.

wave An undulation on the surface of a body of water, resulting in the water surface rising and falling.

wave base The depth corresponding to about one-half wavelength, below which water in unaffected by surface waves.

wave-cut platform A beveled surface that slopes gently seaward; formed by the erosion and retreat of a sea cliff.

wave refraction The bending of waves so that they move nearly parallel to the shoreline.

weathering The physical breakdown and chemical alteration of rocks and minerals at or near Earth's surface.

zone of accumulation The part of a glacier where additions exceed losses and the glacier's surface is perennially covered with snow. Also refers to horizon B in soil where soluble material leached from horizon A accumulates as irregular masses.

zone of aeration The zone above the water table that contains both air and water within the pore spaces of soil, sediment, or rock.

zone of saturation The area below the water table in which all pore spaces are filled with water.

zone of wastage The part of a glacier where losses from melting, sublimation, and calving of icebergs exceed the rate of accumulation.

INDEX

Note: *Italicized* page numbers indicate illustrations.